# Energy Efficient Thermal Management of Data Centers

Yogendra Joshi • Pramod Kumar
Editors

# Energy Efficient Thermal Management of Data Centers

Springer

*Editors*
Yogendra Joshi
G.W. Woodruff School
of Mechanical Engineering
Georgia Institute of Technology
Atlanta, GA, USA

Pramod Kumar
G.W. Woodruff School
of Mechanical Engineering
Georgia Institute of Technology
Atlanta, GA, USA

ISBN 978-1-4419-7123-4 e-ISBN 978-1-4419-7124-1
DOI 10.1007/978-1-4419-7124-1
Springer New York Dordrecht Heidelberg London

Library of Congress Control Number: 2012931944



Printed on acid-free paper

Springer is part of Springer Science+Business Media (www.springer.com)

# Preface

Within the past decade, data centers have taken on an increasingly central role in the progress of two parallel areas of burgeoning global societal interest: Internet delivered information technology (IT) services and telecommunications. The terms data centers or server farms have been widely, and interchangeably, used to describe large facilities, sometimes over 40,000 $m^2$ of floor space, that house computing and storage equipment for IT, and switching equipment for telecommunications. Larger facilities require electrical power delivery in the range of tens of MW, which is converted into heat by the equipment. The need for keeping data centers operational with no downtime for many critical applications, such as banking, electronic commerce, stock transactions, and mobile communications, necessitates deploying significant redundancy in the power delivery and cooling associated with these mission critical facilities. Effective thermal management of data centers is thus essential for their proper operation and has been a key focus since the design of the earliest facilities. As computing capabilities advance, heat dissipations within IT cabinets have increased from a few kW to well over 20 kW within the past decade, particularly in high-performance server cabinets. To enable the successful deployment of this hardware, significant advances in cooling technologies, such as the use of indirect or direct liquid cooling, have also taken place.

The increasing cost of energy has brought a new focus to the design and operation of data centers. A 2007 report to the US Congress showed that the use of electricity by data centers had doubled from 2000 to 2005 in the USA, and such increase was unsustainable. Benchmarking studies undertaken by the US Department of Energy in California's Silicon Valley data centers showed that as much as 30–50% of the overall electricity use of a data center was for cooling. A number of energy saving concepts, such as broadening the environmental temperature and humidity requirements envelope, and using outside air for cooling when appropriate, also termed "free cooling," have since been incorporated in the designs of newer facilities, and recent estimates show a drop-off in the rate of increase in energy usage by data centers.

Several professional societies have been at the forefront of identifying issues relating to the energy management of data centers. Most notably, the American Society for Heating Refrigeration and Air-Conditioning (ASHRAE), in partnership with industry stakeholders, has focused on environmental guidelines for the proper operation of these facilities. Several revisions have been made to allow a broadening of inlet air temperature and humidity requirements, as they have a prominent impact on operational energy costs. Within the past 5 years, ASHRAE has also published a series of books identifying cooling technologies available to the data center designers and industry best practices for a disparate group of professionals, including architects, builders, mechanical engineers, electrical engineers, and IT professionals.

Presently, data centers are managed based on accrued experience or best practices, which often lead to an overly conservative thermal management approach, at the cost of wasted cooling resources. Reducing energy consumption and carbon footprint of data centers, on the other hand, requires a fundamental principles based approach. The data center manager now has to supplement prior experience with conceptual understanding of heat transfer, fluid flow, thermodynamics, computational modeling, metrology, and data acquisition and processing. This book aims to provide such an in-depth understanding by drawing on the expertise of several researchers and industry practitioners. We focus on broader analysis and solution methodologies, rather than specific solutions.

The book focuses on six themes. Understanding, monitoring, and controlling airflows and thermal fields in data centers are discussed in Chap. 2. Power delivery, distribution, and management within the data center are addressed in Chaps. 3 and 4. Energy-efficient operation of a data center requires numerous sensors. The acquisition, processing, and use of such data is an important activity. Quantification of energy efficiency also requires identification of suitable metrics. Another important thermodynamic consideration is that of exergy destruction, which can be minimized through careful design and operation. These concepts are addressed in Chaps. 5–7, 9, and 11. Computational modeling of airflow and temperature patterns within data centers through full field approaches, and rapid simulations are discussed in Chaps. 7, 8, and 10. A number of advances in cooling technologies are being incorporated in data centers to achieve both improved cooling capabilities and energy efficiency. For applications such as high-performance computing, it is clear that air cooling will not be sufficient. A transition to liquid cooling is already taking place, with the added benefit of the possibility of waste heat recovery, as discussed in Chap. 12. A number of other thermal and energy management advances are discussed in Chap. 13.

The editors hope that this book will serve as a useful compilation of in-depth information in this area of growing interest. The book is aimed at the various stakeholders in the data center industry as well as the academic community. The former category includes critical facility designers, cooling, IT, and telecommunications equipment manufacturers, data center end users, and operators.

We hope that established and new academic researchers involved in thermal design, power delivery, and cloud computing will also find the book useful. The Editors are grateful to all the chapter contributors for preparing authoritative compilations of the state of the art and recent progress in their selected topical areas. Finally, we acknowledge the help and guidance extended by the publisher through the editorial assistance of Steven Elliot and Merry Stuber.

Atlanta, GA, USA

Yogendra Joshi
Pramod Kumar

# Contents

# Contributors

**Hrishikesh Amur** Center for Experimental Research in Computer Systems, School of Computer Science, College of Computing, Georgia Institute of Technology, Atlanta, GA, USA

**Cullen E. Bash** Hewlett Packard Laboratories, Palo Alto, CA, USA

**Prajesh Bhattacharya** Lawrence Berkeley National Laboratory, Berkeley, CA, USA

**Siddharth Bhopte** Small Scale System Integration & Packaging Center, Binghamton University–State University of New York, Binghamton, NY, USA

**Van P. Carey** Department of Mechanical Engineering, University of California, Berkeley, CA, USA

**Ada Gavrilovska** Center for Experimental Research in Computer Systems, School of Computer Science, College of Computing, Georgia Institute of Technology, Atlanta, GA, USA

**Hendrik F. Hamann** IBM Thomas J. Watson Research Center, Yorktown Heights, NY, USA

**Ying Hung** Department of Statistics and Biostatistics, Rutgers, The State University of New Jersey, Piscataway, NJ, USA

**Mahmoud Ibrahim** Small Scale System Integration & Packaging Center, Binghamton University–State University of New York, Binghamton, NY, USA

**Yogendra Joshi** G.W. Woodruff School of Mechanical Engineering, Georgia Institute of Technology, Atlanta, GA, USA

**Bhavani Krishnan** Center for Experimental Research in Computer Systems, School of Computer Science, College of Computing, Georgia Institute of Technology, Atlanta, GA, USA

**Pramod Kumar** G.W. Woodruff School of Mechanical Engineering, Georgia Institute of Technology, Atlanta, GA, USA

**Hsien-Hsin S. Lee** School of Electrical and Computer Engineering, Georgia Institute of Technology, Atlanta, GA, USA

**Vanessa López** IBM Thomas J. Watson Research Center, Yorktown Heights, NY, USA

**Jackson B. Marcinichen** Laboratory of Heat and Mass Transfer (LTCM), École Polytechnique Fédérale de Lausanne (EPFL), Lausanne, Switzerland

**Jonathan A. Olivier** Laboratory of Heat and Mass Transfer (LTCM), École Polytechnique Fédérale de Lausanne (EPFL), Lausanne, Switzerland

**Chandrakant D. Patel** Hewlett Packard Laboratories, Palo Alto, CA, USA

**Michael K. Patterson** Eco-Technology Program Office, Intel Corporation, Hillsboro, OR, USA

**Peter Z.G. Qian** Department of Statistics, University of Wisconsin-Madison, Madison, WI, USA

**Emad Samadiani** G.W. Woodruff School of Mechanical Engineering, Georgia Institute of Technology, Atlanta, GA, USA

**Bahgat Sammakia** Small Scale System Integration & Packaging Center, Binghamton University–State University of New York, Binghamton, NY, USA

**Karsten Schwan** Center for Experimental Research in Computer Systems, School of Computer Science, College of Computing, Georgia Institute of Technology, Atlanta, GA, USA

**Amip J. Shah** Hewlett Packard Laboratories, Palo Alto, CA, USA

**Ratnesh K. Sharma** Hewlett Packard Laboratories, Palo Alto, CA, USA

**John R. Thome** Laboratory of Heat and Mass Transfer (LTCM), École Polytechnique Fédérale de Lausanne (EPFL), Lausanne, Switzerland

**Jhenkar Vidyashankar** Center for Experimental Research in Computer Systems, School of Computer Science, College of Computing, Georgia Institute of Technology, Atlanta, GA, USA

**Chengwei Wang** Center for Experimental Research in Computer Systems, School of Computer Science, College of Computing, Georgia Institute of Technology, Atlanta, GA, USA

**Matt Wolf** Center for Experimental Research in Computer Systems, School of Computer Science, College of Computing, Georgia Institute of Technology, Atlanta, GA, USA

**C.F. Jeff Wu** School of Industrial and Systems Engineering, Georgia Institute of Technology, Atlanta, GA, USA

**Sungkap Yeo** School of Electrical and Computer Engineering, Georgia Institute of Technology, Atlanta, GA, USA

# Chapter 1
# Introduction to Data Center Energy Flow and Thermal Management

**Yogendra Joshi and Pramod Kumar**

**Abstract** This chapter provides an introduction to the emerging trends in the growth of data centers. It is seen that projected growth in functionality and performance in information technology and communications equipment is resulting in sharply increasing power dissipation per unit facility footprint. Increased energy usage for powering and thermal management of data centers, which is a potential concern for continued growth of data centers is examined. The guidelines for environmental control of data centers to assure satisfactory equipment performance are discussed. Thermal management of data centers involves multiple length scales. The approaches currently in use and under exploration at various scales are presented. Thermal modeling approaches for data centers are also discussed. Data centers are generally expected to operate continuously, yet there have been documented failure events that have lead to usage interruption. The tier classification of data centers based on redundancy is introduced.

Data centers are computing infrastructure facilities that house large amounts of information technology (IT) equipment used to process, store, and transmit digital information. This equipment is housed within electronic cabinets or racks of standardized dimensions. Data centers also usually contain power conversion and backup equipment to maintain reliable, high-quality power as well as environmental control equipment to maintain the proper temperature and humidity conditions within the facility. An overview of a typical data center facility is provided in Fig. 1.1. A related, but distinct, class of facilities houses equipment devoted to telecommunications, such as telephone exchanges including network equipment.

Y. Joshi (✉) • P. Kumar
G.W. Woodruff School of Mechanical Engineering, Georgia Institute of Technology, Atlanta, GA 30332, USA
e-mail: yogendra.joshi@me.gatech.edu; pramod.kumar@me.gatech.edu

Y. Joshi and P. Kumar (eds.), *Energy Efficient Thermal Management of Data Centers*, DOI 10.1007/978-1-4419-7124-1_1,

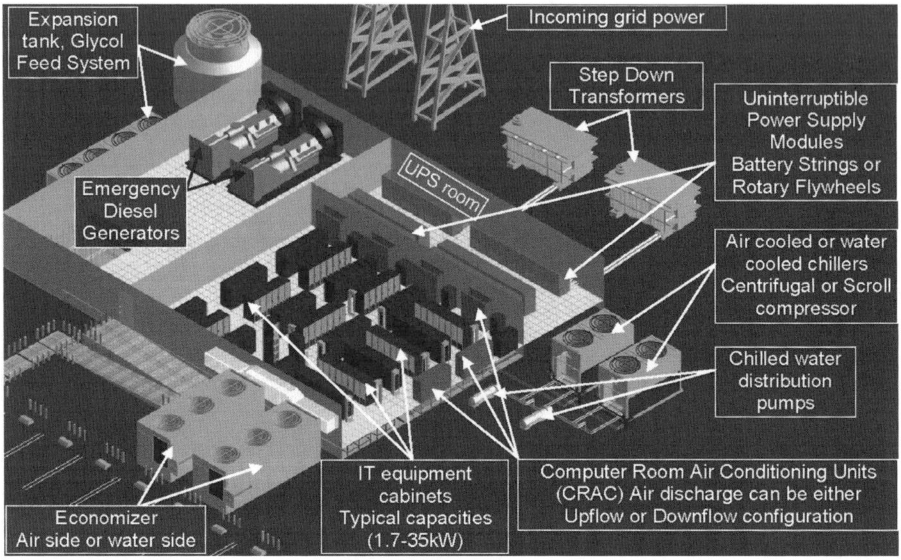


**Fig. 1.1** Typical layout of a data center facility

Data centers are utilized by a broad range of end users, including individuals, Internet service providers, banks, stock exchanges, corporations, educational institutions, government installations, and research laboratories.

## 1.1 Information Technology (IT) Infrastructure Trends

Rapid growth in IT facilities is being fueled by ever-increasing demand for data processing and storage, driven by increased use of electronic transactions in applications such as financial services, electronic commerce, Internet-based communications and entertainment, electronic medical records for health care, satellite navigation, and electronic shipment tracking. The government sector is also seeing an increased usage of these facilities, which can be broadly classified into two categories: owned and operated, versus colocated. The former option is utilized by organizations where one or more of these factors are important: the overall space and power requirements are substantial, which may make it worthwhile to own and operate a dedicated facility, the facility operation requires intensive management, and physical security is a major concern. Colocation, on the other hand is attractive when up-front costs associated with building a data center need to be minimized. The IT infrastructure based on specific network requirements can be rapidly deployed at lower up-front costs. Also, organizational relocations can occur without network downtime or other interruptions.

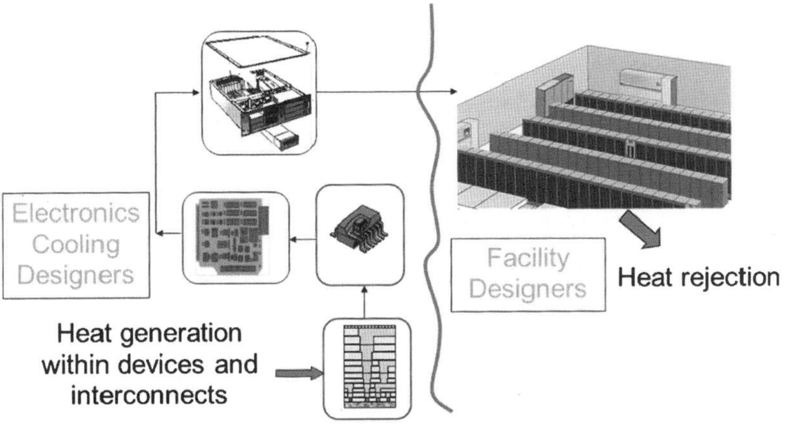


**Fig. 1.2** Multi-scale nature of power and thermal management in data centers. Historically, two separate technical communities have addressed these issues, but increasing power densities and interest in holistic sustainable and energy-efficient design is removing the traditional boundary

### *1.1.1 Increases in Equipment Power Density*

The electrical power consumed by the IT or telecommunications equipment is converted entirely into heat and must ultimately be rejected to the environment. This involves transmission of heat across several decades of length scales from on-chip transistors and interconnects at tens of nanometers, to the data center facility with sizes of hundreds of meters, prior to rejection to the ambient environment, see Fig. 1.2. Historically, the cooling hardware at the scale of electronic systems such as servers or racks has been the focus of microsystems packaging, while facility-level cooling has been addressed by heating ventilation and air-conditioning (HVAC) engineers. With increases in equipment power densities and the need to minimize energy usage for cooling, a holistic consideration of thermal management across the entire length scale hierarchy is essential.

Since the beginning of the past decade, the American Society of Heating Refrigeration and Air-Conditioning (ASHRAE) has played a pioneering role in identifying the trends in power densities, operating environmental guidelines, and cooling of data centers. This has been pursued through the formation of a technical committee (TC 9.9) dedicated to identifying and updating emerging environmental and heat load requirements and best energy management practices for mission critical IT equipment and facilities. The TC 9.9 group has been instrumental in developing a number of guidebooks on data center design and operation, which have been published by ASHRAE as the Datacom Series [1]. An early survey of multiple equipment manufacturers of IT and telecommunications equipment resulted in the historical trends and power density projections, with an updated version seen in Fig. 1.3 [2]. These power densities are several hundred times higher

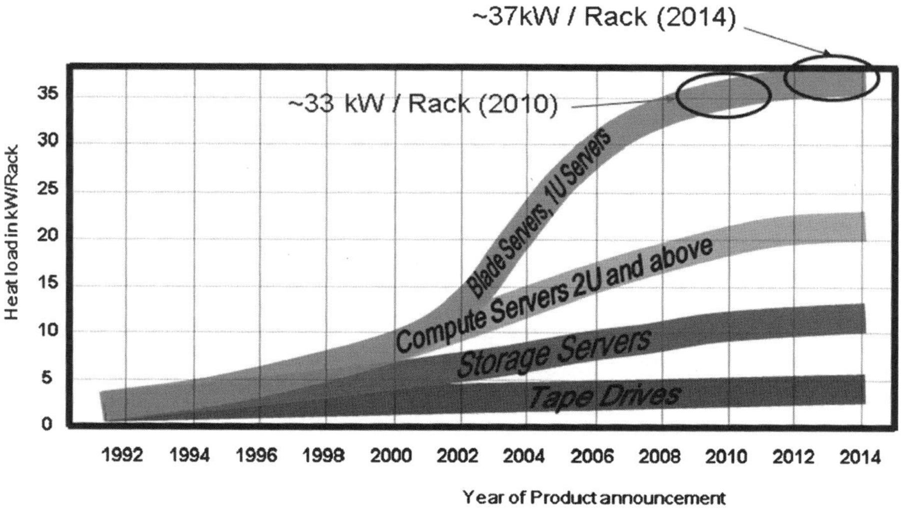


**Fig. 1.3** Power dissipation projections for various IT racks or cabinets; adapted from [2]

than in traditional office buildings and auditoria, and require specialized approaches and guidelines for their handling.

The increases in heat loads per product footprint are being driven by ever-increasing capabilities of IT servers and their dense packaging in racks or cabinets. Increases in demands for power and cooling capabilities are resulting in significant challenges to the growth of existing facilities as well as to the sustainable development of new ones.

### *1.1.2 Chip-Level Power and Packaging Trends*

As illustrated in Fig. 1.2, the energy management of data centers must holistically address all pertinent size scales. Key trends in multi-scale power delivery and management in data centers are discussed in detail in Chap. 3. Here we briefly introduce some of the chip, and cabinet, or rack level, trends. Microprocessor chips are the performance engines of IT data centers. They account for the largest amount of power consumption within compute servers, although power dissipation in other components such as memory is on the rise as well. Microprocessor and memory chip architectures are rapidly emerging, and the challenges associated with their manufacturing and packaging are annually updated by the International Technology Roadmap for Semiconductors (ITRS) [3].

The transition from silicon bipolar to complementary metal oxide semiconductor (CMOS) technologies in the mid-1980s fuelled the rapid growth in microprocessor capabilities. During the 1990s and early part of the past decade the trend was one of increasing levels of functionality via integration, or larger number of transistors on a given chip of roughly 1 cm by 1 cm area. This was possible through the shrinking of feature sizes, currently at 45 nm in production microprocessors, with nearly a billion transistors on a single chip. Clock frequencies of these processors continued to increase, while supply voltages decreased during the same period. These developments sustained the progress of the celebrated Moore's law, which predicted a doubling of microprocessor performance every 18–24 months. The average chip-level heat fluxes in high performance microprocessors increased to approximately 50–75 W/cm$^2$ by 2005 [4]. Since 2005, several challenges associated with continued reduction in feature sizes have resulted in a move away from single core to multi-core microprocessors and a slowing of the Moore's Law scaling. Multi-core processors display average chip-level heat fluxes comparable to single core processors, along with a large degree of spatial variation of heat dissipation on chip. As such, chip-level thermal management is an area of intense current research and development.

Individual servers, each an independent computing and storage entity, are packaged in numerous racks or cabinets within a data center facility. Server powers are increasing and their packaging is constantly evolving to keep pace with new microprocessors [4]. Typical rack dimensions, however, have remained relatively unchanged for the past two decades, with a standard height of 42 U, or 1.86 m, where 1 U = 4.44 cm or 1.75 in. During the same time, the individual servers have become more compact, with some of the smallest form factor ones, called blades, being 1 U thick. The relentless drive towards server miniaturization, which is still housed in standard sized racks, has resulted in significant reductions in the footprint of entire computing machines, at the expense of significantly higher rack level heat dissipation, displayed in Fig. 1.3. In 2002, a 3.7 Terraflop computer, listed in the top 20 fastest machines in the world at the time, consisted of 25 racks consuming a total of 125 kW of power, or an average power dissipation of 5 kW/rack [5]. By 2008, a 3.7 Terraflop machine could be accommodated in a single rack consuming 21 kW [5]. This is a 1/5 total power dissipation achieved in 1/25 of floor footprint for the same computational capability. However, this is at the expense of about four times increased rack power and five times power density based on footprint. Both of these are significant challenges to thermal management, at both the rack and facility levels. Some of the highest power computational racks in the market already dissipate in excess of 30 kW, with projections of continued growth in rack heat loads.

### *1.1.3 Move Towards Virtualization and Cloud Computing*

We explore the rapidly dominating IT concepts of virtualization and cloud computing in Chap. 4. As a brief introduction to these topics, in virtualization the functions enabled by a hardware device are abstracted or segregated from the physical hardware [6]. For example, if a disk provides the service of "data storage" for a server, the virtualization abstraction layer separates the server's access to the "data storage" service from the physical disk on which the data is stored. This allows the disk to be serviced or upgraded to increase capacity without disrupting the application, while data is being stored. The "data storage" service continues even when the physical hardware supporting the service changes. Indeed moving data from one array to another, moving applications across servers, moving network services from one switch to another become feasible in a manner transparent to applications or users.

In cloud computing, applications, computing, and storage resources reside somewhere in the network, or "cloud." Users do not need to be concerned about the location, while rapidly accessing these on demand. Payment models for these services are based on usage, just as they are with water, electricity, or energy services provided by utility companies. Virtualization technology combined with cloud computing enables dynamic provisioning of resources, as applications are not constrained to a specific physical server, or data to a single storage device. Cloud computing over the Internet is commonly called "public cloud computing." When used in a dedicated data center, it is commonly referred to as "private cloud computing." The two models differ in who maintains control and responsibility for servers, storage, and networking infrastructure, and ensures that service level agreements (SLA) are met. In public cloud computing, some or all aspects of operations and management are handled by a third-party "as a service." Users can access an application or computing and storage using the Internet address of the service. Google Gmail is a well-known example of public cloud computing, in which virtualization resides between the Internet connection of an individual user and the data centers delivering the Google service. Companies offering public cloud computing services account for some of the largest growth in data center facilities in the world.

## 1.2 Data Center Energy Usage: The EPA Report and LBNL Benchmarking Studies

According to a report released by the United States Environmental Protection Agency (EPA) [7], in 2006, data centers in the United States consumed about 61 billion kWh, or 1.5% of total US electricity consumption, for a total electricity cost of about \$4.5 billion, see Fig. 1.4. This is equivalent to the electricity consumed by approximately 5.8 million average US households and is estimated to be more than double the amount used in 2000. Telecommunication facilities roughly account for similar additional energy consumption and growth rates. Such sharp increases in energy consumption by data centers and telecommunications facilities are unsustainable over the long term. The EPA report [7] considered a number of possible scenarios that could potentially reduce energy consumption needs of data centers through a variety of technological

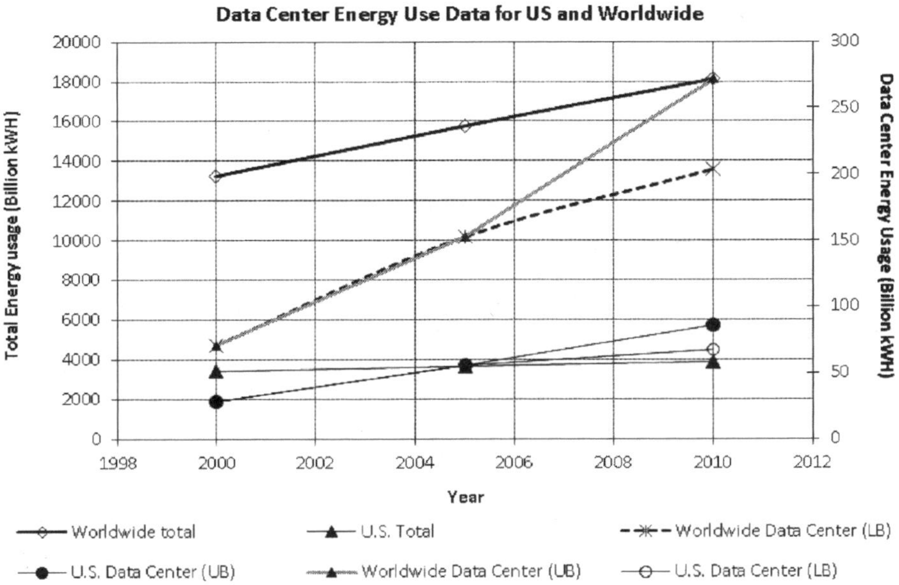


**Fig. 1.4** Energy consumption by data centers in the USA States and Worldwide, based on data reported by J. Koomey in the New York Times, July 31, 2011

advancements. Their possible impact on the future energy consumption by data centers is shown in Fig. 1.4. The EPA report followed a set of energy usage benchmarking studies performed in the Master's thesis of Mitchell-Jackson [8] and by the Department of Energy Lawrence Berkeley National Labs (LBNL) [9] at over 20 data center facilities in the Silicon Valley. These are discussed in some detail in Chap. 3. One of the outcomes of the benchmarking studies was identification of best practices that data center facilities could follow to reduce energy consumption [7].

As seen in Fig. 1.4, employing some of these and other practices has already reduced the data center energy consumption rate from 2005 to 2010, compared to the 2000–2005 period, as reported in a study by J. Koomey in the *New York Times* in July 2011. The data reported by Koomey for 2010 present Upper Bound and Lower Bound estimates for USA and worldwide data center energy usage. In [7], three possible future scenarios that could more aggressively curb energy consumption were also considered: *improved operation*, *best practice*, and *state of the art*, with the last two either flattening the total energy usage or reversing the trend. Each of these was based on improvements targeted at both the IT equipment as well as the facilities.

The state-of-the-art scenario represents the maximum technical potential for energy usage reduction. The best practice scenario is based on efficiency gains that could be realized using existing technologies. The improved operation scenario represents incorporation of low-cost energy-efficiency opportunities.

The *improved operation* for IT equipment requires continued current trends for server consolidation, elimination of unused servers, adopting "energy-efficient" servers to modest level, and enabling power management on all applicable servers. It assumes modest decline in energy use of enterprise storage equipment. For the infrastructure, a 30% energy efficiency improvement through improved airflow management is assumed.

The *best practice* operation requires all measures in the "Improved Operation" scenario. Additionally, server consolidation to moderate extent, aggressive adoption of "energy-efficient" servers, and moderate storage consolidation are also assumed. For the infrastructure, up to 70% improvement in energy efficiency from all measures in "Improved operation" scenario is assumed. Additionally, improved transformers and uninterruptible power supplies, improved efficiency chillers, fans, and pumps, and incorporation of economizers are assumed.

The *state-of-the-art* scenario includes all measures in the "Best Practice" scenario. Additionally, aggressive server and storage consolidation, enabling power management at data center level of applications, servers, and equipment for networking and storage are assumed. For the infrastructure, up to 80% improvement in energy efficiency, due to all measures in "Best practice" scenario is assumed. Additionally, direct liquid cooling of servers, and cogeneration approaches are included. Since the best practice and state-of-the-art scenarios would require significant changes to data centers that may only be feasible during major facility renovations, it was assumed in these predictions that the site infrastructure improvements requiring new capital investments would apply to only 50% of the currently built data centers. For IT equipment, it was assumed that the entire existing stock becomes obsolete within the 5-year forecast period.

A significant fraction of the energy costs associated with the operation of a typical data center can be ascribed to the cooling hardware. As seen in Fig. 1.5, the ratio of the total power input to data center, to the power to the IT equipment dropped from 1.95 to 1.63 during 2003–2005, based on the LBNL benchmarking study cited above [9]. Despite this, energy usage by the cooling equipment, which can often account for 20–50% of the total facility energy consumption continues to be a major concern.

Increasing energy usage in data centers has prompted interest in a variety of solutions targeted towards demand reduction through smart utilization, and use of alternative resources. In the former category are new power management features at the chip level, which may selectively switch off power to certain regions, when not needed. These are discussed further in Chap. 3. At the data center level, similar considerations can be made to the utilization of cooling equipment, which can be switched off or run at reduced capacity, based on IT demands. Such demand-based provisioning is further discussed in Chap. 4. Recent attention has also been focused on the reduction of energy usage through the utilization of air outside the data center for cooling. Depending upon geographical location and season, it may be possible to either bring in outside air directly into the data center (air economizers) or utilize an air-to-liquid heat exchanger for precooling the CRAC coolant (fluid economizers). With the air economizers, there are concerns about introducing particulate or gaseous contamination into the facility. Prototype implementations of both have

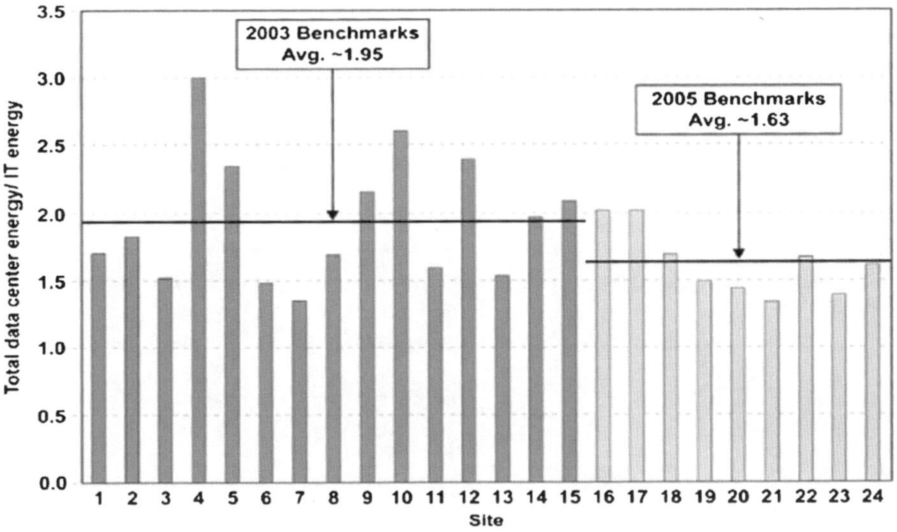


**Fig. 1.5** Benchmarking of ratio of total data center energy usage per unit of IT energy. Implementation of best practices presumably results in improved average benchmarks from 2003 to 2005 [9]. Figure courtesy W. Tschudi

been made. Using air economizers 91% of the time, with 100% air exchange at up to 32.2°C and with no humidity control and minimal air filtration, 67% estimated power savings were reported [10]. Using a fluid economizer to replace chillers part of the year improved the ratio of power used for IT equipment to power used for cooling by 85% [11]. Further discussion of economizers is presented in Chap. 2.

## 1.3 Data Center Energy Flow Chain

Figure 1.6 gives a typical power flow diagram for a data center [12]. The power from the utility grid, or in-house power generator feeds into the Switch Gear. Subsequently, it is divided amongst the IT equipment, the infrastructure facilities, and support systems. The IT power is supplied via the uninterrupted power supply (UPS). The UPS provides power to various IT racks, which contain servers and data storage units, and the networking systems. The UPS is also powered from energy storage media, such as a backup battery, in case of grid power failure. The switch gear also provides for the lighting systems inside the data center. On the facilities end, power lines are routed from the switch gear to the cooling tower, chiller, and chiller pumps.

In Fig. 1.6, the CRAC unit, which is part of the cooling infrastructure is shown powered via the UPS for damage mitigation during a possible power failure. In many facilities, only the IT loads may be connected to backup power sources, while the cooling infrastructure may be connected to the grid power. In such instances, in the event of a grid power failure, while the IT load moves to backup power,

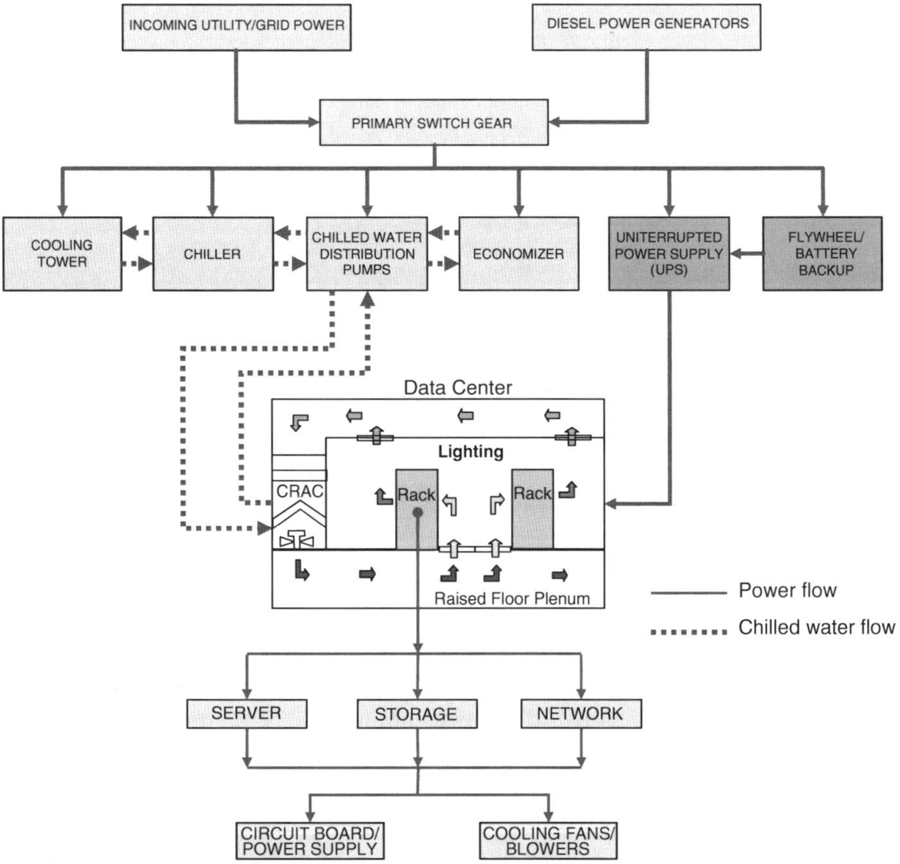


**Fig. 1.6** Power flow diagram in a typical data center. Adapted from [12]

the cooling equipment including the chillers and CRAC units would lag behind until it moves to the backup generator, with the associated start-up time. During this period of transition, while the IT equipment is generating heat, the cooling infrastructure is unavailable, with the possibility of significant air-temperature rise within the facility, which may result in adverse consequences within the facility, such as a deployment of sprinklers, beyond a certain temperature rise. We explore some of the possible transient scenarios in data center thermal management further in Sect. 1.8.

## 1.4 Operating Environment Guidelines for Data Centers

Guidelines for air-cooled data centers specifying dry-bulb air temperature and relative humidity levels at the inlets of the IT equipment have been a focus of the ASHRAE TC 9.9 Committee. These have a profound influence on the cooling

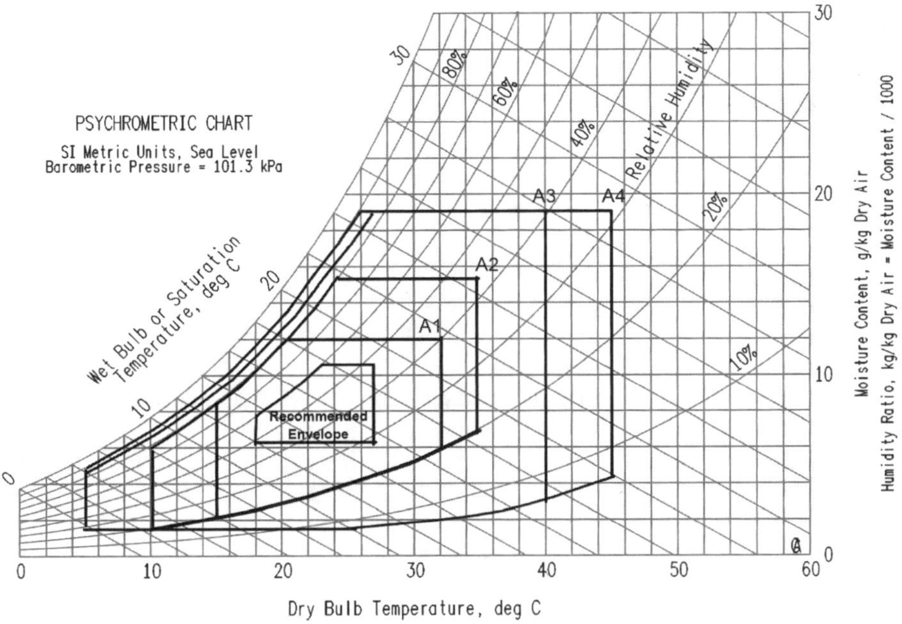


**Fig. 1.7** ASHRAE environment guidelines for the cooling air at the inlet of IT racks (© 2011 ASHRAE, reprinted with permission [14])

energy consumption by the data center, as discussed in Chap. 2. The original guidelines were developed in 2004 [13], and defined "recommended" and "acceptable" regions. The former narrower range was specified for continuous operation and was deemed safe for reliable operation of the IT equipment. The latter broader range was specified for short-term operation, while still insuring safe operation. Subsequent to the publishing of the original guidelines, energy efficiency improvements in data centers have taken on an ever-increasing importance. Since the environmental control constitutes a key part of the data center operating expenses, energy efficiency efforts have focused on reducing these requirements as much as possible, while insuring safe operation of the IT equipment.

In 2011, ASHRAE further revised the guidelines and enlarged both the "recommended" and "acceptable" regions of operation [14]. This expansion allows for reduced energy consumption for cooling as well as extension of conditions for which "free cooling" such as the use of economizers is possible. These regions are shown in Fig. 1.7 for class 1 and 2 data centers, requiring the highest operational reliability. The revised 2011 guidelines also introduced two new classes A3 and A4 to further expand the operating environmental envelopes. The basis for the introduction of new classes A3 and A4 is to provide a wider operating range for a limited class of IT equipment and not mandate all IT equipment to adhere to the wider operating range such as 40°C inlet temperature, which would substantially increase the cost of enterprise and volume servers. The new classes A3 and A4 have been defined to accommodate IT equipment housed in office or lab environment without stringent requirement of environment control. The operating range for classes A3 and A4 varies from 5°C to 45°C. The guidelines also specify appropriate derating factors for operation at higher altitudes. The deratings have also been suitably modified with the intension of keeping the Total Cost of Ownership (TCO) low.

The 2011 ASHRAE guidelines provide a recommended temperature range of 18–27°C (64.4–80.6°F) for dry-bulb temperature, as shown in Fig. 1.7. Since the air density decreases with altitude, the heat removal ability at a given volume rate decreases, and the recommended maximum temperature is derated by 1°C/300 m (1.8°F/984.25 ft) above 1,800 m (5,905.51 ft) for class 1 and 2 facilities. The moisture guidelines at the low end are 5.5°C dew point temperature and at the high end 15°C dew point temperature and 60% relative humidity. The upper moisture guidelines target reliability degradation due to printed wiring board failure mechanisms such as conductive anodic filament growth, which could short neighboring metal lines. High moisture can also cause damage to tape and disk drives. The lower limit extension, on the other hand, is motivated by the desire to extend the hours per year, where humidification and its associated energy costs (see Chap. 2) are not needed. The lower limit of moisture is set by the danger for electrostatic discharge (ESD), which can occur in very dry air.

As per the ASHRAE, the recommended operating range is defined for most reliable and acceptable operation, while achieving reasonable power efficiencies. The guidelines state that it is acceptable to operate outside the recommended envelope for short periods of time without affecting the overall reliability and operation of IT equipment. The recommended operating envelope is NOT the absolute operating environmental limits. Depending on the equipment class and the business and technical needs of the end user, the allowable envelope should be treated as the designed operating limits. Operating within the ASHRAE recommended limits does not assure that the data center will operate at maximum efficiency. This requires consideration of the optimum conditions for the equipment within each facility. Some of the equipment manufacturers in fact specify an even wider window of inlet air temperatures for safe operation of their equipment. Patterson [15] estimates the effect of higher inlet air temperatures in data centers and finds that beyond an optimum value, increasing these may in fact increase the overall power consumption due to increased power dissipation in many parts of the overall multi-scale system, for example from the chips and server fans. In this regard, the 2011 ASHRAE guidelines suggest that increasing the inlet air temperature from 15°C to 35°C could result in 7–20% increase in the IT equipment power consumption due to increased fan speeds. It is estimated that the air requirements of the server could increase 2.5 times over the 10–35°C temperature range. The corresponding increase in air requirement due to higher supply temperatures would ramp up the server fans' speed, thus increasing the data center acoustic noise levels. Empirically, it has been found that acoustic noise is proportional to the fifth power of fan speed. Data center managers should consider the implication of increasing the supply air temperatures on acoustic noise levels in the data center. Many countries have stringent laws on acceptable noise levels in work place.

pioneered the development of the mobile measurement technology (MMT) for sensing [30], which is discussed in Chap. 7. The need for centralized monitoring and archiving of energy usage and its optimization has been addressed through introduction of software tools such as Tivoli [31] and Maximo [32].

Intel has studied data centers from a total cost of ownership (TCO) perspective [33]. They conclude that higher density data centers are a better choice for lowering the total cost of ownership. Densities of workload equivalent to ~10,000 W/m$^2$ (1,000 W/ft$^2$) are achievable with high efficiencies. They also suggest that the design and construction phase of the data center should make greater use of CFD/HT tools to improve airflow management. Intel has also explored monitoring and controlling the various features of a data center in real time [34].

IT power reduction has been a key goal of microprocessor and server companies, such as Intel [35], IBM [36], and Fujitsu [37]. This is being achieved through a combination of chip architecture and packaging advances as well as demand-based thermal management. A number of companies including Microsoft [38], Google [39], and Oracle [40], are using modularized or container data centers, where the power demand for IT and cooling services can grow based on demand.

## 1.6 Multi-scale Thermal Management State-of-the-Art

Thermal management of data centers requires coordinated attention to the multiple length scales seen in Fig. 1.2. Zuo [41] introduced the term *advanced thermal architecture* to address this multi-scale nature of data center cooling. These need to be properly interfaced or coupled, to create the overall thermal management solution. Their systematic consideration increases design freedom, and enables the designer to achieve the highest performance, while being energy efficient. Also, it allows continuous implementation of thermal management advances to keep pace with the growing heat dissipation requirements at various length scales. Samadiani et al. [42] have introduced the concept of open thermal design of data centers to allow continuous scalability in data center designs, while optimizing energy usage. To illustrate the different scales and choices of thermal solutions discussed in their work, a blade server cabinet, shown in Fig. 1.8, in a representative raised-floor data center is considered.

### *1.6.1 Scale 1: Chip Level*

This scale includes different methods to enhance heat dissipation from the chip itself. For example, it includes the design of effective heat sinks and micro heat exchangers attached to the chip for single- or two-phase heat transfer. An example is seen in Fig. 1.9, which utilizes a stacked microchannel heat sink for greater chip temperature uniformity and reduced pressure drop [43].

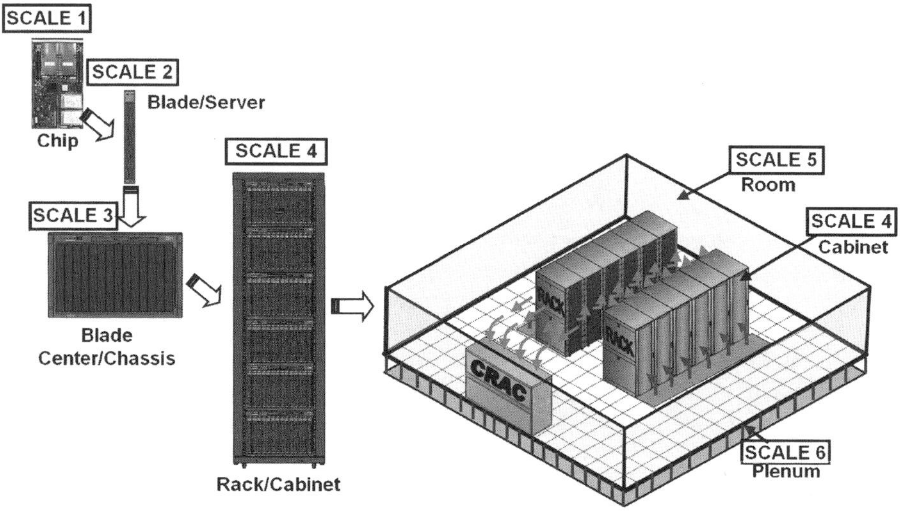


**Fig. 1.8** Multi-scale thermal management of data centers

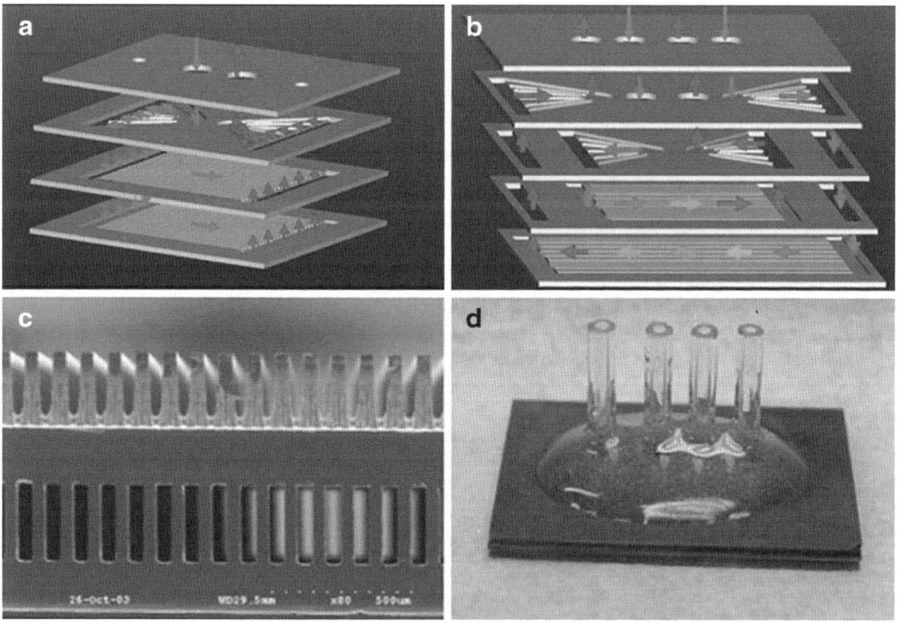


**Fig. 1.9** Chip-level thermal solutions. A chip-integrated stacked microchannel concept explored by Wei and Joshi [43] is shown above to achieve heat flux capability in excess of 200 W/cm$^2$, with moderate pressure drops. A two-layer parallel flow stack (**a**) and a counter-flow stack (**b**) are shown with the flow direction of the coolant. In (**c**) a bonded two-layer stack is shown, and in (**d**) a fully fabricated 1 cm $\times$ 1 cm device with coolant inlet and exit port as well as ports for measuring pressure drop is shown

**Fig. 1.10** Server level indirect cooling using a two-phase thermosyphon. The working fluid evaporates in an enclosure in intimate thermal contact with the processor chip. The vapor rises by natural convection within the central tubing and is condensed within the air-cooled condenser. It returns to the evaporator under gravity

Also, high conductivity thermal interface materials (TIM) development and innovative methods for chip attachment to the heat spreader may be considered at this scale. A review of high performance compact thermal management devices at this scale can be found in Wei and Joshi [44].

### *1.6.2 Scale 2: Server Level*

Solutions at this scale are integrated with the chip carrier or printed circuit board and include liquid cooling using cold plates, in combination with the chip scale solutions. The cold plate liquid circulation may utilize pumped liquid loops, or a buoyancy-driven thermosyphon, as seen in Fig. 1.10. As suggested by Gurrum et al. [45], the innovative solutions at this scale could include providing a two-way heat flow path from the chip, including both top and bottom. The bottom pathway would be facilitated by effectively directing heat from the chip to the substrate and finally to the ultimate ambient. The heat transfer through substrate/board can be done by, for example, using additional solder balls as thermal interconnects, and heat spreaders or substrate-integrated liquid cooling. Examples of thermal management solutions at this scale can be found in publications such as Electronics Cooling [46]. This is an area of continuing technology development. As an example, Wiliams and Roux [47] report on an air-cooled base plate with graphite foam for thermal enhancement.

### 1.6.3 Scale 3: Chassis Level

In current air cooling systems, this scale is used for fan installation and air-delivery pathway to the servers, and could play a role in applying various solutions to the cabinet, especially in combination with two previous scales. For example, this large space can be used by one or more macro heat exchangers, which can transfer the heat from the chips of the servers into the cold air flowed by CRAC units at the cabinet scale (Scale 4). Also, installing a plate at this scale can support components of a compact refrigeration system [48] to maintain sub-ambient chip operating temperature for improved performance.

### 1.6.4 Scale 4: Cabinet or Rack Level

Air cooling is the most popular solution at this scale, due to easy maintenance, low operational cost, and acceptable cooling efficiency. Several approaches currently employed for cabinet cooling are reviewed by ASHRAE [2]. Selected configurations are seen in Fig. 1.11.

Efforts have been made to improve airflow within cabinets [49], and to optimize placement of servers within the cabinet to take advantage of the prevailing airflows [50]. Herrlin [51] introduced the concept of Rack Cooling Index to quantify cooling effectiveness. Air cooling has typically been utilized for rack heat loads up to ~8 kW. At higher heat loads, it needs to be replaced or augmented by other techniques, such as single phase, or phase change liquid cooling, or refrigeration. Liquids, primarily because of their higher density and specific heat, are much more effective in the removal of heat than air. Figure 1.12a shows a typical implementation of a cooling loop that utilizes a liquid-to-chilled water heat exchanger internal to the rack. Typically, the liquid circulating within the rack is maintained above dew point temperature to avoid condensation of ambient moisture. Figure 1.12b depicts a design where the primary liquid loop components are housed outside the rack to permit more space within the rack for electronic components. A detailed discussion of liquid-cooled cabinets is taken up in Chaps. 12 and 13.

A combination of air and liquid cooling, hybrid cooling, is increasingly utilized in a variety of applications. Figure 1.13a shows a liquid loop internal to the rack, where the exchange of heat with the room occurs via a liquid-to-air heat exchanger.

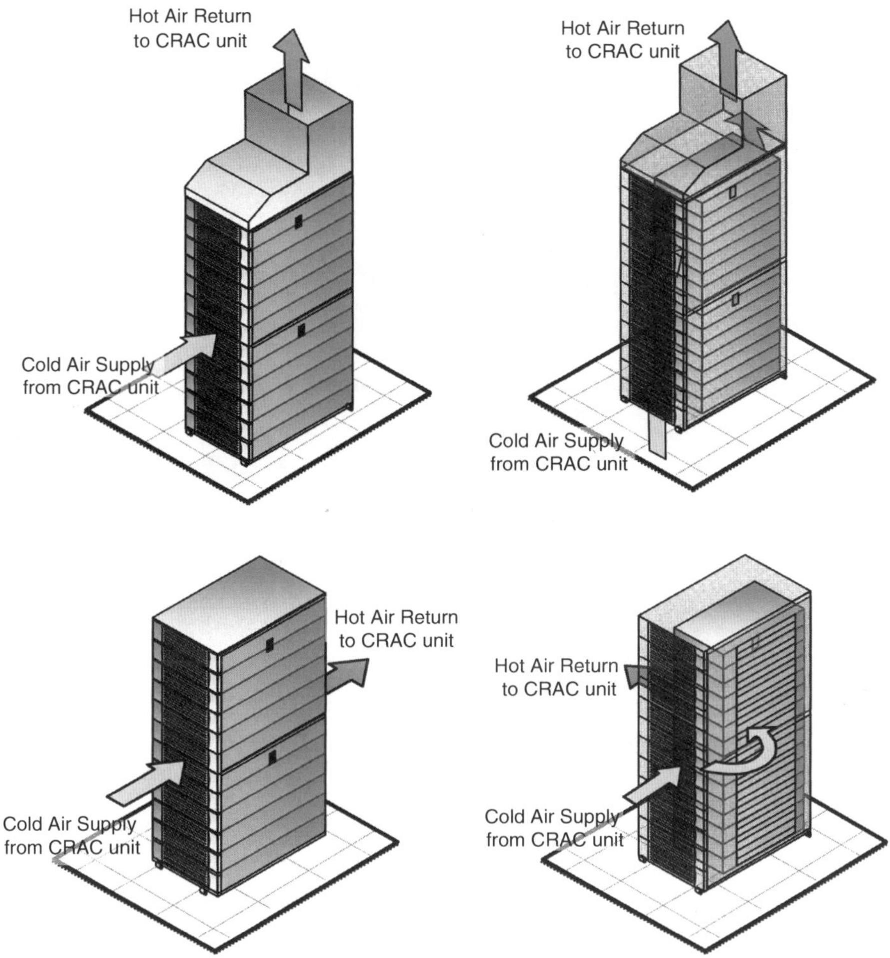


**Fig. 1.11** Rack-level thermal solutions; air-cooled cabinets

The heat generated in the electronics is removed by enclosed air which circulates inside the rack. This cooling configuration is widely used in shipboard cabinets, where the electronic components need to be isolated from the ambient environment. Figure 1.13b shows a schematic of a hybrid air and thermoelectrically cooled cabinet. The air removes the heat from the electronics, rejecting it to the thermoelectric (TEC) modules mounted to the sidewall. The air inside the plenum flows across the hot side of the TEC, driven by the exhaust fan at the outlet of the plenum to remove the heat from the TEC modules. Other examples of hybrid cooling at this scale include rear door water-cooled heat exchanger [29] to cool the cabinet hot air before discharge to the hot aisle of a data center. Also, an air-water or air-refrigerant heat exchanger can be installed on the top or sides of the cabinet.

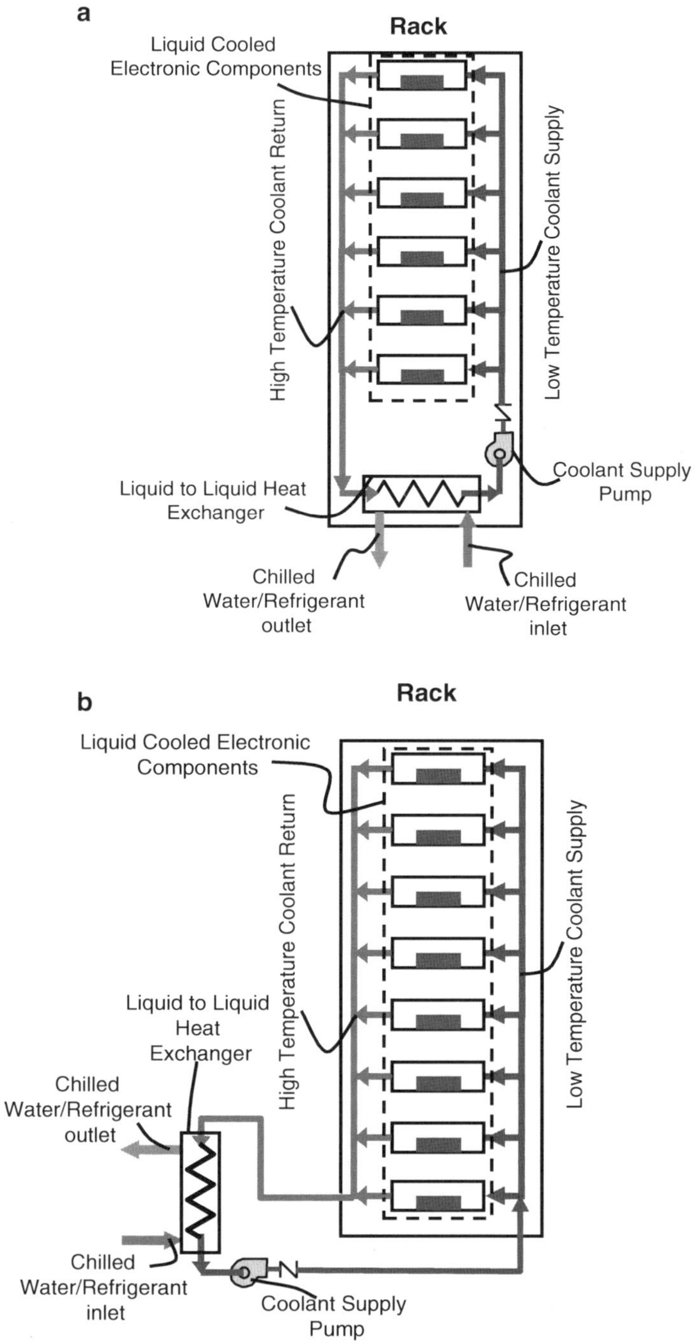


**Fig. 1.12** Rack-level thermal solutions; liquid-cooled cabinets

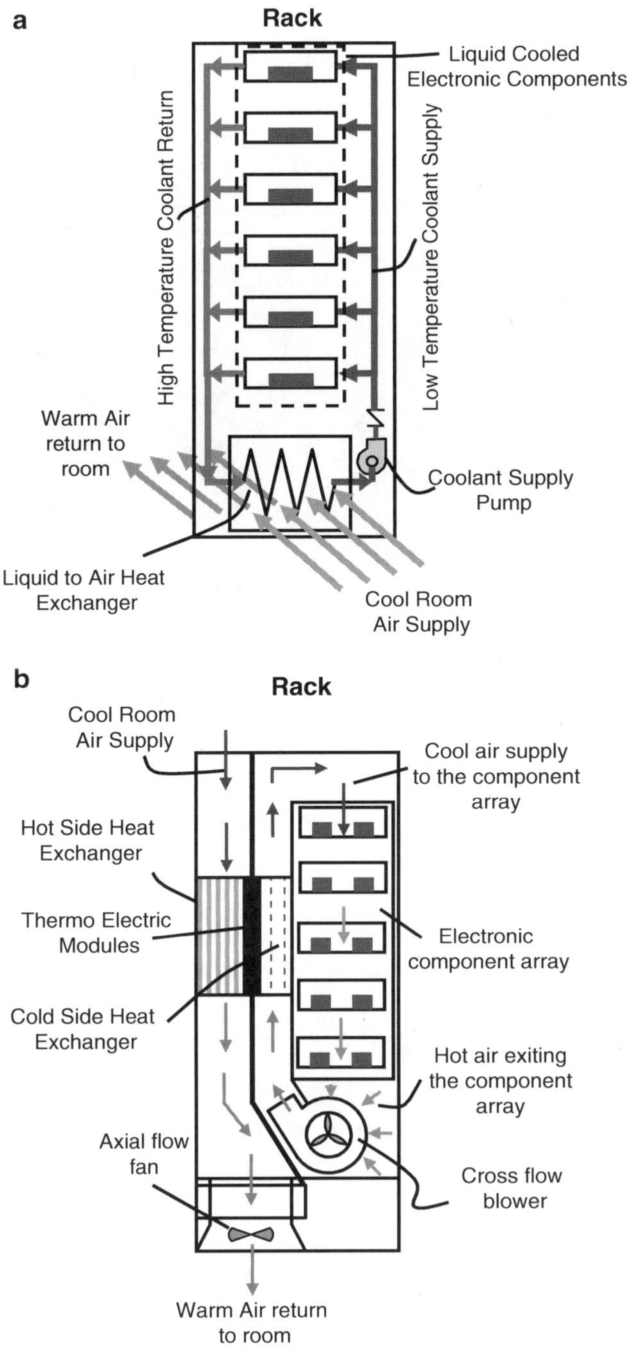


**Fig. 1.13** Rack-level thermal solutions; hybrid cooling approaches

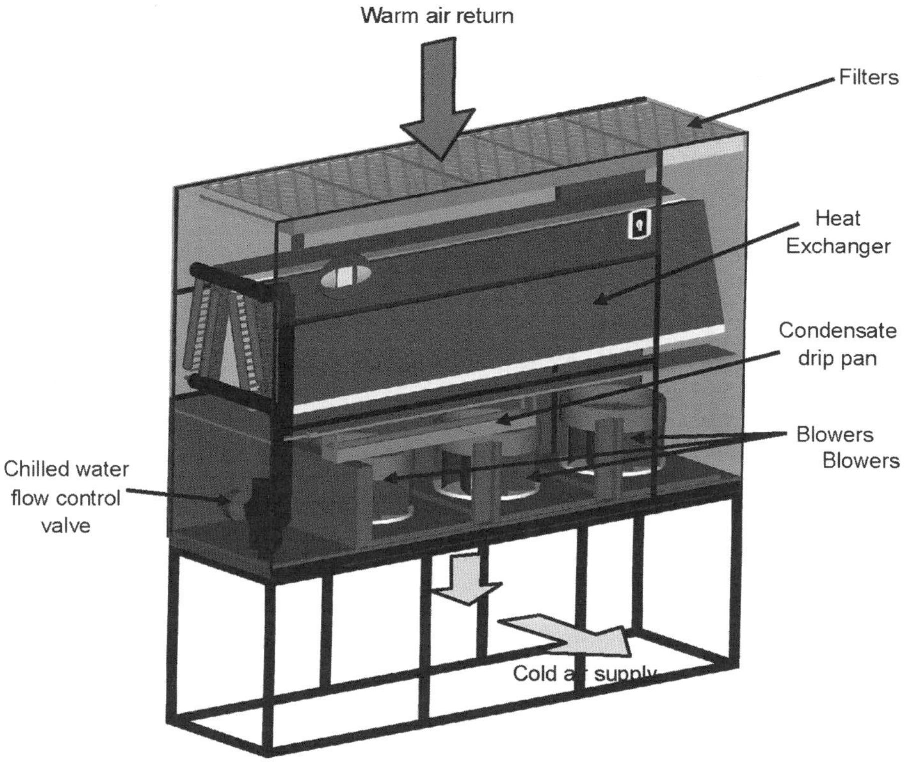


**Fig. 1.14** Schematic of CRAC unit [14]

## *1.6.5 Scale 5: Room Level*

Currently, for a majority of data centers, room-level cooling is achieved through computer room air-conditioning (CRAC or AC) units that deliver cold air to the racks through perforated tiles placed over a raised-floor plenum (RFP). A typical CRAC unit is shown in Fig. 1.14.

In a commonly used room-level cooling configuration seen in Fig. 1.15, perforated tiles are located to deliver the cool supply air to the front of the racks. The hot exhaust air from the rear of the racks is collected and returned from the upper portion of the facility by the CRAC units, completing the airflow loop. Racks are typically arranged in rows with alternating "hot" and "cold" aisles. The perforated tiles are located in the cold aisles and the exhaust air is collected in the hot aisles with solid tiles. This hot aisle–cold aisle (HACA) approach attempts to separate the cold supply from the hot exhaust air, and thus increase the overall efficiency of the air delivery and collection from each rack in the data center.

Several alternate air delivery and return configurations are employed, particularly when a raised floor arrangement is unavailable. Only certain combinations of

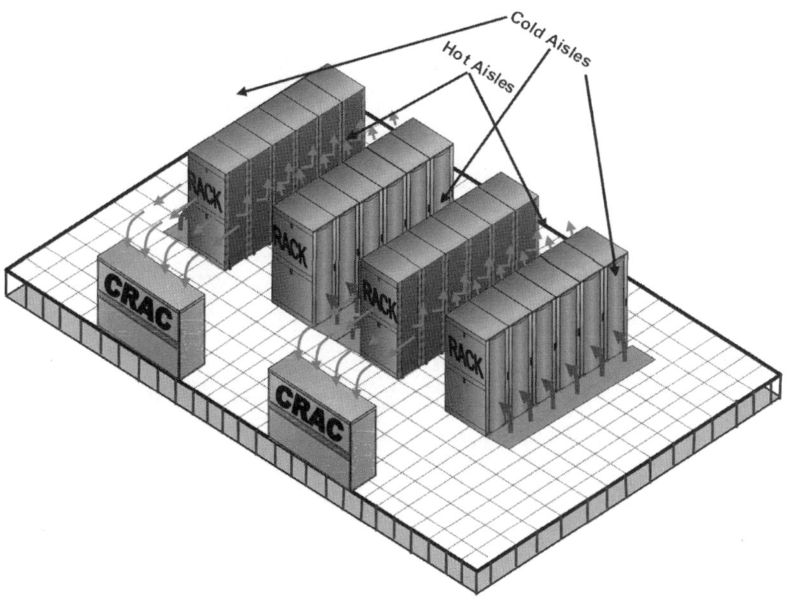


**Fig. 1.15** The hot-aisle/cold-aisle (HACA) layout for room level-cooling of a raised floor plenum data center

CRAC unit supply and return are typically feasible due to mechanical constraints of the CRAC units, without introducing an excessive amount of ductwork. Some of these are seen in Fig. 1.16 [52].

The room scale has been the focus of efforts to enhance the effectiveness of the HACA layout and prevent recirculation. Proposed solutions for avoiding the mixing of cold supply air and hot return air include hot-aisle containment as illustrated in Fig. 1.17a. An alternative or companion concept is cold-aisle containment, as seen in Fig. 1.17b, and computationally explored by Gondipalli et al. [53].

Additional concepts include directed hot-air extraction above the rack, Fig. 1.18, or from the hot aisle, Fig. 1.19. Airflow delivery advancements can be targeted at the room, row, and rack levels, as seen in Fig. 1.20 [54]. Coupled consideration of rack and room airflows to optimize cooling has been considered [55, 56]. Alternative layouts to the HACA to improve air delivery have also been studied, as the pod layout [57], which is described in detail in Chap. 13.

### *1.6.6 Scale 6: Plenum Level*

This scale includes air-delivery plenums, typically below or above the data center room space. Their design includes consideration of flow through the plenum and perforated tiles, which is impacted by the plenum depth, under-floor partitions, and

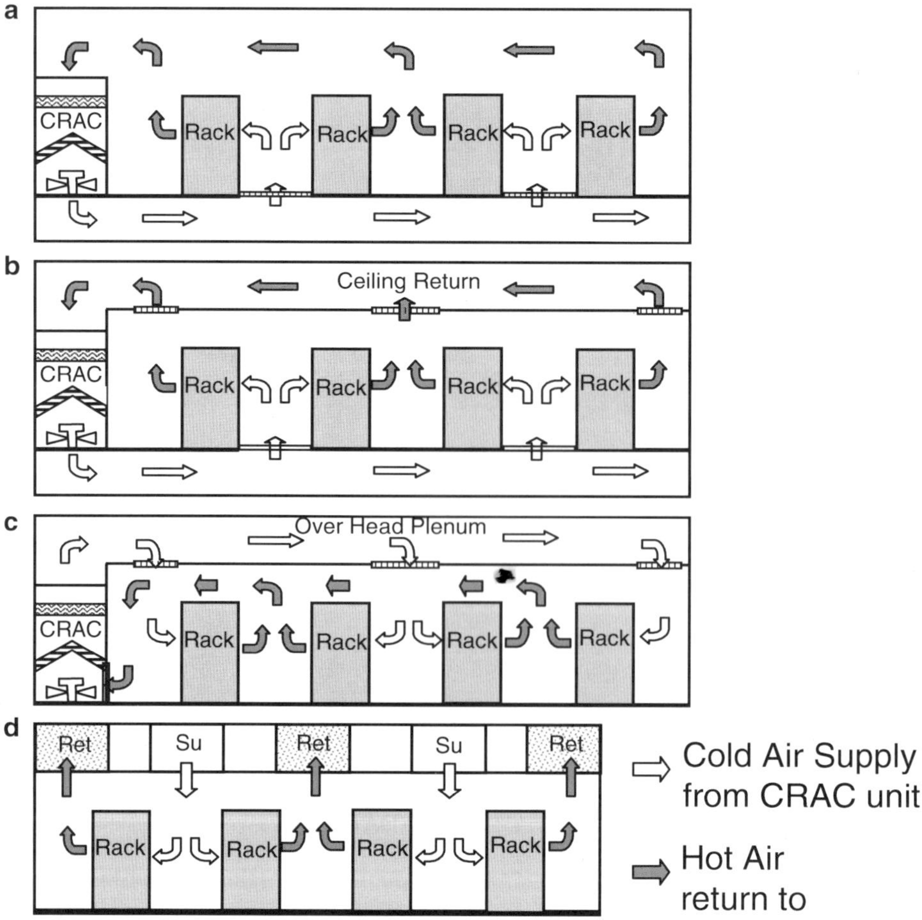


**Fig. 1.16** Other than the conventional raised floor hot-aisle/cold-aisle arrangement (**a**), several other arrangements are employed [52]. Hot air return to the CRAC may be through a ducted passage (**b**). In several existing facilities it may not be possible to provide for a raised floor. Distributed overhead cooling air supply units can be used to create an alternating hot- and cold-aisle arrangements (**c**). Alternately, wall-mounted CRAC units may provide ducted overhead supply of cold air (**d**)

tile specifications. Moreover, because chilled water pipes for the CRACs and any liquid-cooled cabinets as well as electrical cabling often pass through the plenum, it plays an increasingly significant role in cooling future data centers. The effective use of this scale, in combination with the previous ones provides multiple options in configuring the overall multi-scale cooling systems of data centers.

Data center facility infrastructure is typically designed and constructed for a period of 15–20 years, while the IT equipment is more frequently upgraded, sometime in as little as 2 years. In the early years of operation, a data center may

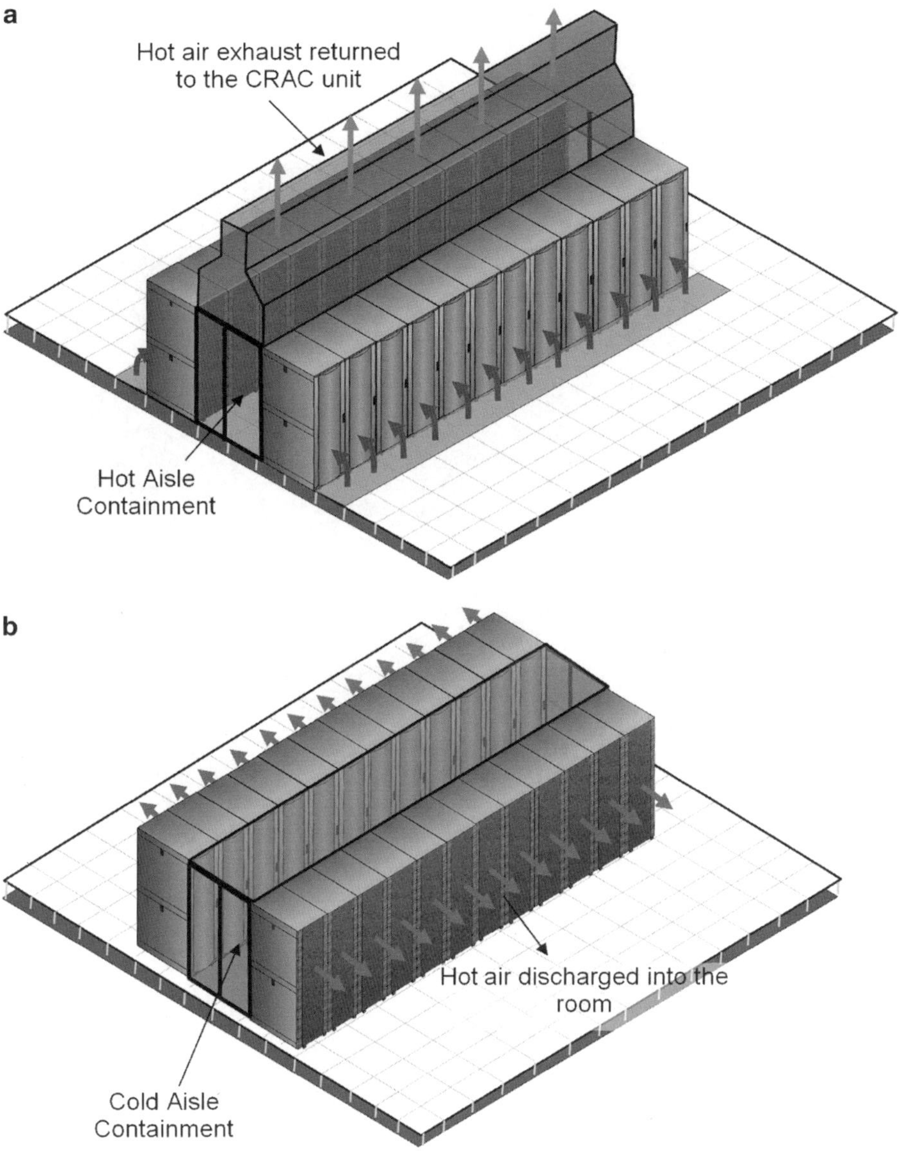


**Fig. 1.17** Recent room-level air cooling advances: (**a**) hot-aisle containment, (**b**) cold-aisle containment

often be underutilized, or over-provisioned, in terms of, space, power, and cooling resources. Through the use of the multi-scale design approach discussed above, it is possible to plan ahead for the systematic growth of thermal management capabilities for the data center facility for its entire design life, while optimizing the usage of energy. This is demonstrated in Chap. 10.

**Fig. 1.18** Recent room-level air cooling advances; over the rack heat collection (reprinted with permission from Emerson Network Power)

## 1.7 Thermal Analysis Approaches for Data Centers

Physical access to operating data centers for the purpose of thermal characterization is usually limited due to security and high reliability constraints. Also, large variations in data center architectures limit the extendibility of measurements across facilities. As a result, the most common characterization approach utilizes computational fluid dynamics and heat transfer (CFD/HT) to predict airflow and

**Fig. 1.19** Recent room-level air cooling advances; hot aisle air extraction (courtesy: www.energyscopethai.com)

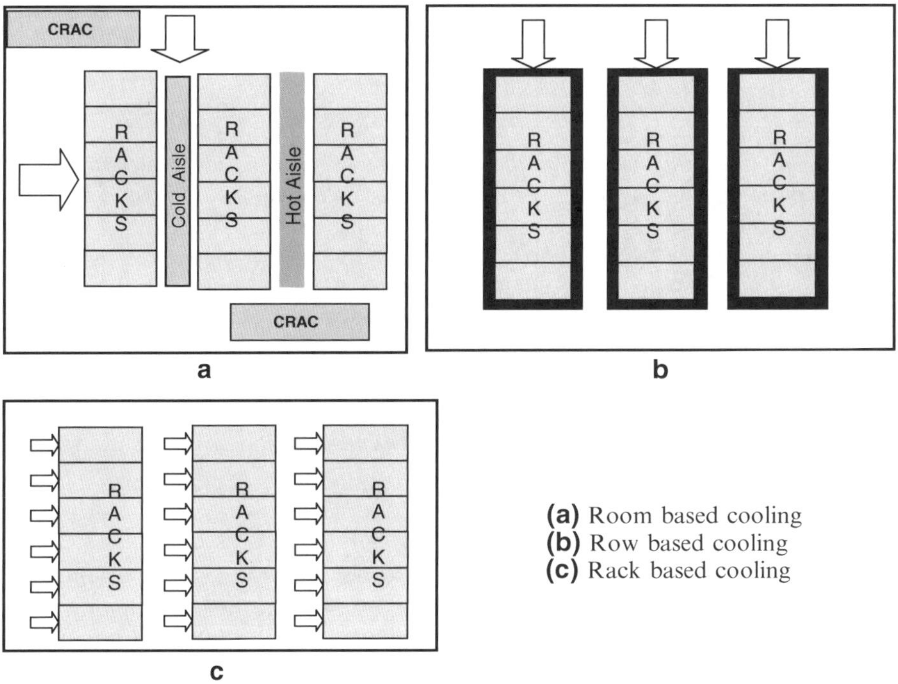


**Fig. 1.20** Recent air cooling advances; various options: (**a**) room-based cooling, (**b**) row-based cooling, (**c**) rack-based cooling

heat transfer. These simulations can enable identification of potentially dangerous local hot spots and provide rapid evaluation of cooling alternatives, as the facility IT loads or equipment change. With continual upgrades, the optimal arrangement of new heterogeneous IT equipment needs to be determined to minimize its effect on neighboring equipment. Additional constraints imposed by the grouping of IT equipment by functionality and cabling requirements often conflict with thermal management strategies, and data center managers need to provide localized supplemental cooling to high-power racks. Thermal simulations are the best way to determine these requirements.

The first published results for data center airflow modeling appeared in 2000 [58]. The various CFD/HT modeling efforts since then have ranged from individual component modeling to rack and power layouts and can be classified into the following main categories [52]:

1. Raised floor plenum (RFP) airflow modeling to predict perforated tile flow rates
2. Thermal implications of CRAC and rack layout and power distribution
3. Alternative airflow supply and return schemes
4. Energy efficiency and thermal performance metrics
5. Rack-level thermal analysis
6. Data center dynamics: control and lifecycle analysis

A detailed discussion of data center CFD/HT modeling is presented in Chap. 8.

### *1.7.1 Reduced Order or Compact Models*

CFD/HT models of data centers, while providing very detailed information on flow and temperatures within data centers, are also highly computationally intensive. Compact models often offer an acceptable tradeoff between modeling details and computational expense. The term "compact model" has received wide use in the thermal modeling literature, without a consensus on the definition. Here, a "compact model" is defined as one that uses a number of internal states to produce predefined output data, given a prescribed set of inputs and parameters [52]. The use of internal states is the key differentiating factor between compact and 'lumped' models. The latter are input–output functions obtained from physical conservation laws, or curve fits to experimental data. For example, fan and resistance models that specify a pressure–velocity relationship are common lumped models that compute the velocity through a surface as a function of the local upstream and downstream pressure.

Compact models in the form of thermal resistance networks have been widely used in thermal design of electronics packaging [59, 60]. The model accuracy in these strongly depends on how the resistance network is constructed and how multi-dimensional effects are approximated with equivalent one-dimensional resistances.

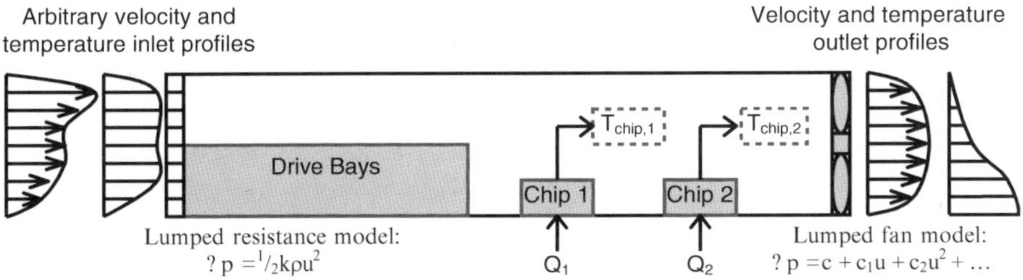


**Fig. 1.21** Server thermal model sketch for compact model development [52]

Compact models use slightly more degrees of freedom (DOF) than lumped models, but provide internal states based on additional details that allow for further examination of the model. Figure 1.21 shows the compact model development procedure for a single server. The main features of the model include some flow obstructions created by drive bays and power supplies that provide a representative system pressure drop. A number of power dissipating components are also modeled with arbitrarily complex details including multiple materials and chip-level thermal management devices such as heat sinks. The model includes an induced-draft fan to pull the air through the server box and may also include a lumped resistance to simulate the pressure drop through screens at the front of the server. The model thermal inputs are inlet temperature and component-wise power dissipation rates. The flow inputs may vary depending on the scope of the compact model, with the simplest being a fixed velocity that also fixes the outlet velocity by continuity. More advanced strategies would use a fan model to drive the flow against the server system resistance accounting for the inlet and outlet static pressures. Since the ultimate goal is for the compact model to be integrated into the full CFD/HT computation of data center facility, detailed velocity, pressure, and temperature profiles are available from the neighboring grid cells in the computational mesh. This means that the compact models can be formulated to accept and output either profiles or area-averaged quantities. The model parameters are the geometry and material properties, which may range over specified values, if the model is constructed to accommodate such variability.

The process of taking a model from a large number of degrees of freedom (DOF), either from detailed numerical simulations or full-field experimental measurements, to a model involving significantly fewer DOF is termed *model reduction* [61]. Figure 1.22 illustrates this taxonomy of efficient modeling procedures by comparing the level of description and model size, measured in DOF.

Figure 1.23 presents a summary of the various types of thermal modeling activities undertaken for data centers [52]. The three main thrusts are evaluation of global cooling schemes, investigating the effects of local and supplemental cooling schemes, and improving energy efficiency at all levels of the data center.

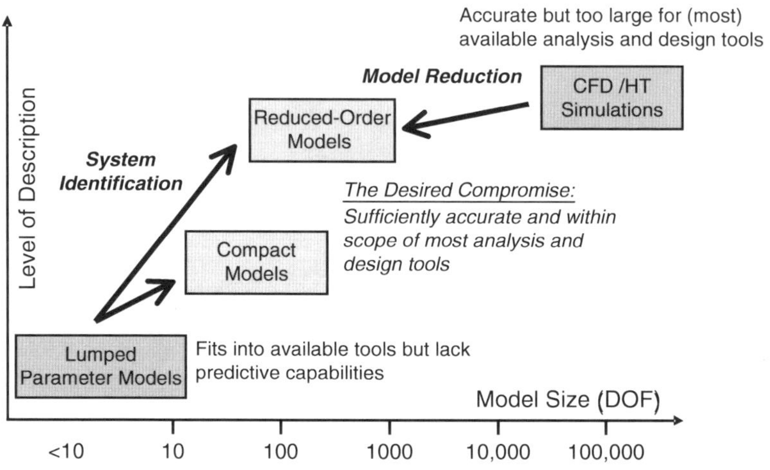


**Fig. 1.22** Modeling description level and DOF taxonomy

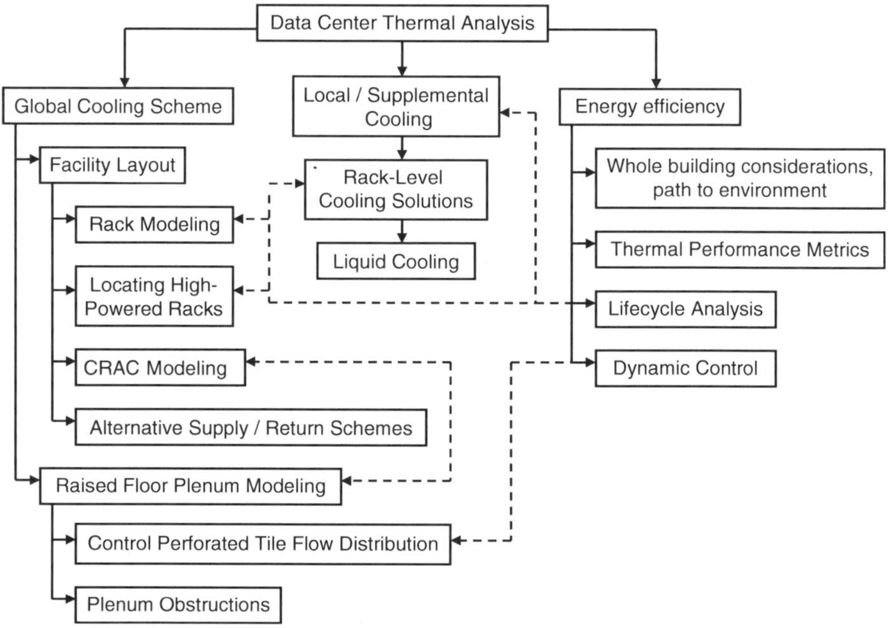


**Fig. 1.23** Data center thermal management organizational chart [52]

## 1.8 Power Failures, Their Consequences, and Mitigation Strategies

Any downtime in the operation of data centers incurs expense to the user, and typically the equipment owner, if a different party, as in colocation data centers. Under the service level agreement (SLA), the owner may have to pay, or forfeit being paid by, the user under such conditions. A company computer may lose information currently being processed if it goes down. A data center controlling an industrial process could extend the process downtime caused by a power failure as the data center is being restarted, and unavailability of Web sites may cause missed opportunity for sales. In general, an IT cluster that loses power unexpectedly may have a complex restart requiring expert IT support. The US EPA cites that 90% of data centers will be affected by a power failure within a 1-year period [62], with outages during peak demand period costing as much as $30 M/min [63]. While some of these power failures are too short in duration to affect facility operation, their percentage is still very high in terms of actual number of impacted facilities. A failure at a Dallas Rackspace data center illustrated that even the best prepared facility can stop operating, given a series of events that repeatedly caused infrastructure to fail [64].

Shields [65] considered a number of failure events in a raised floor air-cooled facility of the type illustrated in Fig. 1.15. Heat flows from the IT equipment to the data center air, and the energy convected by the air is released to the circulating chilled water via an air-to-water heat exchanger within the CRAC, such as in Fig. 1.14. The chiller, located outside the data center, then removes the heat from the warm return water from the CRAC, through coupling with a refrigerant-based vapor compression system. There are various levels of backup power infrastructure that are designed to maintain data center operation during a power failure.

A classification of mission critical facilities developed by the Uptime Institute [66] divides these into categories called Tiers. This classification requires evaluation of other factors besides power and cooling infrastructure, as discussed in detail in Chap. 6. The backup power infrastructure provided, or not provided, to each level of cooling infrastructure determines the most likely transient scenario during a power failure. Several possible scenarios are shown in Fig. 1.24 and described below.

Some data centers do not have any backup power or provide backup power within the rack for the head nodes only, just enough to minimize the amount of labor necessary to bring the IT equipment back online after the power is restored. In this case, cooling of IT equipment is generally not an issue, since the power density of data center drops drastically during the power failure. Such a data center falls in the category of Tier 1.

### *1.8.1 Uninterrupted Power Supply for Compute Equipment Only*

The lowest level of backup power infrastructure would be to provide UPS for IT equipment only. This means that the IT equipment would continue to process data

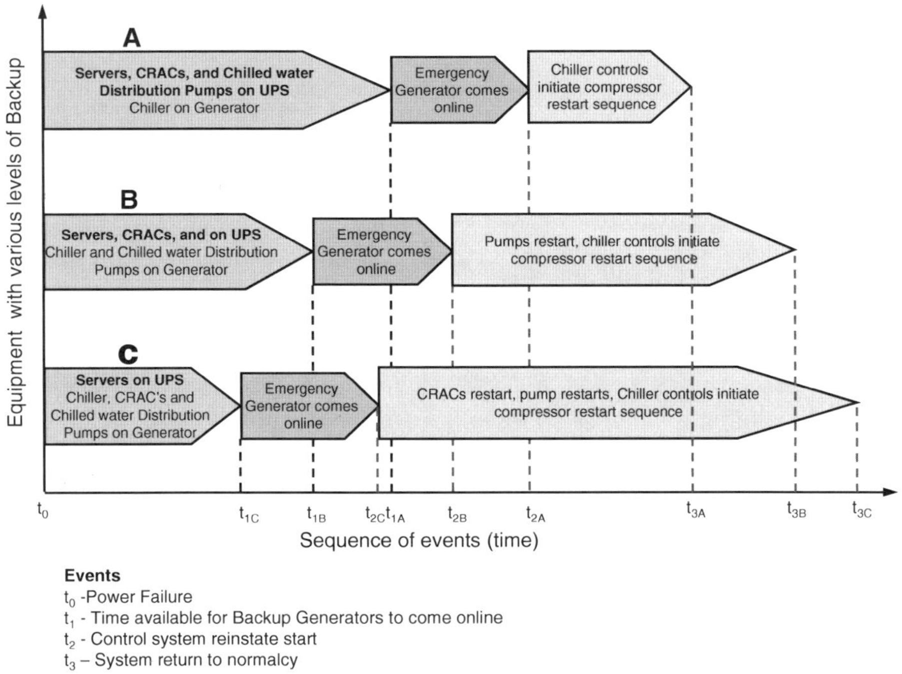


**Fig. 1.24** Power failure scenarios

and also dissipate heat. Running IT equipment without cooling infrastructure is a cause for concern. This design may cause IT equipment air inlet temperatures to rise unacceptably if the power failure disrupts cooling for an extended period, and the problem is compounded as power density increases. It will likely be necessary for this type of data center to shutdown IT equipment automatically or manually in case of an extended power failure. In some cases, a data center with lower power density may be able to continue operation without downtime if an emergency generator is also installed.

### 1.8.2 *Emergency Generator provisioning for Cooling Infrastructure*

The addition of an emergency generator to the backup power infrastructure shortens the maximum theoretical time span of a power failure for any equipment that it supplies. Adding power infrastructure to the emergency generator tends to be substantially less expensive than providing it with UPS. Thus, chillers—the cooling infrastructure that tends to draw the largest amount of power—generally are not provided with UPS, but may receive emergency generator backup power infrastructure. Typically, an emergency generator can restore power within about 30 s after a power failure. At this point, the CRACs and chilled water (CHW) pump with emergency generator backup power infrastructure would start almost immediately, even before the chillers restart. When the CRAC and CHW pump begin working, heat is transferred from the data center air to the CHW, which is still at approximately the chiller evaporator setpoint.

Once the CHW makes a complete circuit around the CHW loop, the CRAC will have a reduced cooling capacity due to lower log mean temperature difference (LMTD) across the HX. With CHW storage, the temperature rise in the CHW loop can be extended proportionally to the amount of CHW stored. Chillers typically require 2–3 min for a restart sequence for each compressor. Thus, a facility provided with emergency generator backup power infrastructure for its IT equipment and cooling infrastructure needs to determine what additional steps are needed to ensure the facility does not reach unacceptable temperatures before the chiller has a chance to come back online.

### *1.8.3 UPS for CRAC Fans*

Without any UPS provided for the CRAC, there is necessarily a delay in air circulation within the data center. With higher power density compute equipment, this delay can lead to recirculation of hot air from IT equipment outlets to inlets. An improvement to the backup power infrastructure, after providing UPS to IT equipment and emergency generator backup power infrastructure to all cooling infrastructure, is to provide UPS for the CRAC. An expected benefit from running the CRAC during the power failure is circulation of the air within the data center, which will maintain the airflow patterns that pressurize the cold aisle and minimize recirculation and hot spots. Another benefit of running the CRAC is the ability of heat transfer to other media besides the room air. This includes the cooling coil within the CRAC and the solid surfaces in the facility, such as the concrete floor, raised floor plenum tiles, walls, and ceiling. Any extension of the time window within acceptable operating temperatures gives the facility more time for power to be restored, either by emergency generator or by the utility service, and cooling infrastructure to come back online.

### *1.8.4 UPS for Chilled Water Pumps and Air Handler Fans*

In many cases, providing backup power infrastructure—especially UPS—to the CRAC may be enough to keep temperatures within acceptable ranges long enough for the emergency generator and chillers to come back online. However, some owners may desire a greater degree of reliability. If power density is considered high enough within the data center that there is concern that server inlet air temperatures may rise too rapidly, or if a greater degree of reliability is desired, the CHW pump can also be placed on UPS backup power infrastructure. Under this scenario, the data center will continue operating at steady state, with no temperature rise, until all the CHW in the piping has made a complete circuit through the AH. With sufficient chilled water storage, the facility can operate with no rise in air temperature for an extended period. This would allow for false starts of both the generator and chiller.

As discussed in Chap. 6, the highest tier ratings for data centers are based not only on backup power infrastructure, or CHW storage, but also on redundancy. Redundancy allows pieces of equipment or systems to fail without disrupting operation of the data center.

## 1.9 Chapter Summary and Book Outline

This chapter has provided an introduction to the energy and thermal management of data centers.

Trends such as sharp increases in chip and cabinet-level power density are discussed, along with computing paradigms such as virtualization and cloud computing. Early energy usage benchmarking studies for data centers, and scenarios for their sustainable growth are identified. The energy flow chain from the grid to the data center facility, which is key to reducing energy usage is introduced. Environmental guidelines for the operation of data center facilities, which dictate the cooling requirements for data centers are discussed. During the past decade, significant advances in energy usage have been achieved by the industry. Best practices based on these experiences are briefly discussed. Thermal management of data centers requires a multi-scale approach. This is illustrated through the adoption of an open approach that allows continued growth of a facility. In order to optimize thermal management, it is essential to have efficient and accurate thermal modeling approaches. An introduction to reduced order modeling is provided. Finally, the dynamic nature of the data center operation is explored through facility provisioning for mitigation of power failures. The following 12 chapters explore many of these ideas in greater depth.

Chapter 2 explores the current state-of-the-art of the mechanical and thermal design of data centers, with a focus on airflow management. In Chap. 3 we explore the important topic of power delivery and losses from the grid to the individual chips that perform the IT operations of the facility. Chapter 4 focuses on emerging trends in the IT component of the overall power consumption. We discuss in detail how emerging usage models of computing are profoundly impacting. Monitoring of facility and IT data and its use in control of data centers is discussed in Chap. 5.

The path to sustainable data center growth requires quantifying and improving the meaningful metrics for energy efficiency. Due to the multitude of energy flow and loss streams, this requires careful consideration, as explained in Chap. 6. Significant advances have been made in sensing of data centers to improve their performance and reduce energy consumption. The use of sensing in concert with modeling offers great promise in improving energy efficiency of data centers. Chapter 7 focuses on metrology tools and their use in data centers, in concert with modeling. In Chap. 8, the significant progress made over the past decade in the thermal and fluid flow modeling in data center spaces, and the design insights obtained through these simulations are discussed. Exergy analysis provides a promising avenue to identify and reduce the thermodynamic inefficiencies in data center facilities. This is the focus of Chap. 9. Computational simulations for design and optimization of data centers must be able to provide rapid responses. Reduced order modeling to achieve this, and its use in improved designs of data centers are discussed in Chap. 10. Handling the large amount of data gathered from data center facilities and using it in a meaningful fashion requires the use of statistical techniques is the focus of Chap. 11. The entire incoming electrical energy into a data center facility is ultimately converted into waste heat. While this waste heat is at relatively low temperatures of 85°C or less, much potential for its possible use exists. Recent work on this topic is presented in Chap. 12. Cooling technology advances impacting energy efficiency in data centers are being rapidly explored worldwide. In Chap. 13, some of these are explored to provide the reader a comprehensive understanding of the challenges and opportunities associated with the explosive growth in Internet and telecom facilities.

## References

1. ASHRAE Datacom Series http://www.ashrae.org/publications/publications_home, February 19, 2011
2. Datacom Equipment Power Trends and Cooling Applications (2005) American Society of Heating Refrigeratingand Air-Conditioning Engineers (ASHRAE), TC 9.9 Committee, Atlanta, GA
3. International Technology Roadmap for Semiconductors http://www.itrs.net/, February 19, 2011
4. Sauciuc I, Prasher R, Chang J-Y, Erturk H, Chrysler G, Chiu C-P, Mahajan R (2005) Thermal performance and key challenges for future cpu cooling technologies. InterPack, San Francisco, CA
5. Patterson MK, Fenwick D (2008) The state of data center cooling: a review of current air and liquid cooling solutions. White Paper, Digital Enterprise Group, Intel Corporation, March 2008
6. Brocade Communications Systems Inc. (2010) Data center industry trends and vision: evolution toward data center virtualization and private clouds, Technical Brief
7. United States Environmental Protection Agency Energy Star Program (2007) Report to congress on server and data center energy efficiency public law 109–431. 2 Aug 2007
8. Mitchell-Jackson J (2001) Energy needs in an internet economy: a closer look at data centers. Masters thesis, University of California, Berkeley
9. Tschudi W, Xu T, Sartor D, Stein J (2003) High performance data centers: a research roadmap. US Department of Energy Lawrence Berkeley National Laboratories
10. Atwood D, Miner JG (2008) Reducing data center cost with an air economizer. Intel Brief Intel Information Technology, August

11. Garday D (2007) Reducing data center energy consumption with wet side economizers. White Paper, Intel Information Technology, Intel Corporation, May
12. Somani A (2008) Advanced thermal management strategies for energy efficient data centers. Master's thesis, Georgia Institute of Technology, December
13. ASHRAE Publication (2004) Thermal guidelines for data centers and other data processing environments, Atlanta, American Society for Heating, Refrigeration and Air Conditioning, Atlanta, Georgia
14. ASHRAE (2011) Thermal guidelines for data processing environments – expanded data center classes and usage guidance, American Society for Heating, Refrigeration and Air Conditioning, Atlanta, Georgia
15. Patterson MK (2008) The effect of data center temperature on energy efficiency. Itherm conference, Orlando, Florida, 28 May to 1 June 2008
16. High Density Data Centers – Case Studies and Best Practices (2008) ASHRAE, Technical Committee 9.9
17. Best Practices for Datacom Facility Energy Efficiency (2009) 2nd edn. ASHRAE, Technical Committee 9.9
18. Sullivan (RF) (2002) Alternating cold and hot aisles provides more reliable cooling for server farms. White Paper, The Uptime Institute, Santa Fe
19. Guidelines for Energy-Efficient Data Centers (2007) The Green Grid. Beaverton
20. Data Center World. http://www.datacenterworld.com/, February 19, 2011
21. 7×24 Exchange. http://www.7x24exchange.org/, February 19, 2011
22. Data Center Dynamics. http://www.datacenterdynamics.com/, February 19, 2011
23. Data Center Alliance. http://www.the-data-center-alliance.com/, February 19, 2011
24. Sharma RK, Bash CE, Patel CD, Friedrich RJ, Chase JS (2003) Balance of Power - Dynamic Thermal Management for Internet Data Centers. Hewlett-Packard White Paper. HPL-2003-5
25. Moore J, Chase J, Farkas K, Ranganathan P (2005) Data center workload monitoring, analysis, and emulation. Eighth workshop on computer architecture evaluation using commercial workloads, February
26. Moore J, Chase JS, Ranganathan P (2006) Weatherman: automated, online, and predictive thermal mapping and management for data centers. Dublin, Ireland: Institute of Electrical and Electronics Engineers Computer Society, Piscataway
27. CoolIT: Coordinating Facility and IT Management for Efficient Datacenters, R. Nathuji, A. Somani, K. Schwan, and Y. Joshi, HotPower Conference, 20009
28. Patterson MK, Meakins M, Nasont D, Pusuluri P, Tschudi W, Bell GC, Schmidt R, Schneebeli K, Brey T, McGraw M, Vinson W, Gloeckner J (2009) Energy-efficiency through the integration of information and communications technology management and facilities controls. Proceedings of the ASME 2009 InterPACK conference IPACK2009, San Francisco, 19–23 July 2009
29. Schmidt RR, Chu R, Ellsworth M, Iyengar M, Porter D, Kamath V, Lehman B (2005) Maintaining datacom rack inlet temperatures with water cooled heat exchanger. In: ASME InterPACK, 2005, San Fransisco, ASME, IPACK2005-73468
30. Hamann HF, Lacey JA, O'Boyle M, Schmidt RR, Iyengar M (2008) Rapid three-dimensional thermal characterization of large-scale computing facilities. In: IEEE transactions on components and packaging technologies, vol 31, no. 2, June 2008
31. IBM Greens the Data Center via Centralized Monitoring. http://www-01.ibm.com/software/tivoli/governance/action/01142010.html, February 19, 2011
32. Maximo Asset Management. http://www-01.ibm.com/software/tivoli/products/maximo-asset-mgmt/, February 19, 2011
33. Patterson MK, Costello DG, Grimm PF, Loeffler M (2007) Data Center TCO: A comparison of high-density and low-density spaces. Intel White Paper, January
34. Filani D, He J, Gao S, Rajappa M, Kumar A, Shah P, Nagappan R (2008) Dynamic data center power management: trends, issues and solutions. Intel Technol J 12(1):59–76
35. Minhas L, Ellison B (2009) Energy efficiency for information technology how to reduce power consumption in servers and data centers, Intel Press, 2009, Hillsboro, Oregon

36. Crippen MJ, Alo RK, Champion D, Clemo RM, Grosser CM, Gruendler NJ, Mansuria MS, Matteson JA, Miller MS, Trumbo BA (2005) BladeCenter packaging, power, and cooling. IBM J Res Dev 49(6)
37. Wei J, Suzuki M (2005) Thermal management of fujitsu high-end unix servers. In: Proceedings of InterPack 2005, San Francisco
38. Microsoft (2008) Best practices for energy efficiency in microsoft data center operations. Microsoft (Fact Sheet), February
39. Barroso LA, Hölzle U (2009) The datacenter as a computer. An introduction to the design of warehouse-scale machines. Synthesis lectures on computer architecture, Morgan & Claypool Publishers
40. Sun Modular Data Center (2011) http://www.sun.com/service/sunmd/, February 19, 2011
41. Zuo J, Hoover LR, Phillips AL (2002) An integrated thermal architecture for thermal management of high power electronics. In: Joshi Y, Garimella S (eds) Thermal challenges in next generation electronic systems. Millpress, Rotterdam
42. Samadiani E, Joshi Y, Mistree F (2007) The thermal design of a next generation data center: a conceptual exposition. Cairo, Egypt: Institute of Electrical and Electronics Engineers Computer Society, Piscataway
43. Wei XJ, Joshi Y (2007) Experimental and numerical study of a stacked microchannel heat sink for liquid cooling of microelectronic devices. ASME J Heat Transfer 129:1432–1444
44. Joshi Y, Wei X-J (2007) Micro and meso scale compact heat exchangers in electronics thermal management-review. Int J Heat Exchangers (Invited Paper), Special supplemental issue on advances in compact heat exchangers, pp 1–32
45. Gurrum SP, Suman SK, Joshi YK, Fedorov AG (2004) Thermal issues in next generation integrated circuits. IEEE Trans Device Mater Reliab 4:709–714
46. Electronics Cooling Magazine. www.electronics-cooling.com, February 19, 2011
47. Williams ZA, Roux JA (2006) Graphite foam thermal management of a high packing density array of power amplifiers. In: AIAA 2006–3609, 9th AIAA/ASME joint thermophysics and heat transfer conference, San Francisco, 5–8 June 2006
48. C.D. Coggins, D. Gerlach, Y. Joshi, and A. Federov. Compact Low Temperature Refrigeration of Microprocessors. in Proceedings of the International Refrigeration and Air Conditioning Conference at Purdue University. 2006, West Lafayette, Indiana
49. Mulay V, Agonafer D, Irwin G, Patell D (2009) Effective thermal management of data centers using efficient cabinet designs, InterPack
50. Rolander N, Rambo J, Joshi Y, Allen JK, Mistree F (2006) An approach for robust design of turbulent convective systems. ASME J Mech Des 128:844–855
51. Herrlin MK (2006), Rack cooling effectiveness in data centers a look at the mathematics, energy and power management. May, pp 20–21
52. Rambo J, Joshi Y (2007) Modeling of data center airflow and heat transfer: state of the art and future trends. Distributed and parallel database, special issue on high density data centers, vol 21, pp 193–225
53. Gondipalli S, Sammakia B, Bhopte S, Schmidt R, Iyengar M, Murray B (2009) Optimization of cold aisle isolation designs for a data center with roofs and doors using slits, InterPack
54. Dunlap K, Rasmussen N (2010) The advantages of row and rack oriented cooling architectures for data centers, White Paper 130, Revision 1, APC by Schneider Electric
55. Schmidt R, Iyengar M (2005) Effect of data center layout on rack inlet air temperatures. In: ASME, IPACK2005-73385, San Francisco, ASME InterPACK
56. Bhopte S, Agonafer D, Schmidt R, Sammakia B. Optimization of data center room layout to minimize rack inlet air temperature. In: ASME InterPACK, San Francisco, ASME, IPACK2005-73027
57. Somani A, Gupta T,Joshi Y (2008) Scalable pods based cabinet arrangement and air delivery for energy efficient data center. International forum on heat transfer, Tokyo, 17–19 September 2008

58. Kang S, Schmidt RR, Kelkar KM, Radmehr A, Patankar SV (2000) A methodology for the design of perforated tiles in a raised floor data center using computational flow analysis. Presented at ITHERM 2000 - Intersociety Conference on Thermal and Thermomechanical Phenomena in Electronic Systems, Las Vegas
59. Bar-Cohen A, Krueger W (1997) Thermal characterization of chip packages - evolutionary design of compact models. In: IEEE transactions on components, packaging, and manufacturing technology - part A, vol 20, pp 399–410
60. Lasance CJM (2004) Highlights from the European thermal project PROFIT. ASME J Electron Pack 126:565–570
61. Shapiro B (2002) creating reduced-order models for electronic systems: an overview and suggested use of existing model reduction and experimental system identification tools. In: Joshi YK, Garimella SV (eds) Thermal challenges in next generation electronics systems. Millpress, Rotterdam, pp 299–306
62. The Role of Distributed Generation and Combined Heat and Power (CHP) Systems in Data Centers (2007) U. S. Environmental Protection Agency Combined Heat and Power Partnership. http://www.epa.gov/chp/documents/datactr_whitepaper.pdf
63. The U.S. Environmental Protection Agency (2008) http://www.epa.gov/chp/documents/datacenter_fs.pdf, updated October 2008.
64. Golson J. How Rackspace Really Went Down. Vallywag. http://valleywag.com/tech/confirmed/how-rackspace-really-went-down-322828.php, February 19, 2011
65. Shields S (2009) Dynamic thermal response of the data center to cooling loss during facility power failure, M.S. thesis, Georgia Institute of Technology
66. Turner W, Seader J, Brill K. Industry standard tier classifications define site infrastructure performance, The Uptime Institute Inc. http://upsite.com/TUIpages/tuiwhite.html, February 19, 2011

# Chapter 2
# Fundamentals of Data Center Airflow Management*

**Pramod Kumar and Yogendra Joshi**

**Abstract** Airflow management is probably the most important aspect of data center thermal management. It is an intricate and challenging process, influenced by many factors. This chapter presents some of the fundamental concepts governing airflows in today's data centers. As such, it provides a foundation necessary for understanding the remaining topics discussed in the book. The chapter begins by introducing the concept of system pressure drop and its influence on the computer room air conditioning (CRAC) unit performance. Various factors contributing to the overall pressure drop, such as plenum design, perforated tile open area, and aisle layouts, are described. Some of the key aspects of room and rack airflows are also discussed. The second part of the chapter highlights the importance of temperature and humidity control in data centers. The basic concepts of psychrometrics are introduced. Specific examples on data center cooling processes, such as sensible cooling, humidification/dehumidification and evaporative cooling are illustrated with the help of the psychrometric chart. The concept of airside and waterside economizers for data center cooling are introduced. The third and final part of the chapter describes an ensemble COP model, for assessing the overall thermal efficiency and performance of the data center.

---

*Disclaimer** The contents of the chapter are based on the compilation of information available in the open literature. Use of the material presented in the chapter for data center design or for other practical applications is offered as suggestions only. The authors do not assume responsibility for the performance or design issues which may arise due to implementation of the suggestions.

P. Kumar • Y. Joshi (✉)
G.W. Woodruff School of Mechanical Engineering,
Georgia Institute of Technology, Atlanta, GA 30032, USA
e-mail: pramod.kumar@me.gatech.edu; yogendra.joshi@me.gatech.edu

Y. Joshi and P. Kumar (eds.), *Energy Efficient Thermal Management of Data Centers*, DOI 10.1007/978-1-4419-7124-1_2,

## 2.1 Data Center Thermal Management: A Historical Perspective

The term data center refers to a facility where Information technology (IT) equipment and the associated supporting infrastructure such as power and cooling needed to operate the IT equipment are located. Success of a data center lies in retrieving information, either being stored or processed in the IT equipment and making it readily available on demand to the end user. The concept of data centers came into existence almost with the birth of the mainframe computer ENIAC at the University of Pennsylvania. Early data centers were mainly limited to space technology, defense, and other government establishments, and had highly restricted access. Only in the past decade or so data centers have become mainstream, due to the rapid evolution of IT and telecommunications products and technologies. Figure 2.1 shows a photograph of a 1960s NOAA's National Climatic Data Center facility housing Honeywell computers.

Two factors contribute heavily to the successful operation of a data center: reliable power and reliable cooling. For many decades, reliable power was a primary concern, and as such emphasis was placed on improving the power quality, availability, and reliability. Today, with the availability of reliable and

**Fig. 2.1** Photograph of the early data center facility (Courtesy: NOAA's National Climatic Data Center facility, reprinted with permission)

**Fig. 2.2** Data center colocated in an office space (Courtesy: NOAA's National Climatic Data Center facility, reprinted with permission)

efficient power conversion products, provisioning adequate power for data centers is a well-understood challenge. Although electrical power is also required to operate the cooling hardware, reliable power still does not ensure reliable cooling, and hence adequately provisioned cooling continues to be a major concern for data center operation.

Data centers designed during the 1960s and 1970s were provisioned to handle average heat loads between 200 and 750 W/m$^2$ with air delivered between 13°C and 17°C. Most of these data centers were co-located with common office spaces. Since the average heat densities were low, the room air conditioners, which were primarily meant for human comfort, could handle the extra heat of the computing equipment. This made cooling a minor concern. Figure 2.2 shows a storage and data entry area within a mainframe computer installation, analogous to a modern co-located data center in an office space.

Over the years, the increase in demand for IT, coupled with the shrinking form factors resulted in unprecedented levels of power and heat densities in data centers. The heat load of IT equipment alone routinely exceeded the entire building heat load. Large chillers and air handlers had to be installed to combat the increase in heat densities of the IT equipment. The noise generated by the fans in the air handlers made the environment inhospitable for human comfort. This led to the isolation of office spaces from computer rooms, thus marking the emergence of stand-alone computer rooms or data centers. Figure 2.3 shows a yesteryear stand-alone data center facility. This facility, a host of IBM Mainframes, was designed to handle large databases, remote computing, and high-throughput multi-programming [1].

**Fig. 2.3** An early stand-alone data center facility (© IBM Corporation, reprinted with permission [1])

Today, modern data centers are warehouse size independent facilities, solely dedicated to housing digital IT equipment and associated cooling and power infrastructure.

## 2.2 Objective of Thermal Management

Poor thermal management can have a number of adverse implications, such as premature failure of severs due to inefficient airflow distribution, increased downtime, poor reliability, all of which result in a significant increase in operating cost. Thermal management is particularly critical to data centers housing financial information. Recent legislations in the USA such as the Sarbanes–Oxley act makes it mandatory for financial companies to disclose information on demand, which makes it imperative that companies not risk down time, or lose data which could jeopardize availability of information in a timely manner [2]. Successful thermal management can significantly reduce operating costs as well. According to a number of recent studies (e.g., Uptime Institute [3]), the cost of power to the server can surpass the cost of the server within ~4 years. Moreover, if the cooling and infrastructure costs are included, the operational cost of a $1,500 server over its lifetime is five times the cost of the server [4]. These statistics suggest that improving the cooling efficiency is a step in the right direction.

What makes data center thermal management challenging is the dynamic and thus unpredictable nature of the IT equipment heat load. In addition, most data centers house racks of various airflow configurations, e.g., front to back, bottom to top, front to top, etc., creating a heterogeneous air distribution environment. These nonuniform airflow directions, coupled with large diversity in equipment from different vendors presents a hurdle in optimizing the air distribution. Unlike power, which is self tuning and easily deliverable through well-defined paths such as cables or overhead bus ways, the absence of well-defined or confined paths for air, often makes it difficult to deliver adequate quantities of cold air, at the inlet of each and every server inside the data center. As an added challenge, provisioning more cooling to one rack could affect the cooling to neighboring racks. ASHRAE [5] has recently recommended a set of thermal guidelines for safe operation of data centers. While these guidelines provide an envelope for safe operation, methods of operating within the safe envelope are application dependent and left to the end user. Presently, most data centers are managed either based on intuition or accumulated experience. The success of data center thermal management lies in providing the right airflow rate, at the right temperature, to all the IT equipment, under all operating conditions.

## 2.3 Overview of Data Center Air Distribution

The primary source of thermodynamic inefficiency in an air-cooled data center is due to intermingling of hot and cold streams. As explained in Chap. 1, the physical separation of hot and cold airstreams is achieved by arranging the IT equipment in alternate rows to form hot and cold aisles as shown in Fig. 1.15. In this arrangement, the racks housing the IT equipment act as physical barriers, isolating the hot and cold airstreams. Once the physical separation is achieved, the next major task is the supply of cold air and extraction of hot air in the most effective manner, expending minimum energy. There are a number of methods of supplying cold air to the IT equipment, classified as room-, row-, and rack-based methods. A comprehensive review of the different methods and the advantages/disadvantages of each method is provided in [6]. In case of room-based cooling, cold air is distributed via an under-floor plenum or using a system of overhead ducts as illustrated in Fig. 1.16a, c. Both these methods are widely used in modern data centers.

In both these systems, large air handlers employing fan coil systems known as computer room air conditioning (CRAC) units are used to force air through the overhead or under-floor plenum. The cold air supplied by the CRAC units in the cold aisle at the inlet of the racks is heated as it passes through the IT equipment and discharged into the hot aisle. In case of an under-floor air distribution (UFAD) system, the hot air collected in the hot aisle is returned to the CRAC units either through the room or by a dedicated ceiling return plenum as illustrated in Fig. 2.4. The hot airstreams returning to the CRAC unit reject heat to the chilled water flowing through the heat exchangers in the CRAC unit. The cool air is again

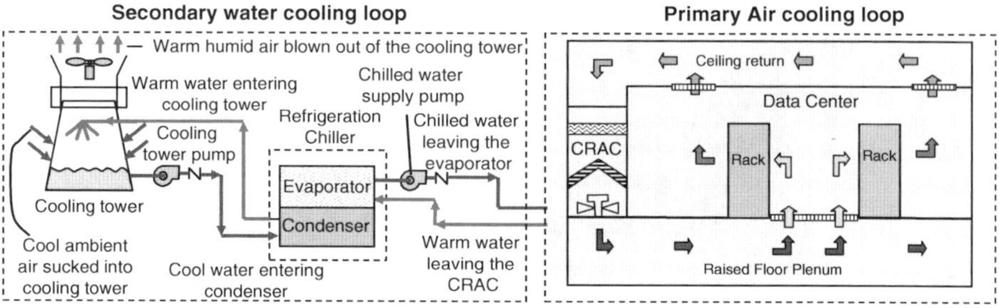


**Fig. 2.4** Primary and Secondary cooling loop in a data center

recirculated back to the IT equipment, thus completing the air cooling loop. The water circulating through the heat exchangers in the CRAC unit is cooled through a secondary loop connected to a chiller or a refrigeration unit, which eventually rejects heat to the ambient environment. The primary air-cooling loop and the secondary chilled-water loop are illustrated in Fig. 2.4.

From Fig. 2.4 one can notice that energy is needed (1) to direct the air to the inlet of the IT equipment and (2) to cool the hot air returning to the CRAC units. The energy needed to deliver the air to the IT equipment is affected by a number of interdependent factors such as, geometry of the room, placement of the CRAC units, type of delivery system used to deliver air to the IT equipment, pressure variation across the room/plenum, and pressure drop across the distribution system, room, and servers. It is the complex interdependency of these factors which makes data center thermal management a challenging task.

## 2.4 Scope of the Chapter

As stated earlier, there is no universal thermal management recipe. A solution that may be very effective in one case may prove to be ineffective, or worse detrimental in another case. Presently, most data centers are managed by following a set of best practices suggested by ASHRAE [5] or other sources. This chapter elucidates some of the key scientific concepts underlying the best practices. This will help the user develop an insight into the fundamental aspects of data center thermal management

The chapter is divided into three sections. The first section highlights the importance of airflow distribution in a data center. Since majority of the data centers today employ a raised air distribution system, emphasis is laid on the factors affecting air distribution in these facilities. The overhead air distribution system is also briefly discussed. The second section focuses on temperature distribution and on humidity control in data centers. Basic psychrometric concepts are used to analyze the various cooling processes. The last section describes a methodology for assessing the data center thermal efficiency using an ensemble COP model.

## 2.5 Fundamental Aspects of Data Center Air Distribution

We know that movement of air is influenced by static pressure variation. Therefore, the first step towards successful air management begins by understanding the static pressure distribution within the data center. To characterize the static pressure distribution, the entire data center comprising of IT equipment and the cooling infrastructure has to be treated as a single unified system.

The following terminology used in the context of data center air distribution is introduced.

(a) Static pressure
Static pressure is defined as the force per unit area. It varies locally along the flow and is obtained either experimentally using a pitot static probe or computationally by solving the fluid flow equations. The CRAC unit must overcome the total static pressure imposed by system to move the air into system. At the inlet to the CRAC unit, the fans create a region of low static pressure resulting in an inflow of air into the CRAC, whereas, at the exit of the CRAC unit, the static pressure is increased due to input of kinetic energy by the CRAC fans.

(b) Dynamic pressure
Dynamic pressure is the pressure along the path of the flow from the CRAC, to the plenum, rack, data center, and back to the CRAC. The dynamic pressure is used to calculate air velocity along the path of flow.

(c) Total pressure
Total pressure is the sum of the static and dynamic pressure. Static pressure and dynamic pressure can change due to acceleration or deceleration of the flow as the airflows though the under-floor plenum (or ducts in case of overhead distribution), perforated tile vents, rack doors, etc. The total pressure is conserved unless lost due to frictional losses.

### *2.5.1 Static Pressure Variation in a Data Center*

Figure 2.5a illustrates a typical raised-floor data center. The facility consists of a CRAC unit comprising of a fan coil system, a raised-floor plenum for distributing the air from the CRAC unit to the cold aisle, a set of perforated tiles for delivering air to the IT equipment and a ceiling return for diverting the hot air back to the CRAC unit. The associated pressure drop across each of the devices is shown alongside in Fig. 2.5b. The corresponding changes in the airstream static pressure from the point of entry into the CRAC unit are illustrated in Fig. 2.5b. This static pressure profile of the data center is useful in illustrating the necessary static pressure rise across the CRAC unit. The CRAC fans increase the air static pressure from room return to a few inches above ambient to pressure at the perforated tile exits are slightly above the atmospheric pressure.

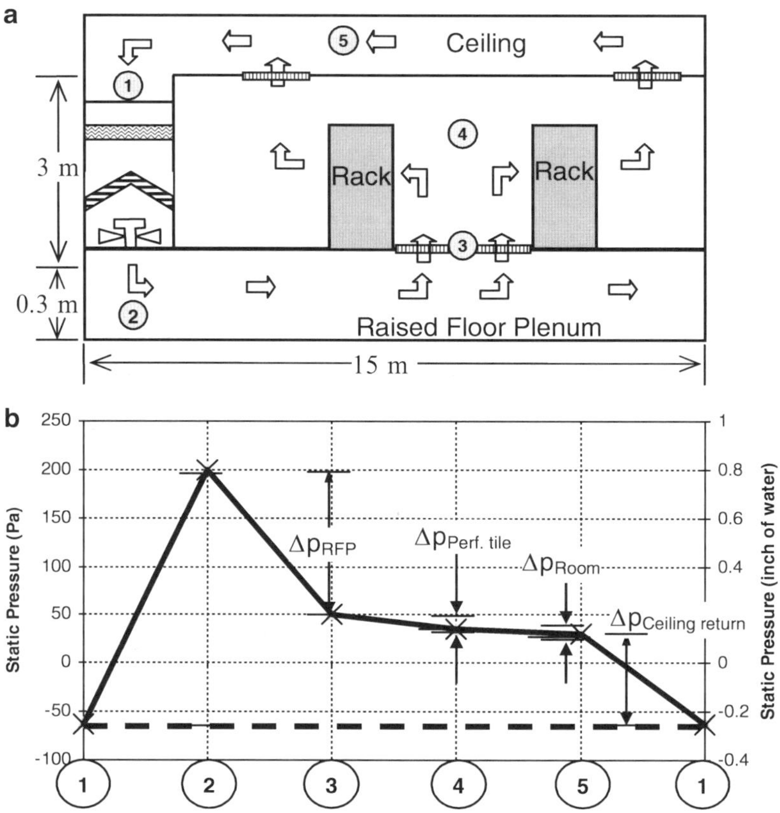


**Fig. 2.5** Pressure drop variation in a data center. (**a**) Schematic of a typical raised-floor facility. (**b**) Pressure drop across various components in the data center

## 2.5.2 *Concept of System Resistance*

System resistance is an obstruction to flow and is used in context with the static pressure loss. In the present discussion, we refer to the entire data center comprising of IT equipment and the cooling infrastructure (air delivery equipment only) as a system [7]. Its flow resistance is a function of the configuration (layout of the CRAC's, aisles IT equipment, etc.), air distribution system employed (over head or under floor), pressure drops across perforated tiles, ceiling grits, rack doors, and CRAC filters. The system resistance is the sum of static pressure losses across all the components participating in air delivery in the data center. It is a function of the flow rate, and is represented as a system resistance curve, as illustrated in Fig. 2.6.

Typically, the system resistance curve is generated using component pressure drop data, either supplied by the manufacturer or generated by using empirical

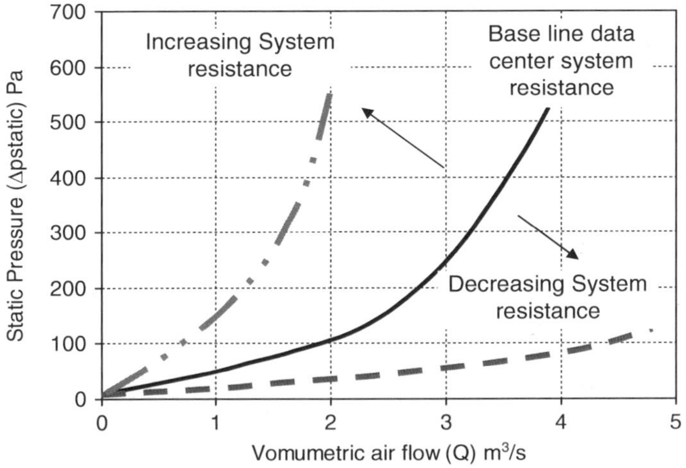


**Fig. 2.6** Typical system resistance curve

measurements or from a CFD model. From the basic principles of fluid mechanics, it is found that the system resistance varies with the square of the volumetric flow rate in laminar flow.

The system resistance has a major role in determining the performance and efficiency of air distribution. Any change in the data center operating conditions will result in a change in the characteristic system resistance curve. For example, if the CRAC filters are clogged, new racks are added, or containment systems are introduced, the system resistance would increase. As a result, the system resistance curve will shift upwards as indicated in Fig. 2.6. Alternatively, if racks are removed, or if the perforated tiles are replaced with higher open area tiles, the system resistance would decrease, resulting in a downward shift in system resistance curve as illustrated in Fig. 2.6.

The system resistance curve provides information on the static pressure head the fans in the CRAC must overcome, in order to deliver desired airflow rate required by the IT equipment. A basic knowledge of the laws governing the operation of the fan is necessary to understand the effect of system resistance on volume of air delivered by the fans in the CRAC unit. These are introduced in the following section.

### *2.5.3 Fan Laws*

All fans operate under a set of laws concerning speed, power, and pressure drop. The fan speed, expressed in revolutions per minute (rpm), is one of the most important operating variables. A change in fan speed (rpm) alters the static pressure rise and fan power necessary to operate at the new speed. The volume of air displaced by the fan blades depends on the rotational speed. The airflow rate increases with increase in fan speed. This relationship is expressed in the *first fan law* as,

$$Q_{N2} = Q_{N1}\left(\frac{N_2}{N_1}\right), \tag{2.1}$$

where $N_1$ is the fan speed (rpm), $N_2$ is the new fan speed (rpm), $Q_{N_1}$ is the volumetric flow rate corresponding to fan speed $N_1$ ($m^3/s$), $Q_{N_2}$ is the volumetric flow rate corresponding to fan speed $N_2$ ($m^3/s$).

The reader may note that a 10% decrease in CRAC fan speed would result in a 10% decrease in the volume of air discharged from the CRAC unit. Thus, changes in the demand of airflow in a data center can be efficiently met by altering the fan speed, within the safe operating range, specified by the manufacturer.

The second fan law relates the static pressure rise across the fan with fan speed. The airstream moving through the fan experiences a static pressure rise due to the mechanical energy expended by the rotating fan blades. As illustrated in Fig. 2.5b, the static pressure at the outlet of the CRAC is always higher than the static pressure at the CRAC inlet. The general equation for calculating the static pressure rise across a fan is provided below.

Total pressure loss = static pressure loss + dynamic pressure loss

$$\Delta p_{\text{Fan,total}} = (p_{\text{faninlet,static}} - p_{\text{fanoutlet,static}}) + \frac{1}{2}\rho_{\text{air}}(V^2_{\text{fan,inlet}} - V^2_{\text{fan,outlet}}), \tag{2.2}$$

where $p_{\text{faninlet,static}}$ is the static pressure at the inlet of the fan ($N/m^2$), $p_{\text{fanoutlet,static}}$ is the static pressure at the outlet of the fan ($N/m^2$), $V_{\text{faninlet}}$ is the air velocity at the inlet of the fan (m/s), $V_{\text{fanoutlet}}$ is the air velocity at the outlet of the fan (m/s), $\rho_{\text{air}}$ is the density of the air ($kg/m^3$).

Since the exit and inlet areas of the CRAC unit are identical, the dynamic pressure loss (or gain) across the CRAC is minimal. The pressure drop across the fan ($\Delta p_{\text{fan}}$) is related to the square of the fan speed as indicated by the *second fan law* expressed below,

$$\Delta p_{N2} = \Delta p_{N1}\left(\frac{N_2}{N_1}\right)^2, \tag{2.3}$$

where $\Delta p_{N1}$ is the static pressure rise across the fan corresponding to $N_1$ fan speed ($N/m^2$), $\Delta p_{N2}$ is the static pressure rise across the fan corresponding to $N_2$ fan speed ($N/m^2$), $N_1$ is the fan speed (rpm), $N_2$ is the new fan speed (rpm).

The fan static pressure rise is usually expressed in inches of water column or in Pa ($N/m^2$). Note that the static pressure rise across the fan rises rapidly with increase in fan operating speed.

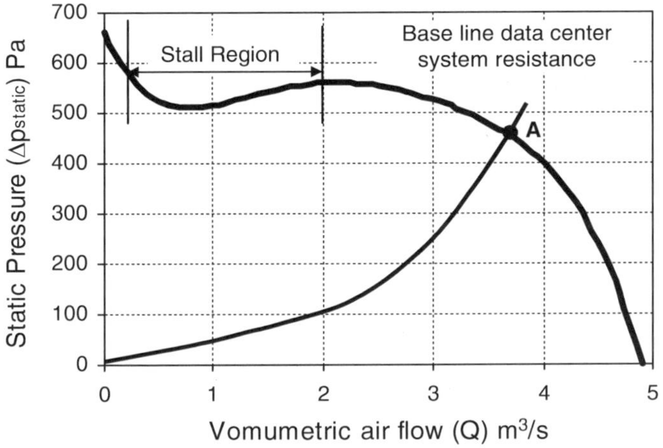


**Fig. 2.7** Characteristic CRAC fan curve

Fan speed variation also affects the fan power consumption. The fan power is proportional to the cube of the fan speed as stated below by the *third fan law* as,

$$W_2 = W_1 \left(\frac{N_2}{N_1}\right)^3, \tag{2.4}$$

where $W_1$ is the fan power corresponding to $N_1$ fan speed (W), $W_2$ is the fan power corresponding to $N_2$ fan speed (W), $N_1$ is the fan speed (rpm), $N_2$ is the new fan speed (rpm).

It is important to recognize that a 10% decrease in fan speed would result in 27% decrease in fan power, suggesting that controlling the fan speed is the most efficient method of flow control.

Note that the above fan laws are valid for geometrically similar fans. The relationship between the volume of air delivered by the fan and the corresponding static pressure rise, for various flow rates at a specific speed, is specified by the manufacturer in the form of a fan performance curve. Figure 2.7 illustrates an example of a characteristic CRAC fan curve for a particular fan speed. The system resistance curve presented in Fig. 2.6 is also overlaid in Fig. 2.7. This data is specific to each CRAC and is based on the fan geometry and operational speed. Different operating speeds will yield different characteristic curves. The intersection of the fan characteristic curve and the system characteristic curve is illustrated by point "A" in Fig. 2.7. With this information, decisions can be made regarding the volumetric flow rate of the CRAC unit, fan operating speed, and the pressure drop of the system.

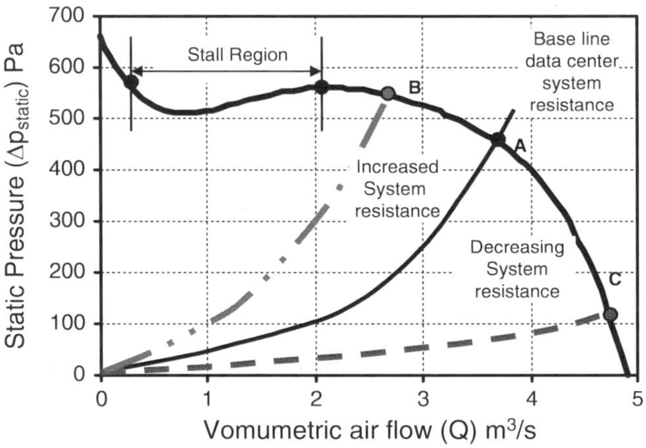


**Fig. 2.8** Effect of change in system resistance on flow rate

## 2.5.4 *Effect of Change in System Resistance*

As long as no changes are made to the data center equipment or layout, the pressure drop across the CRAC will not vary significantly, and the data center will continue to operate at the point "A" illustrated in Fig. 2.7.

This, however, is only an ideal condition and in reality, data centers rarely operate at a fixed operating point. As described previously, the system resistance can change due to a number of normal operational factors, such as clogging of the CRAC filters, and relocation, addition, or removal of IT equipment. Small changes in airflow requirement by the IT equipment due to variation in sever fan speed can also impact the system resistance. The extent of the variation in airflow delivered by the CRAC depends on the characteristic fan curve and the system resistance curve. Some of these effects are highlighted in Fig. 2.8.

Major changes such as introduction of containment systems or introduction of new IT equipment would increase the system resistance, causing the system resistance curve to shift upwards. This increased airflow resistance will shift the operating point to B as illustrated in Fig. 2.8. At point "B," since the static pressure rise across the CRAC is slightly higher compared to point "A," this would result in lower airflow rate from the CRAC. Alternatively, if racks are removed or if the perforated tiles are replaced with higher open area tiles, the system resistance would decrease, resulting in a downward shift in system resistance curve as illustrated in Fig. 2.8. The decrease in system resistance will shift the operating point to "C" that has a slightly reduced static pressure rise across the CRAC fan, resulting in an increase in the airflow rate delivered by the CRAC fan for the same operating fan speed.

Many fans exhibit an unstable behavior if the static pressure exceeds a certain limit. If the static pressure surpasses this limit, the CRAC could slip into the stall region and experience a drastic reduction in volumetric flow. In Fig. 2.8, the stall region is located to the left of point "B." Generally, the CRAC units are designed such that even if the total system resistance approaches the designed static pressure limit, the operating point does not shift beyond "B" on the fan curve. Some of the factors listed above, contributing to a change in the system curve are considered normal, and CRAC fans are designed to accommodate these changes, without appreciably impacting the airflow rate. The flow rate from the CRAC is significantly impacted by a change in room or aisle layout, variation in plenum height, obstruction in the air delivery path, or due to implementation of an improperly designed containment system.

### *2.5.5 Effect of Increasing the CRAC Fan Speed*

Most modern CRAC units are equipped with variable frequency drives (VFDs) to enable variable speed operation of the CRAC fans. Although the initial cost of a VFD-CRAC unit is higher, it provides considerable flexibility in control of airflow. Variable speed operation of the CRAC involves controlling the speed of the fan to efficiently meet the dynamic cooling air requirements of the data center. CRAC fan performance can be predicted at different speeds using the fan laws. Since power input to the fan varies as the cube of the airflow rate, varying the fan speed is the most efficient form of capacity control. If cost is a driving factor in a facility with multiple CRAC units, some units may be fitted with VFDs to accommodate capacity variations. The benefit of variable fan speed can be impaired by the efficiency of the VFD drive and the associated control system. This should be accounted for in the analysis of power consumption.

In many cases, increasing the CRAC fan speed within the manufacturer recommended limits can mitigate the impact of reduction in airflow due to an increase in system resistance. The effect of increase in fan speed on airflow rate from the CRAC unit is shown in Fig. 2.9.

From the first fan law in (2.1), we note that the increase in the airflow rate is directly proportional to the increase in the fan speed. An increase in CRAC fan speed results in a new CRAC fan curve and a new operating point "D" as shown in Fig. 2.9. The new operating point "D" provides increased airflow rate at a higher fan static pressure rise, both of which are higher than the initial conditions specified by point "A." Similarly, a decrease in the fan speed would shift the operating point down to "E" resulting in a decrease in both static pressure and the flow rate. These observations are valid provided there is no change in system resistance. However, if the system resistance decreases, for example, due to installation of high flow tiles, the fan speed could be decreased to maintain identical flow conditions. This condition is represented by operating point "F" in Fig. 2.9. On the other hand, if the system resistance increases, the decrease in airflow, due to increase in static pressure, can be countered by increasing the fan speed as specified by the operating

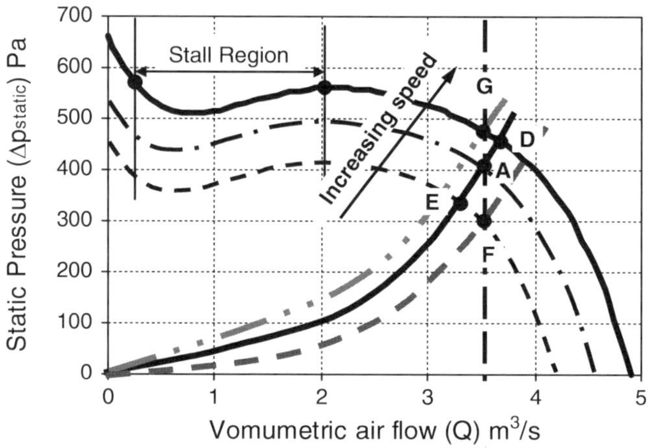


**Fig. 2.9** Effect of change in fan speed on airflow

point "G" in Fig. 2.9. The economic benefits of operating at higher fan speed must be carefully considered, as according to the third fan law (2.3) the power input to the fan varies as the cube of the speed.

### *2.5.6 Minimizing Demand on the CRAC Unit*

The airflow requirement of a data center is based on the cooling demands of the IT equipment, which is either specified by the equipment manufacturer or is determined from supply temperatures and the equipment heat loads. Precise determination of airflow and outlet pressure at the perforated tile (or ceiling grit in case of an overhead system) is important in sizing the CRAC unit. Detailed CFD analysis, as discussed further in Chap. 8, should be carried out to determine the overall system pressure drop. The analysis should account for pressure drop across the plenum, obstructions due to blockages, pressure drop across CRAC filters, perforated tiles, and rack grills. Only when the system flow and pressure requirements are determined, the CRAC unit should be selected. Since, system pressure drop is difficult to predict accurately, often a conservative approach is adopted, with large safety margins. This leads to substantial oversizing of the CRAC units, which operate at flow rates much below their design values, resulting in poor efficiencies.

The subsequent sections of the chapter focus on factors affecting pressure drop across various components and their influence on air distribution in the data center. Both UFAD and OHAD distribution systems are discussed in context with the static pressure drop.

## 2.6 Factors Affecting Under-Floor Air Distribution

Referring back to Fig. 2.5b, we note three distinct gradients in the static pressure curve. The pressure variation across the CRAC unit is the largest, and as stated earlier, is a function of CRAC fan speed. Under normal operating conditions the pressure drop across the CRAC is largely unaffected by static pressure variations across plenum, perforated tiles, or the room. The room pressures (space above the plenum housing the IT equipment) are relatively close to the atmospheric pressure. Since the volume of the air above the raised floor is large compared to the plenum volume, *variations* in room pressure are usually insignificant, compared to the pressure variations across the perforated tiles and the subfloor. On the basis of this assumption, one can neglect the room-level pressure variations relative to the subfloor. This not only simplifies the understanding of flow physics but also provides the flexibility to independently analyze the raised-floor plenum, perforated tiles, and room.

### *2.6.1 Pressure Variation in the Plenum*

The static pressure rise brought about by the CRAC fans is converted to dynamic pressure due to acceleration of the flow at the exit of the CRAC unit. As illustrated in Fig. 2.10, the air discharged from the CRAC enters the plenum vertically downward and turns to proceed horizontally to fill the plenum. Large variations in horizontal velocities in the vicinity of the CRAC unit are incurred, which setup large pressure gradients, giving rise to highly nonuniform flow across the perforated tiles. The decrease in the volume of air due to partial venting through the perforated tiles, results in a gradual decrease in the horizontal velocities. This is accompanied by a corresponding increase in static pressure. The gradual increase in static pressure downstream of the CRAC unit in effect increases the net flow of air from the tiles placed downstream of the CRAC unit as illustrated in Fig. 2.10.

In some cases, the static pressure near the CRAC exhaust can fall below the atmospheric pressure, giving rise to negative flows. The low pressure regions originate due to sudden expansion of the impinging jets of air on the plenum floor,

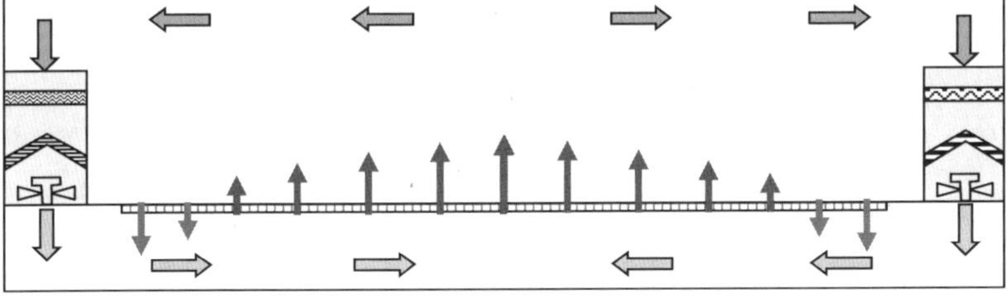

**Fig. 2.10** Distribution of flow across the raised floor

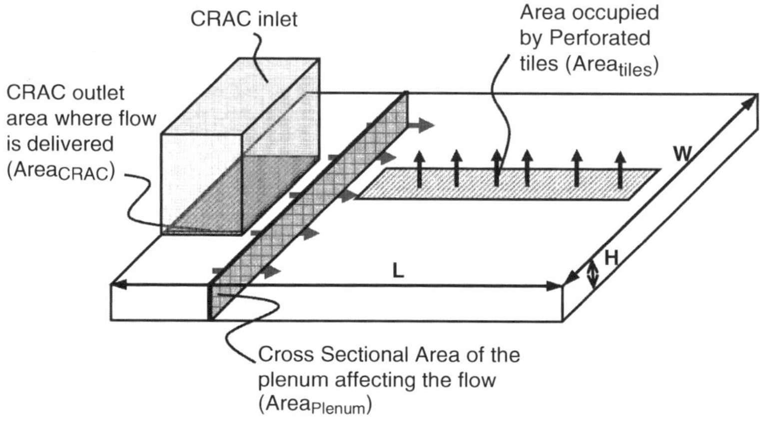


**Fig. 2.11** Cross-section of the plenum affecting the distribution of flow in the plenum

which are then diverted horizontally into the plenum. As illustrated in Fig. 2.10, the high-velocity airstreams in the plenum entrain the room air into the plenum, resulting in reversed flows in the perforated tiles placed near the CRAC unit. This is sometimes referred to as Venturi effect. Reversed flows are generally encountered in areas under the raised floor where horizontal air velocities exceed 162 m/s (530 feet per minute), which is usually within about six tiles off the CRAC unit [1].

### *2.6.2 Factors Affecting Plenum Pressure Distribution*

The variation in the air discharged from the perforated tiles is influenced by the static pressure distribution in the plenum. The static pressure in the plenum is influenced by many design variables such as plenum height, location of the CRAC units, blockages under the perforated floor and open area of the perforated tile. The influence of these factors is discussed in the following sections.

(a) Effect of plenum height
As explained above the primary cause of misdistribution of flow across the perforated tiles is due to the plenum pressure distribution. Changes in local plenum pressure are caused due to variation in velocities in the plenum. The magnitude of variation in the average horizontal velocity is dependent on the cross-sectional area of the plenum (Area$_{Plenum}$) which is normal to the flow as shown in Fig. 2.11.

The cross-sectional area is defined by the plenum height and width (Area$_{Plenum}$ = $H \times W$). For identical CRAC flow rates, shallow plenum depths will result in higher plenum velocities compared to deeper plenums. Lower plenum velocities lead to higher plenum pressures. The magnitude of average horizontal velocities in the

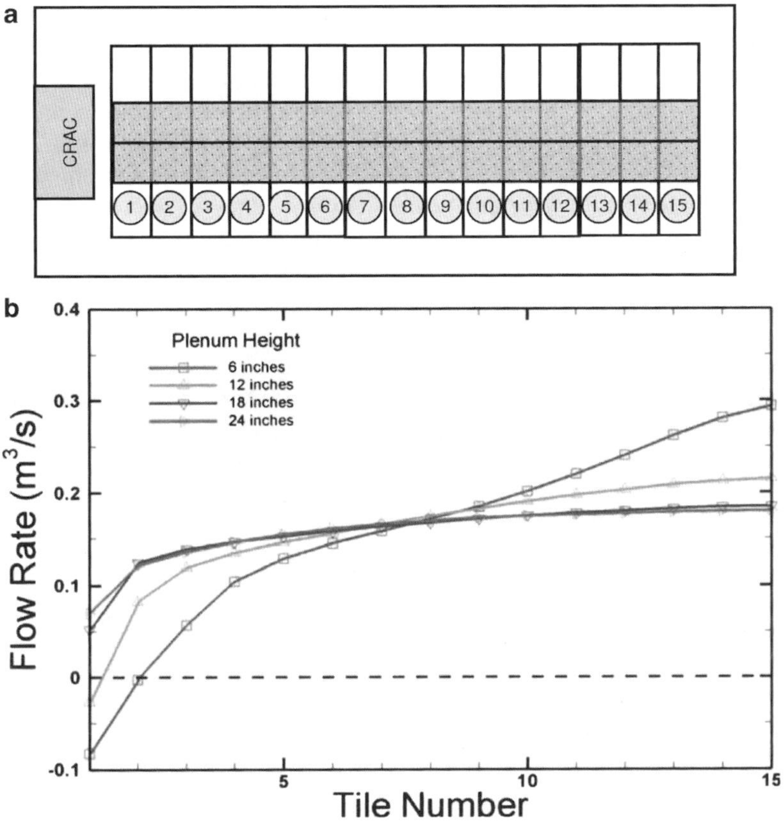


**Fig. 2.12** Effect of plenum height on raised-floor air distribution. (© 2004 ASHRAE, reprinted with permission [8]). (**a**) Tile layout—adapted from [8]. (**b**) Variation in flow rate with tile number for various plenum heights

plenum for a 30-cm plenum depth is three times that for a 90-cm plenum depth. The degree of flow maldistribution is significant only if the variation in plenum pressure is comparable to the pressure drop across the perforated tile. This condition is satisfied if the plenum cross-sectional area is less than or comparable to the total open area of all the perforated tiles ($\text{Area}_{\text{Plenum}} \leq \sum \text{Area}_{\text{tiles}}$). Hence, both the perforated tiles and the plenum height play an important role in plenum pressure distribution. The effect of plenum height on perforated tile flow is illustrated in Fig. 2.12 [8].

The results presented in Fig. 2.12b are obtained for a CRAC flow of 4.72 $m^3/h$ (10,000 cfm) using 25% open area tiles [8, 9]. The data center facility consisted of 30 tiles arranged in two rows to form the cold aisle in front of the CRAC unit as shown in Fig. 2.12a. Figure 2.12b shows the variation of perforated tile flow rate for a simple data center layout for various plenum depths ranging from 15 to 60 cm. As observed from Fig. 2.12b, nonuniformity in flow is found to be significantly higher for a 15-cm plenum compared to a 60-cm plenum depth. The authors also found that the intensity of reversed flow in the perforated tiles placed close to the CRAC unit considerably reduces with increase in plenum depth. The pressure variations in the horizontal plane under the plenum are more pronounced for shallow plenums with large negative pressure zones near the CRAC unit, which is mainly responsible for reversed flows. The pressure variations decrease with increase in plenum depths, resulting in more uniform pressure distributions. For the layout illustrated in Fig. 2.12a, the authors also report that increasing the plenum height beyond 60 cm showed no appreciable change in flow distribution.

(b) Effect of tile open area

The perforated tiles offer substantial resistance to flow of air from the plenum. The flow through the perforated tile is influenced by the open area. A tile with 56% open area offers lower resistance to flow compared to a tile with 25% open area. The discharge velocity from the tile is related to the plenum pressure under the tile via the following relationship. The perforated tile model presented in the following sections is based on the work reported in [10–13].

$$\Delta p_{\mathrm{Perf.tile}} = R|Q|Q \ (\mathrm{Pa}) \tag{2.5}$$

$\Delta p_{\mathrm{Perf.tile}}$ is the difference between the static plenum pressure and room pressure above the raised floor and is defined as

$$\Delta p_{\mathrm{Perf.tile}} = p_{\mathrm{Plenum}} - p_0 \ (\mathrm{Pa}), \tag{2.6}$$

where $p_{\mathrm{Plenum}}$ is the local static plenum pressure under the perforated tile (Pa) and $p_0$ is the room pressure immediately above the tile (Pa). $R$ is the resistance factor and $Q$ is the volumetric flow rate in ($\mathrm{m}^3$/s) defined by

$$Q_{\mathrm{tile}} = A_{\mathrm{Perf.\ tile}} \times V_{\mathrm{Perf.\ tile}} \ (\mathrm{m}^3/\mathrm{s}), \tag{2.7}$$

$A_{\mathrm{Perf.\ tile}}$ is the total tile area ($\mathrm{m}^2$) and $V_{\mathrm{Perf.\ tile}}$ is the discharge velocity of air (m/s). The flow resistance factor is related to the loss coefficient for the tile and is defined as

$$R = \frac{1}{2}\frac{\rho_{\mathrm{air}}}{A_{\mathrm{Perf.\ tile}}^2} K_{\mathrm{Perf.\ tile}}, \tag{2.8}$$

$\rho_{\mathrm{air}}$ the density of air ($\mathrm{kg/m}^3$) and $K_{\mathrm{Perf.\ tile}}$ is the pressure loss due to loss of kinetic energy of the high-velocity jets. $K$ is empirically evaluated using the following relation [14]

$$K_{perf.\,tile} = \frac{1}{f^2}\left(1 + 0.5(1-f)^{0.75} + 1.414(1-f)^{0.375}\right) \tag{2.9}$$

where $f$ fractional open area for the perforated tile.

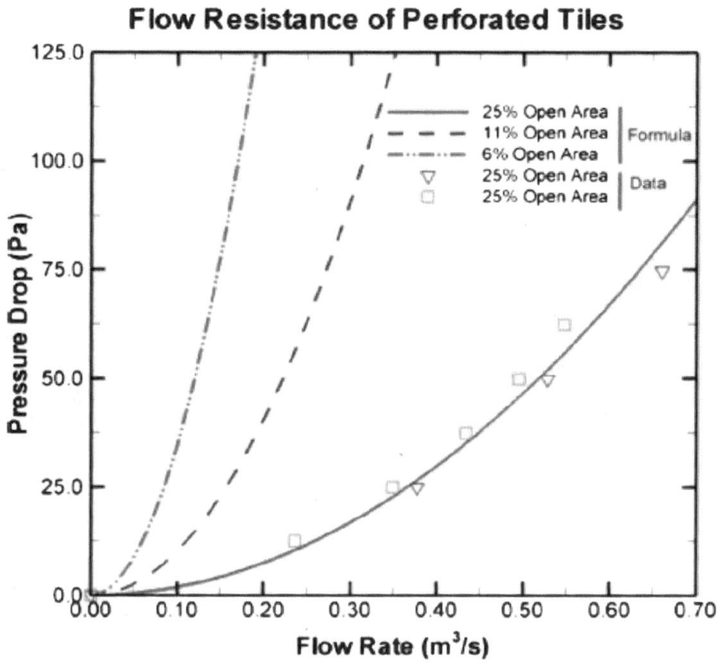


**Fig. 2.13** Effect of perforated tile open area on tile pressure drop (© 2010 ASME, reprinted with permission [9])

The flow direction is determined by the sign of $\Delta p_{\text{Perf.tile}}$. Negative $\Delta p_{\text{Perf.tile}}$ values indicate reverse flows, where the plenum pressure is lower than the room pressure. For the air to flow from the plenum to the room the plenum pressure has to be greater than the room pressure. The plenum air has to overcome the pressure drop in the tile. The $K_{\text{Perf. tile}}$ factor for a 25% open tile is 42. In a physical sense, the $K_{\text{Perf. tile}}$ factor indicates how many velocity heads are lost due to pressure drop across the tile. When the air jets discharged from the tile vents spread into the open space, the pressure change of one velocity head is produced. Since this is small compared to the pressure drop across the perforated tile, the pressure variations in the space above the raised floor can be neglected. Under this assumption, the room is considered to be at uniform atmospheric pressure throughout. The only other regions where the static pressure variations are significant are near the air velocity and CRAC unit. The effect of pressure variation at the rack inlet is investigated later in the chapter.

The pressure drop calculated using the above expression in (2.5) is found to be in good agreement with the experimentally measured data supplied by the manufacturer [9]. Figure 2.13 shows the variation of static pressure drop across a standard 0.6 m $\times$ 0.6 m (2 ft $\times$ 2 ft) tile as a function of tile flow rate.

Usually, a 25% open (standard tile) perforated tile provides approximately 0.24 $m^3/s$ (500 cubic feet per minute (cfm)) at a 5% static pressure drop, while a 56% open (high flow tile) perforated tile provides approximately 0.24 $m^3/s$ (2,000 cfm) [15]. The tile open area plays a major role in maintaining the plenum static pressure. In general, as the tile open area decreases, the flow across the perforated tiles tends to become more uniform. This is because the variations in the horizontal pressure gradients in the plenum under the perforated tile become less significant compared to the pressure drop across the perforated tile. However, the decrease in the perforated tile area is marked by an associated increase in plenum pressure. The increase in plenum pressure leads to loss of air through gaps between perforated tiles and openings in the raised floor provisioned for bringing in electrical cables. As the open area of the perforated tiles decreases, the leakage area becomes comparable to the total perforated tile open area. The flow resistance of the perforated tiles now becomes comparable to the flow resistance of the openings in the raised floor, leading to an increase in plenum air leakage. The effect of plenum air leakage is discussed in the following section.

(c) Effect of blockages in the under-floor plenum

Apart from the primary function of distributing air to the IT equipment, the raised floor is also used for laying chilled water supply and return to the CRAC units, power cables, and sometimes structural beams which would affect the plenum pressure distribution. The blockages impede the plenum flow causing non uniformity in plenum pressures. This results in maldistribution of flow across the perforated tiles. To quantitatively ascertain the extent of flow maldistribution the obstructions have to be included in the CFD model for the plenum. The impact of the flow maldistribution due to under-floor blockages using CFD tools is presented in Chap. 8.

The obstructions reduce the effective area available for the flow of air. The variation in plenum area available for flow causes variation in horizontal velocities in the plenum. Since $\Delta p \propto V^2$, a more significant variation in plenum pressure is observed. Due to retardation of the flow in the upstream side of the blockage, the static pressure increases, resulting in higher perforated tile flow rates in the upstream compared to downstream side of the blockage, where the static pressure is low. In some cases the static pressure can drop below ambient pressure, which would result in reversed flows. Significant variation in flow is usually observed when the dimension of the obstruction normal to the flow is comparable to the plenum height. For example, a 15-cm diameter circular pipe would hardly affect the flow in a 90-cm high plenum. However, this would significantly affect the flow if the plenum height were reduced to 30 cm (12 in.). Figure 2.14a, b [9] illustrate the effect of blockage in a 30-cm high plenum for the layout shown in Fig. 2.12a. The 15-cm diameter pipe is placed parallel to the CRAC unit between tile 5 and tile 6, leaving a clearance of 15 cm above the pipe for the air to flow.

Figure 2.14a shows the pressure and velocity distribution just under the raised floor. Figure 2.14b shows the corresponding perforated tile airflow distribution. Referring to Fig. 2.14a we note that the pipe impedes the airflow from the CRAC

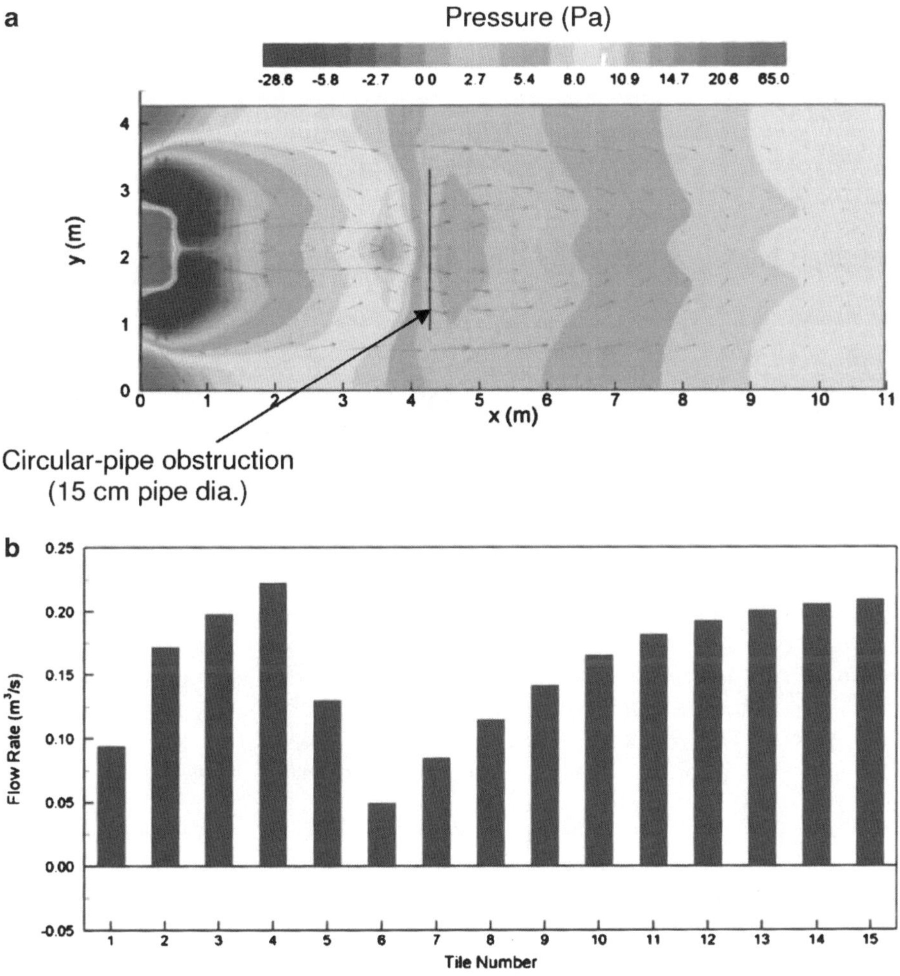


**Fig. 2.14** Effect of blockage on plenum air distribution. (© 2010 ASME, reprinted with permission [9]). (**a**) Pressure distribution and velocity vectors under the raised floor due to presence of a circular pipe. (**b**) Variation in perforated tile flow rate due to presence of a circular pipe

units, resulting in an increase in the static pressure in immediate upstream region of the pipe. The pressure variations in turn affect the perforated tile flow rates as illustrated in Fig. 2.14b. Note that the increase in plenum pressure in the upstream region of the pipe results in higher flow rates in the perforated tiles placed near the CRAC unit.

In general, the obstructions alter the plenum pressure distribution resulting in flow maldistribution across the perforated tiles. The extent of flow maldistribution not only depend on the size of the obstruction but also on the location of the obstruction. Further details on impact of blockages on perforated tile flow rates for various cases are presented in Chap. 8.

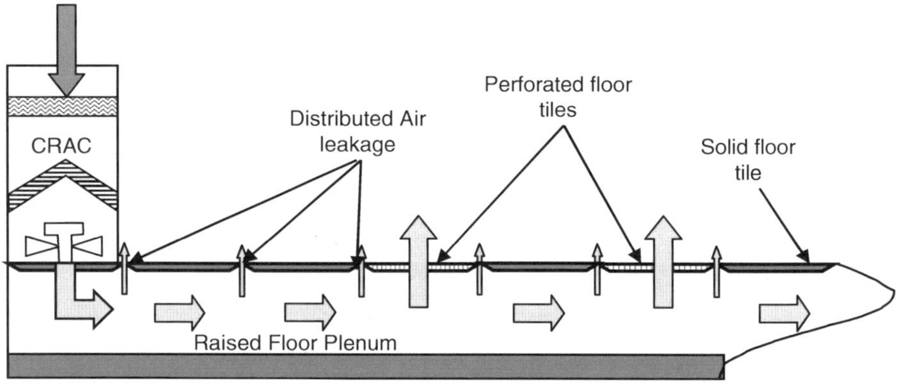


**Fig. 2.15** Distribution leakage between the perforated tile panels

## *2.6.3 Leakage Through the Raised Floor*

Air leakage from the plenum is a major issue. The lost air represents wastage of the precious cooling resource, as it does not contribute to cooling of the compute equipment. If the source of leakage is located in the hot aisle, mixing of the cold and warm airstreams will drop the average hot air return temperature to the CRAC unit. This not only impacts the blower energy use but also impairs the CRAC cooling performance. According to a recent computational study [16] performed on a 1,115-$m^2$ (12,000 $ft^2$) facility, housing 100 server racks, at a heat flux of 100 W/$ft^2$, showed a decrease in average rack inlet temperature of 0.7–0.9°C for every 10% reallocation of leakage flow to a desired rack inlet location.

Air often leaks from the plenum through minute gaps between the perforated tiles and cutouts provisioned for routing the electrical and network cables housed on the raised floor. In addition, air can also leak due to improper sealing in the plenum walls made for accommodating the chilled water pipe lines, and/or due to poor construction quality resulting in an inadequately sealed plenum. The air loss through the panel gaps in the perimeter of the tile between the raised floor and the tiles, as illustrated in Fig. 2.15, are referred to as distribution losses [17, 18].

The leakage area can be in the range of 0.1–0.5% of the exposed floor area and depends on the wear of the tile seals. These losses contribute to 10–15% of the total volumetric flow rate from the perforated units. It is estimated that in a data center operating with 10 cooling units, airflow from one of the CRAC unit is utilized to make up for the air lost due to floor-tile gap leakage. During data center design, distribution losses are accounted as fixed losses and generally based on rules of thumb, or limited experimental measurements. However, in reality these losses are related to variations in plenum pressures, and significantly increase at higher plenum pressures. Decrease in perforated tile area increases the plenum pressure, resulting in an increase in distribution loss. A first-hand estimate of distribution loss can be made by measuring the total airflow rate from all the tiles and comparing it with the CRAC discharge. A model developed in the literature [19] for these losses is described in the following section.

The expression for pressure drop across the tile represented by (2.5) is modified to include the distributed air leakage around the tile perimeter.

$$\Delta p_{\text{Tile total}} = \frac{1}{2}\rho_{\text{air}}[K_{\text{Perf. tile}}V_{\text{Perf. tile}}^2 + \underline{K_{\text{gap}}V_{\text{gap}}^2]}\ (\text{Pa}), \tag{2.10}$$

where $V_{\text{gap}}$ is the velocity of the air escaping through the gap (m/s) and $K_{\text{gap}}$ is the corresponding loss factor associated with the leakage area.

Equation 2.5 can be expressed in terms of perforated tile velocity as

$$\Delta p_{\text{Perf. tile}} = \frac{1}{2}\rho_{\text{air}}K_{\text{Perf. tile}}V_{\text{Perf. tile}}^2\ (\text{Pa}). \tag{2.11}$$

The pressure drop due to distributed leakage is expressed as

$$\Delta p_{\text{gap}} = \frac{1}{2}\rho_{\text{air}}K_{\text{gap}}V_{\text{gap}}^2\ (\text{Pa}). \tag{2.12}$$

Since the plenum pressure is identical we can express the gap air velocity in terms of tile velocity as

$$V_{\text{Perf. tile}} = \left(\frac{K_{\text{gap}}}{K_{\text{Perf. tile}}}\right)^{\frac{1}{2}} V_{gap}\ (\text{m/s}). \tag{2.13}$$

The total quantity of air discharged from the gap and tile is obtained using the following expression:

$$Q_{\text{Tile total}} = A_{\text{Perf. tile}}V_{\text{Perf. tile}} + A_{\text{gap}}V_{\text{gap}}\ (\text{m}^3/\text{s}). \tag{2.14}$$

Using (2.13) and (2.14) we get,

$$Q_{\text{Tile total}} = \left(A_{\text{Perf. tile}}\left[\frac{K_{\text{gap}}}{K_{\text{Perf. tile}}}\right]^{\frac{1}{2}} + A_{\text{gap}}\right)V_{\text{gap}}\ (\text{m}^3/\text{s}). \tag{2.15}$$

Using the above expression the ratio of leakage to total discharge gives

$$\frac{Q_{\text{leakage}}}{Q_{\text{Tile total}}} = \frac{1}{1 + \frac{A_{\text{Perf. tile}}}{A_{\text{gap}}}\left[\frac{K_{\text{gap}}}{K_{\text{Perf. tile}}}\right]^{\frac{1}{2}}}. \tag{2.16}$$

Further, for small values of loss factor $f$ using (2.9) $K$ is approximated as

$$K = \frac{2.9}{f^2}. \tag{2.17}$$

Using the above simplification we get,

$$\frac{Q_{\text{leakage}}}{Q_{\text{Tile total}}} = \frac{1}{1 + \dfrac{A_{\text{Perf. tile}}}{A_{\text{gap}}}\left[\dfrac{K_{\text{gap}}}{K_{\text{Perf. tile}}}\right]^{\frac{1}{2}}} = \frac{1}{1 + \dfrac{A_{\text{tile}} f_{\text{tile}}}{A_{\text{gap}} f_{\text{gap}}}} = \frac{A_{\text{gap}} f_{\text{gap}}}{A_{\text{gap}} f_{\text{gap}} + A_{\text{Perf. tile}} f_{\text{Perf. tile}}}. \tag{2.18}$$

The above expression is valid under the following assumptions:

(a) The plenum pressure is uniform and not affected by the leakages or blockages.
(b) The total perforated tile open area is small compared to the plenum cross-sectional area illustrated in Fig. 2.12.
(c) The distribution of leakage over the entire floor space is uniform.

The pressure drop under the floor becomes more uniform as the number of perforated tiles decrease or if more restrictive tiles are used. As discussed earlier, increasing the plenum height can also result in a more uniform pressure distribution. The above equation (2.18) suggests that the controlling factor for distribution losses is not the individual values of the distribution leakage or the number of tiles but the distribution leakage area ratio. This is an important result suggesting that two data centers with different areas/floor plans and tile porosities will yield identical distribution losses, as long as the distributed area ratios remain identical. The strength of the above expression is exhibited in Fig. 2.16 [17], which is in nondimensional form

Figure 2.16 shows that sealing the intentional openings in the plenum would not only increase the plenum pressure (which leads to an increase in the distribution losses) but also reduce the total open area available for airflow. This in a relative sense would increase the "distributed leakage area/total area of floor openings" ratio; refer to the $x$-axis in the graph in Fig. 2.16. We note that distribution leakage also increases with increase in the distributed leakage area/total area of the floor openings ratio. Hence, sealing the openings in the perforated tiles and plenum walls would only partially increase the available air for cooling.

Typically, the distribution losses are high in large data center facilities operating at partial capacity, as the area occupied by the perforated tiles is small compared to the total floor area. The distribution leakage increases with facility age, due to wear in the tile seals. Usually, the leakage gap between the perforated floor panels is ~0.85 mm. Depending on the wear in the tile seal, the corresponding distributed leakage area can vary between 0.1% and 0.5% of the total exposed area.

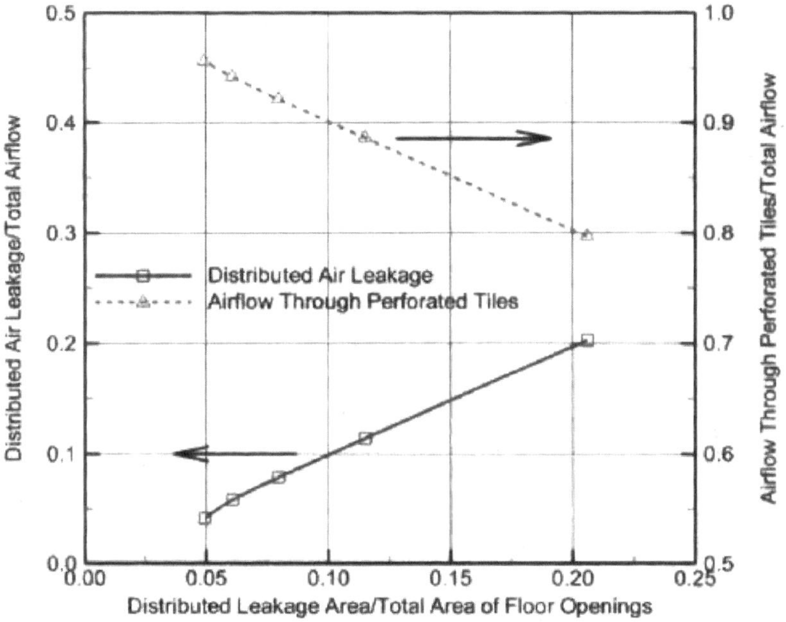


**Fig. 2.16** Variation of nondimensional distributed air leakage and total airflow through perforated tiles with the ratio of distributed leakage area to the total area of the floor openings (© 2007 ASHRAE, reprinted with permission [17])

## 2.6.4 *Controlling the Plenum Pressure Distribution*

Achieving uniformity in flow rates plays a major role in data center floor design. Successful floor layouts exhibit uniform air distribution across all the perforated tiles. Owing to design constraints such as blockages due to obstructions from chilled water pipes, cable trays, and cables it is difficult to achieve this in practice. Furthermore, in many cases the perforated tile flow rate may need to be tuned to rack airflow requirements. Hence, the solution for tuning airflow has to be flexible without involving infrastructural changes such as increasing the raised floor height or installing new CRAC units. There are many ways to balance the flow across the perforated tiles. Two most commonly used methods are discussed below:

(a) Variable open area perforated tiles
The nonuniformity in the flow across various perforated tiles is primarily due the variation in plenum static pressure. As explained in Sect. 2.6.1, the variation in pressure results from a change in flow inertia or momentum effects. From (2.10) we note that the tile pressure drop scales with the square of the tile velocity. The velocity of the air discharged from the tile depends on the open area of the tile. Hence, in principle it is possible to counter the

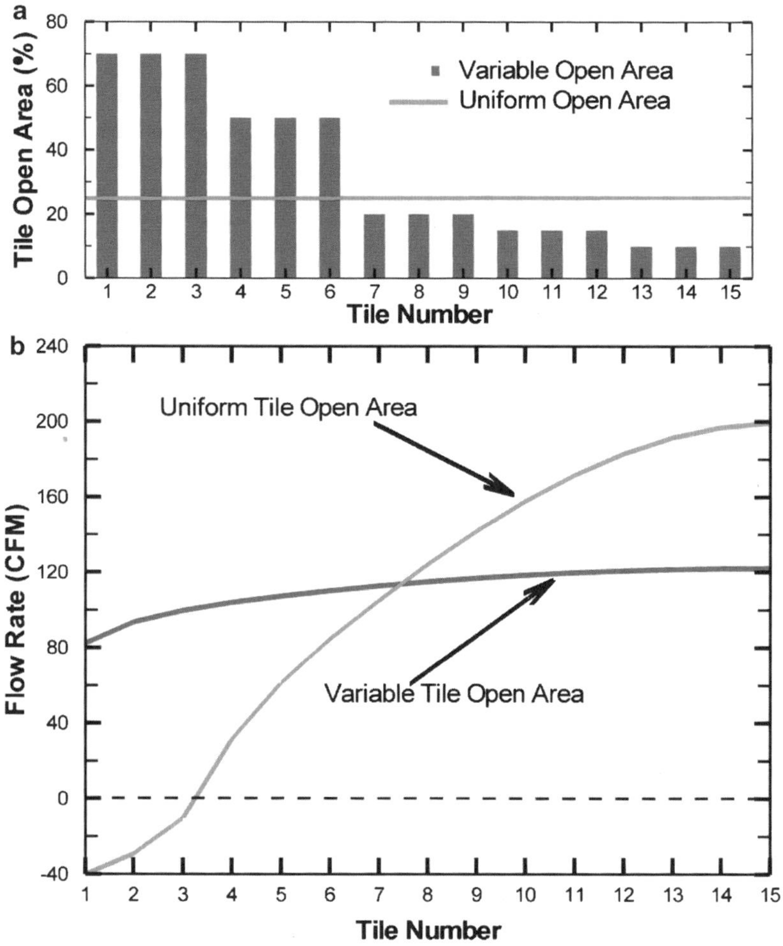


**Fig. 2.17** Effect of variable open area tiles (© 2004 ASHRAE, reprinted with permission [8])

variation in flow across the perforated tile (due to static pressure variation) by controlling the open area of the tile. In general, installing perforated tiles with larger open area will reduce the pressure drop across the tile. This will increase the flow rate from the tile. As such, tiles with larger open area can be installed in areas where the plenum pressures are low such as the vicinity of the CRAC units, downstream side of obstructions, etc. In the regions where the plenum pressures are high such as far end for the aisle away from the CRAC unit or upstream side of an obstruction, etc., installing tiles with reduced open areas will increase the pressure drop and thereby reduce the tile air discharge. Hence, use of variable area tiles help in improving the air distribution. Figure 2.17 [8] illustrates the effect of variable area tile for the layout shown in Fig. 2.12a.

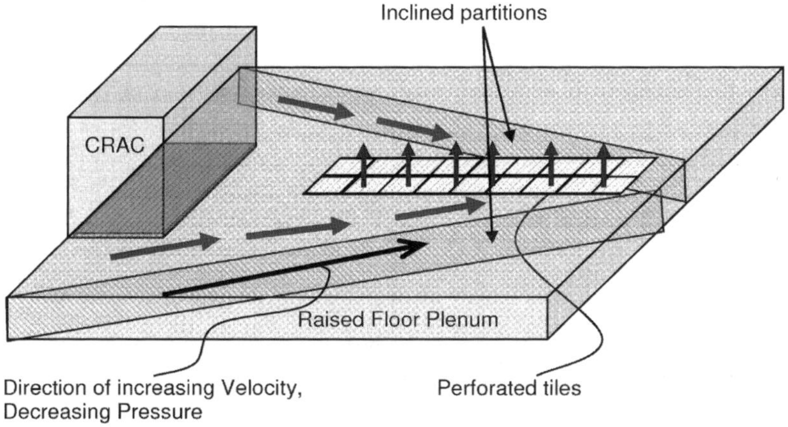


**Fig. 2.18** Use of inclined partitions for controlling plenum flow

(b) Under-floor partitions
As explained above, excessive discharge velocity from the CRAC unit reduces the static pressure through perforated tiles closest to the unit, resulting in inadequate airflow. The static pressure steadily increases, as the high-velocity discharge moves away from the unit, leading to a steady increase in the airflow through the perforated tiles. To counter this effect, vertical barriers or flow deflectors can be used to divert air through the perforated tiles. The use of inclined vertical partitions is illustrated in Fig. 2.18. The inclination of the partition reduces the area available for flow in a linear fashion. For a first estimate, based on a simple analysis using the Bernoulli equation, the variation in pressure along the aisle can be estimated, and accordingly the degree of inclination can be altered to achieve the desired flow distribution from the perforated tiles. In data centers housing multiple units, the use of solid inclined partitions is not recommended as the failure of a particular CRAC unit will result in total loss of air supply to the perforated tiles enclosed within the inclined partitions [9]. In addition, the under-floor partitions could also increase the plenum resistance, thereby resulting in a net reduction in flow discharged from the CRAC unit. From this point of view, using a variable open area perforated tile could be a more suitable solution. Another option is to use adaptive vent tiles [19] or fan-assisted perforated tiles to increase the supply air at locations where the plenum pressure is low, such as immediately downstream of circulation blockages. Fan-assisted tiles can provide 0.0944–0.71 $m^3/s$ (200–1,500 cfm) of supply air.

### 2.6.5 *Effect of Aisle Layout*

In addition to the static pressure variation in the plenum, the distribution of flow across perforated tiles is also influenced by the layout of the aisle. The influence of

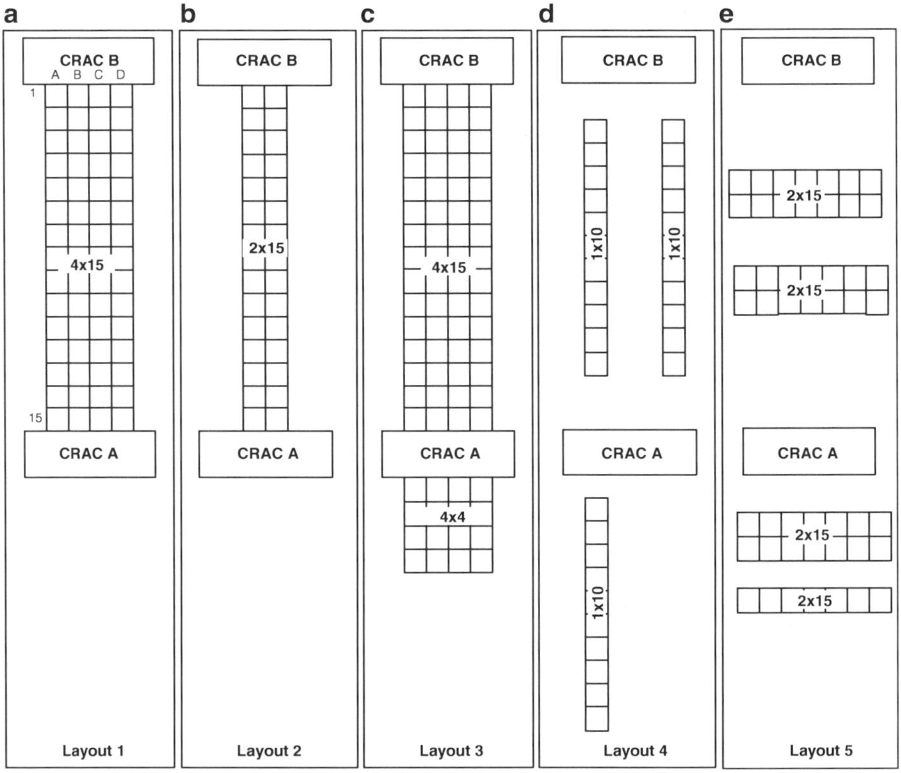


**Fig. 2.19** Effect of CRAC layout on perforated tile flow rates

the aisle layout for a fixed CRAC position is discussed in this section. The results presented are based on the experimental measurements reported in [20, 21]. Experiments were conducted in raised-floor facility with a 30-cm deep under-floor plenum. The standard 0.36 $m^2$ ($2 \times 2$ ft) perforated tiles used in the study provided 19.5% open area to flow. A total of five floor layouts as illustrated in Fig. 2.19 were investigated. The tiles are numbered 1–15 along the length of the aisle and A–D along the width of the aisle. The key observations are summarized here. For more details one may refer to [20, 21].

In all the five cases investigated, when CRAC B was operational, reversed flows were reported in the first two tiles 1A–1D and 2A–2D placed next to CRAC B as illustrated in Fig. 2.19. As explained earlier, negative flows near the CRAC unit result due to large discharge velocities.

(a) Case 1
The layout had an aisle width of $4 \times 15$ tiles as illustrated in Fig. 2.19a. In this case, both CRAC units "A" and "B" were operational. A forward flow deflecting vane referred to as a "scoop" by the authors was mounted on CRAC "A" with the intention to divert the flow towards CRAC B. In this

arrangement, higher flow rates in the perforated tiles closer to CRAC "A" were reported. The highest flow rates were reported in the row of tiles 8A–8D, located midway between CRAC "A" and CRAC "B." Installing a flow deflecting vane was recommended, as it not only reduced the severity of reversed flow but also improved the distribution of air over a larger aisle area. In the same layout when CRAC "A" was turned off, reversed flows were observed in the first three columns (1A–1D till 3A–3D) near CRAC "B." The flow rate monotonically increased towards CRAC "A" which had been turned off.

(b) Case 2
The layout was modified to include a 2 × 15 array of tiles between the CRAC units as shown in Fig. 2.19b. It was observed that decreasing the width of the aisle did not appreciably change the trend of the flow distribution pattern compared to case 1. However, a net increase in the average flow rate per tile, and a reduction in the intensity of reversed flows were reported. The reader may recollect that decreasing the number of tiles effectively decreases the total plenum open area. This results in an increase in the overall plenum pressure leading to a more stable flow. When only CRAC "B" was operational, reversed flows were confined to first two tiles (1A and 1B) located next to CRAC B, as illustrated in Fig. 2.19b. Further, when the aisle width was reduced to one tile, the reversed flows vanished, suggesting that the pressure in the plenum was positive throughout.

(c) Case 3
In this case the layout was modified to include an additional array of 4 × 4 tiles behind CRAC "A," as illustrated in Fig. 2.19c. CRAC "A" was fitted with a deflecting vane as described previously. The flow rate was found to be nearly uniform across all the tiles placed in front and behind CRAC "A." The flow rate gradually decreased in the tiles located towards CRAC "B." When the deflecting vane was removed, a decrease in the flow rate close to CRAC "A" was observed. In both the tests, the tile flow rate increased further away from the CRAC units. Another interesting observation was the emergence of reversed flows in the 4 × 4 array of tiles located behind CRAC "A," after the deflecting vane was removed.

(d) Case 4
The layout was modified with an array of 6 × 2 and 6 × 1 tiles oriented parallel to the CRAC units, as illustrated in Fig. 2.19d. Significantly higher flow rates were observed in perforated tiles placed between the CRAC units in comparison to the perforated tiles placed behind CRAC "A." When CRAC "A" was powered off, reversed flows were reported in perforated tiles adjacent to CRAC "B." The flow was found to be nearly identical in all the other tiles.

(e) Case 5
Figure 2.19e illustrates three isolated rows of perforated tiles, with two rows located in between the CRAC units, and one row to the left of CRAC A. When both CRAC units were operational, higher flow rates were reported in the tiles placed between CRAC units. The flow rates in the tiles behind CRAC "A" were

found to be fairly uniform, whereas an increasing trend in the flow rates close to CRAC "A" were reported in the tiles located in between the two CRAC units. When CRAC "A" was turned off, reversed flows were reported in the tiles next to CRAC "B." Nearly identical flow rates were observed in all other tiles. Compared to the previous cases, substantial increase in the flow rates was observed in the row of tiles placed behind CRAC "A." Further, when CRAC "B" was turned off, reversed flows were reported in all tiles in the row located behind CRAC "A." In addition, a gradual increase in the flow rates in the tiles located in between CRAC "A" and CRAC "B" was reported, with higher flow rates in the tiles located closer to CRAC "B." This example illustrates the influence of perforated tile layout on air distribution in a raised-floor data center. The case study discussed here is specific to a particular plenum height. Hence, the reader is advised not to use these qualitative discussions as general guidelines for planning perforated tile layouts, or placement of CRAC units.

Unlike plenum air distribution which is influenced primarily by the plenum pressure distribution, room-level air management depends on a number of parameters, such as ceiling height, use of containments, distribution of heat loads within the rack and in the aisle, and location of ceiling return vents which are specific to each layout. Understanding the specific influence of each parameter requires detailed CFD modeling, which is discussed in Chap. 8.

## 2.7 Factors Affecting Rack Air Distribution

Successful room-level airflow management originates at the rack inlets. The purpose of the thermal management is not served if the cooling requirements of the racks are not satisfied. This section highlights some of the key parameters affecting rack-level air distribution.

In a front-to-back airflow rack, the cold air discharged from the tile is pulled in by the server fans to cool the components inside the servers. As explained in Sect. 2.5, the quantity of air ingested by the server is dependent on the server fan speed and the pressure drop across the server. A number of external factors, such as perforations on the rack doors, obstruction due to cable management kits, rear door heat exchangers and chimneys induce additional pressure drops, which could significantly affect the mass flow rate of air ingested by the server. Most IT-equipment manufacturers increase the fan speed to account for reduction in quantity of air flowing through the servers due to a "fixed" external static pressure drop created by obstructions.

Apart from physical barriers, the external pressure drop is also affected by the momentum of air delivered by the perforated tile. Since the data center operating environment is not known, it becomes extremely difficult to account for dynamic pressure drop arising due to momentum effects at the rack inlet. Usually, while designing the fan control algorithm, it is assumed that IT equipment will pull air from a quiescent environment. This is only the ideal case of rack air distribution as illustrated in Fig. 2.20a [22, 23].

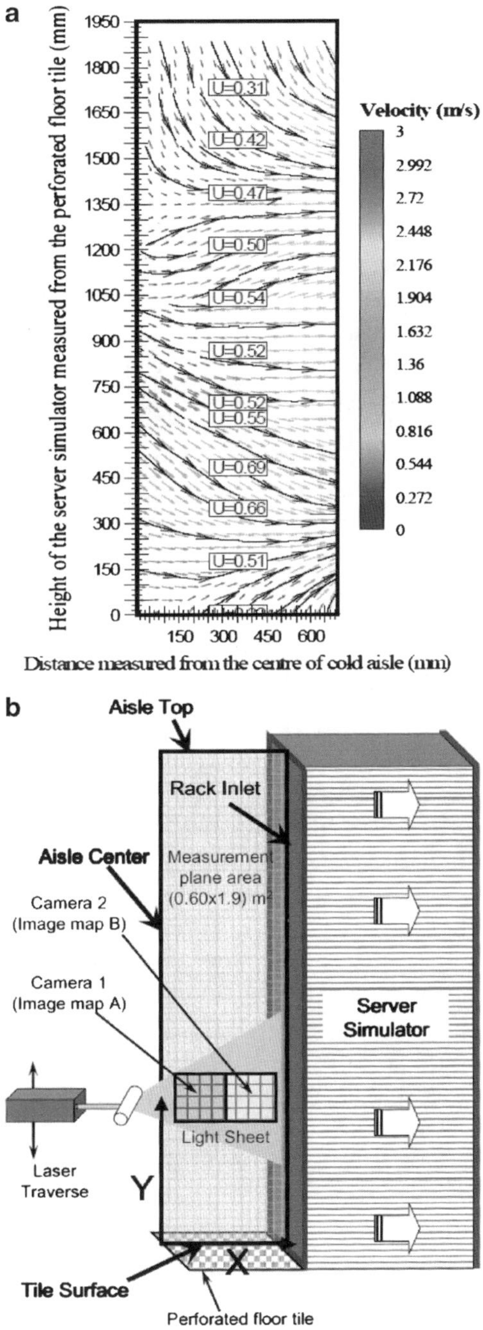


**Fig. 2.20** PIV measurement for ideal rack air distribution (© 2010 IEEE, reprinted with permission [22]). (**a**) Ideal case of rack air distribution. (**b**) PIV measurement plane.

The measurements described in [22, 23] were obtained at the rack inlet using particle image velocimetry (PIV) technique. The technique uses a pulsed laser and a charged coupled device (CCD) camera to capture the flow in the plane illuminated by the laser, as illustrated in Fig. 2.20b. The interrogation area of interest is seeded with neutrally buoyant particles, which scatter light when illuminated by the laser light sheet. The cameras and the laser pulse are synchronized to accurately trace the movement of the particles between two successive light pulses. The captured images are further processed to get information on the flow field. In this technique the seeding of the particle plays a significant role, which determines the accuracy and consistency of the measurements. Since, the particles are neutrally buoyant; it is believed that that trajectory of the particles captured by the camera represents the true velocity of the air in the interrogation area. Further details of PIV technique used for the present measurements can be found in [22, 23].

The PIV measurement plane corresponding to the vector map in Fig. 2.20a is shown in Fig. 2.20b. Referring to Fig. 2.20a, we note that the rack draws air from a quiescent ambient with no momentum-induced effects or adverse pressure gradients prevailing at the server inlet. Figure 2.21 shows a zoomed portion of the flow at the rack inlet showing normal air entry at the rack inlet.

In reality such quiescent conditions rarely exist in a data center, especially in high-density environments with significantly high perforated tile flow rates. The following section describes the effect of perforated tile velocity on rack air distribution.

### 2.7.1 *Effect of Perforated Tile Velocity on Rack Air Distribution*

The cooling of the rack is influenced by both the mass flow rate and momentum of air discharged from the perforated tile. Momentum effects become especially significant at higher flow rates and larger vent sizes, both of which invariably affect rack cooling. A series of experimental PIV investigations reported in [22–26] systematically demonstrate the effect of momentum of the air jet discharged from the perforated tile on rack air distribution.

Figure 2.22a illustrates the velocity field at the exit of perforated tile surface as the air enters the room. The perforated tile is set to discharge 0.754 $m^3/s$ (~1,600 cfm) of air, which is roughly 62% of the rack air requirement based on a $\Delta T_{Rack} = 20°C$ across the rack. As observed from Fig. 2.22b, the flow is highly nonuniform at the perforated tile surface. This is attributed to the geometry of the tile shown in Fig. 2.22c. The sudden expansion of the high-velocity jets emerging from the tile create adverse pressure gradients, leading to reversed flows as shown within the dotted ellipse region in Fig. 2.22a. In this region the air actually flows back into the plenum. The high momentum jets also affect the distribution of air to the rack as shown in Fig. 2.22.

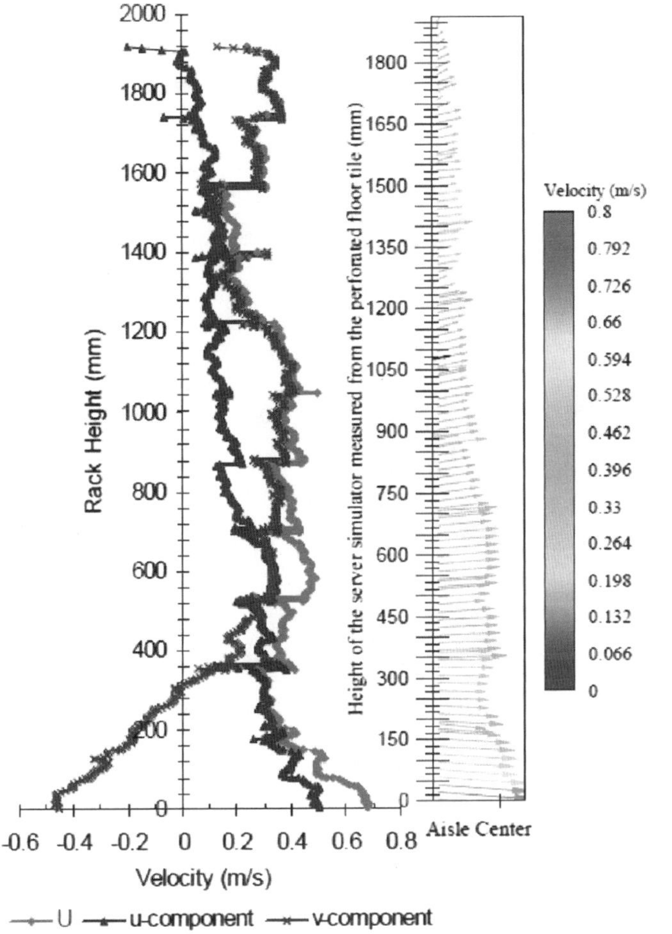


**Fig. 2.21** Shows a zoomed portion of the flow at the rack inlet showing normal air entry at the rack inlet (© 2010 IEEE, reprinted with permission [22])

One can easily comprehend the momentum effect by comparing the flow fields in Figs. 2.20a and 2.22. The high momentum jets of air discharged from the perforated tile wash the face of the rack and escape through the top of the aisle (Fig. 2.24).

Figure 2.24 shows the velocity profile at the top of the rack. It is estimated that 65% of the air supplied by the tile is lost into the aisle.

The deviation in normal air entry to rack is highlighted in Fig. 2.25, which shows a zoomed portion of the flow at the rack inlet. The slanted entry further reduces the net intake of air by the rack. It is important to note that in the region close to the perforated tile, the air actually flows out of the rack into the cold aisle against the preferred direction of flow induced by the server fans. Further analysis of the PIV vector map illustrated in Fig. 2.25 revealed that compared to the ideal case illustrated in Fig. 2.21, there is a 25% reduction in flow to the rack. These observations are further supported

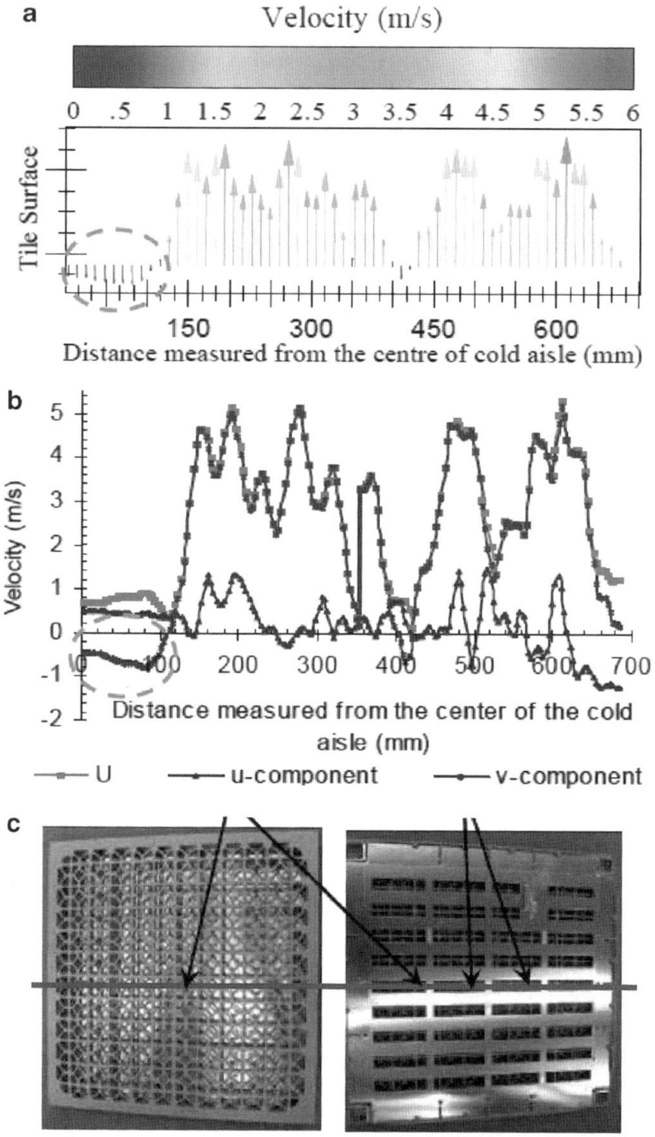


**Fig. 2.22** Details of the flow at the perforated tile surface (© 2010 IEEE, reprinted with permission [22]). (**a**) PIV vector map of the flow at the outlet of perforated tile surface. (**b**) Velocity map corresponding to the flow in (**a**). (**c**) Details of the flow at the perforated tile

by a similar CFD study for a hypothetical data center in [27]. The simulations reported that the high momentum air discharged from the perforated tile can starve the IT equipment placed in the bottom of the rack by reducing the airflow by as much as 15% of the total air needed by the server [27].

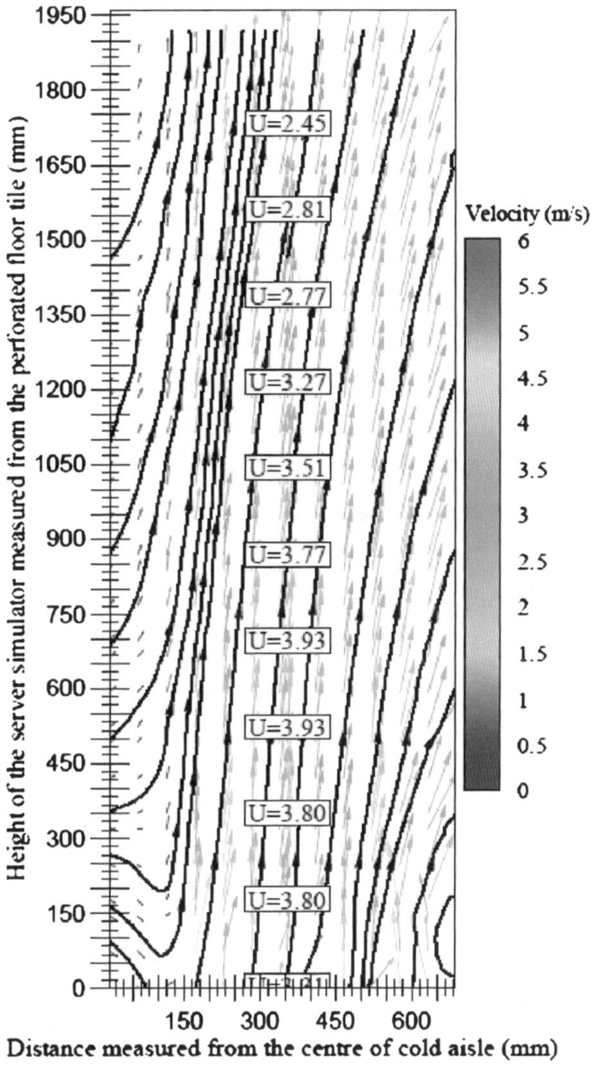


**Fig. 2.23** Air distribution at the rack inlet with perforated tile supply of 0.770 $m^3/s$ (1,415 cfm) (© 2010 IEEE, reprinted with permission [22])

However, as illustrated in Fig. 2.26, if the perforated tile flow is reduced, the rack draws air from the room to satisfy its requirement. This could be either recirculated hot air from the hot aisle or stolen air from neighboring racks in the cold aisle.

The measurements presented in Figs. 2.20–2.26 were obtained using a 22-kW server simulator. Server simulators are essentially a set of heater banks equipped with fans stacked in a standard rack. Since, the server simulator is equipped with actual server fans used in IT equipment; the airflow results presented here are expected to be qualitatively similar to those in a real server. The advantage of using

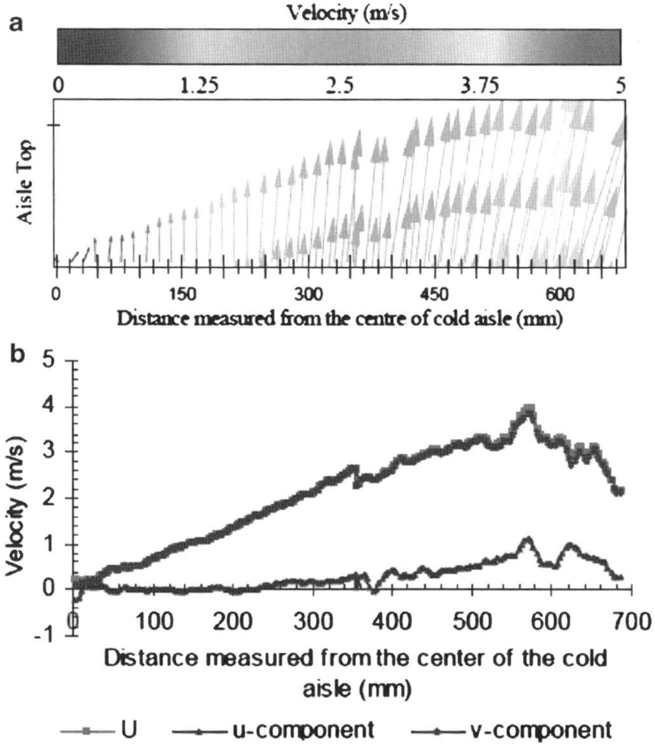


**Fig. 2.24** Details of the airflow at the top of the cold aisle for a perforated tile supply of 0.770 $m^3/s$ (1,415 cfm) (© 2010 IEEE, reprinted with permission [22]). (**a**) PIV vector map at the aisle top. (**b**) Velocity map corresponding to the flow in Fig. 2.24a

a sever simulator is the ability to control the fan speed and heat dissipation, which are difficult in an actual server. In the results presented in Figs. 2.20–2.26 the server simulator fan speeds were set to draw in 1.224 $m^3/s$ (~2,600 cfm) of air.

The observations presented in the previous section bring up the following questions; If the present method of air distribution is crippled with so many issues, why aren't servers in existing data centers failing?, and more importantly how are these servers being cooled? The answers to these questions lie in understanding air distribution to low density server racks, with dissipation <8 kW. Figure 2.27 illustrates the inlet air distribution for a low-density rack corresponding to the PIV measurement plane in Fig. 2.21b.

In this case, the perforated tile discharges 0.31 $m^3/s$ (650 cfm) of air at the server inlet to meet the total rack air demand. From the PIV plot in Fig. 2.26, we note that the entire quantity of air discharged from the perforated tiles is evenly distributed and ingested by the rack. The reader may recollect from (2.1) and (2.5) that perforated tile pressure drop and momentum are both proportional to the square of volumetric flow rate from the perforated tile. Hence, a small reduction in volumetric flow from the perforated tile significantly reduces both pressure drop

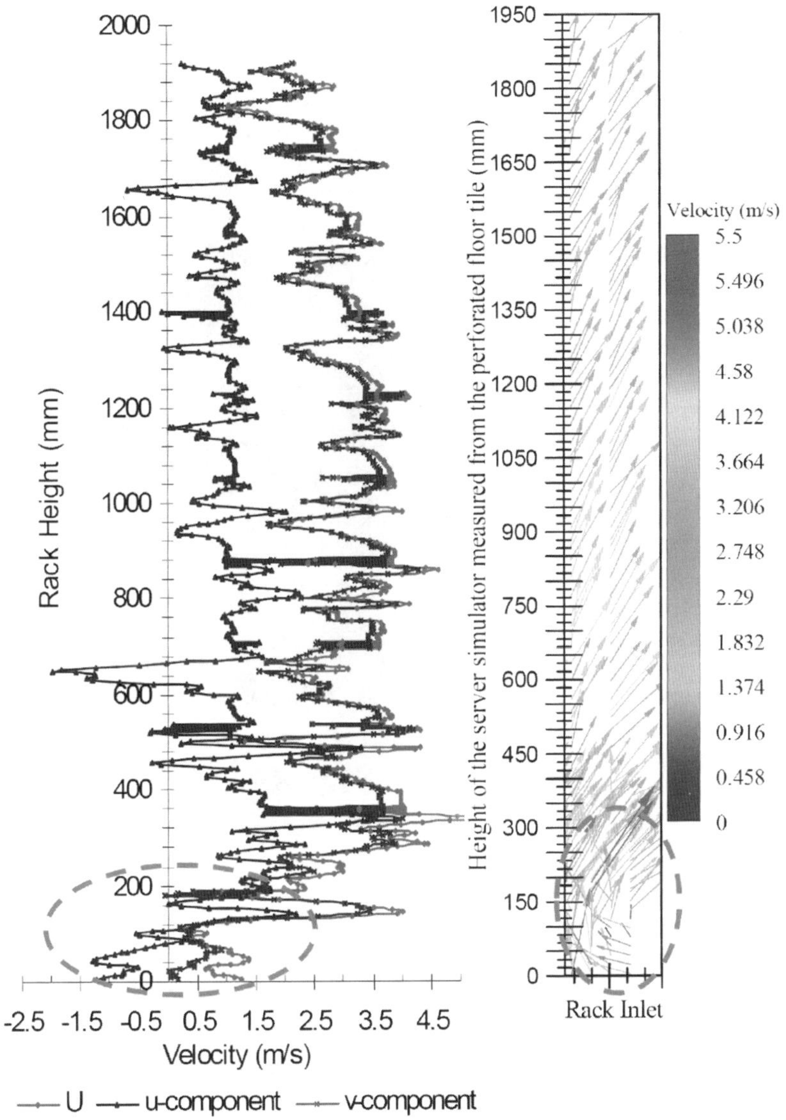


**Fig. 2.25** Zoomed portion of the flow near the rack inlet illustrating slanted air entry for a perforated tile supply of 0.770 $m^3/s$ (1,415 cfm) (© 2010 IEEE, reprinted with permission [22])

and momentum across the tile. As such many of the ill effects of high-density air distribution simply disappear at low cabinet power densities. This is the likely reason for the success of cooling in many of the present data centers.

A further extension of the study involving server-level load migration revealed that the air discharged from the perforated tile had adequate momentum to reach the topmost servers located in the rack. In the 12 cases reported in [26], the airflow

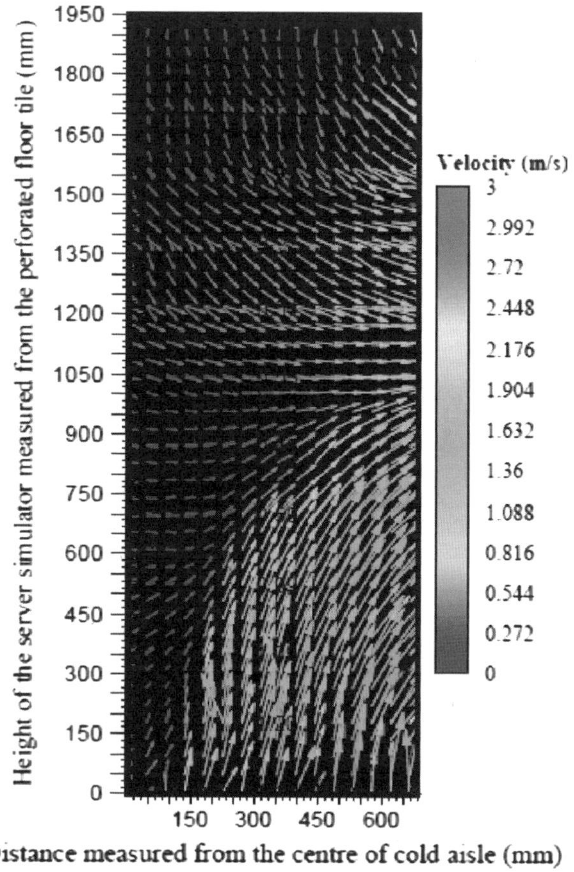


**Fig. 2.26** Air distribution at the rack inlet with perforated tile supply of 0.234 $m^3/s$ (495 cfm) (© 2010 IEEE, reprinted with permission [23])

distribution was found to be uniform irrespective of the location of the server in the rack. The rack air distribution at various server locations is illustrated in Fig. 2.28. As illustrated in Fig. 2.28, at low rack densities, the location of server in the rack is immaterial from the perspective of air distribution. This result is significant, as it helps in understanding the distribution of compute load in a virtualized data center environment. The importance of thermally aware load migration in a virtualized data center is discussed in Chaps. 4 and 10.

Presently, no conclusions can be drawn on the applicability of the above results to higher rack densities; further research is needed to answer this question.

### *2.7.2 Effect of Asymmetrical Aisle Air Distribution*

The mismatch in the perforated tile flow can also affect the air distribution to the rack. Figure 2.29 illustrates the case of symmetrical air distribution at the inlet of

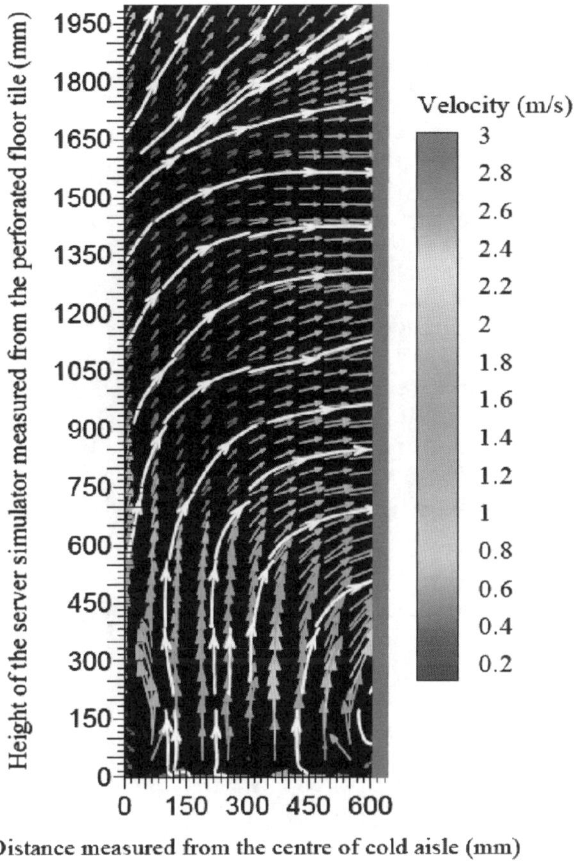


**Fig. 2.27** Inlet air distribution for a low-density rack (© 2011 IEEE, reprinted with permission [26])

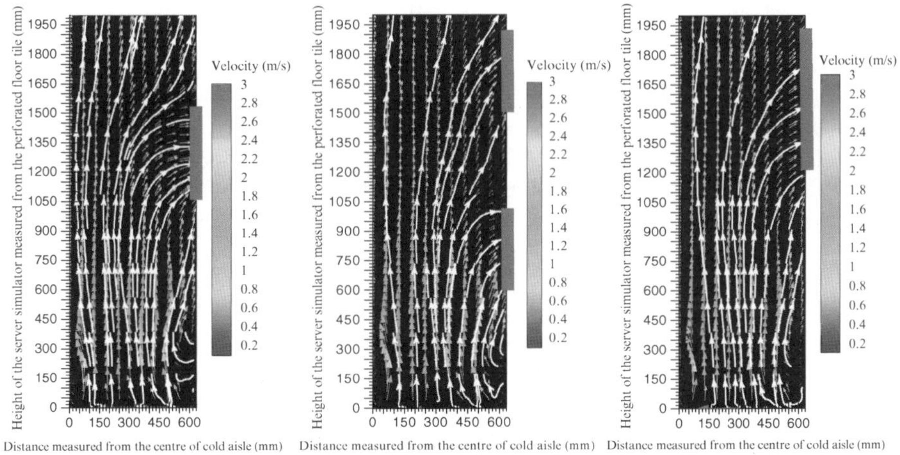


**Fig. 2.28** Effect of server load placement in a low-density rack (© 2011 IEEE, reprinted with permission [26])

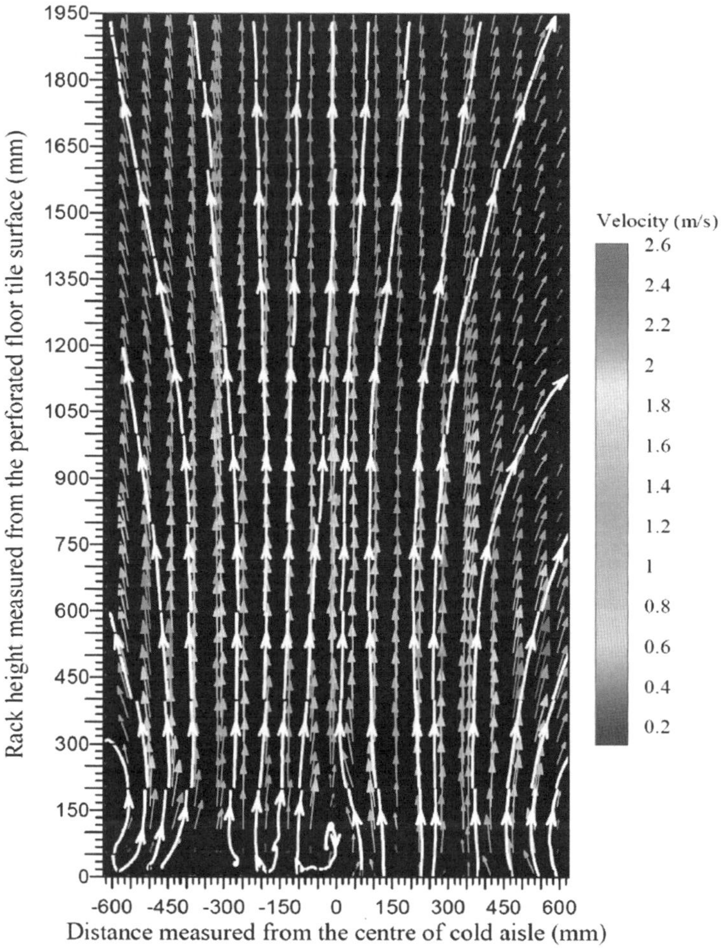


**Fig. 2.29** Uniform aisle air distribution in the aisle (© 2011 ASME, reprinted with permission [24])

two opposing racks located in the center of the cold aisle. In this case, there was no mismatch in flow between the perforated tiles placed in front of the racks. Both the tiles discharged approximately 0.428 $m^3/s$ (~900 cfm) of air. In addition, the racks had identical flow and power ratings. As described in the previous section, in the present case too, the effect of high-momentum jets discharged from the perforated tiles is evident in Fig. 2.29. From Fig. 2.29, and we note that the flow at the inlet of both the racks near the perforated tiles is displaced due to adverse pressure gradients created by the high-velocity jets.

Figure 2.30 shows asymmetrical cold aisle air distribution in two opposing racks.

Asymmetry in flow is created by varying the flow to the left rack, the flow from the left tile is reduced from 0.428 $m^3/s$ (~900 cfm) to 0.226 $m^3/s$ (480 cfm) as shown

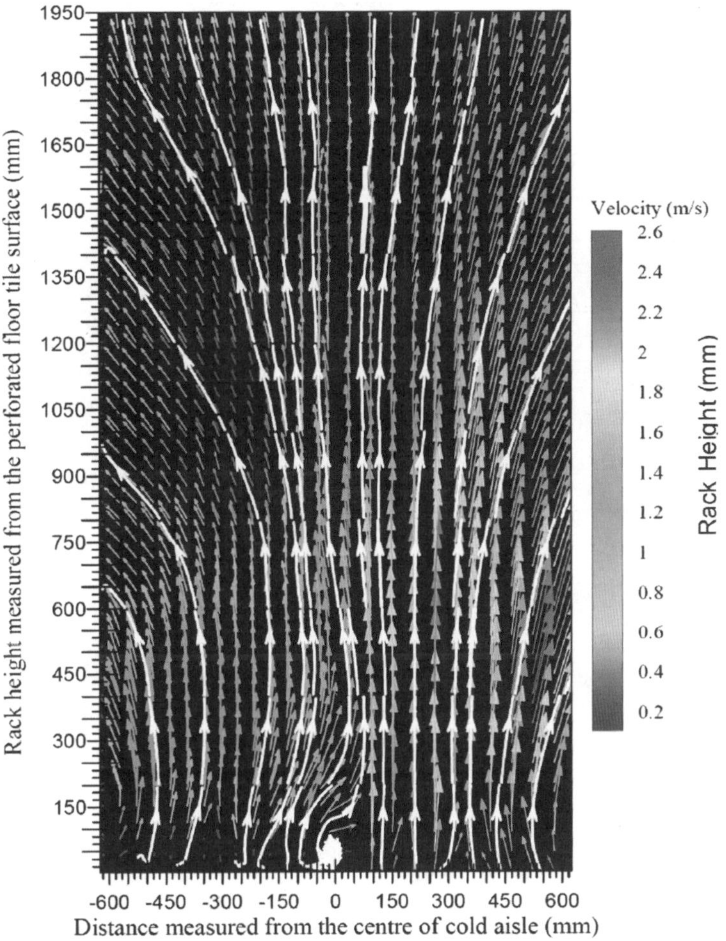


**Fig. 2.30** Nonuniform aisle air distribution (© 2011 IEEE, reprinted with permission [24])

in Fig. 2.30. All other conditions, including the flow rate to the right rack remained identical to the case presented in Fig. 2.29. The disparity in the flow discharged from the two tiles results in a difference in the air velocities emerging from the perforated tile. This mismatch in velocities between the two adjacent layers of air at the aisle center creates an imbalance in static pressure at the center of the aisle. This is schematically illustrated in Fig. 2.31.

The low-velocity airstreams discharged from the left tile are dragged by the adjacent high-velocity airstreams discharged from the right tile. As a result, the high-velocity air discharged from the right tile is retarded while the low velocity air discharged from the left tile is accelerated. The acceleration and deceleration in the flow also creates an imbalance in the static pressure along the aisle center. Due to

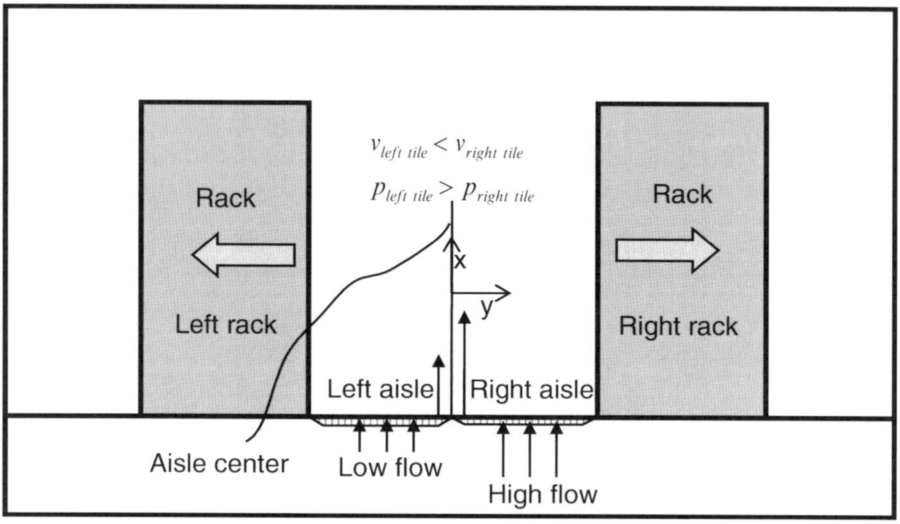


**Fig. 2.31** Static pressure imbalance at the aisle center

high discharge velocities, the local pressure in aisle center corresponding to the right tile is lower than the local pressure in the aisle center corresponding to the left tile as illustrated in Fig. 2.31. The imbalance in static pressure results in entrainment of the air from the left aisle into the right aisle, as indicated by the curved path lines in Fig. 2.30.

It may be noted that the static pressure gradient is created due to the *difference in velocities* between the two adjoining fluid layers and *not the absolute velocities of the fluid streams*. In the extreme case, if the flow from the left tile is completely blocked, the high-velocity jets discharged from the right tile will drag the stagnant column of air in the left aisle resulting in entrainment along the entire aisle height as illustrated in Fig. 2.32. The entrainment region is indicated as a dotted ellipse in Fig. 2.32. The condition represented in Fig. 2.32 can be related to the first or last rack in the aisle where a solid tile is placed immediately next to the perforated tile as illustrated in Fig. 2.33. Air entrainment at the aisle ends are commonly referred to as end effects. In an open aisle data center, the entrained air is usually hot air recirculated from the rear of the racks. The aisle end effects have been well documented and validated through numerous experimental measurements and CFD simulations.

The above examples demonstrate the asymmetrical nature of air distribution between two opposing racks. However, asymmetry can also exist between two adjacent racks placed sidewise. This asymmetry in airflow could affect the net rack air intake, which in turn could affect the rack cooling. This is a topic of ongoing research.

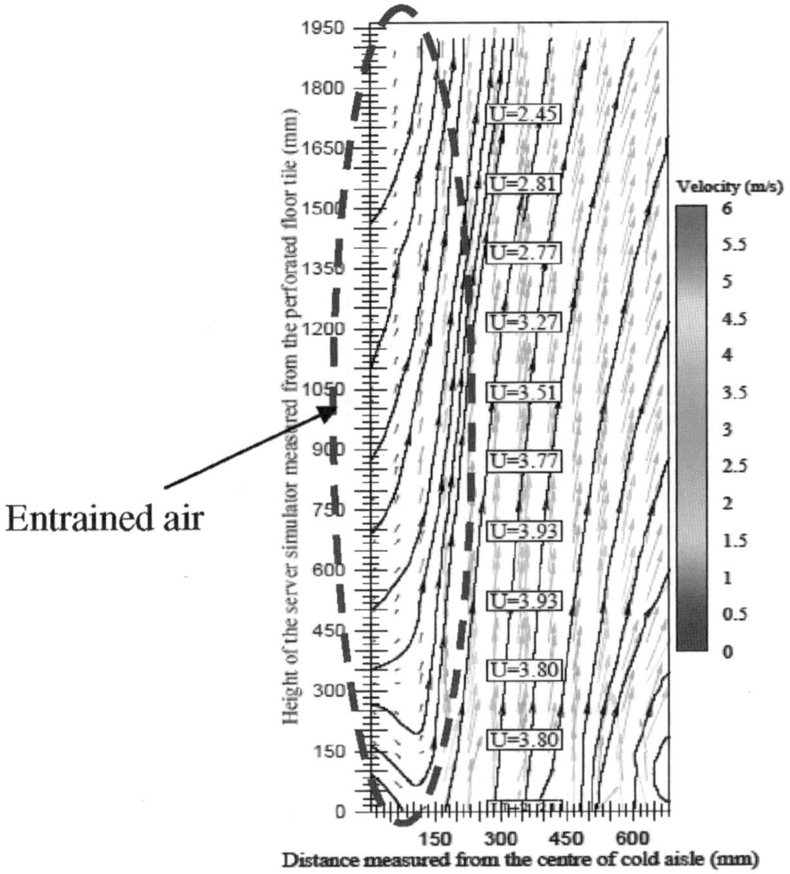


**Fig. 2.32** Entrainment at the aisle center due to mismatch in tile perforated velocities (© 2010 IEEE, reprinted with permission [22])

### *2.7.3 Effect of Supply Air Temperature and Airflow on Rack Cooling*

Till now, the entire discussion revolved around the issue of managing airflow only. Basically, there are two factors that influence cooling efficiency in a data center, (1) the quantity of air supplied and (2) the temperature of the air supplied to the rack. On one hand, increasing the quantity of air supplied to the racks increases the fan energy needed to move the air to the racks, while on the other hand, reducing the supply air temperature increases the chiller energy consumption. Hence, the mandate to manage air distribution must balance both thermal environment and energy savings.

Another important aspect of data center thermal management is that unlike airflow which can be individually tuned to meet the demands of IT equipment, temperature and humidity cannot be specifically tuned to satisfy the individual demands of IT equipment. Hence, changes in temperature or humidity set points have to be done in concert with the air management.

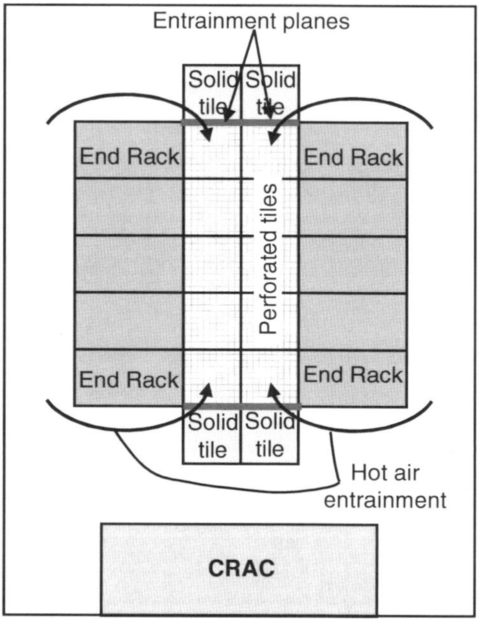


**Fig. 2.33** Entrainment at the aisle ends

The following section discusses the effect of supply air temperature and airflow on rack cooling. The reader may note that the discussions presented in the following section may not be applicable to all situations. The experimental results described in this section were obtained using a 12-kW rack housing 42, 1U (1U-2.45 cm) servers. This data represents performance in a real data center.

(a) Effect of high perforated flow and low supply temperature

Figure 2.34 presents the case for perforated tile flow rate of 0.667 $m^3/s$ (1,413 cfm) and a supply temperature of 12°C measured at the perforated tile surface. The temperature measurements were obtained using a temperature grid consisting of 256 thermocouples. Interested reader is referred to [25] for further details regarding temperature-measurement technique.

Figure 2.34b presents the PIV vector map, and Fig. 2.34a, c, the temperature fields in the plane corresponding to the PIV measurements. The PIV measurements correspond to the plane shown in Fig. 2.20b. A sketch in each figure is provided to aid in identifying the location of the measurement plane corresponding to the reported temperature field.

Referring to Fig. 2.34a, we notice a pocket of hot air at the perforated tile surface near the rack inlet. This is counter-intuitive as one would have expected the temperatures to be lowest in this location, close to the perforated tile surface. This result will no longer be a surprise if the reader recollects the effect of high-velocity jet discussed previously. As explained in Sect. 2.71, the sudden expansion of high-velocity jets discharged from the perforated tile surface creates regions of low pressure at the inlet of the servers placed closer to the perforated tile. Consequently, the hot air from the hot aisle is entrained to balance the pressure in this region, resulting in a hot spot at the rack inlet.

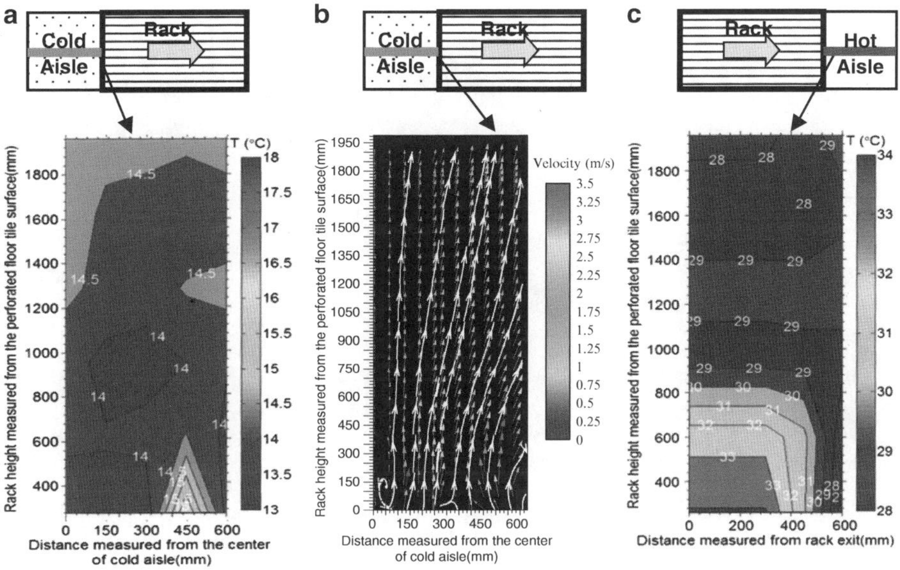


**Fig. 2.34** Rack inlet and exit temperature and flow profiles for supply air temperature of 12°C and perforated tile flow rate 0.667 $m^3/s$ (1,413 cfm) (© 2011 ASME, reprinted with permission [25]). (**a**) Rack inlet temperature profile. (**b**) PIV vector map. (**c**) Rack exit temperature profiles

The reader may note that the impact of the high-velocity jet is much greater than just ingress of hot air. Due to the adverse pressure gradients prevailing at the rack inlet, the server fans are not able to draw in sufficient air required to cool the server. The reduction in cold air supply results in higher exhaust temperatures for the same server heat dissipation. In addition to the reduction in flow, the server also ingests recirculated hot air from the hot pocket shown in Fig. 2.34a. Due to ingress of hot air, the recirculated air participates in cooling the equipment multiple times, resulting in higher exit temperatures. Figure 2.34c shows the temperature variation in the hot aisle at the exit of the rack. As expected, the location of the hot spot at the rack exit coincides with the location of the pocket of hot air at the inlet of the rack, observed in Fig. 2.34a. Note that the temperatures in the hot spot are of the order of 35°C. If the supply air temperature were increased, the hot spot temperature would rise further, which could be deleterious to the server's performance or reliability. In addition, the control system in the CRAC may perceive this local hot spot as a potential cooling deficiency, and increase the operating fan speed or chilled water flow, both of which would lead to an increase in the cooling energy consumption.

(b) Effect of moderate perforated flow and low supply temperature

The effect of a slight decrease in the perforated tile flow on rack cooling is illustrated in Fig. 2.35a–c. The perforated tile flow was decreased from 0.667 $m^3/s$ (1,413 cfm) to 0.523 $m^3/s$ (1,108 cfm). As observed from the flow

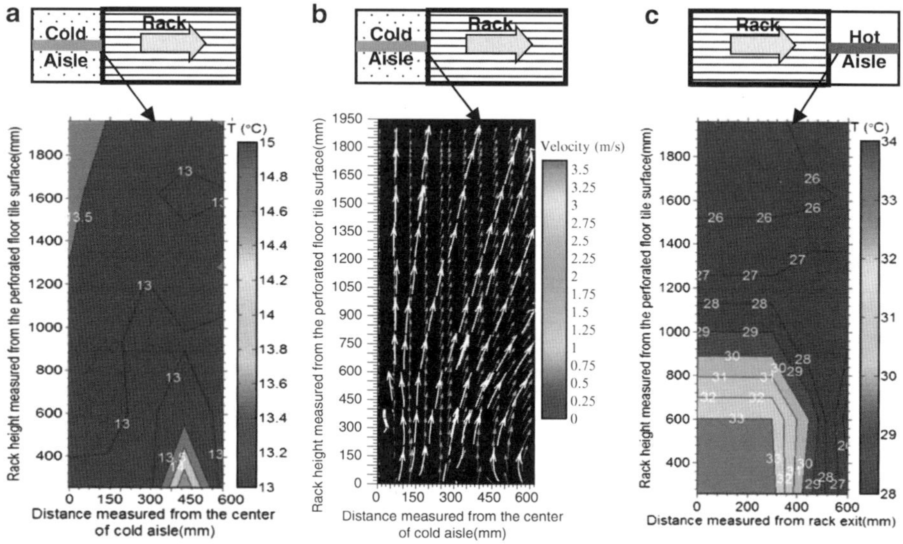


**Fig. 2.35** Rack inlet and exit temperature and flow profiles for supply air temperature of 12°C and perforated tile flow rate 0.523 $m^3/s$ (1,108 cfm) (© 2011 ASME, reprinted with permission [25]). (**a**) Rack inlet temperature profile. (**b**) PIV vector map. (**c**) Rack exit temperature profiles

field in Fig. 2.35a, apart from a decrease in the intensity of reversed flow, the decrease in flow rate does not bring an appreciable change in the overall flow or temperature distribution compared to the previous case presented in Fig. 2.34a. The decrease in perforated tile velocity resulted in minor recirculation in the cold aisle, as observed by the small pocket of hot air at the top of the aisle in Fig. 2.35a.

(c) Effect of low perforated flow with low supply temperature

Next we present the effect of decreasing the perforated tile flow on server inlet and exit temperature distribution. Fig. 2.36a–c presents the results when the perforated tile flow rate is reduced from 0.667 $m^3/s$ (1,108 cfm) to 0.28 $m^3/s$ (586 cfm).

The reduction in flow rate reveals a substantial change in both the temperature and flow field at the rack inlet as illustrated in Fig. 2.36a–c. Referring to Fig. 2.36b we note that the decrease in perforated tile air discharge velocity marks the absence of reversed flows at the perforated tile surface. This suggests that momentum effects no longer create adverse pressure gradients at the rack inlet. The absence of reversed flows is further corroborated by disappearance of the hot air pocket at the inlet of the rack as observed in Fig. 2.36a. Comparing the temperature profile in Fig. 2.35a with Fig. 2.36a, increased stratification in the temperature along the rack height is observed. Higher temperatures in the upper half of the rack suggest substantial recirculation of hot air exhaust into the cold aisle. This is mainly caused by a decrease in the perforated tile flow rate, as the air supplied by the perforated tile does not have enough momentum to reach

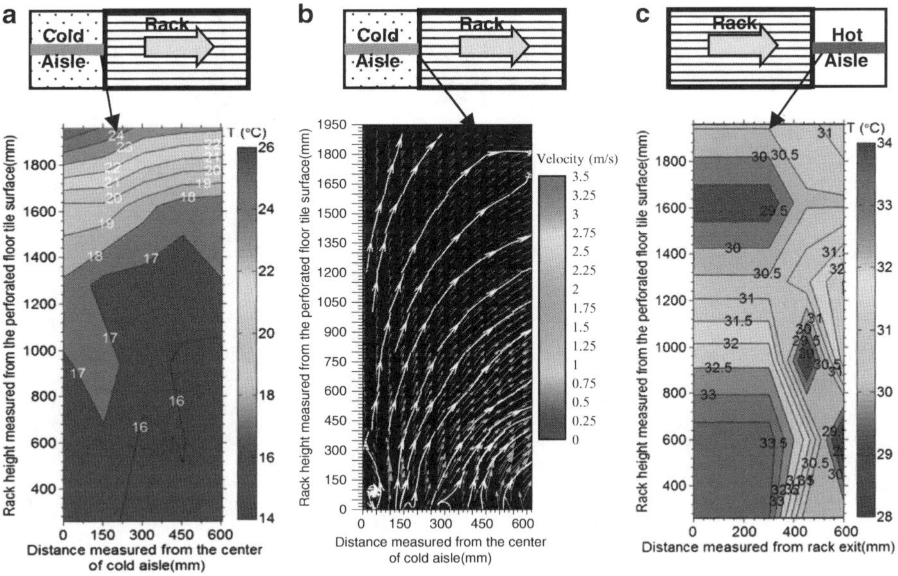


**Figure 2.36** Rack inlet and exit temperature and flow profiles for supply air temperature of 12°C and perforated tile flow rate 0.28 m$^3$/s (1413 cfm) (© 2011 ASME, reprinted with permission [25]). (**a**) Rack inlet temperature profile. (**b**) PIV vector map. (**c**) Rack exit temperature profiles

the top of the aisle. In addition, a sizeable part of the air supplied by the perforated tile is ingested by the servers located in the lower half of the rack, thus resulting in reduced flow to the servers located in the upper half of the rack. It is interesting to observe that even with higher inlet air temperatures the exit air temperatures in the upper half of the rack are relatively lower compared to the rest of the rack inlet. This suggests that the momentum of supply air plays a significant role in the amount of air ingested by the servers. Referring to temperature profiles in Fig. 2.36a, b, we observe that even with higher inlet temperatures the net increase in $\Delta T$ across the rack is only 6–8°C compared to 14–18°C in the lower half of the rack. This suggests that the servers in the upper half of the rack are able to ingest more air due to existence of near quiescent conditions at the inlet. The relatively low tile discharge velocities in the cold aisle are responsible for improved rack air intake efficiencies.

One may note that even with the recirculation at the top, the inlet temperatures in the cold aisle are below the ASHRAE [5] recommended upper limit of 27°C, which leaves ample room for increasing the supply air temperature. The effect of increase in supply air temperature is presented in next.

(d) Effect of low perforated flow with high supply temperature

Figure 2.27a–c illustrates the effect of increasing the supply temperature while keeping the flow rate low. The supply air temperature was increased from 12°C to 22°C.

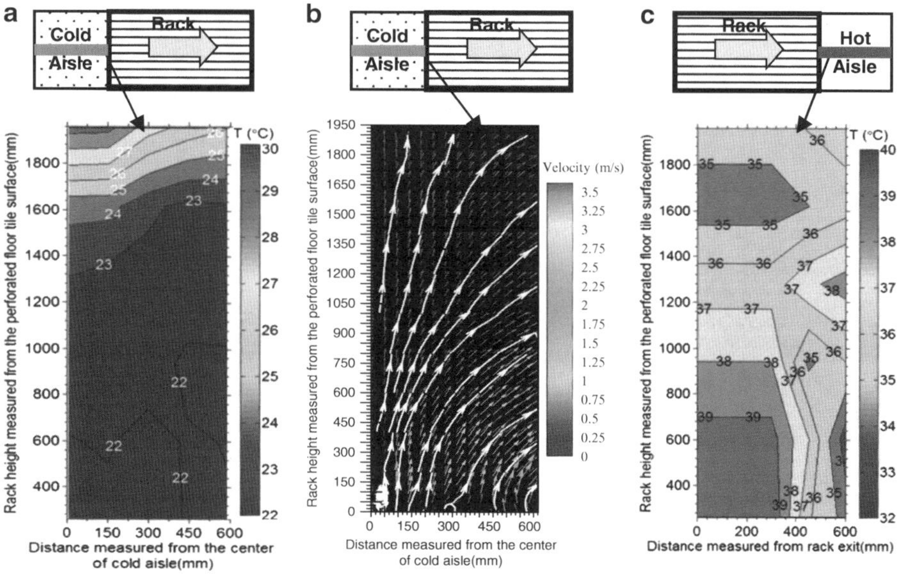


**Fig. 2.37** Rack inlet and exit temperature and flow profiles for supply air temperature of 22°C and perforated tile flow rate 0.28 $m^3/s$ (© 2011 ASME, reprinted with permission [25]). (**a**) Rack inlet temperature profile. (**b**) PIV vector map. (**c**) Rack exit temperature profiles

Comparing the inlet flow fields in Fig. 2.37a, b at the rack inlet, no significant change in the airflow pattern is observed. This suggests that the server fan speed did not vary with a 10°C rise in inlet temperature. This is in agreement with the ASHRAE guidelines [5], which suggest an insignificant change in fan speed for variations in inlet temperatures below 30°C. The increase in the supply air temperature from 12°C to 22°C resuls in a 2.5% variation in air density at standard atmospheric pressure. As such, the cold air in the present case is able to travel higher up into the cold aisle. This is illustrated in Fig. 2.37a. Also, due to higher exit temperatures, the exhaust air tends to rise higher up and settle closer to the drop ceiling. The net effect is a reduction in the ingress of hot air into the cold aisle. This reduction in recirculation is evidenced by the smaller pocket of hot air residing at the top of the cold aisle in Fig. 2.37a, compared to Fig. 2.36a.

Although these observations seem to favor lower tile flow rates, note that the inlet temperature conditions at some locations either exceed or are very close to the ASHRAE-recommended limits of 27°C, thus leaving very little margin for error in control system.

The above observations are further supported by measurements taken in a data center employing raised-floor air distribution system [28]. The study reported higher temperatures at the bottom of the rack. The authors speculated that hot air must have been entrained into the flow from the tile within a few inches above the raised floor. A similar study [29] found that the influx of cold air from the hot aisle

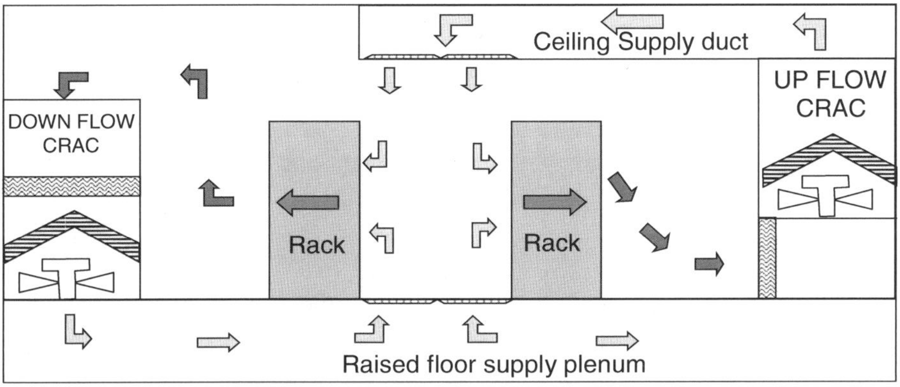


**Fig. 2.38** Dual air supply system

into the cold aisle was greatly reduced with the introduction of blanking panels to close the gap created by the casters between the floor and the rack.

The above investigations on rack air distribution suggest that momentum of supply air can play an important role in high-density cooling. Presently, the average rack densities in legacy data centers range between 5 and 8 kW/rack.[1] With continued replacement of the existing IT equipment with newer high-power racks, these data centers will eventually transit to higher heat densities, thus requiring more airflow. This transition phase has to be handled carefully. As we have observed, merely increasing the perforated tile flow will not resolve the increase in cooling demand. In fact, increasing the perforated tile supply would adversely impact rack cooling. Alternate methods of supplying cold air need to be considered. One option would be to increase the width of the cold aisle to accommodate more perforated tiles. This may drive the average density down but is the most economical solution requiring least infrastructural changes. However, this may not be an attractive proposition where floor space is at a premium. In such cases, the raised floor cooling needs to be supplemented with other cooling methods such as row- or rack-based cooling. These and some other alternate methods have been discussed in Chap. 1. The other option, which is gaining popularity, is dual supply system where air is supplied both from an UFAD system and an OFAD system as shown in Fig. 2.38. In this configuration, in addition to the air supplied by the raised-floor plenum, additional supply air is forced downward in the cold aisle using an overhead air distribution system.

[1] Here average rack density refers to the heat generated by the rack and not the name plate power. It has been observed that even in high density environments with name plate power rack power ratings of ~20 kW the average heat density based on normal operation varies between 6 and 8 kW. However, this is expected to rise in the near future with the virtualization and cloud computing.

## 2.8 Factors Affecting Room Air Distribution

### *2.8.1 Effect of Ceiling Height*

Compelling arguments have been made both in favor and against tall and shallow ceiling heights. The arguments are based on the issue of stratification. Stratification results from a complex interaction of hot air discharge from the IT equipment, aisle, and CRAC layout and location of hot air return ducts which are specific to a particular data center layout. Some data center designers argue that high ceilings allow the hot air to rise higher due to buoyancy, thus minimizing the bypass of hot air into the cold aisle above the racks. While others argue against it stating that tall ceilings are not effective in capturing the hot air, which results in lower CRAC return-air temperatures thus affecting the CRAC cooling performance. The minimum recommended ceiling height is 270 cm (9 ft).

The effect of ceiling height using CFD modeling for various layouts is discussed in Chap. 8. Based on the findings reported in [28, 30–32], it is inconclusive if tall ceilings improve thermal performance of data centers. Hence, presently, no recommendation can be made regarding the height of the ceiling.

### *2.8.2 Effect of Aisle Containment*

Containments are physical barriers used to enhance separation of hot and cold airstreams. They involve enclosing either the hot or cold aisle. The cold aisle containment system (CACS) envelopes the cold aisle by deploying a physical barrier to isolate it from the rest of the room. As such the entire room becomes the hot aisle, resulting in higher room air temperatures. In typical high-density environments with contained cold aisles, the room temperatures can be as high as 47°C. Since the data center houses other systems, such as power distribution transformers and other non-IT equipment outside the cold aisles, adequate provisioning for additional cooling must be provided to these equipments, or the usage of the equipment at higher temperatures needs to be evaluated. There is a limit to the amount of air delivered in a CACS system, which in turn limits the rack cooling capacity. The practical rack heat density limit with CACS system is about 6 kW [33]. However, higher heat densities can be achieved by increasing the quantity of air delivered using fan-assisted floor tiles [33].

In contrast to the CACS system, the Hot Aisle Containment System (HACS) system encloses the hot aisle to collect and isolate the hot exhaust air from the racks, thus allowing the entire room to serve as a cold aisle. In this case, the entire room acts as a reservoir of cold air. By virtue of design HACS embodies the advantages of CACS system, while inherently overcoming some of the draw backs of CACS. Both solutions work well but hot aisle containments are generally considered to be more efficient than cold aisle containment. Successfully designed containment solutions with minimal leakage nearly eliminate all mixing issues.

Although, some manufacturers claim an improvement of 25–30% in cooling efficiency by use of containment [34], it does not necessarily lead to the lowest cooling power consumption. Enclosing the aisles would mean cold air requirement of the IT equipment would have to be exclusively met by CRAC supply. The additional pressure drop imposed by the containment system would drive the CRAC fans to run at higher speeds, increasing the overall cooling infrastructure power requirement. It may be recalled from the fan laws that fan power is proportional to the third power of the fan speed. Hence, an increase in the CRAC fan speed imposes a penalty on the energy savings derived from a containment solution. In order to circumvent the increase in CRAC fan power while enclosing the cold aisle, a bypass branch with a low lift fan has been suggested [35]. An important outcome of the analysis reported in [36] was that enclosing the aisles was beneficial if $\Delta T_{\text{rack}}$ ~20.5°C or higher. The authors further extended the above analysis to servers with variable fan speeds and found an optimum CRAC supply air set point temperature between 16°C and 18°C for an open aisle for minimum power consumption. It was found that enclosing the aisles with higher supply air temperatures increased the server fan speeds leading to higher IT power consumption. In order to satisfy the increased demand for airflow, the CRAC fan speed had to be increased, thus increasing the overall operational cost. For the specific case reported, a 48% reduction in cooling power consumption was noted by maintaining the cold aisle temperature at 22°C instead of 27°C.

Implementing an economizer solution along with a containment system has to be carefully evaluated. Since an active economizer solution offers very little control on the supply air temperature, large fluctuations in supply air temperature would drive the server fans to higher speeds, resulting in higher IT power consumption. These factors have to be carefully considered during the decision process of implementing an active economizer/containment solution.

### 2.8.3 *Effect of Buoyancy*

The temperature gradients in the data center induce buoyant forces aiding in mixing of hot and cold airstreams. It is expected that the buoyancy effects become significant at higher rack densities [37]. The effect of buoyancy is quantified using the Archimedes number (Ar). Archimedes number is a measure of relative magnitude of buoyancy and inertia forces. An Ar ~ O(1) indicates that buoyancy and inertia forces are of the same order, hence, both of them must be included in the CFD model. Large Ar indicates buoyancy as the dominating force for fluid flow, whereas small Ar indicates negligible effect of buoyancy.

The Archimedes number is defined by the following relation:

$$\text{Ar} = \frac{\beta g \Delta T_{\text{Rack}} H_{\text{rack}}}{V^2{}_{\text{Perf. tile}}}, \tag{2.19}$$

where β is the volumetric coefficient of thermal expansion. $\beta = (1/T)$ for a perfect gas, $g$ is the gravitational acceleration (9.8 m/s$^2$), $\Delta T_{\text{Rack}}$ is the temperature difference between the hot and cold aisles, $V_{\text{Perf. tile}}$ is the velocity of air jet from the perforated tile, $H_{\text{Rack}}$ is a *vertical* length scale (usually the height of the rack).

$\Delta T_{\text{Rack}}$ is taken as the temperature rise across the rack. It can be either experimentally measured or estimated thermodynamically, based on rack power draw using the following relation.

$$\Delta T_{\text{Rack}} = \frac{P}{\rho_{\text{air}} C_{p,\text{air}} Q_{\text{tile}}},$$

where $P$ is the rack power dissipation (kW), $Q_{\text{tile}}$ is the volumetric flow rate of air from the perforated tile (m$^3$/s), $\rho$ is the density of air (kg/m$^3$), $C_{p,\text{air}}$ is the specific heat of air at constant pressure (kJ/kg K).

The characteristic velocity is the average velocity of air exiting the perforated tile. Either measured data can be used, or it can be estimated based on net volumetric flow rate from the tile using (2.7).

Using the above expressions, the Ar value for a 20-kW (power draw) rack with a $\Delta T_{\text{rack}} = 20°\text{C}$ for a tile flow rate of 0.38 m$^3$/s (800 cfm) using 25% open area was found to be Ar $= 1.1$. This indicates that natural convection is equally important as forced convection, in some flow regions. Buoyancy not only promotes mixing of hot and cold airstreams but also affects distribution of cold air. Including the buoyancy effects was found to considerably improve the CFD predictions, bringing them closer to measurements. Due to buoyancy, the hot air was found to rise towards the ceiling, thus minimizing the chances of being recirculated back into the cold aisle.

## 2.9 Overhead Air Distribution (OHAD) System

Most data center designers at some point are confronted with the task of selecting an air distribution method for a data center. Selection of any given system depends on numerous design, physical, and financial constraints. Traditionally, OHAD systems have been used for comfort air conditioning applications. The overhead air distribution for a data center application is illustrated in Fig. 1.16b in Chap. 1.

Although both UFAD and OHAD systems are commonly used for both building air conditioning (BAC) and data center applications, the objectives are distinct. The primary objective of the BAC system is to even out the temperature gradients in the room. The air distribution tiles used in BAC systems are designed to entrain surrounding air and quickly diffuse the jet of air discharged from the perforated tile (ceiling or floor). This inherent design of the tile promotes mixing and enhances recirculation, both of which play an important role in diffusing the temperature gradients. These objectives are different from what is desired in a data center, where the primary objective is to maximize the temperature differences by isolating the hot and cold airstreams. The selection of perforated tile becomes particularly important in an open aisle data center employing an OHAD system.

The interested reader is referred to many of the readily available standard industrial ventilation manuals for estimation of the pressure drop in an OHAD system. The higher pressure drop in the ducts makes balancing an OHAD system relatively simple, compared to an UFAD system. However, for an OHAD system the entire process has to be repeated with addition or removal of IT equipment. On the other hand, due to lower plenum pressure drop in a raised floor, small changes in plenum pressures cause large fluctuations in flow. This makes it nearly impossible to completely balance an UFAD system.

Unlike the UFAD system, which offers unlimited flexibility in expansion and layout, the OHAD system requires careful planning for future facility expansion, and accordingly provisions must be made for the accommodating the additional length of the ductwork needed. This is a challenging and complex process, as the designer has to not only account for extra duct work, but also take into consideration the increase in airflow due to continuously increasing rack heat densities. OHAD systems may be preferred over raised floors for data centers with fixed aisle lengths and predictable rack heat densities. This minimizes some of the risks associated with oversizing and the makes the system amenable to optimization. Branch distribution in an OHAD system offers superior airflow management by controlling both pressure and temperature. The absence of branch distribution in an UFAD system makes it difficult to control both temperature and pressure, without intentionally creating partitions in the under-floor plenum. Automatic control of CRAC units in overhead system is not an easy task, particularly if the units need to humidify/dehumidify air, or if the units need to be turned off at part loads.

Typically the fans can consume as much as 40% of the total HVAC system energy in an OHAD system [38]. The authors compared the fan energy costs for the two systems for comfort air-conditioning. The basis for comparison was the use of a centralized air handling unit for distributing cold air into the room. The system is very similar to the traditional data center environment, where the plenum is pressurized using a CRAC unit. Due to absence of ductwork in an UFAD air distribution system the static pressure drop is lower compared to a similar capacity overhead supply system. The analysis showed substantial fan energy savings for UFAD system in comparison to an OHAD system at all operating conditions.

In a data center application, the room airflow and temperature distributions are found to be different for the OHAD and UFAD systems. Unlike the UFAD system, the aisle temperature gradients in an OHAD system are counter-intuitive. One would expect the temperatures to be lowest near the top of the rack. However, contrary to this common conception, the temperatures are lowest near the bottom of the rack closer to the data center floor. This is conceptually explained in Fig. 2.39.

Figure 2.39 shows a cross-section of a rack arranged in a HACA data center facility employing an OHAD system. As illustrated in Fig. 2.39 the rising hot air exhausted from the rack is entrained by the impinging cold airstreams discharged by the ceiling diffuser. As a result, the central core of cold airstream is enveloped by

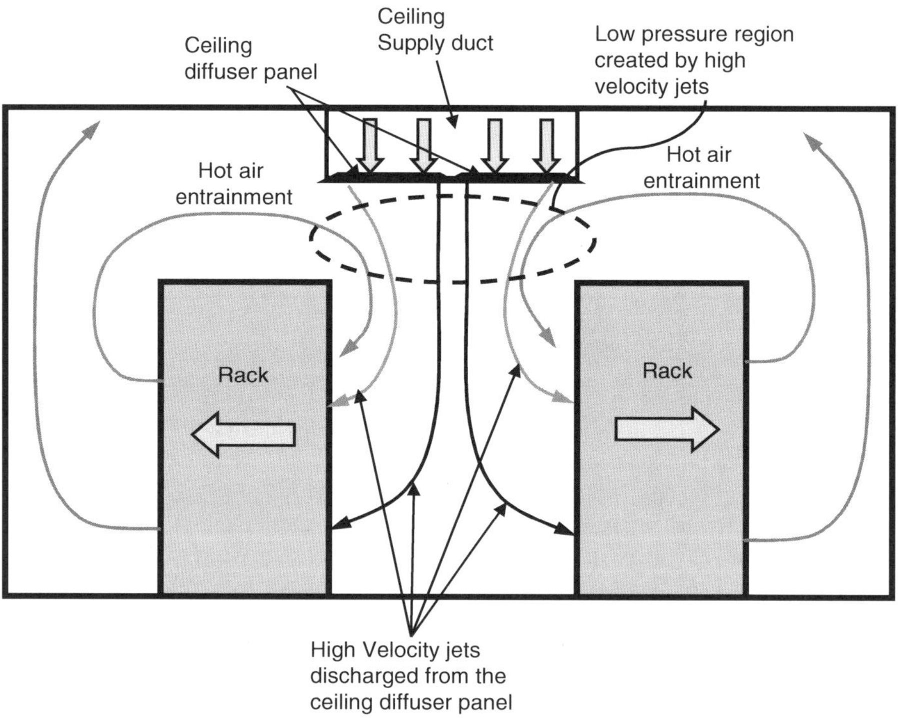


**Fig. 2.39** Typical airflow observed in a data center using an overhead distribution delivery system

a jacket of hot air. As the airstream descends down the aisle, the surrounding hot air envelope is ingested by the servers located in the upper half of the rack, which results in higher inlet temperatures, in the upper half of the aisle.

The quantity of hot air entrained by the descending cold air jets depend on a number of factors.

1. The volume of air ingested by the rack.
2. The temperature of the hot air exhaust responsible for generating sufficient buoyant force for the hot air to rise above the rack.
3. The layout of data center, i.e., the width of the hot aisle and the height of the ceiling, and existence of containment or aisle blanking panels.
4. The geometry of the OHAD system, which defines the discharge velocity and orientation of the descending cold airstream.
5. The geometry of the diffuser tile used for discharging the air. Some tiles promote mixing and entrainment to even out the temperature gradients (a primary requirement for comfort air-conditioning), while others may not.

Careful system analysis using full-scale CFD modeling is needed to ascertain the effect of the above factors. This trend of lower temperature at the bottom of the cold aisle is supported by a number of research articles [30, 31, 35, 39–42].

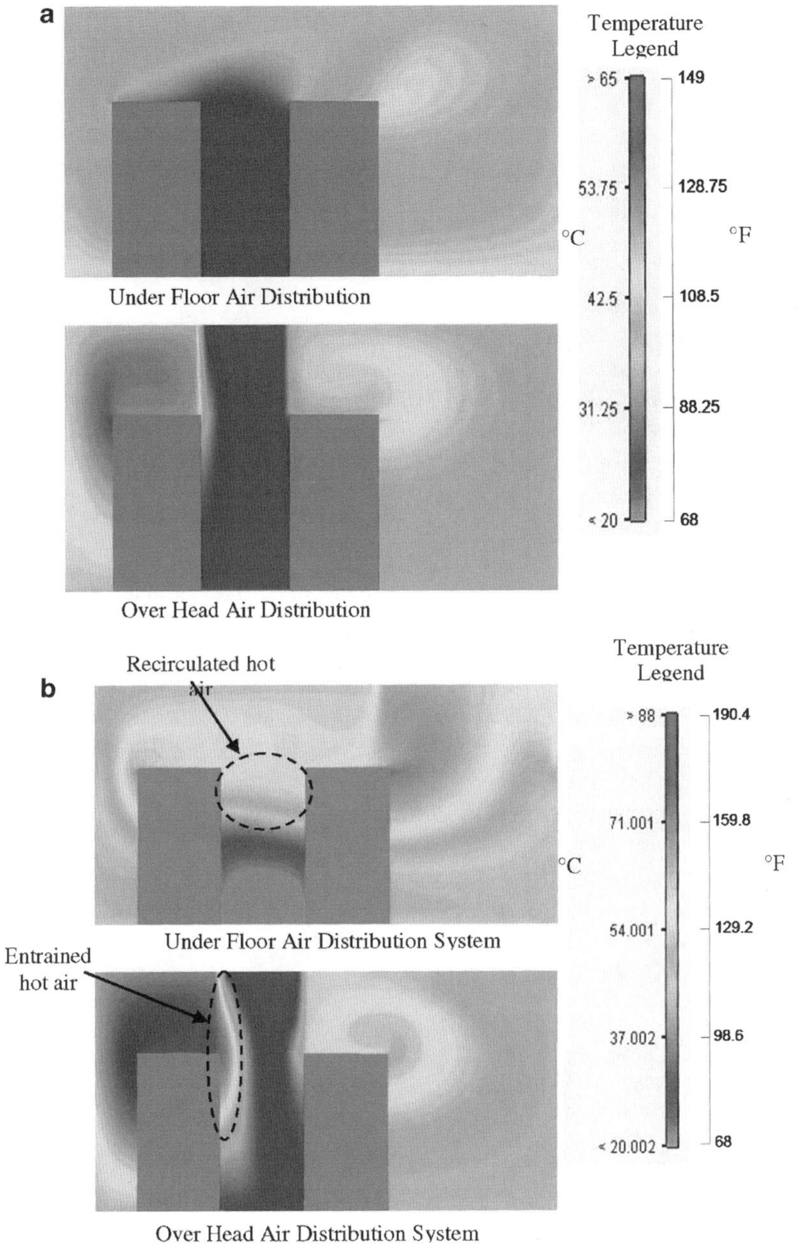


**Fig. 2.40** Comparison of rack inlet temperature contours for OHAD and UFAD system for 60% and 100% air supply. (**a**) 100% supply, (**b**) 60% supply (© 2007 ASHRAE, reprinted with permission [35])

The suitability of raised floor and overhead cooling for high-density applications is investigated in [35, 39]. The temperature profiles in the front of the rack are found to be different for both OHAD and UFAD systems. In both cases, severe recirculation patterns are observed in the end racks located in the aisle, leading to increased temperature along the rack height. The investigations revealed that qualitatively the OHAD system performs better with lower overall rack inlet temperatures compared to UFAD system, when the quantity of air supplied is less than the demand of the IT equipment. The investigations further suggested that UFAD design fared better when 100% of the air is supplied. Some of the observations reported in [35] are illustrated in Fig. 2.40a, b.

For raised-floor designs, lowest temperatures were observed closer to the perforated tile surface and the temperatures gradually increased along the rack height. The magnitude of temperature rise depended on the quantity of air supplied by the perforated tile, and the severity of recirculation. In extreme cases, the temperature at the top of the rack was found to exceed the ASHRAE allowable operating thermal envelope. In general, these inferences are consistent with the other research reported in the literature [30, 39]. However, the findings in [39] suggest that although hot air entrainment is inherent in an OHAD system, the OHAD system is able to maintain inlet temperatures to all the IT equipment, with only 5% excess air supply compared to 20% supply needed for a UFAD system. Hence, from an efficiency stand point the operating cost of an OHAD system is lower compared to UFAD system. Although only 5% excess airflow is required, the biggest drawback with an OHAD system is the limited availability of space in the overhead ceiling, which restricts the size of the duct needed to provide the additional air required by the system.

The above information provided here is generic in nature. It needs to be supplemented with more specific computations to help facility designers and analysts for selecting an air distribution system and assessing its impact on future expansion and operational costs. As stated above, both these systems have their own merits and drawbacks. If properly designed, both UFAD and OHAD systems can perform well for data center applications.

## 2.10 Role of Humidity in Data Center Cooling

In addition to maintaining the inlet air temperatures, data centers also need to adhere to the humidity limits specified by ASHRAE guidelines [5]. While the implications of high room humidity on the IT equipment are fairly well established, the effect of low humidity on performance or reliability of IT equipment is not very clear and is an area of active research [5]. If the data center environment is too humid, condensate can build on components inside the data processing equipment causing corrosion and electrical shorts. In addition, high humidity can cause condensate to form on the heat exchanger cooling coils in the CRAC units, resulting in reduced heat transfer coefficients and increased energy cost due to latent cooling.

On the other hand, it has been found that electro static discharge (ESD) effects exacerbate with decrease in room humidity. ESD not only presents a safety hazard for the people working in the data center, but could also shutdown or damage the IT equipment, leading to loss of productivity. Hence, adequate humidity must be maintained at all times for proper functioning of the IT equipment. The optimal relative humidity (RH) range for a data center environment is 30–60%, which falls within the ASHRAE-recommended envelope for class 1 equipment. Further details on the humidity requirement for various classes of equipment can be found in Chap. 1. In addition to providing sensible cooling, depending on the moisture content in the air, the CRAC unit may need to either humidify or dehumidify the air to maintain the data center environments within the ASHRAE-recommended limits. Dehumidification involves removal of moisture in the air, while humidification involves addition of moisture to the air. The three functions performed by the CRAC unit, viz. sensible cooling, dehumidification, and humidification are explained in the following section. Some data centers may have dedicated equipment for humidification/dehumidification. In such cases the CRAC unit provides only sensible cooling. In any case, the concepts developed here are generally applicable to air-conditioning equipment. The section begins with a brief introduction to basic psychrometric concepts needed for understanding the various cooling functions performed by the CRAC.

### *2.10.1 Principles of Psychrometrics*

The study of moist air or a mixture of air and water vapor is known as psychrometrics. Ambient air is usually a mixture of dry air and water vapor, or moisture. It is the amount of moisture in the mixture which makes the air dry or humid. For the purpose of defining the terminology used to describe the air-conditioning process, consider that the data center room having volume "$V$" to be filled with a mixture of water vapor and dry air (moist air) as shown in Fig. 2.41a. The moist air is at a temperature "$T$" and pressure "$p$." The state of moist air in the data center is schematically represented by point 1 on a temperature–volume diagram in Fig. 2.41b. From the $T$–$v$ diagram we note that the water vapor in the air is in superheated state.

Using the laws of thermodynamics, the following terms are defined:

1. *Mixture pressure* ($p$): Based on mixture laws, the pressure of the moist air mixture is obtained by the summation of the partial pressure of dry air and water vapor as follows:

$$p = p_a + p_v \text{ (Pa)} \tag{2.20}$$

where, $p_a$ is the partial pressure of air and $p_v$ is the partial pressure of water vapor at temperature $T$. The partial pressures of both air and water vapor are calculated using ideal gas law.

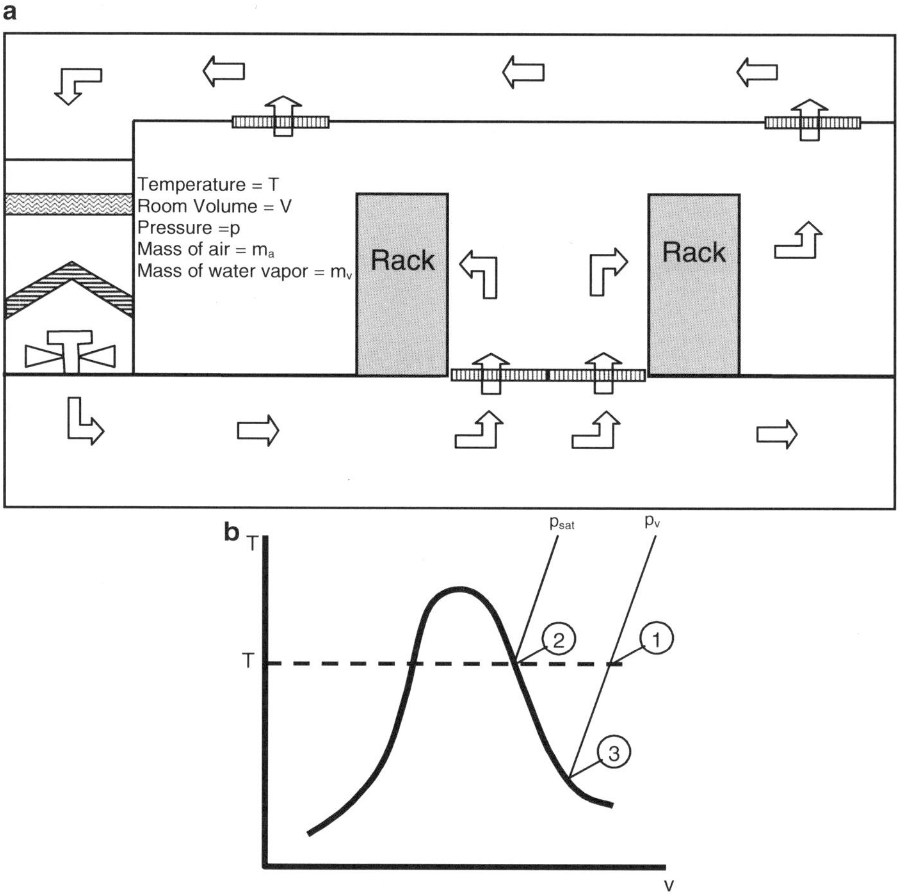


**Fig. 2.41** Thermodynamic representation of the data center conditions. (**a**) Data center control volume showing space conditions. (**b**) Schematic representation of the state of the data center on a $T$–$v$ diagram

$$p_a = \frac{m_a}{M_a}\frac{\overline{R}T}{V}, \quad p_v = \frac{m_v}{M_v}\frac{\overline{R}T}{V} \text{ (Pa)}, \tag{2.21}$$

where $m_a$ and $m_v$ are the masses of water vapor in the room occupying volume "$V$." $M_a = 28.97$ kg/mol and $M_v = 18$ kg/mol are the molecular weights of air and water. $\overline{R} = 8.314$ N m/mol K is the universal gas constant.

2. *Saturated air*: If the water vapor in the moist air mixture is in a saturated state as represented by point 2 in Fig. 2.41, then the air is said to be saturated. The saturation point depends on the temperature and pressure of air. In general, depending on the pressure and temperature, the amount of water vapor in air can vary from zero to maximum (saturated) state.

3. *Humidity ratio or specific humidity* ($\omega$): is defined as the ratio of mass of water vapor to mass of dry air in air–water vapor mixture. This is denoted by $\omega$.

$$\omega = \frac{m_\mathrm{v}}{m_\mathrm{a}} \text{ kg (of vapor)/kg (of dry air),} \tag{2.22}$$

where $m_\mathrm{v}$ is the mass of water vapor in mixture and $m_\mathrm{a}$ is the mass of dry air.

Using ideal gas law and molecular weights of water and dry air, the humidity ratio can be expressed in terms of partial pressures as:

$$\omega = 0.622 \frac{p_\mathrm{v}}{p - p_\mathrm{v}}. \tag{2.23}$$

4. *Relative humidity* ($\phi$) (*RH*): The amount of moisture in air can also be described in terms of ratio of mole fraction of water vapor present in the mixture to the mole faction in saturated mixture at the same mixture pressure and temperature. Using mixture rules, the relative humidity is expressed in terms of partial pressures as:

$$\phi = \left.\frac{p_\mathrm{v}}{p_\mathrm{sat}}\right|_{T,p}. \tag{2.24}$$

A hygrometer is used to measure relative humidity in the room. Relative humidity defines the potential of air to pick up moisture. Air with $\phi = 100\%$ is considered to be saturated and 0% RH is considered to be dry.

5. *Dew-point temperature*: When moist air is continuously cooled at a constant pressure, a point is reached when the mixture becomes saturated. At this saturation state the mixture is said to have reached its dew point. The dew point is represented by point 3 in Fig. 2.41b. If the mixture is further cooled beyond this point, the water vapor in the mixture will begin to condense and separate out as liquid phase. The temperature corresponding to the saturation state of the mixture is called the dew-point temperature. Dew-point temperature is of significant importance in all thermal management applications involving air as the primary coolant.
6. *Dry-bulb temperature* ($T_\mathrm{db}$): Is the temperature of moist air, measured using a standard temperature-measurement device.
7. *Wet-bulb temperature* ($T_\mathrm{wb}$): Is the lowest temperature that can be obtained by evaporating water into the air. Wet-bulb temperature is measured by wrapping a wet wick around the bulb of a standard thermometer. The temperature of water in the wick drops below the dry-bulb temperature due to evaporation of water in the wick. The wick temperature continuously drops till a steady state is reached providing a stable wet-bulb temperature reading. In air-conditioning applications at 1 atm, the wet-bulb temperature is used in place of the adiabatic saturation temperature to calculate the humidity ratio. The vapor pressure $p_\mathrm{v}$ in (2.21) is evaluated at $T_\mathrm{wb}$. This approximation is valid for applications involving moist air in the normal pressure and temperature range of atmospheric air [43].

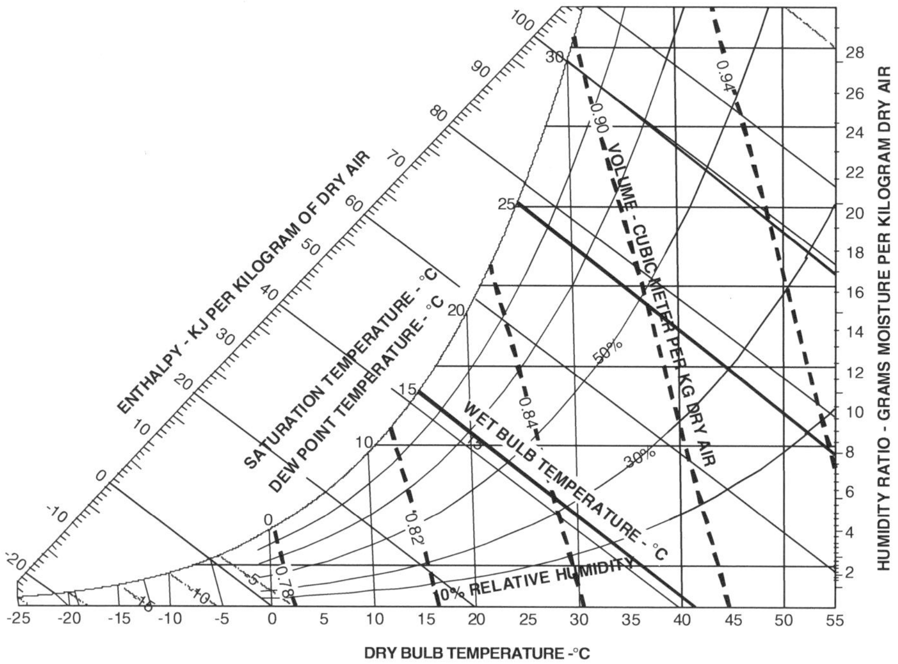


**Fig. 2.42** Illustration of a standard psychrometric chart for 1 atm

A device used to measure both dry- and wet-bulb temperature is known as a psychrometer.

The calculation of the above properties requires thorough knowledge of thermodynamics and property data for water and air at various temperatures and pressures. These calculations are greatly simplified using a psychrometric chart.

## *2.10.2 Description of the Psychrometric Chart*

A psychrometric chart is a graphical representation of the properties of moist air. Air-conditioning systems can be easily analyzed by plotting the dry- and wet-bulb temperatures on a psychrometric chart. The standard psychrometric chart constructed for moist air mixture at a pressure of 1 atm is used for design and analysis of air-conditioning systems. If the mixture pressures vary only slightly from 1 atm the standard chart can still be used with sufficient accuracy for all engineering analyses [43]. In data center applications, since the room pressure varies marginally from 1 atm, the minor pressure variations are ignored and the standard chart can be used to describe the various cooling processes. The standard psychrometric chart for 1 atm is illustrated in Fig. 2.42.

Equation 2.23 simplifies providing a direct relation between humidity ratio and vapor pressure.

$$\omega = 0.622 \frac{p_\mathrm{v}}{1 - p_\mathrm{v}}. \tag{2.25}$$

The dry-bulb temperature is plotted on the $x$-axis and humidity ratio on the $y$-axis. In some charts, the vapor pressure is also plotted along the $y$-axis using the above relation. The curved lines shown in Fig. 2.42 are lines of constant relative humidity. The chart also provides information on the air–water vapor mixture enthalpy per unit mass of dry air in the mixture ($h_\mathrm{a} + \omega h_\mathrm{v}$). The enthalpy of water vapor ($h_\mathrm{v}$) is the vapor enthalpy corresponding to the dry-bulb temperature and is obtained using steam tables. The enthalpy of dry air is obtained using the following relation:

$$h_\mathrm{a} = \int_{T=0^\circ\mathrm{C}}^{T(^\circ\mathrm{C})} C_{p,\mathrm{air}}\,\mathrm{d}T, \tag{2.26}$$

where $C_{p,\mathrm{air}} = 1.005$ kJ/kg K is the specific heat of air at constant pressure at 1 atm. For the corresponding chart in English units the reference temperature is taken as 0°F.

The enthalpy of the mixture is represented on the inclined axis in Fig. 2.42. The projections normal to the enthalpy axis are lines of constant enthalpy. The constant wet-bulb temperature lines approximately coincide with the constant enthalpy lines. The additional set of slanted lines shown dotted in Fig. 2.42 provide information on the mixture volume per unit mass of dry air.

### 2.10.3 Use of Psychrometric Chart for Analyzing Air-Conditioning Processes

The primary use of psychrometric chart for data center applications is in finding the dew-point temperature for a given dry-bulb temperature and calculating the CRAC cooling load. The procedure of obtaining the dew-point and wet-bulb temperatures from the psychrometric chart is illustrated with the help of the following example.

Let's say we would like to find the dew-point temperature of air that is being discharged from the perforated tile at 18°C (60°F) with a relative humidity (RH) of 50%. This point is defined by the intersection of the curved 45% RH line and the 15°C DBT line. The point is marked as "A" in Fig. 2.43. Since, by definition, dew point is the state at which the moisture in the air begins to condense, the dew point (DP) necessarily has to fall on the saturation line or 100% RH line. DP is obtained by horizontally projecting point "A" on the saturation curve as shown in Fig. 2.43. The corresponding temperature is known as the dew-point temperature marked as DP in Fig. 2.43. Similarly, projecting point "A" along the wet-bulb temperature line in the psychrometric chart as illustrated in Fig. 2.43 will give us the corresponding

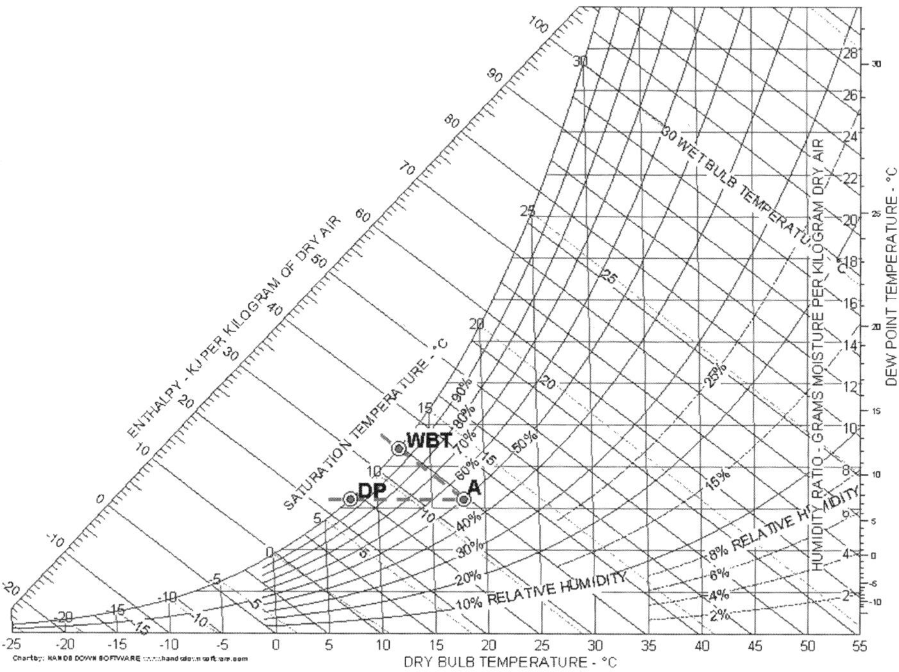


**Fig. 2.43** Psychrometric chart illustrating the procedure for obtaining wet-bulb and dew-point temperature

wet-bulb temperature. This point is marked as WBT in Fig. 2.43. The enthalpy of mixture is obtained by projecting point "A" on the enthalpy axis along the constant enthalpy line.

(a) Sensible cooling process

Sensible cooling is the most common mode of operation of the CRAC unit. There is no humidification or dehumidification involved while operating in this mode. This process is explained with the help of the psychrometric chart using the following example. Let us consider that air discharged from the perforated tile at 18°C with 50% RH is ingested by the servers in the rack. The temperature of the air increases as it passes through servers due to heat gain from the electronic components inside the server and exits the rack, say at 38°C. The procedure for calculating the relative humidity of air as it exits the rack and the amount of sensible heat gained by the air during this process, using the psychrometric chart, is described as follows.

Since, there is no humidification (addition of moisture) or dehumidification (removal of moisture) involved, the humidity ratio ($\omega$) remains constant during this process, therefore $\omega_{\text{Rack inlet}} = \omega_{\text{Rack Exit}}$. Using this condition and the information on the dry-bulb temperatures, the state of air at rack inlet ($\text{DBT}_{\text{inlet}} = 18°\text{C}$ and $\phi = 50\%$) and rack exit ($\text{DBT}_{\text{exit}} = 38°\text{C}$ and $\omega = 6.427$ (const.)) is

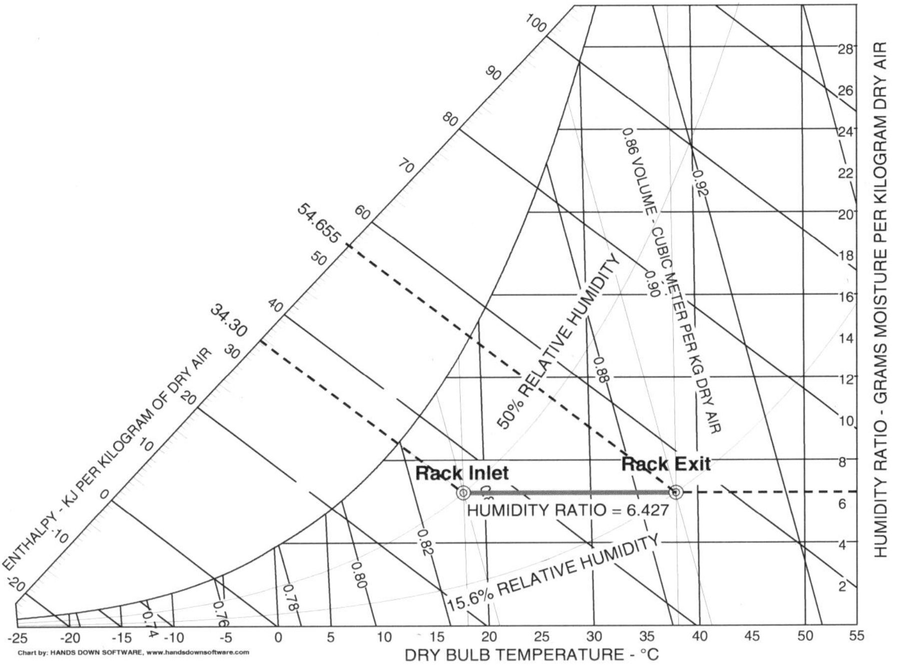


**Fig. 2.44** Psychrometric chart illustrating sensible heating process

determined. This sensible heat gain process is illustrated in the psychrometric chart in Fig. 2.44. The corresponding RH at the rack exit is found to be 15.6%.

Applying the energy balance to the rack, the heat gained by the air is given by the following expression:

$$Q_{\text{air}} = (\dot{m}_{\text{a}} h_{\text{a,inlet}} + \dot{m}_{\text{v}} h_{\text{v,inlet}}) - (\dot{m}_{\text{a}} h_{\text{a,exit}} + \dot{m}_{\text{v}} h_{\text{v,exit}}). \tag{2.27}$$

Using $\dot{m}_{\text{v}} = \omega \dot{m}_{\text{a}}$ from (2.21) and substituting in (2.26) we get

$$Q_{\text{air}} = \dot{m}_{\text{a}} \left[ \underline{(h_{\text{a}} + \omega h_{\text{v}})}_{\text{exit}} - \underline{(h_{\text{a}} + \omega h_{\text{v}})}_{\text{inlet}} \right]. \tag{2.28}$$

The mixture enthalpy per unit mass of dry air as defined in the underlined terms in (2.27) is determined from the psychrometric chart. The values are obtained by projecting the "Rack inlet" and "Rack exit" states along the const. enthalpy line as illustrated in the psychrometric chart in Fig. 2.44.

The enthalpy of air at the rack inlet is obtained as $(h_{\text{a}} + \omega h_{\text{v}})_{\text{inlet}} = 34.30\ \text{kJ/kg}$ and $(h_{\text{a}} + \omega h_{\text{v}})_{\text{exit}} = 54.635\ \text{kJ/kg}$ at the rack exit.

Substituting the values of mixture enthalpies at the rack inlet and exit in (2.27) we get

$Q_{\text{air}} = 20.335\ \text{kJ/kg}$ of air.

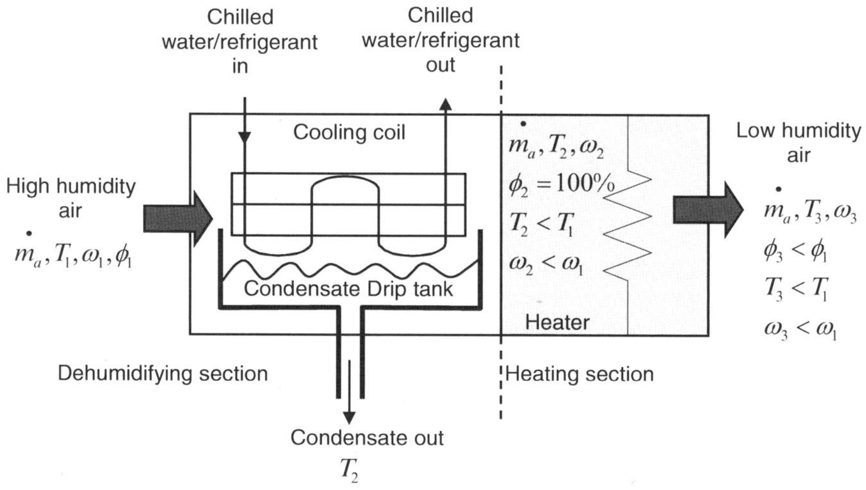


**Fig. 2.45** Schematic of a dehumidification device

From the psychrometric chart the specific volume of air at the rack inlet is determined as 0.833 m$^3$/kg. Using this information the heat gained by the air can be alternatively expressed in terms of the volume of air ingested by the rack as

$Q_{\text{air}} = 24.418\ \text{kJ/m}^3$ of air or $Q_{\text{air}} = 11.5\ \text{W/cfm}$ of air.

Hence, to cool a 20 kW with a 20°C rise in air temperature across the rack the perforated tile would need to supply approximately 0.82 m$^3$/s (1,728 cfm) of air.

The following assumptions were made in calculating the heat gained by the air.

1. The heat gained by the air is a steady state process and the static pressure of the air at the inlet and exit is relatively uniform at 1 atm.
2. Changes in kinetic and potential energy are negligible.
3. All the servers in the rack contribute equally to the heat added to the air and there are no gradients in air temperature within the rack.
4. There is no mixing or interaction of the hot and cold airstreams within the rack or between the inlet and exit of the rack.

The above example highlights an important aspect of data center cooling. In absence of humidification or dehumidification, the cooling or heating processes take place along a horizontal line on the psychrometric chart. Although, the humidity ratio remains unchanged, relative humidity varies significantly during the heating and cooling process. Hence, the placement of the relative humidity sensor in the CRAC or the room is critical for humidity control.

(b) Dehumidification process:

A typical dehumidification device consists of a cooling coil and a heater. Figure 2.45 shows a simplified representation of a dehumidification device. Depending on the design, the CRAC unit may be equipped with a dedicated cooling coil for dehumidification, or it may use the existing cooling coil for both

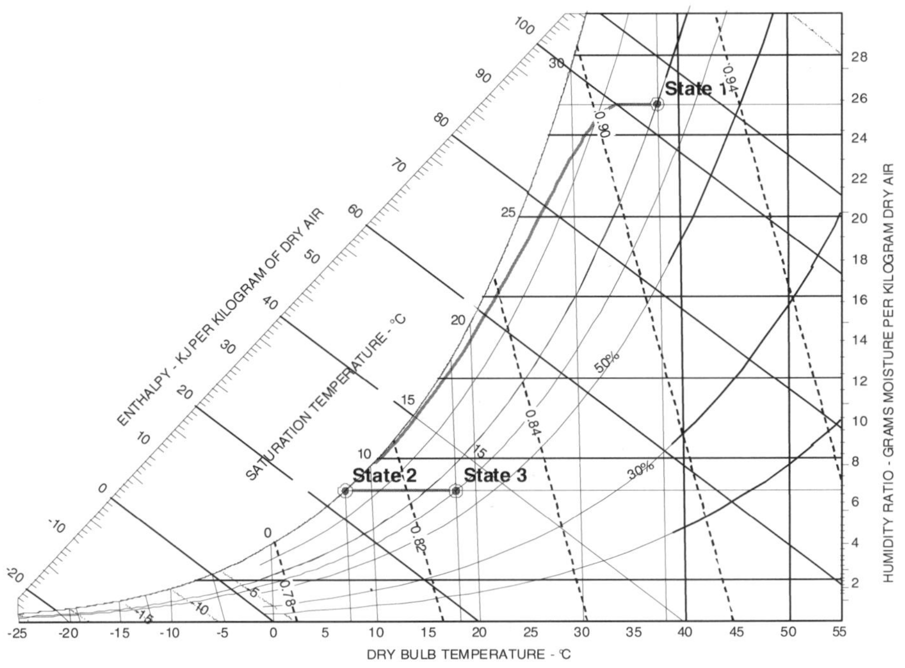


Fig. 2.46 Psychrometric chart illustrating the dehumidification process

sensible cooling and dehumidification. Dehumidification is a two-step process involving cooling as well as heating. The process is explained with the help of the following example.

Let us assume that the hot air from the exit of the racks enters the CRAC at a temperature of $T_1 = 45°C$ with $\phi_1 = 65\%$ RH. This is indicated by state 1 on the accompanying psychrometric chart in Fig. 2.46. The goal of the CRAC is to condition this air and return the air back into the data center in the ASHRAE-recommended range at a temperature $T_3 = 18°C$ with $\phi_3 = 50\%$ RH. This is indicated by state 3 on the psychrometric chart in Fig. 2.46. The humidity ratio corresponding to state 1 and state 3 is determined to be $\omega_1 = 25.54$ g/kg and $\omega_3 = 6.43$ g/kg of dry air. Since, the humidity ratio at state 3 is significantly lower than state 1; the air needs to undergo dehumidification to shed the extra moisture to achieve state 2. As mentioned earlier this is done is two steps.

In the first step, the air is dehumidified by passing it over the heat exchanger cooling coil. As the stream of air flows over the cooling coil, some of the water vapor present in the air condenses due to cooling below the dew-point temperature. The moist air exits the heat exchanger cooling coil as a saturated mixture ($\phi_2 = 100\%$) at a temperature $T_2 = 7.43°C$. This is indicated by state 2 in psychrometric chart shown in Fig. 2.46. Due to condensation, the humidity ratio $\omega_2 = 6.43$ g/kg at state 2 is lower than humidity ratio $\omega_1 = 25.54$ grams/kg at state 1. Although the temperature $T_2 = 7.43°C$ of the moist air at state 2 is lower than the final temperature $T_3 = 18°C$ defined by state 3, it is unsuitable for cooling due to high relative humidity ($\phi = 100\%$). The second step involves decreasing the relative humidity by heating the air. The saturated airstream is passed over a set of heating coils to bring it to the desired relative humidity ($\phi_3 = 50\%$) and temperature ($T_3 = 18°C$), defined by state 3 before being discharged into the room. Note that during the heating process the humidity ratio ($\omega_2 = \omega_3 = 6.43$ g/kg) remains constant between state 2 and state 3, and only the temperature and relative humidity values change.

The amount of energy needed for the entire dehumidification process can be calculated using the following equations, obtained by applying energy balance to the control volume in Fig. 2.45.

*Energy needed for dehumidification*

$$Q_{(1-2)_{\text{Dehumidification}}} = \dot{m}_a\left[(h_a + \omega h_v)_2 - (h_a + \omega h_v)_1 + \underline{(\omega_1 - \omega_2)h_{f,T_2}}\right]. \quad (2.29)$$

*Energy needed for heating*

$$Q_{(2-3)_{\text{Heating}}} = \dot{m}_a\left[(h_a + \omega h_v)_3 - (h_a + \omega h_v)_2\right]. \quad (2.30)$$

The enthalpies at state points 1–3 are readily obtained from the psychrometric chart using the method described previously. However, the specific enthalpy of saturated water, $h_f$ at temperature $T_2$ has to be obtained from the property tables of water. For the example listed above, the following enthalpies are obtained using the psychrometric chart.

$(h_a + \omega h_v)_1 = 102.8$ kJ/kg, $(h_a + \omega h_v)_2 = 22.564$ kJ/kg, $(h_a + \omega h_v)_3 =$ 34.3kJ/kg, $\omega_1 = 0.025545$ kg/kg, $\omega_2 = 0.006428$ kg/kg, $\omega_3 = 0.006428$ kg/kg, $h_{f,T_2}$=31.24 kJ/kg as obtained using standard property table of water at temperature $T_2 = 7.438°C$.

Substituting the above values in (2.28) and (2.29) we get,

$Q_{(1-2)_{\text{Dehumidification}}} = -79.63$ kJ/kg of cooling is required for dehumidification.

$Q_{(2-3)_{\text{Heating}}} = 10.736$ kJ/kg of heating is required for increasing the relative humidity.

As observed from the above example, the contribution of the underlined term in (2.29) is very small compared to the other two terms representing the difference in enthalpies of moist air mixture. It can therefore, be omitted for a first-hand estimate of the energy required for the dehumidification process. The reader is informed that both cooling and heating described by (2.29) and (2.30) require inputs of external energy. These should not be combined, as it provides an incorrect estimate of the energy required for dehumidifying.

(c) Humidification process:
If the air becomes too dry, it may be necessary to increase its moisture content before circulating it back into the data center. Humidification is generally required if outside air is used for cooling. Air can be humidified either by

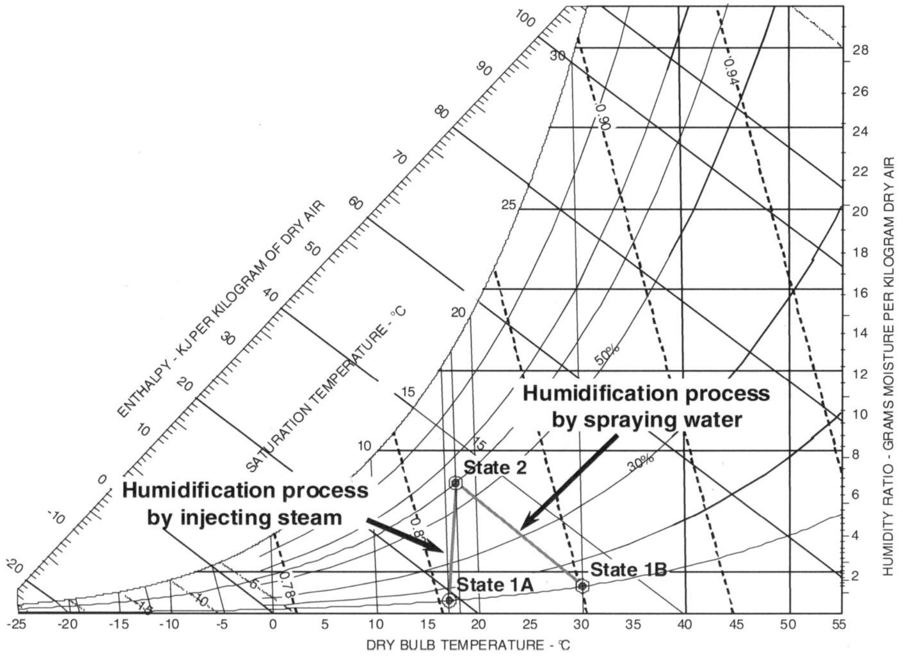


**Fig. 2.47** Psychrometric chart illustrating the humidification process

injecting steam or by spraying water. Depending on the application, both methods are commonly used. These are illustrated schematically in Fig. 2.47. The quantity of steam or water injected and the energy required to humidify can be calculated using the psychrometric chart based on the methodology described previously. The following example illustrates the energy requirement for humidifying using steam injection. Let us assume that air exits the cooling coil in the CRAC at $T_1 = 17°C$ with 5% RH as indicated by the state "A" in Fig. 2.47.

The air is passed through a humidifier where it exits at a temperature $T_2 = 18°C$ and a RH of 50% before being discharged into the data center. The condition of the air leaving the steam humidifier is indicated by state 2 in the psychrometric chart in Fig. 2.48.

We are interested in knowing the condition and quantity of steam to be injected to humidify the air from state 1 to state 2. The mixture enthalpies at state 1 and 2 obtained using the psychrometric chart are listed below:
$(h_a + \omega\, h_v)_1 = 18.521\,\text{kJ/kg}$, $(h_a + \omega h_v)_2 = 34.3\,\text{kJ/kg}$, $\omega_1 = 0.0006$kg (vapor)/ kg$(dryair)$ $\omega_2 = 0.006428$ kg (vapor)/kg (dry air).

Applying energy balance to the steam humidifier we obtain

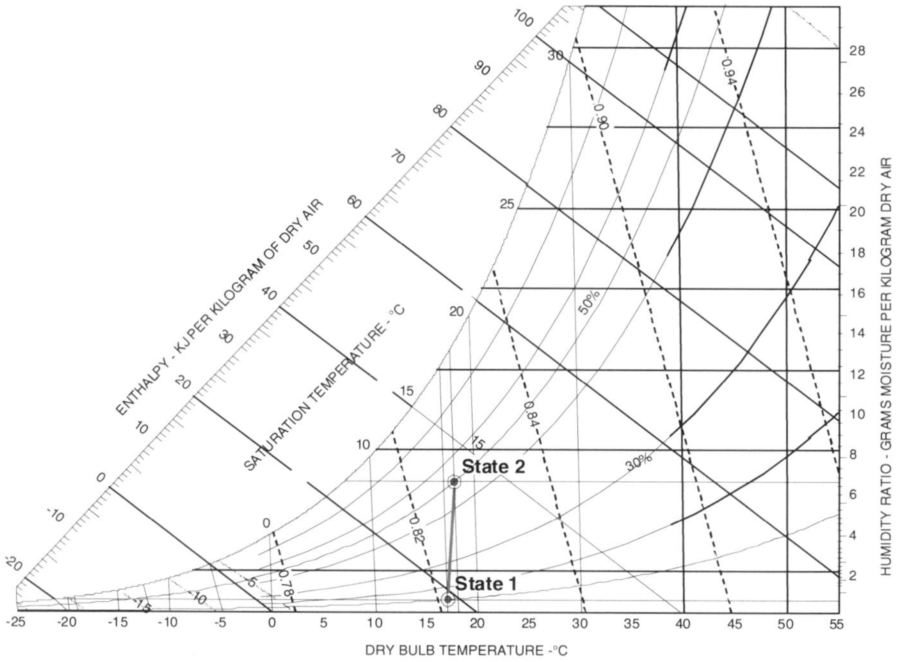


**Fig. 2.48** Psychrometric chart illustrating the humidification process using steam injection

$$h_{\mathrm{v},T_{\mathrm{steam}}} = \frac{(h_{\mathrm{a}} + \omega h_{\mathrm{v}})_2 - (h_{\mathrm{a}} + \omega h_{\mathrm{v}})_1}{(\omega_1 - \omega_2)}. \quad (2.31)$$

Substituting the above values in (2.30), we obtain $h_{\mathrm{v},T_{\mathrm{steam}}} = 2706.52$ kJ/kg.

The corresponding injection temperature of the steam obtained using the steam tables is $T_{\mathrm{steam}} = 120°\mathrm{C}$.

*Quantity of steam required for the humidification*
Applying mass balance across the humidifier
$\dot{m}_{\mathrm{a1}} = \dot{m}_{\mathrm{a2}} = \dot{m}_{\mathrm{a}}$ the mass of dry air in the mixture does not vary,

$$\dot{m}_{\mathrm{v2}} = \dot{m}_{\mathrm{v1}} + \dot{m}_{\mathrm{v_{steam}}}. \quad (2.32)$$

Substituting $\dot{m}_{\mathrm{v1}} = \omega_1 \dot{m}_{\mathrm{a}}$ and $\dot{m}_{\mathrm{v2}} = \omega_2 \dot{m}_{\mathrm{a}}$ in (2.31) we get

$$\dot{m}_{\mathrm{steam}} = \dot{m}_{\mathrm{a}}(\omega_2 - \omega_1) \text{ kg (steam)}. \quad (2.33)$$

The quantity of steam required to humidify the air using (2.32) is found to be $\dot{m}_{\mathrm{steam}} = 5.828$ g/kg of dry air. For the above conditions, a typical CRAC unit discharging 8.02 $m^3$/s (17,000 cfm) of air would require approximately 200 kg/h of steam injection. Generating such high quantities of steam would consume a significant amount of energy, and hence, using steam humidifiers for a data center is not recommended. On the other hand, an adiabatic humidifier can considerably cut down the cost of humidification [44]. Adiabatic humidifiers use evaporative techniques such as water spraying or wetted media for evaporation at constant enthalpy. This process requires significantly less input of external energy, as the energy required to cool and humidify the air is balanced by the release of latent heat of water due to evaporation. However, supplying air to the data center at a prescribed operating temperature would still require cooling, which would add to the cost of energy.

As we notice from the above examples, both dehumidification and humidification processes are energy intensive, and hence, must be avoided to the extent possible. The cost of humidification/dehumidification must be carefully evaluated before using outside-air economizers.

(d) Evaporative cooling

Evaporative or transpiration cooling is a traditional and well-established method of cooling. This technique is particularly effective in hot and dry climates. The method uses the latent heat of evaporation of water to provide cooling. Due to the large latent heat of evaporation for water, (2,257 kJ/kg), this heat transfer mechanism is very effective and leads to cool exit temperatures close to the ambient wet-bulb temperatures. Because, evaporation plays a key role in the heat transfer process, the relative humidity, wet- and dry-bulb temperatures are important parameters to be considered for employing this cooling technique.

Evaporative cooling is achieved by either spraying liquid water into the dry airstream or forcing the dry air through a wetted media. The media is kept damp by constantly replenishing it with water to prevent it from drying. The working of an evaporating cooler is illustrated in Fig. 2.49 and the evaporative cooling process is explained in the psychrometric chart in Fig. 2.50. The dry incoming air at state 1 picks the water from the wetted media and exits at a lower temperature at state 2.

As indicated in Fig. 2.50, both the relative humidity and humidity ratio at state 2 are higher than state 1.

The quantity of water required for cooling the air from $T_1$ to $T_2$ is given by the following relation:

$$\dot{m}_{\mathrm{w}} = \dot{m}_{\mathrm{a}}(\omega_2 - \omega_1)\ (\mathrm{kg/s}), \tag{2.34}$$

where $\dot{m}_{\mathrm{w}}$ (kg/s) is the mass flow rate of water to the wetted media and $\omega_2, \omega_1$ are the humidity ratios at the inlet and the outlet of the evaporative cooler.

*Humidity ratio of the air leaving the evaporative cooler*

Applying energy balance to the control volume in Fig. 2.49, the enthalpy at state 2 is obtained using the following relation.

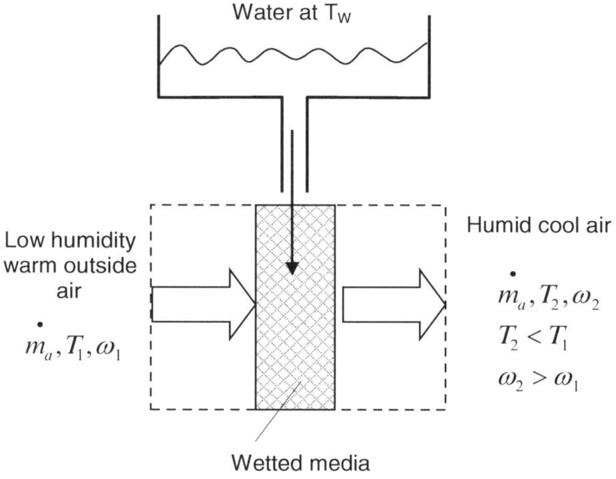


**Fig. 2.49** Schematic of an evaporative cooler

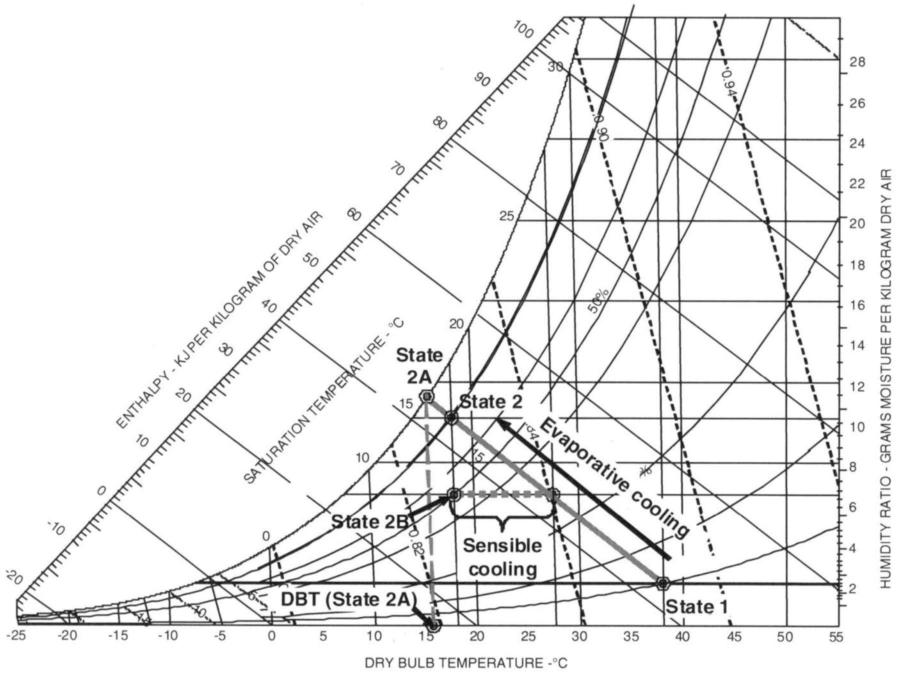


**Fig. 2.50** Psychrometric chart illustrating the evaporative cooling process

$$(h_a + \omega h_v)_2 = (h_a + \omega h_v)_1 + \underline{(\omega_2 - \omega_1) h_{f,T_w}}, \tag{2.35}$$

where $h_f$ is the enthalpy of saturated water entering the wetted media at temperature $T_W$.

Rearranging (2.35) we get

$$\omega_2 = \frac{(h_{a1} - h_{a2}) + \omega_1 (h_{v1} - h_{f,T_w})}{(h_{v2} - h_{f,T_2})} \text{ kg (vapor)/kg (dry air).} \tag{2.36}$$

Using

$$(h_{a1} - h_{a2}) = c_{p,\text{air}}(T_1 - T_2), \tag{2.37}$$

where $c_{p,\text{air}}$ is the specific heat of air at constant pressure,

Substituting (2.36) in (2.35) we get

$$\omega_2 = \frac{c_{p,\text{air}}(T_1 - T_2) + \omega_1 (h_{v1} - h_{f,T_w})}{(h_{v2} - h_{f,T_2})} \text{ kg (vapor)/kg (dry air),} \tag{2.38}$$

$h_{v1}, h_{v2}$ and $h_{f,T_2}$ is obtained using property data of saturated water.

The above equation is based on the assumption that all the water injected into the media evaporates into the airstream and there is no work or heat transfer to the surrounding. The underlined term in (2.35) accounts for the latent energy carried by the water in the wetted media. This term is usually much smaller in comparison to the mixture enthalpies (the remaining terms in (2.35)).

We obtain the following equation by omitting the underlined term in (2.35),

$$(h_a + \omega h_v)_2 \approx (h_a + \omega h_v)_1 \tag{2.39}$$

Based on (2.39) evaporative cooling can be construed as an isenthalpic process, as illustrated on the psychrometric chart in Fig. 2.50. Since the lines of const. mixture enthalpy lines and const. wet-bulb lines are almost parallel on a psychrometric chart, it follows that evaporation takes place at nearly constant wet-bulb temperature. There are some limitations of evaporative cooling. Since it follows a constant enthalpy line, the final state of air mixture must lie on the isenthalpic process corresponding to the initial state of moist air, as illustrated in Fig. 2.50. Theoretically, it is possible to cool the air until it attains the dry-bulb temperature coincident with wet-bulb temperature represented by state 2A in Fig. 2.50. One may also note that the evaporative cooling technique needs to be supplemented with sensible cooling to achieve other states such as 2B, as shown in Fig. 2.50, that do not lie on the isenthalpic line. However, the energy required to achieve state 2B would still be significantly less compared to the process of cooling and humidifying involving steam injection.

The advantage of this method is that it not only provides cooling but also humidifies the air without requiring additional energy. As such, this method of cooling is not only economical but also ideal for data centers located in geographical regions with hot dry climates.

## 2.11 Data Center Cooling Using Economizers

Both air- and waterside economizers have been traditionally used in building air-conditioning applications. However, in the past, stringent environment control coupled with high investment and maintenance costs associated with economizers had discouraged data center designers from implementing economizer-based solutions. It is only recently, that the escalating cooling costs associated with increase in average rack heat density, and subsequent increase in cooling demand have prompted data center operators and designers to seriously consider use of economizers for data center cooling. ASHRAE's Technical Committee 9.9, is now actively advocating the use of economizers for data center cooling. In addition the IT-equipment manufacturers are continuously increasing the acceptable supply air temperatures to encourage greater use of outside air for cooling.

A standard data center cooling system uses a mechanical chiller to remove heat from the data center. The chiller power is a significant component of the total cooling power requirement in a data center. The implementation of an economizer system can either reduce or eliminate the chiller. Depending on the geographical location of the data center, economizers can satisfy a large portion of data center cooling requirements [45]. Economizer systems use outside air, when conditions are favorable, to help meet cooling requirements and provide so-called free cooling cycles for computer rooms and data centers. There are two types of economizer systems; airside and waterside economizers. The working principles of both systems are discussed in the following section.

### *2.11.1 Airside Economizer*

An airside economizer system directly utilizes outside air to partially or fully meet the cooling requirements of the data center. A schematic of a data center utilizing airside economizer is illustrated in Fig. 2.51.

The control logic in the economizer system utilizes a set of sensors and dampers to regulate the volume of air needed to satisfy the data center cooling demand. The sensors continuously measure the ambient and the data center room air conditions. When the outside air temperature and humidity conditions are conducive, the control system regulates the inlet dampers to draw in cool ambient air to replace the hot air exhausted by the IT equipment. The incoming air is filtered to block dust and particulate matter from being introduced into the facility. A set of exhaust air dampers balance the quantity of the airflow into the data center and prevent the facility from becoming overpressurized. In addition, the economizer also has provision for mixing incoming cold and hot airstreams when ambient temperature is below the minimum ASHRAE-recommended temperature of 18°C.

The operation of the outside-air economizer system is illustrated with help of the psychrometric chart in Fig. 2.52.

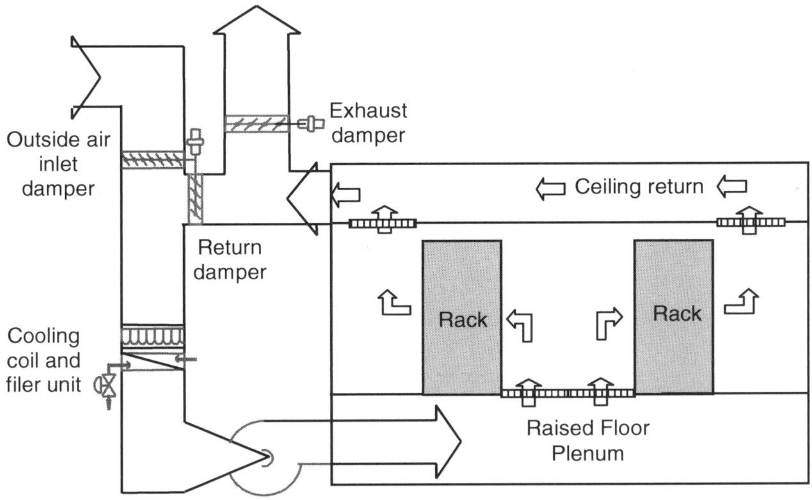


Fig. 2.51 Schematic of an airside economizer system

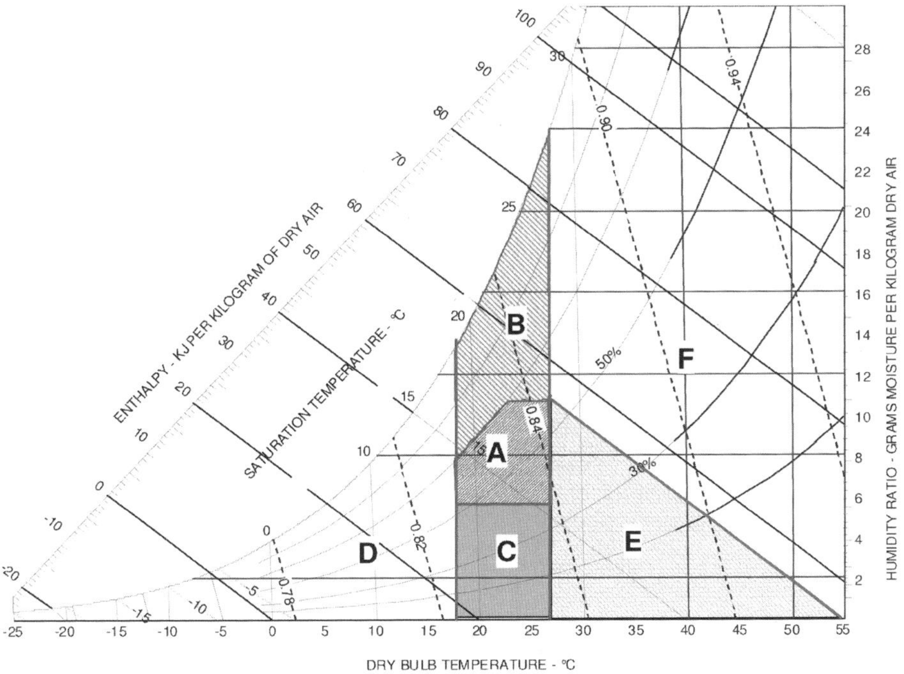


Fig. 2.52 Psychrometric chart showing different regimes of economizer usage

Depending on the ambient temperature and humidity conditions, the psychrometric chart is divided into six regions A–F. The recommended ASHRAE operating range (region A) is also overlaid in the chart. The sequence of damper operation while operating in the various regions is listed below.

- When the outside air temperature and humidity are in the ASHRAE-recommended region "A," the economizer can be operated on 100% outside air. Operating in this region presents highest energy savings as it eliminates the need for both chiller and humidification equipment. While operating in region A both the inlet and the exhaust dampers are 100% opened. The return damper is kept closed.
- If the outside air temperature and humidity fall in region B, outside air still can be used, but would require dehumidification. In this case, the outside air is passed though dehumidification equipment before being discharged into the data center. Depending on the state of the outside air, room air can be blended with outside air or completely exhausted.
- If the ambient air temperature and humidity fall in region C, the air would require humidification to meet the ASHRAE-recommended humidity level. Either steam or adiabatic dehumidifiers can be used for this purpose. In this case also room air can be blended with outside air or completely exhausted based on the condition of the outside air.
- If the ambient air temperature falls in the region D, blending of the outside air with return air is needed to achieve the specified inlet air temperature. In this case the return air is mixed with the outside air so that the temperature falls in the ASHRAE-recommended region. This leads to 100% cooling and chillers can be switched off, but additional humidification or dehumidification may be required. Using the air directly without blending would lead to low room temperatures, which could result in freezing the chilled water pipes in the CRAC units. This might lead to disruptions in the facility, and hence, modulating the outside air and return dampers becomes critical. If carefully designed, operating in region D could also eliminate the usage of chiller.
- Evaporative cooling can be used if the ambient temperature falls in region E. If properly designed, operating in region E could lead to 100% economization using outside air.
- The conditions are not conducive for economization if the ambient conditions fall in region F. For example, when the outside-air temperature is higher than the return-air temperature, no outside air is brought into the data center space In this case, the outside air damper is completely closed resulting in no economization.

(a) Factors affecting the use of airside economizers for data center cooling

The most important factor that must be considered while using outside air is air quality and contamination. Bringing in outside air without proper filtration can introduce fine dust and other airborne particulate or gaseous contaminants into the data center, which could potentially impact the operation and cause damage to the IT equipment. Also, introduction of outside air into the data center makes control of humidity a difficult task. The increased cost of additional infrastructure associated with filtering and humidity control can significantly offset the savings gained from using direct airside economizer system. Some of these issues are specific to the geographical location and need to be addressed in the early design phase of the data center facility.

*Effect of particulate contamination*

As mentioned earlier, the primary concern in introducing external air to the data center is fine particulate matter, which could cause electrical conductor bridging [46]. Ambient concentrations of fine particulate matter in many geographical locations can be higher than the indoor limits established by ASHRAE. Measurements reported in [44] have demonstrated that particulate matter composed of deliquescent salts, especially sulfates can cause failure of electronic equipment due to current leakage arising from particle deposition under high humidity conditions. The dry particles by themselves are not particularly harmful; failure occurs as particles bridging the space between isolated conductors becomes moist and causes leakage currents or shorts. This happens when the relative humidity of the surrounding air rises to a point where the fine hygroscopic dust particles begin to absorb moisture. This process is called deliquescence and occurs at a relative humidity point specific to the composition of the particle. Under these moist conditions, the water-soluble ionic salts in particles can dissociate into cations and anions and become electrically conductive [44]. In summary, the accumulation of hygroscopic dry particle between conductors usually takes place over a larger time scale and is not of primary concern. It is the sudden surge in room humidity that results in deliquescence of deposited particles, leading to failures which can be a rapid event, on the timescale of minutes or seconds.

ASHRAE has recommended a range of filtration techniques and filters for strict control of air quality, while using outside air economization. With 100% recirculating systems, filters with (minimum efficiency reporting value) MERV rating of 8 or 9 (ANSI/ASHRAE52.2-1999; equivalent to 40% efficiency based on the older "dust spot" efficiency rating of ANSI/ASHRAE Standard 52.1-1992) are typically used [47]. These filters are intended to remove only the particulates that are generated due to recirculation or air within the space. When outside air is introduced, it is necessary to increase the MERV rating to 10 or 11 (equivalent to 85% efficient based on dust spot 20 method) so that the filters can extract the increased loading of particulates (i.e., the smaller particles) associated with construction, road highway traffic, industrial processes, and other outdoor pollutants.

Studies on use of outside-air economizer by Pacific Gas and Electric Company and Lawrence Berkeley National Laboratory (LBNL) [48] suggest that the challenges surrounding air contaminants and humidity control can be addressed by use of improved filtration techniques. The study reports slightly higher particulate concentrations (gaseous contaminants were not measured) in data centers with outdoor-air economizer when compared to data centers with 100% recirculation systems. However, by using higher MERV rating filters, it was found that cleanliness of the indoor air, in terms of particulate content, was satisfactory if not better than the cleanliness of a recirculating system with a lower MERV rating. Even with the most conservative particle standards, the particulate concentrations within the

**Table 2.1** Humidification costs for 0.47 $m^3/s$ (1,000) cfm of outside air in New York City (© Liebert Corporation Reprinted with permission [49])

| DB temp. range (°F) | <24 | 25–29 | 30–34 | 35–39 | 40–44 | 45–49 | 50–54 | 55–59 | 60–64 | <64 |
|---|---|---|---|---|---|---|---|---|---|---|
| Moisture to be added* (grains/lb) | 49 | 46.9 | 42.7 | 38.5 | 34.3 | 28.7 | 21 | 12.6 | 2.8 | |
| Hours/year in temp. range | 387 | 344 | 512 | 739 | 791 | 733 | 770 | 722 | 776 | 5,774 |
| Cost to humidify @$0.10/kWh | $457 | $386 | $518 | $667 | $629 | $483 | $367 | $205 | $48 | $3,761 |

* for maintaining 50% RH in the data center

spaces were found to be well within limits. The study concluded that even with the higher fan pressure drop associated with using a filter with a higher MERV rating, the increase in extra energy use was meager (only 1%) in comparison to the savings associated with reduced operation of chiller unit (30–40%).

*Humidification costs associated with use of outside air*
Added cost for an airside economizer results from additional dampers used for closing and opening the economizer windows, improved filters, fans, and humidification–dehumidification units. Though the infrastructure cost is one time, incurred during the initial installation of the airside economizer, the cost of humidification is a recurring expense, which could offset the savings derived from the economizer.

Liebert Corporation evaluated the cost of humidification for a hypothetical data center with airside economizer [49].The results are summarized in Table 2.1. The table shows the cost to humidify 0.47 $m^3/s$ (1,000 cfm) of outside air in New York City based on power cost of $0.10/kWh using a canister-humidifier. The data center space was assumed to be at 22.2°C (72°F) with 50% RH. The study suggests that humidification cost is not trivial and could decrease the energy savings derived from the economizer.

## 2.11.2 Waterside Economizer

A waterside economizer system works in conjunction with a heat rejection loop consisting of either a cooling tower, evaporative cooler or a dry-cooler to meet the data center cooling demand. The system is typically incorporated into the existing chilled water or glycol-based cooling system which delivers chiller water/glycol to the CRAC units in the data center. For economizer operation, the water used in the cooling system passes through an additional set of coils to cool the room air, thus eliminating the need for mechanical cooling. Depending on the system design and specific data center requirements, there are numerous methods of implementing waterside economizers. Discussing the various designs is out of the scope of the chapter and the reader is referred to an economizer design hand

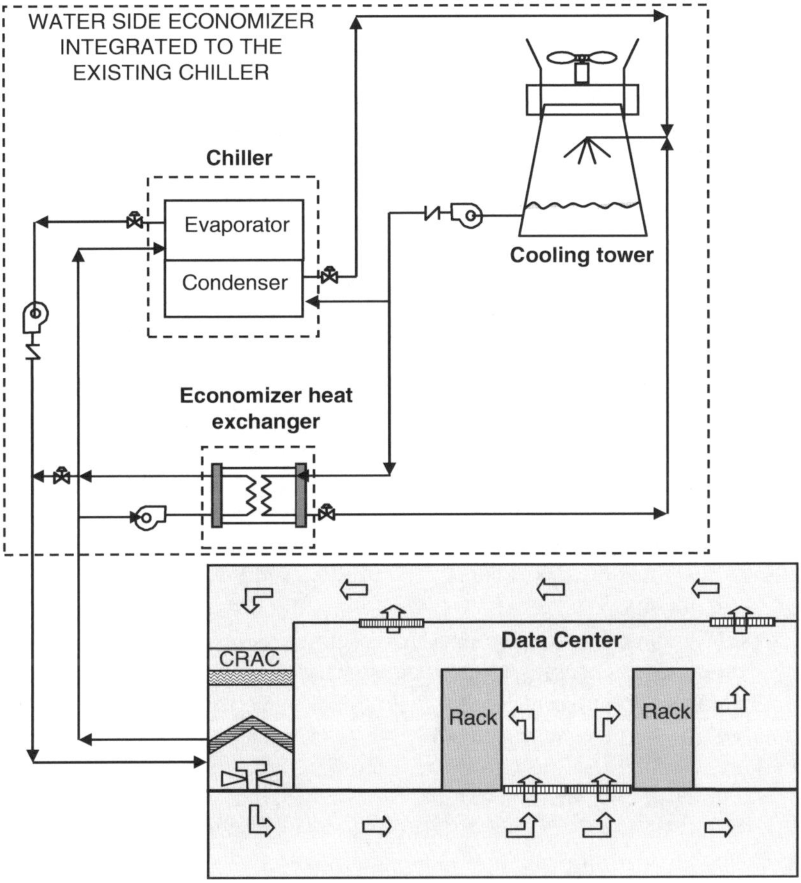


**Fig. 2.53** Schematic of a waterside economizer

book. Figure 2.53 shows a schematic of an integrated waterside economizer utilizing a cooling tower.

The economizer heat exchanger is connected in series with the chiller evaporator and piped in a primary/secondary manner into the chilled water return line from CRAC cooling coils. In this layout, the chilled water entering the economizer heat exchanger is at the highest temperature. Since the chiller and the economizer operate simultaneously, the economizer heat exchanger provides partial pre-cooling, while the chiller provides the rest of the cooling.

During favorable weather conditions or during night times, the data center cooling loads can be served entirely with chilled water produced by the cooling tower alone, by completely bypassing the chiller. The main advantage of waterside economizer is that it can significantly reduce or totally eliminate dependency on compressor-based cooling, without affecting space temperature and humidity set points. In addition, during winter months, waterside economizers also offer an additional level of redundancy by providing backup to the existing mechanical chillers.

### *2.11.3 Practical Considerations in Implementation of Economizer Solutions*

Economization must be engineered into the air handling system. An outdoor-air economizer system is best implemented in the design stage, where any required architectural accommodations can be made at little or no additional cost. The key objective is for all data center air handlers to have access to 100% outside air as well as return air. This is best achieved with a central air handling system. A central air handler and economizer system integrated with an efficient chiller system are ideally suited for large data centers, whereas low cost, mass produced package units integrated with the CRACs are more economical for smaller data centers. Waterside economizers typically use a cooling tower and heat exchanger to generate chilled water, instead of a mechanical chiller. The added costs for a waterside economizer result from increased cooling tower fan power, controls, heat exchangers, and piping. In some cases, the additional costs are incurred due to extra plant floor space needed for installation of additional backup pumps.

Using airside economizer systems in the cold winter months may sound very appealing, but care must be taken to control humidity. Maintaining consistent, acceptable temperature levels can be done with either system (airside economizer or waterside economizer), but maintaining humidity levels, especially with an airside economizer, becomes a significant challenge. The high cost of humidification can offset the potential energy savings derived using outside air. Usually, direct airside economizer would result in 1.5°C increase in IT equipment supply air temperature over the outside-air dry-bulb temperature. For waterside economizer, the data center supply air temperature depends on the temperature of the water flowing in the cooling loop, which in turn depends on the wet-bulb temperature. Another challenge in implementing an economizer system is providing seamless transition into and out of the economizer operational mode. In case of a data center using waterside economizer, the changeover needs to occur with minimal or no disruption of chilled water supply to the CRAC units.

The chilled water supply temperatures to the CRAC units are typically in the range of 8–12°C. In order to supply chilled water at 12°C to a data center facility operating on a waterside economizer, the chilled water leaving the economizer heat exchanger would have to be maintained below 6°C. To achieve this, the maximum wet-bulb temperature would have to be in the vicinity of 4°C. This makes the operation of the economizer unreliable, as a slight increase in the wet-bulb temperature above 4°C (40°F) offsets the capability of the economizer system to generate 12°C (54°F) chilled water, thus requiring the switchover to mechanical cooling. This back and forth transition from economizer mode to mechanical cooling is not recommended and is also not practical, as most chillers require a minimum entering condenser water temperature (typically 15.5–18.3°C) for stable operation.

Before implementing an air/water economizer, it must be ensured that all the IT equipment housed in the data center must be rated to operate in the range of humidity and temperature control that could be achieved by economization. In addition to the temperature and humidity ranges, there are a number of other factors that can result in increased failure rate of IT equipment. The gaseous pollution and particulates can significantly impact the failure rate. These factors have been well documented in [5]. Therefore in addition to DBT and WBT, outside air quality must be carefully considered in selecting the type of economizer.

Despite many practical considerations in the implementation of economizers, the savings in energy can be quite significant. As such, many data centers are resorting to economizer-based cooling solutions over conventional compressor-based chiller systems. The use of economizers is also now required by some building codes. The City of Seattle since 2001 requires the application of economizer systems in computer rooms and data centers. The State of Massachusetts Energy Code for Commercial and High-Rise Residential New Construction, effective in 2001 [50], requires, with some exceptions, air or water economizers for all cooling systems with a total cooling capacity of at least 19 kW (65,000 BTU/h). Pacific Gas and Electric Company in California has published a design guideline scorebook for high-performance data centers [51], addressing various options for improving the data center efficiency, including airside and waterside economizers. They demonstrated a total energy saving of 45% (254,000 kWh/year) on a model data center using airside economizer. Intel has reported large energy savings (40%) by implementing waterside economizers system in data centers [52]. Two cases of waterside economizers were investigated, while designing a high performance data center with a high-density engineering computer server. Liebert [49] has suggested that the use of glycol-based fluid economizer systems could help reduce data center cooling costs and decrease energy usage for a wide range of outdoor temperature and humidity conditions.

## 2.12 General Guidelines on Data Center Best Practices

### *2.12.1 Calculating the CRAC Capacity*

For small data centers, overhead ducted air supply provides total separation of the hot and cold airstreams and hence, no substantial loss of humidity takes place. The cooling occurs along a horizontal line on the psychrometric chart. The air-conditioning systems can operate at maximum efficiency. In large data centers employing raised-floor air distribution systems the air needs to be delivered at substantially lower temperature to overcome the inefficiencies due to mixing of hot and cold airstreams. Lowering the supply air temperatures causes condensation of moisture in the air, resulting in a decrease in relative humidity. The cold air now needs to be replenished with moisture to meet the supply air requirement. This results in an increase in the required CRAC cooling capacity [53]. It is estimated that CRAC oversizing can range from 0% to 30%, depending on the type of air delivery system employed [53].

Unlike building cooling loads which involve both sensible as well as latent cooling, the IT equipment in the data center produces "sensible" heat only. As a result, the bulk of the cooling provided by the CRAC consists of "sensible" cooling. The CRAC cooling capacity is specified in terms of either in kilowatt per hour (kWh) or British Thermal Units per hour (BTU/h) or sometimes in "tons" of refrigeration, where one ton corresponds to a heat absorption rate of 2.5 kWh (12,000 BTU/h) measured at 80°F. These specifications are subjective as they include both "latent" and "sensible" cooling. Therefore, in addition to the above data, CRAC manufacturers also provide cooling capacities in terms of "sensible kWh" (or "sensible BTU/h") at various temperatures and RH values [54]. These numbers can be used to calculate the usable cooling capacity of the CRAC unit. In general, the cooling capacity of the CRAC decreases with decrease in operating room temperature. Since, the data centers usually operate in the ASHRAE-recommended range of 20–25°C (68–77°F) with 40–55% relative humidity (RH), the actual cooling capacity provided by CRAC will be further reduced.

Cooling capacity is sometimes expressed in terms of volumetric flow rate of air discharged by the fan (cfm). As described in Sect. 2.10.3 on psychrometric analysis, the volume of air required to provide cooling depends on the moisture content in the air and the temperature difference between the supply and return air ($\Delta T$) [54]. Using the cfm data, the cooling capacity of the CRAC can be estimated using the following relations.

$$\mathrm{kWh} = \frac{1.08\ \mathrm{cfm} \times \Delta T(^{\circ}F)}{3,412}. \tag{2.40}$$

Or in terms of BTUs/h as

$$\mathrm{BTUs/h} = \frac{1.08\ \mathrm{cfm} \times \Delta T(^{\circ}F)}{\mathrm{BTU/h}}. \tag{2.41}$$

These are empirical estimations only. For all practical applications, appropriate derating due to inefficiencies in air distribution systems and other factors described above must be considered to determine the effective utilization of a particular CRAC unit.

### *2.12.2 Placement of the Aisles*

Typically, the most effective cooling begins about 2.4 m (8 ft.) (measured horizontally) from the CRAC unit [54]. The CRAC capacity, equipment cooling loads, and under-floor conditions (airflow restrictions) will impact the effective cooling range of a CRAC.

### *2.12.3 Aisle Width*

Each rack needs a specific area surrounding the rack to draw in cold air and discharge the warm air. For a front-to-back airflow rack, typically, a width of 120 cm (two floor tiles) is needed in front of the rack and a width of at least 60 cm (one unobstructed floor tile) is needed behind the rack to discharge the warm air and facilitate cable routing. Cold aisles should be a minimum of 4.2 m apart, center-to-center, or seven full tiles [54].

### *2.12.4 Placement of High-Density Racks*

In a data center having IT equipment with varied power densities, it is recommended to disperse the high power rack among low power racks. This would require lower airflow rates to provide cooling to the low power rack. In prior studies conducted [55, 56], it was found that only 31% of the rack flow rate to the high power rack was required to maintain the inlet temperatures within acceptable limits. A further extension of the work found that removing racks adjacent to the high power racks decreased the inlet temperature not only to the high powered rack but also to all the neighboring racks, though the highest reduction in inlet temperature was observed with the removal of one rack only.

### *2.12.5 Data Center Zoning*

The data center IT equipment should be segregated into various zones, depending on the inlet air temperature and humidity requirements. It is recommended that older equipment such as tape and disc storage systems, and printers which require tighter humidity control should be isolated from the rest of the data center. Zone control helps in isolating the few pieces of equipment which require stringent environment control.

### *2.12.6 Placement of the CRAC Units*

The placement of the CRACs should be based on the geometry of the room. The non uniformity in flow is greatly reduced, if all the CRAC units discharge in the same direction. The dead zones beneath are floor are eliminated when the CRAC units deliver air in an opposing manner, as shown in Fig. 2.10. In room layouts that are long and narrow, the CRAC units may be placed at the perimeter, while in large square rooms the CRAC units may be placed at the center, in addition to the perimeter of the room to deliver adequate flow to the racks. In general, CRAC units should be placed to promote even pressure distribution in the plenum. They should be placed perpendicular to the rows of IT equipment (aisles). Case studies using CFD have revealed that aligning CRAC units with the hot aisle generally provides shorter return path for air, thus facilitating higher return-air temperatures to the CRAC units. This leads to better CRAC cooling performance.

### *2.12.7 Effect of Rack Doors*

As described earlier, rack doors contribute to system pressure drop across the CRAC, as well as the server fans. As such, these should be eliminated if possible. If absolutely needed the doors should have a minimum open area of 60% to allow for maximum airflow [57]. Usually, the rear flow out of the rack is obstructed by network and power cables. Improperly managed cables or dense cabling can significantly impede the airflow. In such cases, additional rear door-mounted fans or blowers need to be installed to push the air out of the server.

### *2.12.8 Humidity Control in Data Center*

The reader may recollect that even if the humidity ratio is maintained, relative humidity varies with temperature. This makes humidity control not only expensive, and hard to maintain, but also a point of failure. As a result, many data centers do not operate with humidity control, especially if it involves humidification. In case humidification is necessary, adiabatic or ultrasonic humidifiers should be used, rather than conventional steam humidifiers. Also, due to difficulty in coordinating humidity control among various CRAC units, it is suggested to use dedicated humidification equipment rather than use built-in humidifiers and control programs in the CRAC unit.

These are just a few of the many guidelines and best practices followed in data centers. The list is exhaustive and describing each of them is beyond the scope of the chapter. The interested reader is referred to [3, 5, 6, 54], and in addition, a number of articles are readily available on the Web on best practices followed in data centers.

## 2.13 Ensemble COP of Data Center

Coefficient of performance (COP) is a familiar term for a thermal engineer. For many years it has been the simplest and probably the only means of defining the thermal efficiency of a mechanical system. It combines the information from various energy exchange processes occurring in a cooling system into a single unified number. In addition, COP also provides a basis for comparing two different mechanical systems having a common objective.

This section describes the methodology for calculating the ensemble COP of the entire data center, starting from the individual heat generating chip to the cooling tower, where the ultimate heat rejection to the environment takes place. This model can be used as a tool to evaluate new designs, or tune the operating set points in an existing data center to achieve optimal trade off between thermal efficiency and capacity utilization. The methodology presented here is based on the work reported in [58, 59].

*Definition of COP*

COP is a measure of cooling effectiveness in any thermal system. Higher COP means a more efficient thermal system. In the most basic form, COP of a cooling system is defined as:

$$\mathrm{COP} = \frac{\text{Heat extracted}}{\text{Network input}}. \tag{2.42}$$

With reference to a data center, COP is defined as

$$\mathrm{COP}_{\text{Data center}} = \frac{\text{Heat dissipated by datacenter}}{\text{Energy needed to drive cooling infrastructure}}. \tag{2.43}$$

### 2.13.1 Basis for Development of an Ensemble COP Model

The data center contains different pieces of equipment, often overdesigned for the worst case scenario. The component level overdesign, coupled with improper sizing and selection of the operating parameters, are the primary causes for data center cooling inefficiencies. The ensemble COP model helps in identifying some of these inefficiencies in the early design stage, thus preventing system overdesign. The ensemble model can also be used for broadly characterizing day-to-day operation and answering "what if" questions. The model presented in the following section is based on thermodynamic principles and is not intended to replace or forgo the need for a detailed CFD analysis. The purpose of the model is to understand and assess the influence of the design and operating parameters on the operational efficiency of the data center, and to holistically evaluate the system as a whole rather than evaluate each component individually.

### 2.13.2 Data Center Heat Flow Path

In a data center, the heat exchange process spans across multiple devices and size scales. The source (chip or microprocessor) temperature where the heat originates is much higher compared to the sink (ambient or atmospheric). Thermal inefficiencies

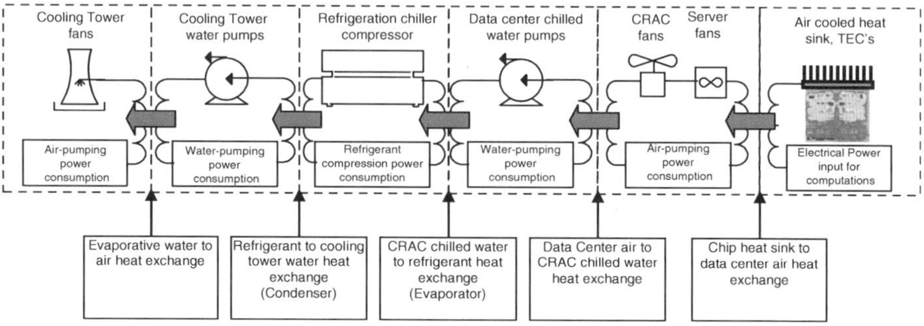


**Fig. 2.54** Schematic of the energy flow in a data center (© 2009 IBM corporation, reprinted with permission [60])

and flow resistances are associated with the intermediate devices in the heat transfer path. A comprehensive understanding of the overall heat transfer path and the associated work involved in extracting the heat is a necessary first step towards development of an ensemble COP model for the data center. The reader is referred to Fig. 2.4 for the energy path in a typical air-cooled data center. Figure 2.54 describes the energy flow associated with Fig. 2.4.

All the electrical energy consumed by the chips, server fans, and the blowers in the CRAC units is eventually converted to heat and released into the data center. The existing practices use air as the primary heat exchange medium to transfer the heat from the chip into the data center and room. This heat is transferred into a chilled water loop in the CRAC units, which is in turn coupled with a refrigerant-water loop to eventually release into the ambient environment. Chilled water from a refrigeration chiller plant is circulated through the air-conditioning units using chilled water pumps. The condenser in the chiller is cooled by circulating water from an air-cooled cooling tower. The warm water entering the cooling tower is cooled due to evaporative cooling by rejecting heat to the environment.

The ensemble COP model traces the available heat transfer path and accounts for additional work required to overcome the inefficiencies and the thermodynamic irreversibilities associated with each device along the heat transfer path. The COP's of the various devices participating in the heat transfer path are described in the following section.

### *2.13.3 COP of the Chip*

The primary source of heat in a data center originates at the chip. This heat is transferred to the heat sink mounted on top of the chip via an intermediate thermal interface material as shown in Fig. 2.55.

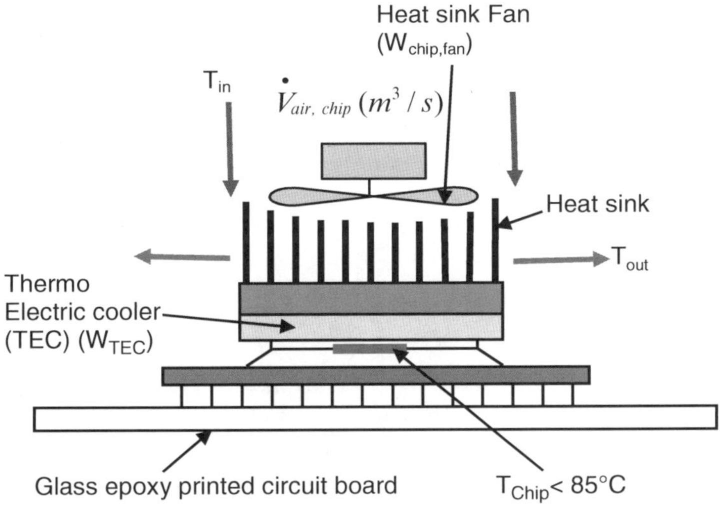


**Fig. 2.55** Schematic of a high-power chip-heat sink system with TEC

Heat transfer devices such as thermo electric coolers (TECs) could possibly be attached to the chip to keep their temperatures below 85°C for Si microprocessors often used in complementary metal oxide semiconductor (CMOS) technology. The TECs offer a negative thermal resistance, thereby lowering the heat transfer resistance to create the necessary temperature difference required to move the heat from processor to the heat sink. Commercially available TEC could consume about 30 W of power with a temperature difference of 15°C between the hot and cold side to dissipate 100 W of heat from the microprocessor [59].

The energy required by the TEC to remove the heat with the ship scale is provided by the following expression:

$$W_{\text{TEC}} = N\{I^2R + I[(S_\text{h} - S_\text{c}) \times (T_\text{h} - T_\text{c})]\}, \tag{2.44}$$

where $N$ is the number of TEC modules, $S$ is the Seebeck coefficient of the material (V/K), $I$ is the current in each element (A), $R$ is the resistance in each element (Ω), $T$ is temperature (K), subscripts h and c denote hot and cold side of the chip.

In addition to TECs, fans or blowers are mounted on top of the heat sink to maintain the temperature within specified limits. The volumetric flow rate of air required to maintain the heat sink temperature depends on the heat dissipated by the processor:

$$Q_{\text{Chip}} = \dot{m}_{\text{air,chip}} C_{p,\text{air}} (T_{\text{out}} - T_{\text{in}})\ (\text{W}), \tag{2.45}$$

$$= \rho \dot{V}_{\text{air,chip}} C_{p,\text{air}} (T_{\text{out}} - T_{\text{in}}), \quad \text{where } \dot{m}_{\text{air,chip}} = \rho \dot{V}_{\text{air,chip}} \tag{2.46}$$

where $Q_{\text{Chip}}$ is the heat dissipated by the chip, $C_{p,\text{air}}$ is the specific heat at constant pressure (kJ/kg K), $T$ is temperature (K), subscripts out and in denote heat sink inlet and exit air temperatures.

The above expression provides the volumetric flow rate of air required for a given temperature rise in air across the heat sink.

From the fan laws in Sect. 2.5.3, we can estimate the fan power required for a given volumetric flow rate using:

$$W_{\text{chip,fan}} = \frac{\dot{V}_{\text{air,chip}} \Delta P_{\text{chip}}}{\eta_{\text{chip,fan}}} \text{ (W)}. \tag{2.47}$$

The $\eta_{\text{chip,fan}}$ at a given speed is usually specified by the fan manufacturer or is obtained by superimposing the efficiency curves on the fan curve, as illustrated in Fig. 2.7 for a given fan speed.

The COP at the chip can be estimated using the above relations as:

$$COP_{\text{Chip}} = \frac{Q_{\text{Chip}}}{W_{\text{Chip}}} = \frac{Q_{\text{Chip}}}{W_{\text{TEC}} + W_{\text{Chip,fan}}} \tag{2.48}$$

The COP of the chip can be optimized by

1. Controlling the supply air temperature, and optimizing the TEC current, which is usually function of TEC geometry and temperature difference.
2. Minimizing the fan power by optimizing the fan speed, pressure drop and volumetric flow rate to operate at maximum fan efficiency.

The above equations are valid for air-cooled processors using TEC and fans. In a similar fashion, the COP of a liquid-cooled processor or heat sink employing heat pipes can be calculated using appropriate relations.

## *2.13.4 COP of the Server*

The heat dissipated by the chip resides in the server enclosure. A server can comprise of multiple chips and other heat generating components such as power supplies, memory and network devices, which might not have active heat dissipation mechanisms such as the microprocessor. The total heat generated within the server enclosure must be expelled into the data center. Similar to a heat sink, servers employ blowers or fans to move the heat into the data center. The fans need to ensure the requisite flow through the servers by overcoming the pressure drop of the server as well as any additional pressure drop offered by the rack doors. The components in the server placed downstream of the flow receive preheated air.

The server fan speeds are predetermined to compensate for the preheating. The COP of the server is given by the following relation:

$$\mathrm{COP}_{\mathrm{Server}} = \frac{Q_{\mathrm{Server}}}{W_{\mathrm{Server}}} = \frac{\sum\left(Q_{\mathrm{Chip}} + Q_{\mathrm{devices}}\right)}{\sum\left(W_{\mathrm{Chip}} + W_{\mathrm{Server,fan}}\right)}$$
$$= \frac{m_{\mathrm{air,server}} C_{p,\mathrm{air}}\left(T_{\mathrm{out,server}} - T_{\mathrm{in,server}}\right)}{W_{\mathrm{Server}}}, \tag{2.49}$$

where $Q_{\mathrm{devices}}$ (W) is the heat dissipated by all other devices in the server such as memory, power supplies, network devices, etc., $T$ is temperature in (K). Subscripts out and in denote heat sink air inlet and exit temperatures.

Once the static pressure drop across the server and volumetric flow rate required to maintain the desired temperature rise are known, the server fan power can be calculated using the characteristic fan curve using (2.47).

From (2.49) we note that the COP of the server can be increased by reducing the fan power. Current research and design practices are aimed at reducing the fan and chip power consumption without sacrificing performance.

### *2.13.5 COP of the Rack*

Usually, a rack does not participate in providing additional cooling. It only serves as an enclosure to physically house multiple servers. The servers directly expel the heat from the chips into the data center. However, a high-power rack may sometimes be equipped with a chimney or a rear door heat exchanger with additional fans to assist movement of air from the servers into the data center room. These additional racks-mounted cooling devices need to be included in determining the rack-level COP. In such cases the rack-mounted COP is determined using the following expression:

$$\mathrm{COP}_{\mathrm{Rack}} = \frac{\sum Q_{\mathrm{Server}}}{\sum W_{\mathrm{Server}} + W_{\mathrm{Rack}}}, \tag{2.50}$$

where $W_{\mathrm{Rack}}$ (W) is the power consumed by the rack-mounted air-assisting devices.

### *2.13.6 COP of the CRAC*

As discussed earlier, an air-cooled data center employs large air handlers or CRAC units coupled with a chilled water distribution system to move the heat dissipated by the racks out from the data center. The hot air exhausted from the racks could be cooled by single or multiple CRAC units and recirculated back into the data center. Depending on the supply air set point in the CRAC unit, cold air supply temperatures vary from 12°C to 22°C. As illustrated in Fig. 2.54, the hot air entering the CRAC unit rejects heat to the chilled water flowing through the cooling coils in the heat exchanger. In some cases, the CRAC units may utilize direct expansion refrigeration instead of a chilled water system.

As emphasized in the previous sections, data center airflows are complex. The highly nonuniform heat load distribution among the IT equipment, coupled with complex air circulation patterns significantly impacts the CRAC utilization and performance. As such, the COP of the CRAC unit depends on the level of capacity utilization and the heat exchange efficiency of the CRAC unit. The capacity utilization of the CRAC units requires knowledge of the airflow, the supply and return air temperatures of air entering and leaving the CRAC unit. These can be obtained through CFD modeling, or using a distributed array of sensors as described later in Chapters 5 and 7.

Typically, the only active power-consuming component in the CRAC unit using a chilled water system is the blower. The power consumed by the CRAC blower can be obtained using the manufacturers data based on blower efficiency and flow rate. Depending on the CRAC utilization capacity and blower power, the COP of a CRAC unit is calculated using the following relation:

$$\mathrm{COP}_{\mathrm{CRAC}} = \frac{Q_{\mathrm{CRAC}} + Q_{\mathrm{CRAC,Blower}}}{W_{\mathrm{CRAC}}}, \tag{2.51}$$

where $W_{\mathrm{CRAC}}$ is the power consumed by the CRAC blower (W), $Q_{\mathrm{CRAC,Blower}}$ is the heat generated by the blower (W), $Q_{\mathrm{CRAC}}$ is the cooling load in the CRAC unit (W), $Q_{\mathrm{CRAC}}$ depends on data center heat load and the CRAC capacity utilization (W).

$Q_{\mathrm{CRAC}}$can vary between individual CRAC units in a data center housing multiple CRAC units. As such, the $\mathrm{COP}_{\mathrm{CRAC}}$ has to be calculated individually for each CRAC on a per-unit basis to assess the impact of utilization variations and benefits of part load efficiencies of the data center air distribution system. Based on first law of thermodynamics, $Q_{\mathrm{CRAC,Blower}}$ refers to the difference between total power consumed by the blower and the flow work.

As described in Sect. 2.5.1, the CRAC fans need to overcome the cumulative impedances in flow offered by the CRAC filters, heat exchanger fins, plenum, room, rack, and the server to provide the requisite airflow needed to cool the IT equipment. Apart from ensuring adequate flow, the CRAC is also responsible for heat exchange between the air and the chilled water. The important design parameters influencing the CRAC COP are heat exchanger coil design, air and water flow rates, and supply air temperature set points. The blower power of the CRAC for a particular operating speed is calculated using (2.47) and the characteristic fan curve data supplied by the manufacturer. If the fan curve is not readily available it can be determined empirically using measured data.

### 2.13.7 COP of Data Center Air Distribution

In addition to the IT equipment, the additional heat generated by the lighting and other ancillary equipment such as uninterrupted power supply (UPS) systems and power distribution units (PDU) must be included in determining the overall heat load in the data center. Also, besides providing cooling, the data center needs to maintain humidity within the ASHRAE-specified limits. This would involve humidification or dehumidification of air requiring additional work, which must be accounted while calculating the COP. The COP of the data center air distribution system is:

$$\mathrm{COP}_{\mathrm{DC}} = \frac{Q_{\mathrm{DC}}}{W_{\mathrm{DC}}} = \frac{\sum_i \left(Q_{\mathrm{CRAC}} + Q_{\mathrm{CRAC,Blower}}\right) + Q_{\mathrm{Ancillary}}}{\sum_i W_{\mathrm{CRAC}} + \sum_j W_{\mathrm{humid-dehumid}}}, \tag{2.52}$$

$Q_{\mathrm{Ancillary}}(\mathrm{W})$ is the heat generated by the ancillary equipment such as PDU's lights, UPS systems, etc. $i = j$, if the CRAC units are equipped with built-in humidification/dehumidification systems.

### 2.13.8 COP of the Chilled Water Distribution Loop

The heat extracted by the data center is transferred to the chilled water flowing through the CRAC unit's heat exchanger. As shown in Fig. 2.54, the chilled water loop comprises of secondary pumps cooling coils, pipes, and flow-regulating valves connected to a central chiller. The chilled water distribution loop is designed to provide required supply of chilled water to the CRAC units in the data center to achieve the preset temperature set points. Additional booster pumps may be required to compensate for the reduction in flow due to resistances offered by the heat exchanger coils in the CRAC, chiller evaporator, the network of pipes, valves, and other miscellaneous devices. The proximity of the data center to the chilled water facility and the design of the hydraulic network determine the pumping power requirement.

The COP of the chilled water distribution loop is given by:

$$\mathrm{COP}_{\mathrm{CWDL}} = \frac{Q_{\mathrm{DC}}}{\sum \left(W_{\mathrm{P}}\right)_{\mathrm{Secondary\ loop}}}, \tag{2.53}$$

where $\sum \left(W_{\mathrm{P}}\right)_{\mathrm{Secondary\ loop}}$ (W) is the power consumed by all the pumps in the secondary loop. The goal here is to utilize minimum pumping power to provide maximum cooling at the CRAC units.

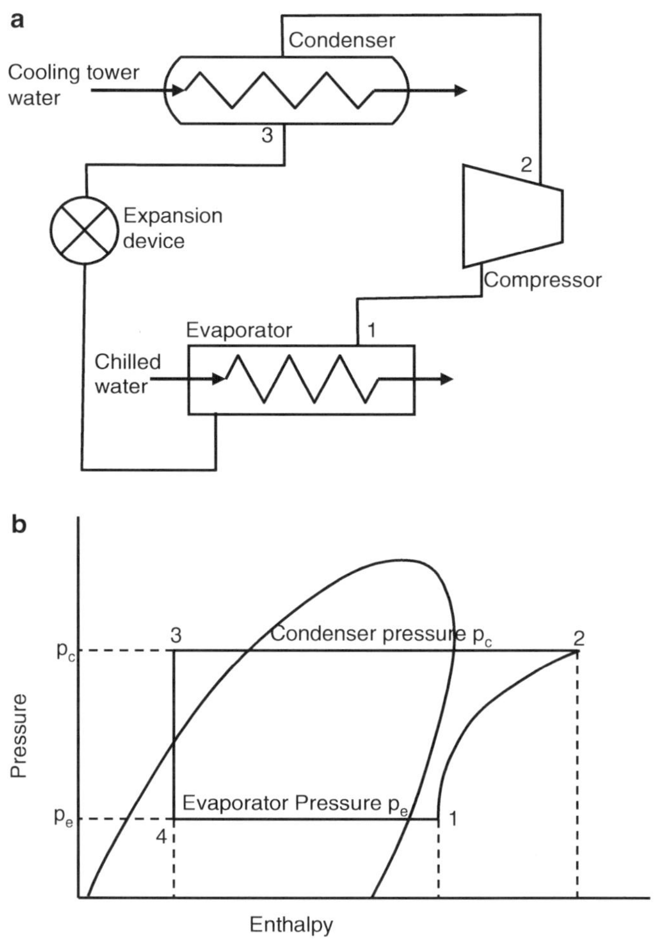


**Fig. 2.56** Operating principle of a chiller (**a**) components of a chiller and (**b**) Basic vapor compression refrigeration cycle

## *2.13.9 COP of the Chiller*

In order to achieve the desired data center supply air temperature, the chilled water to the CRAC unit needs to be at sub-ambient temperatures. A chiller is used for this purpose. It consists of two heat exchangers, a condenser and evaporator, interconnected between the compressor, and an expansion device to form a closed loop circuit as shown in Fig. 2.56a. The chilled water pumps feed the evaporator heat exchanger in the chiller with warm water from the CRAC units. Heat is transferred to the refrigerant circulating through the secondary loop of the evaporator, causing the refrigerant to vaporize. The gaseous refrigerant is compressed to a high pressure–high temperature vapor using a mechanical compressor.

The condenser cools the high pressure vapor refrigerant into saturated liquid. The high temperature–high pressure liquid refrigerant in expanded to a low pressure–low temperature fluid in the evaporator using an expansion device, which completes the vapor compression cycle illustrated in Fig. 2.56b.

In addition to the data center heat load, the chiller is responsible for removing additional loads arising from heat gain due to conduction through data center walls and heat losses in the chilled water distribution system. If these losses are neglected, the chiller heat load is identical to the data center heat load as provided by (2.52). The COP of the chiller is defined as

$$\mathrm{COP}_{\mathrm{Chiller}} = \frac{Q_{\mathrm{Chiller}}}{W_{\mathrm{Chiller}}} = \frac{Q_{\mathrm{Chiller}}}{W_{\mathrm{Chiller,compressor}} + W_{\mathrm{auxillary}}}, \tag{2.54}$$

where $Q_{\mathrm{Chiller}}$ is the total heat extracted by the chiller (W), $W_{\mathrm{auxillary}}$ is the power consumed by auxiliary devices in the chiller such as condenser fans in an air-cooled chiller, or built in primary chilled water pumps in the chiller (W), $W_{\mathrm{Chiller,compressor}}$ is the compressor work (W).

For a positive displacement compressor using a polytropic compression process, the compressor work is defined by the following relation

$$W_{\mathrm{Chiller,compressor}} = \frac{n \times \dot{m}_{\mathrm{ref}} p_{\mathrm{e}} v_{\mathrm{e}}}{\eta_{\mathrm{c}} \eta_{\mathrm{motor}} (n-1)} \left[ \left( \frac{p_{\mathrm{c}}}{p_{\mathrm{e}}} \right)^{\frac{n-1}{n}} - 1 \right], \tag{2.55}$$

where $p_{\mathrm{c}}, p_{\mathrm{e}}$ are the condenser and the evaporator pressures (N/m$^2$), $\dot{m}_{\mathrm{ref}}$ is the mass flow rate of the refrigerant in the compressor (kg/s), $n$ is the polytropic index of the refrigerant, $v_{\mathrm{e}}$ is the volume of the evaporator, $\eta_{\mathrm{c}}, \eta_{\mathrm{motor}}$ are the compressor and motor efficiencies.

The chiller heat load $Q_{\mathrm{Chiller}}$ can be calculated if the chilled water supply and return temperatures and mass flow rate of the water flowing through the chiller are known or measured. The chiller heat load is calculated using the following relation:

$$Q_{\mathrm{Chiller}} = \dot{m}_{\mathrm{water,chiller}} c_{p,\mathrm{water}} (T_{\mathrm{out,ch.water}} - T_{\mathrm{in,ch.water}}), \tag{2.56}$$

where $\dot{m}_{\mathrm{water,chiller}}$ is the mass flow rate of water flowing through the evaporator heat exchanger in the chiller (kg/s), $c_{p,\mathrm{water}}$ is the specific heat of water (kJ/kg K), and $T_{\mathrm{out,ch.water}}, T_{\mathrm{in,ch.water}}$ is the temperature of water leaving and entering the evaporator heat exchanger (K).

There are various analytical models available for evaluating the COP of the chiller. The interested reader is referred to a standard refrigeration design manual. The COP of a typical chiller used for a data center is shown in Fig. 2.57 [58]. The data represents 113 test points based on the manufacturer test data for one of the chillers located at IBM's Poughkeepsie plant. The plot exhibits the characteristic behavior of the chiller, which shows an increasing trend of COP with increasing heat load until the rated capacity is achieved. The chillers operate between 20% and

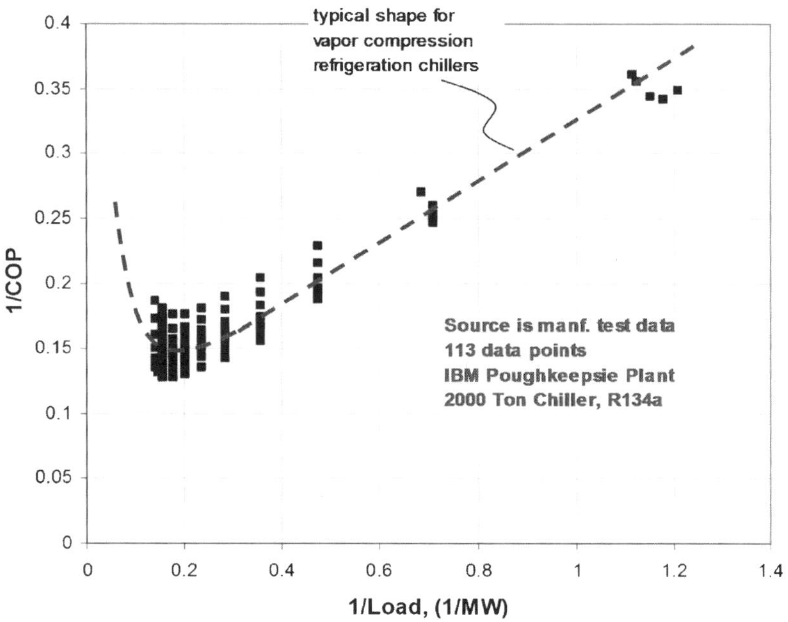


**Fig. 2.57** Typical COP of a data center chiller (© 2009 ASME, reprinted with permission [58])

80% of the rated load. Most chillers operate efficiently in the load range between 60% and 100%. For the specific chiller illustrated in Fig. 2.57, most efficient operating load range is between 70% and 100%, where the COP is maximum.

### *2.13.10 COP of the Cooling Tower*

The cooling tower is the terminal point in the heat transfer path, where the heat extracted from the data center is ultimately rejected to the ambient environment. The warm water from the condenser is sprayed on the fin structures in the cooling tower. Some of the water evaporates, thereby reducing the overall temperature of the remaining water. The temperature of the water leaving the cooling tower is close to the ambient air temperature. The water coming out of the cooling tower needs to be replenished to compensate for the loss due to evaporation. Since, the primary mode of heat transfer is due to evaporation, relative humidity and temperature of ambient air play a major role in the effectiveness of cooling tower. The two main sources of energy consumption in a cooling tower are the cooling tower fans and the supply water pumps. The supply water pumps need to overcome the pressure drop in the piping network, the hydraulic head due to height of the cooling tower, and the pressure drop due the finned structures inside the cooling tower.

The COP of the cooling tower is defined by the following relation

$$\mathrm{COP_{CT}} = \frac{Q_{\mathrm{CT}}}{W_{\mathrm{CT}}} = \frac{Q_{\mathrm{Chiller}} + W_{\mathrm{Chiller,compressor}}}{W_{\mathrm{CT}}}, \tag{2.57}$$

where $W_{\mathrm{CT}}$ (W) is the power consumed by the cooling tower fans and the pumps and $Q_{\mathrm{CT}}$ (W) is the heat removed by the cooling tower.

Alternatively, the heat removed by the cooling tower is provided by the following relation:

$$Q_{\mathrm{CT}} = \dot{m}_{\mathrm{air,CT}}(h_{\mathrm{ao}} - h_{\mathrm{ai}}), \tag{2.58}$$

where $\dot{m}_{\mathrm{air,CT}}$ is the mass flow rate of air entering the cooling tower (kg/s) and $h_{\mathrm{ao}}$ and $h_{\mathrm{ai}}$ are the enthalpy of the air entering and leaving the cooling tower (kJ/kg).

The enthalpies of air entering and leaving the cooling tower can be calculated using the psychrometric chart using the procedure explained in Sect. 2.10.3.

### *2.13.11 Ensemble COP of the Entire Data Center*

The ensemble COP of the overall system is represented by the following relation

$$\mathrm{COP_{system}} = \frac{Q_{\mathrm{Chiller}}}{\begin{array}{l}\sum\limits_k \left(\sum\limits_j \left(\sum\limits_i (W_{\mathrm{Chip}} + W_{\mathrm{server,fan}})\right) + W_{\mathrm{Rack}}\right) \\ + \sum\limits_l W_{\mathrm{CRAC}} + \sum\limits_m (W_P)_{\mathrm{Secondary\ loop}} + W_{\mathrm{Chiller}} + W_{\mathrm{CT}}\end{array}}. \tag{2.59}$$

Rewriting (2.59) in terms of conventional COP defined by

$$\mathrm{COP_{Chiller}} = \frac{Q_{\mathrm{Chiller}}}{W_{\mathrm{Chiller}}}.$$

We get,

$$\mathrm{COP_{system}} = \frac{Q_{\mathrm{Chiller}}/W_{\mathrm{Chiller}}}{\begin{array}{l}1 + \left(\sum\limits_k \left(\sum\limits_j \left(\sum\limits_i (W_{\mathrm{Chip}} + W_{\mathrm{server,fan}})\right) + W_{\mathrm{Rack}}\right)/W_{\mathrm{Chiller}}\right) \\ + \left(\sum\limits_l W_{\mathrm{CRAC}}/W_{\mathrm{Chiller}}\right) + \left(\sum\limits_m (W_{\mathrm{P}})_{\mathrm{Secondary\ loop}}/W_{\mathrm{Chiller}}\right) \\ + (W_{\mathrm{CT}}/W_{\mathrm{Chiller}})\end{array}}, \tag{2.60}$$

**Table 2.2** Typical energy consumption and COP of various cooling components in a data center. (© 2009 IBM corporation, reprinted with permission [60])

| Cooling equipment | Typical % of total data center cooling | Typical % of total data center power consumption | Typical efficiency metric (kW/ton) | Typical COP |
|---|---|---|---|---|
| Server fans | 8 | 3 | 0.07–0.25 | 15–50 |
| CRAC unit | 28 | 15 | 0.15–0.25 | 15–25 |
| Building chilled water pumps | 9 | 3 | 0.05 | 70 |
| Refrigeration system | 46 | 3 | 0.04–1.0 | 3–9 |
| Cooling tower | 9 | 2.5 | 0.1 | 35 |

$$\mathrm{COP}_{\mathrm{Ensemble}} = \frac{Q_{\mathrm{Chiller}}/W_{\mathrm{Chiller}}}{1 + A + B + C + D}, \tag{2.61}$$

where $A$, $B$, $C$, and $D$ are the ratios based on the chiller power consumption. $A$ depends on the power consumed by the cooling infrastructure on the chips, servers, and racks, $B$ depends on the power consumed by the CRAC blowers, $C$ depends on the total power consumed by secondary chilled water distribution pumping system, and $D$ depends on the power consumed by the cooling tower components

The above expression in (2.60) provides a basis to evaluate the ensemble COP with respect to the conventional chiller COP. Inspecting (2.61) we note that the ensemble COP will approach the COP of the chiller if $A$, $B$, $C$, and $D$ are minimized. Usually, $A$ and $B$ are O(1) due to high air demand by the servers leading to high supply by the CRAC to satisfy the server air requirement. In such cases, the ensemble COP drops to less than half of the conventional chiller COP. If the air distribution inefficiencies are considered, the drop can be even higher. The value of "$C$" depends on the proximity of the CRAC to the chiller. It can be significant for large data centers. "$D$" depends on the ambient condition such as dry-bulb temperature and relative humidity based on which the operating speed of the cooling tower fan is selected. Based on these observations, overall COP of even the most efficient data center (as provided by (2.61)) will be less than half of the conventional chiller COP.

Table 2.2 [60] provides an example of the typical COP for the various systems of the ensemble and the relative percentage of total energy consumption in the data center. Referring to Table 2.2 we note that the chiller consumes the maximum energy followed by the computer room air-conditioning units and server fans. Chiller energy accounts for nearly 50% of the total energy consumption in the data center.

The reader is encouraged to refer to [58–60] for specific cases illustrating the application of the ensemble COP model described herein. The use of the ensemble model to assess the impact on the total cost of ownership (TCO) is presented in [59]. In addition to a COP model, there are other methods to evaluate the thermal efficiency of a data center using thermodynamic principles [36, 61–67]. The use of exergy analysis to detect thermal inefficiencies in data centers is described in Chap. 9.

## 2.14 Concluding Remarks and Closure

Data center airflow management is a complex problem, influenced by a number of factors, some of which are dynamic and unpredictable. Presently, data centers are managed based on intuition, or accrued experience, which often lead to overly conservative and energy inefficient solutions. However, with increasing heat densities and the associated energy costs for cooling, there is a growing mandate for reducing the energy consumption and the carbon footprint of data centers. The incumbent data center manager now has to supplement her/his accrued experience with physics-based understanding of air distribution. In this aspect, this chapter provides the necessary foundation needed to understand and appreciate the importance of air management. The chapter focuses on three key themes. The system pressure drop, influenced by the various obstructions in the path of the airflow, determines the airflow rate. The impact of various geometrical features and hardware on the pressure drop are discussed. The general flow features in standard hot-aisle/cold-aisle data centers, which depend on the pressure distribution are also examined. The concepts of temperature and humidity control are central to the effective and energy efficient operation of any data center. The specifications on the inlet environment are examined through the psychrometric charts in this chapter. A number of new data center facilities are embracing the use of "free cooling" by using environmental air for cooling, when the conditions permit. The benefits of airside and waterside economizers are discussed. Operators of data centers are increasingly concerned with improving the facility's operational efficiency. Models for quantifying the coefficient of performance, by considering the multiple size scales involved in the heat removal process, are also presented in this chapter.

A number of subsequent chapters further explore the concepts introduced here. Chapter 5 discusses monitoring and controlling the data center. Metrology techniques are discussed in Chap. 7. Computational modeling of airflows and temperatures in data centers is examined in Chap. 8. Chapter 10 discusses the methods for optimizing data center thermal management hardware, based on reduced order modeling.

## References

1. IBM. Available: http://publib.boulder.ibm.com/infocenter/powersys/v3r1m5/index.jsp?topic=/p7ebe/p7ebegeneralguidelines.htm
2. D. Dunn. (2005 April 18, 12:02 AM ) Step into the future-new technology, security, and reliability requirements are changing the data-center infrastructure. *InformationWeek*. 1–3. Available: http://www.informationweek.com/news/160901346
3. *Uptime Institute*. Available: http://uptimeinstitute.org/index.php?option=com_content&task=view&id=363&Itemid=327
4. Stahl L, Belady C (2001) "Designing an alternative to conventional room cooling," in 23rd international telecommunications energy conference, Edinburgh, 14–18 October 2001, p 109–115

5. ASHRAE T (2011) 2011 Thermal guidelines for data processing environments – expanded data center classes and usage guidance. 1–45. Available: http://tc99.ashraetcs.org/documents/ASHRAE%20Whitepaper%20-%202011%20Thermal%20Guidelines%20for%20Data%20Processing%20 Environments.pdf
6. ASHRAE T. Datacom equipment power trends and cooling applications [Online]. Available: http://www.techstreet.com/cgi-bin/detail?product_id=1703614&ashrae_auth_token=
7. [Online]. Available: http://emt-india.com/BEE-Exam/GuideBooks/3Ch5.pdf
8. Patankar SV, Karki KC (2004) "Distribution of cooling airflow in a raised-floor data center," in 2004 Annual meeting–technical and symposium papers, american society of heating, refrigerating and air-conditioning engineers, 26–29 June 2004, Nashville, p 599–603
9. Patankar SV (2010) Airflow and cooling in a data center. J Heat Transfer 132:073001
10. Karki KC, Patankar SV (2006) Airflow distribution through perforated tiles in raised-floor data centers. Build Environ 41:734–744
11. Patankar SV et al. (2004) "Distribution of cooling airflow in a raised-floor data center," in Technical and symposium papers–2004 annual meeting of the american society of heating, refrigerating and air-conditioning engineers, 26–30 June, 2004, Nashville, p 629–634
12. Karki KC et al (2003) Use of computational fluid dynamics for calculating flow rates through perforated tiles in raised-floor data centers. HVAC R Res 9:153–166
13. Karki KC et al. (2003) "Techniques for controlling airflow distribution in raised-floor data centers," in 2003 International electronic packaging technical conference and exhibition, 6–11 July, 2003, Maui, p 621–628
14. Idelchik IE (1994) Handbook of hydraulic resistance. Begell House, Orlando
15. Available: http://publib.boulder.ibm.com/infocenter/powersys/v3r1m5/index.jsp?topic=/p7ebe/p7ebegeneralguidelines.htm
16. Hamann H et al. (2008) "The impact of air flow leakage on server inlet air temperature in a raised floor data center," in 2008 11th intersociety conference on thermal and thermomechanical phenomena in electronic systems (ITHERM '08), 28–31 May 2008, Piscataway, p 1153–1160
17. Karki KC et al. (2007) "Prediction of distributed air leakage in raised-floor data centers," in 2007 Winter meeting of the american society of heating, refrigerating and air-conditioning engineers, 27–31 January 2007, Dallas, p 219–226
18. Radmehr A et al. (2005) "Distributed leakage flow in raised-floor data centers," in ASME/Pacific rim technical conference and exhibition on integration and packaging of MEMS, NEMS, and electronic systems: advances in electronic packaging 2005, 17–22 July, 2005, San Francisco, p 401–408
19. Wang Z et al. (2010) "Kratos: amanagement of cooling capacity in data centers with adaptive vent tiles," in ASME 2009 international mechanical engineering congress and exposition, IMECE2009, 13–19 November 2009, Lake Buena Vista, p 269–278
20. Schmidt R et al. (2004) "Raised-floor data center: perforated tile flow rates for various tile layouts," in ITherm 2004–Ninth intersociety conference on thermal and thermomechanical phenomena in electronic systems, 1–4 June 2004, Las Vegas, p 571–578
21. Schmidt RR et al. (2001) "Measurements and predictions of the flow distribution through perforated tiles in raised-floor data centers," in Pacific rim/international, intersociety electronic packaging technical/business conference and exhibition, 8–13 July 2001, Kauai, p 905–914
22. Kumar P et al. (2010) "Dynamics of cold aisle air distribution in a raised floor data center," in 2010 3rd international conference on thermal issues in emerging technologies, theory and applications, ThETA3 2010, 19–22 December 2010, Cairo, p 95–102
23. Kumar P, Joshi T (2010) "Experimental investigations on the effect of perforated tile air jet velocity on server air distribution in a high density data center," in 2010 12th IEEE intersociety conference on thermal and thermomechanical phenomena in electronic systems, ITherm 2010, 2–5 June 2010, Las Vegas
24. Kumar P et al. (2011) Cold aisle air distribution in a raised floor data center with heterogeneous opposing orientation racks. In: INTERPACK2011, Portland, p 1–8

25. Kumar P et al. (2011) Effect of supply air temperature on rack cooling in a high density raised floor data center facility. In: IMECE 2011, Denver
26. Kumar P et al. (2011) Effect of server load variation on rack air flow distribution in a raised floor data center. In: 27th Annual IEEE semiconductor thermal measurement and management, SEMI-THERM 27 2011, 20–24 March 2011, San Jose, p 90–96
27. Radmehr A et al. (2007) Analysis of airflow distribution across a front-to-rear server rack. In: ASME InterPACK Conference 2007. IPACK2007, 8–12 July 2007, New York, p 837–843
28. Sorell V et al. (2006) An analysis of the effects of ceiling height on air distribution in data centers. In: 2006 Winter meeting of the American Society of heating, refrigerating and air-conditioning engineers, ASHRAE, 21–25 January 2006, Chicago, p 623–631
29. Mulay V et al. (2010) Effective thermal management of data centers using efficient cabinet designs. In: 2009 ASME interpack conference, IPACK2009, 19–23 July 2009, San Francisco, p 993–999
30. Udakeri R et al. (2008) Comparison of overhead supply and underfloor supply with rear heat exchanger in high density data center clusters. In: SEMI-THERM 08. 2008 24th annual IEEE semiconductor thermal measurement and management symposium, 16–20 March 2008, Piscataway, p 165–172
31. Shrivastava S et al. (2005) Comparative analysis of different data center airflow management configurations. In: ASME/Pacific rim technical conference and exhibition on integration and packaging of MEMS, NEMS, and electronic systems: advances in electronic packaging 2005, 17–22 July 2005, San Francisco, p 329–336
32. Shrivastava S et al. (2005) Significance levels of factors for different airflow management configurations of data centers. In 2005 ASME international mechanical engineering congress and exposition, IMECE 2005, 5–11 November 2005, Orlando, p. 99–106
33. Hot aisle vs. cold ailse containment. Available: http://www.apcmedia.com/salestools/DBOY-7EDLE8_R0_EN.pdf
34. Combining cold aisle containment with intelligent control to optimize data center cooling efficiency. Available: http://shared.liebert.com/SharedDocuments/LiebertFiles/SL-24640_Rev12-09_FINrev3lo.pdf
35. Schmidt RR, Iyengar M (2007) Comparison between underfloor supply and overhead supply ventilation designs for data center high-density clusters. In: 2007 Winter meeting of the American Society of heating, refrigerating and air-conditioning engineers, 27–31 January 2007, Dallas, p 115–125
36. Khalifa HE, Demetriou DW (2010) Energy optimization of air-cooled data centers. J Thermal Sci Eng Appl 2, 2010. p 041005-1-13
37. Abdelmaksoud WA et al. (2010) Improved CFD modeling of a small data center test cell. In: 2010 12th IEEE intersociety conference on thermal and thermomechanical phenomena in electronic systems, ITherm 2010, 2–5 June 2010, Las Vegas
38. Supply Fan Energy Use in Pressurized Underfloor Air Distribution Systems [Online]. Available: http://www.cbe.berkeley.edu/research/pdf_files/SR_fanpower2000.pdf
39. Sorell V et al. (2005) Comparison of overhead and underfloor air delivery systems in a data center environment using CFD modeling. In: American Society of heating, refrigerating and air-conditioning engineers, ASHRAE 2005 annual meeting, 25–29 June 2005, Denver, p 756–764
40. Schmidt R et al (2005) Data centers: meeting data center temperature requirements. ASHRAE J 47:44–49
41. Nevins RG, Ward ED (1968) Room air distribution with air distributing ceiling. ASHRAE Trans 74:2–1
42. Ho SH et al. (2011) Comparison of underfloor and overhead air distribution systems in an office environment. Building and Environment, 46:1415–1427
43. Moran MJ, Shapirio HN (2000) Fundamentals of engineering thermodynamics, 4th edn. Wiley
44. Weschler CJ, Shields HC (2003) Experiments probing the influence of air exchange rates on secondary organic aerosols derived from indoor chemistry. Atmos Environ 37:5621–5631
45. Rumsey P (2007). Using airside economizers to chill data center colling bills. Available: http://www.greenercomputing.com/columns_third.cfm?NewsID=35825

46. Chivers KJ (1965) Semi-automatic apparatus for the study of temperature-dependent phase separations. J Sci Instrum 42:708
47. Air filter materials and building related symptoms in the BASE study. Available: http://eetd.lbl.gov/ie/pdf/LBNL-59663.pdf
48. Data center economizer contamination and humidity study. Available: http://hightech.lbl.gov/documents/data_centers/economizerdemoreport-3-13.pdf
49. Utilizing economizers effectively in the data center. Available: http://www.kelly.net/pdf/liebert-i.pdf
50. *energycodes.gov*. Available: http://www.energycodes.gov/implement/tech_assist_reports.stm
51. High performance data centers. Available: http://hightech.lbl.gov/documents/DATA_CENTERS/06_DataCenters-PGE.pdf
52. Reducing data center energy consumption with wet side economizers. Available: http://www.intel.com/it/pdf/reducing-dc-energy-consumption-with-wet-side-economizers.pdf
53. Calculating total cooling requirements for data centers. Available: http://www.lamdahellix.com/%5CUserFiles%5CFile%5Cdownloads%5C25_whitepaper.pdf
54. Optimizing facility operation in high density data center environments. Available: http://h20000.www2.hp.com/bc/docs/support/SupportManual/c00064724/c00064724.pdf
55. Schmidt R, Cruz E (2003) Raised floor computer data center: effect on rack inlet temperatures when adjacent racks are removed. In: ASME 2003 international electronic packaging technical conference and exhibition, InterPACK2003, 6–11 July 2003, Maui, p 481–493
56. Schmidt R, Cruz E (2002) Raised floor computer data center: effect on rack inlet temperatures when high powered racks are situated amongst lower powered racks. In: 2002 ASME international mechanical engineering congress and exposition, 17–22 November 2002, New Orleans, p 297–309.
57. Effective computer room cooling. Available: http://www.thefreelibrary.com/Effective+computer+room+cooling.+%28Infrastructure%29.-a0106646787
58. Iyengar M, Schmidt R (2009) Analytical modeling for thermodynamic characterization of data center cooling systems. J Electron Packaging 131:021009
59. Patel CD et al. (2006) Energy flow in the information technology stack: Introducing the coefficient of performance of the ensemble. In: 2006 ASME international mechanical engineering congress and exposition, IMECE2006, 5–10 November, Chicago
60. Schmidt R, Iyengar M (2009) Thermodynamics of information technology data centers. IBM J Res Dev 53:449–463
61. Shah AJ, Patel CD (2010) Exergo-thermo-volumes: an approach for environmentally sustainable thermal management of energy conversion devices. J Energy Resour Technol 132:021002
62. Shah AJ et al (2008) Exergy analysis of data center thermal management systems. J Heat Transfer 130:021401
63. Shah AJ et al (2006) An exergy-based figure-of-merit for electronic packages. J Electron Packaging 128:360
64. Shah AJ et al. (2005) Exergy-based optimization strategies for multi-component data center thermal management: part I, analysis. In: ASME/Pacific rim technical conference and exhibition on integration and packaging of MEMS, NEMS, and electronic systems: advances in electronic packaging 2005, 17–22 July 2005, San Francisco, p 205–214
65. Shah AJ et al. (2004) An exergy-based control strategy for computer room air-conditioning units in data centers. In: 2004 ASME international mechanical engineering congress and exposition, IMECE, 13–19 November 2004, Anaheim, p 61–66
66. Sharma RK et al. (2008) On building next generation data centers: energy flow in the information technology stack. In: 1st ACM Bangalore annual conference, COMPUTE 2008, 18–20 January 2008, Bangalore, p ACM Bangalore chapter
67. Guggari S et al. (2003) A hybrid methodology for the optimization of data center room layout. In: 2003 International electronic packaging technical conference and exhibition, 6–11 July 2003, Maui, p 605–612

# Chapter 3
# Peeling the Power Onion of Data Centers

**Sungkap Yeo and Hsien-Hsin S. Lee**

**Abstract** As the concept of cloud computing is gaining popularity, more data centers are built to support the needs. The data centers, which have consumed 1.5% of the total electrical energy generated in the USA in 2006, are paying the majority of their maintenance cost to the electricity bills. Reducing power consumption in the data centers is now a must not only for seizing sustainable development but also for preserving our planet green. Along the effort of building power-efficient data centers, this chapter will start by answering the ultimate question—where did the power go? By taking a top–down approach from the data center level all the way down to the microarchitectural level, this chapter visualizes the power breakdowns and discusses the power optimization techniques for each layer.

## 3.1 Introduction

The concept of cloud computing has emerged as the de-facto future computing model for all types of computing. Ideally, moving computing to the cloud relieves much responsibility from the users, providing higher reliability and availability for data computation and management. With this transformational paradigm shift, the main computing power and resources will be provided by the cloud service providers who maintain and operate a complete infrastructure, solution platforms, and a plethora of applications in the so-called *data centers*. These data centers accommodate computing nodes and peripherals that consume electrical power for computing and cooling facility in the unit of megawatt. For example, the world's largest online game, World of Warcraft, developed by Blizzard Entertainment,

S. Yeo (✉) • H.-H.S. Lee
School of Electrical and Computer Engineering, Georgia Institute of Technology, Atlanta, GA 30332, USA
e-mail: sungkap@gatech.edu; leehs@gatech.edu

Y. Joshi and P. Kumar (eds.), *Energy Efficient Thermal Management of Data Centers*, DOI 10.1007/978-1-4419-7124-1_3,

currently requires more than 13,000 systems with more than 75,000 processing cores for their online services. The utility bills for these data centers can reach several million dollars a month. As seen in Chap. 1, the power consumed by data centers was estimated to account for 1.5% of the total electrical energy consumed in the USA in 2006. Such increased power usage is not only an economical concern for service providers, data center operators, and end users; to generate these power also inevitably emits more carbon dioxide accelerating the environmental contamination. Therefore, operating these data centers at the maximal power efficiency has become a first-class research thrust across multidisciplinary areas for scientists, engineers, and policy makers. However, before any effort to be dedicated, researchers and policy makers need to fully understand the entire power delivery and distribution system, in other words, to be able to answer the question—where did the power go?

In this article, we take a holistic view of power dissipation for the entire data center and analyze them layer-by-layer (thus, peeling the power onion) from the data center level all the way down to the microarchitectural level using data available in the public domain. In addition to the power breakdown analysis, we also review and discuss representative power optimization techniques for each layer. These techniques are either already implemented in the machines shipped by vendors or research ideas and/or prototypes proposed and built by research institutes.

## 3.2 Power Breakdown at Data Center Level

In this section, we will analyze the power distribution from the perspective of the highest level, i.e., the data center level, by taking published data from actual data centers.

### *3.2.1 Data Center Power Distribution*

When the electrical power from a power plant is delivered to a data center, it is consumed to operate two main facilities. Firstly, it powers up all the computing equipments and hosts the computing services. Secondly, as these computing devices convert the supplied power into useful computation and dissipate heat, the data center has to arrange additional power to remove the heat generated from the facility. These computing nodes and their cooling system are two major power consumers in a typical data center. In addition, a data center also uses power for their power delivery infrastructures including the uninterrupted power supply (UPS) units or other supplementary infrastructures, e.g., lighting. In addition to the utility power, modern data centers typically build their own power generators along with UPS systems in order to guarantee a stable power supply system.

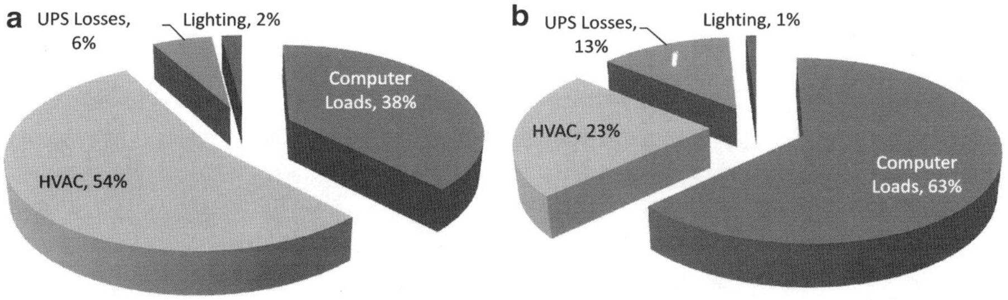


Data Center 8.1 (total 578kW) Data Center 8.2 (total 1681kW)

**Fig. 3.1** Power breakdown of two different data centers [32] (**a**) Data Center 8.1 (total 578 kW). (**b**) Data Center 8.2 (total 1681 kW)

The breakdown of power usage of these components at the infrastructure level is illustrated in Fig. 3.1. The data were taken from a case study performed by the Lawrence Berkeley National Lab [32]. In Fig. 3.1, the portion, "Computer Loads," accounts for the power drawn from the UPS for non-HVAC (heating, ventilating, and air conditioning) purpose. This includes not only the power drawn from actual machines or network switches but also the loss from the power distribution units (PDUs) or power supply units (PSUs). According to the investigation on two different types of data centers, each data center demonstrated rather different characteristics in power usage. For data center facility 8.1, 54% of its available power was consumed in the HVAC while only 38% was for computer loads. In contrast, the data center facility 8.2 spent the majority of its power, 63%, in the computer loads and only 23% for the HVAC. One reason for the difference is that the HVAC of facility 8.1 was running on its full power regardless of the utilization of their computing nodes. In other words, facility 8.1 will continue to dissipate power for the HVAC even if the computer loads are low. Other potential reasons for facility 8.2's higher power efficiency for computing, although not revealed in the original report, could be attributed to different ambient temperatures, different sizes of the facilities, different designs of air flow, etc.

To emphasize the importance of efficient HVAC for maintaining a data center, we can perform simple math to see what if the data center 8.1 could achieve the HVAC efficiency of data center 8.2. If the data center 8.1 can reduce its power consumed by the HVAC down to 23% of its total power as in the data center 8.2, the amount of power for the HVAC will be reduced from 312 kW ( $\sim$ 578 kW $\times$ 0. 54) to 79 kW $\left(\sim 578\,\text{kW} \times 0.46 \times \frac{23}{100-23}\right)$. The difference in power, 233 kW, will become 2,041 MWH ( $\sim$ 233 kW $\times$ 365 days $\times$ 24 h) per year. When this energy saving is converted into dollars by applying roughly $0.1/1 kW H, 2,041 MWH will turn into 204 thousand dollars. This shows why power-efficient cooling is critical in data centers. Figure 3.2 shows the power consumed by the HVAC for a variety of data centers investigated in [34], the power portion by the HVAC alone spans from as low as 20% up to more than 50%.

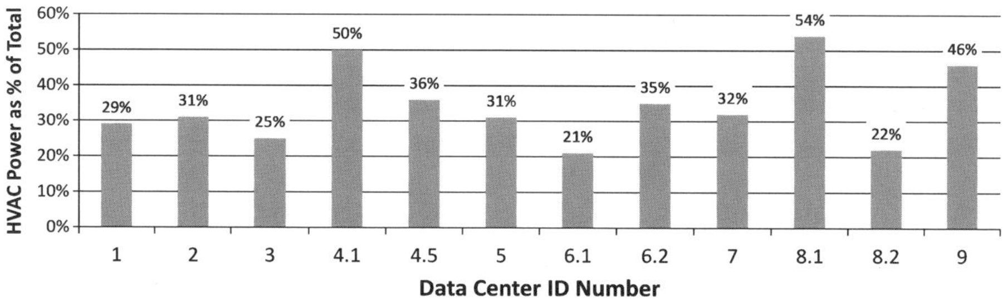


**Fig. 3.2** Power consumed by HVAC out of total power in data centers [34]

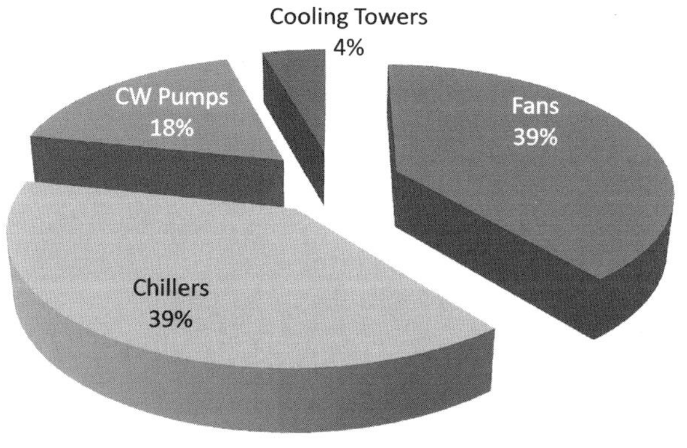


**Fig. 3.3** HVAC power breakdown

Furthermore, the power breakdown of the HVAC itself for data center 8.2 is shown in Fig. 3.3 based on data collected in [32]. According to this study, there are three major power consumers in an HVAC: fans (39%), chillers (39%), and cooling water pumps (18%). In a typical data center with raised floor, fans designed to circulate the air in the server room are connected to two water pipes, one for inlet of cool water and the other one for outlet of warm water. Pumps in Fig. 3.3 are for the water flow while chillers are for cooling down the warm water.

### *3.2.2 Efficiency of Power Delivery Infrastructure*

In addition to HVAC, another major contributor of power consumption is power delivery infrastructure, which is depicted in Fig. 3.4. As shown in Fig. 3.4, power delivery begins with three-phase 480 V AC from power plants. Although three-phase 480 V AC is the most common connection type in the US, other countries may use different types. For example, Canada is known to use three-phase 600 V AC as an input and output current of the UPS. By using higher voltage current before arriving PDU, they can reduce electrical losses from the wires. However,

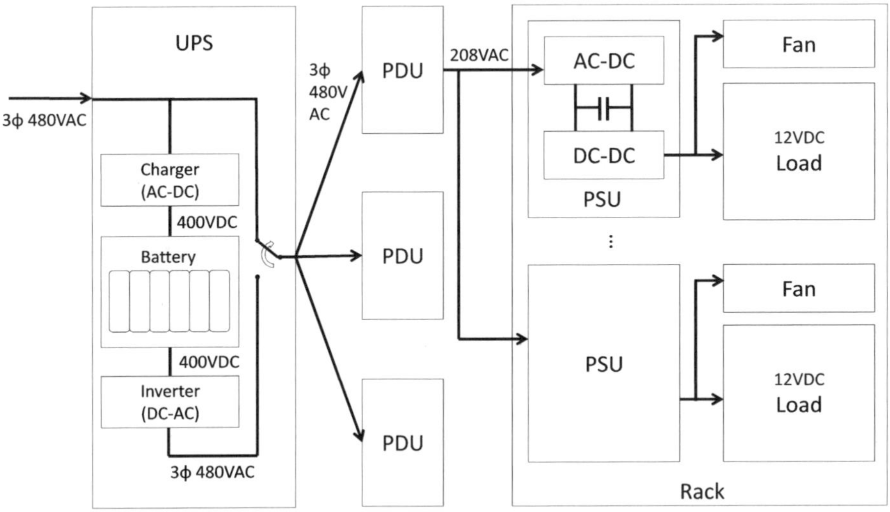


**Fig. 3.4** Power delivery infrastructure

since wires are not the major source of electrical losses in a data center, using higher voltages in the power delivery infrastructure is no longer an issue to many data centers in the United States. On the contrary, three-phase 480 V AC is recently gaining more popularity because a data center can significantly improve the overall efficiency by this. The details are covered at the later pages of this chapter. If we continue on Fig. 3.4, one can find that all the power delivered to the computing nodes come from the UPS. A UPS is designed to provide seamless power supply even in the event of power plant failure. When the power plant fails to supply power, the data center will start up their own backup power generators. Since these power generators take time (tens of seconds to a few minutes) to be fully functional, the UPS has to continue the power supply to avoid any power disruption and service availability during this time period. As the UPS must keep enough energy for its role in the data centers, it is necessary for the UPS to convert AC to DC for charging their battery packs. Because of the overheads such as AC to DC conversion and/or heat dissipated from the batteries while charging, a UPS has relatively lower efficiency than other components in the power delivery system such as the PDU or the PSU shown in Fig. 3.4. In [32], authors showed that the UPS efficiency in their data center facility 8.2 ranges from 68% in the worst case to 87% in the best case. In addition, they also revealed that the main reason of this efficiency deviation was due to the low electrical loads to the UPS. For example, a UPS with an average load of 314 kW reaches a power efficiency of 87% while that with an average load of 79 kW is 68%. The reason for this low efficiency is closely related to the capacity of the UPS. The UPS used in the facility 8.2 was exactly the same UPS with 1,100 kVA energy capacity regardless of its actual loads. Because UPSs are designed to store energy up to its full capacity (1,100 kVA), low power demand directly indicates high overhead.

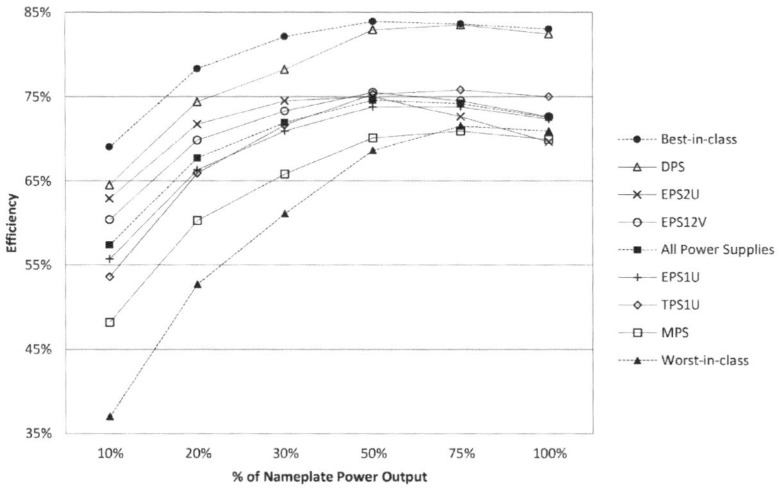


**Fig. 3.5** The average measured efficiency of power supply [13]

When the UPS delivers power to a PDU, the PDU converts the three-phase 480 V power to a single-phase 208 V power to supply power to multiple racks. Because the PDUs are not designed for converting AC to DC, their efficiency are relatively high compared to that of the UPS or the PSU. A typical PDU can sustain more than 90% of the efficiency. A highly efficient PDU can achieve up to 99% at a higher one-time cost [3].

The 208 V power will finally be delivered to the PSU. The design goal of a PSU is to convert 208 V AC to several different types of DC currents, e.g., 12 V, 5 V, and 3.3 V. This process includes both AC to DC conversion and DC to DC conversion, the reason why PSUs experience relatively lower efficiency than the PDUs. Generally speaking, the PSUs are located in each computing node supplying power for each individual machine. For example, 42 PSUs would be used to supply the total power needed for a rack of 42 U space that is fully occupied by 42 1 U machines. However, due to the industrial demand for higher-density server machines, some blade server systems are designed to share the PSUs among a number of blade servers. For instance, IBM offers an 8 U blade server system that contains 14 blade servers that share four PSUs. The idea of sharing the PSU is not only just for saving space but also for increasing the power supply efficiency. The PSU is known for its low efficiency when it is lightly loaded as shown in Fig. 3.5 [13]. In this figure, the *x*-axis represents the load induced to the PSUs in the percentage of the nameplate power while the *y*-axis plots the efficiency with respect to the electric power (W). Each line in the figure represents different form factors. They include EPS12 V, a derivative of ATX standard, and EPS1 U/2 U, a part of Intel's Server System Infrastructure (SSI) for 1 U or 2 U machines. In addition, the line in the center with solid square denotes the average power efficiency of all PSUs tested. For example, given a PSU with 500 W nameplate power, on average, this PSU will obtain around 57% efficiency when load is 50 W, i.e., 10% of the

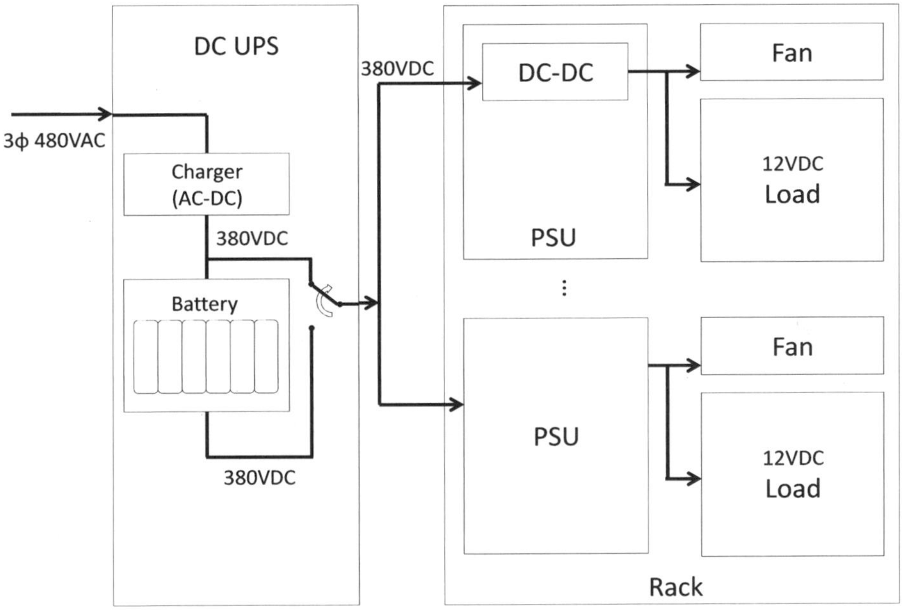


**Fig. 3.6** Power delivery infrastructure in DC-powered data centers

nameplate power. The same PSU will reach 75% efficiency when the load is 50% or 250 W of its nameplate power. Followed by this PSU efficiency trend, the aggregate power consumed by two PSUs with 10% load each will be much higher than one PSU with 20% load. For these reasons, sharing a few PSUs among several servers will attain a better power utilization while having the additional benefit of saving physical space.

All in all, the UPS and PSU in Fig. 3.4 are the two major components that suffer from low efficiency in power delivery simply because they need to convert AC to DC or DC to AC. To alleviate these problems, some data centers proposed to use a new power delivery concept that uses DC power inside the facility to minimize the lost efficiency during AC to DC conversion. For example, when three-phase 480 V AC arrives to a typical AC-powered data center, a UPS typically converts this to 400 V DC for storing energy to its battery packs. This 400 V DC has to be converted to three-phase 480 V AC before delivered to a PDU, which converts 480 V AC to 208 V AC to supply a PSU. In addition, the PSU typically converts 208 V AC to 400 V DC before using DC–DC converters to generate the 12 V DC supply. In short, AC to DC conversion occurs three times in a typical AC-powered data center, each losing power efficiency. In contrast, a DC-powered data center converts AC to DC once at the UPS and uses DC for distributing power all over the facility as shown in Fig. 3.6. In these type of data centers, a specially designed UPS and PSU are used. The UPS does not convert DC back to AC, and the PSU only has a DC–DC converter. These DC-powered data centers were measured to be 4.7%–7.3% more efficient than the most efficient AC-powered data centers [36].

However, some studies suggested that the efficiency gained by adopting a DC distribution infrastructure is less than 1% when it is compared to the efficiency of an AC-based distribution with higher voltage systems [31, 30]. Especially in the data centers of North America, PDUs with step-down transformers are commonly used to convert three-phase 480 V AC to single-phase 208 V and 120 V AC. This transformation was necessary when the IT equipment only accepts 208 V and 120 V AC. However, since almost all IT equipment manufactured today is ready to accept higher voltages [30], PDUs with step-down transformers can either be replaced to the autotransformers with higher efficiency or even omitted in the power delivery infrastructure. For example, if a PSU can accept single-phase 277 V AC then this PSU can directly be connected to the three-phase 480 V AC without additional transformation because line-to-neutral voltage of three-phase 480 V AC is 277 V AC. In this case, a PDU does not perform step-down transformation, and this significantly increases its efficiency. In an AC-powered data center without a step-down transformation in their power delivery infrastructure, the sources of electrical loss are limited to the UPS, PSU, and wires. Because wires show the same efficiency for both AC and DC-powered data centers, comparison has to be made between the DC-UPS and AC-UPS or the DC-PSU and AC-PSU. In a detailed study, DC-UPS showed almost the same efficiency with AC-UPS while the DC-PSU showed 1.5% higher efficiency than the AC-PSU [31]. In summary, even though DC-powered data centers could achieve 1.25% higher efficiency over AC-powered data centers in their power delivery infrastructure, the overall savings in the data center-level power consumption was less than 1%.

### *3.2.3 Power Density*

On the other hand, data center designers have been trying to boost up the computing power within a unit square feet of a facility to improve the overall space efficiency. It was achieved at two different levels: the per-component level and the per-system level. At the component level, data center administrators keep upgrading the underlying processors of each node to improve per-computing-node's performance, which contributes to the overall power increase. For example, the first superscalar microprocessor from Intel, the Pentium Processor, has its thermal design power (TDP) around 15 W while many of the high-end processors today can easily consume more than 100 W. At the same time, the die size was not proportionally enlarged, therefore, the chip power density is exacerbated. When it comes to the system level, the design of server system continues to accommodate more servers for a given space. For instance, the smallest server form factor during late 1990s was the 1 U machine. Now it is common to see two machines are placed in 1 U space, effectively making one machine a half-U unit. Furthermore, some blade systems can accommodate tens of servers in a few 1 U slots. In other words, from system's perspective, the power density is increasing as well. By taking these two levels into account, the overall power density of a data center continues to increase,

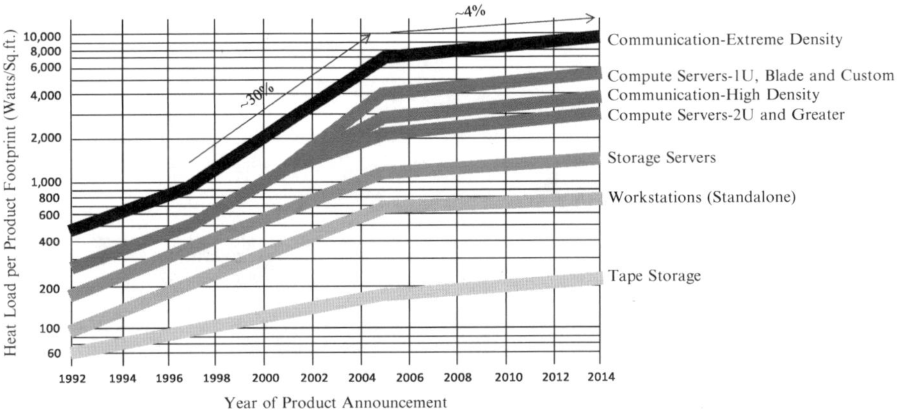


**Fig. 3.7** Power density trends of common equipments in data centers [4]

leading to potential issues in cooling system design and operation cost. Figure 3.7 shows the power density trend of common equipments in a data center from 1992 to present days. In this figure, the power density is clearly increasing during 1992–2005 but the growth rate starts to slow down since 2005. Again, this abrupt turn can be analyzed from the per-component level and the per-system level. At the component level, there was a major paradigm shift in high-performance microprocessor design. When Intel announced their last Pentium 4 Processor family, Prescott, in 2004, people nicknamed it PresHot, making a parody of its power consumption and the heat dissipation mechanism required to cool it down. This old design fashion of scaling up frequency eventually makes a system more susceptible to reliability issues, worse yet, it is also economically infeasible. As demonstrated in Fig. 3.8, which we surveyed Intel's processors based on TDP and published die areas from 1993 to 2010, the power density, prior to Pentium 4, had exponentially increased over different process generations and microarchitectural designs. The former Intel Fellow Fred Pollack had actually warned about this trend in his keynote speech at the International Symposium on Microarchitecture (MICRO-32) at Haifa, Israel, in 1999. In light of this, the slope of the power density (Watts/cm$^2$) across recent processor generations based on Intel's Core microarchitecture (post 2005) has been leveled off or even reduced as also shown in Fig. 3.8. This trend also highly correlates to the power density trend of common equipments in a data center shown in Fig. 3.7.

At the system level, data center designers and operators are well aware of the escalating power density issue, which not only poses design challenges and cost implications of the cooling mechanism but also lowers the overall power efficiency. The latter is known as the paradox of power density. When a data center replaces their older machines with higher power density ones, it must have enough power budget required by the new machines. If not, the power delivery infrastructure needs to be redesigned and additional space may be required. Even if the data center has enough room for power infrastructure for the new machines, this does not mean

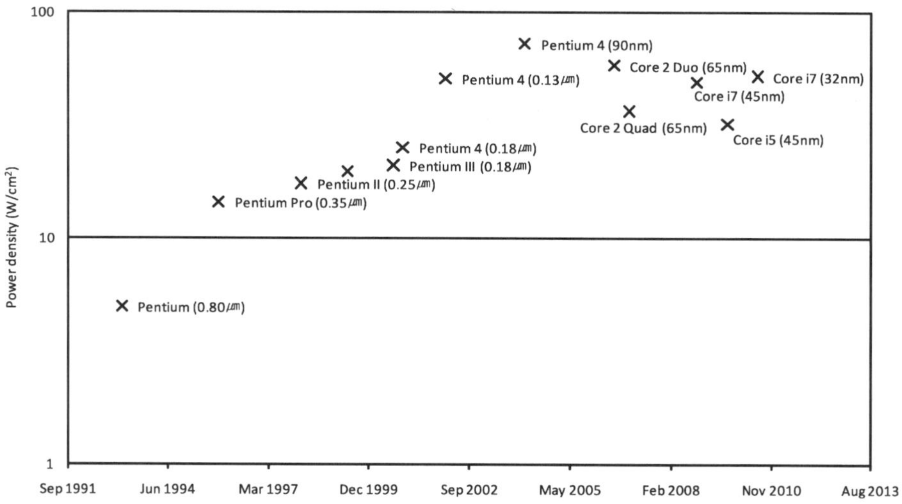


**Fig. 3.8** Power density of Intel's IA32 processors in Watts/$cm^2$

that they can be free to consume all the available power, because the data center also needs to have enough power to remove the additional heat. In summary, putting additional 1 kW on a fully occupied data center will require additional power delivery infrastructure as well as additional cooling capacity for satisfying the expansion. Because both require space in the data center, it will become infeasible to increase the power density of a data center at some certain point. As a result, the need for equipments with higher power density has largely been alleviated, which is another reason for the slowdown of the slope after 2005 in Fig. 3.7. By combining these modern considerations of components and systems, it is observed that the increase trend of the power density at the data center level has been much slowed down.

## 3.3 Power Breakdown of a Single Computing System

As illustrated in Fig. 3.4, power is finally delivered to the components such as the CPUs, PCI slots, memory, motherboards, and disks, of a system. Before we detail the per-system power breakdown, it is important to understand why we need to estimate the actual power consumption of a system instead of the power described in a user manual, i.e., the nameplate power. When calculating the nameplate power, the vendor has to be as conservative as possible to prevent their products from malfunctioning in the face of power deficiency. As a result, the total nameplate power is usually estimated by summing up the worst-case power consumption of all components in a system. In most of the cases, however, not all of the system components will operate with its maximal power simultaneously. Even if all of

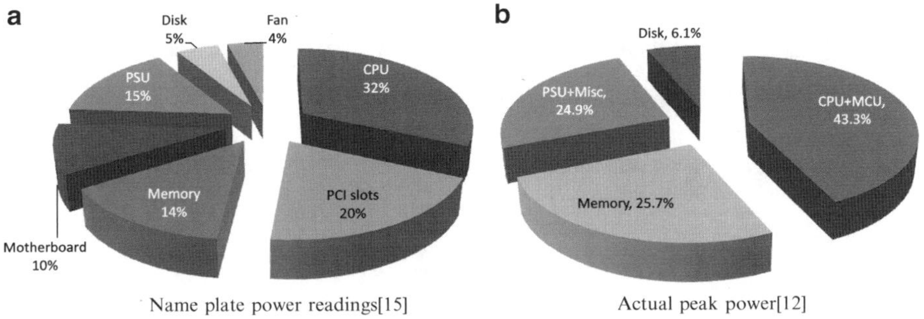


**Fig. 3.9** Power breakdown of a server. (**a**) Nameplate power readings [15]. (**b**) Actual peak power [12]

them are busy at the same time, a system will not reach manufacturer's nameplate power as it is oftentimes overestimated intentionally. In a data center environment, this discrepancy between the nameplate power and the actual measured peak power can cause significant inefficiency in the power delivery infrastructure. As seen in previous figures, a system has to be placed in a rack that typically accommodates tens of servers. Given that the power for a rack is limited by the PDU (e.g., 2.5 kW per rack [15]), the number of systems in a rack is fixed based on either the nameplate power or the measured peak power of a server. For example, if a data center deploys servers based on the nameplate, 213 W, a rack of 2.5 kW will accommodate 11 servers while the actual aggregated peak power of 11 servers is less than 1.6 kW [15]. Under such circumstances, the data center will pay more on the power delivery infrastructure for supporting the nameplate power that can never be reached.

Because the nameplate power is different from the actual power consumption, so does the power breakdown of a server is. According to the nameplate power readings in Fig. 3.9a, a CPU accounts for around one third of the total power of a system followed by 20% for the PCI slots, 14% for the memory, and 10% for the motherboard. On the other hand, Fig. 3.9b shows the actual power consumption of components in a typical blade server using a 2.2 GHz AMD Turion processor. Different from the nameplate power readings, the CPU with an on-die MCU consumes 43% of the total actual power while the memory accounts for a quarter of the total. By comparing these two figures, it is apparent that in the actual deployment, the CPU and memory are the most power-consuming components in a system.

Figure 3.10 also identified that the CPU and memory are the two major power consuming components in a system, accounting for more than half in all three samples that corroborates the data points shown in Figure 3.9. In the worst case, the IBM p670, 67% of the total power were consumed by the CPU and memory. In addition to this fact, it is also interesting to find that Google spends more power on the CPU and I/O devices than the others. This is simply because their main applications are the web search, email, and document services [15]. For the web

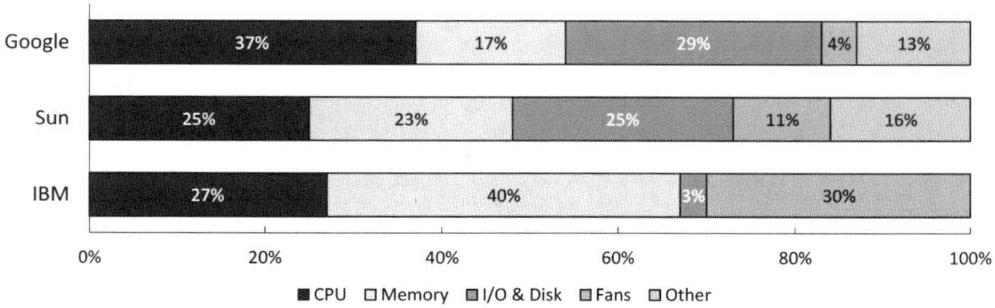


**Fig. 3.10** Per-system power breakdown by company [24]

search service, many computing nodes in the back-end have to sort and index web pages while the front-end nodes have to parse queries. Many of these operations are CPU intensive. On the other hand, email services require a large number of database accesses and file downloads which are primarily I/O operations. Moreover, even though the source [15] did not mention YouTube service or similar types of workloads, it is obvious that these streaming services will demand much more on the I/O side. In summary, the most power-consuming components in a real data center are the CPU and memory, however, depending on the services that a system provides, the power breakdown can be vastly different.

## 3.4 Power Breakdown of a CPU

As the CPU is one of the most power hungry components in a system, it is imperative to examine and understand the power distribution within the CPU. Not only can power reduction in each CPU collectively reduce the overall power consumption of all computing nodes but it also cuts the cost of thermal management hardware, such as the sizes of the heat sinks and cooling fans and the center-level cooling strategy. As a part of this effort, we will cover the power breakdown of a CPU in two different aspects. First, the power breakdown by functional modules such as the register file, fetch logic, or ALU will be included. Second, we will further analyze the power breakdown of a CPU based on different types such as active dynamic power, subthreshold conduction, and gate leakage.

### *3.4.1 Per-CPU Power Breakdown by Modules*

Although there is a scarcity of public literature that breaks down the power distribution of a modern out-of-order (OoO) microprocessor, there were some attempts from both academia and industry that analyzed, modeled, and simulated the power consumption of sophisticated processors at the microarchitectural level. Figures 3.11 and 3.12, both based on the DEC Alpha processor, detail and illustrate

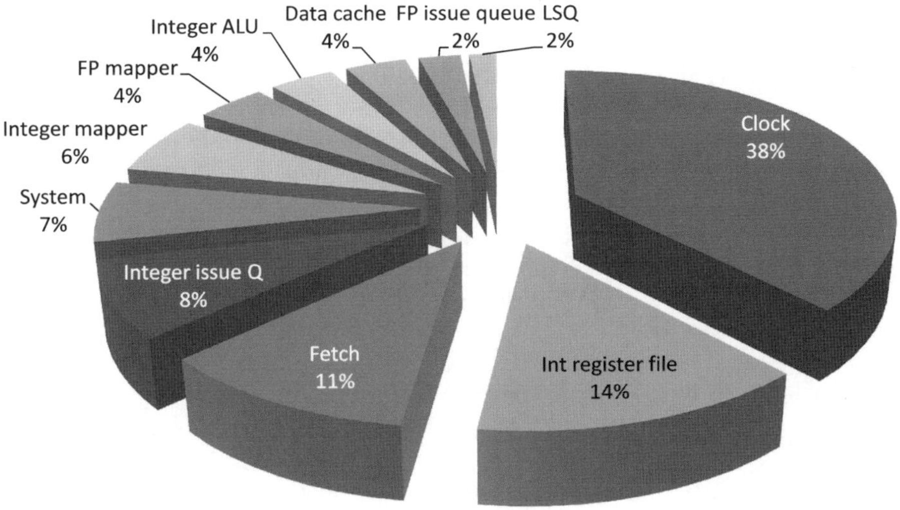


**Fig. 3.11** Power breakdown of Alpha 21264 [27]

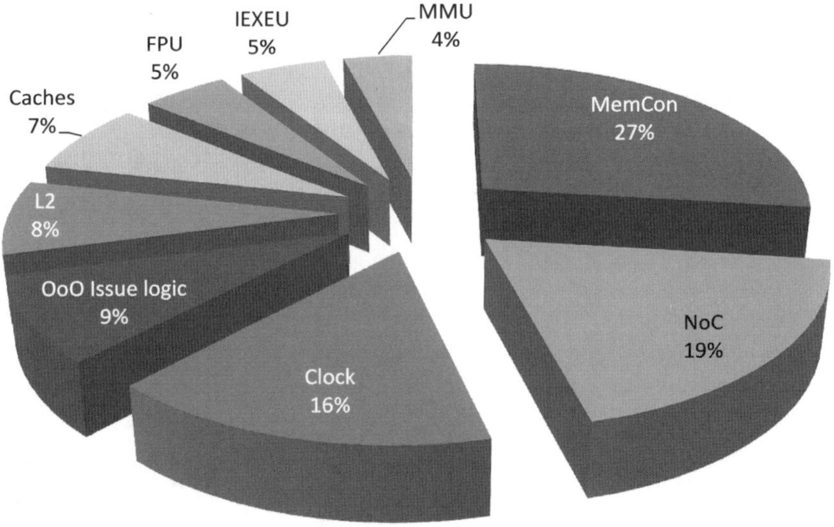


**Fig. 3.12** Power breakdown of Alpha 21364 [23]

such power distribution. Figure 3.11 shows the power breakdown of an Alpha 21264 processor running *gzip* at 600 MHz. These numbers were generated using the microarchitectural Wattch power model integrated with the cycle-level Alpha-sim simulator [27]. Given that the Alpha 21264 processor is a four-wide superscalar microprocessor with OoO execution, speculative execution, and large instruction queues for both integer and floating-point instructions, the power breakdown obtained by modeling this microprocessor will be a good representative for today's high-performance processors. From Fig. 3.11, one can easily find that the clock tree actually accounts for more than one third of the total power dissipation. Note that, the clock signal itself is the fastest switching part of the entire chip, and this has to be done regardless of the modular utilization in the CPU. For example, the clock signal would change the logical state of the floating-point functional unit every cycle even if only an integer application is being executed. Such unnecessary power waste can be eliminated if more advanced circuit techniques such as unit-level, fine-grained clock-gating or dynamic voltage–frequency scaling (DVFS) are applied. We will discuss more of these techniques in subsequent sections. To elaborate more about the clock distribution, it is worth mentioning that the Alpha 21264 processor uses a metal grid that covers the entire die area for distributing the clock signal. A metal grid for clock distribution is known to be the most effective (but not necessarily the most efficient) way of distributing clock signal with minimum clock skew to all the parts of the chip [5]. As a result, this lets a CPU run at a higher operating frequency than other types of clock distribution network such as H-tree for IBM S390 or length-matched serpentine structure for Intel P6. However, this clock distribution network, a metal grid, has a main drawback that it consumes more power than other alternatives due to its large capacitance. Next to the clock signal, the integer register file accounts for 14% of the total power. Because these numbers are generated by running *gzip*, an integer application, the integer register file is heavily used. The accumulated OoO logic accounts for 20% of the total power consumption: 8% for the integer issue queue, 6% for the integer mapper (for register renaming in integer registers), 2% for the floating-point issue queue, and 4% for the floating-point mapper. In exchange for higher performance by exploiting instruction-level parallelism, the power portion of the OoO-related logic is larger than those of the data cache (4%) and the functional units (4%).

On the other hand, Fig. 3.12 shows the power breakdown of Alpha 21364 microprocessor generated by an integrated framework called McPAT that models power, area, and timing done by HP Labs. The Alpha 21364 processor is the successor of Alpha 21264 with minor changes on the core design with major differences on other supplementary logic including an on-die memory controller ("MemCon" in Fig. 3.12), L2 cache, and network on chip controller ("NoC"). The design philosophy of Alpha 21364 was to improve bandwidth of the memory subsystem as well as maintaining scalability for future many-socket systems. With this objective, the memory controller and network on chip controller have become the most power-consuming components—accounting for almost half (46%) of the entire chip power budget. For the rest of the chip, the clock distribution accounts for 16% while the OoO issue logic is about 9%.

### 3.4.2 Per-CPU Power Breakdown by Sources

Figure 3.13 illustrates the power breakdown of a CPU by sources such as active power, subthreshold conduction (subthreshold leakage), or gate leakage across

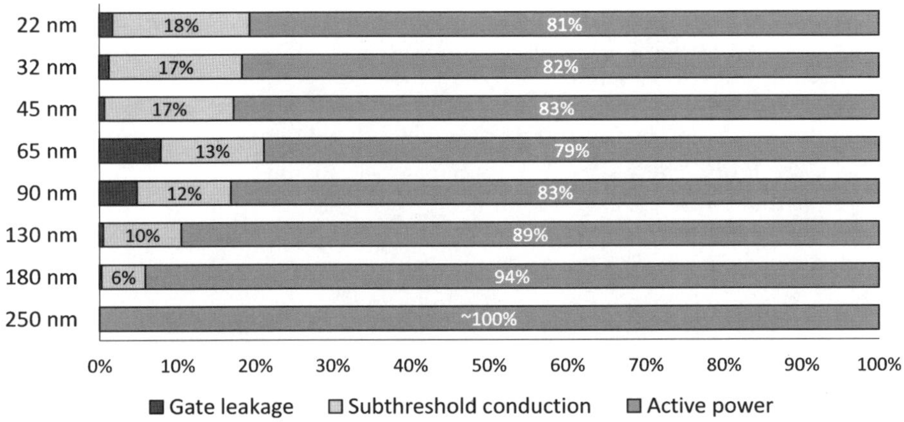


**Fig. 3.13** CMOS leak power trend by fabrication process technologies [6, 23, 33]

different fabrication process technologies. We collected these data from multiple sources [6, 23, 33]. As the feature size shrinks, as shown in Fig. 3.13, the portion of the subthreshold conduction continues to increase and reaches almost 20% of the total power in the 22 nm technology node. This increasing trend is a tradeoff for reducing the active power. To lower the power of a processor, designers employ lower supply voltage ($V_{dd}$) as the active power of a CMOS device is proportional to $V_{dd}^2$. When $V_{dd}$ was high (e.g., 5 V), CMOS gates can be operated at relatively high threshold voltages (e.g., $V_{th} = 700$ mV). Due to the high threshold voltage, subthreshold leakage current were negligible as shown in the following formula where $I_{off}$ is the subthreshold leakage current and $s$ is the subthreshold swing in mV/decade [26].

$$I_{off} \propto 10 - \frac{V_{th}}{s}. \tag{3.1}$$

According to (3.1), for a given subthreshold swing, the subthreshold leakage current is exponentially and negatively proportional to the threshold voltage. Meanwhile, $V_{dd}$ has been lowered from 5 V to sub-1V today, $V_{th}$ was also scaled down to 200 mV. For a subthreshold swing of 100 mV/decade, every 100 mV drop in $V_{th}$ will cause ten times more subthreshold leakage current. On the other hand, gate leakage is also exacerbated as the technology node advances. The increasing trend was because of the fact that with technology scaling, the capacitance of the gate oxide material in a MOSFET also scaled down. Equation (3.2) shows the relationship of capacitance ($C$) with the dielectric constant (*kappa*), area ($A$), permittivity of free space ($\varepsilon_0$), and insulator thickness ($t$).

$$C = \frac{\kappa \epsilon_0 A}{t}. \tag{3.2}$$

Since smaller fabrication process technology reduces area ($A$) of the gate oxide, the overall capacitance of the gate oxide becomes smaller, which increases the gate leakage current. As an alternative method for increasing the capacitance of the gate oxide material, material with higher κ value has been used since 45 nm fabrication process technology, e.g., Intel's high-k metal gate technology revolution. As a result, with the "High-κ" material, the gate leakage has almost disappeared in Fig. 3.13 since 45 nm.

## 3.5 Hierarchical Power Optimization Techniques

Power optimization is one of the most active research areas in several engineering disciplines for the last decade. Moore's Law continues to drive more and more transistors to be integrated on a single chip that consume exponentially higher dynamic power. On the other hand, device miniaturization increases the operating frequency at the expense of higher dynamic power at the same time worsens the leakage power. Technologies at the device level (e.g., Intel's high-k metal gate in their 45 nm process) all the way up to the design of a data center all aim at minimizing power consumption. Note that, even a small percentage of improvement could help curtail millions of dollars paid for energy. For example, data center operators such as Google are willing to hire hundreds of engineers to just optimize the power usage of their date centers. If successful, the savings achieved will be much more than enough to cover these engineers' compensation. In this section, we provide some representative samples of such thrusts from a hierarchical perspective starting from the infrastructure, and then system, and finally the microarchitecture level.

### *3.5.1 Infrastructure Level*

#### 3.5.1.1 Energy-Proportional Computing

First of all, we will start our discussion from the infrastructure-related issues. In typical data centers, the average utilization is known to be as low as 20–30% [8]. One reason for this low utilization is that since data centers are prepared to serve the highest demand of a day or a week, their computing power is over-provisioned to satisfy the worst-case scenario even when the number of requests is low. For example, we can expect more searching queries at noon than midnight; however, search engine providers have to maintain computing nodes in their data centers for the demand of noon. Given the low utilization of a data center by its nature, the need for energy-proportional computing [7] has risen. The basic concept of the energy-proportional computing is that when the utilization of a computing node is under 100%, say 50%, the power consumption of the computing node should be half the power of 100% utilization. To apply this concept to a data center, a energy-proportional data center with 30% utilization should consume only 30% of its peak power. However, the energy-proportional computing concept is not ready to the vast majority of today's equipments. A power model for today's common computing node shows that the computing node consumes almost half of its peak power when it is completely idle (0% utilization) and consumes about 75% of its peak power when utilization is 50% [7]. This is a far worse result than common sense. To alleviate this problem, a new idea has been proposed for data centers with common equipments [35]. In this work, by considering that even common equipments have a (near-)energy-proportional characteristic at high utilization, some computing nodes are suggested to be turned off to keep the others busy. For example, when ten computing nodes of the same type are around 5% utilization, the idea suggests to turn nine machines off but keeping only one node up and running. In the ideal situation of this technique, the aggregate power consumption can be very close to the utilization even with non-energy-proportional machines. However, there are a few major drawbacks that keep some services from adopting this technique. First, the utilization of a data center should be precisely predictable. Because turning on the machines takes time, a poor prediction of utilization will lead to serious performance degradation. Second, to turn off a machine, a job or a virtual machine must be safely migrated to other machines for execution, which also consumes additional power and network resources. Third, storage on the machines cannot be used when they are turned off. Google pointed out that although utilization is low for their data centers, their servers cannot be turned off as they are seldom completely idle [7]. This is because of the distributed file system [17] that distributes data storage to different physical hard disk drives (HDDs). For seamless service of their file system, the entire HDD has to be migrated to some other nodes if the node has to be turned off. Fourth, the system memory can not be accessed when the machines are down. For some applications such as in-memory search indexing, the system memory is used for storing data structures. Evicting these index values to a disk or to another node will increase search latency a lot. For these reasons, achieving higher power efficiency by turning off some machines may not be practical for certain types of services.

#### 3.5.1.2 PowerNap

As some services are not ready to turn off machines, PowerNap [24] has been proposed for eliminating the server idle power. The basic idea of PowerNap starts from the fact that once a server becomes idle, it is measured that the average time the server spends in this state (average idle time) is around 100 ms for most of the services while some other services (DNS or scientific computing cluster) have longer average idle time up to one or several seconds. For these reasons, if a server can be turned off and brought back in a few milliseconds, the server can effectively be turned off during its idle period. For this fast transition between full performance and *nap* modes, PowerNap suggests to use the "S3" sleep state (also known as Standby state) of the CPUs, self-refresh for dynamic random access memories

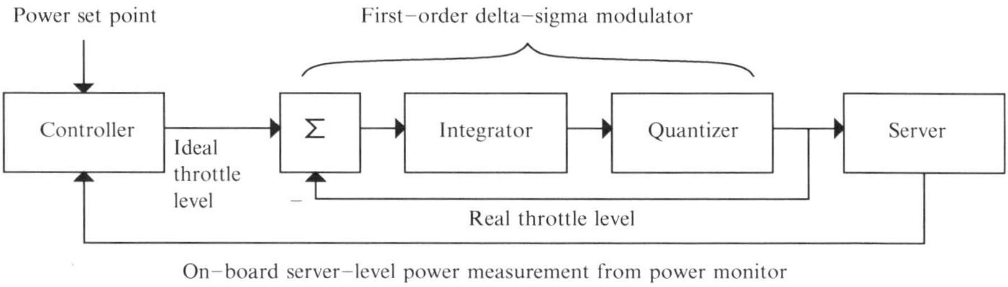


**Fig. 3.14** System diagram for power capping controller [22]

(DRAM), solid state disks (SSDs) for the storage, and wake-on-LAN for the network interface card (NIC). By using these features, a typical blade can change its power state from full performance mode to the *nap* mode in 300 μs and vice versa. With the penalty of less than 1 ms transition time, a typical server that consumes 270 W when idle and 450 W when active can save significant power while in the *nap* mode because it consumes only 10 W during the nap mode. Further comparison between PowerNap and DVFS technique showed that PowerNap technique with less than 10 ms of transition time always outperforms DVFS in terms of response time and power scaling. As a result, PowerNap yields a steep power reduction up to 70% for Web 2.0 servers.

#### 3.5.1.3 Power Capping

On the other hand, power capping [22] is a system-level power-control technique that guarantees the power consumed by a server to be confined within a given power envelope, i.e., the capped value. For example, if a server with power capping capability is set to 200 W, the power controller inside the server will keep the power consumption of this server below 200 W. To achieve this design goal, the controller throttles performance by using DVFS technique when it consumes more power than the capping value. Figure 3.14 illustrates the closed-loop feedback controller for the power capping feature. First, the controller is set to a certain value representing the maximum allowed power budget for this server. The controller calculates the ideal throttle level based on the set point and the measured power consumption. Second, the actuator that in fact is a first-order delta-sigma modulator calculates the target throttle level based on the ideal and real throttle level retrieved from other sources. By using this extra controller on top of the conventional server power supply design, a server can safely be under-provisioned, the key to enhance efficiency of the power delivery infrastructure. For example, consider a server that draws less than 200 W for more than 99% of its lifetime while consumes 250 W under very rare circumstances. When such servers are deployed in a rack of 2 kW, only eight servers can be placed based on the worst case scenario. However, if these servers have the power capping feature and are set to 200 W, a rack of 2 kW can accommodate ten servers with minor performance degradation.

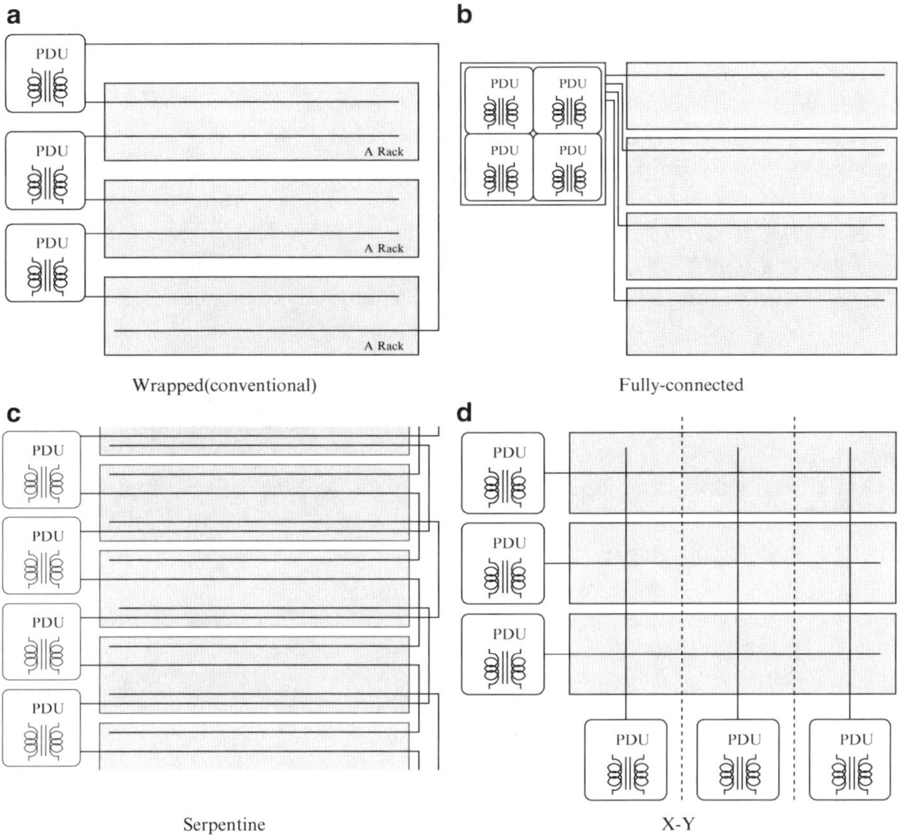


**Fig. 3.15** Power distribution topologies for power routing [28]. (**a**) Wrapped (conventional), (**b**) Fully-connected, (**c**) Serpentine, and (**d**) *X–Y*

#### 3.5.1.4 Power Routing

Power Routing [28] is a technique for reducing redundant power delivery infrastructure. In a high-availability data center, more than one PDU are used for supporting a server cluster to reduce the risk of PDU failure. In the event of PDU failure, other PDUs take over the duty of the broken PDU to support uninterrupted service. Hence, high availability and reliability in data centers can be achieved via such over-provisioning to provide reserved capacity. The amount of the reserved capacity that causes overhead in power delivery infrastructure highly depends on the topology used by the data center. For example, a wrapped topology in Fig. 3.15a shows that two PDUs can be brought in to recover a single PDU failure. In other words, each PDU in the wrapped topology needs to have 50% of the reserved capacity for recovering a single PDU failure. On the other hand, when it comes to a single PDU failure, an example of a fully-connected topology as shown in Fig. 3.15b can be used to have three additional PDUs for replacing one failed PDU. In this case, the amount of redundant capacity that each PDU must have is 33% of the peak power a rack can draw.

The bottom line is that depending on the connectivity among PDUs and server clusters in a data center, the reserved capacity can vary for recovering a PDU failure. Because more reserved or redundant capacity in PDUs directly indicates that more money is to be spent on power delivery infrastructure, it is important to choose a better routing topology with less redundancy and better scheduling ability. Power Routing is one representative of such techniques. Power Routing comprises two parts. First, this idea introduced many different topologies between PDUs and server clusters such as serpentine topology in Fig. 3.15c or *X–Y* topology in Fig. 3.15d. Second, Power Routing introduced a heuristic scheduling algorithm for assigning a power line to servers while balancing loads. As this power feed assignment is an NP-Complete problem, authors first let the servers be fractionally assigned to the power feeds by using standard linear programming methods. From this fractional solution, the real problem will be solved approximately. When the approximate solution fails to meet the requirements from PDU specs or fails to balance between AC phases, they repeat the second step. By applying real data center power traces to this idea, Power Routing could save 5%–10% of the required power capacity for conventional data centers and 22%–28% for the energy-proportional servers.

#### 3.5.1.5 Disk and Storage Power Strategies

For the last few decades, HDDs are the main storage devices for server storage systems. Common magnetic HDDs contain several mechanical parts such as the actuator arms and spinning disks, which consume nontrivial power for the following reasons. First, to increase the performance of the HDDs, they employ higher rounds per minute (RPM) internal disks that consumes more power. For example, a typical 7200 RPM HDD consumes around 10 W [20], while a commercial 15 k RPM HDD consumes more than 16 W on average [1]. Second, the capacity of a HDD is directly proportional to the number of internal disks in the HDD. As the number of physical disks in the HDD continues to increase, they add more mass to spin resulting in higher power consumption. A prior study [11] showed that an oracle algorithm that puts a disk into the power-saving mode only if it is absolutely necessary can save the disk power by up to 50% without any performance penalty. In addition, in the same study, a realistic algorithm that uses only previous history for deciding when to put a HDD into the power-saving mode can save about 40% of the disk power with negligible performance penalty.

Even though HDDs still account for the majority of the storage for server systems, it is expected that SSDs may replace the magnetic HDDs in the near future. Compared to the power consumed by HDDs, SDDs are more than ten times power efficient as these devices consume less than 0.15W even in the active mode [10]. One main drawback of SDDs is the price per capacity. From today's

**Table 3.1** Today's storage solutions and total cost of ownership (TCO) for 10 years in a data center

| *Name* | *Power* | *Price per capacity* | *TCO for 10 years (1 TB)* |
|---|---|---|---|
| Commodity level HDD | 10 W | \$0.1/GB | \$280–\$380 |
| High-performance HDD | 16 W | \$1/GB | \$1,288–\$1,450 |
| SSD | 0.15 W | \$3/GB | \$3,000 |

market, a commodity HDD costs 10 cents per gigabyte while high-performance HDDs and SSDs cost \$1 and \$3, respectively, as listed in Table 3.1. Even though the gap between these two different solutions seems too big to have any advantage of using SSDs, the math for data centers is different from that of normal consumers. For data centers, there is a rule of thumb for exchanging power savings with the dollars; for every watt in a data center, it costs 10–20 dollars for power building cost [15] and around eight dollars for 10-year energy bill [15]. If an SSD that consumes less than 1 W has the same performance of a 10 k RPM HDD that consumes more than 20 W, a data center can value the SSD for an extra \$200–\$400 over a 10 k RPM HDD for power savings. In Table 3.1, even with 20 $\times$ difference in price per gigabytes, the gap is not as wide as 20 $\times$ when the power savings is also considered.

Although SSDs are more power efficient than the legacy HDDs, the solution itself suffers from a major limitation—write endurance, in other words, the storage device, based on NAND flash, can only tolerate a limited number of write operations before it wears out. Depending on the device types, one can write to the device cell around 10 thousands to one million times. To decelerate cell failure and extend the lifetime of these memories, a wear leveling scheme is typically employed to evenly distribute writes across the entire memory space. The basic operations of wear leveling is described as follows. First, the controller of the drive internally maintains a map that links the OS' addresses with their corresponding physical NAND flash memory addresses. Second, for every write operation, the controller assigns a new empty NAND flash page and links the new page to the original address. By using this policy, even if a user writes to the same file over and over, these write operations are evenly spread out across the device. This wear leveling scheme was good enough to ensure a reasonable lifespan for the typical use. However, researchers soon realized that this wear leveling scheme (also known as dynamic wear leveling) cannot secure a reasonable lifetime for SSDs for the following reasons. First, because SSDs are no longer a secondary storage device but a primary storage device, the expected writing rates for SSDs will be much higher than a portable thumb drive. Second, when some NAND flash pages are occupied by the files that do not change over time, those pages will never be empty and cannot be reused and participate in the dynamic wear leveling. To overcome these shortcomings in dynamic wear leveling, static wear leveling schemes are commonly used in SSDs. In the static wear leveling, infrequently updated static files are relocated to a new NAND flash page so that all the NAND flash pages are written almost the same number of times. These movements are triggered and performed using a given wear-leveling threshold by a garbage collector inside the SSD

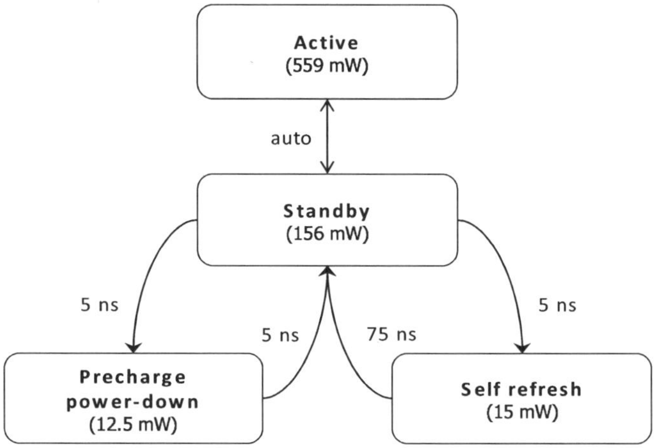


**Fig. 3.16** Operating modes for DDR DRAM [25]

controller. However, this scheme may incur additional writing operations due to moving static files on top of the original write request. Because this characteristic exacerbates *write amplification* [18] of SSDs, the writing performance is degraded in the static wear leveling when compared to its dynamic counterpart.

## *3.5.2 System Level*

### 3.5.2.1 DRAM Power Management

The main memory made of DRAM is a power hog as demonstrated in Fig. 3.10. To save DRAM power, modern DRAM supports up to six different power states for RDRAM [29] or four different power states for double data rate (DDR) DRAM [25]. More specifically, a DRAM controller can put an entire rank[1] of the main memory into the low power state if the rank has not been used for a given period of time. However, when a rank is in the low power state, there will be nonnegligible delay before it becomes ready to be read or written again. Figure 3.16 illustrates this cycle. There are four power modes implemented in current DDR DRAM [25]. When a rank is in the standby mode, it is automatically moved to the active mode when a read or write request arrives. On the contrary, a transition to

[1] In DRAM, a rank is uniquely addressable 64 bits or 72 bits (when supporting 8 bits error correction code) data area. In a dual rank memory module, for example, memory controller uses chip select signal to choose what rank to access. In other words, the memory controller can access only half of the entire memory space in a cycle.

the other two modes, self-refresh or power-down mode, is done manually by the memory controller. The power-down mode starts when the memory controller lowers the clock enable signal (CKE) to the idle DDR DRAM rank, and the self-refresh mode starts when CKE is lowered as well as the auto-refresh signal is sent. These two low power modes are essentially similar in terms of power savings; however, they are different in terms of allowed interval in each mode. For the power-down mode, a rank cannot be in this mode more than maximum refresh interval, because no refresh signal is sent to a rank in this mode. In contrast, a rank can be in the self-refresh mode without time limit, because the on-chip timer in DRAM generates periodic refresh signal for a rank in this mode. This is why the self-refresh mode has longer transition delay and requires slightly more power than the power-down mode. To make use of these different power states for saving power in DRAM, Hur et al. in [19] proposed a simple power-down policy. First, each rank of the main memory has a counter that resets upon every read or write request and increases upon every idle cycle for bookkeeping the number of idle cycles for the rank. Second, when the counter reaches a threshold value, the memory controller checks the internal queue to verify whether there is a read or write request for this rank. If a rank has been idle for more than the threshold time and there is no read or write request in the queue, the memory controller puts the rank into the power-down mode. This policy is reported to increase DRAM energy efficiency by 11%–43% for different benchmark programs.

#### 3.5.2.2 Dynamic Voltage–Frequency Scaling (DVFS)

Dynamic voltage–frequency scaling is a technique for reducing the dynamic active power by lowering the operating voltage and/or frequency of a microprocessor. The following power equation indicates that the active power of a CMOS circuit is linearly and quadratically proportional to the frequency and the operating voltage, respectively.

$$\text{Active Power} \propto C \cdot V_{dd}^2 \cdot f \tag{3.3}$$

Therefore, for certain instances such as when the utilization of a processor is low, when the response time is insensitive, or when the running tasks are not critical, etc., a system with DVFS technique can reduce its operating voltage and frequency on the fly with minimal impact to the quality of service. Although the voltage and frequency can be controlled independently in a typical microprocessor, it is more common to use a lower voltage for lowering the frequency. This is because when using a lower operating voltage, the time for charging any given capacitor takes longer than when using a higher operating voltage, which leads to a slower operation or slower operating frequency. The main drawback of this technique is that lower voltage and frequency can inadvertently penalize the performance.

#### 3.5.2.3 Clock Gating and Power Gating

Distributing the clock signal across the entire die area in synchronous circuits requires more than one third of the total chip power. It gets worse if a chip uses a metal grid clock distribution network for minimizing the clock skew as discussed earlier. For reducing the active power for the clock distribution network, the most commonly used technique is clock gating. The basic idea of clock gating is to cut off the clock signal for the regions that are not used. When the clock signal does not enter a particular region of a circuit, it avoids the switching activities of its flip-flops and clock buffer tree, thereby saving power. To achieve this goal, two types of solutions are employed: a latch-free clock gating and a latch-based clock gating. In the latch-free clock gating design, a simple two-input AND gate is used to enable or disable the clock signal while the latch-based design uses a level-sensitive latch for holding the enable signal. Whenever the enable signal is off, the delivery of the clock signal is cut off. The main drawback of this clock gating is that the additional combinational logic will likely elongate the propagation delay in delivering clock signal to all corners of a chip. Due to this extra propagation time that exacerbates the clock skew, a circuit with clock gating may reduce the operating frequency.

Although clock gating can help reduce the active power of unexercised circuits, this cannot save leakage power. As the leakage power continues to worsen when the feature sizes shrink due to lowered threshold voltage (as shown in Fig. 3.13), power gating is introduced to disconnect the unused circuits from the power source using a sleep transistor with a high threshold voltage to eliminate the leakage current. Figure 3.17a illustrates an example of a sleep transistor that gates off the power supply path via $V_{ss}$ of an SRAM cell. This more aggressive power-saving technique faces several drawbacks if not used wisely. First, power-gating a circuitry, from active to inactive or vice versa, takes time in order to stabilize the circuit operation. Depending on the scale of the circuit block, the circuit may need to be switched off in multiple steps to keep the ground bounce noise under safety margin. Hence, it could affect the overall performance. Second, switching the states consumes extra power. For these reasons, when and where to power off must be chosen carefully. In other words, power gating should be performed only when the penalty in power and time for turning on and off is significantly less than the power that can be saved.

### *3.5.3 Microarchitectural Level*

Microarchitectural power-reduction techniques have been an active research area among processor architects. A majority of these studies focus on on-chip memories, i.e., caches. Some techniques combine circuit and microarchitectural optimization techniques to reduce power. In the subsequent sections, we review some major tasks toward these efforts.

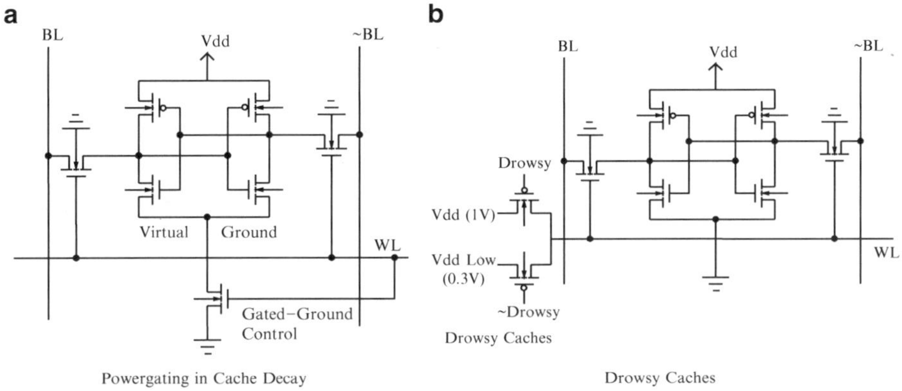


**Fig. 3.17** Circuit diagram for Cache decay and Drowsy cache. (**a**) Power gating in Cache decay. (**b**) Drowsy caches

#### 3.5.3.1 Reconfigurable Caches

*Selective cache ways* is one of the earliest architectural techniques proposed for reducing power consumption in caches of a processor. It selectively turns off a subset of cache ways for an associative cache at run-time. The idea starts from the fact that large on-chip caches are usually partitioned into several subarrays for reducing latencies. Because each subarray effectively stores one data cache way, it can readily be turned off at the hardware level. The mechanism can be supported with minimal additional hardware—a Cache Way Select Register (CWSR), to store which cache ways to use, and special instructions for reading and writing the CWSR. An application can disable selected cache ways during the period of modest cache activities without much performance impact. As shown in [2], this on-demand cache resource allocation mechanism saves 40% in overall cache energy dissipation in a four-way set associative cache with less than average 2% performance penalty.

#### 3.5.3.2 Cache Decay

Given the trend of integrating larger and larger on-die caches continues, researchers have studied and proposed various techniques to control the leakage power of these components. *Cache decay* [21] is one of such techniques that combine power-supply-gating shown in Fig. 3.17a with dynamic architectural behavior for controlling the leakage power of each individual cache line. It was motivated by the observation that cache lines are "dead" for more than 70% of the time. The dead time of a cache line is defined as the time of its last access and the time it is evicted. To avoid leakage power consumed during the dead time, if one can predict a cache line is dead, the line can be evicted and powered off earlier than the actual replacement taking place. The prediction is achieved by employing a decay counter for each cache line to book-keep the idleness of the line. When the down-counter reaches zero indicating the line is not being accessed for a given threshold, the line will be early evicted and enter the power-off state using power-gating to save leakage energy.

One drawback of the cache decay technique is the potential performance loss due to the fact that early power gating loses cache data, which may causes additional cache misses. Therefore, "when to decay a cache line" becomes critical. This work experimented different decay intervals from 1 k cycles to 512 k cycles and showed that a decay interval of 8 k cycles showed the best saving result with a 70% reduction of the leakage power.

#### 3.5.3.3 Drowsy Caches

*Drowsy cache* [16] was proposed to ameliorate the performance issue due to data loss of cache decay. In a drowsy cache, a cache line can choose between two different supply voltages, a normal voltage ($V_{\mathrm{dd}}$) for regular cache lines, and a lowered one ($V_{\mathrm{dd-low}}$) for drowsy cache lines. When a line is put into the drowsy mode, the data content is preserved although it has to pay a slight penalty (one to two cycles) to reinstate the line back to normal operated voltage before it can be re-accessed. Cache lines with the scaled down supply voltage can significantly reduce the leakage current by $6\times-10\times$ due to short-channel effects. For the drowsy cache technique, there are additional hardware overheads. First, a drowsy bit is added to each cache line to indicate whether the cache line is in drowsy mode or not. Second, a voltage controller is added as illustrated in Fig. 3.17b to supply a normal voltage for active state cache lines and a lowered voltage for drowsy state cache lines. Third, the word line gating circuit is added to prevent direct access to drowsy cache lines. With these additional hardware overheads, cache lines periodically change its state to the lower power one, and the line is woken up in the penalty of one cycle when it has to be accessed. Due to the overhead of additional cycle to wake up a cache line, performance could be degraded as much as 2% with an average of less than 1%. With this small impact on performance, the total energy (including static and dynamic) consumed in cache lines were reduced by 75%.

#### 3.5.3.4 Razor

Razor [14], a combination of microarchitectural and circuit level techniques, can substantially reduce the power consumption of a microprocessor by aggressively adopting low (subcritical) voltage in the pipeline. Similar to DVFS, Razor dynamically lowers the operating voltage (i.e., DVS) to significantly reduce power consumption. However, even with the DVFS technique, there are voltage margins to be obeyed to avoid any execution error in the processor. For example, there have to be a process margin to consider manufacturing variations, an ambient

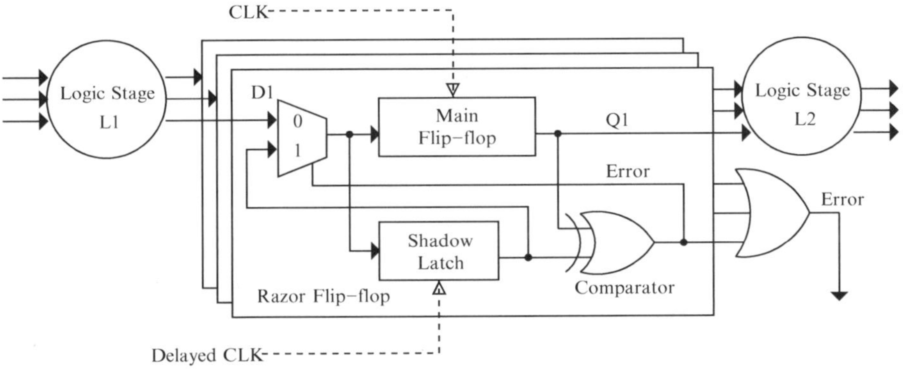


**Fig. 3.18** Razor flip-flop—A shadow latch with delayed clock

margin to prevent processors from malfunctioning due to high temperature, and a noise margin to tolerate various unknown noise sources. Without these voltage margins, a processor could generate incorrect computation results due mostly to timing failure in the slow latches. In many cases, these margins are overestimated to guarantee a reasonably large guardband for correctness. The design rationale of Razor challenged this worst-case design constraint and proposed to aggressively and dynamically scale down the operating voltage until an error is detected. Once an error is detected, a recovering mechanism will be triggered to correct these errors dynamically. As such, Razor can approach the minimal power consumption by lowering the supply voltage to the lowest possible value. The error detection mechanism is achieved by employing a *shadow latch* with a delayed clock to each normal flip-flop. As shown in Fig. 3.18, the shadow latch with the delayed clock is designed to ensure to latch the correct incoming data while the normal flip-flop could fail due to too aggressive DVS. Whenever the value of the shadow latch mismatches the value in the DVS-ed flip-flop, a timing error is indicated. Then, a pipeline flush and replay similar to branch misprediction recovery will follow with incrementally increased supply voltage. This supply voltage feedback control system will eventually reach the optimal operating voltage for a specific processor that runs a specific application. As shown in [14], the error-detecting circuits with aggressive DVS can reduce the power consumption of a processor by 64% with 3% performance impact. This technique is currently being implemented and integrated into future ARM products [9].

## 3.6 Challenges in Power Delivery and Optimization

Although the power consumption of targeted components can be improved with the power optimization techniques we discussed in the previous section, it does not guarantee an overall saving when they are applied altogether. Under certain circumstances, the savings of individual optimization techniques are not additive, worse yet, they could cancel each other out. Moreover, these techniques may substantially reduce the overall cost-effectiveness, making them economically impractical. Therefore, whenever a new power optimization is being considered, it must be thoroughly evaluated together with all existing solutions applied to the data center. Here we list some of the caveats and challenges in power delivery and optimization for data centers.

- One common pitfall in power optimization is so-called *balloon effect*. In the balloon effect, suppressing one corner of a balloon may inadvertently inflate the other side. Similarly, saving power on one particular component may increase power consumption of others in the system. For example, fans sometimes consume more than one fourth of the entire system power as shown in Fig. 3.10. The administrator could lower the fan speed to conserve power given it will not trigger the fail-safe mechanism due to thermal emergency. Even though this can effectively reduce power consumed by the fans, the optimization may induce more leakage power consumption in the processors as it causes higher average inner temperature that results in higher leakage current in the processors. Without proper tradeoff evaluation prior to such optimization, the overall system power may end up being increased rather than reduced.
- Considering the tradeoff between *power* and *energy* is another challenge. In several cases, optimizing power consumption can come with a performance penalty which induces additional computation time and may be energy. For example, the power capping technique lets a computing node operated under a given power budget; this can be used to cap the peak power of a data center. However, whenever the computing node requires more power than the power capping value, the performance penalty cannot be avoided. Of course, this performance penalty will induce additional energy consumption. Because many data centers pay their electricity bills based on their peak power draw as well as the total energy consumed in a given period, designers must be more careful in making such tradeoff during power optimization.
- Reducing power consumption can drastically increase *maintenance cost*. For instance, a higher average temperature of a data center may reduce power consumption in HVAC units; however, it will also result in a higher failure rate, reducing their availability. Replacing failed components requires human resource as well as additional down time of the data center. As another example, using DC over AC on the power delivery infrastructure of a data center cannot be a good solution when the data center has to pay more on buying nonstandard DC-powered products such as DC-UPS and DC-PSU.
- Some techniques focused on reducing the *average* power consumption of a data center; however, these can increase the *peak* power draw which results in a higher cost. For example, a scheduling system for deciding where to migrate the virtual machines or what virtual machine to migrate can reduce the average power consumption of a data center at the cost of additional power for the scheduling system itself. However, when the data center is running at the highest

load, or all the physical nodes are busy, there is not much room for such a scheduler to choose a good migration policy. In this case, the data center will experience higher peak power draw which can diminish the overall benefit of the scheduling system.

- In designing the power delivery infrastructure, the tradeoff between efficiency and human safety must be carefully considered. Whenever a more efficient solution is to be adopted, the potential risk to human operators and public safety must be accessed For example, using higher voltage current induces higher electrical efficiency in most of the components in a data center, such as wires or PSUs. Wires with higher voltages have less loss than those with lower voltages because the latter has to convey more current than the former for delivering the same power. In addition, PSUs experience higher efficiency with higher input voltages. However, it is apparent the use of higher voltage may increase the risk of operations and require extra emergency response plans.
- Last but not least, a working solution for a data center may not be a panacea for all other data centers. As mentioned earlier, energy-proportional computing is one of the major goals for most of the power-saving ideas, and powering off the physical machines after migrating virtual machines is known to closely achieve the goal. However, for certain data centers it is simply infeasible to turn off the machines for many reasons. Thus, data center administrators have to understand the purpose and implications of their data center before applying any new idea for saving power.

## 3.7 Summary

As future computing paradigm is in the cloud, all types of computing including our day-to-day work, entertainment, large-scale data processing and management, social networking, etc., will be performed by the invisible machines in the data centers maintained by cloud service providers. As discussed in this article, the electricity and the power consumed by these data centers, including that used to power up the facility as well as to remove the heat generated by the facility, have already accounted for a nontrivial portion of the overall energy infrastructure, and the trend is only continuously escalating. Not only is this a serious cost issue for every one but it also indirectly leads to disastrous consequences of the environments due to carbon emission. Hence, minimizing the power consumption of data centers has become a major mission as we rely on them more and more to accomplish computing tasks.

To address the emerging power crisis in these computing platforms, researchers first need to understand the sources and causes of these power consumption and their associated components in various stages of the power delivery infrastructure. This article aims at providing a holistic view of the power distribution in a data center, from the top infrastructure level down to the intra-chip level. The layer-peeling breakdown analysis in this article was performed to improve the understandings of the significance of power consumption in each layer. For example, at the top layer, certain data center designs would waste more than half of the electrical power in the cooling facility whereas more efficient data centers will spend only 14%. Power loss in the UPS, for another example, is in fact the second largest power overhead for supporting computing loads. At the system level, the processors account for close to half of the power consumption while the system memories consume around one quarter. At the innermost level, we presented simulated data collected by the research communities and showed the power breakdown based on functional modules inside a processor. Finally, we review representative power optimization techniques at each layer, either implemented in real systems or studied by researchers. These techniques were shown to effectively improve power delivery efficiency, conserve power whenever possible, and reduce both dynamic and leakage power for each module. These techniques are orthogonal and can be applied simultaneously across different layers to reach synergistic reduction in the overall power.

## References


1. Seagate Cheetah 15K 3.5-inch Hard Drives. http://www.seagate.com/www/en-us/products/enterprise-hard-drives/cheetah-15k, 2010.
2. D. H. Albonesi. Selective Cache Ways: On-Demand Cache Resource Allocation. In *International Symposium on Microarchitecture*, pages 248–259, 1999.
3. American Power Conversion (APC) Corp. PDPM288G6H (300MM RACK, 266kW, Auto Transformer, 72 poles, Modular Distribution). http://www.apcmedia.com/salestools/MTAI-7P4S9D_R0_EN.pdf, 2009.
4. American Society of Heating, Refrigerating and Air-Conditioning Engineers, Inc. Datacom Equipment Power Trends and Cooling Applications. Chapter 3, Fig 3.10, 2005.
5. Anand Lal Shimpi. Intel's Atom Architecture: The Journey Begins. http://www.anandtech.com/show/2493, 2008.
6. C. Auth, A. Cappellani, J. Chun, A. Dalis, A. Davis, T. Ghani, G. Glass, T. Glassman, M. Harper, M. Hattendorf, et al. 45nm High-k+ metal gate strain-enhanced transistors. In *2008 Symposium on VLSI Technology*, pages 128–129, 2008.
7. L. Barroso and U. Holzle. The Case for Energy-proportional Computing. *IEEE COMPUTER*, 40(12):33, 2007.
8. P. Bohrer, E. Elnozahy, T. Keller, M. Kistler, C. Lefurgy, C. McDowell, and R. Rajamony. The Case for Power Management in Web Servers. *Power Aware Computing*, 62, 2002.
9. D. Bull, S. Das, K. Shivshankar, G. Dasika, K. Flautner, and D. Blaauw. A Power-efficient 32b ARM ISA Processor Using Timing-error Detection and Correction for Transient-error Tolerance and Adaptation to PVT Variation. In *Digest of Tedhnical Papers in the International Solid-State Circuits Conference (ISSCC)*, pages 284–285, 2010.
10. D. DeVetter and D. Buchholz. Improving the Mobile Experience with Solid-State Drives. *Intel Whitepaper*, http://download.intel.com/it/pdf/Improving_the_Mobile_Experience_with_S%olid_State_Drives_2009.pdf.
11. F. Douglis, P. Krishnan, and B. Marsh. Thwarting the Power-hungry Disk. In *Proceedings of the USENIX Winter 1994 Technical Conference on USENIX Winter 1994 Technical Conference*, page 23. USENIX Association, 1994.

12. D. Economou, S. Rivoire, C. Kozyrakis, and P. Ranganathan. Full-system Power Analysis and Modeling for Server Environments. In *Workshop on Modeling, Benchmarking, and Simulation (MoBS)*, 2006.
13. Ecos Consulting and EPRI Solutions. High Performance Buildings: Data Centers - Server Power Supplies. Lawrence Berkeley National Laboratory (http://hightech.lbl.gov/dctraining/reading-room.html), 2005.
14. D. Ernst, N. S. Kim, S. Das, S. Pant, R. Rao, T. Pham, C. Ziesler, D. Blaauw, T. Austin, K. Flautner, and T. Mudge. Razor: a Low-power Pipeline based on Circuit-level Timing Speculation. In *Proceedings of the 36th Annual IEEE/ACM International Symposium on Microarchitecture*, pages 7–18, 2003.
15. X. Fan, W.-D. Weber, and L. A. Barroso. Power provisioning for a warehouse-sized computer. In *Proceedings of the 34th annual International Symposium on Computer Architecture*, pages 13–23, New York, 2007.
16. K. Flautner, N. Kim, S. Martin, D. Blaauw, and T. Mudge. Drowsy caches: simple techniques for reducing leakage power. In *Proceedings of the 29th Annual International Symposium on Computer Architecture*, pages 148–157, 2002.
17. S. Ghemawat, H. Gobioff, and S. Leung. The Google file system. *ACM SIGOPS Operating Systems Review*, 37(5):29–43, 2003.
18. X.-Y. Hu, E. Eleftheriou, R. Haas, I. Iliadis, and R. Pletka. Write Amplification Analysis in Flash-based Solid State Drives. In *Proceedings of SYSTOR 2009: The Israeli Experimental Systems Conference*, SYSTOR '09, pages 10:1–10:9, New York, 2009.
19. I. Hur and C. Lin. A Comprehensive Approach to DRAM Power Management. In *Proceedings of the 14th International Symposium on High Performance Computer Architecture*, pages 305–316. IEEE, 2008.
20. A. Karabuto. HDD Diet: Power Consumption and Heat Dissipation. http://ixbtlabs.com/articles2/storage/hddpower.html, 2005.
21. S. Kaxiras, Z. Hu, and M. Martonosi. Cache Decay: Exploiting Generational Behavior to Reduce Cache Leakage Power. In *Proceedings of the 28 th annual International Symposium on Computer Architecture*, volume 29, pages 240–251, 2001.
22. C. Lefurgy, X. Wang, and M. Ware. Power Capping: a Prelude to Power Shifting. *Cluster Computing*, 11(2):183–195, 2008.
23. S. Li, J. Ahn, R. Strong, J. Brockman, D. Tullsen, and N. Jouppi. McPAT: an integrated power, area, and timing modeling framework for multicore and manycore architectures. In *Proceedings of the 42nd Annual IEEE/ACM International Symposium on Microarchitecture*, pages 469–480, 2009.
24. D. Meisner, B. T. Gold, and T. F. Wenisch. PowerNap: Eliminating Server Idle Power. In *Proceeding of the 14th International Conference on Architectural Support for Programming Languages and Operating Systems*, ASPLOS '09, pages 205–216, New York, 2009.
25. Micron. 512Mb DDR SDRAM (x4, x8, x16) Component Data Sheet. http://download.micron.com/pdf/datasheets/dram/ddr/512MBDDRx4x8x16.pdf, 2000.
26. S. Narendra and A. Chandrakasan. *Leakage in nanometer CMOS technologies*. Springer-Verlag New York Inc, 2006.
27. K. Natarajan, H. Hanson, S. Keckler, C. Moore, and D. Burger. Microprocessor Pipeline Energy Analysis. In *Proceedings of the 2003 International Symposium on Low Power Electronics and Design*, pages 282–287, 2003.
28. S. Pelley, D. Meisner, P. Zandevakili, T. F. Wenisch, and J. Underwood. Power routing: Dynamic power provisioning in the data center. In *ASPLOS '10: Proceedings of the fifteenth edition of ASPLOS on Architectural Support for Programming Languages and Operating Systems*, pages 231–242, New York, 2010.
29. Rambus Inc. 128/144-MBit Direct RDRAM Data Sheet, May 1999.
30. N. Rasmussen. Increasing Data Center Efficiency by Using Improved High Density Power Distribution. *White paper, American Power Conversion (APC) Corp*, 128, 2008.

31. N. Rasmussen and J. Spitaels. A Quantitative Comparison of High Efficiency AC vs. DC Power Distribution for Data Centers. *White paper, American Power Conversion (APC) Corp*, 127, 2008.
32. RUMSEY Engineers, Inc. Data Center Energy Benchmarking Case Study — Facility 8. Lawrence Berkeley National Laboratory (http://hightech.lbl.gov/dctraining/reading-room.html), 2003.
33. Semiconductor Industries Association. Model for Assessment of CMOS Technologies and Roadmaps (MASTAR). http://www.itrs.net/models.html, 2007.
34. G. Shamshoian, M. Blazek, P. Naughton, R. S. Seese, E. Mills, and W. Tschudi. High-Tech Means High-Efficiency: The Business Case for Energy Management in High-Tech Industries. Lawrence Berkeley National Laboratory (http://hightech.lbl.gov/dctraining/reading-room.html), 2005.
35. N. Tolia, Z. Wang, M. Marwah, C. Bash, P. Ranganathan, and X. Zhu. Delivering Energy Proportionality with Non Energy-proportional Systems: Optimizing the Ensemble. In *HotPower '08 Workshop on Power Aware Computing and Systems*. USENIX Association, 2008.
36. M. Ton, B. Fortenbery, and W. Tschudi. DC Power for Improved Data Center Efficiency. *Lawrence Berkeley National Laboratory, EPRI Solutions, Ecos Consulting, Tech. Rep*, 2008.

# Chapter 4
# Understanding and Managing IT Power Consumption: A Measurement-Based Approach

**Ada Gavrilovska, Karsten Schwan, Hrishikesh Amur, Bhavani Krishnan, Jhenkar Vidyashankar, Chengwei Wang, and Matt Wolf**

**Abstract** The continuing, unsustainable increase in datacenter power consumption is causing researchers in industry and academia to be heavily invested in addressing power management challenges. This chapter presents the basic elements of a measurement-based approach toward managing distributed datacenter and cloud computing systems to meet both application and end-user needs and to obtain improved efficiency and sustainability in their operation. The main components of the approach presented include (1) continuous online monitoring, measurement and assessment of systems and applications behaviors and power consumption, including for online estimation of the power usage of virtual machines running application components in these virtualized systems; (2) the ability to perform these tasks efficiently at scale, so as to deal with the ever-increasing sizes and complexity of modern datacenter infrastructures; and (3) the importance of "coordinated" management methods that operate across multiple levels of abstraction and multiple layers of the management stack in an orchestrated manner.

## 4.1 Introduction

The continuing, unsustainable increases in datacenter power consumption are causing researchers in academia and in industry to be heavily invested in addressing the issue [1–4]. Efforts to develop power-aware datacenter management techniques range from multi-scale methods for managing IT system power usage [5], to power capping to deal with increased server densities [6–8], to integrating into the

---

A. Gavrilovska (✉) • K. Schwan • H. Amur • B. Krishnan • J. Vidyashankar • C. Wang • M. Wolf
Center for Experimental Research in Computer Systems, School of Computer Science, College of Computing, Georgia Institute of Technology, Atlanta, GA, USA
e-mail: ada@cc.gatech.edu; schwan@cc.gatech.edu; amur@cc.gatech.edu; bhavani.krishnan@gatech.edu; jhenkar.vidyashankar@gatech.edu; flinter@cc.gatech.edu; mwolf@cc.gatech.edu

Y. Joshi and P. Kumar (eds.), *Energy Efficient Thermal Management of Data Centers*, DOI 10.1007/978-1-4419-7124-1_4,

**Table 4.1** Dynamic power for different VMs on a Core2 Xeon platform

| | CPU Utilization | Power |
|---|---|---|
| VM1 | 50% | 6 W |
| VM2 | 100% | 12 W |
| VM3 | 100% | 42 W |

management processes information regarding the datacenter cooling infrastructure, the latter aimed at improving facility-level metrics like Power Usage Effectiveness (PUE) [3, 4, 9].

A common element of power-aware datacenter management is its exploitation of virtualization technology for server consolidation and for dynamic load distribution and/or load balancing. With virtualization, applications are represented as sets of self-contained virtual machines (VMs) that can be moved between different datacenter machines, even while they are running—termed VM migration. Used in both private datacenters as well as in emerging cloud computing systems, virtualization involves a broad range of technologies developed by companies like IBM, VMWare, and Citrix, and supported by hardware extensions for platforms (e.g., from Intel and AMD), for networks (e.g., by Cisco), and for the large-scale storage systems used in datacenters (e.g., products from IBM or NetApp).

Virtualization offers increased degrees of freedom concerning the mapping of IT loads to machines—through dynamic VM migration—and with this freedom comes the need for new methods for managing IT systems and workloads, including to manage the migration actions being taken. One important goal of IT system management, for instance, is to achieve some desired datacenter power state with a limited number of reconfigurations and VM migrations [10]. Furthermore, consider the data presented in Table 4.1 below, which shows the power utilization for three VMs. When load migration is needed, a decision to migrate VM1 may be insufficient for meeting some lower target power cap. This means that in order to make appropriate decisions, there must be precise estimates of VMs' power consumption. Unfortunately, the common practice of using a VM's current average CPU utilization for this purpose [4] does not offer much precision. This is because CPU utilization values include the time processors spend waiting for memory and thus, do not accurately reflect either actual CPU usage or CPU power consumption. For instance, VM2 and VM3 in Table 4.1 both show 100% CPU utilization, but VM3 is memory bound whereas VM2 is CPU bound. As a result, despite having the same CPU utilization, the power contributions of the two VMs are quite different.

While the data in Table 4.1 illustrates limitations in using overly simplistic models of per-VM power usage, the scenario nonetheless shows that there is substantial value in having per-VM information: (1) power-aware management methods can benefit in terms of making appropriate decisions about the VM migrations or other reconfiguration actions required to achieve some given power state [10, 11]; (2) datacenter administrators can use it to develop customer-facing billing or chargeback policies [12]; and (3) environmentally responsible consumers of datacenter and cloud resources can employ it to minimize their workloads' carbon footprints [13].

Assessing a VM's power usage is difficult for reasons beyond those discussed above. In datacenter systems, for instance, there will be generational differences between the many computational platforms being used in each facility, and in addition, there will be large runtime variations in the workloads imposed on utility or cloud computing applications. Moreover, VM-level information about the applications being run is typically not available to the operators who are responsible for managing the datacenter/cloud infrastructures. "No one tells the operations team what applications are being run" is a common message we have heard from the datacenter operators with whom we have interacted. What is needed, therefore, are techniques that automatically determine VM and application behaviors, including a VM's power usage, and these techniques should not rely on applications to provide them with (potentially faulty) information. Instead, there should be "black-box" methods that continuously monitor and assess datacenter applications and systems, using the platform-level monitoring support available in modern hardware and systems, and operating with degrees of accuracy similar to what past work has shown possible for statically developed VM energy profiles [1].

Datacenter compute platforms, hardware, surrounding infrastructure (e.g., cooling subsystems), and the applications for which they are used are complex. At the same time, as stated above, there will be limited knowledge about applications, their behaviors and needs, the algorithms and methods they use, and the workloads imposed on them and on the underlying datacenters. In response, there has been increasing interest in the development and use of measurement-based approaches to understanding applications, systems, and their current behaviors. Stated succinctly, such approaches seek to manage datacenter systems by (1) dynamically and continually extracting measurement data from infrastructure, hardware, systems, and applications, (2) using such measurements to better understand current system behavior based on the runtime analysis of such data, and (3) using the insights gained in this fashion to manage infrastructure, systems, and applications, where (4) management goals are driven by metrics capturing both the goals of IT system providers and users [14].

This chapter articulates the basic elements of a measurement-based approach to datacenter management. It begins by first providing a brief introduction to virtualization technology, its use in modern datacenters, and particularly, in emerging cloud platforms. Next, using a concrete case study based on the measurement results shown earlier in the chapter, we describe a simple power-metering methodology using black-box monitoring methods developed for instrumented virtualized computing platforms. Far from being complete or conclusive, the study's main purpose is to demonstrate opportunities and challenges for future work. Second, we review the state of the art in datacenter monitoring, identifying challenges and suggesting future work, followed by a discussion of measurement and analysis methods, the latter focused on black-box techniques that do not require end users or system providers to divulge detail about the applications or systems being run, or the actual cooling technologies being used. Third, we briefly outline some recent advances in management methods and algorithms, again focusing on those that are based on measurement and runtime understanding rather than static modeling or profiling. We conclude by discussing alternative management goals and ways to deal with them.

## 4.2 Overview of Virtualization Technology and Compute Clouds

A brief overview of virtualization technology and emerging virtualized compute cloud platforms appears below.

### *4.2.1 Virtualization Technology*

The use of virtualization technology has become commonplace in the large datacenters that power today's Internet services. Using this technology, the platform hardware is "virtualized" to create what appear to be multiple "virtual" instances of each single physical platform. This is achieved by multiplexing these virtual instances across the underlying physical machine resources. The software run on a virtual platform is called a "Virtual Machine" (VM) or "domain," and it is typically comprised of its own operating system (OS) managing the virtual instance's resources and of applications being run on top of the OS.

The ability to share a single physical platform across multiple virtual machines is a key benefit of virtualization, as it makes it possible to *consolidate* multiple virtual machines, each running potentially different operating systems and applications, onto one physical machine, i.e., one physical server. This can significantly reduce the operational expenditures of an enterprise, by improving the utilization of its IT platforms and/or reducing the total number of machines required by the enterprise's applications. Such reductions can also lead to reduced operational and management costs, because fewer hardware nodes need to be controlled, monitored, and serviced. A second benefit of virtualization is that it enables dynamic, at runtime, "migration" of VMs from one physical platform to another. Combined with consolidation, this means that the number of machines currently turned "on" in a datacenter can be made directly proportional to actual current IT workloads, termed "power proportionality," rather than being an artifact of how certain subsystems have been configured. This is because VMs can be migrated to less loaded physical nodes and load can be easily redistributed to deal with peak demands. Virtualization also simplifies the creation of methods for reliability or fault tolerance—since a VM can be migrated from a node that is determined to be misbehaving to a valid one, with few or no long-term impacts on VM performance [15] (Fig. 4.1).

Delving a little deeper into virtualization technology, support for virtual machines is provided both by modern hardware and by a "thin" layer of software resident on each physical machine. This software, termed the Virtual Machine Monitor (VMM) or hypervisor, makes available to each VM the resources it requires. This involves presenting a VM with some subset of hardware resources dedicated for its own use, for instance, a portion of the machine's RAM memory and some disk space, and it involves multiplexing other hardware resources across multiple VMs, such as the CPUs, network devices, and disks, present in the system.

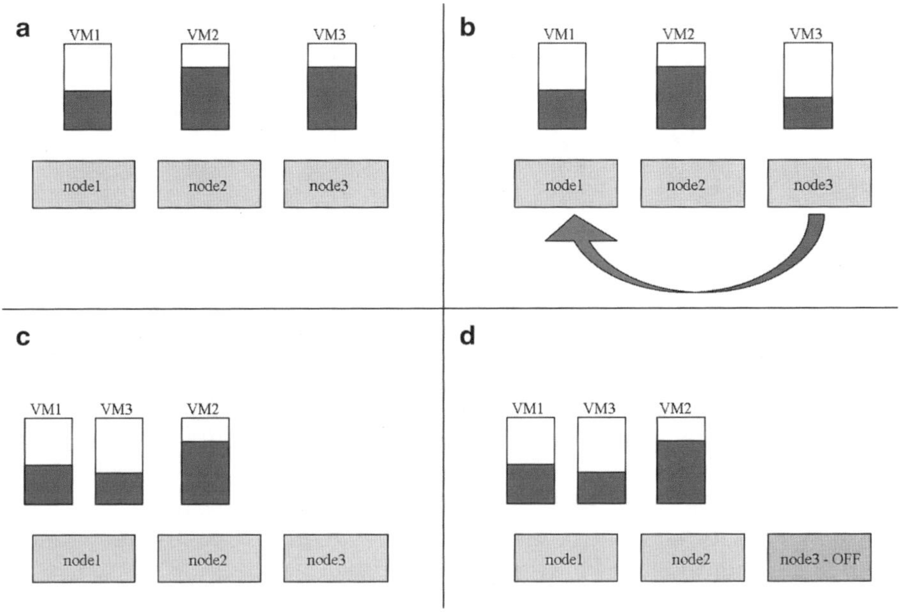


**Fig. 4.1** Dynamic VM migration: (**a**) Initial load configuration. Red bars in VMs indicate their resource (e.g., CPU) utilization. (**b**) The resource utilization of VM3 drops. (**c**) VM3 is migrated to node1, where it is collocated with VM1. Node3 is now idle. (**d**) Node3 is powered off

In all such cases, the hypervisor controls resource allocations and sharing, and it does so in ways that give each VM the illusion of running on what appears to be its own machine. Examples of popular hypervisors include VMware's ESX [16], Microsoft's HyperV [17], or Xen, available as open source [18] or as a commercial product through Citrix [19].

In some hypervisor products, e.g., in Xen and HyperV, the lowest level of software residing between a virtual machine's operating system and the physical hardware is responsible only for the most critical and time-sensitive platform management operations, such as CPU scheduling, memory protection, etc. All other functionality needed to multiplex the full platform's resources is resident in a special "privileged" domain, called control domain, or Service VM, or, in Xen, referred to as "dom0." We mention this fact only because the presence of such domains with privileged access to underlying hardware also makes it possible to dynamically monitor and measure virtual machines and their actions, without explicitly involving them and without the need to update or change their software in any fashion. This constitutes "black-box" monitoring since at this level of abstraction, there is no explicit knowledge about VM internals, including whether they run Unix or Windows operating systems, whether they run performance-critical or other applications, etc. Instead, there is visibility into which, how many, and how much virtual machines use the resources the hypervisor has assigned to them. Examples include amounts of memory or disk used, extents of

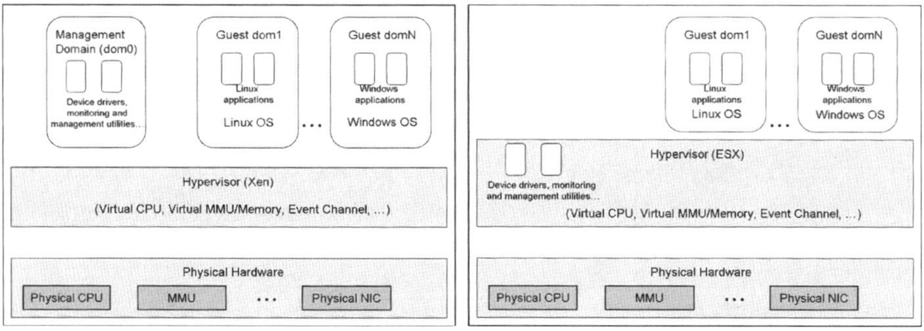


**Fig. 4.2** Virtualization architecture for two popular hypervisors, Xen and VMware's ESX. In Xen, the management functionality is part of the privileged domain dom0. In ESX, it is integrated into the hypervisor layer

CPU utilization, network packet rates (e.g., Ethernet frame rates), numbers of low-level requests to disks, etc. Black-box monitoring, then, uses such information, it is possible to infer VMs' dynamic behavior and then control their resource allocations, without their involvement or knowledge.

A dom0 as used in Xen is shown on the left side in Fig. 4.2. An alternative organization of a hypervisor, as used in VMware's ESX, is depicted on the right side of Fig. 4.2, where the same "privileged" functionality is integrated into the hypervisor layer. For convenience, the remainder of this chapter assumes the Xen model of privileged domains when explaining how and which measurements and controls can be realized in virtualized environments.

### *4.2.2 Virtualized Datacenters and Compute Clouds*

Cloud computing takes a step beyond simply running multiple virtual machines on some set of machines owned by a single organization or company and resident in a single facility. Specifically, extending beyond such "private clouds," more general cloud computing infrastructures seek to deal with peak loads or aim to curtail companies' IT costs, by providing to them "public cloud" systems that are accessible to cloud users whenever they need them. Moreover, but still limited by certain implementation challenges, there are "hybrid clouds," where an organization's private cloud computing infrastructure includes public cloud components, expanding into and using the public cloud whenever needed. Another dimension of cloud technology is the multiple levels of cloud services they can make available. The most common of these are fixed sets of software services—Software as a Service (SaaS)—offered to clients, such as email or office productivity services, that are run on the datacenter systems operated by the cloud provider, rather than on facilities owned by the user. More recently, providers like Amazon have generalized such offerings to also permit end users to run arbitrary applications on the provider's

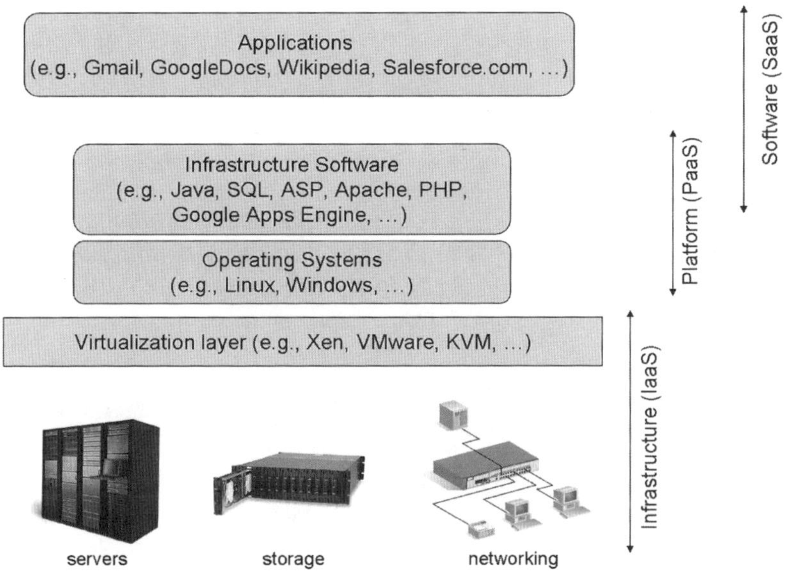


**Fig. 4.3** Clouds may provide Infrastructure-as-a-Service (IaaS), Platform-as-a-Service (PaaS), or Software-as-a-Service (SaaS) functionality

datacenter machines, using proscribed Web-based APIs for accessing and using cloud resources—termed Infrastructure as a Service (IaaS) [20], and there are additional efforts to give end users more control over the kinds of machines on which their cloud applications are run—Platform as a Service (PaaS) [21]. Fig. 4.3 illustrates these three models.

This chapter's purpose is not to delve deeply into cloud computing systems. Instead, we note the opportunities and challenges presented by these technologies for future datacenter systems and their management:

- Computations are becoming increasingly "fungible," meaning that computations are no longer bound to individual machines configured for them and that computation to machine bindings can be changed when needed and appropriate.
- Fungibility can be used to increase the levels of efficiency at which datacenters operate, increasing machine utilization beyond the 20% commonly seen at "typical" times in traditional corporate datacenters.
- There are entirely new opportunities for sharing computational resources and thus, more effectively using them to satisfy end user or application needs. These range from making it easier for a company to deal with sudden increases in IT load, to shifting computation to sites not affected by local disasters, to moving IT loads to where energy is currently abundant (or cheap), to creating a sustainable global IT ecosystem in which computations are run when and where most appropriate for meeting $CO_2$ emission limits or complying with companies' goals for energy use and cost.

These opportunities, however, raise technical challenges that will drive IT industry efforts for many years to come. One such challenge is that of effective system management:

> *How to manage distributed datacenter and cloud computing systems both to meet application and end user needs and to obtain efficiency and sustainability in their operation.*

The remainder of this chapter projects a research path toward attaining this goal: through continuous online system and application monitoring, measurement, assessment, and management of—a *measurement-based approach* to—utility datacenter and cloud computing. We next describe the problem in taking even the first step in this ambitious goal: VM-level power metering, its feasibility, and challenges.

## 4.3 VM-level Power Metering: Feasibility and Challenges

A study of VM-level power metering is an important step toward managing datacenter and cloud computing systems. This is because it is the application IT loads—embedded in VMs—rather than "what's running" on some physical machine that can be meaningfully moved across underlying datacenter machines and facilities. This means, of course, that it must be known which VMs belong to which application, a fact that cannot be trivially assumed in virtualized systems, but may need to be determined by black-box monitoring methods like those described in [22]. Nonetheless, given that the set of VMs constituting an application—commonly termed a VM ensemble—is known, we can then dynamically place and move applications or suitable application subsets, as per their power needs and in accordance with the sustainability policies sought by datacenter operators. Doing so requires us to assess each VM's power usage, and the purpose of this chapter is to describe a measurement-based approach for doing so.

### *4.3.1 A Case Study*

Multiple issues make it difficult to estimate the power a VM will consume: (1) the dynamic behavior of the workload imposed on the applications run by the VM, (2) the fact that VM performance is affected by the other VMs with which it shares the physical platform on which it runs, and (3) that VM performance and power consumption will differ across platforms, subject to memory or cache availability, processor speed, and many other factors. Another issue is that the methods for continually monitoring VM power usage themselves are subject to constraints that include (1) independence of specific underlying hardware or system features,

(2) low cost in terms of low delays in running them and few resources used for doing so, and (3) generality in terms of their ability to deal with a wide variety of underlying hardware and system platforms.

We advocate a measurement-based approach to estimating VM power consumption. Toward this end, power models are constructed by continually correlating a VM's usage of specific types of resources to its system-level power consumption, both of which are easily measured at hypervisor level, using commonly available system monitoring tools and hardware sensors. Such measurements can record all resources used by VMs, including memory, the latter particularly important because of the increasing contributions of memory to system power usage and the increasingly complex memory hierarchies present on current and next-generation hardware [23].

Using a broad range of workloads exhibiting different CPU and memory usage patterns, derived from a synthetic benchmark and from the SPEC benchmark suite, we next demonstrate the ability to accurately estimate an individual VM's contributions to platform power consumption. In fact, the data gathered during our power model construction and experimental analysis indicates that estimation requires consideration of the VM's CPU and its memory resource utilization. The accuracy of such estimates, however, is strongly dependent on additional insights into the VM's usage of the memory system, such as their utilization of different levels of the cache hierarchy or its attained memory-level parallelism. In the absence of such additional information (due to increasing overheads of finer-grain monitoring of low-level hardware counters or need for additional application instrumentation), we demonstrate that it is still feasible to establish bounds on a VM's power usage. In addition, we experimentally demonstrate that even when using black-box methods, it is possible to estimate the per-VM power usage, contrary to arguments made in [1, 24].

Finally, we demonstrate that model construction and the runtime measurements gathered for online power usage estimation must consider certain architectural features, including the design of the platform's cache architecture. It is precisely this fact, namely, that precise models may require detailed architectural information, which leads us to advocate our measurement-based rather than purely modeling-based approach to assessing VM power usage. This is because without continual online measurements, there will be too many architectural details and issues that must be known and taken into account for static power models to be sufficiently precise, and to do so in the dynamic, ever-changing, multigenerational datacenters commonly used in cloud and utility computing systems is not likely feasible or reasonable. The creation of such static models is even further complicated by the complex ways in which collocated VMs can interfere with one another, and the architecture-specific way in which collocation with different VMs, running different types of applications, affects their performance and resource consumption (e.g., overall slowdown due to increased number of cache misses or I/O contention).

#### 4.3.1.1 Methodology

Our VM power metering approach first establishes a power model for the various system resources present on a given platform, specifically targeting those resources that have significant contributions to overall system power. It does so by correlating the utilization level of each specific resource to the overall system power, when other types of resources are maintained at extremely low utilization levels. Next, at runtime and using lightweight monitoring tools, we measure the per-VM utilization of various resources. Our current implementation targets platforms virtualized with the Xen hypervisor, and relies on tools such as Xenoprof [25] for dynamic performance profiling and access to hardware counters. The hardware counters record various events—such as cache misses, rate of instruction execution, etc., and based on these values, we can then determine the utilization of the corresponding resources. For instance, each cache miss results in access to main memory, so the cache miss rate counter is used to determine the memory utilization. Similarly, the counter corresponding to the number of executed instructions is used to determine the CPU utilization. Finally, the VM's power usage is estimated based on the power levels corresponding to the appropriate resource utilization, as derived in the resource' power models, and few additional factors, such as the number of active cores or sockets, to include the often significant transition costs from activating a system component.

The total system power consumed by a physical server machine can be written as follows:

$$\text{Dynamic } P_{\text{server}} = P_{\text{idle}} + P_{\text{cpu}} + P_{\text{mem}} + P_{\text{disk}} + P_{\text{io}} \tag{4.1}$$

Idle power is measured when all cores and the memory subsystem are idle. While on modern platforms (e.g., Intel's core i7 generation and beyond) the idle power level continues to be reduced, it still represents a significant component of the power consumption of datacenter hardware. Disk usage has been modeled successfully in previous work [1, 2] and hence, we do not focus on it in great detail below. We also presently ignore $P_{\text{io}}$, network power usage [26]. For the workloads and platforms currently used in our work network contributes to small fraction of the overall power consumption. If needed, such information can easily be added to this model.

Since one of our goals is low cost—lightweight—VM power metering, we limit the amount of information monitored to only few types of events. Specifically, we use hardware counter values for instructions retired per second (inst_ret/s) and last-level cache misses per second (llc_miss/s). As a result, a single type of event may represent more than one resource utilization state. For instance, the same value for inst_ret/s may correspond to certain CPU utilization and its accompanying power level, or, in the event of a cache-bound workload, it may represent cache usage plus some different CPU utilization value, resulting in different total power usage. Similarly, the same llc_miss/s value may correspond to different levels of memory utilization depending on memory-level parallelism and memory overlap resulting from a cache miss.

In order to deal with this while avoiding the development of fine-grain models for every single platform element, and the subsequent issues with scaling such fine-grain models to datacenter systems, we explore the possibility of establishing bounded models. The goal is to limit the required monitoring state and runtime overheads, while at the same time understanding the level of estimation accuracy.

#### 4.3.1.2 Modeling

As an illustrative example, we establish the CPU and memory power models for two different platforms, a dual-socket quadcore Nehalem Core i7-based system, and a quadcore Core 2-based Xeon system, termed Nehalem Core i7 and Xeon Core 2 in the remainder of this chapter, respectively. We demonstrate the validity of the models using a set of simple experiments. The Nehalem Core i7 has 12 GB RAM and 8 MB last-level (L3) cache. The Xeon Core 2 has a 6 MB last-level (L2) cache.

CPU Model

To accurately model the power consumed by the CPU subsystem, which includes processor and cache power, we measure for each VM the number of executed instructions, represented via the CPU's counter "instructions retired per second" (inst_ret/s), which we then correlate with the measured value for dynamic server power. The mapping from instructions retired per second to CPU power usage is complicated by the fact that memory references that hit in different levels of the cache consume different amounts of power. For instance, we observe that with a measurement of x instructions retired per second, there will be much less power consumed if each of those instructions executed on the CPU hits the L1 cache as compared to the case where each of their memory references hits only in the last level cache (LLC). As shown in Fig. 4.4, this may explain the varying values for power usage for equal levels of CPU utilization observed in [1]. In fact, if a majority of the accesses hit in L1, we see larger values for inst_ret/s than if the accesses hit in LLC for a given power value.

The problem with the cache-related observation noted above is that it implies that relatively small changes in processor architecture (e.g., differences in cache size) and/or changes in program behavior (e.g., differences in cache footprint) can change the power consumption observed for the same values of CPU utilization. In order to limit the need to monitor the utilization of each cache level present in the architecture, we explore the utility of building a CPU power model using the inst_ret counter only, the intent being to provide bounds on CPU power for any executing workload by considering the extreme cases. To establish these bounds we consider to extreme cases: an L1-bound workload, which only accesses

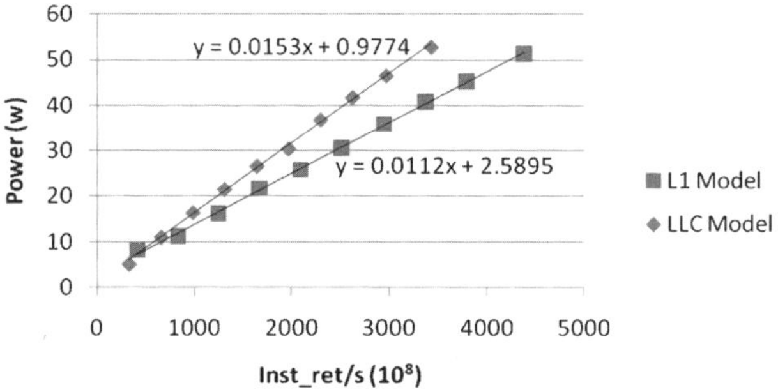


**Fig. 4.4** CPU model for L1 and LLC

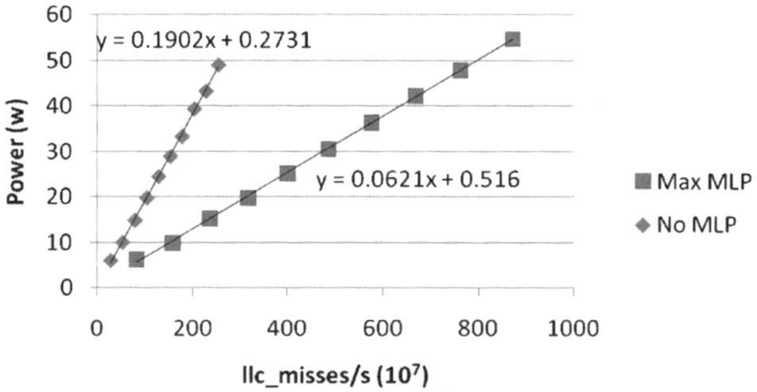


**Fig. 4.5** Memory model for MAX MLP and NO MLP

the first-level cache during execution, and an LLC-bound workload, for which every instruction accesses the last-level cache available on the given platform. Referring the reader to [27] for detail, using an artificial benchmark, we determine that with 100% CPU utilization for an LLC-bound process, we see 22% fewer inst_ret/s than for a 100% CPU utilization for an L1-bound process. This means that reality will lie between the two extreme case models that represent all hits going to the first level of cache vs. the last level of cache, respectively. This provides us with the bounds shown in Fig. 4.4.

### Memory Model

For a baseline memory model, we consider the last-level cache misses per second (llc_misses/s) only. Since each cache miss must result in access to main memory (i.e., if the data is not present in the cache, it must be fetched from memory), we believe the llc_misses/s counter can be used to understand the VM's memory behavior, and build the memory power model by correlating llc_misses/s and the dynamic server power. There are complications, however, due to differences in processors' memory architectures. This again results in bounds on memory power usage depicted in Fig. 4.5.

Both models also include the package wake-up costs, as well as costs for activating a second socket. Modern platforms, such as the Intel Core i7, have deep power states, and waking up a core from a deep idle state can itself consume power. For the hardware used in our research, these wake-up "costs" correspond to 41 W. Fully loading a core on the same socket adds 11 W to this power level, whereas fully loading a core on the second socket adds additional 11 + 2 W. Similar results can be obtained for the Core 2 platform evaluated in our work.

#### Validation and Experimental Results

We do not explain in detail the validation of the bounded models shown above, but we note that experimentation establishes that the CPU and memory power consumed by a VM is additive in terms of the number of cores used. Further, there are issues with the Core 2 architecture due to its cache design, which can cause an L1 cache to be "turned on" (i.e., it consumes power) although the processor using it is idle. This information must be considered during the initial phase of establishing the baseline resource utilization—to—power relationships, which are later used during online power metering.

Experimental evaluations show that the approach to VM power metering shown above can misestimate the power consumption of CPU-intensive applications by 10–20% and that of memory-bound applications up to 7% of measured values, with typical measurement errors under conditions in which multiple VMs share the use of a single platform being around 5%. At the same time, monitoring overheads remain small, with a 3% maximum overhead (in terms of application slowdown) in the worst case. Rather than showing such results in detail, here, we simply note that experimental results are obtained by using power measurement equipment only when results are validated, whereas the measurement approach utilizes only standard hardware and hypervisor metrics and does not require neither a VM's participation nor any a priori knowledge about a VM.

### *4.3.2 Other Approaches*

There are other approaches to VM-level power metering, including those described in [2] and [28]. They are similar to what we have presented in that they seek to utilize the events gathered by hardware counters present on modern hardware platforms to dynamically, at runtime, extract the resource utilization of given software components, i.e., VMs, but they do not properly model the increasingly important memory power usage of modern machines. An alternative approach described in [1] argues against the online instrumentation of applications or VMs, as such overheads may be prohibitive, and instead, suggests the use of a priori established VM profiles. This may not be a suitable choice for next generation datacenter and cloud computing systems and their increasingly dynamic applications.

## 4.4 Datacenter Management

We now move from discussing the prerequisites to measuring the power consumption of datacenter applications, i.e., VM power metering, to describing the state of the art in datacenter-wide and utility cloud monitoring and management. Concerning existing software, there are many partial or subsystem-level solutions to systems management. At one end of the spectrum, there are rich all-encompassing commercial monitoring and management solutions such as HP System's Insight Manager, IBM's Tivoli, and VMware's vCenter (and vCloud extensions) for datacenter environments. These systems perform centralized data collection and analysis and provide some support for script-based triggering mechanisms for taking pre-defined runtime actions when certain conditions arise. There is also hardware-level support—beyond that for each single machine—focused on certain physical subsystems, such as HP's iLO or IBM's Director solutions for blade centers. Finally, supporting these systems, there are numerous industry standards and tools for representing monitoring information so that it can be transported efficiently and interpreted and read by any tool that needs to do so [29].

Unfortunately, none of the these solutions currently scale to the sizes needed for next generation datacenter systems, which start with 1,000's of machines and are expected to reach multimillion core sizes in the near future. Further, none can deal with the distributed nature of the multiple datacenters operated by Web companies by like Yahoo, Google, or Microsoft, or by IT providers like IBM or HP.

Potential solution approaches using the measurement-based approach to power management advocated in this chapter can take advantage of open source tools for collecting monitoring data and for cluster-level monitoring. The popular tool Ganglia, for instance, uses a hierarchical approach to monitoring in which attributes are replicated within clusters using multicast methods and aggregated via a tree structure [30]. Alternative aggregation structures are evaluated in several related efforts, including [31–33]. Reference [34] supports on demand but not complex queries. Reference [35] constructed on top of Hadoop provides monitoring and analysis for large data-intensive codes and systems, focused on large volumes logs for failure diagnosis. Some systems use gossiping techniques or data structures like DHTs [36] to access distributed monitoring data, and peer to peer infrastructures like Pastry [37] coupled with management and integration software Scribe [38] can be used to construct and manage aggregation structures like those cited above. The outcome is scalable data aggregation and distribution, supporting continuous or one-shot queries via custom or general, hierarchical or peer-to-peer topologies. Not addressed by such efforts, however, are the methods used to analyze and evaluate monitoring data, and the integration of such methods into monitoring systems. Such integration, however, is precisely what is needed for the measurement-based approach to datacenter management described in this chapter.

A measurement-based approach requires us to combine online monitoring with online analytics—termed "monalytics"—leveraging the robustness and scale properties of the methods listed above, but then enhancing them to also (1) combine data collection and aggregation with arbitrary analysis tasks, (2) permit dynamic deployment and reconfiguration of monalytics graphs and operators wherever and whenever they are needed, (3) provide scalability through data local analysis, i.e., by analyzing data at the points where they are captured, whenever possible, and finally, (4) extend such analysis with additional local management functions. An example of the latter is the immediate movement of the VM when a power cap is exceeded, triggered by management software located on the node that detects that violation. Clearly, the earlier discussions of per-VM power metering can aid in such decision making, perhaps by moving the minimum number of VMs (and the right ones) to meet the power cap goal, but there are additional issues that must be taken into account when managing power consumption at datacenter scale, as discussed next.

### *4.4.1 Managing Power Consumption at Datacenter Scale*

In Chaps. 1 and 2 of this book, the datacenter environment is described as a complex electromechanical system in which management goes beyond the IT resources it contains to also including power generation and distribution, multiple IT and facility cooling technologies, and facility aspects like heating or lighting. A management system, therefore, should be extensible to include all of these manageable entities, despite the fact that they exist in different management silos. One approach is to integrate the multiple management silos—e.g., cooling system with IT system management—into one entity. However, this is neither commercially viable, due to market forces, nor is it practical when dealing with such complex systems. There are many other reasons for this fact, involving lack of standards concerning system interactions, the ability to make independent progress with how to better manage individual subsystems, and industry competition and intellectual property in such subsystems or in management methods. The solution we advocate is one that adopts a "coordination-based" approach, in which different subsystems (1) each manage their own resources, but also (2) share select information about internal states relevant to management, and (3) provide external interfaces to receive inputs from other management systems about suitable management actions. An example of such an approach is the CoolIT [3] system design for managing both datacenter cooling and IT resources. CoolIT principles include those of the vManage infrastructure, which uses coordination methods to deal with different management goals, such as those that aim to reduce total power usage vs. those that maintain certain performance levels [10]. Similar work at IBM [11] has created coordination methods between IBM's Director and Tivoli management systems, including the recent addition of datacenter environmental models as a basis for cooling management [5]. As it will be discussed in more detail in Chap. 6, a common metric for such management has been to reduce the energy-centric PUE metric, which represents the energy efficiency of a datacenter, i.e., what fraction of its total energy consumption is directed toward useful work performed by its IT

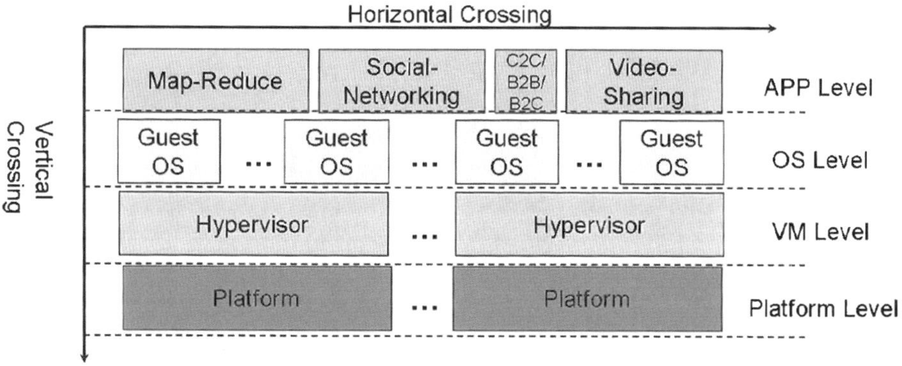


**Fig. 4.6** The functional view of a virtualized datacenter or a utility cloud (© 2010 IEEE, reprinted with permission [33])

resources (e.g., compute, networking, and storage systems) vs. other tasks (e.g., cooling overheads, heating, etc.).

We next further explain the kinds of cross-layer coordination necessary for effective datacenter resource management, based in part on our interactions with datacenter operators.

## *4.4.2 Cross-Layer Coordination*

Figure 4.6 depicts a sample "software stack" present in the large-scale datacenters supporting Web applications. Here, interactive software systems like those used for social networking share resources with applications that analyze Web data and trends in order to improve search results, with search engines and Web portals. Such application software is embedded in sets of VMs each running appropriate software configurations, such as certain versions of operating systems coupled with database software, etc. Supporting these, there are hypervisors or virtualization and management infrastructures like VMWare's ESXServer and VirtualCenter [16] coupled with additional hardware-centric software provided by vendors like HP or IBM [39]. Datacenter software, therefore, spans many machines and operates at multiple levels of abstraction (e.g., OS vs. application level) and consequently, system management involves coordination both across different machines—one might think of this as the "length" scale—and different levels of abstraction—the "time" scale, with monitoring and management actions customized to both. For a rapidly changing subsystem, for instance, such as a Web server that handles millions of requests, it may be desirable to quickly respond to issues like undue response times, in order to prevent clients from perceiving them and/or reduce the spread of problems into other subsystems. For subsystems that fail commonly, such as disks, redundancy methods like those that duplicate valuable data will quickly deal with

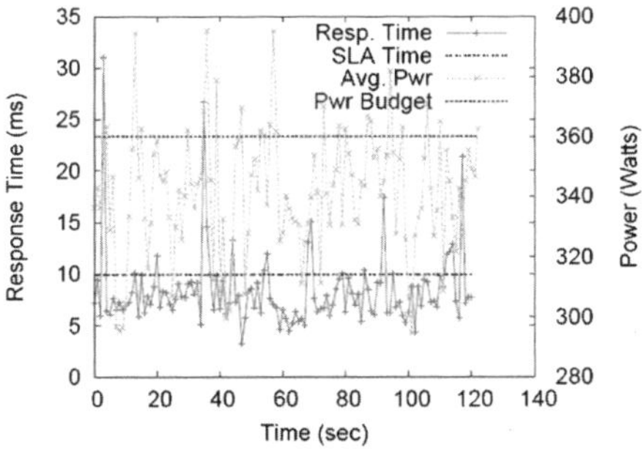


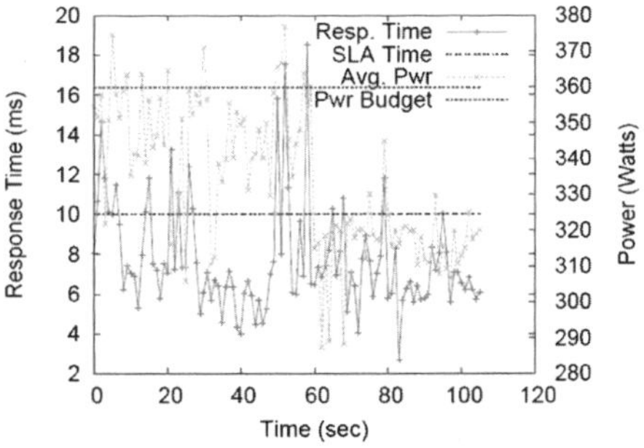


**Fig. 4.7** Need for coordination for improved efficiency of power management algorithms. In the graph on the left, power budget violations are managed by reducing the CPU speed, thereby resulting in increased response time, i.e., increased number of SLA violations. SLA violations are managed by increasing the CPU speeds, thereby triggering new power budget violations. In the graph on the right, the two management entities coordinate, and these oscillations are greatly diminished (© 2009 IEEE, reprinted with permission [10])

losses, and for highly critical systems, there may even be replication across multiple datacenters, to deal with brownouts or catastrophic failures.

The multi-machine, multilayer, and sometimes, even multi-datacenter nature of modern IT systems constitute one reason for a coordination-based approach to management, but another important reason is that there are often multiple coexisting management criteria and metrics. An example reproduced from [10] described below is a case in which a system is managed both to meet power caps and to meet application requirements stated as SLAs (Service-Level Agreements), the latter stated as minimum response times required by the application.

#### 4.4.2.1 Case Study: vManage Management Infrastructure and Methods

As shown in Fig. 4.7, when different entities (and their internal management algorithms) independently try to enforce power caps and response time limits, respectively, oscillatory behavior (see the left part of Fig. 4.7) results, prompting the need for coordination techniques that attempt to correct for this behavior (see the reduced number of threshold violations on the right). In this case, the coordination methods used try to predict whether the target machine to which a VM is moved from an overloaded source machine will continue to have available power (or "performance" to meet response time needs) for some time in the future, to avoid situations in which a VM is moved only to find itself in an overload situation soon thereafter. We refer the reader to [10] for a detailed discussion of the case study and solution methods, but we conclude from these discussions that there is a need for multi-machine or multi-site, and multilevel management methods that coordinate the potentially conflicting actions taken in different management silos. Ongoing research is addressing this need.

## 4.5 Scalable Runtime Analysis of System Behavior

We now turn to a practical and necessary prerequisite for effective online management of datacenter systems, such as the simple example shown above. The topic is the flexible and scalable analysis, where there is a need for both infrastructure—systems and software for carrying out monitoring and analysis—and for methods—techniques and their implementations that analyze the data being captured and aggregated while the system is running—data analytics. We term the combination of both—monitoring and analytics—"monalytics," and we next describe first some simple methods for understanding monitoring data—based on visual inspection—followed by an outline of the state of the art seen in automated methods for understanding program and system behavior. The descriptions are motivated with an example drawn from the following typical hardware layout found in modern datacenter systems, where machines are arranged as a cloud comprised of multiple datacenters, each datacenter consisting of multiple containers, with containers full of racks, enclosures, nodes, then sockets (i.e., into which are plugged memory and processor units), then cores (i.e., processors), and finally, the aforementioned software stack ending in VMs running on those cores.

In the physical hierarchy of a modern virtualized datacenter or a utility cloud illustrated in Fig. 4.8, the numbers are typical quantities of components at each level. The hierarchical relationship between hardware components and virtual environment—datacenter, container, rack, enclosure, node, socket, core, and virtual machines (VMs for short)—is typically configured statically at hardware, but software configuration and deployment (e.g., VM placement and migration) are dynamic. Such systems have the following properties:

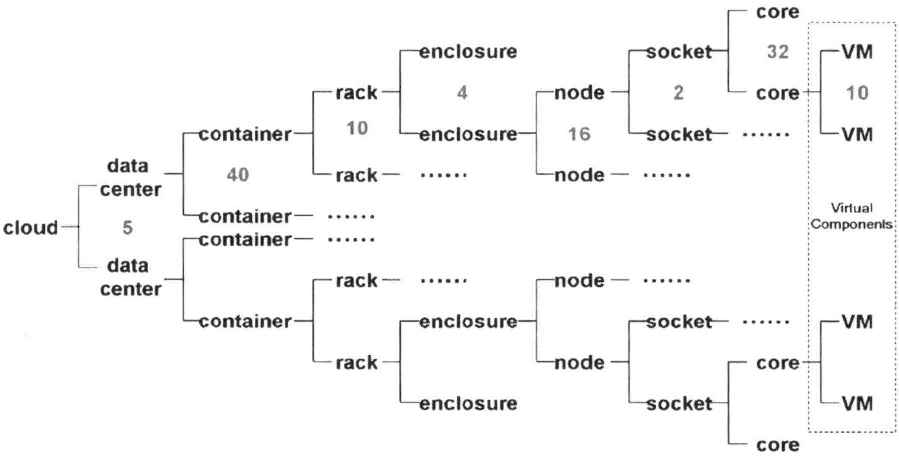


**Fig. 4.8** A cloud hierarchy where the numbers are typical quantities of components (cores, sockets, racks, etc.) to be expected in future clouds. These typical quantities would lead to $5 \times 40 \times 10 \times 4 \times 16 \times 2 \times 32 = 8{,}192{,}000$ physical cores or 81,920,000 virtual machines (© 2010 IEEE, reprinted with permission [33])

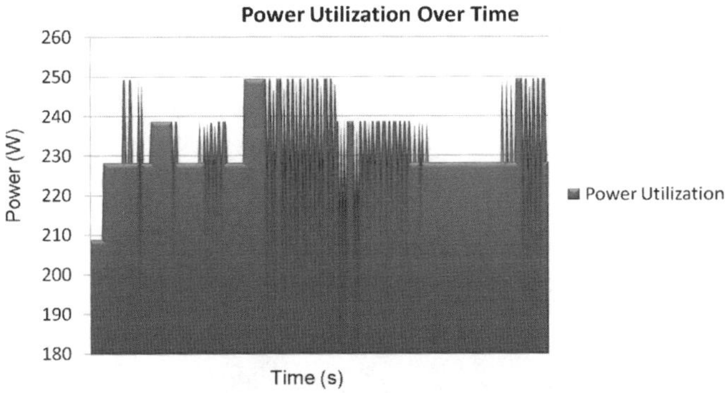


**Fig. 4.9** Power utilization over time for a server running a RUBiS workload

1. *Exascale*. For up to 10M physical cores, there may be up to 10 virtual machines per node (or per core). Given the non-trivial number of sensors per node and additional sensors for each level of physical hierarchy, the total number of metrics present in such systems can reach exascale, $10^{18}$.
2. *Dynamism*. Web applications and more so, future utility clouds support heterogeneous applications that include, but are not limited to, analytical jobs implemented via the Map-Reduce paradigm, "front" facing software interacting with end users including for social networking, supported by multitiered applications with database backends, coexisting with entirely different sets of codes such as those used for data or video streaming, photo sharing and manipulation, and others. While running simultaneously, individual applications will have their own and sometimes unique workload/request patterns, and with modern cloud infrastructures, there will be dynamic application arrivals and departures (e.g., as a company moves software "into" or "out of" the cloud to deal with peak loads).

Dynamism can lead to substantial changes in the power usage for even a single cloud application, as demonstrated by the power measurements for a single Web application depicted in Fig. 4.9 above. Here, we measure the power consumption of a three-tier Web application, termed RUBiS, on a standard server system. Changes in power usage are due to dynamic variations in the workload imposed on the application, driven by varying numbers and behaviors of Web clients. Workloads are meant to emulate the behavior seen on Web sites like eBay [40].

Figure 4.10 shows the concurrently obtained measurements from performance sensors associated with the applications running on the platform, demonstrating the strong correlation between the power usage and request rates seen on the platform (e.g., higher request rates imply larger power usage).

### *4.5.1 Measurement Interfaces and Standards*

More important than the changes observed in the picture is the way in which these metrics are obtained. In response to the diversity of hardware technologies,

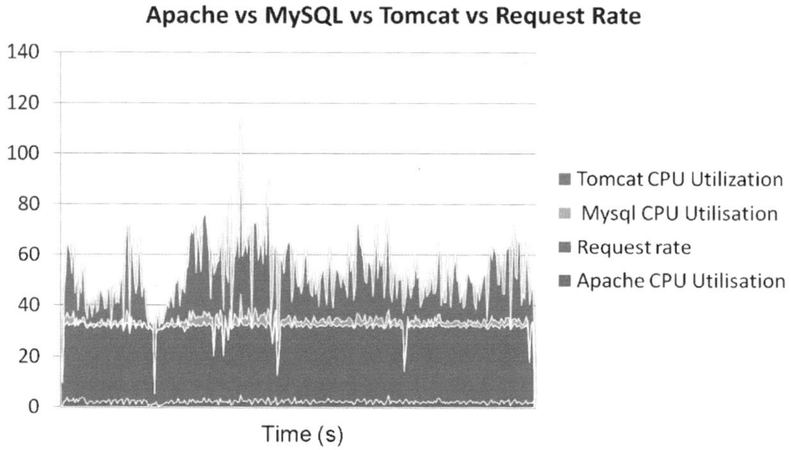


**Fig. 4.10** Variation of request rate (in requests per second) and CPU resource utilization (in percent) for the various RUBiS components

hardware and software vendors, software stacks, management tools, etc., present in modern datacenters, industry has adopted several de facto standard interfaces and protocols to be used for monitoring and management of datacenter platforms. In this section, we describe two such examples used in our work: IPMI (Intelligent Platform Management Interface) and SNMP (Simple Network Management Protocol).

#### 4.5.1.1 IPMI

Modern motherboards, typically for server platforms, embed so-called Baseboard Management Controllers (BMC) that interact with various sensors and gathers various events about the platform state—e.g., temperature, fan speeds, operating system state, etc. The BMC manages the interface between system management software and the platform management hardware, and provides autonomous monitoring, event logging, and recovery control.

In order to simplify the interoperability of different management software stacks with the array of vendor-specific BMCs, Intel led a standardization effort which resulted in the IMPI (Intelligent Platform Management Interface) standard. IPMI is a hardware-level interface specification that is "management software neutral" providing monitoring and control functions that can be exposed through standard management software interfaces, such as DMI, WMI, CIM, SNMP, etc. As a hardware-level interface, it sits at the bottom of a typical management software stack, as illustrated in Fig. 4.11 below. IPMI is best used in conjunction with system management software running under the operating system.

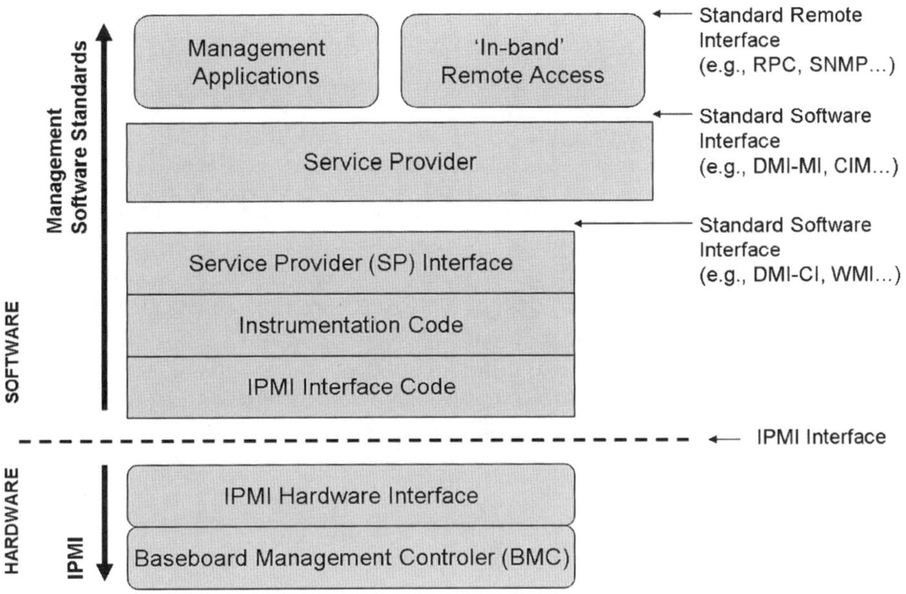


**Fig. 4.11** IPMI management stack

This provides an enhanced level of manageability by providing in-band access to the IPMI management information and integrating IPMI with the additional management functions provided by management applications and the OS. System management software and the OS can provide a more sophisticated control, error handling, and alerting, than can be directly provided by the platform management subsystem.

IPMI supports the extension of platform management by connecting additional management controllers to the system via a "management bus" IPMB (Intelligent Platform Management Bus). IPMB is used for communication to and between management controllers and provides a standardized way of integrating chassis features with the baseboard. Because the additional management controllers are typically distributed on other boards within the system, away from the "central" BMC, they are sometimes referred to as satellite controllers. By standardizing IPMB, i.e., the interconnect, a baseboard can be readily integrated into a variety of chassis that have different management features. IPMI's support for multiple management controllers also means that the architecture is scalable. A complex, multi-board set in an enterprise-class server can use multiple management controllers for monitoring the different subsystems such as redundant power supplies, hot-swap RAID drive slots, expansion I/O, etc., while an entry-level system can have all management functions integrated into the BMC.

Access to monitored information, such as temperatures and voltages, fan status, etc., is provided via the IPMI *Sensor Model*. Instead of providing direct access to the monitoring hardware IPMI provides access by abstracted sensor commands, such as the *Get Sensor Reading* command, implemented via a management controller. This approach isolates software from changes in the platform management hardware implementation. Sensors are classified according to the type of readings they provide and/or the type of events they generate. A sensor can return either an analog or discrete reading. Sensor events can be discrete or threshold based.

The different event types, sensor types, and monitored entities are represented using numeric codes defined in the IPMI specification. IPMI avoids reliance on strings for management information. Using numeric codes facilitates internationalization, automated handling by higher-level software, and reduces management controller code and data space requirements. Tables 4.2 and 4.3 provide examples of the IPMI data gathered for two different entities—fan speed and power supply.

On our platforms, Intel's BaseBoard Management Controller (BMC) offers logging capabilities through system event logs (SEL) that are then cast into the standards-based metrics reported through IPMI. The same interface can also be used to issue commands for clearing, adding, and deleting of events from the SEL. Log records are time-stamped to differentiate the specific data polling interval and are obtained at an average of 1.5sec, which provides sufficient granularity at the scale of management for VM migration. Finally, for remote access to this data, we use the SNMP protocol, which is described next.

#### 4.5.1.2 SNMP

The Simple Network Management Protocol (SNMP) is by far the dominant protocol in network management. It was designed to be an easily implementable basic network management tool that could be used to meet network management needs. SNMP is more generally a protocol for management of network-attached devices on IP networks, including the Internet, and is defined by the Internet Engineering Task Force (IETF). In addition, SNMP is also widely used for management of datacenter network and their interconnected entities, including routers and switches, servers and network appliances, storage, printers, etc.

A key reason for its widespread acceptance, besides being the main Internet standard for network management, is its relative simplicity. The protocol defines management entities—SNMP managers and agents, which interact to manage the various devices on the IP network. The multiple versions of the protocol are all based on few simple operations, like GET and SET, which are used to monitor and modify the state of the managed entity (e.g., server node).

SNMP does not itself define the exact variables used by the management processes. Instead, this information is encoded via Management Information Bases (MIBs). MIB is a data structure that describes all data for which a device can report the status of, or even set its value. MIBs are hierarchical—e.g., a top-level MIB can correspond to a server rack, with internal elements corresponding to different servers, power supplies, and other rack components.

**Table 4.2** Sensor data for 1 of the 16 fans on a Dell server rack

| | |
|---|---|
| Sensor ID | FAN MOD 1A RPM (0 × 30) |
| Entity ID | 7.1 |
| Sensor type (analog) | Fan |
| Sensor reading | 8,475 (±75) RPM |
| Status | ok |
| Lower non-recoverable | na |
| Lower critical | 4275.000 |
| Lower non-critical | na |
| Upper non-critical | na |
| Upper critical | na |
| Upper non-recoverable | na |
| Assertion events | |
| Assertions enabled | cr- |
| Deassertions enabled | cr- |

**Table 4.3** Sensor data record for the power supply

| | |
|---|---|
| Sensor ID | Status (0 x 64) |
| Entity ID | 10.1 (Power supply) |
| Sensor type (discrete) | Power supply |
| States asserted | Power supply [Presence detected] |
| Assertion events | Power supply [Presence detected] |
| Assertions enabled | Power supply [Presence detected] [Failure detected] [Predictive failure] [Power supply AC lost] |
| Deassertions enabled | Power supply [Presence detected] [Failure detected] [Predictive failure] [Power supply AC lost] |

### *4.5.2 State of the Art in Datacenter Analytics*

Visual inspection, as shown above, is a technique commonly used in today's datacenters, enhanced with the threshold-based approaches discussed next. For future systems, however, visual inspection and threshold-based techniques must be enhanced with automated methods that (1) detect problems or unusual behaviors and (2) mitigate or correct them. We complete this chapter with a brief summary of the state of the-art in techniques for monitoring and analysis of behavior of large-scale datacenter systems.

#### 4.5.2.1 Threshold-Based Approaches

Threshold-based methods are pervasively leveraged in industry monitoring products [30, 41–44]. Using their knowledge about datacenter applications and their SLAs, operators first set up upper/lower bounds for each metric being monitored. These can be set statically or dynamically. Whenever any of the metric observations violates a threshold, an alarm is triggered. Providing a moderate volume of metrics to operation teams with highly trained expertise, threshold-based methods are widely used, with advantages of simplicity and ease of visual presentation. They have several shortcomings for modern datacenter systems and cloud applications:

1. *Incremental False Alarm Rate (FAR).* Consider a threshold-based method monitoring $n$ metrics: $m_1, m_2 \ldots m_n$. For each metric (e.g., per node power reading, per node CPU load, per-VM memory utilization, etc.) $m_i$, if the false alarm rate FAR is $r_i$, then the overall false alarm rate of this method is the sum of all $r_i$'s. Thus, this means that when monitoring 50 metrics with FAR 1/250 each (i.e., 1 false alarm every 250 samples), there will be 50/250 = 1/5, i.e., 1 false alarm every 5 samples! Thus, the false alarm rate in threshold-based technique grows too fast with an increase in the number of monitoring metrics.
2. *Detection after the Fact.* Consider 100 Web Application Servers (WAS) running the same service, with a common deficiency in their code, e.g., *memory leaks*. The memory utilization metrics from all those WASes may stay below their thresholds for a period of time as memory use is slowly increasing. Thus, no anomaly is detected. However, when one of the WASs raises an alarm because it crosses the threshold, it is likely that all other 99 WASs will raise alarms soon thereafter, thereby causing a reporting avalanche and likely, leading to disastrous consequences for the running application.
3. *Poor Scalability.* It is obvious that as we approach exascale volumes in monitoring metrics, it is no longer efficient to monitor metrics individually.
4. *Generality.* Generally useful methods should operate for many of the metrics present in datacenter systems or applications, at least in terms of the first step being taken to distinguish normal from abnormal operation and behavior. Methods specific to certain metrics or scenarios can then follow such general techniques.

The lesson we learn from threshold-based approaches is that detecting anomalies by visual inspection of individual metric values or by simple threshold testing may not work well for future systems. Needed are new detection gauges and novel ways of aggregating and analyzing metrics, typically termed "anomaly detection" methods. We note however, that there remains a substantial difference between detecting some anomalous system behavior, i.e., being able to say that "something unusual is occurring," and being able to act on that behavior. This is because an anomaly can be due to many causes, only some of which may correspond to actual system failures that can be corrected or mitigated [45]. There is a need, therefore, to better couple anomaly detection with failure (performance or reliability failures) detection and correction techniques.

Table 4.4 Typical statistical approaches and their features. H-crossing means aggregating metrics among peer distributed components. V-crossing means aggregating metrics cross-layer in the hardware-software implementation stack (© 2010 IEEE, reprinted with permission [33])

| Technique | Function | Scalability | Online | Black-box | H-crossing | V-crossing |
|---|---|---|---|---|---|---|
| SLIC [46] | Problem determination | n | n | n | y | y |
| Nesting/ Convolution [47] | Performance debugging | n | n | y | y | n |
| Pinpoint [48] | Problem determination | n | n | n | y | n |
| Magpie [49] | Performance debugging | n | n | y | y | n |
| Pranaali [50] | SLA management | y | y | n | y | n |
| E2EProf [51] | Performance management | n | n | y | n | n |
| SysProf [52] | Performance management | n | n | y | y | n |
| Sherklock [53] | Problem determination | n | n | y | y | n |
| EbAT [33] | Anomaly detection | y | y | y | y | y |

#### 4.5.2.2 Statistical Methods

There exist many promising methods for anomaly detection, typically based on statistical techniques. However, only a few of them can deal with the scale of future cloud computing systems and/or the need for online detection, because they use statistical algorithms with high computing overheads and/or onerously mine immense amounts of raw metric data without first aggregating it. In addition, they often require prior knowledge about application service-level objectives (SLOs), service implementations, request semantics, or they solve specific problems at specific levels of abstraction (i.e., metric levels). A summary of features of well-known statistical methods appears in Table 4.4. In the table, we also compare those methods with a novel statistical technique based on entropic methods that has better scalability properties, termed EbAT and described in [33].

Discussing these methods, Cohen et al. [46] developed an approach in the SLIC project that statistically clusters metrics with respect to SLOs to create system signatures. Chen et al. [48] proposed Pinpoint, which uses clustering/correlation analysis for problem determination. Magpie [49] is a request extraction and workload modeling tool. Aguilera et al's [47] nesting/convolution algorithms are black-box methods to find causal paths between service components. Arpaci-Dusseau et al. [54] develop gray-box techniques that use information about the internal states maintained in certain operating system components, e.g., to control file layout on top of FFS-like file systems [55]. Concerning datacenter management, Agarwala et al. [51, 52] propose profiling tools, E2EProf and SysProf, which can capture monitoring information at different levels of granularity. The methods shown focus on identifying relationships among VMs [22] rather than anomalies. Kumar et al. [50] proposed Pranaali, a state-space approach for SLA management of distributed software components. Bahl et al. [53] developed Sherlock using an inference graph model to auto detect/localize internet services problems. In comparison, our EbAT (Entropy-based Anomaly Testing) method [33] aims to address the scalability needs of future systems by providing an online lightweight technique that can operate in a black-box manner across multiple horizontal and vertical metrics. We leverage entropy-based analysis technique that has been used in the past for network monitoring [56, 57], but we adapt them for datacenter monitoring operating at and across different levels of abstraction, including applications, middleware, operating systems, virtual machines, and hardware.

## 4.6 Chapter Summary

This chapter describes the basic elements of a measurement-based approach to power management of virtualized datacenter and cloud systems. First, what is needed are lightweight and efficient mechanisms to dynamically estimate the power usage of individual VMs running on such platforms, as well as entire sets of VM executing on behalf of a single client or application. Second, this information is then used in distributed methods for load deployment and redistribution, needed to meet required power caps and power distribution properties across the entire datacenter, e.g., so as to control the thermal properties in the datacenter facility and ensure efficiency of the cooling infrastructure. Such methods for datacenter- and cloud-scale power management rely on continuous monitoring and analysis of hardware- and software-level events across the entire system. Toward this end, third, we also discuss the state of the art in datacenter monitoring, along with a discussion of measurement and analysis methods that can help automate the detection of anomalous or critical behaviors across arbitrary sets of datacenter hardware and software components, such as those which can lead to failures or sudden spikes in power usage. In addition, the chapter also includes a brief survey of related efforts in industry and academia targeting similar problems.

## References

1. Kansal A, Kothari N, Bhattacharya A (2010) Virtual machine power metering and provisioning. In: ACM symposium on cloud computing SOCC, 2010
2. Bohra A, Chaudhary V (2010) VMeter: power modelling for virtualized clouds. In: Workshop on high-performance, power-aware computing (HPPAC), in conjunction with the international parallel and distributed processing symposium, 2010

3. Nathuji R, Somani A, Schwan K, Joshi Y (2008) CoolIT: coordinating facility and IT management for efficient datacenters. In: HotPower, in conjunction with the USENIX annual technical conference, 2008
4. IBM, Syracuse University New York State to build one of the world's most energy-efficient datacenters, http://www-03.ibm.com/press/us/en/pressrelease/27612.wss, 2009
5. Das R, Kephart JO, Lenchner J, Hamann H (2010) Utility-function-driven energy-efficient cooling in datacenters. In: International conference on autonomic computing, June, 2010
6. HP datacenter management solution reduces costs by 34 percent, http://www.hp.com/hpinfo/newsroom/press/2007/070625xa.html, 2007
7. IBM helps clients "meter" datacenter power usage to help lower energy costs, http://www-03.ibm.com/press/us/en/pressrelease/19695.wss, 2006
8. Nathuji R, Schwan K (2008) VPM tokens: virtual machine-aware power budgeting in datacenters. In: Proceedings of high performance and distributed computing HPDC, 2008
9. Samadiani E, Amur H, Krishnan B, and Schwan K. Coordinated Optimization of Cooling and IT Power in Datacenters. ASME Journal of Electronics Packaging, 2010
10. Kumar S, Talwar V, Kumar V, Ranganathan P, Schwan K (2009) vManage: loosely-coupled platform and virtualization management in datacenters. In: International conference on autonomic computing ICAC, 2009
11. Kusic D, Kephart J, Hanson J, Kandasamy N, Jiang G (2008) Power and performance management of virtualized computing environments via lookahead control. In: International conference on autonomic computing ICAC, 2008
12. Agarwala S, Routray R, Uttamchandani S (2008) ChargeView: an integrated tool for implementing chargeback in IT systems. In: Network operations and management symposium NOMS, 2008
13. Corrigan K, Shah A, Patel C (2010). Estimating environmental costs. In: SustainIT – workshop on sustainable information technology, 2010
14. Ferrero RW, Shahidehpour SM, Ramesh VC (1997) Transactional analysis in deregulated power systems using game theory. IEEE Transactions on power systems 12(3):1340–1347
15. Clark C, Fraser K, Hand S, Hasen JG, Jul E, Limpach C, Pratt I, Warfield A (2005). Live migration of virtual machines. In: Proceedings of the symposium on networked systems design and implementation (NSDI), May 2005
16. The VMware ESX Server. http://www.vmware.com/products/esx/
17. Microsoft Hyper-V Server. http://www.microsoft.com/hyper-v-server/
18. Xen Open Source Hypervisor. http://www.xen.org
19. Citrix XenServer. http://www.citrix.com/xenserver
20. Amazon Elastic Compute Cloud (EC2). http://aws.amazon.com/ec2/
21. Open Cirrus HP/Intel/Yahoo open cloud computing research testbed, http://opencirrus.org
22. Apte R, Hu L, Schwan K, Ghosh A (2010) Look who's talking: discovering dependencies between virtual machines using CPU utilization. In: HotCloud, in conjunction with USENIX annual technical conference, 2010
23. Carter J (2010) Looming challenges in server and datacenter energy efficiency. One researcher's perspective. In: CERCS energy workshop, 2010
24. Koller R, Verma A, Neogi A (2010) WattApp: an application aware power meter for shared datacenters. In: International conference on autonomic computing, 2010
25. Menon A, Santos JR, Turner Y, Janakiraman G, Zwaenepoel W (2005) Diagnosing performance overheads in the Xen virtual machine environment. In: USENIX annual technical conference, 2005
26. Mahadevan P, Sharma P, Banerjee S, Ranganathan P (2009) A power benchmarking framework for network devices. In: Networking, 2009
27. Krishnan B, Amur H, Gavrilovska A, Schwan K (2010) VM power metering: feasibility and challenges. In: GreenMetrics, 2010
28. Lim MY, Porterfield A, Fowler R (2010) SoftPower: fine-grain power estimation using performance counter. In: High performance and distributed computing HPDC, 2010

29. Common information model distributed management task force CIM-DMTF. http://www.dmtf.org/standards/cim
30. Massie ML, Chun BN, Culler DE (2004) The ganglia distributed monitoring system: design, implementation and experience. Parallel Computing 30:817–840
31. Renesse RV, Birman KP, Vogels W (2003) Astrolabe: a robust and scalable technology for distributed system monitoring, management and data mining. ACM Transactions on Computer Systems 21(2):164–206
32. Strom R, Banavar G, Chandra T, Kaplan M, Miller K, Mukherjee B, Sturman D, Ward M (1998) Gryphon: an information flow based approach to message brokering. In: International symposium on software reliability engineering, 1998
33. Wang C, Talwar V, Schwan K, Ranganathan P (2010) Online detection of utility cloud anomalies using metric distributions. In: Network operations and management symposium NOMS, 2010
34. Liang J, Ko SY, Gupta I, Nahrstedt K (2005) MON: on-demand overlays for distributed systems management. In: Workshop on real, large distributed sytems, 2005
35. Boulon J, Konwinski A, Qi R, Rabkin A, Yang E, Yang M (2008) Chukwa: a large scale monitoring system. In: Cloud computing and its applications, 2008
36. Ko SY, Yalagandula P, Gupta I, Talwar V, Milojicic D, Iyer S (2008) Moara: flexible and scalable group based querying systems. In: International conference on middleware, 2008
37. Druschel P, Rowstron A (2001) Pastry: scalable, distributed object location and routing for large-scale peer-to-peer systems. In: International conference on distributed systems platforms (Middleware), 2001
38. Rowstron A, Kermarrec A.-M, Castro M, Druschel P (2001) SCRIBE: the design of a large-scale event notification infrastructure. In: International workshop on networked group communication, 2001
39. HP integrated lights out (iLO). http://www.hp.com/go/ilo
40. Cecchet E, Marguerite J, Zwaenepoel W (2002) Performance and scalability of EJB applications. In: International symposium on object oriented programming, systems, languages and applications OOPSLA, 2002
41. HP systems insight manager, http://www.hp.com/go/sim
42. IBM Tivoli, http://www.ibm.com/tivoli
43. Nagios – industry standard in open source monitoring, http://www.nagios.org/
44. Nimsoft, http://www.nimsoft.com
45. Candea G, Kawamoto S, Fujiki Y, Friedman G, Fox A (2004) Microreboot – a technique for cheap recovery. In: Symposium on operating systems design and implementation OSDI, 2004
46. Cohen I, Goldszmidt M, Kelly T, Symons J, Chase JS (2004) Correlating instrumentation data to system states: a building block for automated diagnosis and control. In: Symposium on operating systems design and implementation *OSDI*, 2004
47. Aguilera MK, Mogul JC, Wiener JL, Reynolds P, Muthitacharoen A (2003) Performance debugging for distributed systems of black boxes. In: Symposium on operating systems principles SOSP'03, 2003
48. Chen MY, Kiciman E, Fratkin E, Fox A, Brewer E (2002) Pinpoint: problem determination in large, dynamic internet services. In: Dependable systems and networks DSN, 2002
49. Barham P, Donnelly A, Isaacs R, Mortier R (2004) Using magpie for request extraction and workload modelling. In: Symposium on operating systems design and implementation OSDI'04, 2004
50. Kumar V, Schwan K, Iyer S, Chen Y, Sahai A (2008) A state-space approach to SLA-based management. In: Network operations and management symposium NOMS, 2008
51. Agarwala S, Alegre F, Schwan K, Mehalingham J (2007) E2EProf: automated end-to-end performance management for enterprise systems. In: Dependable systems and network DSN, 2007

52. Agarwala S, Schwan K (2006) SysProf: online distributed behavior diagnosis through fine-grain system monitoring. In: International conference on distributed computing systems ICDCS, 2006
53. Bahl P, Chandra R, Greenberg A, Kandula S, Maltz DA, Zhang M (2007) Towards highly reliable enterprise network services via inference of multi-level dependencies. In: SIGCOMM, 2007
54. Arpaci-Dusseau AC, Arpaci-Dusseau RH (2001) Information and control in gray-box systems. In: Symposium on operating systems principles SOSP, 2001
55. Nugent JA, Arpaci-Dusseau AC, Arpaci-Dusseau RH (2003) Controlling your place in the file system with gray-box techniques. In: USENIX annual technical conference, general track, 2003, pp. 311–323
56. Lakhina A, Crovella M, Diot C (2005) Mining anomalies using traffic feature distributions. In: SIGCOMM, 2005
57. Lakhina A, Crovella M, Diot C (2004) Diagnosing network-wide traffic anomalies, In: SIGCOMM, 2004

# Chapter 5
# Data Center Monitoring

**Prajesh Bhattacharya**

**Abstract** Since the early days of data centers, the data center operators have been challenged with device placement, capacity planning, equipment maintenance, and failure & downtime. Lately, things have become even harder for them due to the spotlight on low energy efficiency and low utilization of available capacity. The idea is that more needs to be done with less, while maintaining the same level of service. While piecemeal and inadequate attempts have been made from time to time, to address some of these challenges, the key to solving all of these problems lies in combining IT and facility monitoring, or even just facility monitoring. Therefore, in this chapter, first we briefly discuss some of the aforementioned issues. Then we get into the details of what physical quantities should be monitored in a data center and what value that can bring. We focus on the cooling distribution chain and the power distribution chain. Since power monitoring is a very popular topic and a major portion of it deals with power quality, a separate section has been dedicated on brief explanations of different power quality issues. Finally, we discuss the requirements for the software infrastructure that is needed to support this type of holistic monitoring.

## 5.1 Introduction

While most of the other chapters in this book delve into the fundamental theories behind the respective topics, this chapter deals with the implementation side. Interested readers can refer to ref. [1] for the fundamentals on this topic. Monitoring on the Information Technology (IT) side of the data center is a well-developed and well-implemented field. Large companies such as IBM, HP, and Computer

P. Bhattacharya, PhD (✉)
Lawrence Berkeley National Laboratory, One Cyclotron Road, MS 90R3111, Berkeley, CA 94720-8134, USA
e-mail: pbhattacharya@lbl.gov

Y. Joshi and P. Kumar (eds.), *Energy Efficient Thermal Management of Data Centers*, DOI 10.1007/978-1-4419-7124-1_5,

Associates have extensive product suites that address IT monitoring. On the same token there are well-known and widely used open source tools as well, such as Nagios [2] and Ganglia [3]. Although IT is the reason behind the mere existence of the data center, non-IT assets and operations comprise a large portion of the data center assets and operations. It is well understood by the data center owners, operators, and vendors why IT monitoring is important. However, the rationale behind the monitoring of the facilities side is still not well understood by the community. Not to mention, the value of combining IT monitoring with facility monitoring is far from being explored [4]–[7]. Historically, data center operators have had several difficulties that can be solved by combining IT and facility monitoring, or even just by facility monitoring. Therefore, in this chapter, first we will briefly discuss some of the issues that can be addressed by either only facility monitoring or a combination of IT and facility monitoring. Then we will get into the details of what should be measured in a data center and what value that can bring. Since power monitoring is a very popular topic and a major portion of it deals with power quality, a separate section has been dedicated on brief explanations of different power quality issues. Finally, we discuss the type of software infrastructure that will be needed for this.

## 5.2 Why Monitor

Let us look at some of the challenges that data centers have been plagued with.

### *5.2.1 Device Placement*

Device placement will probably be the first one up the rank. Where should we place the new servers or new racks? There has to be enough space, enough power, enough cooling, and not to mention ease of software administration and hardware maintenance. There are Configuration Management Database (CMDB) software tools that can be used to manually keep track of locations. Then there are the new generation of tools that provide information on power consumption/availability of circuits and temperature prevalent in specific areas. There is another type of tools that helps understand the cooling behavior in a what-if scenario. But these tools, individually, do not solve the actual problem; instead they add extra work for the data center staff. The capabilities of these tools should be combined so that with minimal manual intervention, the data center staff can obtain an optimal solution for device placement. The device placement methodology should minimize Total Cost of Ownership (TCO) while satisfying the constraints, some of which were described above.

### 5.2.2 Capacity Planning

If device placement ranks 1st, capacity planning would probably rank 2nd in the list of most challenging and lingering issues in the data center industry. With increasing power density of the servers and racks, power capacity and cooling capacity have taken the front seat in the challenge of capacity planning. However, capacity planning should start from the capacity planning of the applications that are hosted in a data center. Today's customary solution for capacity planning is to either start with a lot of capacity and then slowly use that capacity or to run out of capacity and not able to expand in time. Only a handful of visionaries explore the concept of modular increment in capacity on an as needed basis. This requires careful forecasting and results in reduction in unutilized asset cost, yet, uninterrupted service.

### 5.2.3 Equipment Maintenance

It is a well-known fact in any industry dealing with expensive and mission critical equipment that fault prognosis and Condition Based Maintenance (CBM) of equipment is a much more cost-effective and low-risk maintenance policy than either a wait-for-the-failure approach or schedule-based preventive maintenance. The first one leads to costly and disastrous downtimes and the second one leads to unnecessary expenditure on maintenance or sometimes unexpected early failure that results in costly downtime. CBM minimizes maintenance cost and almost eliminates the possibility of equipment failure.

### 5.2.4 Capacity Utilization

Most data center operators tend to maintain more headroom for the data centers' power and cooling capacity than that is necessary. And the methodology to arrive at the buffer level is not very scientific. For example, the peak power consumption used in the calculation might have come from manual readings off of the power meter taken at 1:30 PM on the last Wednesdays of the month over the last year. First of all, the data used in calculations should be more scientific, and also, the calculation methodologies should be more scientific. It should involve trend analysis, cause-effect analysis, and finally forecasting. This leads to (a) greater utilization of existing capacity, resulting in more revenue; (b) reduced risk of capacity overshoot, meaning avoidance of downtime cost; and (c) deferred capital investment that would have otherwise been required for premature capacity expansion.

### 5.2.5 Failure and Downtime

The first order check against equipment failure and downtime is to send notifications about critical conditions to relevant people. This is already done in the industry, but based on only real-time data. The algorithm behind the notification trigger should include rigorous analysis and proper forecasting in order to avoid false positives. False positives reduce the importance of alarms. The second level of action against failures will involve automated failure prevention or graceful exit. This second method not only avoids or minimizes the effect of failure but also minimizes manual intervention. Costs associated with both failure and manual labor are significant, which can be avoided with sophisticated algorithms. However, in case of eventual failure, automated and accurate root-cause analysis should be performed scientifically as opposed to manually guessing the cause. This can help in avoiding future failures.

### 5.2.6 Energy Efficiency

The customary approach for data center cooling has been to overcool the room 24 $\times$ 7 to cover the worst case cooling situation for a given piece of IT equipment. A slightly better approach is to perform on-demand cooling. A few vendors are currently planning advanced controls for just CRAH or CRAC units. However, what we need is an algorithm that optimizes the data center as a single system for the lowest energy cost. That would involve considering multiple *inputs* such as ambient conditions, power, and temperature reported directly from the IT equipment and the IT workload; and several *setpoints* such as CRAH setpoint, chilled water setpoint, and cooling tower setpoint for every piece of equipment in each category.

### 5.2.7 Future Purchasing Decision

The state of the art in data center equipment purchase decision making is vendor brochure/catalog, prior experience with the vendor, and personal relationship with the salesperson. The lack of scientific approach results in poor uptake of better technologies, vendor lock-in, and other undesirable situations. Instead, decisions should be made based on a comparative study of the candidate equipment under consideration, which should, in turn, be based on real performance data for each of the candidates. Similar situation is true for the design process of data centers. The design, instead, should be based on performance data of the technology and/or equipment considered.

Solutions to all of the above challenges require reasonable quality measured data. And the measured data can come only from continuous monitoring.

## 5.3 Candidates for Monitoring

### *5.3.1 Cooling System*

The cooling system in today's data centers requires a comprehensive monitoring solution to ensure energy optimization, as well as performance, reliability, and security. The combination of real-time measurements and historic information, along with the ability to analyze, filter, and visualize, is required for data center cooling optimization. *Historic data* allows users to understand the baseline capacity loading, temperature, and power consumption profiles and their variability during normal operation. Rather than using a brief snapshot during a random survey or a 30-min average from a limited number of assets, the ability to capture all of this data and store it for analysis provides the user with a detailed profile of asset utilization history and variability. Once the normal baseline capacity and performance information are understood, *real-time data* can be compared to the "normal" baseline using statistical analyses to determine trends and provide early warning whenever any of these items is out of its normal range. In addition, with data center staff increasingly tasked to do more with less, automation and logic are needed to ensure the equipment runs reliably instead of searching for problems and the ability to predict a failure in advance. This section describes the importance of historic data in efficient and reliable operation of the data center cooling system. Figure 5.1

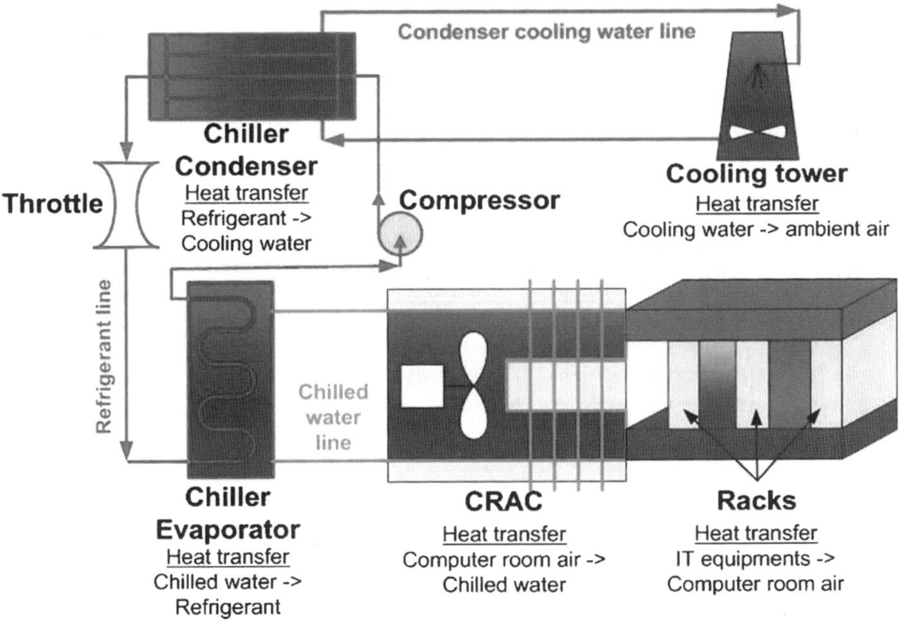


**Fig. 5.1** Data center cooling system schematic diagram

shows a top level schematic view of a data center cooling system. The reader may recollect the primary and secondary loop described in Chap. 2.

(a) Chiller data

As discussed in Chaps. 1 and 2, chillers represent a substantial capital investment and are a major contributor to operating costs for most data centers. For many organizations, chillers are the largest energy users. Therefore, even a slight improvement in the operating efficiency of the chiller translates into substantial energy savings and consequently cost savings. Chiller design efficiencies have improved steadily over the past decade due to advances in controls, refrigerants, and equipment design. Therefore, the main source of chiller inefficiency for the newer chillers is inefficient operation of the chillers. Regular measurement of various chiller-related quantities and keeping a history of that measured data have become extremely important to enhance the chiller operational efficiency [8].

*Quantities to measure*

- Power consumed by the chiller
- Cooling provided by the chiller
- Condenser pressure
- Evaporator pressure
- Water quality on the condenser side
- Water quality on the evaporator side
- Water and refrigerant temperatures at the entrance and at the exit of the evaporator
- Water and refrigerant temperatures at the entrance and at the exit of the condenser
- Purge unit [9] run-time (low-pressure chillers)
- Moisture accumulation at the purge unit (low-pressure chillers) [9]
- Refrigerant charge level
- Pressure drop across the water filters
- Pressure drop across the oil filters

*How to use the data*

Every chiller has a design efficiency, the ratio of the electrical power input to the generated cooling effect (kW/ton), provided by the manufacturer. If the historic data shows that this value is consistently higher than the design value, then one can conclude that the chiller is under-performing. Since the chiller unit consists of several components such as the condenser, evaporator, throttle, compressor, refrigerant, cooling water, chilled water, piping, and the filters, it is important to figure out which component or components are causing this inefficiency. Many times a problem with one component would surface out as an inefficiency of another component.

Refrigerant Approach Temperature (RAT) is the first quantity to be checked for heat exchanger (either evaporator or condenser) inefficiency. The RAT is determined by calculating the difference between the leaving fluid (water) and the saturated temperature of the refrigerant being heated (evaporator) or cooled (condenser). As can be seen in Fig. 5.2, $\mathrm{RAT}_{\mathrm{condenser}} = T_{\mathrm{c}} - T_{\mathrm{cooling,out}}$ and

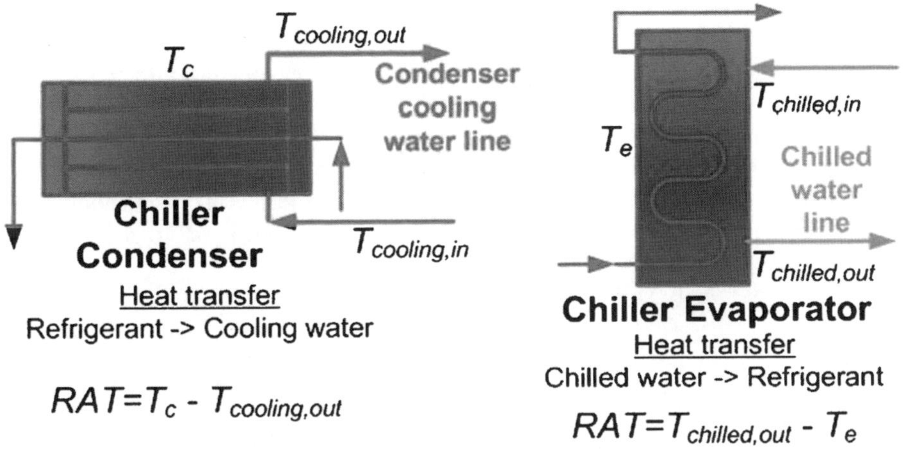


**Fig. 5.2** Definition of Refrigerant Approach Temperature (RAT)

$\mathrm{RAT}_{\mathrm{evaporator}} = T_{\mathrm{chilled,out}} - T_{\mathrm{e}}$, where $T_{\mathrm{c}}$ is the refrigerant temperature inside the condenser, $T_{\mathrm{e}}$ is the refrigerant temperature inside the evaporator, $T_{\mathrm{cooling,out}}$ is the temperature of the cooling water leaving the condenser, and $T_{\mathrm{chilled,out}}$ is the temperature of the chilled water leaving the evaporator.

Every chiller has a design RAT. When it is exceeded on a regular basis (which can be known only from the historic data), one can conclude that there is a problem with the heat exchange process in the chiller. Possible causes of high RAT include the following.

- Low refrigerant level
- Noncondensable gasses
- Low/high flow rates
- Part loads at low Entering Condenser Water Temperature (ECWT)
- A scaled or fouled heat exchanger

For a high-pressure chiller system, refrigerant can leak out of the system. This is hazardous to the environment and would lead to lower chiller efficiency. If the refrigerant charge level decreases over time, then one can conclude that there is a leakage in the system. Otherwise, if the evaporator pressure is less than the design value or if the evaporator temperature is less than the design value, then one should check for leakage in the system.

For a low-pressure chiller, moisture and air are noncondensables that can get into and then remain trapped inside the refrigerant system. Both lead to increased condenser pressure, increased compressor power and reduced chiller efficiency. Trapped moisture can also create acid inside the system that can corrode different chiller components. Moisture creates rust on the inside walls of the components, which reduces heat transfer coefficient and consequently the chiller efficiency. The fine rust particles flowing through the system at a high velocity can impinge on the component wall to permanently damage the components. Therefore, if the measured condenser pressure or temperature are consistently greater than the design value, one should check for noncondensables and fouling inside the condenser tubes. Another indication of the existence of moisture inside the refrigerant system is the purge-unit run-time. The purge-unit purges out trapped moisture from time to time. If it runs too frequently (check the total run-time over the last week, last month, last 3 months, last 6 months, etc.), then one should look for moisture leakage into the system.

Also, typically chiller efficiency is lower at lighter loads. Most data centers have redundant chillers that operate all the time at light loads. It is always advisable to operate fewer chillers at higher loads and to turn on the redundant ones only when it is required. If one has a history of the chiller load, then this redundant chiller start-up does not have to be at a critical time, rather it can very well be scheduled. Thus, historic data helps achieve greater energy efficiency while avoiding any emergency situation.

The fouling of the heat exchanger tubes results in reduced heat transfer coefficient which in turn results in lower chiller efficiency. Also, extreme fouling can lead to permanent damage to the tubes. The rate of waterside fouling depends on the water quality and the operating temperature. If the amount of dissolved chemicals or particles in the water is consistently greater than the recommended high limit, then one should check the water treatment system.

Filters for condenser-water systems are very effective at maintaining clean water, if properly maintained. Maintenance of strainers and filters limits chiller-tube erosion caused by sand or other small particles moving at high velocities. Erosion and tube pitting decreases overall heat-transfer effectiveness and decreases efficiency. If uncorrected, these conditions can lead to plugged tubes or catastrophic tube failure. Greater pressure drop across a water filter indicates accumulation of debris at the filter. A consistent increase in the pressure drop across the water filter indicates greater particle density of the supply water. The filter should be changed or cleaned if that pressure drop exceeds the recommended number for the system.

Greater pressure drop across the oil filter indicates change in the oil characteristics. The change can occur due to various reasons such as oil-moisture contamination, oil-refrigerant contamination, unacceptable compressor wear etc. High moisture content in the oil can signal problems with the purge unit. Oil-refrigerant contamination points to mixing between oil and refrigerant. All these things lead to inferior chiller performance and eventual chiller failure. Therefore, it is very important to monitor the pressure drop across the oil filter on a continuous basis and look for reasons when it starts to increase.

*Stacking* [10] is the abnormal accumulation of refrigerant in the condenser, commonly caused by a decrease in the difference between the condenser pressure and the evaporator pressure. This reduced pressure drop prevents the refrigerant's ability to physically flow back to the evaporator and maintain a normal refrigerant cycle. *Carryover* is the presence of liquid refrigerant droplets in the cold, low-pressure vapor produced in the evaporator. If excessive, carryover can be detrimental to the performance and reliability of a chiller. When this occurs, liquid refrigerant droplets travel from the evaporator to the compressor inlet. As these liquid droplets enter the compressor system they vaporize on the metal internals, stripping away lubricant. Ultimately this oil stripped from the compressor becomes entrained in the hot, high-pressure refrigerant vapor.

Stacking and carryover both affect chiller performance and are not easily detected from a single sensor. *Keeping detailed records* [10] of oil additions between oil changes and checking for leaks when additions are made can make a difference. If no leaks exist, foaming in the evaporator refrigerant sight gauge (manual data logging) can be a sign of high oil levels in the refrigerant. Refrigerant boiling in the compressor oil sight gauge (manual data logging) may also be a sign of stacking and/or carryover.

(b) IT equipment inlet and outlet temperature and air flow data

It is of utmost importance to ensure adequate cooling of the IT equipment. Otherwise, equipment failure is unavoidable during extended operation. Sometimes one might need to increase the cooling at a particular portion of the data center when required. Figure 5.3 shows a schematic of the hot aisle/cold aisle arrangement as described in Chap. 1. In this case, IT equipment inlet air temperature is equivalent to the cold-aisle temperature and the IT equipment outlet temperature is equivalent to the hot aisle temperature. On the other hand, in a data center room with liquid-cooled racks, aisles are not used for cooling. The inlet and outlet temperatures in that case are measured inside the racks.

Nevertheless, cooling is expensive due to the amount of energy it consumes. On many occasions cooling equipment serving a particular segment of a data center continues to operate at the same level even when the power drawn by the IT equipment and consequently the heat dissipation by the same at those areas have gonc down.

The hot air leaving the hot aisle from the top of the aisle is supposed to go to the air handling facility (viz. the CRAC/CRAH unit). As discussed earlier in Chap. 2, the reader may recollect that due to various reasons, a portion of this hot air gets inside the adjacent cold aisle from the top of the aisle and reenters the IT equipment from the front side resulting in recirculation. Since this circulated air is hot, it fails to cool the IT equipment at the top of the rack resulting in possible equipment failure and damage.

If there is a blockage in the air path across the rack, cold air would not be able to flow to the IT equipment. In a controlled environment, the increase in the IT equipment temperature might trigger more air flow from the air handler serving that affected location, resulting in unnecessary power consumption. On the other hand, in an uncontrolled environment (i.e., CRAC/CRAH control is not based on the temperature of the affected area), the temperature rise of the IT equipment would go unnoticed, and it would only be the eventual equipment failure that would draw people's attention.

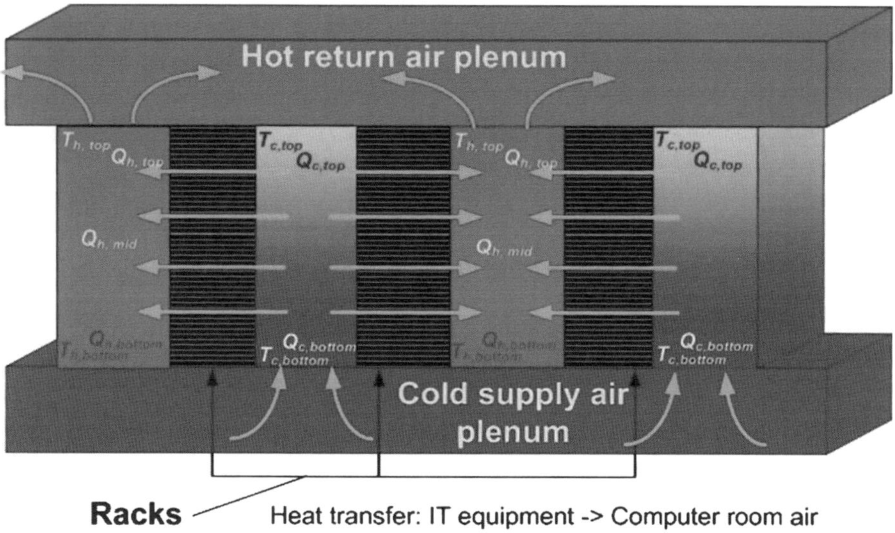


**Fig. 5.3** Schematic view of the air flow inside the computer room

*Quantities to measure*

- Air temperatures at the top and at the bottom of each cold aisle ($T_{\text{c,top}}$ and $T_{\text{c,bottom}}$ in Fig. 5.3) at a minimum, while temperature readings can be obtained from inside each server for better resolution
- Air temperatures at the top and at the bottom of each hot aisle ($T_{\text{h,top}}$ and $T_{\text{h,bottom}}$ in Fig. 5.3) at a minimum, while temperature readings can be obtained from inside each server for better resolution
- Air flow at the top, middle, and at the bottom of each hot aisle ($Q_{\text{h,top}}$, $Q_{\text{h,mid}}$ and $Q_{\text{h,bottom}}$ in Fig. 5.3)
- Air flow at the top and at the bottom of each cold aisle ($Q_{\text{c,top}}$ and $Q_{\text{c,bottom}}$ in Fig. 5.3)

*How to use the data*

Based on power consumption of the IT equipment, one can estimate the rate of heat to be removed from the IT equipment at different operational levels. One can achieve that heat transfer target using different combinations of the supply air temperature (adjacent cold aisle temperature) and the air flow through the IT equipment. If the supply air (cold aisle) temperature is too high, then either the required heat removal cannot be achieved (resulting in equipment failure) or a greater air flow will be required. On the other hand, a low return air (hot air) temperature means one is putting more load either on the chiller or on the CRAC/CRAH unit than is required, resulting in energy wastage. Since the IT load and consequently the heat dissipated by the IT equipment are dynamic in nature, one needs to perform these calculations based on historic data instead of instantaneous values. That way one would not end up with "crazy" control system behavior.

If in a cold aisle the top temperature is consistently greater than the bottom temperature, then one needs to look for possible hot air recirculation at the top of the adjacent racks. The cold aisle air flow is supposed to decrease from the maximum at the bottom to zero at the top of the rack. If there is a persistent cold air flow at the top of the cold aisle, then the air flow path across the rack needs to be checked for air blockage across the rack or lack of driving pressure across the rack (IT equipment fans not working properly). The air flow data measured at the middle of a hot aisle can help detect the location of such problems. If the value of this quantity is consistently less than what it should be, then one can conclude that the problem is at a level below the sensor level.

(c) Humidity control

As discussed previously in Chaps. 1 and 2, the reader may recollect that humidity in a computer room cannot be excessively low because of increased risk of electrostatic discharge (ESD). At the same time, a very high humidity could cause corrosion; electrical short-circuits on the Printed Circuit Board (PCB); and delamination of the PCB among other things. Thus, both low and high humidity are unacceptable to computer rooms and hence the humidity needs to be controlled.

There are four types of humidifiers—spray pad, electric steam, ultrasound, and infrared [11, 12]. Each type has its relative advantages and disadvantages. Humidifiers can be local and attached to the individual CRAC units, or they can be centralized. Despite the different mechanism and configuration options, the basic concept behind a humidifier is the same. Some amount of water is consumed; it is either evaporated or boiled off into steam and then thrown into the moving air; and this process consumes some amount of energy.

As we have seen in Chap. 2, dehumidification, involves subcooling of the computer room air below the dew point so that some of the moisture contained in the air will get condensed on the cooling coil and then a possible reheating of it to take it to the desired supply air temperature. In order to subcool the air, either the chilled water temperature has to be lower or the chilled water flow has to be more relative to if the dehumidification were not required. Either of these means greater load on the chiller/chilled water system, resulting in higher power consumption. Thus, dehumidification consumes energy at two levels—first, at the chiller/chilled water system and second, at the reheater.

Therefore, it is always advisable to run humidification/dehumidification as little as possible, which would mean it should be controlled over a "set-range" rather than a setpoint. Also, it is important to know why the humidification or the dehumidification was required during a particular time frame. Was there an introduction of outside air during that time or was there nothing special that happened? If nothing special happened, then the reason for humidification might have been prior dehumidification and vice versa.

*Quantities to measure*

- Absolute humidity of the CRAC/CRAH supply air
- Absolute humidity of the CRAC/CRAH return air

- RH of the outside air
- Makeup air flow rate
- Absolute humidity of the central humidity control supply air (if applicable)
- Water consumed by the humidifier
- Power consumption of the humidifier
- Time of operation of the humidifier
- Water condensed on the cooling coils (dehumidification)
- Power consumption by the reheater
- Time of dehumidification

*How to use the data*

By having the historic absolute humidity data of the CRAC/CRAH supply air, one can conclude if high/low humidity was one of the reasons for a device failure. If there is a consistent difference between the absolute humidity of the CRAC/CRAH supply air and CRAC/CRAH return air, one has to check if there is any air leakage or if there is any source of humidity in the air path. One can estimate the amount of humidification or dehumidification required based on the makeup air flow rate, moisture content of the outside air, temperature of the outside air, and the moisture content and the temperature of the CRAC/CRAH supply air. If this estimated number is consistently different from the actual number (which is either the amount of water consumed by the humidifier in case of humidification or the amount of water condensed on the cooling coils in case of dehumidification), then one needs to look for the existence of humidifying or dehumidifying agents in the air path.

(d) CRAC/CRAH data

CRAH units (CUs) are air handlers with slightly different purpose and design specifications than air handlers used in traditional building air-conditioning. CUs of smaller capacity used at smaller computer rooms can reject heat directly to the chiller evaporator in which case the chiller would have a roof-top type condenser. However, in most cases the CU rejects heat into the chilled water system which in turn rejects heat to the chiller evaporator. The fan and the humidity control unit are the only power-consuming components of the CU. Therefore, efficient operation of these two is important for the efficient operation of the CU. There are different factors because of which the CU can become inefficient such as improperly controlled or uncontrolled air flow and the improperly controlled humidity control. Since the topic of humidity has already been discussed in Section 5.3.1c, we will skip it in this subsection.

*Quantities to measure*

- Return air temperature
- Supply air temperature
- Fan speed
- Air flow (CFM)
- Fan power
- Chilled water inlet temperature

- Chilled water outlet temperature
- Chilled water mass flow rate
- Air temperature upstream of the humidity control station

*How to use the data*
Most data centers operate the CUs (most of the times overdesigned) at a constant speed at all times only to overcool the room. One can save considerable energy by operating the CU in such a way that it provides only the CFM required to keep the IT equipment temperature below the prescribed limit. This would need either real-time controlling or scheduling of the CRAC fans. Because of the dynamic nature of the IT load and consequently the IT power, a control system based only on instantaneous data would make the system unstable. The control system needs to perform its calculations based on sufficient amount of historic data (predictions based on machine learning). The other option, scheduling, also requires knowledge about the history of the IT load and the IT power consumption.

Based on the chilled water inlet and outlet temperatures and the mass flow rate, one can calculate the heat absorbed by the chilled water. On the other hand, the CRAC inlet air temperature, air temperature upstream of the humidity control station, and the fan CFM will provide one with the heat rejected by the air. Ideally, the two numbers should match. However, if the chiller number is consistently greater than the air heat rejection number, one can conclude that there is some heat loss in the CRAC.

One can calculate the CU efficiency from the measured data over a period of time and then compare it with the vendor-provided efficiency. Also, historic data on the CFM and the fan power would enable one to verify the fan curve provided by the supplier and decide on the regime in which the CRAC fan should be operated. That would reveal if the CU is operating as per the supplier's promise.

(e) Cooling tower data
The performance of a cooling tower degrades when the efficiency of the heat transfer process declines. Some of the common causes of this degradation include [10] scale deposition on the evaporation surface, clogged cooling water spray nozzles, poor air flow, poor pump performance, etc.

*Quantities to measure*

- Water treatment chemical residual
- Pressure drop across the strainer in the basin
- Makeup water flow
- Vibration in pumps and fans
- Pressure drop along the cooling water line
- Pump power
- Fan power

*How to use the data*
If the concentration of the water treatment chemical residual is consistently in excess of its design value, one should address the matter with seriousness,

because these chemicals can corrode the heat transfer surfaces and can also cause scaling [13]. Thus, historic data can help in preventive maintenance and also in enhancing the heat transfer efficiency leading to improved energy efficiency. If the pressure drop across the strainer increases consistently over time, one can conclude that the supply water has a lot of debris in it. As this pressure drop increases, the water flow would become slower impacting cooling tower performance. Therefore, a threshold value for this pressure drop should trigger a strainer-cleaning activity. Related to this is the problem of persistent high pressure drop along the cooling water line, which points to the existence of debris and particles in the cooling water. Either these particles were too small to get filtered at the basin strainer or the strainer itself is damaged. In any case, the straining operation would need attention.

A cooling tower loses a portion of its cooling water due to evaporation and that is the way the remaining water is cooled. Therefore, the water loss can be calculated based on the cooling load on the cooling tower. The makeup water is the water that replenishes this lost water. Therefore, if the makeup water flow is consistently more than what it should be, one can conclude that water is being lost due to some reason other than the cooling tower evaporation. Typically, that reason would be a leakage in the water path. Thus, historic data can help in cooling water leak detection.

All pumps and fans come with manufacturer-specified vibration numbers (amplitude and frequency). If the measured value is persistently outside the manufacturer-specified acceptable limit for a specific piece of equipment, then the equipment has to be checked. Also, having such data helps in performing condition-based preventive maintenance.

(f) Ambient air for cooling—air-side and water-side economizers

Cooling constitutes as much as 40–50% of the data center power consumption. Therefore, even the slightest percentage savings on that translates into big savings in the absolute terms. There are many locations where the outside air is sufficiently cold and dry for 3–6 months during the year so that using the outside air for cooling is much more energy efficient and economical, compared to using the cooling system to do the same. However, there are a few factors that have to be taken into account.

*Quantities to measure*

- Outside air temperature
- Relative humidity of the outside air
- Pressure drop across the air filter
- Absolute humidity of the CRAH supply air
- Absolute humidity of the CRAH return air
- Absolute humidity of the central humidity control supply air (if applicable)
- Water consumed by the humidifier
- Power consumption of the humidifier
- Time of operation of the humidifier
- Water condensed on the cooling coils (dehumidification)

- Power consumption by the reheater
- Time of dehumidification

*How to use the data*

If the outside air is brought into the computer room for cooling, then it is called air-side economizing. Air-side economizing can obviate the chiller and the cooling tower, saving considerable energy. While performing air-side economizing, the humidity, the dust particle content and other contamination present in the outside air have to be monitored very carefully. The computer room air quality control system has to be monitored carefully, because this system has to treat the incoming outside air continuously, without any failure. Sometimes bringing in outside air into the computer room can be difficult from an infrastructure point of view.

When the outside temperature and the relative humidity are sufficiently low, the warmer chilled water coming from the computer room can be cooled directly by the cooling water from the cooling tower, bypassing the need of the chiller. This is called water-side economizing. A heat exchanger is used for the heat transfer between the two water streams. When compared to the air-side economizing, this employs one extra energy-consuming component in the form of the cooling tower, but presents a hassle-free operation as far as aggressive maintenance of air quality is concerned.

### 5.3.2 Power Generation and Distribution System

The power distribution system in today's data centers requires a comprehensive monitoring solution to ensure energy optimization. The enhancement of performance, reliability, and security also requires a comprehensive monitoring solution. The combination of real-time and historic information along with the ability to analyze, filter, and visualize are required for data center energy optimization. *Historic data* allows users to understand the baseline capacity loading, temperature, and power consumption profiles and their variability during normal operation. Rather than using a brief snapshot during a random survey or a 30-min average from a limited number of assets, the ability to capture all of this data and store it for analysis provides the user with a detailed profile of asset utilization history and variability. Once the normal baseline capacity and performance information are understood, *real-time data* can be compared to the "normal" baseline using statistical analyses to determine trends and provide early warning whenever any of these items is out of its normal range. In addition, with data center staff increasingly tasked to do more with less, automation and logic are needed to ensure reliable operation of equipment and the ability to predict a failure in advance instead of searching for causes of problems, after they occur. Figure 5.4 shows a typical power distribution topology inside a data center and also the logical locations where power-related quantities should be measured.

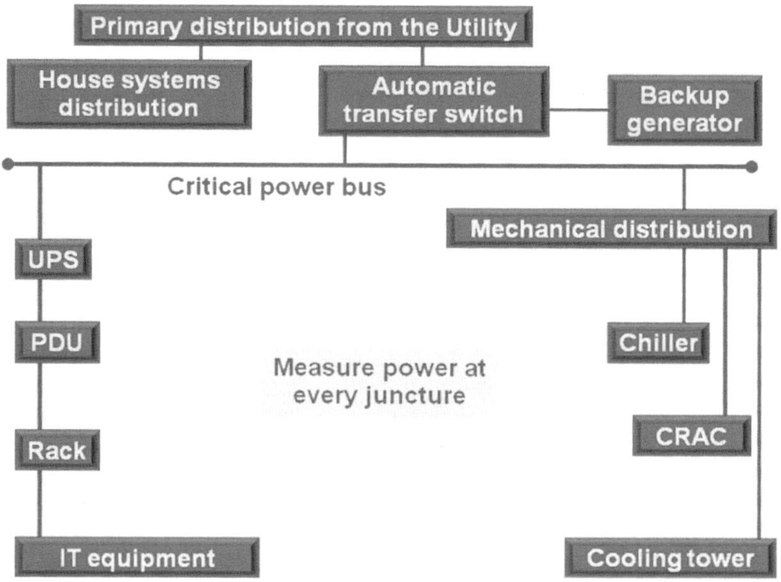


**Fig. 5.4** Typical power distribution topology and power measurement locations

In section 5.3.2 we discuss what quantities should be measured at each equipment, starting from the building entry point for the power. Then we discuss the value of having that data. We have excluded the mechanical equipment from this section because it has been discussed in detail in Section 5.3.1.

### *5.3.3 Building Meter and Automatic/Static Transfer Switch Data*

The building meter is the entry point for the power from the utility company into the building. The data center pays the utility company based on the reading of this meter. Therefore, it is important to keep track of what and how much energy is being consumed.

*Quantities to measure*

- Utility voltage at the building meter/transfer switch for each phase
- Utility frequency at the building meter/transfer switch for each phase
- Utility active power at the building meter/transfer switch for each phase
- Utility reactive power at the building meter/transfer switch for each phase
- Generator voltage at the transfer switch for each phase
- Generator frequency at the transfer switch for each phase

*How to use the data*

Supply voltage and supply frequency constitute the quality of power. Please see Section 5.3.8 for more details on the nature, associated problems, and possible solutions of different power quality issues. History of power quality from the utility helps in understanding how well the power conditioning equipment (such as the UPS) is performing. It also helps in root-cause analysis of problems with equipment performance by performing trend analysis and correlation analysis. Having power quality data can serve as a proof if one wants to challenge the utility company for poor power quality.

The integral of active power over time is energy and that is what we pay for. One immediate benefit of tracking active power and then calculating energy is the ability to compare this energy number with the energy consumption number provided by the utility company.

Capacitive and inductive loads go through a cycle of energy dissipation and energy storage every half-cycle of the alternating current. Consequently, this energy is wasted. Moreover, this charge–discharge cycle gives rise to heat dissipation from these elements and more active power is used to remove that heat. Therefore, it is important to try to reduce the total reactive power loss for the whole data center and track that reduction.

Since in most cases, the generator in a data center runs at a time when there is no other backup, it is of utmost importance to ensure that it is in good health. Typically, when the data center is running on the generator, it is an emergency situation and the data center staff might not have time to look at the generator data as such. However, once out of that emergency situation, it is extremely important to look at the generator power quality data from the last run. If the power quality was poor then the generator needs to be checked to ensure that the generator does not fail to perform when the next emergency comes.

### *5.3.4 UPS Data*

All of the power to the IT equipment in a data center goes through the UPSs. We can safely assume that 40% –80% of the total power consumption of the whole data center passes through the UPS. The range is fairly wide because of the varied operational efficiency of different data centers. There is always some power loss in the UPS. The extent of this loss depends on the design efficiency of the UPS and the operational condition of the same. For an example, in a 5 MW data center with a PUE = 2 and a UPS efficiency of 90%, the data center is losing more than 275 kW of power just at the UPS!

*Quantities to measure for each AC phase*

- Input voltage to the UPS
- Input current to the UPS
- Input power to the UPS

- Frequency of the input voltage to the UPS
- Output voltage from the UPS
- Output current from the UPS
- Output power from the UPS
- Frequency of the output voltage from the UPS

*Quantities to measure for the UPS battery*

- Battery actual voltage
- Battery replace indicator
- Battery run-time remaining
- Battery temperature

*How to use the data*
Every UPS has specific efficiency versus load characteristics and typically, greater the operating load, higher the efficiency. It is important to keep track of the load and the corresponding efficiency of the UPS to check if it is following the vendor-supplied characteristics.

It is advisable to operate one UPS at a higher load rather than operating multiple UPSs at lighter loads. IT service load is dynamic in nature and consequently the IT power load is also dynamic to some extent. Therefore, it might not be beneficial to control the operation of UPSs based on just the last snapshot of the real-time data, because that might lead to switching load from one UPS to the other at a frequent rate. Instead, scheduling the IT power load to bypass a particular UPS for a certain period of time would be a more robust approach to run UPSs at higher loads. This will be possible only if one has the history of the UPS power draw.

Real-time data for the battery replace indicator helps in taking action about replacing the battery. On the other hand, the historic data for the same quantity can reveal how frequently the indicator is coming on and how it is related to certain other things such as the battery temperature, ambient temperature, and the total time in use.

### 5.3.5 PDU Data

The Power Distribution Unit (PDU) typically consists of two major components—a step-down transformer and a power panel. Figure 5.5 shows the schematic view of a $\Delta$-$Y$ step-down transformer that can reside inside a PDU.

Transformer losses are broadly categorized into copper loss and iron loss. Copper loss takes place in the windings (the lines around the cores in Fig. 5.5) and iron loss takes place in the magnetic circuit (the cores themselves). The different types of energy losses in transformers include winding resistance loss, hysteresis loss, eddy current loss, magnetostriction loss, mechanical loss, and other stray losses.

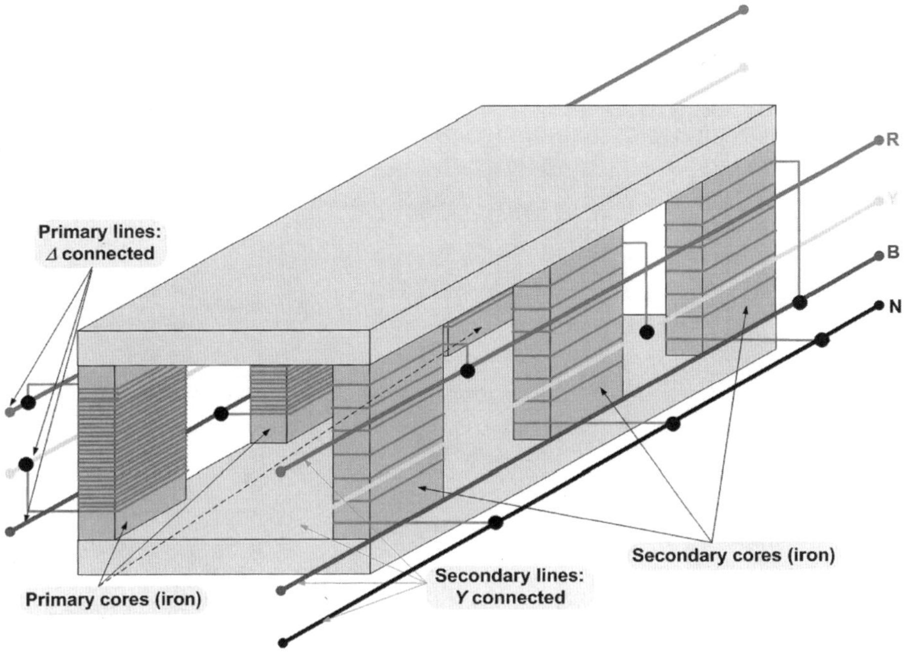


**Fig. 5.5** Schematic view of a Δ-$Y$ step-down transformer that can reside inside a PDU

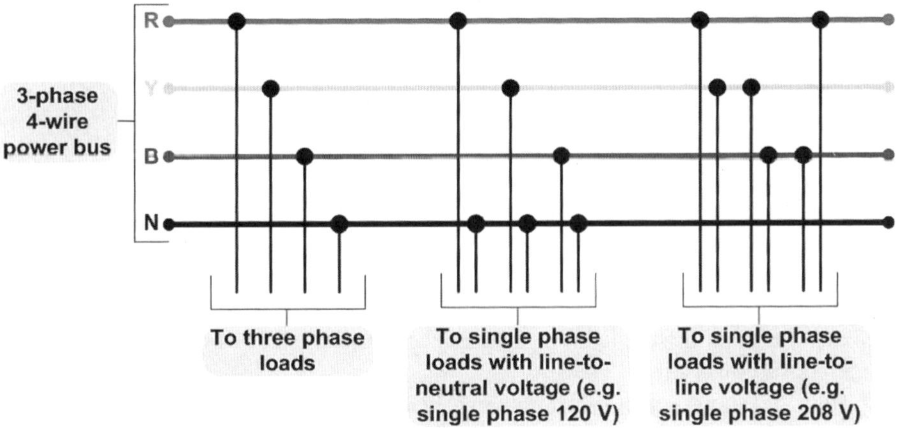


**Fig. 5.6** Three ways to connect load to a three-phase power bus

The output of the transformer (the secondary) arrives at the power panel in the form of a three-phase power bus and it is then fanned out into several circuit breakers. Therefore, another name for the power panel is breaker panel. The branch circuits coming out of these breakers take the power to the IT racks. Figure 5.6

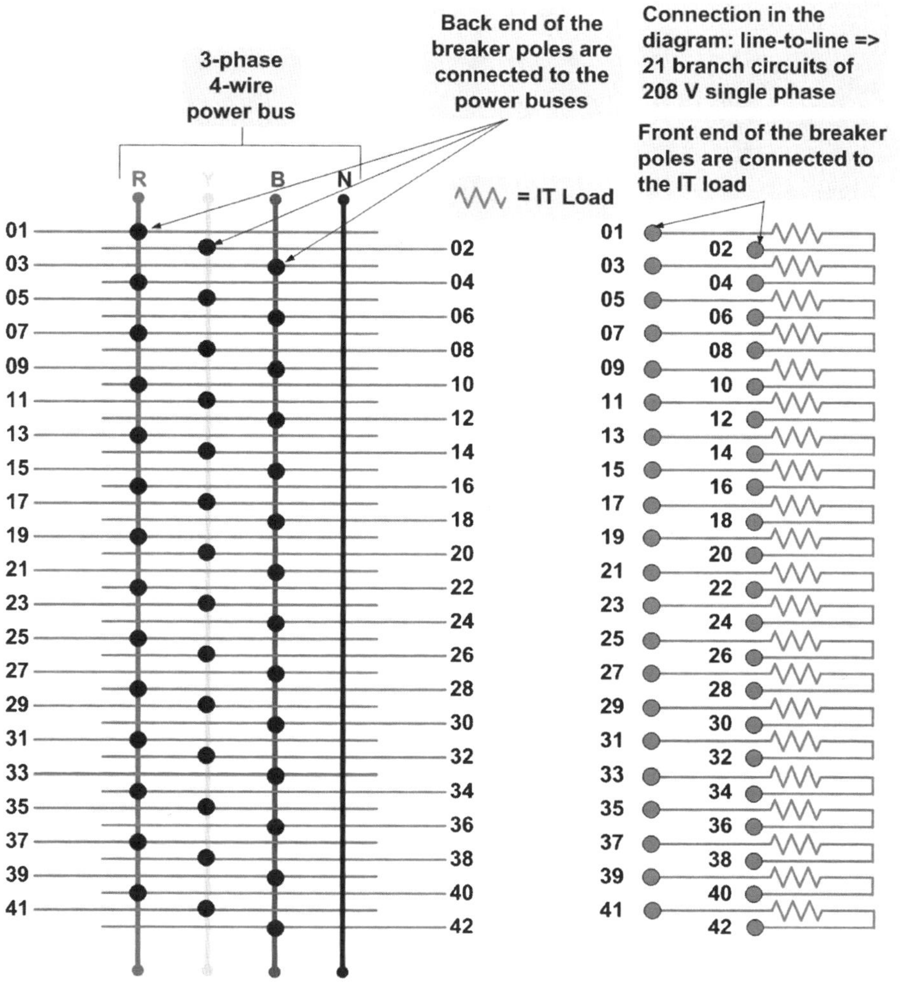


**Fig. 5.7** Schematic view of a breaker panel

depicts the three different ways in which electrical loads can be connected to a three-phase power bus.

Figure 5.7 shows a sample power panel electrical connection where the three-phase power bus is a continuation of the secondary lines shown in Fig. 5.5. The poles are connected in the third way portrayed in Fig. 5.6, i.e., line-to-line.

There are several quantities that one should monitor inside a PDU to save energy, avoid downtime, and to extend its life.

*Quantities to measure*

- Input voltage for each phase
- Input current for each phase

- Input active power for each phase
- Input reactive power for each phase
- Output voltage of the transformer for each phase
- Output current of the transformer for each phase
- Output active power of the transformer for each phase
- Output reactive power of the transformer for each phase
- Temperature of the primary cores of the transformer
- Temperature of the secondary cores of the transformer
- Temperature of the coolant of the transformer
- Dimensions of the cores of the transformer
- Vibration of the cores of the transformer
- Voltage on each branch circuit
- Current on each branch circuit
- Active power on each branch circuit
- Reactive power on each branch circuit

*How to use the data*

By measuring and comparing the PDU input active power and the transformer output active power one can obtain the total loss at the transformer. Historic data on the temperatures of the cores and the coolants and the currents in the windings can help one understand the correlation among those quantities. A consistently high coolant temperature is a sign of impending danger.

Increased current in windings results in higher winding temperature, leading to higher temperature of the coolant. A higher temperature of the winding would mean a greater resistance of the winding, resulting in a positive feedback effect. Prior knowledge of when to expect a greater load can help in scheduling extra cooling effort.

The hysteresis loss is dependent on the core material and the frequency of the voltage. Although the core material cannot be changed very easily, keeping track of the frequency (UPS output) is a good way to mitigate the hysteresis loss. The core temperatures are indicators of the resistive heat loss from the cores due to eddy current loss, which is dependent on the frequency. The relation can be figured out by correlating the two.

The change in the magnetic flux causes the cores to expand and contract during each power cycle. This phenomenon causes frictional heat loss, also termed as magnetostriction loss. By analyzing the historic data on the dimensions of the cores one can estimate the extent of this loss.

The historic data on the vibration amplitude and frequency can enable one to estimate the mechanical losses in the transformer.

As mentioned above, some of the losses can be mitigated by operational improvements, but mitigation of the others would need component replacement. Based on the relative magnitudes of the different loss types, one can decide on the component replacement approach to save additional energy. Having historic data on these quantities also help in convincing the supplier of the equipment in case of Service Level Agreement (SLA) violation.

Monitoring and keeping the history of the voltage across each branch circuit enable one to check for voltage fluctuation and to troubleshoot problems. Also, one can use the voltage data to calculate and verify the power factor of the circuit.

The history of current data and active power data for each branch circuit helps in understanding how the current capacity of the circuit is being used. It also enables one to identify the most power-consuming racks and possibly find out why they consume that much power. If individual branch circuits are not shared among multiple customers, then the historic data for branch circuit active power will be a straightforward way to bill back the customer.

Monitoring the reactive power on a branch circuit helps one in pinning down the source and the type (capacitive or inductive) of the reactive impedance. Once the values for all the branch circuits are available, one can start fixing them, possibly starting with the worst ones. Having the historic data will also help in evaluating the measures taken to mitigate reactive power loss on an individual branch circuit basis.

### 5.3.6 IT Equipment Power Data

The older servers do not have the capability of operating at different power consumption levels based on the computational load they serve. New generations of servers do have this capability. Therefore, it becomes important to track the IT load and the power consumed, in order to understand if the equipment is being utilized to achieve energy efficient operation. The following data can be collected either from a "smart power strip" or directly from the individual pieces of equipment using the *Intelligent Platform Management Interface* (IPMI). Further details on IPMI can be found in Chap. 4.

*Quantities to measure across individual pieces of equipment*

- Voltage
- Current
- Active power

*How to use the data*

Ideally the voltage supplied by the individual outlet of the power strip will be the same across all the outlets and should remain constant over time. Each computer application has its own metrics of how busy it is from a compute service standpoint. If the history of those metrics and that of the active power for a specific piece of IT equipment are compared, the power consumption relative to the compute performance will become evident for that particular piece of equipment. This same analysis can be rolled up from the individual server level to all the servers catering to a specific service. That will help one understand if a specific service is prone to less energy efficient operation. The power-per-service performance can be enhanced by implementing server virtualization and consolidation of servers. The reader may refer to Chap. 4 for more details on server work load distribution in virtualized environments.

### 5.3.7 Power Consumption and Cost of Power

Tracking compute load, power consumption, and the cost of power and acting on that historic data can provide significant savings in operational costs of data centers. Typically both the IT load and the cost of power are high during the day and low during the night. However, the time of the day in one region is different from another as we go around the globe. If a load distribution scheme is in place such that the IT load on a data center is high when the power cost it incurs is low, and the IT load is low when the power cost it incurs is high, the data center can significantly reduce its operational energy expenditure. This can also enable a data center to "make money" by selling their demand at the time of critical peak demand times.

*Quantities to measure*
Demand cost, usage cost, compute load for different services.

### 5.3.8 Generator Data

The generator in a data center, for most part, is a piece of backup equipment which is started and operated only during a time of emergency. Therefore, even though an operator has more important things to care about during the time a generator runs, it is of utmost importance for the operator to come back and have a look at the generator data once the data center is out of the emergency situation, in order to ensure that the generator starts and runs fine when the next emergency takes place. Also, the data collected during the test-run period should be analyzed for the same purpose.

*Quantities to measure*

- Intake air temperature
- Intake air flow
- Exhaust gas temperature
- Exhaust gas flow rate
- Carbon emission
- Total run-time (daily, weekly, monthly, annual)
- Fuel tank level
- Chemical characteristics of the fuel stored in the tank
- Flue gas excess air content
- Pressure drop across the air filter
- Pressure drop across the fuel filter
- Jacket water (coolant fluid) level
- Jacket water (coolant fluid) inlet temperature
- Jacket water (coolant fluid) outlet temperature
- Cooling air intake temperature
- Cooling air exhaust temperature
- Cooling fan power

- Cooling air flow
- Lubricant level in the lubricant-oil sump (makeup reservoir)
- Lubricant-oil inlet temperature
- Temperature of the battery driving the starter motor
- Charge level of the battery driving the starter motor
- Engine speed
- At each start-up: time taken to start, time spent idling, time taken to get to sync speed and time taken to breaker closure
- Frequency of the AC output
- Voltage of each phase of the AC output upstream of the voltage regulator
- Voltage of each phase of the AC output downstream of the voltage regulator
- Active power on each phase
- Reactive power on each phase
- Noise level
- Vibration level

*How to use the data*
Each generator has a mandatory stipulation on the total operating time during a fixed time period, usually a month. If that limit is exceeded, the data center is heavily penalized by the local law enforcement authority. Therefore, monitoring and retaining a record of the total run-time (daily, weekly, monthly, annual) of a generator is very important. The best way to do that is by archiving real-time data on the start time and the end time of each "run" the generator undergoes.

Fuel tank level should be monitored to ensure adequate availability of fuel at the time of emergency. Also, having historic data on fuel tank level gives a perfect measure of the energy input into the generator over time. One can evaluate the efficiency of the generator over time based on that.

Typically, generator fuel stays in the tank more than it is used. The chemical characteristics of the fuel stored in the tank change over time. In order to maintain the chemical properties of the fuel, chemical treatment is required. Therefore, it is worthwhile to monitor the fuel property on a continuous basis and perform chemical treatment based on a predefined condition. Having the history of the fuel property can also help in diagnosing certain behaviors of the generator from the combustion standpoint.

When a generator is running, the combustion makes the engine crank. However, when a generator starts, the engine needs to crank for the first combustion to take place. And that is achieved by the use of the starter motor. The starter motor is the first component of the generator that has to work when a generator has to start. This motor is battery powered. Charge is lost from the battery at the time of operation and also when the battery is sitting idle. Therefore the charge level of the battery driving the starter motor should be monitored very closely. Having the historic data on this charge level will help in troubleshooting engine start-up problems. Charging a battery can cause the battery temperature to increase. Also, the battery temperature can increase due to other external reasons. Increase in the temperature can hinder battery operation, especially charging. Analysis of historic data on the temperature of the battery driving the starter motor will help in understanding charging problems with the battery and can prolong the battery life.

It is important to analyze the historic data on the time taken to start, time spent idling, tame taken to get to the sync speed, and time taken for breaker closure for every generator start-up. If there has been any abnormality in any of the above steps, it should be addressed at the earliest opportunity, because that same problem can aggravate and can cause the generator to fail to start during the next emergency.

A higher intake air temperature enhances the efficiency of the generator. Historic data on the intake air temperature can provide insight into the performance of the effort made in increasing the intake air temperature (by recovering heat from a heat source such as the exhaust gas).

The intake air flow dictates the generator efficiency. The flue gas excess air content can be monitored to decide on whether to increase or decrease the intake air flow. Thus, historic data on those two quantities can help one understand and improve the generator efficiency.

The exhaust flue gas temperature and the exhaust gas flow rate decide the amount of heat energy being lost from the generator. The first step to reduce this loss is to monitor these quantities. Once a measure has been taken to reduce this loss, historic data can verify the performance of that measure. Associated with this history is the record of carbon emission. By having high fidelity historic data on carbon emission, one will be able to trade carbon credit.

The pressure drop across the air filter tells one if it is time to clean or change the filter. Historic data on that can help one understand how it affects the generator efficiency. Similarly, the pressure drop across the fuel filter can be used to decide on the time when it should be cleaned or changed. If the pressure drop reaches the high limit fairly quickly after every cleaning, then one needs to check for sources of contamination in the fuel.

The jacket water level in the reservoir should be monitored to ensure adequate supply of the fluid. Historic data on the jacket water level in the reservoir helps one understand the rate at which it is disappearing. Under normal circumstance, this rate would be slow and steady corresponding to the evaporation during operation. However, a fast decrease in level can indicate a leakage.

The inlet and the outlet temperatures of the jacket water provide a way to calculate heat loss through cooling. If a technique is established to recover this heat, its effectiveness can be verified using the historic data. Similarly, the inlet and the outlet temperatures of the jacket water cooling air can be used to understand the heat loss. Also, measuring all the four temperatures will enable one to calculate the efficiency of the heat exchange process between the jacket water and the cooling air. When the generator is running at low load, the jacket water outlet temperature will be lower requiring less effort in cooling it down to the inlet temperature. Under that circumstance one can use a lower fan speed and consequently reduce the power consumption of the cooling air fan. Historic data on cooling fan power and cooling air flow will help in analyzing the performance of this energy-saving technique.

The lubricant level in the lubricant-oil sump (makeup reservoir) should be monitored to ensure adequate supply of the fluid. Historic data on the lubricant level in the sump helps one understand the rate at which it is being consumed. Under normal circumstance, this rate would correspond to the evaporation during operation. However, a rapid decrease in level can indicate a leakage. The lubricant-oil inlet temperature should be monitored closely to ensure it is below the high limit. Historic information on this quantity helps in understanding the rate at which the lubricant level in the sump went down and also helps in troubleshooting any problem that might have been caused by lubricant overheating.

Frequency of the AC output constitutes one part of the quality of power delivered by the generator. Historic data on the frequency, engine speed, active power drawn, and the reactive power drawn are required in order to know the reason behind the frequency variation. These historic data streams, in association with historic data on the voltages upstream and downstream of the voltage regulator allow one to understand the power load at different points in time and how the generator reacts to that load variation.

Historic data on the noise level outside the facility enables the data center manager to prove that the operation of the generator is not disturbing the neighbors. Historic data on the noise level outside the facility, together with that near the generator, can show how effective the noise shielding has been and how it can be improved. Historic data on vibration near or at the generator body proves very important to understand gradual loosening or degradation of the mounting elements.

## 5.4 Power Quality

This section provides brief explanations of different types of power quality issues. For more in-depth analysis on this topic, readers should refer to books dedicated on power quality. It is intended for those who are not power quality experts but need to make a sense out of their power measurements and take action on that.

### *5.4.1 Impulsive Transients*

It can happen due to various reasons such as electrostatic discharges, lightning strikes, etc. Although its duration could be very short ($<5$ ns), its impact on the electrical components could be devastating due to its very high magnitude. In order to protect the electrical components, one needs to use Transient Voltage Surge Suppressor (TVSS). TVSS either absorbs the extra energy of the transient or short circuits it to the ground. If one measures and stores the voltages upstream and downstream of the TVSS, then it would be possible to know how frequently an impulsive transient takes place, what the possible reasons of those transients, how one can reduce the occurrences of such transients, how the TVSS performs to protect the electrical components, etc. (Fig. 5.8).

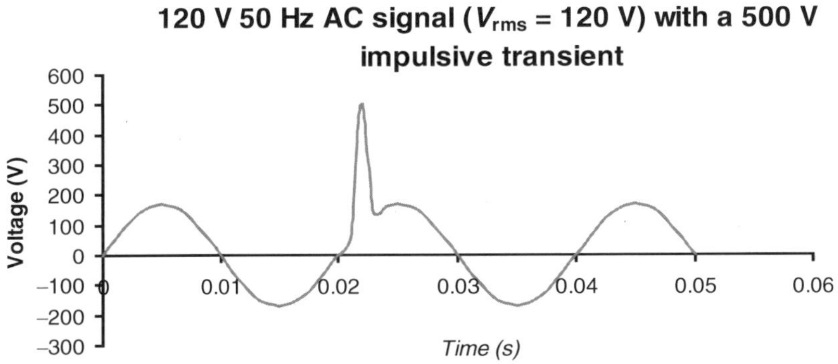


**Fig. 5.8** Impulsive transients

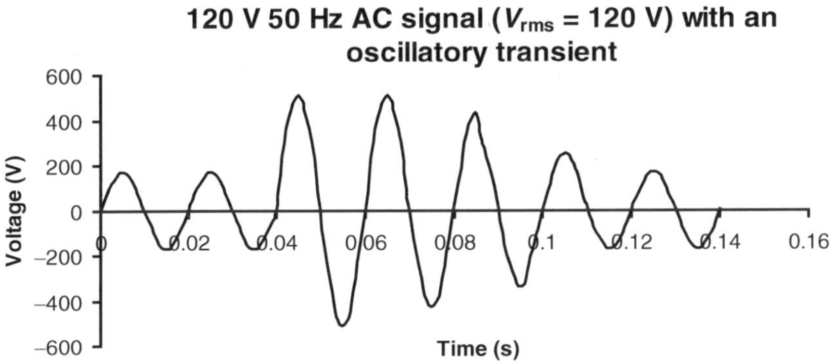


**Fig. 5.9** Oscillatory transients

## *5.4.2 Oscillatory Transients*

Heavy electrical loads turning on and off can cause oscillatory transients. The voltage amplitude suddenly rises above the nominal level and then slowly comes down to its nominal value over several cycles. If one measures and stores the voltages at every relevant point at the facility, then it would be possible to determine how often an oscillatory transient takes place, why the transient takes place, how it can be prevented from occurring, etc. (Fig. 5.9).

## *5.4.3 Interruptions*

As the name suggests, interruptions are nothing but temporary break in the power 0supply. Interruptions are classified in four different categories based on

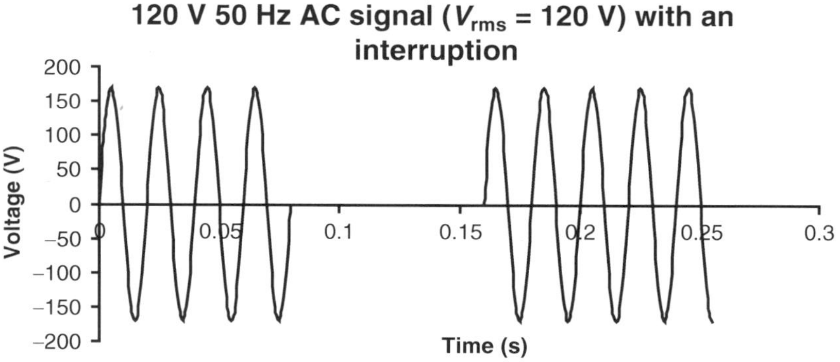


**Fig. 5.10** Interruptions

their durations—instantaneous, momentary, temporary, and sustained. As we all know, Uninterruptible Power Supplies (UPS) are used to supply power during an interruption. It is very important to keep track of the frequency and the duration of the power interruptions so that one can discuss the matter with the utility provider, decide on the right on-site back-up generator, decide on the right UPS, evaluate the possibility of building on-site power plant, etc. In order to keep track of interruptions, one needs to measure the voltage upstream of the UPS (Fig. 5.10).

### *5.4.4 Sag/dip*

A sag is a reduction in the AC voltage for a duration of greater than 0.5 cycle and less than 1 min while the AC frequency remains unaffected. Sags are caused by system faults, switching on of equipment drawing high current during start-up etc. Power line conditioners and UPSs can compensate for sags. If one measures and stores the voltages both upstream and downstream of these sag compensating equipment, then it would be possible to determine how often a sag happens, why it happens, how the occurrence of the sag can be avoided, how the equipment is performing to compensate for the sag, etc (Fig. 5.11).

### *5.4.5 Undervoltage*

Undervoltage is a reduction in the AC voltage for a sustained period of time (>1 min) while the AC frequency remains unaffected. Undervoltage is a long-term result of the problems that create sags. It can create overheating in many electromechanical equipment and can lead to the failure of nonlinear loads such as a

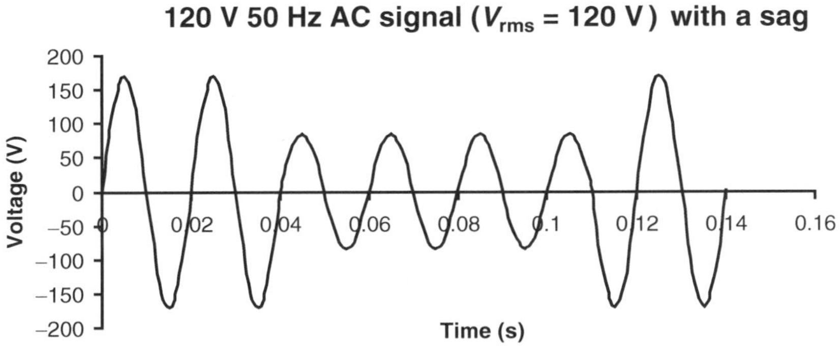


Fig. 5.11 Sag/dip

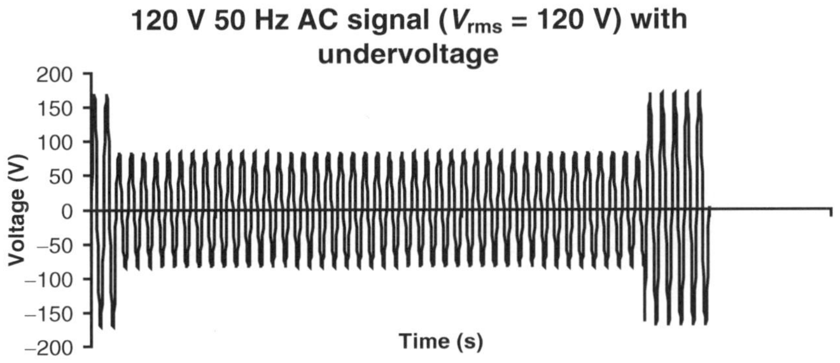


Fig. 5.12 Undervoltage

computer power supply. As in the case of sag, power line conditioners and UPSs can compensate for undervoltage. It is of utmost importance to keep track of undervoltage (Fig. 5.12).

### *5.4.6 Swell/Surge (Opposite of Sag)*

A swell is an increase in the AC voltage for a duration of greater than 0.5 cycle and less than 1 min while the AC frequency remains unaffected. The common causes of swells are high impedance neutral connections, sudden load reductions, single phase fault on a 3-phase system, and when a large load is suddenly turned off. Power line conditioners and UPSs can compensate for swells. If one measures and stores the voltages both upstream and downstream of these swell-compensating equipment, then it would be possible to determine how often a swell occurs, why it occurs, how the occurrence of the swell can be avoided, how the equipment is performing to compensate for the swell, etc (Fig. 5.13).

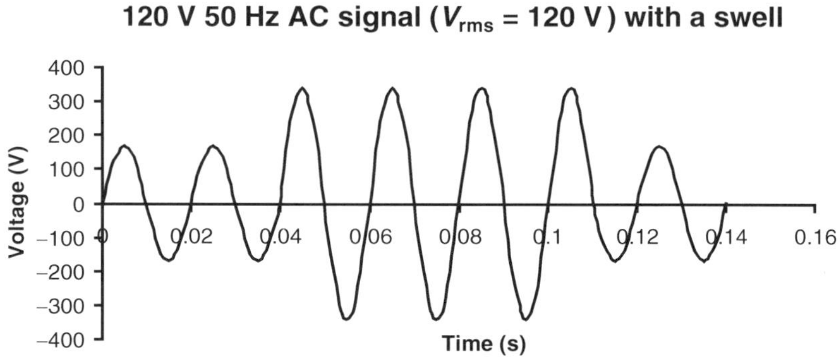


**Fig. 5.13** Swell/surge (opposite of sag)

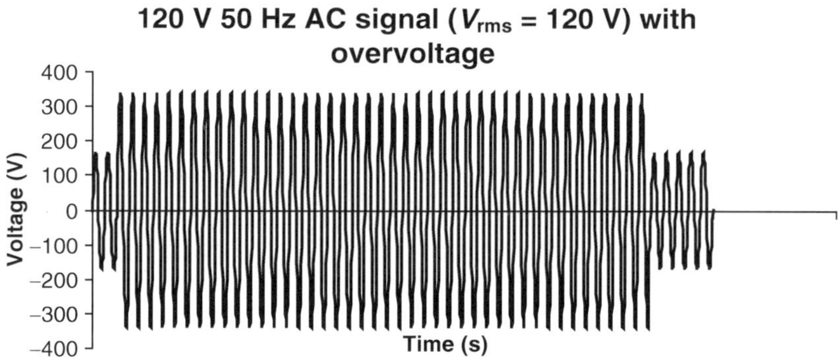


**Fig. 5.14** Overvoltage

### *5.4.7 Overvoltage*

An overvoltage is an increase in the AC voltage for a duration greater than a few seconds with the AC frequency remaining unaffected. Overvoltage is caused where the supply transformer tap settings are incorrectly set and where loads have been reduced but the commercial power systems continue to compensate for the load changes that are no longer necessary. Overvoltage condition can result in high current draw and unnecessary tripping of downstream circuit-breakers as well as overheating and stress on equipment. Power line conditioners and UPSs can compensate for overvoltage. It is of utmost importance to keep track of overvoltage (Fig. 5.14).

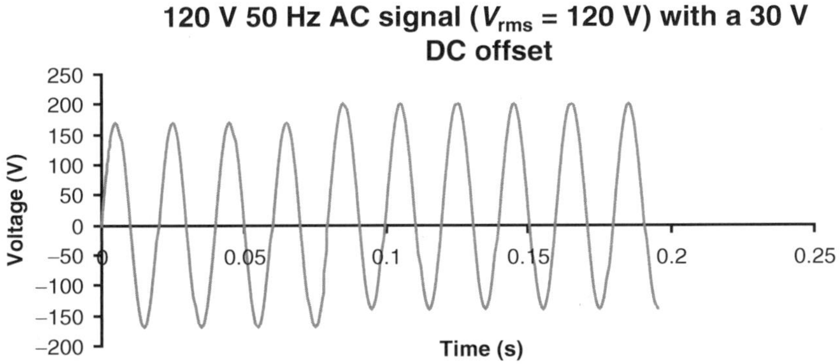


**Fig. 5.15** DC Offset

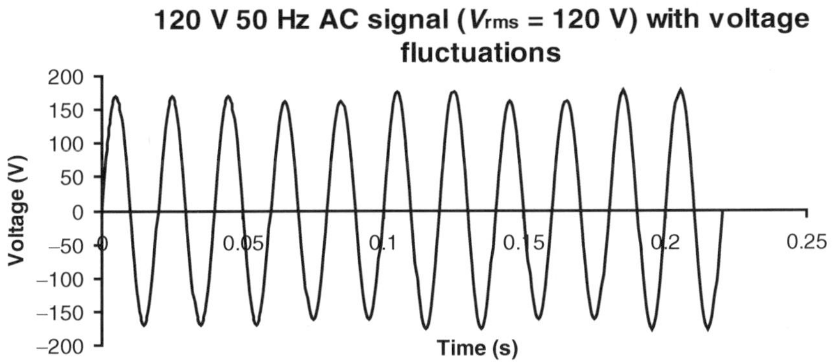


**Fig. 5.16** Voltage fluctuation

## *5.4.8 DC Offset*

DC offset happens when a direct current is added to an AC power source. A DC offset can overheat and damage electrical equipment (Fig. 5.15).

## *5.4.9 Voltage Fluctuation*

Voltage fluctuation is either a systematic variation of the voltage waveform or a series of random voltage changes of small dimensions (95–105% of the nominal). Typically, the frequencies of these variations are less than 25 Hz. Power line conditioners and UPSs can compensate for voltage fluctuation (Fig. 5.16).

### 5.4.10 Harmonic Distortion

The power system voltage can depart from the ideal sinusoidal pattern in several respects. Harmonic distortion is the name for a departure in which every cycle of the waveform is distorted equally. Harmonics appear on the power distribution system as a harmonic current. Equipment such as computers and other electronic components draw current in a non-sinusoidal fashion. These patterns are actually superimposition of different sinusoids—one having the original supply frequency (say 60 Hz) and the others having frequencies that are multiples of that original frequency (such as 180 Hz, 300 Hz, 420 Hz, etc.). These higher frequencies are called the harmonics of the fundamental frequency. Since the existence of these harmonics gives rise to the distortion in the voltage waveform, the distortion is called harmonic distortion.

Rotating equipment such as motors are badly affected by harmonic current. They get overheated. Power factor correction in the presence of harmonics is a confusing subject. When no harmonics is present we can write the following.

$$I_{\text{total}}^2 = I_{\text{power}}^2 + I_{\text{reactive}}^2$$

However, in the presence of harmonics, the above equation becomes [14]:

$$I_{\text{total}}^2 = I_{\text{power}}^2 + I_{\text{reactive}}^2 + I_{\text{harmonics}}^2$$

In many cases, e.g., with computer installations, $I_{\text{reactive}}$ is close to zero, but $I_{\text{harmonic}}$ is large and the power factor is less than 1 [14]. If such a customer installs power factor correction capacitors, then $I_{\text{reactive}}$ increases due to the capacitor current, further increasing $I_{\text{total}}$, worsening the power factor.

There are several measures of the harmonic distortion, including the level of the voltage at each harmonic. Two of those measures are Total Harmonic Distortion (THD) and notch depth. The first is important for long-term thermal effects, the second for equipment malfunction.

Unlike most other types of supply problems, harmonics can go unnoticed for many years unless equipment temperature or the voltage waveform is monitored on a continuous basis. Equipment not containing harmonic correction feature should be isolated from a circuit having harmonics.

## 5.5 Data Collection and Management

Implementation of monitoring involves various tasks. Understanding of what needs to be measured, selection of hardware, and finally the installation of the hardware are all challenging tasks. Often, this challenge becomes a barrier to the implementation of monitoring.

As far as software is concerned, certain monitoring tools can collect facilities data, but not IT data and there are others that can collect IT data but not facilities data. A few monitoring tools can collect both facilities data and IT data, but are restricted to specific platforms (OS). Therefore, choosing the right software tool for monitoring can be challenging; and under certain circumstances a suitable tool might not even be available.

In most cases, the out-of-the-box system is far from being a comprehensive solution. There are a few solutions that collect data, but do not produce any meaningful result with it, in which case custom applications need to be built at every site. Other solutions collect data from a single system, say a power meter or a power strip, and provide a solution based only on that data, in which case the big picture scenario is not captured.

Data center is a dynamic environment where rack positions are changed and servers are moved around. Monitoring of such environments requires (preferably automatic) change registration at the data collection level and then fully automated propagation of that change through the rest of the monitoring and application system. Unfortunately, no monitoring tool offers this type of change-friendliness out of the box. Thus, either the monitoring system has to be maintained manually by the in-house staff or custom application has to be built on top of the monitoring tool. Either of the two options inflates the cost of the monitoring system.

Therefore, implementing and maintaining a data center monitoring system is a challenge. Reference [1] is a great resource in learning about how to choose the hardware for data center monitoring. In this section, we discuss the requirements for the software part.

### 5.5.1 Features

In this subsection we discuss the different features or aspects that a monitoring software tool should have. One can use this portion as a guide for choosing a monitoring software product.

(a) Platform and browser support
It's probably not necessary for the core components (the middleware and the database) of the software to support both Windows and Linux, but it would be nice. However, it is absolutely necessary for it to support all major Web browsers.

(b) Database
The system should be as standards oriented as possible so that the support ecosystem for the system is large. That guides us to have a system that uses SQL standard-based database. Ideally, it should be compatible with all major SQL database systems such as Oracle, PostgreSQL, DB2, MySQL, and MS SQL Server.

(c) DataType support
The system should support at least the following data types.

- 64 bit floating point (double precision)
- Binary Large Object (BLOB)—this data type is required to store images both for metadata and for time series coming out of monitoring camera.
- String

(d) Data collection
There should be file download and file unzip capabilities where file parsing is concerned. There should be backfill mode of data collection to recover history. The tool should support the following protocols for data collection:

- XML file parsing
- Structured text file parsing
- SQL-based database query
- BACNet
- Modbus
- DDE (Dynamic Data Exchange)
- DNP 3
- XML SOAP
- SNMP
- IPMI

(e) Time stamp

- Timestamp should be portable across time zones.
- Difference between device clock and server clock should be identified and correction should be applied automatically.

(f) Metadata

*Source of metadata*

- Provision for manual entry should be there. However:
- It should nicely integrate with the CMDB to pick up information contained in there.
- It should communicate with the electronic design document of the data center to automatically gather information on structural, mechanical, electrical, plumbing, ducting, and IT.
- For the facilities side, which calls for compatibility with Building Information Model (BIM), specifically Industry Foundation Classes (IFC); while in order to be compatible with the IT side, the metadata system should comply with Common Information Model (CIM).

*Navigation*
Navigation through all the points should be easy and intuitive. It should not expect the user to remember what BAC.OAK.CC.1 means. That was just an example from typical practice of Building Automation System (BAS) System

Integrators (SI). The point is that the metadata should be extensive but lucid. More details on this can be found in Section 5.5.1).

*Data model*
The data model of the metadata should contain context (for example, the location that this data pertains to), physical quantity, data source type (measured, model1, model2, etc.), any asset associated with the location, asset history of the location (if assets were moved out or replaced), and history of a single asset (for example, maintenance history).

(g) Security
There should be a per-user or per-group based security for both metadata and time series data. This way, a particular piece of information in the database can be made accessible to one user without making it visible to another user. There may be a need, especially for collocation providers, to make the data for a time-series point available to a customer only for a specific time window, and at times making it exclusively available to that customer. In order to accomplish that, there should be a provision for time-sliced security for a single time-series point.

(h) Data query
In addition to querying for raw data, users typically need to query for summary data or features of a time-series point, such as mean, max, and min over a period of time. There should be options for that. Also, there should be provisions for data filtering based on user-defined filtering rules and data interpolation. Often, it is discovered that the calibration of a sensor was not correct during a time period. Therefore, in order to use the data coming out of that sensor during that time period, a data processing rule has to be applied to the raw data. And this rule can change over time for the same sensor. There has to be a way for the user to store these rules for the specific sensors and optionally apply them during data query.

(i) Horizontal Scalability
It should be scalable in terms of *number of points*. There should not be any limit as long as we can provide commodity grade hardware (NOT a $500,000 cluster). Essentially, it should be able to scale out easily (horizontal scalability). At the same time, it would be nice to have the result of a benchmark test to know the approximate number of points that a specific hardware can handle with reasonable performance.

It should be scalable in terms of *data throughput*. Consider, collecting 600,000 points every 30 s (We do not necessarily need to store all of them.). That's 20,000 points/s distributed over the time period. The system should be able to scale just by the addition of commodity grade hardware into the system.

(j) User Interface

- The visualization should be Web based.
- It should have 2D and 3D views; zoom and pan capabilities; its search for location, asset, physical quantity, point, etc should be like google maps, i.e., easy but informative; there should be capability to superimpose time-series

data on the metadata (location, asset, etc.); not to mention plain and simple time-series displays such as trend and dial gauge should be there.
- There has to be a way to upload graphic and then to superimpose numbers on that. This could be very useful in putting numbers on a floor layout or campus map in the absence of design data or even in other conditions.
- There should be buttons on the screens for sending manual signals to controllers or analysis engines.

(k) Extensibility

*Analytics*
The system should be extensible, that is, it should have an API so that anybody can build his/her own analysis that interacts with this system. The native API can be faster and can be used by applications developed in the native programming environment. However, there should be a standards-based API, such as XML Web service based or JSON based, so that nonnative applications can use this system as well.

*Visualization*
Similar is true for displays, that the tools and components should be available so that users can build new display screens easily.

(l) Data compression
The tool should have lossless data compression mechanism. We want to keep everything forever, but in an efficient manner. Some tools average the data as the data gets old. That prevents retroactive analysis. The data compression should be only on the streaming data, so that data retrieved from the database at any time would have repeatability.

(m) Ease of deployment
The time required for configuring the system for a particular site should be as little as possible. Built-in templates for various equipment and auto-discovery of IP-enabled devices help reduce deployment time.

(n) Redundancy
The system should support redundancy, which is necessary for high availability and load balancing. We can assume that at a minimum, a monitoring system will have a Web server for serving visualization, a database server for storing and serving the data and data collection engines for communication with the data sources. There should be redundancy at the Web server level, the database level, and the data collection level.

(o) Forecast data
Forecast data is different from actual historic data in terms of timestamp. In historic data, there is only one timestamp associated with a value. In forecast data, there is a timestamp for when the forecast was made and a second timestamp for the time for which the forecast was made. For example, weather forecast for Sunday is made on Thursday, Friday, and Saturday. The system should capture both notions of time for the sake of clarity, correctness, and possible analysis of the performance of the forecast. The system should be able to handle forecast data.

(p) Change management
It should have the capability to propagate changes made at one part of the system to all the parts of the system that are affected by that change.

## 5.6 Conclusions

As we have seen in this chapter, having historic data for all relevant measurable quantities in a data center cooling system helps improve energy efficiency. It also helps in performing preventive maintenance on equipment. Reliability of operation depends on the reliability of the equipment. As we noticed, historic data helps in maintaining the reliability of various pieces of equipment such as chiller components, cooling tower components, and CRAC unit components.

We discussed about how having historic data for all relevant measurable quantities in a data center power distribution system helps improve energy efficiency of the data center. It also helps in performing capacity planning and preventive maintenance on equipment. We discussed about what data to collect from the building meter, transfer switch, generator, UPS, PDU, and rackPDU and/or IT equipment. We also discussed how one can analyze the historic data to achieve better operational efficiency. Decisions related to data center power need to be based on analysis. From this chapter, we can conclude that this analysis should be based on the historic data on all the power-related quantities.

Finally, we discussed about what technical aspects to look for while choosing an appropriate monitoring software for the data center. However, there are always aspects other than the technical ones that also play roles in the decision making, such as initial cost of software procurement, initial cost of software deployment, support and maintenance cost of the software, proprietary or open source, and the existing support ecosystem of the software.

## References

1. ASHRAE TC9.9 and The Green Grid (2010) Real-time energy consumption measurements in data centers
2. http://www.nagios.org/
3. http://ganglia.sourceforge.net/
4. EPA report to the congress, Aug 2007
5. Staff writer, Gartner notes that Data centers account for 2% of global $CO_2$ emissions, Data Center Journal, 10 Oct 2007
6. Partridge E (2007) The ten steps towards a greener Data center. The Data Center Journal, 17 Oct 2007
7. Otterson M (2007) Where facilities and IT meet. The Enterprise Data Center, 2007
8. Graham KM (2004) 5 Ways to chiller efficiency. Maintenance Solutions, FacilitiesNet, Oct 2004

9. Iowa State University – University Extension, Coordinate by CIRUS, 2005, Energy-related best practices: a sourcebook for the chemical industry, Chapter 6, p 80
10. Clarke D (2006) Refrigerant stacking and carryover in water-cooled chillers. Sustainable Facility, 30 Oct 2006
11. Evans T (2009) Humidification strategies for data centers and network rooms. APC by Schneider Electric, White Paper 58 Revision 2
12. All Climate HVAC, http://www.allclimatehvac.com/humidification.html
13. Western Area Power Administration Energy Services, Technical Brief WSUEEP98013, Optimizing Cooling Tower Performance, Rev 2/98 http://www.wapa.gov/es/pubs/techbrf/cooling.htm
14. Harmonic distortion in the electric supply system, Integral Energy Power Quality Center, Technical Note No 3, March 2000

# Chapter 6
# Energy Efficiency Metrics

**Michael K. Patterson**

**Abstract** In this chapter, metrics for measuring and improving data center efficiencies are explored. Metrics at varying levels from the infrastructure components to the entire data center are reviewed. The primary data center efficiency metric, PUE is discussed at length as well as variants of PUE. The challenges of defining a metric around computing output or data center useful work are also considered. The chapter includes discussions on a variety of other related topics such as codes, standards, and rating systems. Through a thorough review of the chapter, the reader will also gain a strong insight into some of the fundamental issues in data center design and operations.

## 6.1 Metrics and Data Center Issues

Data Centers are at the core of today's information economy. Whether it's the high-performance computing designing the next airliner, order placement for an on-line retailer, or even the payroll for the local school district; all of these activities are taking place on servers in a data center of some size. As seen in Chap. 1, these critical pieces of the economy use a significant amount of energy (61 billion kiloWatt hours (kWh) or 1.5% of the US electricity costing $4.5 billion, based on estimates from the EPA [1]) and it will continue to grow for the foreseeable future. To ensure continued operation with the lowest total cost of ownership (TCO), IT managers must be able to operate their data centers with high-energy efficiency, in order to get the highest computational output for their energy cost input. Metrics are required to achieve this objective.

The data center has evolved significantly in the past decade. Previously, the focus was on mainframe computers in large rooms. Energy was a minor issue as the

M.K. Patterson PhD, PE, DCEP (✉)
Eco-Technology Program Office, Intel Corporation, Hillsboro, OR 97124, USA
e-mail: michael.k.patterson@intel.com

Y. Joshi and P. Kumar (eds.), *Energy Efficient Thermal Management of Data Centers*, DOI 10.1007/978-1-4419-7124-1_6,

computing used the primary amount of energy. The space was large and the overall heat density was manageable. The trend towards reduced cost and higher-powered servers has caused the shift from a few mainframes to servers, where the near commoditization of the IT building blocks has created high-density, high-performance racks of servers. As seen in Chap. 3, the energy density of the IT equipment has shown significant growth, and at first glance this has been seen as a problem. However, the performance and output of the IT equipment has increased at an even greater rate; so performance per Watt is growing at a Moore's Law—type growth rate. This is good news. The challenge has become data center design to support these new IT configurations. The engineering associated with the infrastructure now is on par with the engineering required for the IT side. Metrics have become a key tool in this balance.

Another challenge has been the fast pace of innovation from the IT community. Servers may have a useful life of 3–5 years (before the next generation's performance becomes so compelling that an IT refresh is warranted) while the data center is generally in a building with an expected life span of 15–25 years. The number of refresh cycles in the data center can causes mismatches in the IT technology and the supporting infrastructure. Here again metrics can play a fundamental role in the process, in both determining when to refresh and what facilities changes may be in order to ensure the best TCO and continued successful operations.

Innovation in the data center is also occurring. There is a wide range of new power and cooling architectures. Examples of these are high voltage (on the order of 380 V DC) DC power distribution and liquid cooling. In these and the full range of other innovations, the metrics below still apply. Care must be taken to ensure all inefficiencies are still considered and understood. There have been many claims about the overall "green" benefit from varying technologies but the best tools to ascertain the efficiency are still the metrics discussed herein.

The number of metrics in the industry is ever-increasing. This chapter does not attempt to cover all of them, many are redundant or very specialized and of marginal value. Here, we cover the most widely used metrics as well as those that will provide the data center designer, operator, owner, or regulator with the best opportunity for reducing energy while allowing the continued growth in the information economy.

## 6.2 Glossary

AC—Alternating current
CADE—Corporate average data center efficiency
CEF—Carbon emissions factor
CFM—Cubic feet per minute, volumetric airflow measurement
COP—Coefficient of performance
CRAC*—Computer room air conditioner
CRAH*—Computer room air handler

DC—Direct current
DCeP—Data center energy productivity
DCiE—Data center infrastructure efficiency
ERE—Energy reuse effectiveness
ERF—Energy reuse factor
EWIF—Energy water intensity factor
GHG—Greenhouse gases
HPC—High-performance computing
HVAC—Heating, ventilating, and air-conditioning
IT—Information technology
itEUE—IT energy usage effectiveness
LEED—Leadership in energy and environmental design
PDU—Power distribution unit
pPUE—Partial power usage effectiveness
PSU—Power supply unit
PUE—Power usage effectiveness
UPS—Uninterruptable power supply
VR—Voltage regulators
TDP—Thermal design power

*Note on CRAC and CRAH usage: CRAC, in its most common usage refers to any modular cooling unit, typically around the perimeter of a data center, providing cooling air to a section of the data center. In a more strict and far less common usage, CRAC (computer room air *conditioner*) refers only to those cooling units with compressors for air-conditioning, while CRAH (computer room air *handler*) refers to those cooling units which usually contain a cooling coil but getting the cooling fluid from a building chilled water system, so their primary function is air handling. In this book, the more common usage is in place. It is incumbent upon the reader to ensure that they recognize the difference and apply the characteristics of each appropriately in their usage.

## 6.3 Example Data Center

To help demonstrate the metrics in this chapter, their use, and their issues, a hypothetical data center will be considered. This data center will be based upon actual server data as well as industry-typical values for infrastructure (Fig. 6.1).

The server data is from an EPA report [2]. The data center is initially configured as shown in Table 6.1.

In the above example, the site receives electrical power from the utility; the power feeds the UPS as well as the chiller plant. The chiller provides chilled water to the CRAC, which in turn provides cool air to the data center to cool the IT equipment. The UPS feeds power to the CRAC as well as into the data center where the PDU distributes it to the racks of IT equipment.

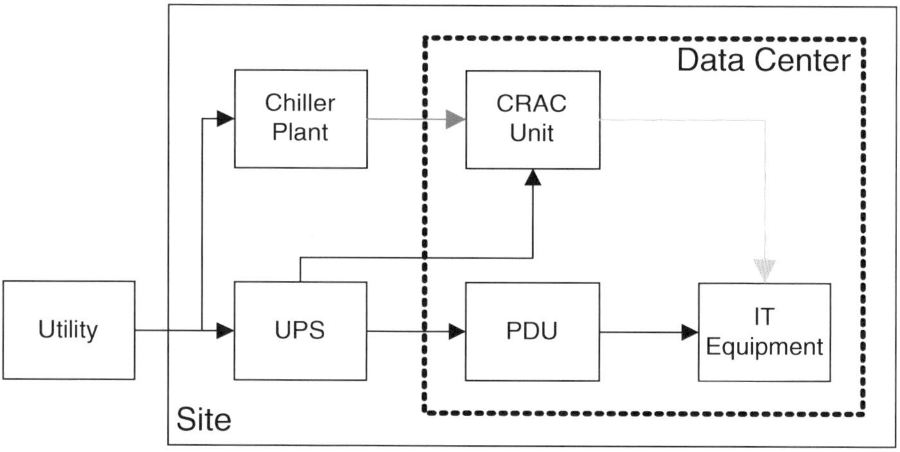


**Fig. 6.1** Simplified schematic of example data center

**Table 6.1** Hypothetical data center for metrics discussion

| | |
|---|---|
| Area (sq ft) | 2,000 |
| Number of racks | 500 |
| Servers per rack | 20 |
| Total servers | 1,000 |
| *Server vintage (Year)* | *2006* |
| Average server power[a] (W) | 273 |
| Average data center connected load (total of all IT) (kW) | 273 |
| Peak server power[b] (W) | 427 |
| Annual IT energy use (kWh) | 2,391,480 |
| Annual energy loss in UPS to IT (kWh) | 144,000 |
| Annual energy loss in PDU (kWh) | 96,000 |
| Annual energy use in lighting/security (kWh) | 60,000 |
| Annual energy use in chiller plant (kWh) | 450,000 |
| Annual energy use in CRAC units (kWh) | 700,000 |
| Annual energy loss in UPS to cooling plant (kWh) | 50,000 |
| Total energy use (crossing the data center boundary) (kWh) | 3,891,480 |
| Total IT energy use (kWh) | 2,391,480 |

[a]Assuming 30% annual average utilization
[b]Assuming 100% CPU utilization

The values and schematic above are intended for demonstration purposes only and should not be misconstrued as targets or as representing any particular facility or recommended design. The differences between average and peak server power are dealt with in ref. [3]. The issues of system sizing are beyond the scope of this chapter.

## 6.4 Metrics Background

### *6.4.1 Metrics Use*

There are a range of possible uses for a metric and this use should be considered in their development and application.

*Self-improvement over time*—Measuring a given data center for a metric, then tracking it over time, generally year to year, will provide a path towards improvement of that data center. In addition, a metric's use before and after a change to the system design or operations will provide insight into the impact of the change.

*Comparisons with other data centers*—This could be internal to a company that has multiple sites, or a comparison of other data centers in the same region or with the same workload profile (HPC vs. enterprise vs. collocation vs. Internet portal data center). In addition, comparisons can be made to obtain a place within a rating system. However, appropriate caution and consideration must be given to any metric used to compare data centers. A high availability data center in a warm humid climate SHOULD have a different energy efficiency performance than a less critical data center located in a cool dry climate. Attempting to drive one data center to achieve the same level of energy efficiency through as specific metric could have negative consequences.

*Site selection/design*—Any new data center should have targets in place for energy efficiency, cost, sustainability, as well as an understanding of any code or rating system compliance that may be required or desired. The range of metrics in this chapter and selection of the most appropriate will assist in the site selection and the design process. As mentioned earlier, a given site may not have the environmental conditions, space, or power (or water) necessary for the stated goals of the new data center called out by the chosen metrics.

*Operations*—Several of the metrics in the chapter also lend themselves well to operational improvements in the data center. While some are defined as annual metrics, trending data and the resultant temporal value of the metrics can provide the operations staff with useful insight to the health of the ongoing operation. Problems can often be detected early and avoided if a metric is trending in a direction that has no attributable cause.

*Collocation selection*—Finally, metrics can assist in selection of a hosting data center. In the case of an IT organization, outsourcing their IT equipment to a collocation type facility metrics can help with the selection of the best site or host. Reviewing the metrics of the potential hosts can provide the IT owners with insight into the energy efficiency, sustainability, availability, and expected costs associated with such a business decision.

## 6.4.2 *Range of Coverage*

Metrics covered in this chapter range from component specific (e.g., a server fan) up to the full data center suite of IT and infrastructure combined. The metrics are generally applicable to energy efficiency but the coverage does extend to a few sustainability metrics as well as performance metrics.

In addition to the metrics reviewed here, appropriate codes and standards are also referred to when they have direct applicability to the efficiency of data centers. These will certainly remain a fluid list of codes and standards and the reader is encouraged to explore beyond this chapter for the latest applicable codes, standards, and versions of the same.

The chapter reviews a variety of data center rating systems. While not specifically metrics, these rating systems can help in developing an understanding of the appropriate metrics. Further, these rating systems and other rules-of-thumb can be useful in the initial understanding of a data center or also in the conceptual planning phase of new data center.

## 6.4.3 *Metrics Goals*

If the goal of a specific metric is kept in mind, its development and application can be much more fruitful. In general, the goal for any metric is to measure and improve some value function or parameter. Whether that goal is met lies largely in the answer to the following questions about the metric.

### 6.4.3.1 Metrics Questions

*Does it drive a measureable activity?* Once a metric is established and tracked and trended, there should be an action that can be put in place to improve the metric. If not, the only reason to record such a metric may be for environmental compliance or required reporting.

*Will improving the performance to the metric bring measurable gains?* Improving the data centers performance against a metric should result in a higher efficiency, greater computing output, lower costs, higher availability, or a lower carbon or water footprint. If the metric does not drive these results, its value should be analyzed.

*Can its components be readily measured? If not, can they be estimated?* The greater the accuracy of the measurements, the stronger the case for improvements the metric calls out. But even when the parameter must be estimated (vs. measured), it is generally better than not doing the metric. For example, PUE (which will be discussed later) is the primary infrastructure efficiency metric in the industry. It should include lighting in the data center and support spaces. If this branch circuit is not metered, it would be better to sum the number of fixtures times their bulb's name plate Wattage and use that value than to not calculate PUE at all. The reader may refer to Chap. 5 for power measurement techniques available for data centers.

*Will the measurement of the metric cost more than it will save?* For small data centers or server closets, the installation of real-time energy measurement for the power and cooling systems as well as for the IT equipment itself may cost more than the annual energy costs to operate the data center. Value-appropriate measurements and metrics need to be considered. Weekly recording of power levels may be sufficient in such an installation.

*Is the metric independent of hardware?* A metric that would, e.g., only be able to be used with a specific manufacturers IT equipment or with a specific airflow management strategy will be limiting.

*Is the metric independent of architecture?* Will the metric work for different workloads; e.g., HPC or Web-based services? Air or liquid cooling? The broader the applicability, the wider spread will be the adoption.

*Is the metric aligned to a specific topic?* There have been metrics suggested that combine a wide range of metrics to determine an overall "green" score. However, when multiple disparate metrics are combined in such a fashion the results and comparisons become less meaningful. Comparison to another data center would be of little value as each underlying metric could have varying values to each site. Comparison to a single data center over time, showing improvement, would be of little use as much greater visibility to the data center operations will be had looking at the individual metrics that make up the proposed metric.

*Is the metric at the right level?* If the goal is to improve airflow management in a section of the data center, using an overall data center efficiency metric that included power distribution, chiller performance, and lighting would not provide the best resolution. However, it would be needed if the goal was working towards an overall infrastructure energy cost reduction.

*How easily can the metric be gamed?* A metric intended to measure server internal power efficiency that just looked at AC–DC conversion efficiency (PSU efficiency) could be easily gamed by the manufacturer spending more on the PSU but keeping their total cost even by reducing costs on voltage regulators.

*Can the measurements be taken without impacting operations?* Data center operations are critical to the business of the parent organization. Causing the data center to run a benchmark or a measureable non-output-bearing workload would, in almost all cases, be a significant detriment to that business. One exception to this would be the HPC space where specific benchmarks are run periodically to establish a place on industry wide ranking schemes. One important point here is that generally these benchmarks do not represent well the computers intended usage nor does the measurement typically occur over a period of time that the energy use is representative of any annual IT or infrastructure basis.

## 6.5 Trends and Rules-of-Thumb

There are historical trends and experiences around data centers that have provided the initial basis for data center comparisons and evaluations. These have led to the extensive set of metrics covered in this chapter and are worth a brief review as they help set the stage for a full understanding of the data center efficiency challenges. These values below should not be used directly in planning or sizing of data centers as the actual numbers will be site and need specific but these can be informational as a starting point. These trends are discussed extensively by Koomey [4].

*Area/area*—As the computational side of the equation follows a Moore's law [5] improvement for performance per Watt, the infrastructure side is actually trending the opposite direction. As the IT becomes more dense, the energy per unit volume of the IT increases. The advantage is that the computational output per unit volume is increasing at an even greater rate. The disadvantage of this is that the infrastructure to support it becomes more complicated and requires more space. While the level of availability required and the sophistication of the energy efficiency features built into the data center infrastructure can vary this significantly; in 2011 the area required for the data center infrastructure is roughly equivalent to the area required for the IT equipment. This value of 1:1 is trending towards the infrastructure required area to be larger and the IT space smaller. For the example data center of Table 6.1, there would be a need for roughly 2,000 sq ft of space for power distribution and cooling equipment.

*Cost/cost*—Similarly the ratio of the cost of the IT equipment to the cost of the data center is trending in the same direction. Historically, the IT equipment had been the larger of the two by a significant margin. Currently, it is not infrequent that the capital cost of the data center is approximately equal to the capital cost of the IT equipment.

*Percentage of cost*—The cost of the data center itself is composed of three main areas: the civil/structural/architectural (CSA), power, and cooling. Again the values are site specific and variable but each of these costs on an order of magnitude scale are equal. In general, power costs are trending up in comparison to the CSA. Cooling costs are becoming more variable based upon the extent of the sophistication of the cooling systems. Some simple free-cooling type designs are driving this portion lower, but highly available data centers with extensive efficiency features in the cooling systems are also trending higher than CSA costs.

*Cost/cost*—Evidence of the need for energy efficiency and metrics to drive the industry there are found in Belady's work [6] where he points out that the costs (driven largely by energy) to operate the IT equipment and the supporting infrastructure are roughly on par with the costs to buy the IT equipment itself. Clearly, this ratio is based on the expected life of the IT equipment as well as the overall efficiency of the data center, but the fact that on order of magnitude these two costs are similar is reason enough to increase focus on energy efficiency.

*Area/rack*—The final rule-of-thumb is space density. Current best practices are that each rack in a standard data center will use between 25 and 40 sq ft of total data center area. This number is greatly affected by data center layout, number of aisles, and maintenance/move-in space but can serve as a first pass metric on data center layout efficiency. This ratio is shrinking (<25 sq ft per rack) in the case of containerized data centers and other high-density strategies.

## 6.6 Basic Data Center Metrics

These basic data center metrics have been in use for an extensive length of time. While they are the simplest metrics, they also can be the source of significant problems when used incorrectly. The discussion and understanding of the attributes of each is imperative to not use them improperly.

*Power/area*—(W/sq ft or W/m$^2$) The most frequently used metric in the description of data centers is also the one whose applicability has fallen the most. This metric gives an overall power density for the data center but it has very little value in sizing the specific cooling or power architecture and the efficiency inherent in that architecture. The only benefit for such a metric is on the large scale where the utility or central cooling plant sizing can be done. For example, in our data center of Table 6.1, the power density is 136.5 W/sq ft. Since the total square footage is known, the total connected IT load can be determined. Knowing something about the cooling efficiency and power distribution (see PUE later in this chapter), a preliminary overall utility feed and chiller plant sizing can be made. That is where the use of the metric should stop. Data center power density, even in an individual data center, can vary greatly across the room. It is also a function of layout efficiency. These two variables preclude its use in designing specific cooling or power architectures. Instead, the more appropriate metric is discussed next.

*Power/rack*—(kW/rack) The ability to efficiently cool or power IT equipment has more to do with local density than average data center efficiency. For this reason, a more important power density metric is at the rack level. This is further discussed by Herrlin [7]. Racks of 2 kW or of 20 kW can be in the same data center and in close proximity to each other. Providing power or cooling to these racks may require very different solutions. An often misunderstood aspect of this metric is any assumption of implied efficiency. The lower-density rack is easier (i.e., requires less engineering) to power and cool, but that has no direct correlation to efficiency. In fact, the higher density and required care in design and operations will often lead to a more efficient (and less costly) data center. This is further discussed by Patterson [8]. W/sq ft does not provide any visibility to these critical aspects. In our example data center (Table 6.1), we have a very homogenous rack loading and have a peak connected load of nearly 7 kW/rack and an average 5.4 kW/rack. These values allow the engineer visibility to what the infrastructure must be able to do.

**Table 6.2** Uptime Institute cost model

| Tier | \$/kW[a] | \$/sqft |
|---|---|---|
| I | \$11,500 | |
| II | \$12,500 | |
| III | \$23,000 | |
| IV | \$25,000 | |
| All | | \$300 |

[a]kW of redundant UPS capacity for IT

*Cost/area*—(\$/sq ft, €/m$^2$) As with the power per area, this metric can be misleading. In fact, unlike the power/area which has some use, cost/area is of no use at all. As demonstrated above with the percentage of costs, the CSA costs will likely be 1/3 or less of the total cost of the data center, with power and cooling representing larger fractions. The data center power density can vary greatly and total square footage is independent of this value and as such will invariably lead to poor results. The most problematic example of this is the using \$/sq ft to estimate the cost of the data center. This type estimate can work well for general office or industrial space but is of little value in the data center. Consider two data centers; one (a) with two hundred 2 kW racks, the other (b) with twenty 20 kW racks. The floor space of (a) will be 10× of (b); however, the total connected IT load of (a) and (b) are equal. At first order, the power and cooling infrastructure costs can be considered roughly equal between the two (actually it may be slightly less for (b) due to less distribution of power and cooling), while the floor space for (a) is the only aspect that is 10×. Using an estimate of \$/sq ft could lead to very serious problems. The Uptime Institute [9] discusses this in detail.

*Cost/power*—As described above, the majority of the data center cost is in the power and cooling infrastructure. Therefore, a better metric is that of cost/kW of connected IT load. This metric takes out of the equation, the uncertainties (and the ability to falsely skew the metric) of layout efficiency, and rack or room power density.

Compounding the issue is the concept of availability. (See rating systems later in this chapter—Sect. 6.10.5, but very briefly Tier I represents the minimal configuration for a data center in terms of availability, while Tier IV represents the highest availability with redundant infrastructure components and distribution systems; higher availability → more hardware → higher costs) These more robust designs will certainly cost more in operations as well. This is why the Uptime Institute [9] has gone beyond either of these metrics and proposed a combination of the two. Their recommendation for cost estimation is shown in Table 6.2.

The model is a combination of costs, cost/power plus cost/area to get a total capital construction cost. An estimate of the cost for the example data center would be based upon Tier level or the extent of the availability built into the infrastructure. Assuming (e.g., purposes) that the data center needs to be a Tier III data center, the capital cost for the data center (this does not include the IT hardware) is based upon the connected IT load and the area of the data center.

$$\text{Cost} = \$12\frac{500}{\text{kW}} \times 273\text{kW} + \frac{\$300}{\text{sq ft}} \times 2,000\,\text{sq ft} = \$4,012,500.$$

The Uptime cost model estimates the data center to cost just over $4 million dollars. It is instructional to point out that the costs associated with square footage was only $600 K with the costs associated with the power and cooling (kW of connected load) was over $3 M.

## 6.7 Infrastructure Efficiency Metrics

### *6.7.1 PUE*

The primary data center efficiency metric is PUE or power usage effectiveness. The metric is defined as:

$$\text{PUE} = \frac{\text{Total data center annual energy}}{\text{Total IT annual energy}}.$$

PUE, in its most basic use, provides visibility to the burden or overhead that the data center infrastructure causes on the site. For example, a PUE of 2.0 means that the total energy used by the data center is 2× that used by the IT equipment or that the infrastructure energy use is equal to that of the IT energy. See the subsequent section on rating systems for data on some actual performance data in the industry.

The total data center energy should include all energy needed to support the data center; primarily (but not limited to) the energy lost in power conversion and distribution from the utility to the IT equipment (transformers, UPS, and PDU), energy used to cool the data center and support areas (chiller plant, CRAC/H, and/or other cooling hardware), lighting, security systems, and the IT equipment itself. The IT annual energy should include the servers, network, and storage equipment in the data center.

The Green Grid has a number of references available to look at the requirements further, see ref. [10], for example. In addition, the definition of PUE has been the subject of cross-industry collaboration among The Green Grid, ASHRAE, the DOE, the EPA, and others [11]. The use of PUE is increasing and is being included in rating systems developed by these organizations.

PUE, in its most rigorous form, would be calculated on an annual basis with on-line instruments and actual measurements tracking all of the energy inputs. This should, for example, include down to the level of the diesel fuel used to operate emergency back-up generators periodically to keep them available. It should also include the HVAC of the equipment rooms where the UPS and chillers are housed. If there is a service elevator for the data center, it should include that energy as well.

**Table 6.3** Data center energy use summary

| | |
|---|---|
| Annual IT energy use (kWh) | 2,391,480 |
| Annual energy loss in UPS to IT (kWh) | 144,000 |
| Annual energy loss in PDU (kWh) | 96,000 |
| Annual energy use in lighting/security (kWh) | 60,000 |
| Annual energy use in chiller plant (kWh) | 450,000 |
| Annual energy use in CRAC units (kWh) | 700,000 |
| Annual energy loss in UPS to cooling plant (kWh) | 50,000 |
| *(Sum of all the above infrastructure and IT energies)* | |
| Total energy use (crossing the data center boundary) (kWh) | 3,891,480 |

The above is not meant to be comprehensive but to define by example the range of energies that need to be accounted for. In the simplest example, a stand-alone data center, running purely on grid electricity (i.e., no diesel, gas, steam, chilled water, etc. is fed to the site), the PUE would simply be the annual energy use purchased from the utility by the site divided by the annual energy used by IT equipment. Rarely the calculation of this is straightforward but its consideration demonstrates the intent of the PUE metric.

The calculation of PUE for the example data center of Table 6.1 would be as follows (Table 6.3).

$$\mathrm{PUE} = \frac{\text{Total data center annual energy}}{\text{Total IT annual energy}} = \frac{3,891,480\ \mathrm{kWh}}{2,391,480\ \mathrm{kWh}} = 1.63.$$

Mixed-use facilities are those that are not stand-alone data centers, such as a large office or manufacturing building which also houses a data center supporting that operation. The PUE for these facilities requires more analysis and likely an allocation of central services, partially used by the data center.

An illustrative example would be a large campus with a central chilled water plant (chillers, cooling towers, distribution pumps, etc.). That chilled water plant supplies the data center CRACs with chilled water and provides the path for rejection of heat to the outdoor ambient. To accurately calculate PUE, the cooling energy must be known and properly attributed to the data center and other campus uses. The cooling system energy use is the energy used by the chillers, the cooling towers, chilled and condenser water loop pumps, and even including the HVAC and lighting in the space housing the central plant. Once that total annual energy use is known, the data center portion must also be calculated or apportioned. The simplest method would be to apportion the energy based upon the flow rate of the central plant that goes to the data center CRACs, e.g., if they get 35% of the total flow, then 35% of the total chiller plant energy should be included in the numerator of the PUE calculation. A better estimate would be to include the energy carried by each flow stream to the different portions of the campus. The simplest process would be to consider the supply return chilled water differential temperature and a product of the actual flow rate to each zone served by the central plant, the zones with a higher differential temperature using a greater portion of the plant's capacity.

The above methodology is quiet involved when, to get the annual energy use, the differential temperatures and flow rates must be integrated over the entire year to determine PUE.

The challenge of the determining PUE to the finest degree possible may preclude its accomplishment. However, PUE can and should then be calculated using an estimate or simpler methodology to gain the benefit of tracking the metric. The data center operator must understand the rigor under which their calculation must be performed. If they are submitting their data center into a specific rating system for national recognition, there will likely be fairly rigorous requirements. The operator should consult the specifics of the local authority. On the other hand, if PUE is being used to provide a measure for internal continuous improvement, a monthly "snapshot" measurement and calculation by the site mechanical engineer would more than likely suffice to obtain sufficient resolution in the metric to track the desired improvement.

An alternate method to the above would be to assign a source-/site-based multiplier (as discussed in a later section, see Sect. 6.7.1.3) that includes an estimate of the general efficiency of the production of chilled water and multiply that by the energy content of the chilled water crossing the data center boundary (based on $\Delta T$ and mass flow rate). While this process may be suitable for submission to the EPA who has published the values, it would be of little use to the site if they were actively working towards a central chilling plant efficiency improvement plan. In summary, the method and specifics of data collection for calculating PUE must be goal-acceptable. The purpose for tracking PUE must be considered when the level of effort in measuring it is decided.

#### 6.7.1.1 pPUE

Because PUE is so broadly used as an energy efficiency metric in data center discussions, it can be misapplied, and as such further refinement of its definition is an ongoing process. One example of such misuse comes when PUE is applied to only a section of a data center's infrastructure. An example of this would be the use of a containerized solution for housing IT equipment. This strategy has a number of advantages, such as a quick deployment, high density, and an integrated, single-supplier IT and infrastructure solution. These containerized solutions generally come preloaded with IT equipment and merely need chilled water, network, and power connections. The problem becomes using PUE to define the efficiency of the container. There will likely be some existing site-based infrastructure that supports the container solution and its energy use must be considered to understand the true efficiency of the installation. For example, consider a container based on the example data center of Table 6.1. For discussion's sake, assume that the IT equipment is housed in containers that also house energy equivalent equipment in place of the CRAC, PDUs, and lighting. On site, there was an existing chiller for a chilled water source and a UPS for high-quality reliable 480 V AC power (Table 6.4).

**Table 6.4** Site and container energy use

| | |
|---|---|
| Annual IT energy use (kWh) | 2,391,480 |
| Annual energy loss in PDU (kWh) | 96,000 |
| Annual energy use in lighting/security (kWh) | 60,000 |
| Annual energy use in CRAC units (kWh) | 700,000 |
| *(Sum of all the above infrastructure and IT energies)* | |
| Total container energy use (including IT energy) | 3,247,480 |
| Annual energy use in chiller plant (kWh) | 450,000 |
| Annual energy loss in UPS to container (kWh) | 144,000 |
| Annual energy loss in UPS to cooling plant (kWh) | 50,000 |
| Total site energy use (external to container) | 596,000 |

Using a physical boundary of the container, a "PUE" could be calculated of total container energy (3,247,480 kWh) divided by the IT energy of 2,391,480 kWh. The "container PUE" would then be equal to 1.36. Quotes are used as these values are not proper use of the definition of PUE, but at the same time could be a beneficial metric, e.g., to track the container efficiency over time if improvements are made to it. To provide this PUE-like metric capability, The Green Grid has defined another metric, pPUE. The above math demonstrates the calculation of pPUE. Its value can be in optimizing a subsection of a data center or container over time, or, for example, of comparing two containers that are supported by a common central infrastructure. For this example, all of the energy losses remain the same as the based case but in this one the boundaries are what define the pPUE and the PUE. The pPUE is 1.36 and the PUE remains 1.63.

### 6.7.1.2 Power or Energy?

PUE was initially conceived of as a very simple planning metric. What is the connected IT load in comparison to the connected total load? From the very early beginnings, its value to the user has increased as its definition has been filled out, with energy/energy use being the better use in terms of energy efficiency. Energy is what costs the data center money through the monthly utility bill, and energy will also drive the carbon or water footprint of the data center. At the same time, an instantaneous PUE will have merit when used properly. For instance, determining a data center's instantaneous PUE (perhaps in various free-cooling modes) will indicate how well parts of the system are designed for the varying environmental conditions. Similarly, knowing the highest PUE (lowest efficiency) does give some indication of the power and cooling infrastructure that is needed to handle peak loads. It can also help in optimization of systems. For instance, if an operator recorded his PUE for a given month, the following year (assuming environmental conditions are similar) re-calculating PUE for that month will be of benefit to determine which way efficiency is trending. The strict definition of PUE, being an annual average cannot do that with as much precision. PUE can be beneficial as power or energy (short and long term). The key to its use is to also be very clear about the application of the PUE number stated.

#### 6.7.1.3 Site or Source?

Another aspect of PUE that needs to be made clear in its discussion and use is that of site vs. source energy. PUE, per the consensus of the groups listed in ref. [11], is a source energy term. This implies that the energy consideration must include the generating source of the energy. This will provide a better accounting for the inefficiencies of that generation. The specific use of PUE may or may not benefit from this inclusion; however, as described in ref. [11], the distinction between site and source is generally less profound than would be imagined for PUE.

PUE when comparing two data centers in different regions of the country for their overall impact on the electric grid, or perhaps their carbon footprint, a source-based PUE would have benefit. For a single data center with a goal of improving their local site operations by tracking and working to lower their PUE, a site-based metric is sufficient and a source-based metric would simply create additional work.

The simplification comes from the fact that as a ratio if the data center is essentially fully driven by grid-based electricity, the site-based and source-based PUE are identical. This is not the case when the site brings in natural gas, or district chilled water, or any other energy stream. In these cases, EPA [11] provides factors to multiply the energy content by; thereby converting the entire metric to a source-based value. But as the majority of data center energy is electricity, the additional math is fairly limited.

In the discussion above around a mixed-use facility and the example discussed attempting to allocate the energy used to create the chilled water, a simplification can be had using the source-based multipliers. Any time the central plant shares an energy stream (chilled water, diesel, natural gas, steam, etc.) with the data center and other portions of the campus, a reasonable PUE may be calculated by using the multipliers to put the central plant as a "source" rather than on the site. Inherent in these multipliers are equalizing factors for energy generation for varying fuels or types of energy so they can be used when other allocation methods may be too expensive or complicated. The main concern in this method is that using PUE for optimizing the process at the "source" (in the example above, this would be the central chiller) becomes impossible because it assumes a given efficiency for that process and as such is not measured.

It is expected that the source-based multipliers may be modified as data continues to be collected and more granular.

#### 6.7.1.4 PUE Gaps

Understanding PUE is fundamental to its appropriate use. That understanding should also include knowing when PUE should not be used. One example of this would be in the case of an IT refresh. In our example data center, we are using the 2006 vintage servers of ref. [2]. Table 6.5 shows the revised values if the data center would replace all of the servers with the 2009 vintage.

**Table 6.5** Data center with modernized IT

| | |
|---|---|
| Area (sq ft) | 2,000 |
| Number of racks | 500 |
| Servers per rack | 20 |
| Total servers | 1,000 |
| *Server vintage (Year)* | *2009* |
| Average server power[a] (W) | 194 |
| Average data center connected load (total of all IT) (kW) | 194 |
| Peak server power[b] (W) | 307 |
| Annual IT energy use (kWh) | 1,699,440 |
| Annual energy loss in UPS to IT (kWh) | 102,330 |
| Annual energy loss in PDU (kWh) | 68,220 |
| Annual energy use in lighting/security (kWh) | 60,000 |
| Annual energy use in chiller plant (kWh) | 319,780 |
| Annual energy use in CRAC units (kWh) | 700,000 |
| Annual energy loss in UPS to cooling plant (kWh) | 35,531 |
| Total energy use (crossing the data center boundary) (kWh) | 2,985,301 |
| Total IT energy use (kWh) | 1,699,440 |

[a]Assuming 30% annual average utilization
[b]Assuming 100% CPU utilization

It is assumed that the energy loss or use in the UPS, PDU, and chiller all scale with load. This assumption is somewhat suspect as generally these systems are more efficient fully loaded and less efficient lightly loaded, but assuming a constant efficiency is suitable for our demonstration purposes. We also assume that the lighting and security loads remain constant, as does the CRAC unit energy.

The new PUE with newer servers is 1.76, an increase from 1.63. Two conclusions should be taken away from this. First, any update of IT equipment should also cause a review of the base infrastructure to ensure it is still sized appropriately. Second, while PUE is a powerful tool, it does not handle improvements in IT efficiency properly and its use should be focused primarily on the infrastructure and its improvements. From ref [2], the new servers show a performance increase of roughly 2.8× (at 30% utilization, similar numbers for other utilizations) and a power decrease of 29% (again at 30% utilization). For a significant improvement in performance and reduction in energy use, PUE went up. Understanding PUE and its basis is critical to the proper application to the energy efficiency improvements.

### *6.7.2 DCiE*

*DCiE* was a metric conceived of at the same time as PUE and is simply its inverse.

$$\mathrm{DCiE} = \frac{1}{\mathrm{PUE}}.$$

Each metric had its proponents. DCiE was considered the more technically appealing by engineers who appreciated that it was an efficiency measure, and the ideal value of 1 (or 100%) was approached from below. PUE, on the other hand, was preferred by operation staffs as it gave them better visibility to the "tax" they had to pay for their infrastructure. Anything over 1.0 (1.0 = ideal—all energy used solely for the IT equipment) was simply a burden on their energy bill and cost them money. With PUE, the infrastructure burden was easily seen with respect to the IT load. PUE is optimized by approaching 1.0 from above.

Both metrics represent the same thing equally; they are just of a different format. However, the common usage has been so overwhelmingly that of PUE that DCiE has fallen out of favor, even to the point that The Green Grid (who originally defined both) has dropped DCiE from its vernacular.

### *6.7.3 ERE*

As discussed in previous sections, with such widespread use of PUE, misuse of the metric is an unfortunate but not unexpected occurrence. One prime example of this are reports of a PUE below 1.0. With PUE being defined as total energy over IT energy, this value is impossible. The claims were being made in the context of waste energy from the data center being beneficially used elsewhere. While the concept is admirable, the metric is not amenable to this variation. To resolve this, The Green Grid defined a new metric [12], energy reuse effectiveness (ERE), to do just that; to account for energy reuse.

$$\mathrm{ERE} = \frac{\text{Total energy} - \text{Reused energy}}{\text{IT energy}}.$$

There is an alternate construct for ERE.

$$\mathrm{ERE} = (1 - \mathrm{ERF}) \times \mathrm{PUE},$$

where ERF or energy reuse factor is defined as

$$\mathrm{ERF} = \frac{\text{Reuse energy}}{\text{Total energy}}.$$

One benefit of the form of ERE is that it has the same denominator as PUE; total IT energy. This allows the comparison and contrast of the PUE and ERE values.

ERE has a theoretical ideal value of 0. This would imply that all of the energy brought to the data center (IT and infrastructure) was reused beneficially somewhere else. While this is, from a practical purpose, very unlikely, the intent here is more to draw comparison to PUE where the value cannot go below 1.0. ERF has a theoretically ideal value of 1.0 which states that 100% of the energy is reused after the data center.

Note that ERE, when no energy is reused simply, is the same as PUE. It could be argued then that a single metric, ERE could suffice. However, there is benefit in having both. Consider a very efficient data center with a PUE of 1.2 and a decent ERF of 0.25 (25% of the total energy gets reused). The ERE would be 0.9. Conversely, a relatively inefficient data center with a PUE of 2.0 may be doing very well with energy reuse and be reusing 55% of its total energy (ERF = 0.55). In this case, the ERE is again 0.9. However, both data centers have very different profiles and should be each working on different operational parameters of their data center. One may separately want to lower its PUE, while the other considering improving its ERF. In both cases, ERE would benefit, but as a single stand-alone metric it does not provide visibility on what to focus on.

### *6.7.4 CADE*

*CADE*—Corporate average data center efficiency was introduced by the Uptime Institute and Mckinsey & Company [13].

$$\begin{aligned}\text{CADE} = {}& \text{Infrastructure utilization} \times \text{Infrastructure energy efficiency}\\ & \times \text{IT utilization} \times \text{IT energy efficiency}.\end{aligned}$$

When first presented, several of the terms had not been resolved and CADE remained a proposed metric that could be used when the four components were properly resolved. Because of this, CADE is not frequently referred to, but its consideration does yield some informational value and insight.

The first concept to be dealt with is that of utilization. This can best be discussed through an example. The sample data center above has a peak connected load of 427 kW (one thousand 427 W servers) (assuming this peak will ever occur would generally be an overly conservative estimate; in addition 427 W is the peak power consumption, not the nameplate rating associated with the PSU. Nameplate ratings would not occur in any normal operations (see ref. [3]). However, we will assume an HPC type workload where reaching the 427 W value is more plausible—all servers at peak power consumption would depend on workload and data center type. For the 427 kW to be operated, the data center would need a power distribution system and a cooling system. Those systems would likely be somewhat bigger than the connected load based on a combination of having a safe buffer, potential growth, and the digital nature of adding increments of capacity. For example, the cooling units may come in 150 kW capacities. Four of these would yield a 600 kW cooling capacity. In addition, assume the power system has a 700 kW capacity. What would be the utilization value? In the ideal world, the two capacity values (power and cooling) would be equal. However, this is rarely the case.

Consider choosing the cooling for the utilization evaluation. In this case, the utilization is 0.71. This value could then be used in the CADE calculation. However, if utilization is to be improved, using the better (higher) value may be misleading, clearly the power system is further underutilized than the cooling system and may be the right area to focus on improving utilization. But if the power system were selected to represent utilization, it can be seen that it may never be fully utilized. If the power system is modified, it could be that the cooling system could eventually be the subsystem that is limiting growth as its utilization may near the acceptable upper limit. The recommended method is to use the lower of the infrastructure utilizations. But at the same time it is useful to trend all of them to know if any are nearing 100% utilization (reaching full capacity), then the focus should turn towards balance instead of utilization improvement. Belady discusses the use of a spider chart to assist in this analysis [14]. The example above was oversimplified, looking solely at power and cooling. A more thorough utilization analysis should include a wider range of facility parameters. Cooling, for instance, should include both the central chiller plant (total heat removal capacity) and volumetric airflow. It is certainly possible that these two are not precisely balanced (particularly, in the case of chilled water-based CRACs), and their utilizations are different. Volumetric space, floor (area) space, and weight are also parameters that could limit a data center. And while it more strongly borders the IT side of the equation, the network's capacity (to/from the data center) could be a utilization limiter.

The second component of CADE is that of infrastructure energy efficiency. Fortunately, this metric currently exists and is simply DCiE or 1/PUE.

The third component of the metric is IT utilization. McKinsey originally suggested processor utilization; however, this makes only slightly more sense than suggesting airflow as the way to measure infrastructure utilization. One reason for this could be the frequently made incorrect assumption that CPU power represents a majority of server power, when in fact it is typically in the range of 25–40% of the total server power. Processor utilization is often measured and can represent the weak link in the chain for IT utilization, but any server could have different levels of capacity, from the CPU, to the amount of memory, to the local storage, to the I/O at the server level. The ideal server has a balance of each of these and as such no one component is always the utilization limiter. But beyond the servers, the IT system could be bound by total storage or network capacity. As with the infrastructure utilization, the full range of parameters needs to be tracked and understood to properly improve the overall utilization.

The fourth component in CADE was that of IT energy efficiency and was left as a future metric to be developed. However, the form of this metric (IT hardware energy efficiency) is in the same form as PUE and invites a PUE-like metric which is discussed in the next section.

The theoretically Ideal value of CADE is 1.0 or 100%. This value cannot be achieved, but is included to enhance the reader's understanding of the math associated with CADE. CADE values can actually be very low. Consider our example system. The PUE is 1.63. The lowest infrastructure utilization is 0.61. The IT utilization has been taken as 30% (and this is better than typical).

itEUE is not known but a rational estimate for a well-designed modern server could be 1.2. CADE then is:

$$\text{CADE} = \text{Infrastructure utilization} \times \text{Infrastructure energy efficiency} \\ \times \text{IT utilization} \times \text{IT energy efficiency}.$$

Using PUE values for energy efficiency:

$$\text{CADE} = \text{Infrastructure utilization} \times \frac{1}{\text{PUE}} \times \text{IT utilization} \times \frac{1}{\text{itEUE}}.$$

And solving:

$$\text{CADE} = 0.61 \times \left(\frac{1}{1}\right) 0.63 \times 0.3 \times \left(\frac{1}{1}\right) 0.2 = 0.09.$$

This value is not atypical of the current state of the art for data centers. The best way to improve CADE would be to work on the four individual components separately; tracking each to reduce each sector's contribution to inefficiency.

Corporate average data center efficiency is one of a few metrics that are composed of other metrics that could be used to make changes. Generally, combined metrics tend to defuse the problem area and make it difficult to focus on the issue that most needs addressing. A closer evaluation of CADE shows that it does represent a unique combination of capital expenditure (both infrastructure and IT) utilization and operational expenditure inefficiency (how much of the energy purchased does NOT go to doing computations) in a single number. The above description of utilization challenges and metrics that have yet to be fully established (itEUE) make CADE a very difficult number to calculate once. Tracking it over time would likely be beyond any temporal value. But it's calculation as a one-time snapshot could instead alter the business direction. In the example above, a CADE of 0.09 was calculated. In an abstracted manner, it could be argued that only 9% of the combined capital (infrastructure and IT) and energy expenditures are being used to do actual computations. This value may actually motivate the owners to either improve efficiency or perhaps move their data center utilization to external cloud providers where only time and compute cycles are purchased and it could be expected that the large cloud service providers would have optimized their data centers for higher utilizations and efficiencies. The best way for some data centers to get to higher efficiencies may be through a change in business processes vs. changes to their specific data center operations.

### *6.7.5 itEUE*

itEUE as noted above could be the fourth metric in the CADE calculation. If PUE is total energy divided by IT energy, itEUE would be total IT energy divided by computational energy.

$$\text{itEUE} = \frac{\text{Total energy into the IT equipment}}{\text{Total energy into the compute components}}.$$

As PUE identifies the infrastructure "tax" on the IT equipment, itEUE would identify the IT internal "tax" on computing. This tax would be composed of the IT internal fans, power supplies, and voltage regulators. The compute components would be the CPU, memory, and storage. itEUE can practically only be a theoretical construct or perhaps measured in the laboratory. The challenge for anything more than this would include adding significant cost and measuring points and IT output for the additional data. Complications stem from the extensive variability in IT configuration as well as the variability in the energy use of the different components at different workloads and with the IT equipment in different environmental conditions. Identifying and locking these down (workload, inlet temperature) would drive itEUE to a laboratory-only measurement. Based on this, it is expected that itEUE will remain a theoretical construct or an estimated value. As discussed above, it lends itself well to CADE estimates but the true value of itEUE is more likely in the discussions around more advanced and more integrated infrastructure solutions. Difficulties with the simple concept of PUE come about when the line between infrastructure and IT are not clear. For example, many large supercomputers come with an integrated cooling system. Some components of these systems would generally be part of a data center room infrastructure in a more standard situation, but in these large specialized systems the standard IT servers, storage, and network are also not as easily distinguishable to assist in drawing the line between infrastructure and IT. itEUE, as well as the concept of pPUE, can be useful tools in being able to better compare different data centers. Again cooling provides an approachable example. Generally, there are fans in both the room and the IT equipment. Consider technology advancements such that the IT equipment fans could be deleted and simply the room fans drove all the cooling. The IT energy use would go down and the PUE would go up, even with perhaps a reduction in total power. Another example application could go the other way. In this case, the room fans are deleted and the IT equipment fans will do all of the air movement. Total energy goes down, IT energy goes up, and PUE is reduced. In the first example, the PUE went up, in the second the PUE went down. Both could use less total energy but PUE would suggest the second is a better choice, which may not be the case. In fact, based on the fan affinity laws (which state that fewer larger fans will be more efficient than more smaller fans) moving to eliminate the smaller server fans should be the more efficient direction, but PUE indicates the opposite. If the question were looked at in terms of the theoretical construct of a combined itEUE and PUE, then the metric would drive the correct behavior.

## 6.8 Compute Efficiency Metrics

The data centers purpose is to do useful computational, transactional, or archival work. The ultimate goal for energy efficiency would then be work divided by the energy required to produce it. PUE and the other infrastructure metrics have not covered the IT output aspect of the data center and have only covered the energies associated with them.

In this context, PUE simply uses the IT energy consumption as a proxy for IT output. It assumes that if there is a fixed quantity of energy available to the data center, then the same amount of work could occur if PUE is comparing the data center this year to last year or it is comparing two different data centers in different regions. In both cases, with IT Energy as the numerator there is an implicit assumption that the amount of work produced by either could be the same. With an understanding of this assumption, PUE has been shown to be a very effective tool for infrastructure improvements, but ineffective when the actual IT capability is included. For example, in the preceding section on PUE, it was shown that replacing older servers with newer servers that were both more capable and used less energy, the PUE went up.

The ability to define useful work in the data center and then divide that by the total energy consumed would provide the most complete metric for truly assessing data center efficiency. Unfortunately, that metric may never come. The primary issue is that defining useful work is problematic. Even for a single data center useful work can change over time. Data centers are extremely dynamic and the server loading and the specific data center tasks regularly change. Over any given month, the balance of work tasks in a data center will shift, so measuring a consistently defined "useful work" is the problem.

Useful work for an enterprise class server may be the number of emails that have been processed through the data center, while an HPC data center useful work could be the number of iterations completed on a given CFD analysis. Additional servers may be added to the enterprise-based data center, so it is now processing emails as well as the payroll for the corporation. The useful work remains a difficult concept.

Attempts to look at some specific computational metric have merit but will generally lead to an overly precise workload type. For example, an Internet search engine site could use search queries answered as the right useful work, but clearly only valid for that type data center. The potential to use information packets transmitted into and out of the data center could be a way to get to a common numerator in the metric. However, the network traffic as an indicator for productivity on one workload type is not the same as on another.

Server CPU utilization could possibly be a measure of output. But as seen in the section above, newer more efficient servers had a 2.8× improvement in output as compared to the older generation at the same CPU utilization. In addition, the software itself may be variable in terms of its efficiency to produce work in the same time and/or for the same energy input.

No clear path forward to a single metric is currently seen. The Green Grid and other organizations such as SPEC [15] are continuing to work on these aspects but for the interim, alternate methodologies must be adopted to gauge efficiency.

Another challenge to the compute efficiency metric is the premise that the data center cannot or should not have its ongoing work stopped and a benchmark be loaded across the entire IT suite and run to gain some measured output to use as a figure of merit. The only exception is HPC; this was discussed earlier briefly and

**Table 6.6** Proxy recommendations of The Green Grid [17]

| Proxy # | Method |
|---|---|
| 1 | Useful work self-assessment and reporting |
| 2 | DCeP subset by productivity link |
| 3 | DCep subset by sample workload |
| 4 | Bits per kiloWatt hour |
| 5 | Weighted CPU utilization—$\text{SPEC}_{\text{int\_rate}}$ |
| 6 | Weighted CPU utilization—$\text{SPEC}_{\text{power}}$ |
| 7 | Compute units per second trend curve |
| 8 | Operating system workload efficiency |

will be discussed further in a later section. But beyond that application it would not be a useful endeavor to attempt to get a specific measurable IT result from a test or benchmark instead of running the data center for its intended purpose.

### *6.8.1 DCeP*

The Green Grid's first attempt at a metric for data center compute efficiency metric was DCeP [16]. This metric is defined as:

$$\text{DCeP} = \frac{\text{Useful work produced in a data center}}{\text{Total data center energy consumed to produce that work}}.$$

While the metric is elegant in its simplicity, it is difficult to actually implement in a data center and is also limited to measuring a single data center against itself at a later time. Inherent in this process, is the assumption that over that time period (first measure of DCeP to the second measure of DCeP) the mission of the data center and the definition of useful work is consistent.

### *6.8.2 Proxies*

The limitations of the DCeP metric have caused The Green Grid to look into alternate methods that could be used to compare across data centers with different mixes of IT equipment and missions. The Green Grid has published recommendations [17] to use various proxies for productivity or useful work that could be more readily measured or assessed. These eight suggested proxies are listed in Table 6.6.

Each of these proxies has advantages and disadvantages, and none are ideal. The methods each have their use case where they would be the most appropriate. Detailed study of the proposed proxies and their pro's and con's is required before any are chosen to attempt to measure the computing energy efficiency of a data center.

In this chapter, the values for performance come from the SPECpower metric [18] and in essence the proxy method number 6 has been loosely applied. Further information can be found at ref. [2].

Beyond these proxy-based methods, work towards an actual performance per Watt type metric may possibly come from a combination of software (workload based) counters with IT and infrastructure energy use. Software development kits are available that can embed counters in the code that, when applied to useful work output units and progress, can give a "useful work" measure that is transportable with the application between data centers and across different IT equipment and operating systems. Much work remains to get industry agreement on the use of such counters but this area does seem potentially fruitful for the next step of a performance per unit energy metric.

### *6.8.3 The HPC Case*

HPC represents a special class of data centers that has some further work accomplished in this space; however, even here this does not appear to be a single metric that provides a true efficiency. Twice a year the Top500 organization publishes a list of how the top HPC supercomputers in the world perform on a tightly specified benchmark; LINPACK [19]. To be rated on the Top500 list, a LINPACK result must be submitted. LINPACK solves a dense set of linear equations. The advantages of LINPACK are that it can or has been run on a majority of HPC machines and it also provides a good stress test as the benchmark causes the cluster to run near the IT equipment's TDP, providing a challenge to the power and cooling infrastructure. The disadvantage of LINPACK is that it is an artificial workload and its performance measurement only resembles a narrow set of HPC workloads. There are a full range of HPC benchmarks and these can be run on various HPC machines to determine the best performance of a proposed cluster for a given capital expense and energy input. Subsequent to the Top 500, another rating system, the Green 500, has come into being [20]. This system uses the same performance benchmark (LINPACK) as the Top 500, but then divides the performance result by the energy used to obtain it. While this seems desirable in its simplicity and worthwhile in that it gives output divided by energy input, it should only be considered in those as guidance in those cases where the actual workload is very similar to the LINPACK operation. Unfortunately (because it requires the most work), the best way to ensure the highest efficiency for a given energy input is to run the specific intended workload on the cluster or a subset of similar compute nodes and to measure performance/energy. If this cannot be accomplished, the second alternative is to look at more specific application benchmarks that have been run on those systems (e.g., CFD, Monte-Carlo, Oil and Gas, Climate, etc.) that are similar to the workload machines intended use. These benchmarks will yield a more efficient choice than the Green 500 list.

One other aspect of the Green 500 list is that the energy consumption is solely for the machine itself and does not include any data center infrastructure energy use, so it really is not a data center efficiency metric but a compute metric. This points out another subtlety in the HPC case that needs to be considered. Invariably the useful work output in HPC is based on a specific supercomputer or cluster. This cluster is in a data center. The issue is that only in certain cases is that cluster alone in the data center. In most cases there are other IT equipment in the same space so while it may be possible to measure the cluster's work output, measuring energy input to the data center (vs. just the cluster) will measure energy consumption for more than just the specified work output.

## 6.9 Component Metrics

There are a wide range of component and subsystem metrics that define some aspect of the overall efficiency chain. Generally, these are not metrics that can be worked towards improvement as part of the data center operations, but instead represent the efficiency of the component as installed. They are included to be informational and to provide a broader understanding of the overall efficiency chain. These are broken down into infrastructure component level metrics and IT component level metrics. Theses metrics focus on a specific piece of equipment and can drive efficiency in those components. However, there is no direct way to build an overall data center efficiency model or result from the individual components. Certainly use of, and optimization of, these components will improve the chance for a good data center efficiency, but the overall design of the infrastructure and IT system will be the fundamental driver of the total efficiency. The most efficient components in the world, put together poorly, will still yield an inefficient system.

### *6.9.1 Infrastructure*

*Chillers*—Power/power—(kW/ton) For the evaluation of the chillers. Generally, defined as a COP for the chiller, this is a fairly typical mechanical engineering metric for HVAC applications. A higher COP indicates that the chiller system can move more heat from the data center for a lower energy input. COP would generally be a procurement focused metric or operationally for the facilities staff in monitoring the chiller system itself.

*Room air distribution*—Airflow/power (CFM/kW) The data center airflow units (CRACs or CRAHs) deliver airflow through the data center to IT equipment. The ability to do that efficiently will contribute to the overall data center efficiency. A measure of the airflow divided by the power to move it provides a basis for CRAC or CRAH efficiency. As with any metric, there is risk of not capturing true performance unless other aspects are understood or held constant. For example, one way to improve airflow/power would be a lower-grade filter. Air would flow more easily, improving the metric; however, the unit has been made to sacrifice performance in other areas.

*Rack cooling index—RCI*—The RCI metric [21] provides a very simple method to score airflow distribution within the data center, particularly at a rack by rack level. Ideal airflow management would provide the same optimized temperature to every server within the rack. However, when problems exist in the data center, it is common to find that the base supply airflow is too cold and that the airflow management is imperfect and that recirculation brings warm server exhaust air to the front of some servers that do not receive the right amount of cooler supply air. $RCI_{HI}$ is simply the percentage of servers in a given rack that are below the recommended maximum temperature (above the maximum they are likely suffering recirculation issues). $RCI_{LO}$ is simply the percentage of servers in a given rack that are above the minimum recommended temperatures. For both $RCI_{HI}$ and $RCI_{LO}$, the optimum value would be 100%.

*Airflow/airflow*—(CFM/CFM) A good indicator of total room airflow efficiency from a design or capacity perspective is the ratio of total room supplied airflow to the total IT required airflow. This metric can be quite varied and dynamic as well. The theoretical ideal would be a value of 1.0, indicating that every CFM of airflow provided by the CRACs or other air-movers was used by the IT equipment and that no excess was provided. In a very rigorous airflow containment or localized cooling solution, this arrangement could provide a ratio of just above 1.0 (actual best case performance likely in the 1.1–1.2 range).

In data centers with poorly managed airflow, the ratio could range from well below 1.0 to high numbers such as 5 or more. Each case is worth a brief review. In the case of a value below 1.0, the room airflow rate is less than the IT flow required. This implies that there is recirculation occurring. One minus the ratio's value is very roughly the same as the number of servers that would be getting recirculated air. Leakage in the raised floor or supply ducting would compound the problem. To overcome the inadequate airflow, the cooling system temperature would need to be set otherwise unnecessarily low to provide the entire room with acceptable cooling. While this data center may save energy in room-level fans, it will use excess energy in the chiller system, and as such is not a recommended approach. Values well above 1.0 indicate that the excess air is being provided to the room. The room-cooling layout will dictate what value as a minimum can be made to work. A room with no orderly thermal arrangement will generally require a very large volume of air to ensure adequate cooling for each rack. The simplest thermally driven layout of hot-aisle/cold-aisle can be operated with a more reasonable ratio, saving significant amounts of fan energy. The most advanced aisle segregation methods or chimney cabinets can have values nearing 1.0.

The ratio may also be very low in the case of liquid cooling at the rack level through rear doors or fully enclosed liquid cooled racks. For these architectures, the ratio may actually be 0 or 0.1 (little to no room airflow for cooling). These systems can add to the efficiency by reducing CRAC or room fan energy but will carry a trade-off of additional rack-based fans and liquid-loop pumps. The overall efficiency must be reviewed using PUE and variants thereof.

*Return temperature index—RTI*—Herrlin provides an analogous metric to the above airflow ratio [22] but from a differential temperature basis. The ideal values and ranges are identical, but the RTI uses a ratio of differential temperatures between the IT equipment and the CRAC unit. Herrlin also goes into a detailed discussion on the range of the metric, pointing out that values below 1.0 will result in recirculation. Looking at the room air distribution using RTI as well as the volumetric ratio and RCI will give the best perspective on the health of the airflow management.

## 6.9.2 IT

*Fan—Airflow/power (CFM/kw servers) or power/power*—Similar to the room air-handling units, efficiency at the server level can be measured by looking at the flow rate of the internal fans compared to the power dissipation (the cooling load) of the server. Another way to consider this is the fan power divided by the total platform power. These two numbers have been trending downward as server design practices become more focused on energy consumption and the energy used in the fans. The two metrics listed are generally used for different purposes; even though they are linked by the efficiency of the fan itself (fan flow vs. fan input power). But in server design and data center efficiency, the value of flow/power is a good indicator of the quality of the server thermal management system. Lower required airflow for a more powerful server represents a better thermal design. Power/power yields a similar result but is more generally used in evaluating the overall power efficiency chain; particularly, if the entire cooling burden (IT through the infrastructure) is being considered.

*Power supply/voltage regulators (AC/DC/DC power)*—IT power conversion efficiencies inside the platform are based primarily on PSU and VR efficiencies. Inefficiencies generally come from power conversion performance. There are several rating systems for power supplies that are discussed later in this chapter. The IT equipment will be fed a medium (100) to high voltage (480) AC or DC power stream that must be typically converted to a 12, 5, and 1 V DC power inside the IT equipment. The efficiency of those power conversions plays a role in the overall data center efficiency. The metric can be applied to each step to understand that step's efficiency, but the overall power step down efficiency needs to consider all of them together to look at an overall power efficiency. PSU efficiencies are discussed further in Sect. 6.10. Server metrics have been proposed that is primarily AC power into the IT divided by DC power used by the IT. This metric is essentially power supply efficiency and as such has never caught on as an IT level metric.

## 6.10 Rating Systems

There are a numerous rating systems that deal with energy efficiency and other data center aspects that the practitioner needs to be aware of. These are regularly being updated so they should be consulted on a frequent basis. They cover the range from power supplies to data centers and the components within. The intent here is simply to provide awareness rather than a detailed technical explanation.

### *6.10.1 Energy Star*

The US Environmental Protection Agency runs the Energy Star program and it rates a wide array of products [23]. At a very high level, Energy Star is awarded to products or buildings that rank in the top 25% of the measured class. The EPA, in 2009, completed the first version of Energy Star for Servers [24]. Initially, this rating system included no performance efficiency as there was no consensus metric for this. In 2010 and 2011, there were significant efforts to define the SERT metric [15] for incorporation into the Energy Star program. In 2010, the EPA implemented an Energy Star program for Data Centers, and in 2011 focused on Energy Star for Storage (large-scale data storage for data centers).

### *6.10.2 EU CoC*

The European Union Code of Conduct for data centers [25] provides a resource for European data centers to submit and compare efficiency data and well as providing an extensive best-practices document on energy efficiency, from software through the data center infrastructure.

### *6.10.3 PUE Results*

Both the Energy Star program and the EU Code of Conduct use PUE (or DCiE) in their ratings systems. The 2010 results are worth a point of discussion. The Energy Star program showed an average PUE of around 1.9 and the EU CoC showed an average of around 1.8. These values should not be considered as industry averages. Consider that the participation in both programs in 2010 is voluntary. The participants can safely be assumed to be those whom are particularly interested in energy efficiency, and have likely had some level of attention to their PUEs and energy efficiency improvements. Based on this, the industry average PUE remains well above 2.0.

**Table 6.7** Uptime Institute fundamental requirements for various Tier Ratings [26]

| | |
|---|---|
| Tier I | Nonredundant capacity components and a single, nonredundant distribution path serving the computer equipment |
| Tier II | Redundant capacity components and a single, nonredundant distribution path serving the computer equipment |
| Tier III | Redundant capacity components and multiple independent distribution paths serving the computer equipment. Typically, one distribution path serves the computer equipment at any time. All IT equipment is dual powered and installed properly to be compatible with the topology of the site's architecture |
| Tier IV | Multiple, independent, physically isolated systems that each have redundant capacity components and multiple, independent, diverse, active distribution paths serving the computer equipment. All IT equipment is dual powered and installed properly to be compatible with the topology of the site's architecture |

### *6.10.4 LEED*

The US Breen Building Council administers the LEED program for Green Building Certification [26]. While some buildings with data centers and some stand-alone data centers have participated, the LEED program is not specific enough to be useful for data center energy efficiency rating. The current program includes data centers with other types of commercial buildings and does not provide enough focus on the unique high-energy use of data centers. While it provides a tool for evaluating the "greenness" of general buildings, its future benefit for data center energy efficiency work remains unclear.

### *6.10.5 Tier Ratings*

The Uptime Institute has published [27] a Tier Rating system where the specific Tiers I, II, III, and IV are precisely defined and certified. The full Tier concept is beyond the scope of this chapter but a basic understanding can help in understanding data center efficiency.

Uptime states [27] that tier levels have the following fundamental requirements (Table 6.7).

The Uptime Institute recommends that the data center design Tier level should be based upon the purpose of the data center. In ref. [27], Uptime provides the following examples (Table 6.8).

See ref. [28] for the full descriptions and other examples.

As seen above, the concept of Tier Ratings is very complex, yet powerful. The Uptime Institute should be consulted for anyone interested in a full implementation of the program.

**Table 6.8** Uptime Institute examples of typical applications [27]

| | |
|---|---|
| Tier I | Small businesses where information technology primarily enhances business process |
| Tier II | Scientific research, e.g., chip design, oil exploration, seismic processing, or long-term weather modeling, which typically does not have on-line or real-time service delivery obligations |
| Tier III | Companies that support internal and external clients 24 × 7. Such as service centers and help desks, but can accept short periods with limited service due to a site failure |
| Tier IV | Businesses based on E-commerce, market transactions, or financial settlement processes |

In terms of energy efficiency, availability, and cost, all three are interrelated. As seen in an earlier segment, the Uptime Institutes cost model shows that the Tier level has a direct impact on data center cost. Similarly, the Tier level would be expected to have an impact on energy efficiency. As redundant components and redundant distribution systems are added to higher tier levels, the efficiency would be expected to go down without additional funding supporting offsetting efficiency enhancements.

The three factors can be envisioned as a triangle; energy efficiency, availability, and cost all affecting each other. Any improvement in one will have an adverse effect on one or both of the other legs of the triangle. It then becomes a business decision for the data center owner to best optimize the cost to build, operate, the energy efficiency, the service level required by the customer, and the sustainability of the data center. Each data center is unique and will require a similar optimization.

### *6.10.6 Power Supply Unit*

Power supply efficiency is an important aspect of the IT platform energy efficiency. Power supplies have been an issue in the past as low-cost parts with poor efficiency could find their way into servers where first cost was the primary decision driver. The issue was further complicated by the limited number of PSU capacity ratings that were being carried by manufacturers. In the worst-case scenario, an oversized, poor efficiency, low-cost PSU was sold into the servers for the data center. The problem can then be compounded by poor data center designs, basing infrastructure sizing on PSU nameplate data, driving an oversized, poor efficiency data center. See ref. [3] for more information on sizing data center systems.

Two programs are now in place that are driving PSU efficiency. Climate Savers Computing Initiative [29] and 80-Plus [30] provide rating systems for the highly efficient PSU. Climate Savers is also engaged in IT equipment power management activities.

### *6.10.7 Codes and Guidelines*

The American Society of Heating, Refrigerating, and Air-conditioning Engineers (ASHRAE) is a technical society that provides extensive information and support for data center design and operation, as well as administering standards applicable to energy efficiency of the building and some infrastructure components.

ASHRAE 90.1 [31] versions before 2010 specifically excluded data centers as process cooling. Due to the increased focused on energy use in data centers, ASHRAE has dropped this exclusion and 90.1-2010 now covers data centers.

ASHRAE Standard 127 [32] covers CRAC and CRAH unit testing and efficiency.

ASHRAE TC 9.9 is the technical committee that deals with mission critical facilities. Through ASHRAE, the IT manufacturers have collaborated and agreed to a common set of recommended environmental conditions (temperature and humidity) [33] for data centers. Adherence to these guidelines allows the data center design to be optimized while ensuring that the operating conditions do not adversely affect the IT equipment. TC 9.9 has published a full series of books [34] on the data center that provides extensive information for design and operations, contamination issues, liquid cooling, and real-time energy measurement.

## 6.11 Sustainability Metrics

Much of the attention in data center engineering and improvements has been on the energy efficiency of the data center and the IT equipment. As the challenges of the overall energy use and its impact on the environment are better understood, other metrics will become more important. Sustainability will become a greater part of the data center evaluation and operation. There are metrics now available in this area but they are relatively immature and will continue to evolve.

### *6.11.1 Carbon*

Carbon will become an increasing important aspect of data center operations. Cap-and-Trade or Carbon taxes could become a part of site selection and operational focuses for data centers. Some regions have already begun taxing energy users a carbon fee. Carbon generation will come from two sources, primarily from the energy procured for the data center operations and carbon emissions from on-site energy generation. These are essentially Scope 1 and Scope 2 level emissions.

Scope level descriptions from The Greenhouse Gas Protocol Initiative [35]

- Scope 1: All direct GHG emissions.
- Scope 2: Indirect GHG emissions from consumption of purchased electricity, heat, or steam.

- Scope 3: Other indirect emissions, such as the extraction and production of purchased materials and fuels, transport-related activities in vehicles not owned or controlled by the reporting entity, electricity-related activities (e.g., T&D losses) not covered in Scope 2, outsourced activities, waste disposal, etc.

The Green Grid defined a metric for carbon; carbon usage effectiveness (CUE). This metric [36] is in the family of metrics that includes PUE; all with a common denominator of total annual IT energy usage (in kWh). Carbon (and other GHGs) are in units of equivalent kilograms of $CO_2$ [37, 38].

$$\mathrm{CUE} = \frac{\text{Total CO}_2\text{ emissions caused by the total data center energy}}{\text{IT equipment energy}}.$$

Total $CO_2$ emissions are the sum of the Scope 1 and Scope 2 emissions based upon the operations of the data center. Scope 3 emissions are excluded. In addition, for the mixed-use case (data center as part of a larger complex), the data center-dependent emissions must be separated from the total site Scope 1 and 2 emissions.

A better understanding of the CUE (and WUE discussed below) in comparison to PUE can be had by considering their limits. The theoretically ideal value of PUE is 1.0 as discussed earlier. In this case, 100% of the power goes to IT. For CUE (and WUE), the theoretically ideal value would be when zero carbon (or water for WUE) is emitted (or used) through the operation of the data center.

To make the methodology more approachable, the calculation for CUE can be taken back to a factor, CEF, and PUE. CEF is further discussed in ref. [36].

$$\mathrm{CUE} = \mathrm{CEF} \times \mathrm{PUE},$$

$$\mathrm{CUE} = \frac{\text{CO}_2\text{ emitted (kg CO}_2\text{ eq)}}{\text{Unit of energy (kWh)}} \times \frac{\text{Total data center energy}}{\text{IT equipment energy}}.$$

As an example, consider a site selection activity. The EIA [39] lists the CEF for Oregon and Washington as 0.147 and California as 0.350. If we assume that the PUEs are similar for the data center (previously calculated as 1.63), the CUE for the Northwest-based data center would be 0.24 vs. 0.57 for California. The owner must weigh many factors for site selection but if carbon were a priority, the values of CUE would be of interest.

In the event of on-site generation, a combination of source (grid)-based carbon emissions (Scope 1 and Scope 2) and site (local generation)-based carbon emissions (Scope 1) are used based on the ratio of annual total energy from each.

### *6.11.2 Water*

Similar to carbon and CUE, The Green Grid has defined two water sustainability metrics [40]. Water usage effectiveness (WUE) and $\mathrm{WUE}_{\mathrm{source}}$ are both important for the overall understanding of the data center on the local water shed. These metrics are defined as:

$$\mathrm{WUE} = \frac{\text{Annual water usage}}{\text{IT equipment energy}}$$

and

$$\mathrm{WUE}_{\mathrm{source}} = \frac{\text{Annual source energy water usage } + \text{Annual site water usage}}{\text{IT equipment energy}}.$$

The units of WUE are L/kWh.

The annual water usage in WUE is all water used directly in or for the data center. This primarily will include water evaporated in the cooling towers (if they are part of the cooling system) and water evaporated into the data center for humidity control. These waters are analogous to Scope 1 emissions in CUE.

In $\mathrm{WUE}_{\mathrm{source}}$, the water use is analogous to a combined Scope 1 and Scope 2 emission.

The metrics have different practical uses. WUE is the better metric for optimizing ongoing operations at a site or data center. Reducing water use directly will show an improved WUE. $\mathrm{WUE}_{\mathrm{source}}$ is more appropriate for use in site selection or site design activities. Different energy generation schemes use different amounts of water. The EWIF represents the water use as a function of generation type [41]. To get the total source energy water usage, the total site energy usage must be used. This value multiplied by the EWIF will give source water usage.

Then $\mathrm{WUE}_{\mathrm{source}}$ can also be calculated as:

$$\mathrm{WUE}_{\mathrm{source}} = (\mathrm{EWIF} \times \mathrm{PUE}) + \frac{\text{Annual site water usage}}{\text{IT equipment energy}}.$$

Water usage at the site is generally a balance between energy and water treatment (generally chemistry). Reductions in water use can first be gained by ensuring that system operations are as designed, past that there will likely be trade-offs with using more energy or using more treatment chemistry. Note that in the case of more energy use, the water use may simply just be shifted to the electrical energy generating plant. This is why WUE and $\mathrm{WUE}_{\mathrm{source}}$ both need to be considered. The local site conditions and constraints (water vs. energy vs. chemistry) need to be considered before large-scale changes are implemented.

## 6.12 Conclusion

Metrics in the data center industry will continue to develop and will provide the tools for improvements in efficiency and sustainability. The best metrics are those that track an important and actionable parameter or a set of parameters; that will drive an improvement for or by the designer, builder, owner, operator, customer, or end user. In addition, the components of the metric should preferably be easily measured or, as a minimum, able to be estimated with a reasonable accuracy. The industry will, through natural selection, develop metrics that are best suited for the needed improvements. The good ones will become pervasive, while the weak ones will fall out of use. Interestingly, the measure of "good" will be the results driven by the metric, far more so than the scientific or mathematical strength of the metric.

The data center and the IT industry is the heart of the information economy and tracking and driving its improvement over time will ensure that it can continue to grow. Metrics are vital to that task.

## References


1. EPA (2007) Report to Congress on server and data center energy efficiency, Public Law 109–431, http://www.energystar.gov/ia/partners/prod_development/downloads/EPA_Datacenter_Report_Congress_Final1.pdf
2. EPA (2010) Energy savings from energy star-qualified servers
3. The Green Grid (2009) Proper sizing of it power and cooling loads, White Paper #23
4. Koomey JG et al. (2009) Assessing trends over time in performance, costs, and energy use for servers, http://www.intel.com/assets/pdf/general/servertrendsreleasecomplete-v25.pdf
5. Intel (2011) What is Moore's Law?, http://www.intel.com/about/companyinfo/museum/exhibits/moore.htm
6. Belady CL (2007) In the data center, power and cooling costs more than the IT equipment it supports. In: Electronics cooling, February, pp. 24–27, http://www.microsoft.com/presspass/features/2009/apr09/04-02Greendatacenters.mspx
7. Herrlin MK, Patterson MK (2009) Energy-efficient air cooling of data centers at 2000 W/ft$^2$, http://www.ancis.us/images/IPACK2009-89224_FINAL.pdf, InterPACK'09
8. Patterson MK et al. (2007) Data center TCO; A comparison of high-density and low-density spaces, http://www.intel.com/technology/eep/datacenter.pdf
9. Turner PW (2010) Dollars per kW plus Dollars per square foot of computer floor, The Uptime Institute, http://uptimeinstitute.org/wp_pdf/(TUI3029A)CostModelDollarsperkWPlusDollars.pdf
10. The Green Grid (2009) Usage and public reporting guidelines for The Green Grid's infrastructure metrics PUE/DCiE - WP #22, http://thegreengrid.org/en/Global/Content/white-papers/Usage%20and%20Public%20Reporting%20Guidelines%20for%20PUE%20DCiE
11. EPA (2010) Recommendations for measuring and reporting overall data center efficiency, http://www.energystar.gov/ia/partners/prod_development/downloads/Data_Center_Metrics_Task_Force_Recommendations.pdf
12. The Green Grid (2010) ERE: A metric for measuring the benefit of reuse energy from a data center, White Paper #29 http://www.thegreengrid.org/~/media/WhitePapers/ERE_WP_101510_v2.ashx?lang=en
13. Kaplan JM et al. (2008) Revolutionizing data center energy efficiency, McKinsey and Co., http://uptimeinstitute.org/content/view/168/57
14. The Green Grid (2009) The Green Grid productivity indicator, WP #15, http://www.thegreengrid.org/~/media/WhitePapers/White_Paper_15_-_TGG_Productivity_Indicator_063008.ashx?lang=en
15. SPEC (2011) Server efficiency rating tool, http://www.spec.org/sert/, Standard Performance Evaluation Corporation
16. The Green Grid (2008) A Framework for Data Center Productivity, WP #13, http://www.thegreengrid.org/~/media/WhitePapers/WhitePaper13FrameworkforDataCenterEnergyProductivity5908.ashx?lang =en

17. The Green Grid (2009) Proxy proposals for measuring data center productivity, WP #17, http://www.thegreengrid.org/~/media/WhitePapers/White%20Paper%2017%20-%20Proxies%20Proposals%20for%20Measuring%20Data%20Center%20Efficiencyv2.ashx?lang=en
18. SPECpower Committee (2011) SPECpower rating tool, http://www.spec.org/power_ssj2008/
19. The Top 500 (2011) http://www.top500.org/
20. The Green 500 (2011) http://www.green500.org/
21. Herrlin MK (2005) Rack cooling effectiveness in data centers and telecom central offices: the rack cooling index (RCI). ASHRAE transactions, Vol 111, Part 2. American Society of Heating, Refrigerating and Air-Conditioning Engineers, Inc., Atlanta, GA
22. Herrlin MK (2008) Airflow and cooling performance of data centers: two performance metrics. ASHRAE transactions, Vol 114, Part 2. American Society of Heating, Refrigerating and Air-Conditioning Engineers, Inc., Atlanta, GA
23. EPA (2011) Energy star for data centers, http://www.energystar.gov/index.cfm?c=prod_development.server_efficiency
24. http://www.energystar.gov/ia/partners/prod_development/downloads/Data_Center_Rating_Development_Results_Nov12.pdf
25. European Union (2011) European codes of conduct for ICT, http://re.jrc.ec.europa.eu/energyefficiency/html/standby_initiative_data_centers.htm
26. US Green Building Council (2011) LEED, http://www.usgbc.org/DisplayPage.aspx?CategoryID=19
27. The Uptime Institute, Resource Center, Understanding Tiers, http://professionalservices.uptimeinstitute.com/understand_tier.htm, 2011
28. Turner WP et al. (2008) Tier classifications define site infrastructure performance, http://uptimeinstitute.org/wp_pdf/(TUI3026E)TierClassificationsDefineSiteInfrastructure.pdf, Uptime Institute, 2008
29. Climate savers computing (2011) http://www.climatesaverscomputing.org/
30. ECOS (2011) Plug load solutions, http://www.plugloadsolutions.com/80PlusPowerSupplies.aspx
31. ASHRAE, Standard 90.1-2010 (I-P Edition) – Energy standard for buildings except low-rise residential buildings (ANSI Approved; IESNA Co-Sponsored), ASHRAE, 2010
32. ASHRAE, Standard 127–2007 – Method of testing for rating computer and data processing room unitary air conditioners (ANSI Approved), ASHRAE, 2007
33. ASHRAE TC 9.9, Thermal guidelines for data processing environments, 2nd Edn., http://tc99.ashraetcs.org/, ASHRAE, 2010
34. ASHRAE TC 9.9, Datacom Series, http://www.ashrae.org/publications/page/1900, ASHRAE, 2011
35. The Greenhouse Gas Protocol Initiative (2011), Frequently asked questions, http://www.ghgprotocol.org/calculation-tools/faq
36. The Green Grid, *Carbon Usage Effectiveness (CUE): A Green Grid Data Center Sustainability Metric* - WP#32, http://www.thegreengrid.com/en/Global/Content/white-papers/Carbon_Usage_Effectiveness_White_Paper, 2010
37. United States Environmental Protection Agency (2005) Metrics for expressing greenhouse gas emissions: carbon equivalents and carbon dioxide equivalents http://www.epa.gov/oms/climate/420f05002.htm
38. US Energy Information Administration http://www.eia.doe.gov/environment.html;
39. U.S. Department of Energy, Energy Information Administration Form EIA-1605 (2007), Appendix F. Electricity Emission Factors, http://www.eia.doe.gov/oiaf/1605/pdf/Appendix%20F_r071023.pdf
40. The Green Grid (2011) Water usage effectiveness (WUE): a Green Grid data center sustainability metric – WP#33
41. Torcellini P, Long N, Judkoff R, Consumptive water use for U.S. power production, NREL/TP-550-33905, 2003, Golden, CO

# Chapter 7
# Data Center Metrology and Measurement-Based Modeling Methods

**Hendrik F. Hamann and Vanessa López**

**Abstract** This chapter describes data center measurement systems and supporting modeling methods. The first part concerns techniques and systems for taking relevant physical measurements to characterize the environmental conditions in data centers. This includes a brief discussion about how design choices, for example, sensor placement, sensor density, and measurement frequency, depend on the supporting modeling approach. Wireless and wired sensing solutions and the role of internal and external sensors are addressed as well. The second part of the chapter deals with how these measurements can be utilized as inputs for subsequent heat transfer modeling in data centers. Two different modeling approaches are discussed, namely a simplified physics-based model (Laplacian model), where the measurements are used to provide the required boundary data, and a reduced order modeling approach using proper-orthogonal decomposition. Case studies for these two different techniques are presented.

## 7.1 Data Center Measurements and Monitoring

Monitoring and measurement technologies play a central role in the management of the physical infrastructure of a data center (DC) for several reasons. First of all, DCs are highly complex with millions of processes running simultaneously. For example, as described in Chaps. 1, 2, and 8, the temperature distributions in DCs are given by a complex set of physics equations (i.e., the Navier–Stokes equations for fluid flow coupled with a heat transfer equation) where boundary conditions for the momentum equations are determined by the operation of large-scale fans, while the source terms for the energy equation are dominated by the power dissipation of the information technology (IT) services running on the different pieces of equipment.

H.F. Hamann (✉) • V. López
IBM Thomas J. Watson Research Center, Yorktown Heights, NY, USA
e-mail: hendrikh@us.ibm.com; lopezva@us.ibm.com

Y. Joshi and P. Kumar (eds.), *Energy Efficient Thermal Management of Data Centers*, DOI 10.1007/978-1-4419-7124-1_7,

Second, DCs are mission-critical facilities and consequently the physical infrastructure should be monitored and managed so that reliability of the operation is warranted. Evidently, without a measurement system which can provide optimum visibility into the operations of the DC, any changes (including energy efficiency improvements) are seen as risky and thus unlikely to occur.

Finally, an important aspect of a monitoring technology relates to energy efficiency metrics discussed in Chap. 6. Although DC energy efficiency can be assessed with only a few monitoring points using metrics such as DCIE (DC infrastructure efficiency = total energy usage/energy consumption for all IT equipment) [1], such metric is quite coarse and depends on various other factors, which are often not under immediate control of the operator (such as weather conditions, equipment mix, legacy, cooling systems, etc.). Because of this, energy efficiency improvements are often not verifiable and thus investments in new, more energy efficient technologies lack a clear business case (and associated charge back mechanisms). This is in particular problematic because facilities (responsible for energy consumption) and IT operations are often organizationally separated. Clearly, a comprehensive monitoring technology, which enables quantitative measurements on a more granular level is a critical component for making additional advances in DC energy efficiency. Although the importance of DC monitoring has been recognized, it can be challenging to deploy comprehensive measurement systems due to the following reasons:

1. The DC is made up of very different technology components, which are segmented in different parts of the DC: facility or building, power or cooling infrastructure, and information technology. Typically, each of these segments has at least one monitoring system, often with a propriety architecture and interface where little data can be shared. The lack of integration between the different existing measurement systems, as well as organizational issues with a separation between facilities and IT, often prevents an effective monitoring of the entire facility.
2. It is an underappreciated fact that despite the trend toward standardization, most DCs are unique as it relates to equipment mix, infrastructure, IT services, business requirements, etc. Consequently, monitoring solutions need to have some level of customization, which complicates the deployment of such technologies.
3. Because of this customization aspect and the fact that such monitoring technologies often only apply to a subset of DCs, the commercial market for such solutions has been small and is fragmented and consequently the respective technologies have been not scalable, thus preventing a cost-effective deployment to the larger market.
4. In many cases, it has been difficult to justify the business case for a monitoring system. The main reason for this is that most monitoring systems are not coupled to analytical or modeling tools, which would provide DC managers with smarter decision support. In fact, most monitoring systems are being used just as simple reporting tools with little data processing and analytics, where then the operators use their own technical background and skills to translate the data into improvements of the facility. Evidently, such an approach will provide suboptimal results.

### 7.1.1 Relevant Physical Measurements

In the following sections, we discuss briefly different measurements which are relevant to characterize the environmental conditions in a DC. Here, we limit the discussion to the physical measurements which are required as inputs to the supporting models presented in the second part of the chapter. This includes measurements of

(a) Temperature
(b) Pressure and airflow/velocity
(c) Power
(d) DC layout and dimensions

We also describe briefly some other important measurements and emerging sensing technologies such as relative humidity and corrosion sensing. We note that the following discussion is not intended to be complete, but rather to provide a brief overview of different measurement technologies and systems.

#### 7.1.1.1 Temperature

Temperature measurements are clearly very important for characterizing a DC environment. There are numerous ways to monitor temperatures, for example, with bimetallic, resistive, expansion, bandgap, and infrared sensors, as well as thermocouples. Many of these sensors can be obtained as small digital chips, which are already integrated with a readout circuitry. The accuracy of these sensors can be obtained from specifications of the sensor manufacturer and will generally range from 0.05°C to 1.0°C at a low bandwidth (~1 Hz). The signal-to-noise is generally governed by the intrinsic properties of the sensor rather than the data collection. Specifically, with proper signal conditioning and considering the small measurement range (10–40°C), a 16-bit A/D converter will be able to read a signal with $30°\text{C}/2^{16} = 0.000046°\text{C}$ accuracy, which is generally far more precision than what the sensor can provide.

*Air Temperature Measurements*: Because DC operators (often) manage temperatures and relative humidity based on the ASHRAE (American Society of Heating, Refrigerating and Air-conditioning Engineers) guidelines [2], air temperatures to the inlet of the servers are of interest. Due to the low heat transfer coefficient for air ($h \sim 10$–$100$ W/m$^2$ K), special attention is required because the mounting or the connection of the sensor might interfere with the actual measurement. More specifically, Netwon's law governs the heat transfer from the air to the sensor, where the thermal resistance between the air and sensor can be described by

$$R_s = \frac{1}{h \times A_s} \tag{7.1}$$

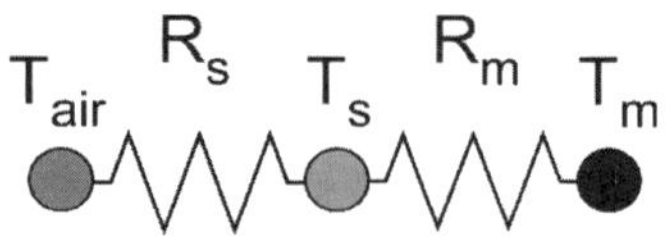


**Fig. 7.1** Thermal model for estimating a systematic error for air temperature measurements

with $h$ denoting the heat transfer coefficient and $A_s$ the sensor area. We note that the following discussion is not generally applicable for estimating the impact of the mounting. In fact, we recommend to model the actual sensing geometry accurately using (standard) thermal modeling software. However, for the purpose of discussion, we describe the thermal resistance between a sensor and its mounting point by

$$R_m = \frac{l_m}{k \times A_m}, \tag{7.2}$$

where $k$ is the thermal conductivity, $A_m$ the cross-sectional area of a wire connecting the sensor to a mounting point, and $l_m$ the length of this wire. It should always be checked that $R_m$ is significantly (~10×) larger than $R_s$. For example, let us assume an air temperature $T_{air} = 21°C$ and a sensor surface area $A_s = 1\ mm^2$ with a heat transfer coefficient of $h = 20\ W/m^2\ K$. The sensor is mounted via a 36-gauge wire ($k = 100$ W/m K, ~0.127 mm) to a circuit board, which is at $T_m = 25°C$. The wire length is $l_m = 10$ cm. The respective thermal resistances are $R_S = 5.0 \times 10^4$ K/W and $R_m = 6.2 \times 10^4$ K/W. The total heat flow $Q$ from the mounting point to the sensor is then given by

$$Q = (T_m - T_{air})/(R_s + R_m), \tag{7.3}$$

(here $Q \sim 36\ \mu W$). We can estimate the temperature of the sensor as

$$T_s = T_m - R_m \times Q, \tag{7.4}$$

which yields 22.8°C (instead of the true air temperature of 21.0°C). The *systematic error* ε is then given by the difference in temperature between the sensor and air,

$$\varepsilon = T_s - T_{air}, \tag{7.5}$$

which here is ~1.8°C. Equations (7.1) – (7.5) can be rewritten to yield a formula for the systematic error of the thermal measurement Fig. 7.1, namely,

$$\varepsilon = \frac{R_s}{R_m}(T_m - T_s). \tag{7.6}$$

One popular choice for measuring air temperatures are thermocouples, because, due to their small size, it is relatively easy to decouple them from the mounting environment. Gauge 40 thermocouples are available, which can provide good thermal isolation from the environment. Because thermocouples rely on Seebeck coefficients of dissimilar metals at the junction (a typical Seebeck coefficient is a few μV per degree Kelvin), it is important to pay attention to the electrical connection of the thermocouple to avoid the generation of additional thermocouple junctions, which could interfere with the measurements.

*Infrared Measurements*: At a first glance, it seems that infrared imaging and sensing would be a viable approach for measuring DC temperatures. While infrared technologies can give a fast and qualitative view of temperatures as well as hotspots, it is important to be aware that such an approach has some drawbacks. Because the technology relies on the detection of infrared radiation from heated objects, the actual air temperatures cannot be measured (because air is basically transparent in the infrared wavelength range). Since each material/surface has a different emissivity (which is the ability of the material to emit energy by radiation) the measurements have to be calibrated. Such calibrations are cumbersome and rely on careful reference measurements where each material is heated to a known temperature and its radiation is being measured.

*Transient Responses*: The transient response of a thermal measurement system can be complex and it should be generally modeled and/or measured. A measurement could involve the exposure of the thermal sensor to a heat pulse (which could be applied via Joule or laser heating) and then measure the actual time response. Besides the actual thermal response of the measurement, we note that the bandwidth of the data acquisition system should also be characterized, especially when measuring small voltages from thermocouples.

An important concept to consider when designing a measurement system is the thermal diffusivity, which is a measure of the transient thermal response of a material to a change in temperature:

$$\kappa = \frac{k}{\rho c_p}, \tag{7.7}$$

(with $\rho$ as the density and $c_p$ as the specific heat). The thermal diffusivity can be related to a thermal response time $\tau$ over a distance $l$ via

$$\tau = \frac{l^2}{\kappa}. \tag{7.8}$$

Equation (7.8) shows a nonlinear (i.e., quadratic) relation between sensor size (e.g., junction size of a thermocouple) and its transient response, which is important to keep in mind when designing the measurement system. We reiterate that actual thermal response time of a thermal measurement system requires detailed calculations or experiments.

#### 7.1.1.2 Airflow

Other important measurements in a DC relate to airflow or air velocity. There are many different techniques available, which allow monitoring air movement using mechanical, ultrasonic, pressure-based, optical, and thermal mass methods.

Commonly, one will find two different types of measuring devices: balometers, which measure the *total airflow* (integrated air velocity) through a defined surface such as a perforated tile, and anemometers, which measure the *local air velocity*. Generally, balometers require capturing the airflow, which means that the measurement is more robust but "intrusive," while anemometers are typically "nonintrusive" but often less comprehensive. For DC application, it is in particular important to measure the airflow through the perforated tiles or vents as well as the air conditioning units (ACUs). These applications are described first before discussing methods for obtaining airflow measurements for the IT equipment.

*Perforated Tiles/Vents*: Balometers are often used in DCs to measure the airflow through a perforated tile. An example is a flow hood [3], which consists of an airflow capturing device (this is often a fabric hood mounted over a collapsible frame that collects the airflow) and a sensing element (which is typically a static Pitot tube). A typical Pitot tube allows measuring the dynamic pressure $p_{\mathrm{d}}$ (difference between total and static pressure) by providing access ports parallel and perpendicular to the flow direction. Using Bernoulli's law the air velocity $v$ at the Pitot tube can be determined as

$$v = \sqrt{\frac{2p_{\mathrm{d}}}{\rho}}. \tag{7.9}$$

Equation (7.9) typically includes an additional correction factor, which we have omitted here. Such factor depends on the design of the Pitot tube. By knowing the surface area $A_{\mathrm{t}}$ from which the air was collected (typically the area of one tile), the measured flow $f_{\mathrm{measured}}$ can be readily calculated as

$$f_{\mathrm{measured}} = A_{\mathrm{t}} \times v. \tag{7.10}$$

Next, the flow hood measurement has to be corrected for the influence of the flow hood itself (as it restricts the airflow somewhat). A correction factor c can be calculated, following [4], as

$$c = \frac{f_{\mathrm{corr}}}{f_{\mathrm{measured}}} = \sqrt{\frac{R_{\mathrm{t}} + R_{\mathrm{hood}}}{R_{\mathrm{t}}}}, \tag{7.11}$$

where $R_{\mathrm{t}}$ and $R_{\mathrm{hood}}$ are, respectively, the flow impedances of the tile and flow hood and $f_{\mathrm{corr}}$ and $f_{\mathrm{measured}}$ denote, respectively, the corrected (or "true") and measured airflow. Equation (7.11) can be readily derived by relating the flow measurements to the pressure via the respective flow impedances:

$$\Delta p = R_{\mathrm{t}} \times f_{\mathrm{corr}}^2 \quad \text{and} \quad \Delta p = (R_{\mathrm{t}} + R_{\mathrm{hood}}) \times f_{\mathrm{measured}}^2. \tag{7.12}$$

Flow impedances for perforated tiles and the hood should be determined independently, for example, in a flow chamber experiment or by referring to the equipment manufacturer.

The flow impedance of a perforated tile can also be estimated based on the perforation of the tile [5]. More specifically, the perforated tile impedance $R_t$ is given by the equation

$$R_t = \frac{1}{2} \times \frac{\rho}{A_t^2} \times K, \tag{7.13}$$

where $A_t$ is the area of the tile and $K$ a loss coefficient. For a standard US tile with an area of $A_t = 4\ \text{ft}^2 = 0.371612\ \text{m}^2$ and an air density of $\rho = 1.184\ \text{kg/m}^3$, we can rewrite this equation as

$$R_t = 9.54 \times 10^{-7}\ \text{Pa/cfm}^2 \times K = 4.28\ \text{Pa/(m}^3\text{/s)}^2 \times K. \tag{7.14}$$

The loss coefficient $K$ can be estimated [6] as

$$K = \frac{1}{o^2}(1 + 0.5 \times (1 - o)^{0.75}) + (1.414(1 - o)^{0.375}), \tag{7.15}$$

where $o$ denotes the fractional opening or perforation of the tile. As an example, let us assume a hood impedance of $R_{\text{hood}} = 2 \times 10^{-5}\ \text{Pa/cfm}^2 = 89.79\ \text{Pa/(m}^3\text{/s)}^2$. For a 10% perforated tile we obtain a loss coefficient of $K = 282$, a tile impedance of $R_t = 2.7 \times 10^{-4}\ \text{Pa/cfm}^2 = 1{,}212.21\ \text{Pa/(m}^3\text{/s)}^2$, and a correction factor of $c = 1.036$ (3.6%). For a 20% perforated tile the equations yield a loss coefficient of $K = 68$, a tile impedance of $R_t = 6.5 \times 10^{-5}\ \text{Pa/cfm}^2 = 291.83\ \text{Pa/(m}^3\text{/s)}^2$, and a correction factor of $c = 1.143$ (14.3%). As illustrated with this example, correction factors are becoming more important for higher throughput tiles because the impact of the hood itself becomes more and more dominant.

*Air Conditioning Units*: While there are many robust balometer-based solutions for tile and vent flow measurements available, it is much more difficult to measure the flow for ACUs. One approach includes making relative (or noncalibrated) measurements $f^i_{\text{ACU-NC}}$ for the different ACUs using a standard flow sensor such as an anemometer or flow hood. For example, a flow hood would be placed in a repeatable manner over the different sections of the ACU intake. In cases where different types of ACUs are being deployed, the surface area of the ACU air intake should be included in the analysis to derive relative flow values. An alternative approach entails using just the nominal flow capacities from the manufacturer to obtain $f^i_{\text{ACU-NC}}$. In case that the ACUs are equipped with variable frequency drives (VFDs), the nominal flow capacity is given by the blower settings $\gamma^i$(0–1)

$$f^i_{\text{ACU-NC}} = f^i_{\text{ACU-NC},o} \times \gamma^i \tag{7.16}$$

with $f^i_{\text{ACU-NC},o}$ representing the nominal flow capacity at 100% ($\gamma^i = 1$).

Absolute airflow values $f^i_{\mathrm{ACU}}$ for each ACU can be obtained using the principle of energy balance [7]. The total airflow for all active ACUs in the DC is given by

$$f^{\mathrm{total}}_{\mathrm{ACU}} \approx \frac{P_{\mathrm{RF}}}{\rho c_{\mathrm{p}}} \sum_{i=1}^{\#\mathrm{ACU}} \frac{1}{\Delta T^i_{\mathrm{ACU}}}, \tag{7.17}$$

where $P_{\mathrm{RF}}$ denotes the total heat load (which is dissipated in the DC room and removed by the ACUs) and $\Delta T^i_{\mathrm{ACU}}$ denotes the temperature differential between the return and discharge of the temperature of each ACU. The actual airflow for each ACU is then given by

$$f^i_{\mathrm{ACU}} = \frac{f^i_{\mathrm{ACU-NC}}\, f^{\mathrm{total}}_{\mathrm{ACU}}}{\sum_{i=1}^{\#\mathrm{ACU}} f^i_{\mathrm{ACU-NC}}}. \tag{7.18}$$

In some DCs, monitoring solutions allow measuring the total heat load, for example, by measuring the temperature differential and flow of the chilled water loop or by electrical metering. Here, we provide a simple method for estimating the total dissipated power in the DC room using

$$P_{\mathrm{RF}} \approx P_{\mathrm{IT}} + P_{\mathrm{light}} + P_{\mathrm{ACU}} + P_{\mathrm{PDU}} + P_{\mathrm{misc}}, \tag{7.19}$$

where $P_{\mathrm{PDU}}$ is the power dissipation due to losses associated with the power distribution units (PDUs) (which can be sometimes approximated by $P_{\mathrm{PDU}} \approx 0.1\, P_{\mathrm{IT}}$) and $P_{\mathrm{misc}}$ is miscellaneous power consumption that is not captured by the PDU measurements. $P_{\mathrm{IT}}$ is by far the largest term in (7.19) and can be (conveniently) obtained from measuring the power consumption level $P^j_{\mathrm{PDU}}$ at the PDU level for each PDU via the relationship

$$P_{\mathrm{IT}} = \sum_{j=1}^{\#\mathrm{PDUs}} P^j_{\mathrm{PDU}}. \tag{7.20}$$

Details regarding PDU power measurements are discussed further below. $P_{\mathrm{light}}$ is the power for lighting and can be often approximated by $P_{\mathrm{light}} \approx 1.5\ \mathrm{W/ft^2}\, A_{\mathrm{DC}} = 16.1\ \mathrm{W/m^2}\, A_{\mathrm{DC}}$, where $A_{\mathrm{DC}}$ is the DC area. In cases where water cooling is deployed, (7.19) has to be modified by subtracting the power removed by the water cooling systems. $P_{\mathrm{ACU}}$ is the power dissipation of the ACUs and is mainly given by the blower (or fan) power of the ACUs ($P^i_{\mathrm{blower}}$) as

$$P_{\mathrm{ACU}} \approx \sum_{i=1}^{\#\mathrm{ACU}} P^i_{\mathrm{blower}}. \tag{7.21}$$

We have assumed here that the ACUs are located in the DC. The actual ACU power consumption can be substantially higher than what (7.21) yields because it neglects other operations of the ACU (such as dehumidification), which consume power. For ACUs equipped with VFDs, the blower power is determined by the blower settings $\gamma^i$(0–1) and can be described by

$$P^i_{\text{blower}} = P^i_{\text{blower},o} \times \gamma^{i\ \text{nb}}, \tag{7.22}$$

with $P^i_{\text{blower},o}$ denoting the blower power at 100% ($\gamma^i = 1$) and nb the exponent of the dependence of the blower power as a function of the settings. In general, nb can vary depending on the details but can be as high as nb $= 3$ for efficient VFDs.

*IT equipment*: It is an unfortunate fact that exact airflow values for the different pieces of the IT equipment are often not known. While it is already difficult to measure airflow for the different types of equipment in a DC, the situation is further complicated because often fans within the servers are controlled by the inlet air temperature, which makes these inputs variable. One approach is to refer to equipment manufacturers and use nominal values. In cases where such information is not available, a similar approach can be used as we discussed for the ACU airflow measurements. In essence, the temperature differentials $\Delta T^s_{\text{IT}}$ and relative airflow or velocity measurements $f^s_{\text{IT-NC}}$ across the equipment can be used to obtain the IT equipment airflow $f^s_{\text{IT}}$ by

$$f^{\text{total}}_{\text{IT}} \approx \frac{P_{\text{IT}}}{\rho c_{\text{p}}} \sum_{s=1}^{\#\text{IT}} \frac{1}{\Delta T^s_{\text{IT}}} \quad \text{and} \quad f^s_{\text{IT}} = \frac{f^s_{\text{IT-NC}}\, f^{\text{total}}_{\text{IT}}}{\sum_{s=1}^{\#\text{IT}} f^s_{\text{IT-NC}}}. \tag{7.23}$$

Alternatively, one can consider an approach where power measurements are used as a "proxy" to gauge how much airflow a respective IT equipment is requiring. Although large differences might exist between the different types of equipment (storage, network, server, tape, etc.) as well as between the manufactures, a simple equation might be used to estimate the airflow as

$$f^s_{\text{IT}} \approx \frac{P^s_{\text{IT}}}{\rho c_{\text{p}} \Delta T^s_{\text{IT}}}, \tag{7.24}$$

where $P^s_{\text{IT}}$ denotes the power consumption of the IT equipment. Generally, it is recommended to determine $\Delta T^s_{\text{IT}}$ with actual temperature measurements. Experiments have shown that in a typical DC the value of $\Delta T^s_{\text{IT}}$ is on average only ~2.5°C (~4°F) but it can be as large as 12°C (~21°F) for high-performance servers and blade systems.

*Anemometer*: Anemometers measure air velocity in the DC in a particular direction. Common approaches use mechanical systems. Hot wire anemometry measures the heat transfer of a heated wire to a downstream sensor to monitor air velocity. Although there are several other viable anemometer technologies available, including ultrasound, the challenge for DC applications has been that detailed and spatially resolved air velocity measurements (in all three dimensions) are required to extract really useful information from such an approach. Currently, such techniques are not established. Besides reducing cost for such an approach, it will require additional research and development to understand which sensing technology and type of implementation would be most viable for DC applications. One example for this research includes the application of particle image velocimetry (PIV) for DCs [8]. PIV allows for full 2D and 3D visualization of the air velocity.

#### 7.1.1.3 Pressure

We have already discussed pressure sensing in the context of measuring airflow with a Pitot tube in a flow hood. Another typical application for pressure monitoring in DCs involves the sensing of the pressure differential between the plenum and raised floor with the goal of determining airflow from the perforated tiles. In contrast to the flow hood measurements (discussed in Sect. 7.1.1.2), this approach allows for real-time monitoring of the flow values from each perforated tile without disturbing the airflow.

We note that the pressure differentials between the plenum and raised floor are typically quite small, which requires some attention while taking these measurements. Typically, static pressure sensors do not have enough sensitivity. Generally, differential sensors with high flow impedance, a linear response over the pressure range, and a sensitivity of less than 50 Pa are required. Such differential pressure sensors have two access points, where one is used to connect a tube to the plenum space while the other connects to the raised floor. Pressure sensors should be placed in areas of low flow with the tube pointing perpendicular to the main direction of the flow to avoid picking up dynamic pressure.

For illustration purposes of such a pressure monitoring system, we consider the deployment of four pressure sensors in a small DC (97.55 $m^2$ (1,050 $ft^2$)). The DC has two types of tiles (with 10% and 20% perforation) and two ACUs with VFDs, which allow changing the plenum pressure. Using 25 different combinations of settings for the VFDs of the two ACUs, 25 different pressure distributions were generated. In order to obtain the actual pressure at the tile locations (vs. the pressure sensor location), a simple inverse distance interpolation algorithm was used [9].

In an inverse distance approach, the weight $w_{jk}$ of each data point on an interpolation point is calculated based on the distance $r_{jk}$ between these two data points. For a 2D approach, the following equations can be applied:

$$w_{jk} = \frac{1}{(r_{jk} + c)^b} \exp(-\mu \times r_{jk}) \times im_j, \tag{7.25}$$

$$r_{jk} = \sqrt{(x_j - x_k)^2 + (y_j + y_k)^2}. \tag{7.26}$$

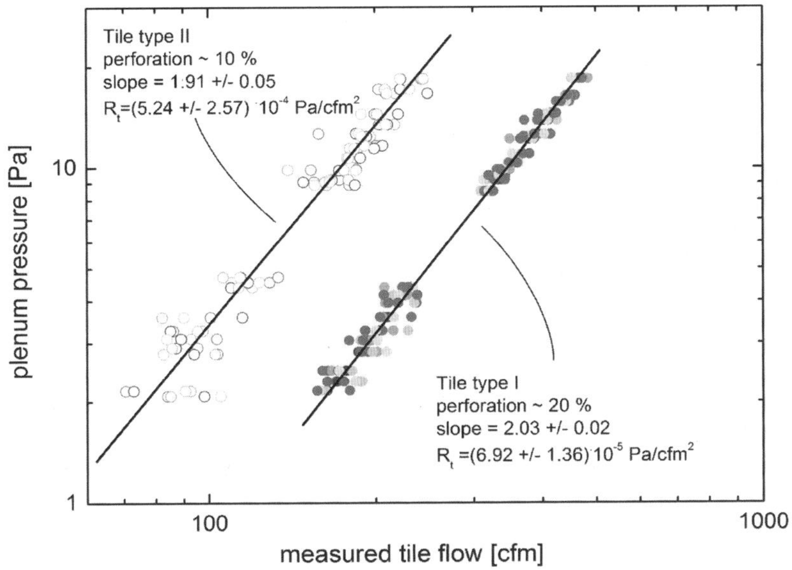


**Fig. 7.2** Measured plenum pressure differentials as a function of measured tile flow for two different types of tiles. Details of this DC are described in [19, 36]. Figure reprinted, with permission, from [19] (Hamann HF, López V, Stepanchuk A. Thermal zones for more efficient data center energy management. Proceedings of ITherm2010 (© 2010 IEEE))

In (7.25) and (7.26) $x$ and $y$ are the spatial coordinates, $c$ is a smooth parameter (default $= 0$), $b$ and $\mu$ determine the attenuation distance, $im$ is a weighting factor for a particular sensor (default $= 1$), and $j$ and $k$ denote the data and interpolation point index, respectively. The interpolated values at any interpolation location are then given by

$$z_k(x, y) = \sum_{j=0}^{n-1} \frac{w_{jk} z_j}{\sum_{j=0}^{n-1} w_{jk}}, \tag{7.27}$$

with $z$ as the value of the data point and $n$ the total number of data points.

For each of the 25 pressure distributions, the airflow through each tile was measured as well using the previously described flow hood method. Each flow hood measurement was corrected as discussed in Sect. 7.1.1.2. The resulting data, plotted in Fig. 7.2, show the quadratic relationship between the airflow and pressure differential (refer to (7.12)). The results indicate that the airflow for each tile can be measured with reasonable accuracy in real time by monitoring the plenum pressure field. Using the standard deviation for the higher impedance tile and considering very low pressures (where the data shows more noise), we can estimate an error in the flow measurements of ~25%. The results are significantly improved for the higher throughput tiles to ~10% accuracy in the flow measurements.

#### 7.1.1.4 Power

It is very important to measure the power usage of the different components in a DC. While power measurements are important at all levels (including the chiller plant, etc.), we restrict the discussion here to power measurements of the IT equipment. Unfortunately, often operators have little insights in the actual power distribution within the DC. While it is relatively easy to estimate and measure the total power dissipated in the DC (e.g., see (7.19)), it is a much more complicated task to assess or measure the power dissipation of every piece of equipment ($P^s_{\mathrm{IT}}$). One common approach includes the "nameplate" power of the equipment manufacturers ($P^s_{\mathrm{IT-NP}}$). This information is often also found in the respective asset management system of the DC. It is important to understand that the purpose of the nameplate power is to indicate the *limits for maximum power draw*. By no means does such information give the actual power draw during normal operation. Generally, the power draw will change over time and is usually significantly less (~50–60%) than the nameplate power. One solution involves using the nameplate power for relative allocation of power among the IT equipment and then leveraging (possibly real time) power measurements for the entire IT equipment to normalize it using

$$P^s_{\mathrm{IT}} = \frac{P^s_{\mathrm{IT-NP}} P_{\mathrm{IT}}}{\sum_{s=1}^{\#\mathrm{IT}} P^s_{\mathrm{IT-NP}}}, \tag{7.28}$$

where the total IT power $P_{\mathrm{IT}}$ can be obtained, for example, from (7.20).

An alternative and superior approach leverages instrumented PDUs. Today, many power delivery and distribution units such as PDUs or uninterruptible power supply (UPS) systems are equipped with appropriate sensing technologies. The data can be often obtained using Web-based protocols interfacing with the unit. One illustrative example of various sensor points of a PDU is shown in Fig. 7.3.

In overview, the power from the various UPS systems is distributed to different PDUs throughout the DCs. Any large-scale DC (>4.7k m$^2$ (50k ft$^2$)) will often have more than 50 of such PDU systems (each at ~400 A). Commonly, each PDU is fed with three phases with a voltage between the lines of ~480 V. The input is transformed to lower voltage. Most modern PDUs measure the real power of the PDU, $P^j_{\mathrm{PDU}}$, as well as the apparent power $S^j_{\mathrm{PDU}}$. The ratio defines the power factor $\mathrm{PF}^j$ for the entire PDU by

$$\mathrm{PF}^j = \frac{P^j_{\mathrm{PDU}}}{S^j_{\mathrm{PDU}}}. \tag{7.29}$$

The output voltages from the transformer between the three phases ($V^j_{\mathrm{ABout}}$, $V^j_{\mathrm{CAout}}$, $V^j_{\mathrm{BCout}}$, typically ~208 V) and the neutral line ($V^j_{\mathrm{ANout}}$, $V^j_{\mathrm{BNout}}$, $V^j_{\mathrm{CNout}}$, typically ~120 V) are often measured. The total output current for each phase from

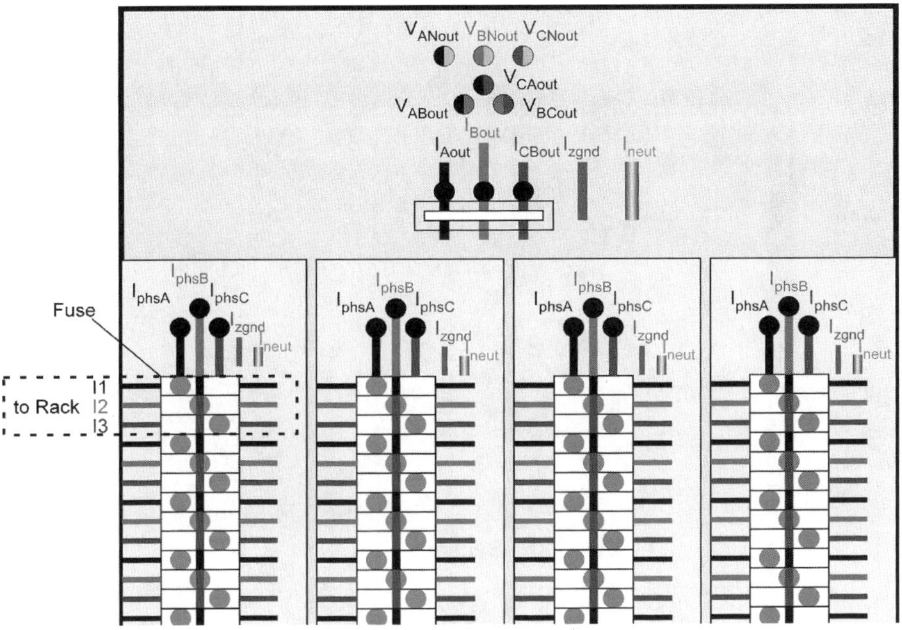


**Fig. 7.3** An example of different monitoring points of a power distribution unit. Three phases are supplied to four panels, which can then be branched into 42 single phase circuits (here we only show the first branches)

the transformer is sensed ($I^j_{\text{Aout}}, I^j_{\text{Bout}}, I^j_{\text{Cout}}$) as well any stray current in the ground and neutral line ($I^j_{\text{zgnd}}, I^j_{\text{neut}}$). The power of the PDU is split into four panels ($p = 1, \ldots, 4$). For each of the four panels the currents of the three phases ($I^{j,p}_{\text{A}}, I^{j,p}_{\text{B}}, I^{j,p}_{\text{C}}$) as well as the currents in the ground and neutral line ($I^{j,p}_{\text{zgnd}}, I^{j,p}_{\text{neut}}$) are monitored. Each panel supports 42 (=3 × 14) branches ($b = 1, \ldots, 42$), which allow connecting 14 three-phase circuits to the IT equipment. The currents for each line are monitored ($I^{j,p,b}$). Using these measurements, as depicted in Fig. 7.3, the power $P^s_{\text{IT-NC}}$ to each piece of IT equipment can be estimated. For example, let us assume that a server is connected to PDU#42 ($j = 42$) on panel#1 ($p = 1$) using the branches 4, 5, and 6 ($b = 4$–6). Then, the (noncalibrated) power can be obtained for this particular case as

$$P^s_{\text{IT-NC}} = \frac{I^{42,1,4}V^{42}_{\text{ABout}} + I^{42,1,5}V^{42}_{\text{CAout}} + I^{42,1,6}V^{42}_{\text{BCout}}}{\sqrt{3} \times \text{PF}^{42}}. \tag{7.30}$$

In some cases, PDUs also report the power factor on an individual panel and branch although this is generally costly. The power factor PF for newer equipment will be around 0.9 and higher, while for legacy equipment it could be as low as 0.5. In the absence of a power factor value for each branch, it is recommended to use

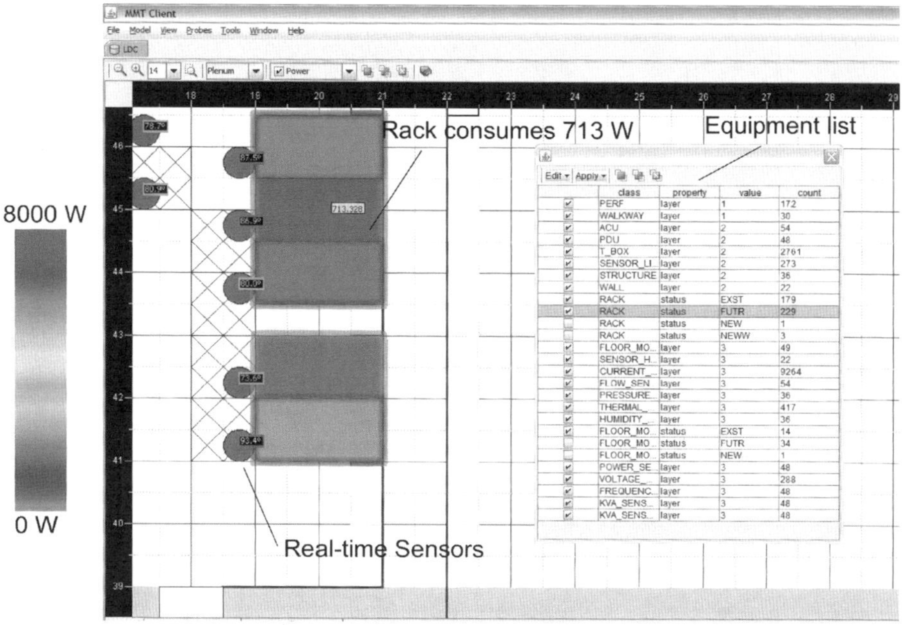


**Fig. 7.4** Real-time power map using the described PDU power monitoring system. The racks are shaded with a color map, which is shown on the *left*. For example, the second rack from the top consumes ~713 W of power

the power factor for the entire PDU (see (7.29)). Although the power estimation using (7.30) will be quite accurate, it is recommended to use them as relative power levels and then normalize it using the power of the entire IT equipment,

$$P_{\mathrm{IT}}^{s} = \frac{P_{\mathrm{IT-NC}}^{s} P_{\mathrm{IT}}}{\sum_{s=1}^{\#\mathrm{IT}} P_{\mathrm{IT-NC}}^{s}}. \tag{7.31}$$

Figure 7.4 shows an example where the described approach was used to characterize the full power distribution in a DC.

We note that a legacy DC does not typically have active PDU monitoring. A preferred choice for retrofitting legacy DCs includes current clamps or Hall sensors, which pick up the current-induced magnetic field. Both Hall sensors and current clamps have the advantage that they can be retrofitted without disconnecting existing circuits.

### 7.1.1.5 Assets

Any physics-based modeling approach will require some detailed information (e.g., dimensions, specifications, etc.) about the assets in the DC. A tight integration of the measurement system with the asset management system is therefore important. Such asset management system should also be tied to a life-cycle management system. The asset management system is supported by a database, where typically all information with respect to the assets is being managed. Such database will generally include power, location, equipment, customer information, and much more. The equipment information typically includes dimensions of the racks and servers, which can be used to define the modeling domains. Technologies which monitor the location of a given asset using radio frequency identification (RF-ID) are emerging. In some cases, robots equipped with laser scanners and/or ultrasonic sensors are used to map the physical dimensions of the DC space [10].

#### 7.1.1.6 Other Sensing Technologies

*Relative Humidity*: As seen in Chap. 2, humidity control is another important aspect in DC energy management. Unfortunately, accurate humidity measurements are actually difficult to obtain. Relative humidity is defined as follows:

$$\mathrm{RH}(\%) = \frac{P_{\mathrm{H_2O}}}{P_{\mathrm{H_2O}}^{\mathrm{vapor}}} \times 100, \tag{7.32}$$

where $P_{\mathrm{H_2O}}$ is the partial water pressure and $P_{\mathrm{H_2O}}^{\mathrm{vapor}}$ is the vapor pressure of water (maximum amount of water before condensation at a given temperature). As discussed in Chap. 2, the classical way to measure relative humidity involves measuring the dry $T_{\mathrm{DB}}$ and wet $T_{\mathrm{WB}}$ bulb temperature. $T_{\mathrm{DB}}$ can be readily measured with a standard temperature sensor, while $T_{\mathrm{WB}}$ is measured when the sensor is kept wet with water and a continuous airflow evaporates the water from the sensor. A psychrometric chart at a given barometric pressure can then be used to obtain the relative humidity.

There are also commercially available digital relative humidity sensors. Typically, these sensors use a thin film capacitor where the dielectric is a polymer, which absorbs or desorbs water depending on the relative humidity. The change in capacitance is then measured to gauge the relative humidity. Such capacitive sensors are convenient to use but often the accuracy and long-term stability are less than what traditional methods yield.

The vapor pressure $P_{\mathrm{H_2O}}^{\mathrm{vapor}}$ is a function of temperature, described by the Clausius–Clapeyron [11] as follows:

$$P_{\mathrm{H_2O}}^{\mathrm{vapor}}(\mathrm{mbar}) \sim 6.11 \times 10^{\frac{7.5T(^\circ\mathrm{C})}{237.7+T(^\circ\mathrm{C})}}. \tag{7.33}$$

Note that (7.33) yields ~1,000 mbar at $T = 100^\circ\mathrm{C}$, which is the boiling point of water at atmospheric pressure.

For DC applications the dew point, which is the temperature at which water would condense at a given humidity, is of particular interest. Equations (7.32) and

(7.33) calculate the partial water pressure ($P_{H_2O}$). Then (7.33) can be used again to obtain the corresponding temperature, letting $P_{H_2O}$ denotes the vapor pressure. An approximate equation for the dew point can be found as long as the relative humidity is high (>50%):

$$T_D(^\circ C) \sim T_{DB}(^\circ C) - \frac{(100 - RH(\%))}{5}. \tag{7.34}$$

Equation (7.34) describes a common rule of thumb, where for each 5% of relative humidity decrease/increase the dew point decreases/increases by 1°C. For example, for a DC operating at 60% relative humidity for 25°C, the dew point will be 17°C (i.e., any equipment below 17°C will have condensation risk).

*Corrosion*: Another emerging environmental measurement in DCs includes the corrosion level. The challenge here lies in the required sensitivity. Corrosion in DCs can be a concern even at rates as low as 10 Å/day. In order to manage appropriate filter technologies and the daily use of free cooling, it is desirable to measure the corrosion impact on a daily basis. Corrosion sensors include mass balance and resistive sensing, but current state-of-the-art corrosion monitors cannot detect such low corrosion rates reliably. In addition, corrosion is a highly complex mechanism which depends on many different parameters including humidity, temperature, chemical contents, surface properties, etc. It is expected that DC corrosion monitoring and management will be an active research area as free cooling becomes more pervasive.

### *7.1.2 Measurement: Modeling Systems*

The design of the measurement system is a very important aspect of the entire monitoring solution. Clearly, there are tradeoffs between (a) density of sensors (how many sensors), (b) location of sensors (where to place sensors), and (c) the measurement frequency (how often to read the sensors) as well as other parameters (such as business needs, reliability requirements, etc.). Unfortunately, design choices are often made without understanding the dependencies with regard to underlying modeling technologies. Any viable DC measurement system needs to be supported by some modeling technology which allows interpreting the measurement values because *a sensor only provides information at one given point in space and time*. This is particularly important for temperature sensing because thermal gradients in DCs can be quite large. For example, assuming that a sensor positioned at the return of an ACU measures the "return temperature" or assuming that a sensor positioned at the corner of a rack server measures the "inlet temperature," can lead to very flawed conclusions. One should always keep in mind that strong temperature gradients can exist in DCs and depending on the details of the DC one might find that temperatures across the intake of the ACU or a server inlet can vary by several degrees. Supplemental sensors on ACU returns will often

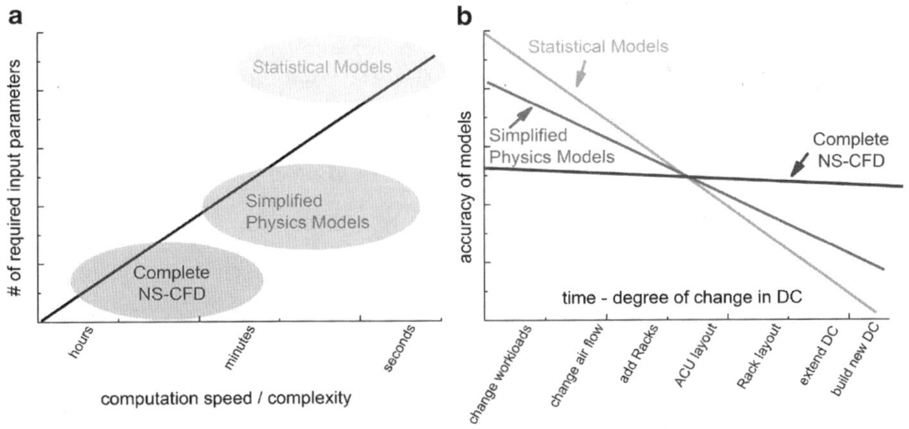


**Fig. 7.5** Dependencies of different modeling technologies on number of measurements and computation speed (**a**), and accuracy with different levels changes in a DC (**b**)

provide different values than what the manufacturers show. Both sensors can be right but measure the "return temperature" at different points. Only *detailed modeling* of the full temperature distribution using the discrete set of measured points obtained from sensors can provide reliable temperature distributions and meaningful insights from these real-time measurements. Even more importantly, as the DC undergoes changes the modeling technology needs to be able to take such changes into account as it affects the relation of the sensors to the complete temperature field.

#### 7.1.2.1 Models

Some of the important dependencies to consider when modeling heat transfer in DCs are illustrated in Fig. 7.5. We have separated here the different modeling technologies into three groups:

(a) Traditional computational fluid dynamics (CFD) models solving the Navier–Stokes equations for fluid flow coupled with a heat transfer equation (Chap. 8).
(b) Simplified physics-based models, such as the Laplacian model (LM) discussed in the second part of this chapter.
(c) Statistical or reduced order models (SM/ROM), such as proper-orthogonal decomposition (POD-) based models, which we describe in more detail further below (also, see Chap. 10).

In Fig. 7.5a, we have plotted the level of description (or number of required measurements) for these different modeling approaches as a function of computation time, which will govern the measurement frequency. On one side, traditional CFD requires only few measurements but comes with longer computation times (often hours), while on the other side of the spectrum SM/ROM requires many more measurements but can provide results much faster (often within seconds).

While Fig. 7.5a might suggest that SM and ROM with many supporting sensor points might be the most advantageous approach, Fig. 7.5b illustrates another important aspect, specifically how these different models maintain accuracy as the DC undergoes changes. For discussion purposes, we have assumed in Fig. 7.5b that statistical and reduced order models may provide the best accuracy at the beginning—simply because typically these models can readily leverage the largest number of measurement points and thus are easily calibrated with larger data sets as available from thermal assessment technologies. We note that in the *absence* of such initialization with larger detailed data sets (which is unfortunately a common practice) it is very unlikely that these models would be accurate at all because little physics is being leveraged. Figure 7.5b illustrates that such models are likely to fail and will become more and more erroneous as the DC undergoes significant changes, and initial data sets (which reference real-time sensors to the complete temperature field and/or calibrate these models) are less and less accurate.

We note that Fig. 7.5 is designed to facilitate the discussion of advantages and disadvantages of the different models but clearly the details in Fig. 7.5 have not been fully proven nor entirely investigated. Figure 7.5 is useful to develop suitable architectures, which would support different data sources and supporting models. It is also quite insightful to discuss the implications of Fig. 7.5 as it relates to the measurement system including sensor density, measurement frequency, and sensor placement.

*Sensor Density, Measurement Frequency, and Sensor Placement*: The right sensor density and read frequency will depend on the modeling technology. For example, if the preferred model technology is SM/ROM, a higher density of sensors is required (possibly 4–5 per rack), ideally in combination with high resolution thermal assessment data. Sensor read frequency could be as high as every 10 s because the modeling technology will allow rapid feedback. On the other hand, if the supporting modeling technology includes physics-based models (such as CFD or LM) the number of sensors can be significantly lower (possibly one per rack or even less). The modeling technology would not require full assessment data to get calibrated. However, sensor read frequency should be lower because the computation time will be longer. In practice, time-averaged values would be used to feed such models.

Generally, the placement of sensors is less critical for CFD or LM than for statistical and reduced order models. Typically, the inlet locations of the most critical servers should be targeted. In the case that SM/ROM is the supporting modeling technology, it might be useful to place sensors at the outlet of the servers as well. In the case a single sensor per rack is used, the sensor should be placed more toward the top of the rack than the bottom (in order to pick up recirculation effects). If the racks are aligned into long aisles, sensors should be placed higher at the "corner" racks. If the sensors cannot be mounted in the middle of these corner racks, they should be on the "outer" side to pick up recirculation. Sensors for racks

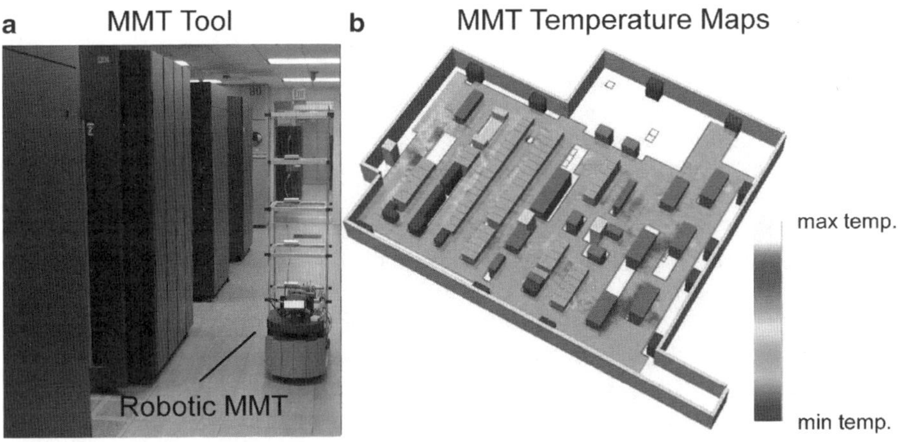


**Fig. 7.6** IBM's measurement and management technologies (MMT). (**a**) Depicts a robotic version of the MMT tool while (**b**) shows example data for a given height of a particular DC

in the middle of a long aisle can be placed lower. All ACUs should be equipped with temperature (for return and discharge) and airflow sensors. If the ACUs are in galleries or a larger central air handling unit is used, the air inlets and outlet ducts to the DC room should be monitored (return, and inlet temperatures as well as airflows).

#### 7.1.2.2 High Resolution and Real-Time Measurement Systems

*High resolution*: One example for a technology, which allows for a very detailed assessment of DCs, is IBM's Measurement and Management Technologies (MMT), depicted in Fig. 7.6. In essence the technology (commonly referred to as MMT 1.0) rapidly "digitizes" the facility using a 3D scanning process, whereby relevant environmental parameters such as temperature, flow, humidity, and physical dimensions of the rooms and racks are gathered [12]. The tool includes a supporting frame loaded with sensors.

Details of this technology can be found in [12], where the accuracy of this sensing solution has been investigated in more detail. Specifically, thermal time constants (5–95% of steady state) were measured to be less than 5 s. Absolute errors of the measured steady-state temperatures for different mounting options as well as different directions of airflow (i.e., sideways, from below, etc.) were always less than 2.5°C at 50°C.

Exemplary data such as a 3D temperature map for a given height is shown in Fig. 7.6, which was directly obtained from the MMT tool. The 3D data is being complemented with detailed airflow measurements from each ACU and perforated tile as well as the asset information. The MMT tool has also been used for mapping the dimensions of the facility leveraging laser scanners. A robotic version of this tool has been built as well [10]. The data sets are automatically postprocessed yielding specific energy savings recommendations [7]. Although the scan can be quite fast (e.g., for a large DC with ~5,575 $m^2$ (60k $ft^2$) a scan takes less than 10 h) often faster feedback is required—especially, if the measurement system is being used for active controls with the facility.

*Real-time Sensing*: The static representation derived from the spatially dense thermal distributions obtained with MMT 1.0 provides a very accurate and detailed "snapshot" of the conditions within a DC at the time of measurement. Although each MMT 1.0 measurement has a timestamp and can be referenced to the changing conditions in the DC, it is practically difficult to permanently scan the facility. However, over time the configuration of the equipment, including networking and storage devices, as well as the operational conditions thereof, together with the airflow and associated cooling system, can change. Stationary sensors can be queried in near real time and thus can provide faster insights. Naturally, the spatial density of stationary sensors is less than what a scanning technology such as MMT can provide and thus sensor placement and the spatial density of sensors are important issues.

#### 7.1.2.3 Wired and Wireless Systems

Another important aspect of the monitoring solution is whether the sensors should be wired or wireless. Wireless sensing solutions have been deployed in DCs [13], mostly using radio frequency, and bring the advantage of the ease of installation. Important issues of wireless systems include cost, battery lifetime, reliability, and possibly interference of the radio frequency signals with the IT equipment. On all of those fronts wireless sensing solutions have made steady progress and clearly wireless sensors are today a viable approach as a DC monitoring system.

An often overlooked challenge of a sensing solution is the fact that any sensor has to be tracked. At a minimum, the sensor value has to be related back to the position of the sensor in the DC. Toward that end, wired solutions might have some advantage as it is less likely that a rack gets moved with a sensor attached without updating the position information. One wired solution uses a grid-like network with a one-wire protocol [14], where in a regular distance (typically one or two tiles) access points are located allowing for the addition, removal, and movement of sensors; see Fig. 7.7a. Figure 7.7a also depicts how sensor height is being altered with sensors being placed somewhat higher at the corners of an aisle to pick up recirculation effects. As shown in Fig. 7.7b access points are generally located in the hot aisle if the grid network is being installed in the plenum. As shown in Fig. 7.7b the grid network infrastructure includes areas of the DC which have not been populated yet so future growth can be managed without deploying any additional access points. From each access point one or more sensors can be teed off as needed.

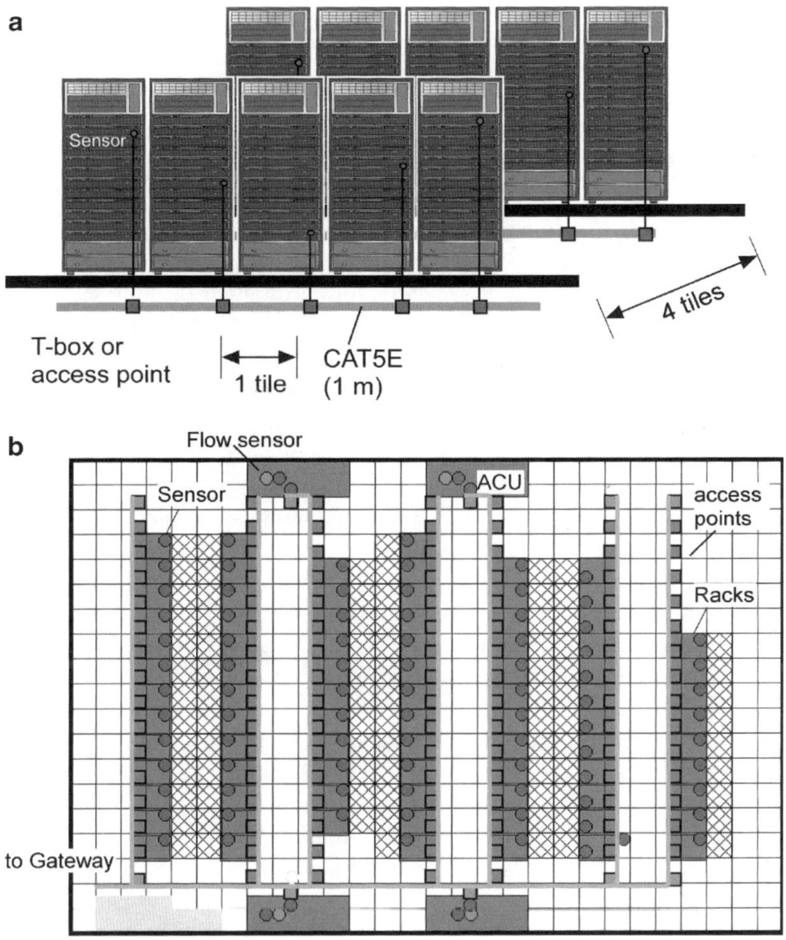


**Fig. 7.7** Example of a wired sensor network using a grid-like infrastructure in a vertical (**a**) and horizontal view (**b**)

In Fig. 7.7 each ACU is being monitored with two temperature sensors (at the discharge and return) and a flow sensor. These sensors are connected to the grid as well. In a typical install pressure, humidity and other sensors would be added to the grid network, which is not shown in Fig. 7.7. The grid network is connected to a gateway Web server, which interfaces between the one-wire sensor network and TCP/IP (Transmission Control Protocol/Internet Protocol). Each Web server can support more than 400 sensors. Software queries the Web servers at a desired rate (as fast as every 300 ms) to gather the latest data. The software also relates each sensor (identified with a unique id) to a data model entry, which contains position information.

This technology will support temperature, power/current, humidity, differential pressure, airflow, and corrosion rate monitors—*all on the same one-wire grid network.*

#### 7.1.2.4 Internal and External Sensors

Most IT equipment already contains several sensors which are mounted inside the enclosure and which can be used for environmental monitoring of DCs. In addition, IT equipment includes often other relevant monitoring capabilities such as microprocessor utilization. However, most DC environmental monitoring systems today prefer external sensors for various reasons, which include network and accessibility challenges, as well as the fact that external sensors are IT vendor agnostic.

In addition, since the need for environmental monitoring with internal sensors is relatively new, there are no clear standards about what the exact measurements mean. In fact, the measurement has to be related back to the specifications of the manufacturer, which can make it difficult to use such data in a consistent way. It is expected, though, that in the long term more and more DC monitoring systems will complement external sensors with internal ones. However, granted the fact that many DCs still have a lot of legacy equipment, it is unlikely that all external sensors will be replaced by internal ones any time soon. It is also clear that more research in understanding the placement of internal sensors, as well as more standard work, is required.

One example, where internal and external sensors are being leveraged at the same time, is shown in Fig. 7.8. In this demonstration over 100 external sensors have been installed. In addition, IBM's Systems Director Active Energy Manager (AEM), which measures IT, power, and thermal components built into IBM systems, is used to obtain the values of internal sensors—here for a subset of the IT equipment.

#### 7.1.2.5 Protocols

The topic of protocols is highly complex. Most communication for data center applications will go over the Internet and uses protocols from the TCP/IP suite, which includes more than 1,000 protocols on different layers (data link layer, network layer, transport layer, session layer, application layer, routing, tunneling, and security). Each of the protocols in the TCP/IP stack perform different functionalities. For example, ModBUS (Modicon Communication Bus), which is commonly used in building automation, is an application layer messaging protocol that provides client/server communication between devices connected on different types of buses or networks. More specifically, ModBUS TCP uses TCP as a transport layer protocol. SNMP (Simple Network Management Protocol) is another application protocol of the TCP/IP suite, which allows network components to communicate and to be managed by a global network

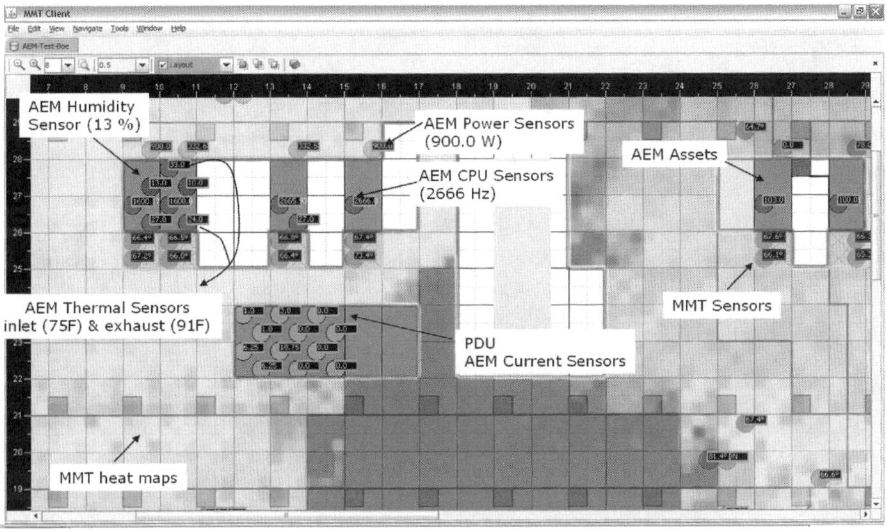


**Fig. 7.8** Example of how internal and external environmental sensors can be integrated into the same platform for DC monitoring

management architecture. An example of yet another protocol is IPMI (Intelligent Platform Management Interface), which includes link, transport and session layers and that enables remote monitoring and control of servers, networking equipment and other assets regardless of their operating system or hardware platform. IPMI resides on an $I^2C$ (inter-integrated circuit) physical layer but can be interfaced to the TCP/IP suite [15]. A further discussion of the topic of protocols is beyond the scope of this chapter. We refer the reader to the following reference [16] for more details.

## 7.2 Measurement-Based Modeling Approaches

In what follows, we discuss more concretely how the measurement data described in the first part of this chapter can be used to model heat transfer in DCs. Traditional modeling approaches of heat transfer in DCs rely on CFD calculations solving the Navier–Stokes equations including turbulence effects, as discussed in Chap. 8. While this approach has been successfully used and is the most accurate modeling technology, especially for planning purposes, it comes with long calculation times. Recently, there has been focus on the development of faster physics-based models, as well as reduced order models, leveraging the fact that the availability of measurement data may be "traded" against the complexity of the model description. In this section, we review recent work on the subject and present case studies employing

**Table 7.1** Units used in the case studies

| Physical quantity | Unit |
|---|---|
| Length | Meters (m) (feet (ft)) |
| Time | Seconds (s) |
| Energy | Joules (J) |
| Temperature | Degree Celsius (°C) or Kelvin (K) |
| Potential function | $m^2/s$ ($ft^2/s$) |
| Air velocity | m/s (ft/s) |
| Air density | 1.205 $kg/m^3$ (0.034 $kg/ft^3$) |
| Thermal conductivity of air | 0.026 W/m K (0.0079 W/ft K) |
| Specific heat of air | 1005 J/kg K |

such measurement-based modeling techniques. We begin with a Laplacian model for DCs, based on potential flow theory, and discuss how (real-time) sensor measurements can be used to define the required input information to the model, mainly in the form of boundary values for the partial differential equations (PDEs) appearing in the model. Afterward, an approach using POD models and their use in DC energy management will be discussed. See also Chap. 10 for a discussion of the application of the POD approach for design of energy efficient DCs.

### *7.2.1 Laplacian Model for Data Centers*

Studies on the use of potential flow theory for DC energy management are documented in the literature. For instance, it has been used to develop and analyze rack-cooling performance metrics [17, 18], improve air conditioning utilization [19], and in heat transfer simulations [20–22, 45]. Advantages of using a potential flow model include applicability of the principle of superposition for linear PDEs (i.e., a linear combination of solutions is again a solution of the PDE) and the availability of fast and robust numerical methods for solving the Laplace equation (which is at the heart of the model). For thorough treatments on potential flow theory, and, more generally, fluid dynamics and the use of PDEs to model physical phenomena, we refer the reader to [23, 24]. Next we present only what is needed to describe the proposed air and heat transfer model for DC energy management use. The mathematical definition of the model, along with a discussion on how to use measured data (e.g., airflow and temperature readings from sensors) to supply the required input information to the model, appears in Sect. 7.2.1.1. Information on the numerical solution of the boundary value problems comprising the model is given in Sect. 7.2.1.2, followed by applications of the model, which are presented in Sect. 7.2.1.3. Finally, it is important to note that in order for the results to be physically meaningful, one must use consistent units when working with the model. The units used in the discussions that follow are as indicated in Table 7.1.

#### 7.2.1.1 Model Description

Several assumptions are being made which allow us to work with a model based on potential flow theory [23, 24, 45], rather than one based on the Navier–Stokes equations for fluid flow, namely:

1. The flow is considered inviscid, that is, the effects of viscosity on the fluid flow are considered negligible.
2. The effects of turbulence are not taken into account.
3. The fluid is incompressible, or, in other words, constant air density is assumed.
4. The fluid flow is irrotational, that is, the curl of the velocity $v$ is zero ($\nabla \times v = 0$) at each point in space and time.

The condition for irrotational flow is satisfied by assuming that the velocity field $v$ is given by the gradient of a potential function $\phi$, that is,[1,2]

$$v = -\nabla\phi. \tag{7.35}$$

The mass transport equation is then obtained by substituting (7.35) into the continuity equation

$$\frac{\partial \rho}{\partial t} + \nabla \cdot (\rho v) = 0. \tag{7.36}$$

Since the density $\rho$ is assumed constant, one has, after factoring out $\rho$,

$$-\nabla \cdot (\nabla\phi) = -\nabla^2\phi = -\left(\frac{\partial^2\phi}{\partial x^2} + \frac{\partial^2\phi}{\partial y^2} + \frac{\partial^2\phi}{\partial z^2}\right) = 0, \tag{7.37}$$

which is the Laplace equation. We thus refer to (7.37), coupled with suitable boundary conditions, as the *DC Laplacian model*.

A typical DC layout lends itself to the use of Neumann boundary conditions for specifying air velocity at the boundaries of the problem domain. That is, sources and sinks of air located on boundaries of the domain can be represented with the use of the Neumann boundary condition

$$n \cdot \nabla\phi = g_{\mathrm{N}}, \tag{7.38}$$

where $n$ denotes the unit outward normal vector at a point on the boundary of the solution domain and $g_{\mathrm{N}}$ represents the value of the air velocity in the direction

[1] $\nabla$ denotes the gradient operator; $\nabla \cdot$ denotes the divergence operator.

[2] The minus sign in (7.35) has been chosen here simply as a convention; use of both a plus and minus sign appear in the literature. What is important then is to use signs in accordance with the convention adopted when modeling sources and sinks of air, for example, when using (7.42) or a non-zero right-hand side in (7.37).

normal to the boundary at the point. For instance, boundary surfaces for specifying inflow or outflow of air in a three-dimensional domain are naturally defined by including the outer surface of objects such as ACUs, equipment racks, and furniture as part of the boundary of the PDE solution domain. Clearly, the floor, walls, and ceiling in the room also correspond to domain boundaries. If air is supplied into the room via the perforated tiles, while being recycled back through the ACUs intake locations, then perforated tiles and ACUs intake locations correspond to surfaces where there is (nonzero) inflow and outflow of air, respectively. Inflow and outflow of air is also generated from equipment such as servers. Surfaces where leakage occurs, for example, cable cut-out locations, may also be used to represent air inflow or outflow locations. For other boundary surfaces, like walls and furniture, zero-flow boundary conditions will apply (i.e., $g_{\mathrm{N}} = 0$ in (7.38)).

One can take advantage of the availability of (real-time) sensor measurements to supply the value of the function

$$n \cdot \nabla\phi \tag{7.39}$$

at given points on the boundary of the problem domain, where $n$ denotes the unit outward normal vector at a point. This will be required when numerically solving the DC Laplacian model (7.37) coupled with the Neumann boundary condition (7.38). For example, perforated tile airflow can be estimated from real-time pressure sensor data [19]. As for the server racks inlet/outlet sides and ACUs, if sensor data is unavailable, airflow values can be estimated from the nameplate power of the IT equipment. Details about such airflow estimations are provided in Sect. 7.1.1.2.

For the purpose of discussion, we denote by $f_{\mathrm{measured}}$ the measured value of air flow (in units of m$^3$/s (ft$^3$/s)) through a given surface $S$, for example a perforated tile, which is a part of the PDE domain boundary. Then one has that

$$\iint_S n \cdot \nabla\phi \, \mathrm{d}S = F, \tag{7.40}$$

where[3]

$$F = \begin{cases} f_{\mathrm{measured}} & \text{if } S \text{ is an air inflow (source) surface} \\ -f_{\mathrm{measured}} & \text{if } S \text{ is an air outflow (sink) surface} \end{cases}. \tag{7.41}$$

Hence, the distribution of the air velocity in the direction normal to the surface $S$ should be done such that (7.40) holds.

In the case study discussed in Sect. 7.2.1.3, Three-Dimensional Heat Transfer Simulations, the air velocity is assumed to be uniformly distributed, that is,

---

[3] The signs are chosen in accordance with equations (7.35) and (7.37) so that sources and sinks are properly defined.

$$(n \cdot \nabla\phi)_i = \frac{F}{A_S}, \tag{7.42}$$

at each point $i \in S$, where $(n \cdot \nabla\phi)_i$ denotes the value of $n \cdot \nabla\phi$ at a point $i \in S$ (cf. (7.38)) and $A_S$ is the surface area of $S$ (e.g., the area of a perforated tile). It is then easily verified that the requirement (7.40) holds. Note also that, as the measured value $f_{\text{measured}}$ of air flow is in units of $m^3/s$ ($ft^3/s$), the function $n \cdot \nabla\phi$ in (1.42) (7.42) is in the appropriate units of m/s (ft/s) (refer to Sect. 7.2.1, Table 7.1). We remark that the number of points at which (7.42) is to be evaluated depends on the size of the grid or mesh used in the numerical calculations. Specifically, in our simulations, where the finite element method was used, the number of such points depended on the number of elements falling on the corresponding surface (e.g., the area of a perforated tile). (Similarly for the function evaluations in (7.43), (7.46), (7.54), and (7.55).)

The DC Laplacian model could also incorporate airflow sources and sinks at nonboundary locations of the PDE solution domain with a nonzero right-hand side $f$ in (7.37). An example of such a problem is provided by the case study in Sect. 7.2.1.3. In this situation, the right-hand side can be specified by setting

$$f_i = \frac{F}{V_{\text{object}}}, \tag{7.43}$$

along points $i$ of the region $V$ occupied by the object acting as a source or sink of air within the domain, where $V_{\text{object}}$ denotes the volume of $V$, $f_i$ denotes the value of $f$ at a point $i \in V$, and $F$ is as defined in (7.41). Note that in accordance with (7.37), $f$ is in units of $s^{-1}$.

Since the solution of the boundary value problem (7.37)–(7.38) for airflow is defined only up to a constant, the Laplacian model is further coupled with a Dirichlet boundary condition

$$\phi = g_D, \tag{7.44}$$

where $g_D$ denotes the value of the potential function at given points on the boundary. For the numerical solution of the problem, it suffices to specify (7.44) at one grid point. In our numerical simulations we set $g_D = 0$ at such point. Finally, before proceeding to discuss the boundary value problem for temperature distribution, we point out that in a potential flow model for incompressible flow the relationship between velocity and pressure is provided by the Bernoulli equation

$$\frac{\partial\phi}{\partial t} + \frac{\|v\|^2}{2} + \frac{p}{\rho} = \text{constant},$$

where $p$ denotes pressure and $\|\cdot\|$ the Euclidean norm. The latter relationship is not required as part of the case studies presented in Sect. 7.2.1.3. For further details on the Bernoulli equation, the reader is referred to [24].

The case study presented in Sect. 7.2.1.3 considers a steady-state temperature distribution, in which the energy equation is represented via the convection–diffusion equation

$$\rho c_p v \cdot \nabla T - \nabla \cdot (k \nabla T) = h, \tag{7.45}$$

where $T$ denotes temperature, $\rho$, $c_p$, and $k$ correspond, respectively, to density, specific heat, and thermal conductivity of air, $v$ is the air velocity, and $h$ represents sources or sinks of heat (at nonboundary locations). Analogous to the case (7.43) for sources or sinks of air in the interior of the domain, evaluation of the function $h$ in (7.45) can be done by setting

$$h_i = \frac{P}{V_{\text{object}}}, \tag{7.46}$$

where $P$ is the power being generated by the source (in which case $h_i$ has a positive sign) or removed by the sink (in which case $h_i$ has a negative sign). Since $P$ is in units of watts (W), $h$ is in the required units of W/m$^3$ (W/ft$^3$).

For a complete specification of the model, (7.45) must be coupled with appropriate boundary conditions prescribing temperature and/or heat flow at the domain boundaries. Precisely, a Dirichlet boundary condition

$$T = u_{\mathrm{D}}, \tag{7.47}$$

where $u_{\mathrm{D}}$ represents the value of $T$ at given boundary points, is used to prescribe temperature. For this purpose, measurements from temperature sensors can be directly utilized. Thus, the boundary condition (7.47) is easily handled. Heat flow at boundaries of the solution domain, for example, as that generated by equipment such as servers, can be represented via the boundary condition

$$- n \cdot (k \nabla T) = q + \alpha T, \tag{7.48}$$

where the coefficient function $q$ is defined via measurements (or estimates) of the heat generated or removed by the equipment and $\alpha$ depends on the air velocity through the surface of such equipment; precise expressions for these coefficients are derived below using standard methods for this purpose. Note that with the coefficient $\alpha = 0$, (7.48) corresponds to a Neumann boundary condition (analogous to that in (7.38)), while the case $\alpha \neq 0$ defines (what is typically referred to as) a Robin boundary condition.

To derive expressions for the coefficients $q$ and $\alpha$ in the heat flow boundary condition (7.48), consider a piece of heat generating equipment in a DC, such as a server. Recall that the outer surface of equipment in the room defines in part the boundaries of the solution domain of the PDE. That is, the space occupied by the equipment is excluded from the solution domain of the PDEs. Integrating (7.44) over the volume $V$ occupied by the piece of equipment and using the Divergence Theorem,[4] one has that

$$-\iint_S n_{\mathrm{E}} \cdot (k \nabla T)\, \mathrm{d}S = \iiint_V h\, \mathrm{d}V - \iiint_V \rho c_{\mathrm{p}} v \cdot \nabla T\, \mathrm{d}V, \tag{7.49}$$

[4] Divergence Theorem: $\iiint_V \nabla \cdot F\, \mathrm{d}V = \iint_S n \cdot F\, \mathrm{d}S$, where $F$ is a continuous and differentiable vector field.

where $S$ corresponds to the equipment surface area and $n_{\mathrm{E}}$ denotes the unit normal vector pointing from the surface of the equipment into the room, i.e., $n_{\mathrm{E}} = -n$, where $n$ is as defined in (7.38). Since the temperature $T$ is a scalar function, the air velocity $v$ a vector, and we consider the density $\rho$ and specific heat $c_p$ to be constant,[5]

$$\begin{aligned} \nabla \cdot (\rho c_p v T) &= \rho c_p v \cdot \nabla T + \rho c_p T (\nabla \cdot v) \\ &= \rho c_p v \cdot \nabla T, \end{aligned} \tag{7.50}$$

where the last equality results from the assumption $\nabla \cdot v = 0$ (per equations (7.35) and (7.37)). Therefore the integrand in the last term of the expression in (7.49) can be replaced by the function $\nabla \cdot (\rho c_p v T)$. Furthermore, by the Divergence Theorem, one has

$$\iiint_V h \, \mathrm{d}V = \iint_S n_{\mathrm{E}} \cdot \tilde{q} \, \mathrm{d}S = \mathrm{P}, \tag{7.51}$$

where, as in (7.46), P denotes the power being generated by the piece of equipment and $\tilde{q}$ is the corresponding heat flow density (i.e., $\nabla \cdot \tilde{q} = h$). Then equation (7.49) becomes

$$-\iint_S n_{\mathrm{E}} \cdot (k \nabla T) \, \mathrm{d}S = \iint_S n_{\mathrm{E}} \cdot \tilde{q} \, \mathrm{d}S - \iint_S n_{\mathrm{E}} \cdot (\rho c_p v T) \, \mathrm{d}S \tag{7.52}$$

or, equivalently,[6]

$$\iint_S n \cdot (k \nabla T) \, \mathrm{d}S = -\iint_S n \cdot \tilde{q} \, \mathrm{d}S + \iint_S n \cdot (\rho c_p v T) \, \mathrm{d}S. \tag{7.53}$$

This gives a means for prescribing boundary conditions of the type (7.48) (from a comparison between the expressions in (7.48) and (7.53)). For example, the numerical simulations discussed in Sect. 7.2.1.3 use[7]

$$q_i = (n \cdot \sigma_1 \tilde{q})_i = -\sigma_1 \frac{P}{A_{\mathrm{S}}} \tag{7.54}$$

and

$$\alpha_i = -(n \cdot \sigma_2 \rho c_p v)_i \tag{7.55}$$

where $\sigma_1$ and $\sigma_2$ are two dimensionless weight coefficients, the subscript $i$ is again used to denote evaluation of the functions at a point and the remaining symbols are

[5] The identity $\nabla \cdot (fF) = f \nabla \cdot F + F \cdot \nabla f$, where $f$ is a scalar field and $F$ a vector field, is used here.

[6] Recall that $n_{\mathrm{E}} = -n$.

[7] Distributions of power and air velocity other than the uniform ones adopted may also be appropriate, although such are not treated here.

as previously defined. Note that $q$ and $\alpha$ are, respectively, in the appropriate units of W/m$^2$ (W/ft$^2$) and W/m$^2$ C (W/ft$^2$ C). The coefficients $\sigma_1$ and $\sigma_2$ were introduced so that the right-hand side $(q + \alpha T)_i$ of (7.48) is comparable in magnitude to that of physically meaningful values of heat flow through the boundary. Hence they depend on the (expected) order of magnitude of $T$ and $\nabla T$, as well as on the order of magnitude of the values used in the computations for the remaining physical parameters in (7.48). Finally, by (7.55), for surfaces through which there is no flow of air (i.e., $n \cdot \rho c_p v = 0$), the boundary condition (7.48) is of the Neumann type. This would describe, for example, heat flow generated by a server. For equipment having airflowing through (sections of) its surface, for example server inlets and outlets, one has a Robin boundary condition. The latter could also apply to other surfaces with nonzero airflow, like the location of perforated tiles, in which case $q$ can be set to zero.

#### 7.2.1.2 Numerical Solution of the Boundary Value Problems

Application of the airflow and heat transfer models for use in DC energy management will be considered in the upcoming sections. Details specific to the numerical solution of the boundary value problems for each of the case studies presented will be provided in the particular sections; here, we discuss general aspects of the numerical solution process. A detailed discussion on the numerical solution of PDEs, and more generally the topic of scientific computing, is outside the scope of this chapter, thus we refer the reader to, for example, [25–27] and references therein for thorough treatments. The C programming language [28] was used to develop finite element solvers for the numerical solution of the boundary value problems. Of course, programming languages and methods different than those adopted here could be utilized. For instance, the reader is referred to [29, 30] as examples of PDE (or CFD) programming environments.

The boundary value problems (7.37)–(7.38), (7.44) and (7.45), (7.47)–(7.48) were solved using the Galerkin finite element method [25, 26, 31]. This method was chosen based on its flexibility and widespread use. Although it is common practice to use an unstructured set of grid points when working with the finite element method, and some problem geometries indeed require it, the layout of a typical DC lends itself to the use of a structured grid and, thus, was the approach adopted in the numerical simulations. As noted in Sect. 7.2.1.1, the outer surface of objects such as ACUs, server racks, and furniture, along with the floor, walls, and ceiling in the room, are used to define the boundaries of the solution domain of the PDEs. Illustrations and details of the domains and corresponding finite elements meshes for the case studies appear in Sect. 7.2.1.3.

Since the mass transport equation (7.37) is independent of the temperature $T$, the solution of (7.37) and (7.45) can be decoupled. Clearly, the energy equation (7.45) must be solved after the mass transport equation (7.37), as it is dependent on the velocity field (7.35). Under the assumption that the parameters $\rho$, $c_p$, and $k$

(corresponding to density, specific heat, and thermal conductivity of air) are independent of $T$, the equations are linear in the unknowns $\phi$ and $T$. Let[8]

$$\phi(x,y,z) \approx \sum_{j=1}^{N} \tilde{\phi}_j \eta_j(x,y,z) \tag{7.56}$$

denote an approximation to the solution $\phi$ of the mass transport equation (7.37) and

$$T(x,y,z) \approx \sum_{j=1}^{N} \tilde{T}_j \eta_j(x,y,z) \tag{7.57}$$

that of the solution $T$ of the energy equation (7.45), where $N$ denotes the number of nodes in the finite elements mesh and $\eta_j$, $j = 1, \ldots, N$, are piecewise linear basis functions satisfying

$$\eta_j(x,y,z)_i = \begin{cases} 1 & \text{if } j = i \\ 0 & \text{otherwise} \end{cases}, \tag{7.58}$$

where $i = 1, \ldots, N$ and $(x,y,z)_i$ denotes the spatial coordinate of the $i$-th node. Note that the unknowns are now the coefficients $\tilde{\phi}_j$ and $\tilde{T}_j$ in the linear expansions (7.56) and (7.57) and that, by virtue of the choice of basis functions, $\tilde{\phi}_j$ and $\tilde{T}_j$ correspond, respectively, to the approximate value of $\phi(x,y,z)$ and $T(x,y,z)$ at the $j$th node, for $j = 1, \ldots, N$. Application of the Galerkin finite element method to the boundary value problem (7.37)–(7.38), (7.44) results in a system of linear algebraic equations

$$A\tilde{\phi} = b, \tag{7.59}$$

the solution of which gives the coefficients $\tilde{\phi}_j$ in the linear expansion (7.56) and thus an approximation to the potential function $\phi$ at the nodes (i.e., grid points) of the finite elements mesh. An approximation to the velocity (7.35) is then computed inside each element of the mesh by differentiating the finite element approximation (7.56) to the potential function $\phi$, as it is required for computing the numerical approximation to the temperature $T$ in (7.45). Application of the Galerkin finite element method to the boundary value problem (7.45), (7.47)–(7.48) for the temperature $T$ likewise results in a system of linear algebraic equations

$$B\tilde{T} = c, \tag{7.60}$$

[8] A detailed account of the finite element method is outside the scope of this text. The reader is referred to [26, 27, 31] and references therein for a thorough treatment (including theory and examples) of the method.

whose solution $\tilde{T}$ provides the coefficients $\tilde{T}_j$ in the linear expansion (7.57) and hence an approximation to the temperature $T$ at the nodes of the finite elements mesh.

The implementation of the finite element solvers [45] was done building upon that presented in [31]. The classical artificial diffusion stabilization technique [26] was incorporated as part of the solver for (7.45) to deal with numerical instability issues arising when solving convection-dominated problems. The CLAPACK routines [32] were used for the solution of the resulting systems of linear algebraic equations for the thermal zones case study presented in Sect. 7.2.1.3, which is a problem formulated in two spatial dimensions. To better handle the larger systems resulting from the three-dimensional simulations presented in Sect. 7.2.1.3, the UMFPACK software [33] was used. Details on performance (i.e., computational time) are provided in each of the case study sections, along with suggestions for improvement, like taking advantage of reusability of the mesh, fast numerical solution techniques, and the superposition principle for PDEs, among others.

#### 7.2.1.3 Case Studies

The Laplacian model (7.35) – (7.38), (7.44) has been the object of several studies geared toward evaluating alternatives to more costly CFD simulations with the Navier–Stokes equations. Although the latter provide more accurate numerical approximations to actual fluid flow behavior, especially when considering small-scale behavior, experimental studies suggest that, on a large-scale, the potential flow model may still provide reasonable approximations to airflow distribution in DCs [17, 20, 22, 28]. As such, we discuss next two applications of the Laplacian model for DCs; the first deals with optimization of cooling resources within DCs, while the second concerns three-dimensional simulations of heat transfer. We emphasize that these approaches to DC energy management are still in their experimental stages, yet the initial results, along with those from previous and concurrent studies, are encouraging.

Thermal Zones

The optimization of cooling resources within DCs, which includes the localized provisioning of cooling power with respect to the heat dissipation, can lead to significant reduction of energy consumption [34]. While such matching is rather straightforward for an entire DC, on a more granular level, however, the large range of heat densities of today's IT equipment has made this task much more challenging. One complication comes from the fact that there is little operational information, for example, which physical areas (or zones) are supplied by the different ACUs so that local cooling demands can be met [35].

Although best practices have been widely publicized and have contributed to more efficient usage of cooling resources, it remains challenging to provision the right amount of cooling. In particular, the optimization of the individual ACUs, for which utilization levels $\upsilon^i$ can be defined by

$$\upsilon^i = \frac{P^i_{\mathrm{COOL}}}{P^i_{\mathrm{CAP}}}, \tag{7.61}$$

where $P^i_{\mathrm{COOL}}$ denotes the actual power removed (or cooled) by a particular ACU (the $i$th ACU) in the room and $P^i_{\mathrm{CAP}}$ its respective cooling capacity [13], can be difficult. $P^i_{\mathrm{COOL}}$ can be readily estimated by

$$P^i_{\mathrm{COOL}} \approx \rho c_p f^i_{\mathrm{ACU}} \Delta T^i_{\mathrm{ACU}}.$$

Although different areas within a DC often require very different amounts of cooling, it is not uncommon to find DCs with more than twice the amount of active cooling than actual heat load just to be able to meet the cooling demands of a few high power density regions within the facility. For example, a DC might require $N + 1$ ACUs which, with $N = 7$, should result in a cooling utilization of approximately 85%. However, in reality average ACU utilization levels are often as low as 40%. One reason for this is the lack of operational insights and information on both the utilization of each ACU, as well as the respective regions or zones which the individual ACUs are serving.

There has been recent work focused on developing techniques for determining and visualizing such *thermal zones*. For example, thermal zone mapping has been used to optimize equipment placement and to investigate failure scenarios and corrective actions [35]. Here, we present a study [19] aimed at defining such thermal zones based on real-time data in an effort to provide operators with information for optimizing ACU utilization in a DC. Although the thermal zones should be viewed as a concept or framework and further validation is needed, the case study shows how this methodology can be used to provide valuable decision support to effectively increase ACU utilization within a DC, thereby improving its energy efficiency. Evidently, thermal zones are governed by many parameters such as the location of ACUs, the airflow produced by each ACU, placement of perforated tiles, tile types, and room dimensions, among other factors. It is not unusual for a DC to have more than 50 ACUs somewhat randomly distributed across the DC space with more than 1,000 perforated tiles. In order to be able to calculate these zones quickly so that operational decisions can be based on such information, simplifications in the airflow model were deliberately sought, while still trying to capture the governing physics. Therefore, the Laplacian model (7.35)–(7.38), (7.44) was used, with the model inputs obtained from real-time measurements. The thermal zones were determined from the resulting velocity field following streamlines, as outlined below. We note that the concept of streamlines has previously been used to help operators improve DC energy efficiency and refer the reader to [18] for details on such study.

Once the air velocity field has been calculated, an efficient algorithm is employed to trace the air from each area of the DC back to the originating ACU. The traces are determined using an algorithm for computing streamlines. These traces connect a given ACU with its corresponding zone. Specifically, the algorithm simulates the actual trajectory of airflow starting at a certain location in the DC. The step size used in the algorithm can be varied and depends on the magnitude of the velocity vector. This is done to avoid long calculation times in areas with low air movement. The trajectory is calculated until it either intersects with an ACU location or with a location which has already been assigned to an ACU. The origin point of that trajectory is then assigned to that particular ACU.

Having determined the thermal zones, each zone can be attributed with the utilization $\upsilon^i$ (see (7.61)) or with a coefficient of performance (COP), which is defined as [13]

$$\mathrm{COP}^i = \frac{P^i_{\mathrm{COOL}}}{P^i_{\mathrm{BLOWER}}}, \tag{7.62}$$

where $P^i_{\mathrm{BLOWER}}$ represents the blower power consumption of the $i$th ACU. The cooling power $P^i_{\mathrm{COOL}}$ in the metrics (7.61) and (7.62) is derived from real-time temperature and airflow (and/or water flow for waterside ACUs) measurements. Finally, for presentation purposes the discussion that follows is limited to a raised floor DC, where the zones are based on the cooled air supply from the plenum, modeled as a problem in two spatial dimensions. However, we note that the same concepts and algorithms can be applied to other types of DCs as well as for determining three-dimensional zones, which are based on discharge and return airflows of each ACU.

The case study was carried out for a raised floor DC (603.87 $\mathrm{m}^2$ (6,500 $\mathrm{ft}^2$)) with an average heat density of 818 $\mathrm{W/m}^2$ (76 $\mathrm{W/ft}^2$). The height of the plenum is 0.6096 m (2 ft). A subview of the general layout is depicted in Fig. 7.9. The DC has 4 ACUs (labeled #1–4) with 4 hp blowers ($P_{\mathrm{CAP}} = 70$ kW), 4 ACUs (#5–8) with 1.2 hp blowers ($P_{\mathrm{CAP}} = 25$ kW), as well as one large-scale central ACU (#9) with $P_{\mathrm{BLOWER}} = 33$ kW and $P_{\mathrm{CAP}} = 570$ kW, which is ducted from the basement into the plenum. The inlets to the plenum of this central ACU are distributed along the south wall of the DC while the return air is ducted through the ceiling back to the central ACU.

The DC is being monitored in real time using IBM's MMT [12, 13]. Each server rack and ACU has at least two thermal sensors (identified with circles in the layout in Fig. 7.9). Rack sensors are placed at the inlet side(s) of the equipment, while for each ACU temperatures at the inlet (return) and outlet (discharge) are measured. In total, 172 thermal sensors were deployed, with some additional thermal sensors located in the plenum and ceiling, 8 sensors for measuring pressure differentials, 9 flow sensors monitoring the ACU flow outputs, and 9 humidity sensors. The real-time data are being gathered from each sensor every 60 s and fed via a software application into a database.

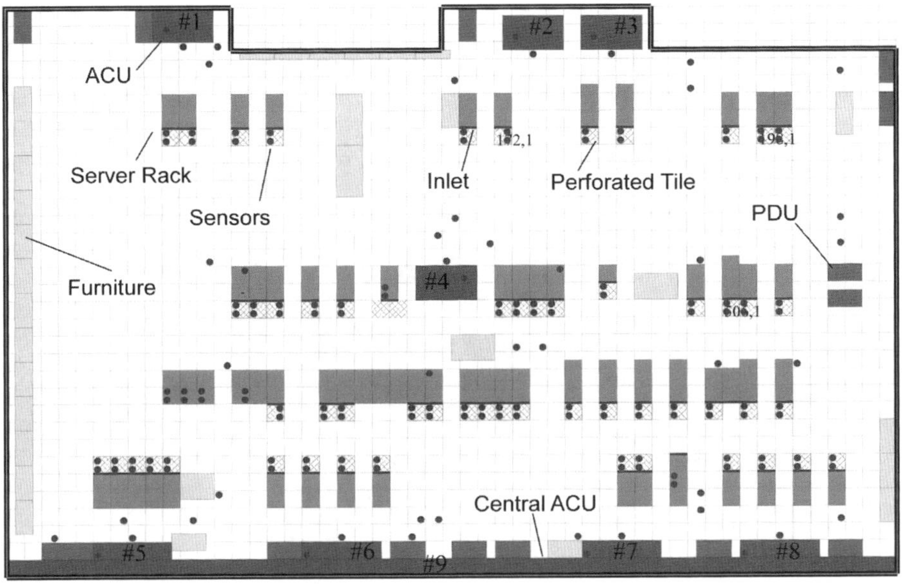


**Fig. 7.9** DC layout for the thermal zones case study. (Figure adapted, with permission, from [19]. © 2010 IEEE)

Pressure differentials are measured at eight locations in the plenum (plenum height = 0.6096 m (2 ft)). Little variations (<15%) between the values from the different locations are observed, which shows that the pressure distribution is quite uniform (<±15%). To obtain a complete pressure field from these eight real-time sensors, an inverse distance interpolation algorithm is used to estimate the pressure at any given location. Refer to Sect. 7.1.1.3 for a description of such algorithm. Estimates $f_{\text{estimate}}$ of airflow through the perforated tiles and ducts are computed using the pressure differential $\Delta p$ and flow impedance $R_{\text{TILE}}$ (at the corresponding tile or duct location) via the relationship (refer to (7.12))

$$f_{\text{estimate}} = \sqrt{\frac{\Delta p}{R_{\text{TILE}}}}. \tag{7.63}$$

The impedance values used in this case study were determined experimentally, as described previously in Sect. 7.1.1.

The values of airflow through perforated tiles and ducts used in the case study, estimated with the aforementioned procedure, as well as those for the ACUs, are depicted in Fig. 7.10. The height of the bars indicates the airflow values for the perforated tiles, which range from 0 to 0.3233 m$^3$/s (685 cfm). The airflow values for the ACUs were determined by calibration of the data collected from the sensors monitoring the ACU flow outputs in the DC, using the procedure described in Sect. 7.1.1.2.

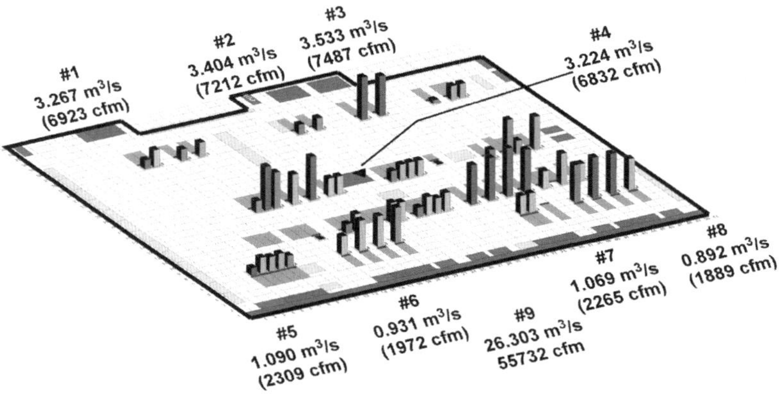


**Fig. 7.10** Airflow values in the case study DC. The height of the bars indicates the airflow value for the perforated tiles, which ranges from 0 to 0.3233 $m^3$/s (685 cfm). The ACUs are labeled with their identification number, along with the corresponding airflow. (Figure adapted, with permission, from [19]. © 2010 IEEE)

As previously mentioned, the thermal zones in the case study were determined using the air velocity field for the plenum, which was modeled in two spatial dimensions. The boundary of the solution domain of the PDE, identified by dark lines in Fig. 7.9, was thus defined by the outer walls of the DC. The perforated tiles and inlets to the plenum of the ACUs are located inside the PDE solution domain and, hence, the corresponding measured airflow for these objects was represented in the Laplacian model via a nonzero right-hand side in (7.37). Air leakage through cable cut-outs and other locations was neglected in the study so in order to satisfy the principle of conservation of mass the ACU airflow values were adjusted to match the total (measured) airflow through the perforated tiles. Since there is no flow of air through the solution domain boundary (i.e., the walls), one has $g_N = 0$ for the Neumann boundary condition (7.38) at the wall boundaries. Finally, as indicated in Sect. 7.2.1.1, it is sufficient to select one node to prescribe the boundary condition (7.44). The first node in the mesh was selected for this purpose, with the value $g_D = 0$ in (7.44).

To generate the finite elements mesh, a node was placed at every 0.3048 m (1 ft) within the domain and then the elements, in this case triangles, were constructed connecting the nodes. All equipment and assets (perforated tiles, ACUs, building pillars, etc.) were "mapped" to this one foot grid. Mesh elements falling within regions occupied by ACU inlets to the plenum define locations of air sources in the Laplacian model, whereas elements overlapping with perforated tiles define locations of sinks (refer to (7.43)). With the chosen grid spacing, the mesh consisted of only 6,646 nodes and 12,944 elements. It was established that finer meshes did not significantly change the results of the calculation. There were 776 elements corresponding to source terms and 464 elements corresponding to sinks. Figure 7.11 visualizes the flow potential $\phi$, along with the normalized velocity vectors, resulting from the numerical solution of the Laplacian model (7.35)–(7.38), (7.44) with input

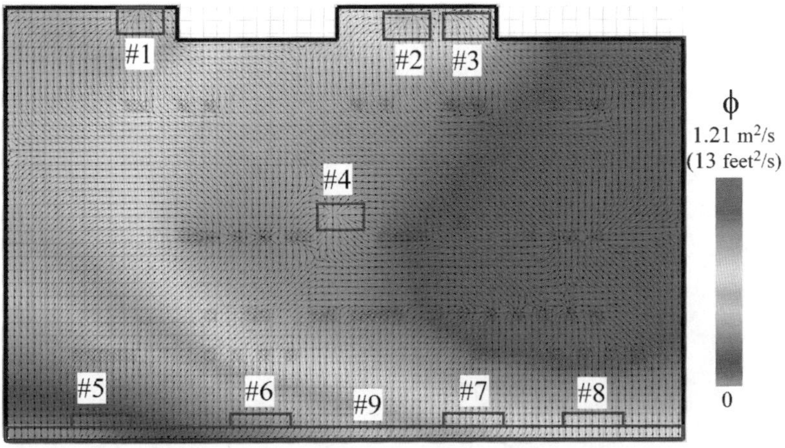


**Fig. 7.11** Plenum flow potential $\phi$ with normalized air velocity vectors (plenum height = 0.6096 m). (Figure adapted, with permission, from [19]. © 2010 IEEE)

airflow data as in Fig. 7.10. The dark boxes show the sources (ACUs), while the hatched regions depict the perforated tiles, or sinks.

Figure 7.12 shows the resulting thermal zones for two different cases, where in case (a) all nine ACUs were running while in case (b) ACUs #1, 5, and 6 were turned off (and sealed). The units were turned off based on the respective utilization levels from case (a). In the figures, the ACUs are shown as dark boxes, while the tiles are depicted using hatched squares. Figure 7.12a, b is interesting if combined with the respective utilization levels of each ACU, shown in Fig. 7.13a, b. In case (b) three underutilized ACUs were turned off. As a result, the remaining active ACUs serve a larger physical area, remove more power and thus run at higher utilization, as evident from Figs. 7.12 and 7.13. The average utilization (weighted by the cooling capacity of each unit) increases from ~50% to ~68%, saving ~7 kW (which includes the power savings at the chiller plant due to the reduced load in the DC). As discussed in previous publications [7], increased utilization almost always decreases the average plenum temperatures (here by 0.72°C (1.3°F)), which can compensate entirely for the increase in temperatures due to less airflow (here, 0.67°C (1.2°F)). Finally, the MMT sensor network was used throughout the experiments to ensure that inlet temperature requirements were met.

An illustration of how this methodology could be used to optimize the ACU utilization within a DC, thereby improving its energy efficiency, is provided by Figs. 7.12 and 7.13. In case, conditions are changing, for example, tiles are getting moved to work on cabling, different tile types are deployed, or an ACU gets turned off, one can readily repeat the calculations and rezone accordingly. The computation time on a laptop (2 GHz dual core) computer was less than 25 s, which includes mesh generation, model setup, and zoning. In case the heat load changes, the ACU utilization levels get updated in real time. In this particular case study, the heat load variations throughout the day were less than 7% and did not affect the optimization procedure.

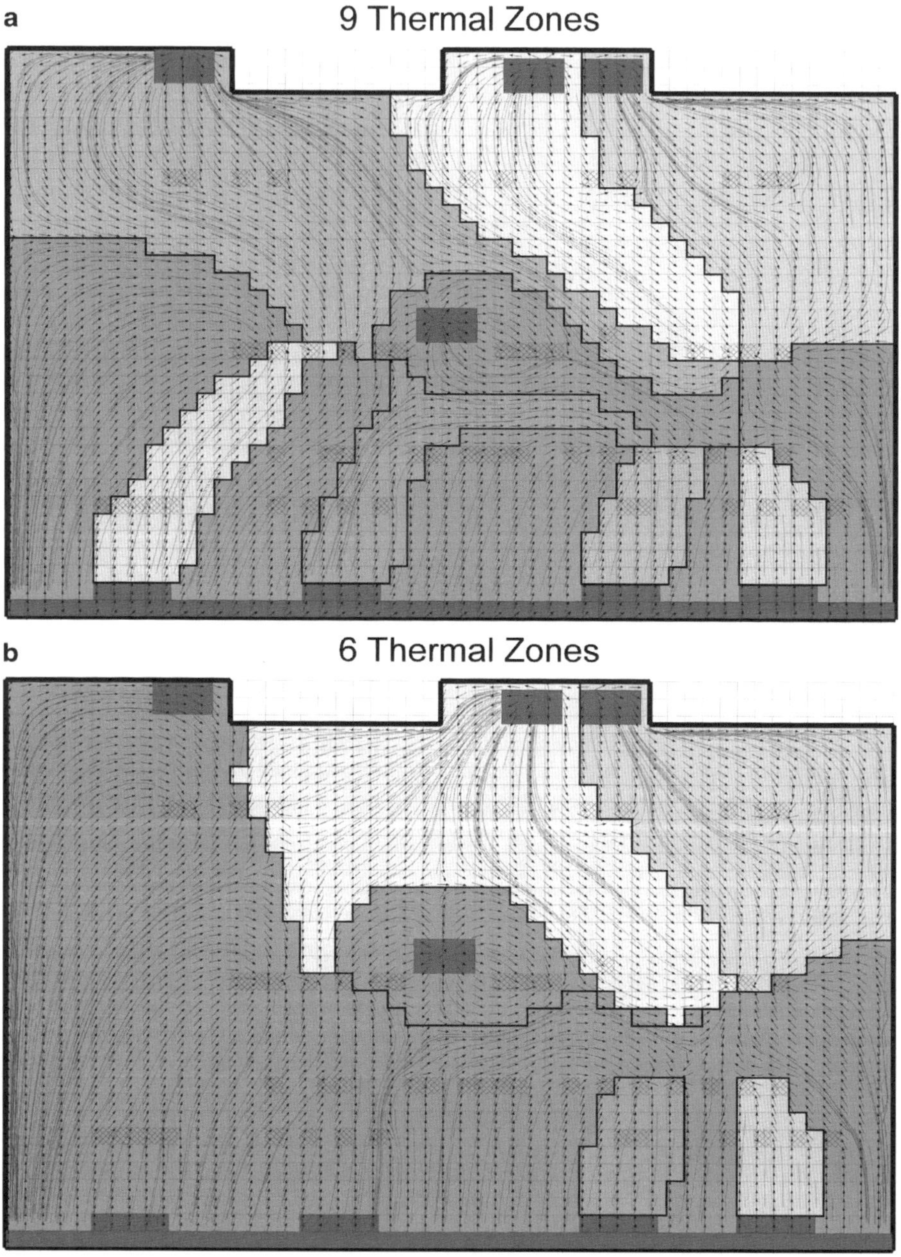


**Fig. 7.12** Plenum thermal zones shown as polygons in different colors for (**a**) nine and (**b**) six ACUs being active. The streamlines are shown as lines (plenum height = 0.6096 m). (Figure adapted, with permission, from [19]. © 2010 IEEE)

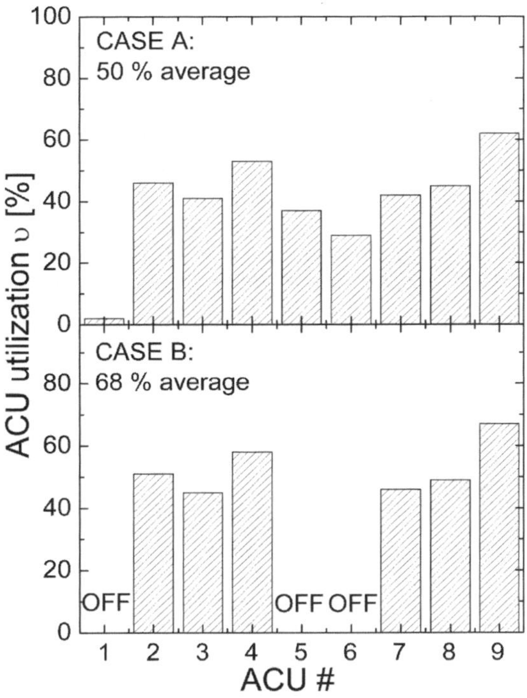


**Fig. 7.13** Corresponding ACU utilization levels for each zone for the two cases depicted in Fig. 7.12. (Figure reprinted, with permission, from [19]. © 2010 IEEE)

At this point it is worth noting that, as long as such changes in the DC do not require modifications to the finite elements mesh, repeating the calculations will require significantly less computing time since the nodes and elements do not have to be regenerated each time. Furthermore, if a direct solver for linear systems is used to solve the system resulting from the discretization of (7.37), as in this case study, savings in computational time are also possible as long as the coefficient matrix in the linear system remains unchanged (which is the case if the conditions changing correspond only to modifications in the measured flow at the sources and sinks, as this results only in a modified right-hand side of the system). Since a direct solver typically employs two phases, numerical factorization of the coefficient matrix followed by the solution of the system with the factored matrix, the numerical factorization of the coefficient matrix need to be done only once, as long as the matrix remains unchanged. The numerical factorization is the most time consuming of the two phases, so doing the factorization only when strictly needed can result in considerable savings in computational time.

We also note that since the PDE is linear, the algorithms can include the exploitation of the superposition principle by taking a linear combination of two solutions of the model (or, generally, any number of solutions) to obtain another solution, as long as the solutions being added correspond to the same domain (geometry) and the resulting solution is physically meaningful. For example, say $\phi_1$ is a solution of $-\nabla^2\phi = f$ obtained with only ACU #1 turned on (at a given fan speed setting), while all the other ACUs were off, and $\phi_2$ is a solution of

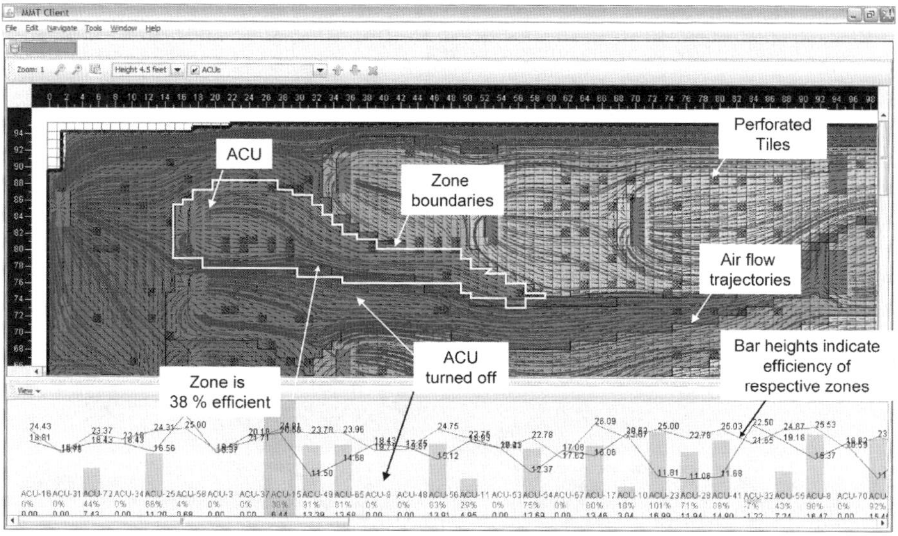


**Fig. 7.14** MMT client software application for real-time monitoring of thermal zones and associated efficiencies (utilization levels and COPs). (Figure reprinted, with permission, from [19]. © 2010 IEEE)

the PDE obtained with only ACU #2 on (at a given fan speed setting), while all the other ACUs were off. The velocity field for the scenario with only ACU #1 on is $v_1 = \nabla\phi_1$, while that for the scenario with only ACU #2 on is $v_2 = \nabla\phi_2$. Then $\phi_3 = a_1\phi_1 + a_2\phi_2$, where $a_1$ and $a_2$ are arbitrary constants, is also a solution of the PDE and $v_3 = a_1v_1 + a_2v_2$ corresponds to a velocity field for the scenario with ACU #1 and ACU #2 on (at the corresponding fan speeds for which the original scenarios were obtained, scaled by the constants in the linear combination), while all other ACUs are off. This may provide ways to compute the respective airflow patterns faster for varying conditions by avoiding redundant calculations and could be in particular important for DCs with VFDs where one can have (basically) an infinite number of combinations of different ACU flow settings.

Finally, because the presented zoning methodology is aimed at operational decision support, a JAVA-based software application was developed, which allows for real-time monitoring and optimization of ACU settings in DCs. Some of the capabilities of this application are shown in Fig. 7.14. For example, the bar chart shows the respective real-time utilization of each ACU. The actual thermal zones are shown as polygons and highlighted. The tool also allows, among other capabilities, interactively controlling the layout, physical data representation, and real-time data management. Although the concept of thermal zones presented here is at an experimental stage and additional benchmarking is needed to understand more accurately the actual physical relevance, the case study demonstrates how this technique, in combination with ACU utilization and performance metrics, can provide valuable operational decision support to optimize energy efficiency.

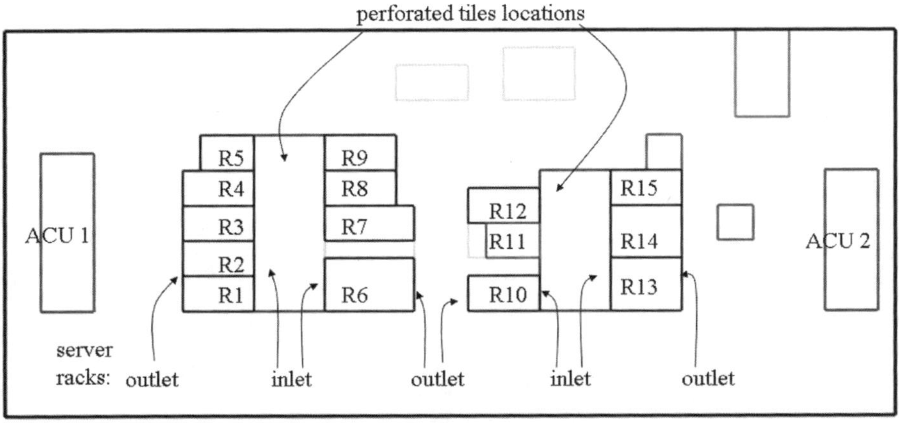


**Fig. 7.15** DC layout for the 3D simulations. Objects in the room, such as equipment and furniture, are represented as *boxes*. ACUs intake locations and server racks inlet sides correspond to air outflow locations, since air is flowing out of the problem domain through these surfaces, while perforated tiles and server racks outlet sides correspond to air inflow locations. The DC has 15 racks, labeled R1–R15, and two ACUs. The ACU intake locations are at the top of each ACU. The remaining unlabeled objects are furniture and other types of equipment. The perforated tiles locations correspond to cold aisles, while the hot aisles are located in between the ACUs and the racks. Each rack has space for seven servers (refer to Tables 7.2 and 7.3 for further server details). (Figure adapted, with permission, from [21]. © 2010 IEEE)

## Three-Dimensional Heat Transfer Simulations

The Laplacian model (7.35)–(7.38), (7.44) has been the focus of several studies geared toward evaluating its suitability (when coupled with an equation for heat transfer) for generating numerical approximations to temperature distributions in DCs. For instance, the study in [22] considers a convective heat transfer model, along with numerical solution techniques. The reader is referred to [22] for specific details. In what follows we consider the convection–diffusion equation (7.45) and discuss numerical experiments done using the steady-state heat transfer model (7.45), (7.47)–(7.48), following up on the work presented in [21, 45].

The required input data for the model were generated using 3D experimental data obtained via IBM's MMT [13, 36]. Such measurements were collected for a DC occupying a space of approximately 238 $m^3$ (8,400 $ft^3$), with a physical layout as depicted in Fig. 7.15.[9] A spacing of 0.1524 m (0.5 ft) was used to generate the finite elements mesh.[10] This value was chosen experimentally, as slightly larger and smaller spacing yielded qualitatively similar numerical results, and it is comparable

[9] Figures for the 3D simulations were generated using the Mayavi2 visualization package (http://code.enthought.com/projects/mayavi).

[10] The 3D finite elements mesh was generated using Matlab (http://www.mathworks.com/products/matlab/).

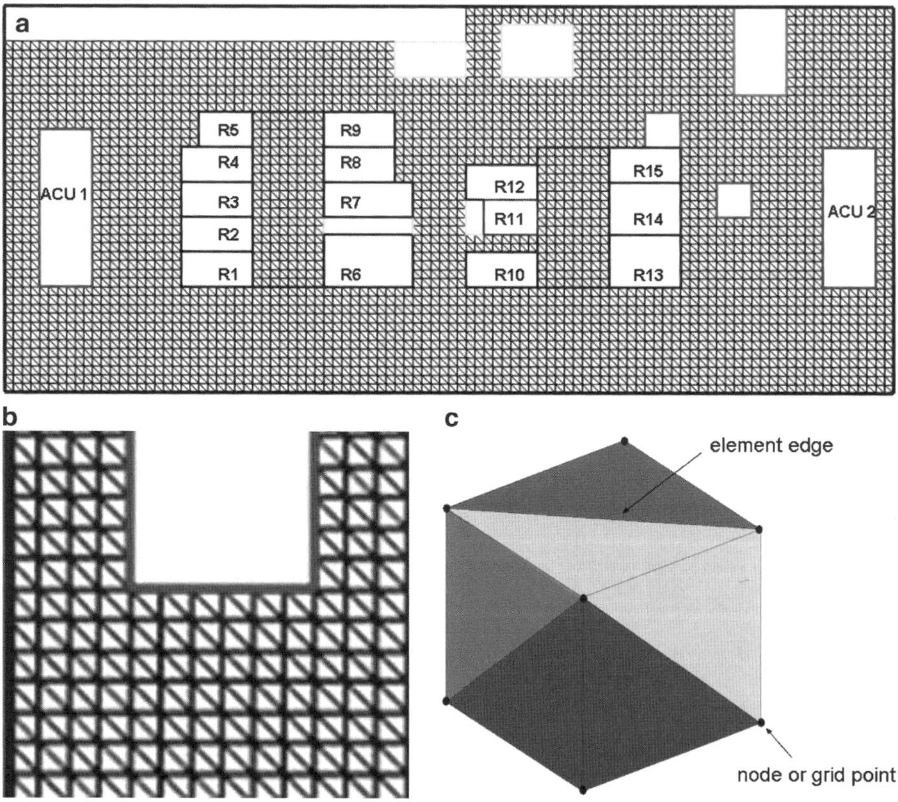


**Fig. 7.16** (**a**) 2D view of the finite elements mesh at a height of 1.37 m (4.5 ft) from the floor of the 3D domain of the DC layout from Fig. 7.15. (**b**) Magnification of the 2D view of the mesh around the lower left corner of ACU 1. (**c**) Mesh components: elements of the mesh, in this case tetrahedra, have edges and nodes, or grid points, in common. (Figure adapted, with permission, from [21]. © 2010 IEEE)

to what has been used previously in the literature in CFD simulations for DCs [5]. A view of the mesh from the top of the room appears in Fig. 7.16. With such spacing, the 3D finite elements mesh had a total number of 63,705 nodes and 327,360 elements. Although the finite elements solver used (refer to Sect. 7.2.1.2) was not optimized for speed, the execution times on a Linux desktop computer were very reasonable. For example, with the aforementioned mesh, assembling a linear system took approximately 8 s, while the execution time for solving the linear system using the UMFPACK software [33] was around 20 s: 1 s for the symbolic factorization of the coefficient matrix, 18 s for the numerical factorization, and 1 s to solve the factored system. As previously mentioned, such fast execution times, especially for solving the factored system, can offer significant reductions in the total computation time whenever the matrix factorization can be reused (i.e., instead of computing it repeatedly).

**Table 7.2** Server power values used in the numerical simulations

| | S1 | S2 | S3 | S4 | S5 | S6 | S7 | Total (W) |
|---|---|---|---|---|---|---|---|---|
| R1 | 110 | 563 | 563 | 839 | 0 | 0 | 0 | 2,075 |
| R2 | 110 | 110 | 563 | 563 | 563 | 422 | 422 | 2,753 |
| R3 | 243 | 422 | 422 | 422 | 503 | 0 | 0 | 2,012 |
| R4 | 422 | 422 | 422 | 519 | 110 | 422 | 422 | 2,739 |
| R5 | 0 | 0 | 0 | 0 | 0 | 0 | 0 | 0 |
| R6 | 0 | 0 | 0 | 0 | 0 | 0 | 0 | 0 |
| R7 | 1,053 | 1,053 | 1,053 | 1,053 | 1,053 | 1,053 | 1,053 | 7,371 |
| R8 | 390 | 390 | 390 | 390 | 390 | 390 | 390 | 2,730 |
| R9 | 390 | 390 | 390 | 390 | 390 | 390 | 390 | 2,730 |
| R10 | 1,614 | 1,614 | 1,615 | 1,614 | 1,614 | 1,614 | 1,614 | 11,299 |
| R11 | 0 | 0 | 0 | 0 | 0 | 0 | 0 | 0 |
| R12 | 221 | 221 | 221 | 221 | 221 | 214 | 169 | 1,488 |
| R13 | 1,053 | 1,053 | 1,053 | 1,053 | 1,053 | 1,053 | 1,053 | 7,371 |
| R14 | 1,279 | 1,279 | 1,279 | 1,279 | 1,279 | 1,279 | 1,279 | 8,953 |
| R15 | 243 | 243 | 0 | 0 | 0 | 0 | 0 | 486 |

There are 15 racks (R1–R15), each one represented by 7 servers (S1–S7). (Table reprinted, with permission, from [45]. © 2011 Elsevier)

The set of MMT data used in the numerical simulations had a total measured airflow through the perforated tiles of approximately 2.99 $m^3/s$ (6,340 cfm). This measured airflow was obtained using the procedure described previously (Sect. 7.1.1.2). Air leakage (e.g., through cable cut-outs) was neglected and the ACUs intake airflow adjusted (in proportion to the measured ACU airflow) to match the total measured airflow for the perforated tiles. For each server, the flow of air through the inlet side was set to match that through the outlet side. In this way conservation of mass in the airflow model was ensured. As airflow and power measurements for the servers were unavailable, the server airflow was estimated from the nameplate power of the IT equipment, which was normalized to the total DC power consumption (52 kW) as discussed in Sect. 7.1.1.4. A constant temperature differential of 11.67°C (21°F) between each inlet and outlet was assumed. The server airflow was then computed using (7.24). The resulting power and airflow values for the servers, which should be viewed as estimates, are listed in Tables 7.2 and 7.3.

As for the energy equation (7.45), two different alternatives were considered for selecting nodes (i.e., grid points) where temperature was to be specified at the boundary, with the idea of simulating different placement of temperature sensors at the server racks inlet sides. The first corresponds to a scenario where temperature was specified at a large number of boundary nodes, namely 1,676 nodes. This represents approximately 9% of the total number of boundary nodes. In the second case, a sensor was assumed to be located at the center of each server inlet side; refer to Fig. 7.17 for further details. The number of boundary nodes for which temperature was specified for the second scenario was 230, or approximately 1.3% of the total number of boundary nodes. The latter case corresponds to a more practical situation where a limited number of temperature sensors would be in place.

**Table 7.3** Server airflow values (in $m^3/s$) used in the numerical simulations

| | S1 | S2 | S3 | S4 | S5 | S6 | S7 | Total ($m^3/s$) |
|---|---|---|---|---|---|---|---|---|
| R1 | 0.0080 | 0.0401 | 0.0401 | 0.0595 | 0 | 0 | 0 | 0.1477 |
| R2 | 0.0080 | 0.0080 | 0.0401 | 0.0401 | 0.0401 | 0.0302 | 0.0302 | 0.1968 |
| R3 | 0.0175 | 0.0302 | 0.0302 | 0.0302 | 0.0359 | 0 | 0 | 0.1439 |
| R4 | 0.0302 | 0.0302 | 0.0302 | 0.0368 | 0.0080 | 0.0302 | 0.0302 | 0.1959 |
| R5 | 0 | 0 | 0 | 0 | 0 | 0 | 0 | 0 |
| R6 | 0 | 0 | 0 | 0 | 0 | 0 | 0 | 0 |
| R7 | 0.0746 | 0.0746 | 0.0746 | 0.0746 | 0.0746 | 0.0746 | 0.0746 | 0.5220 |
| R8 | 0.0278 | 0.0278 | 0.0278 | 0.0278 | 0.0278 | 0.0278 | 0.0278 | 0.1949 |
| R9 | 0.0278 | 0.0278 | 0.0278 | 0.0278 | 0.0278 | 0.0278 | 0.0278 | 0.1949 |
| R10 | 0.1147 | 0.1147 | 0.1147 | 0.1147 | 0.1147 | 0.1147 | 0.1147 | 0.8028 |
| R11 | 0 | 0 | 0 | 0 | 0 | 0 | 0 | 0 |
| R12 | 0.0156 | 0.0156 | 0.0156 | 0.0156 | 0.0156 | 0.0151 | 0.0118 | 0.1048 |
| R13 | 0.0746 | 0.0746 | 0.0746 | 0.0746 | 0.0746 | 0.0746 | 0.0746 | 0.5220 |
| R14 | 0.0906 | 0.0906 | 0.0906 | 0.0906 | 0.0906 | 0.0906 | 0.0906 | 0.6343 |
| R15 | 0.0175 | 0.0175 | 0 | 0 | 0 | 0 | 0 | 0.0349 |

There are 15 racks (R1–R15), each one represented by 7 servers (S1–S7). (Table reprinted, with permission, from [45]. © 2011 Elsevier)

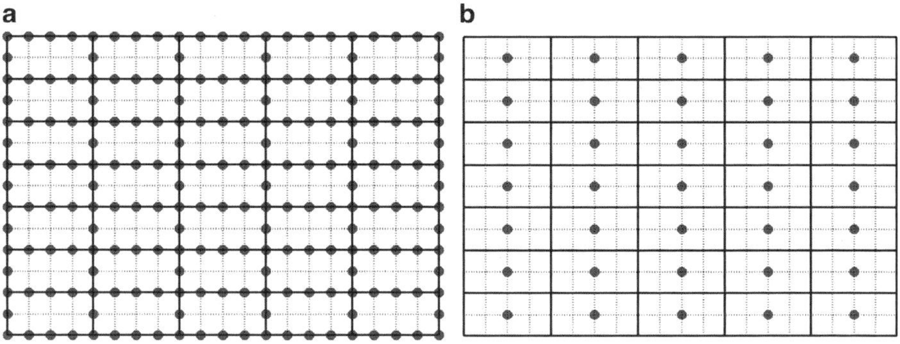


**Fig. 7.17** Two different boundary data patterns for the numerical simulations: (**a**) each server has a dense set of temperature sensors on its inlet side and (**b**) each server has a temperature sensor at the center of its inlet side. Sensor locations (marked with a *circle*) correspond to boundary nodes for which temperature was specified as part of the formulation of the boundary value problem (7.44), (7.46)–(7.47). For the remaining nodes, located at the intersection of the grid lines, temperature was obtained via the numerical solution of the boundary value problem. (Figure adapted, with permission, from [21]. © 2010 IEEE)

Again, the value of the temperature at the simulated sensor locations (i.e., the boundary nodes where temperature was specified) was taken from the aforementioned data set of measurements collected by the MMT. Temperature was also specified at some of the nodes falling on perforated tiles locations, ACUs intake sides, and server racks outlet sides. The boundary nodes where temperature was specified for these surfaces followed the same pattern as for the server racks inlet sides (see Fig. 7.17). In the case of the Robin boundary condition (7.48), the finite element solver required evaluation of (7.54) and (7.55) at the center of each element face falling on the surface of the servers inlets and outlets, ACUs intake sides, and perforated tiles locations. These function evaluations were done as per the discussion in Sect. 7.2.1.1.

The airflow and heat transfer models had not been calibrated, for example, by using multiple sets of measured data to solve inverse problems in an effort to estimate parameter values which the model equations depend on. Hence values for the parameters typically used in experiments, namely, $\rho = 1.205\ \mathrm{kg/m^3}$ ($0.034\ \mathrm{kg/ft^3}$) for the air density, $k = 0.026\ \mathrm{W/mK}$ (0.0079 W/ft K) for the thermal conductivity, and $c_p = 1,005\ \mathrm{J/kg\ K}$ for the specific heat were employed in the numerical simulations. A listing of the units used in the simulations appears in Sect. 7.2.1, Table 7.1. Finally, without having performed calibrations on the models, it would be premature to try to draw definite conclusions on the predictive capabilities of the models based on detailed comparisons between the measured temperature data and the numerical results. Nevertheless, the outcome of the simulations, presented next, suggest that even without calibration the models can produce results which, in qualitative terms, compare favorably to the measured data. This is an indication that the models are quite capable of providing meaningful information for use in DC energy management and that more research, including further calibration and detailed quantitative analyses, is worth pursuing.

An illustration of the air velocity field resulting from the numerical solution of the Laplacian model (7.35)–(7.38), (7.44) is shown in Fig. 7.18. The first plot in this figure corresponds to a two-dimensional slice of the velocity field at a height of 0.0609 m (0.2 ft). As such, one can observe airflow into the server inlets from the perforated tiles. The second plot is for a height of 1.9812 m (6.5 ft), where flow of air into the ACUs intake sides can be seen (refer to Fig. 7.15 for a description of the DC layout). Two-dimensional slices of the temperature distribution at different heights in the DC for the measured and the numerically computed temperature appear in Figs. 7.19–7.21. Although a node-by-node comparison of the measured and numerically computed temperatures shows noticeable differences, at each height one can observe that the numerical results approximate quite well the distribution of temperature into sections that are colder or warmer (hotter) in the room. Refer to the figure captions for more detailed information.

For the purpose of discussion the metric of *absolute error at a node* is adopted, with a focus on the server racks inlet sides. Here, the absolute error is defined as

$$\text{Absolute error } = T_{\text{measured}} - T_{\text{numerical}}, \tag{7.64}$$

where $T_{\text{measured}}$ denotes measured temperature at a node (or grid point) and $T_{\text{numerical}}$ corresponds to the numerical approximation of temperature at a node. Both $T_{\text{measured}}$ and $T_{\text{numerical}}$ are given in degrees Celcius, and therefore so is the absolute error. Additional metrics include the root-mean-squared error (RMSE), defined as

$$\text{RMSE} = \sqrt{\frac{1}{N}\sum_{i=1}^{N} a_i^2}, \tag{7.65}$$

**Fig. 7.18** Numerical approximation of the air velocity field: (**a**) 2D slice at a height of 0.06 m (0.2 ft); (**b**) 2D slice at a height of 1.98 m (6.5 ft). The color coding of the velocity vectors indicates the magnitude of the vector; the color bars range from 0 m/s at the *bottom* to 1.15 m/s (3.77 ft/s) at the *top*. Basic features of the flow, like flow of air into the room through the perforated tiles (**a**) and into the ACUs intake locations (**b**), as well as into/out of the server racks inlet/outlet sides, can be observed (refer to Fig. 7.15 for a description of the data center layout)

where $N$ corresponds to the number of nodes on the surface of the server racks inlet sides for which temperature was estimated numerically and $a_i$ denotes the absolute error at the $i$th node, and the average absolute error,

$$\text{AVG AE} = \frac{1}{N}\sum_{i=1}^{N}|a_i|. \tag{7.66}$$

We note that these metrics do not take into account measurement uncertainties, that is, in this case study the measured data $T_{\text{measured}}$ are considered the "true" value with which the numerical results are to be compared.

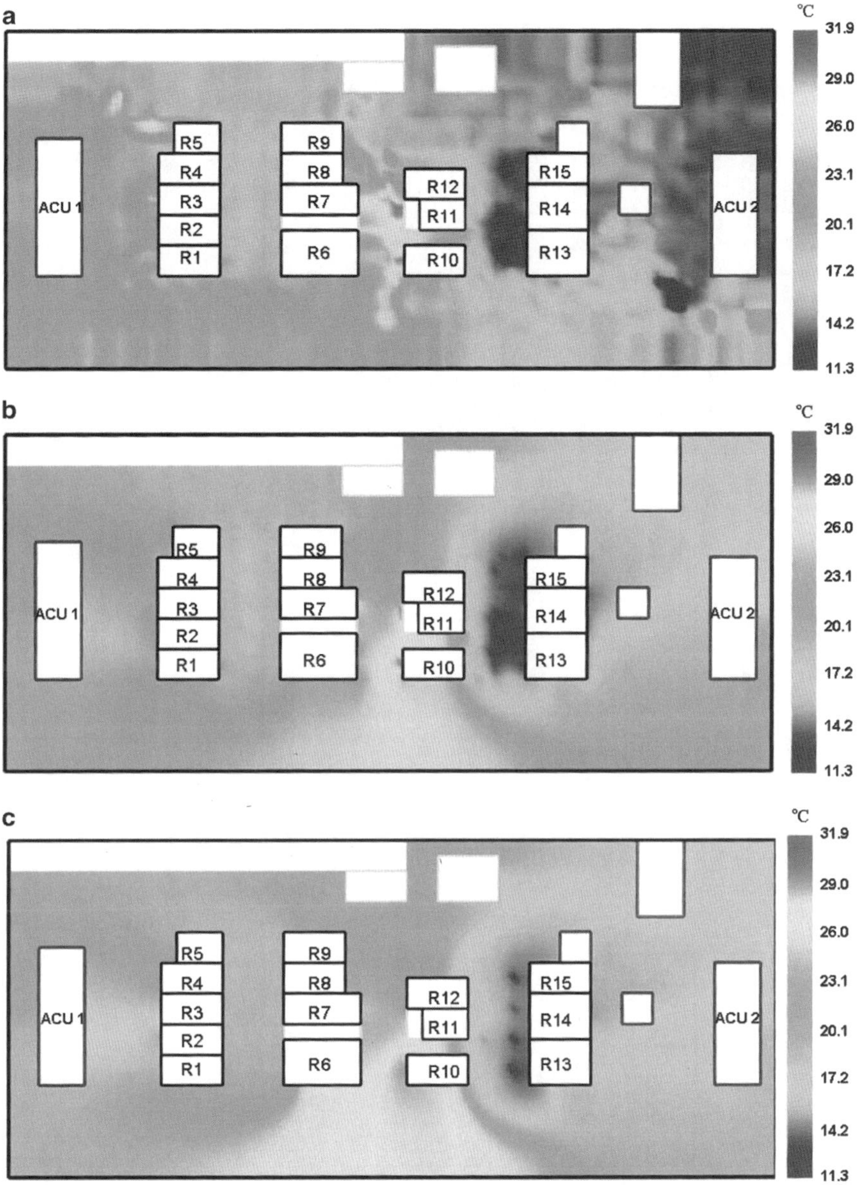


**Fig. 7.19** Temperature distribution at a height of 0 m: (**a**) measured temperature; (**b**) numerical solution resulting from the use of the boundary data pattern (**a**) described in Fig. 7.17; and (**c**) numerical solution resulting from the use of the boundary data pattern (**b**) described in Fig. 7.17. The color bars range from 11.3°C (52.34°F) at the *bottom* to 31.9°C (89.42°F) at the *top*. The sectioning between colder and warmer/hotter areas is fairly well approximated by the numerical results. Improvements may be sought, for example, through inclusion of cable cut-outs on the floor to account for leakage and/or with the availability of measured temperature information at other locations, in addition to the perforated tiles areas. (Figure adapted, with permission, from [45]. © 2011 Elsevier)

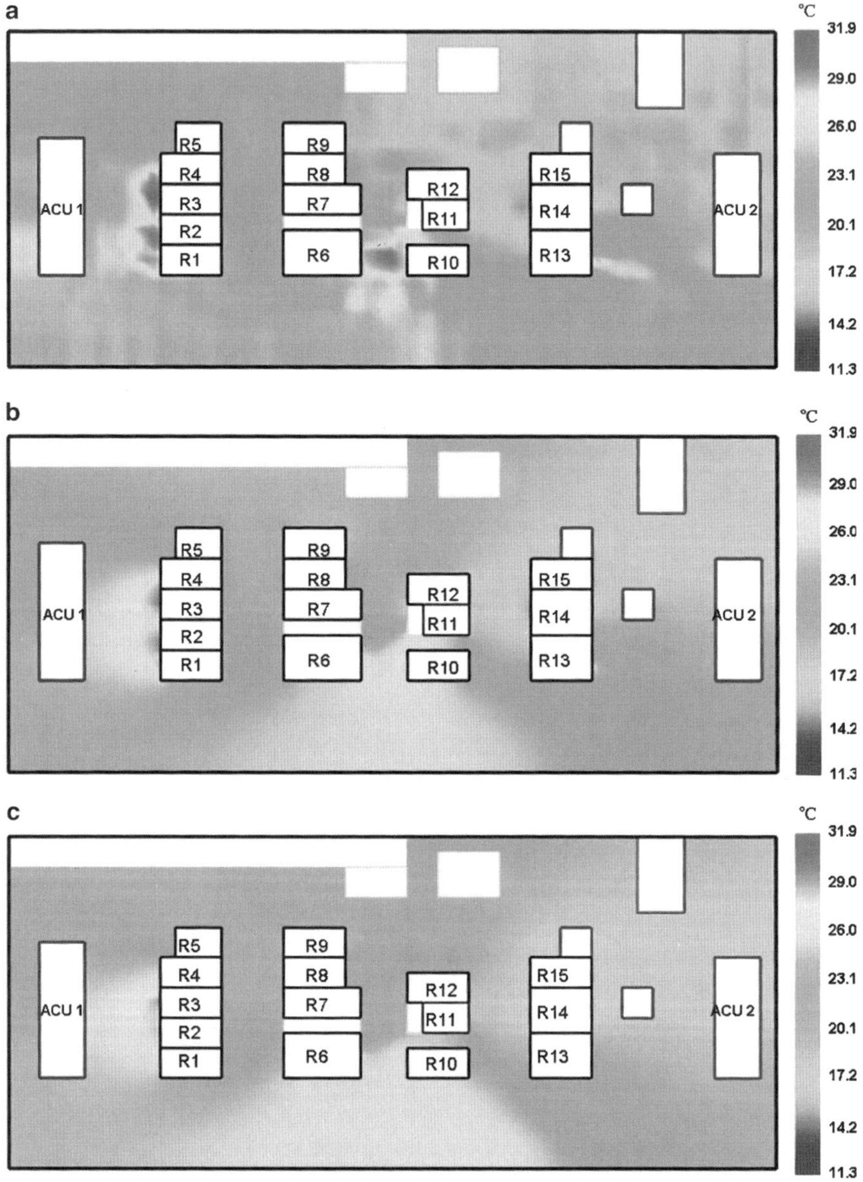


**Fig. 7.20** Temperature distribution at a height of 1.524 m (5 ft): (**a**) measured temperature; (**b**) numerical solution resulting from the use of the boundary data pattern (**a**) described in Fig. 7.17; and (**c**) numerical solution resulting from the use of the boundary data pattern (**b**) described in Fig. 7.17. The color bars range from 11.3°C (52.34°F) at the bottom to 31.9°C (89.42°F) at the top. The sectioning between colder and warmer/hotter areas is fairly well approximated by the numerical results. Improvements may be sought, for example, with a different distribution of power (for the Robin boundary condition) along the rack surfaces and/or with the availability of measured temperature information at additional boundary nodes. (Figure adapted, with permission, from [45]. © 2011 Elsevier)

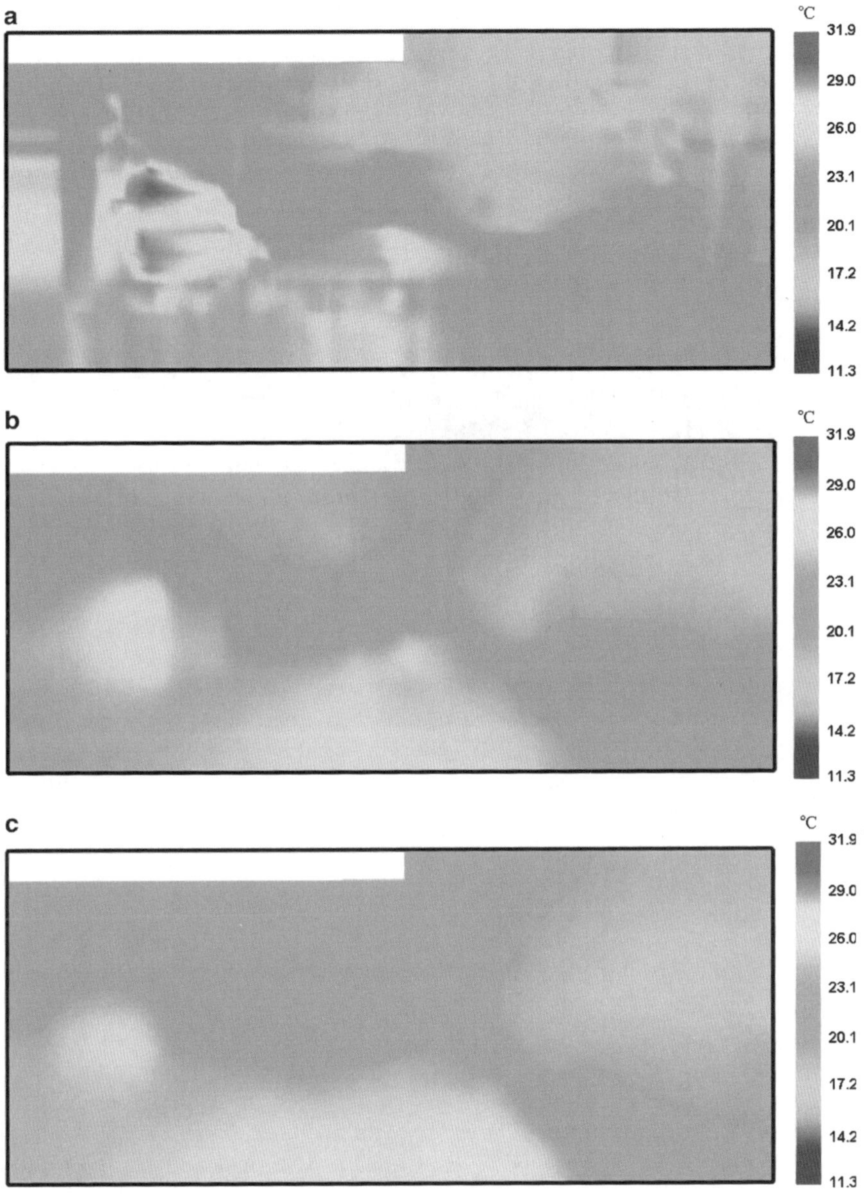


**Fig. 7.21** Temperature distribution at a height of 2.286 m (7.5 ft): (**a**) measured temperature; (**b**) numerical solution resulting from the use of the boundary data pattern (**a**) described in Fig. 7.17; and (**c**) numerical solution resulting from the use of the boundary data pattern (**b**) described in Fig. 7.17. The color bars range from 11.3°C (52.34°F) at the bottom to 31.9°C (89.42°F) at the top. The sectioning between colder and warmer/hotter areas is fairly well approximated by the numerical results. Improvements may be sought, for example, with a different distribution of power (for the Robin boundary condition) along the top of the racks and/or with the availability of measured temperature information at additional boundary nodes (e.g., on the ceiling or top of the racks). (Figure adapted, with permission, from [45]. © 2011 Elsevier)

As one would expect, the value of these error metrics increased as the number of boundary nodes at which temperature was specified decreased. To illustrate this, surface plots of the temperature for the inlet sides of racks 13–15 are displayed in Fig. 7.22. For the boundary data pattern (a) described in Fig. 7.17, which prescribed temperature at a larger number of boundary nodes, the absolute error on the inlet sides of the racks ranged in the interval [−2.01, 2.27]°C ([−3.62, 4.09]°F), while for the boundary data pattern (b) the range was [−2.40, 4.84]°C ([−4.32, 8.71]°F). Note that by (7.53), a negative value for the absolute error indicates that the numerical result overestimated the measured temperature at the node, whereas a positive value for the absolute error denotes a node for which the numerical result underestimated the measured temperature. The range of the numerically computed temperature at the inlets of racks 13–15 in case (b) was [13.8, 22.0]°C ([56.84, 71.6]°F), although the color bars for the temperature surface plots run all on the same scale of [12.1, 22.6]°C ([53.78, 72.68]°F), corresponding to that for the measured temperature as well as for the numerical solution from case (a). In spite of the range of absolute errors, the surface plots for temperature show good qualitative agreement in both cases. Finally, histograms of the absolute error, grouped by rack numbers, appear in Figs. 7.23 and 7.24. The values of the RMSE and average absolute error metrics are listed in Table 7.4, while the standard deviation and mean of the measured and numerically computed temperature at the server racks inlet sides are listed in Table 7.5.

To summarize, although the proposed Laplacian model (7.35)–(7.38), (7.44) for DCs is simpler than those based on the numerical solution of the Navier–Stokes equations for fluid flow, since it is also physics-based, it can still provide meaningful information for use in DC energy management. Support for this statement is provided by the (preliminary) numerical simulations discussed here, in addition to those presented in [19–22]. Improvements should be sought, for example, by (a) using multiple sets of measured data to calibrate the model via, say, the solution of inverse problems to estimate model parameters; (b) exploration of different ways to account for leakage in the airflow model; (c) experimentation with other alternatives for distributing the measured data when providing the Robin boundary conditions for heat flow (this could also be posed as an inverse problem for the model); and (d) use of different numerical stabilization techniques for the solution of the convection–diffusion equation [26, 37] for estimating temperature, among other options. With the availability of real-time sensor data to provide input information, mainly in the form of boundary data, such improvements could result in an airflow and heat transfer model suitable for operational and real-time usage. In addition to day-to-day operational usage, other applications could include, for example, use of the models to determine optimal placement of sensors within a DC.

Since the model equations for fluid flow and temperature are linear in the unknowns, taking advantage of the availability of fast methods for solving systems of linear equations and/or the principle of superposition for PDEs can result in significant reductions in computational time. In addition, reductions in computational time could also be achieved, for example, by reuse of the mesh and reuse of the numerical factorization of the coefficient matrices of the linear

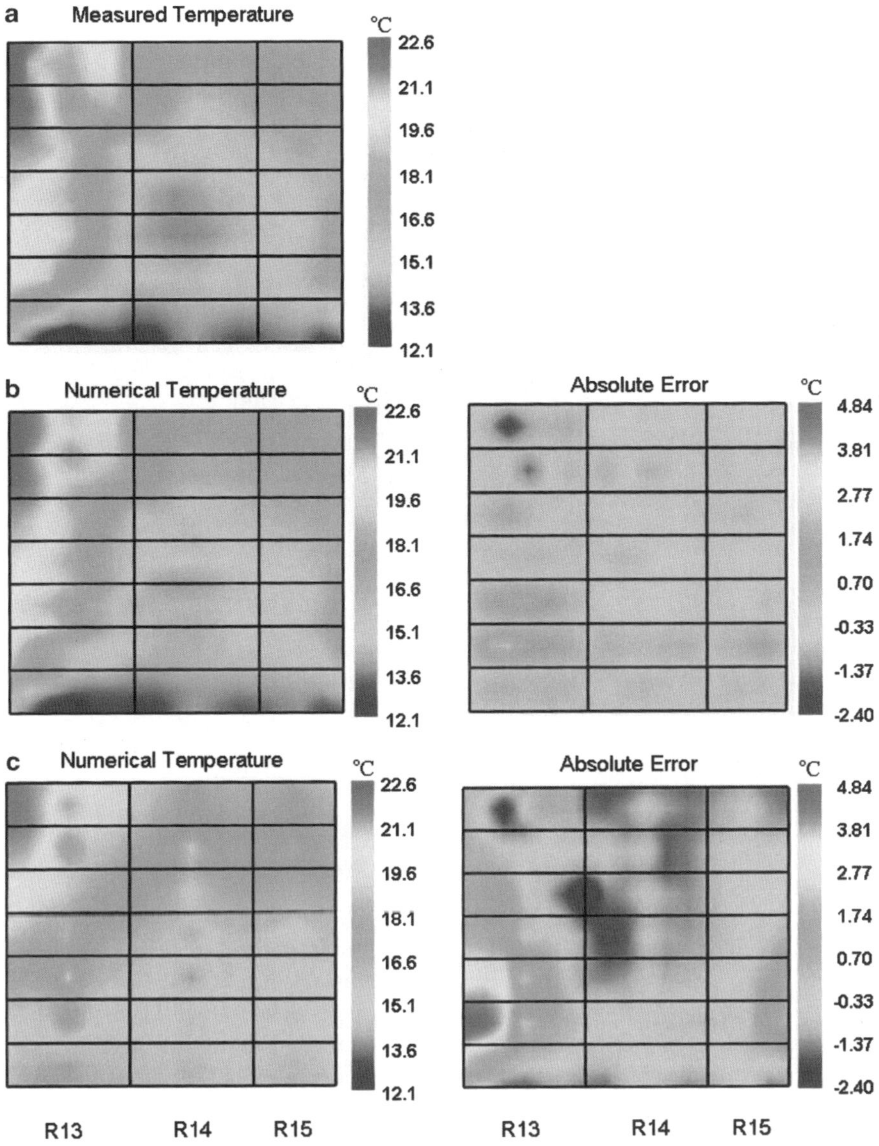


**Fig. 7.22** Surface plots of the temperature (*left*) and absolute error (*right*) at the inlet sides for racks 13–15: (**a**) measured temperature; (**b**) numerical solution resulting from the use of the boundary data pattern (**a**) described in Fig. 7.17; and (**c**) numerical solution resulting from the use of the boundary data pattern (**b**) described in Fig. 7.17. In each surface plot, rack boundaries are defined by solid black vertical lines, with racks 13, 14, and 15 represented, respectively, by the first, second, and third column defined by the *black vertical lines* (refer to the labeling under figure (**c**)), while server locations are delimited by *horizontal lines* (recall that each rack has space for seven servers). The color bars for the temperature range from 12.1°C (53.78°F) at the bottom to 22.6°C (72.68°F) at the top. The color bars for the absolute error range from −2.40°C (−4.32°F) at the bottom to 4.84°C (8.71°F) at the top. The absolute error at a node is defined by (7.63).

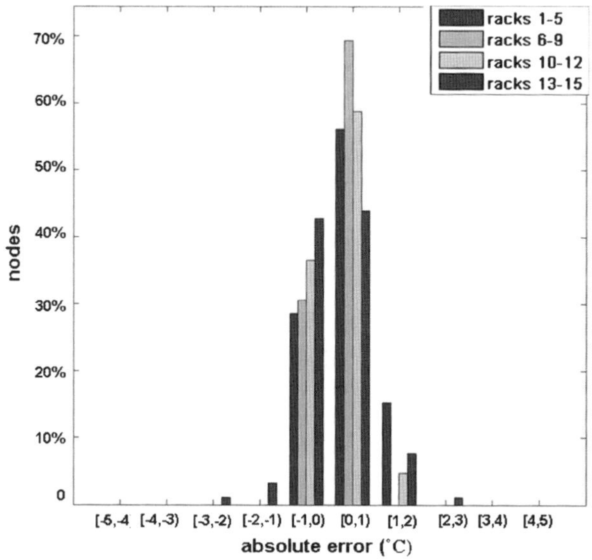


**Fig. 7.23** Histograms of the absolute error, as defined by (7.63), resulting from the use of the boundary data pattern (**a**) described in Fig. 7.17. The nodes on the surface of the server racks inlet sides have been grouped by rack number, per the DC layout (refer to Fig. 7.15). The *vertical axis* denotes the number of nodes in each bin, as a percentage of the total number of nodes on the racks inlet sides for which temperature was estimated numerically. The *horizontal axis* indicates the range of the absolute error, in °C, for each bin. All nodes for racks 6–9 had an error in the range [−1, 1]°C ([−1.8, 1.8]°F), while racks 1–5, 10–12, and 13–15 had, respectively, 85%, 95%, and 87% of their nodes in this range. The remaining 15% for racks 1–5, 5% for racks 10–12, and 13% for racks 13–15 had errors between 1°C and 3°C (1.8°F and 5.4°F), in magnitude. (Figure adapted, with permission, from [45]. © 2011 Elsevier)

systems (in instances when a direct solver for linear systems is used), whenever these components remain unchanged. Such reductions in computational time would be particularly advantageous whenever multiple runs are performed, for example, as part of a time-dependent simulation. Time dependence could be incorporated into the model, as the required inputs to the boundary value problems (i.e., time-dependent boundary data) can be taken from a sequence of sensor readings in the time interval of the simulation.

## 7.2.2 *Reduced Order Model for Data Centers*

Let us briefly recap our previous modeling discussions. We touched on CFD models, which solve the complete Navier–Stokes equations for fluid flow coupled with a heat transfer equation. CFD is the most advanced modeling technology, which—by its nature—is computationally intense with long calculation times.

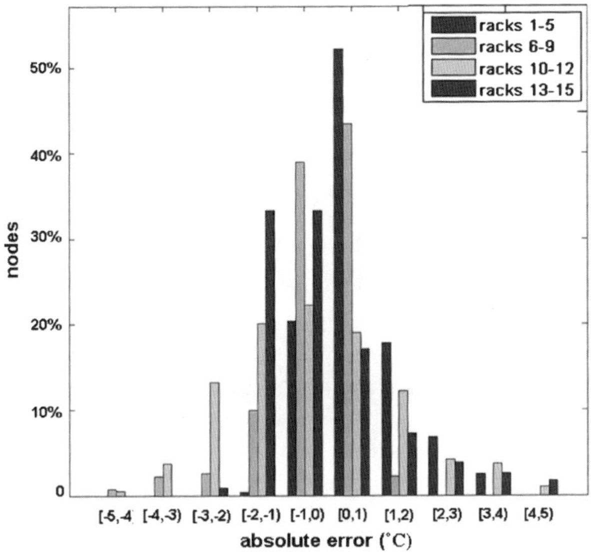


**Fig. 7.24** Histograms of the absolute error, as defined by (7.63), resulting from the use of the boundary data pattern (**b**) described in Fig. 7.17. The nodes on the surface of the server racks inlet sides have been grouped by rack number, per the DC layout (refer to Fig. 7.15). The *vertical axis* denotes the number of nodes in each bin, as a percentage of the total number of nodes on the racks inlet sides for which temperature was estimated numerically. The *horizontal axis* indicates the range of the absolute error, in °C, for each bin. The percentage of nodes having an error in the range [−1, 1]°C ([−1.8, 1.8]°F) was 73%, 82%, 41%, and 50%, respectively, for each of the rack groups 1–5, 6–9, 10–12, and 13–15. The percentages for the error range [−2, 2]°C ([−3.6, 3.6]°F) were 90%, 94%, 74%, and 91%, respectively, for each of the rack groups 1–5, 6–9, 10–12, and 13–15. The remaining 10% for racks 1–5, 6% for racks 6–9, 26% for racks 10–12, and 9% for racks 13–15 had errors between 2°C and 5°C (3.6°F and 9°F), in magnitude. (Figure adapted, with permission, from [45]. © 2011 Elsevier)

**Table 7.4** RMSE and average absolute error (in °C), as defined by (7.64) and (7.65), at the server racks inlet sides for the two boundary data patterns described in Fig. 7.17

| Nodes with temperature prescribed | RMSE | Average absolute error |
|---|---|---|
| Case (A): 1,676 | 0.5842 | 0.4364 |
| Case (B): 230 | 1.2879 | 0.9478 |

As the number of boundary nodes with temperature prescribed decreased, the value of the metrics increased. Recall that in the present case study prescribing temperature at boundary nodes simulated placement of temperature sensors at different locations (refer to Fig. 7.15), so the metrics indicate dependence on sensor placement (as one would expect)

Second, we presented a simpler, still physics-based Laplacian model (LM), which can leverage readily measured data as boundary conditions. We demonstrated that LM can predict temperature fields quite accurately with only few measured data points. LM is physics-based and should provide good robustness in a changing DC

**Table 7.5** Standard deviation and mean of the numerically computed temperature $T_{numerical}$ and measured temperature $T_{measured}$ (in °C) at the server racks inlet sides for the problems defined by the two different cases from Fig. 7.17

| Case | SD $T_{numerical}$ | Mean $T_{numerical}$ | SD $T_{measured}$ | Mean $T_{measured}$ |
|---|---|---|---|---|
| (A) | 2.3615 | 18.9012 | 2.5343 | 19.0997 |
| (B) | 2.3679 | 19.2029 | 2.5620 | 19.2040 |

Note that the mean of the measured temperature is slightly different for each of the cases. This is due to the fact that, for each test case, the mean was computed omitting the set of nodes at which temperature was prescribed as part of the input data to the boundary value problem, since temperature at such nodes was therefore not an unknown

environment. In this section, we discuss an even higher level model from the spectrum depicted in Fig. 7.5a.

Specifically, we present a reduced order model which uses POD. In this model, the physics knowledge/learning is generated from previous observations, which can include both measurements (as described in Sect. 7.1) and/or calculations from models such as CFD or LM. Because of this, such model requires "initialization" data. Enough observations have to be made to cover the simulation range of interest. In contrast, because of their physics-based nature, such initialization data are typically not required for CFD or LM. Main advantages of the reduced order or statistical models are the superior computational speed. Typical applications for POD-based reduced order models include design optimization and/or finding an optimum control point within the range of observations, which would be infeasible with CFD. The ability to predict the system's properties beyond the range of observations is most likely to be limited. Consequently, as the DC undergoes some changes, the model will lose accuracy. The application of POD-based reduced order models to DCs has been pioneered by Joshi et al. [36, 38, 39].

The POD has several properties that make it well suited for turbulent flows, which fundamentally determine the complex temperature field in DCs: First, because the basis functions are determined empirically, the method is naturally ideal for nonlinear problems. Secondly, it has been shown that this approach captures turbulent effects better than any other alternative linear decomposition method [40]. Finally, experiments have demonstrated that such low-dimensional models can capture quite well the role of coherent structures, which are responsible for turbulence generation [40]. POD has already been used to create reduced order models of transient temperature fields using a Galerkin projection of the system POD modes onto the governing equations [41]. Here, we restrict the discussion to steady-state application.

#### 7.2.2.1 Model Description

The POD, also known as the Karhunen–Loeve decomposition, is a model reduction technique that expands a set of observations on empirically determined basis functions for modal decomposition. Detailed descriptions of POD can be found elsewhere [42–44]. Here, we only give a short summary. See also Chap. 10 for additional details.

Suppose that there is a set of $N$ observations, which were obtained from either numerical simulations or detailed experimental measurements. Each observation corresponds to $N$ different states of the design (or control) parameter of the system (e.g., ACU airflow rates). In a POD method, the temperature field is expanded into $M$ basis functions (or POD modes) $\psi_k$,

$$T = T_o + \sum_{k=1}^{M} b_k \psi_k, \tag{7.67}$$

where $b_k$ denotes the POD coefficient of the $k$th POD mode, $T_o$ represents a source term (reference temperature field), and $M$ is the number of modes kept in the decomposition, $M$ can range from 1 up to $N$. We note that for transient problems the coefficients $b_k$ in (7.67) are time-dependent and are (typically) determined via application of a Galerkin projection, which converts the governing PDEs into a finite, lower dimensional system of ordinary differential equations. In this case, the solution approach is part of the spectral-Galerkin methods. In the steady-state case, the coefficients $b_k$ in (7.67) may be determined using an interpolation procedure, as outlined below.

There are two ways for calculating the POD modes, either using the snapshot and/or the direct method. Details of these methods can be found elsewhere [40]. The method of snapshots has the advantage that it replaces the need to compute an autocorrelation matrix. Each POD mode $\psi_k$ (for $k = 1, \ldots, N$) can be expressed as a linear combination

$$\psi_k = \sum_{n=1}^{N} a_{k,n}(T_{\text{obs},n} - T_o), \tag{7.68}$$

where $T_{\text{obs}}$ is a matrix of which each column $T_{\text{obs},n}$ (for $n = 1, \ldots, N$) is a complete temperature field corresponding to the $n$th observation. Each component of the reference temperature field $T_o$ in (7.67) and (7.68) is defined as the average of all observations at a given point. The weight coefficients $a_k$ in (7.68) are obtained by solving an $N \times N$ eigenvalue problem

$$Ra = \lambda a, \tag{7.69}$$

with $R = (T_{\text{obs}} - T_o)^* \otimes (T_{\text{obs}} - T_o)/M$ (recall that $M$ denotes the number of modes retained in (7.67) and $N$ the number of observations). The solution of (7.69) results in a set of eigenvalues $\lambda_k$ (for $k = 1, \ldots, N$) along with the corresponding eigenvectors $a_k$. The components $a_{k,n}$ (for $n = 1, \ldots, N$) of the $k$th eigenvector $a_k$ give the weight coefficients $a_{k,n}$ of the $k$th POD mode $\psi_k$ in (7.68). This allows computing the $N$ POD modes in (7.68). The "energy" captured by each POD mode is proportional to the magnitude of the corresponding eigenvalue. Typically, the eigenvalues are being sorted in descending order so the first POD mode $\psi_1$ captures the most "energy" (is most important), while the last POD mode $\psi_N$ captures the least "energy" (is least important). Finally, the POD coefficients $b_k$ in (7.67) for a new situation (i.e., not corresponding to the set of observations) can be obtained via an interpolation process.

#### 7.2.2.2 Specification of Model Inputs

As discussed the POD model requires initial observations (typically large-scale data sets), which can come from CFD calculations and/or high resolution assessment data as discussed in Sect. 7.1.2.3. First, the reference temperature field $T_o$ is calculated as the average for a given point for all observations. Second, using the observation matrix $T_{obs}$ and solving the eigenvalue problem of (7.69) respective POD modes can be determined using (7.68). Precisely, for each observation $T_{obs,n}$ (for $n = 1, \ldots, N$) the POD coefficients required to reconstruct the observed temperature field (refer to the expansion (7.67)) are obtained by projecting each POD mode onto the observation $T_{obs,n}$. That is, for $n = 1, \ldots, N$,

$$b_{k,obs,n} = (T_{obs,n} - T_o) \cdot \psi_i, \quad \text{for } k = 1, \ldots, M. \tag{7.70}$$

We note that there are more convenient ways to compute a complete coefficient matrix, in which each column is a coefficient vector $b_{obs,n}$. Third, the POD coefficients $b_k$ in the expansion (7.67) for a new (i.e., not observed) scenario can be obtained via an interpolation between POD coefficients at the observed design parameter values (e.g., ACU flow rates) to match a desired design (or control) value (e.g., the desired ACU flow rate for each ACU).

#### 7.2.2.3 Case Study

Following [36], POD modeling was applied to the same DC facility described in Sect. 7.2.1.3. As depicted in Fig. 7.15 the DC has two ACUs, where one (ACU#1) was turned off for this case study. Seven observations were generated by changing the blower setting of ACU#2 to 96%, 92%, 88%, 80%, 76%, 72%, and 65% of its nominal flow capacity (5.8521 $m^3/s$ (12,400 cfm)) and then measuring the complete temperature field using a MMT 1.0 tool (see Sect. 7.1.2.2). Each MMT measurement generated 10,584 measurements throughout the facility. Thus, the observation matrix $T_{obs}$ contained 10,584 rows and 7 columns. Figure 7.25 displays two of the seven observations at a height of 1.0668 m (3.5 ft) showing that the temperatures increase as expected by reducing the airflow from 96% to 65%. As explained previously, the seven observations allow computing seven POD modes, each with its unique POD coefficient $b_k$ (refer to (7.70)), as shown in Table 7.6.

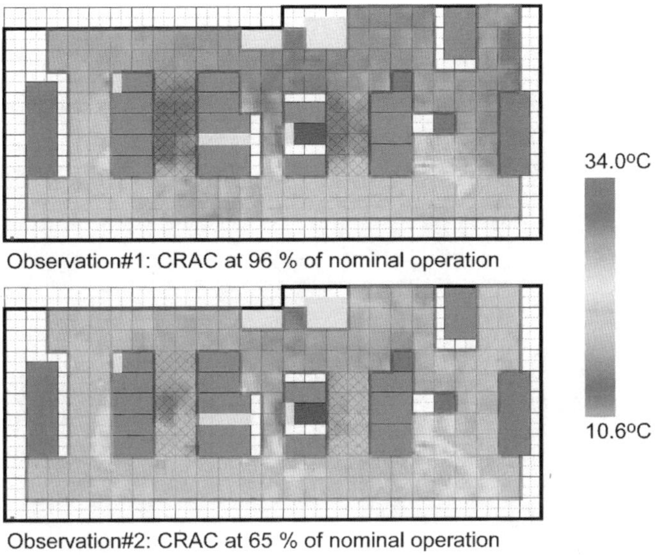


**Fig. 7.25** Two of the seven MMT observations used for POD model. (Figure reprinted, with permission, from [36]. © 2009 ASME)

**Table 7.6** POD coefficients (table reprinted, with permission, from [36] © 2009 ASME)

| | Observation number | | | | | | |
|---|---|---|---|---|---|---|---|
| Mode number | 1 | 2 | 3 | 4 | 5 | 6 | 7 |
| 1 | −108.1 | −95.9 | −61.4 | −0.2 | 24.2 | 71.9 | 169.5 |
| 2 | 22.1 | 32.1 | −18.5 | −64.3 | −25.1 | 43.1 | 10.7 |
| 3 | −29.8 | −9.4 | 2.1 | 12.9 | 13.3 | 63.2 | −52.3 |
| 4 | 27.0 | 5.9 | −64.5 | 36.7 | −6.7 | 6.0 | −4.4 |
| 5 | −2.8 | 8.8 | −21.4 | −24.5 | 60.9 | −13.4 | −7.6 |
| 6 | −39.5 | 47.8 | −6.4 | 10.4 | −8.1 | −8.3 | 4.2 |
| 7 | 0.0 | 0.0 | 0.0 | 0.0 | 0.0 | 0.0 | 0.0 |

In the next step, the blower settings are changed to 84% and 68% and new weight coefficients were calculated using interpolation to the new blower settings, as indicated in Sect. 7.2.2.1. These new weight coefficients (here linearly interpolated to the new settings) allowed in combination with POD modes to rapidly predict the temperature field for these two new conditions. The POD generated temperature field was obtained in less than 2 s on a ×86 desktop computer with 2.8 GHz CPU and 2.75 GB of RAM.

In order to understand the accuracy of the POD modeling, the MMT tool was used to actually measure the temperature distribution for these two new cases so that the prediction could be compared to the actual measurements.

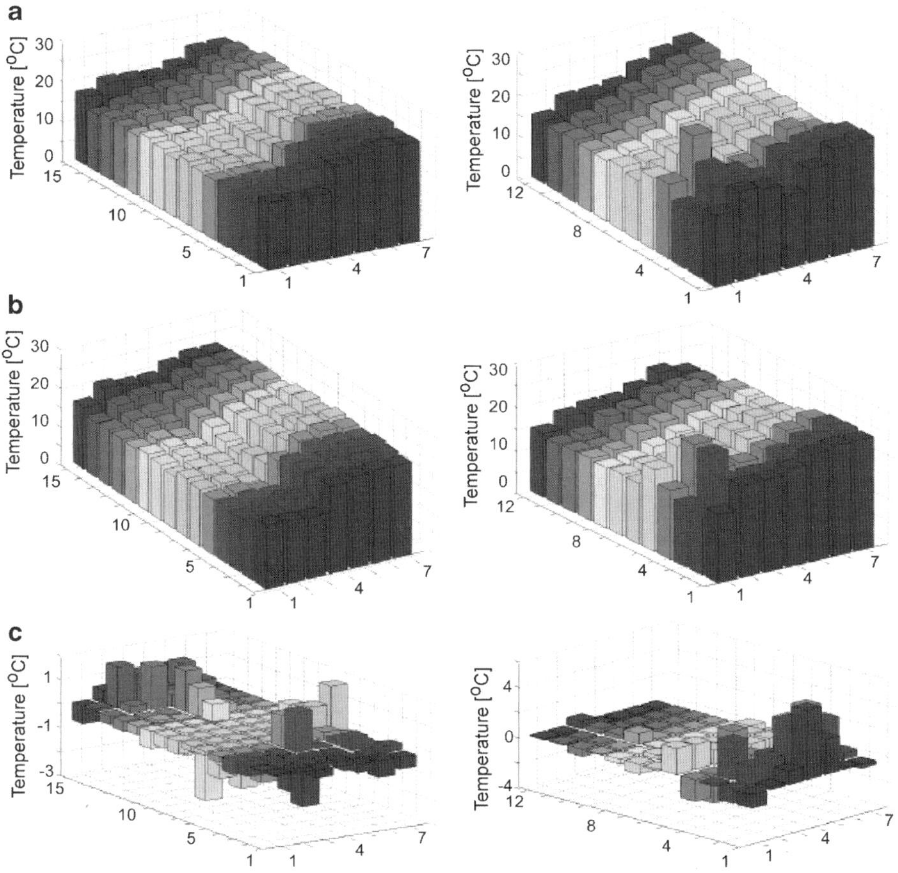


**Fig. 7.26** (**a**) Measured temperatures, (**b**) POD generated temperatures, and (**c**) temperature errors for 68% of ACU operation at the inlets of racks R1–R5 (*first column*) and racks R10–R12 (*second column*). (Figure reprinted, with permission, from [36]. © 2009 ASME)

In Figs. 7.26 and 7.27, we compare the POD predictions with the MMT measurements at the inlets of the racks R1–R5 and R10–R12 for the 68% and 84% case. There are 15 temperature readings on the horizontal axis for R1–R5 (first column) and 12 for R10–R12 (second column). In the vertical direction we have seven readings, from 0.1524 m (0.5 ft) to 2.286 m (7.5 ft).

Figures 7.26 and 7.27 demonstrate that POD can rapidly and accurately predict the temperature distribution. For 84% operation of the ACU unit, the mean error in the domain, 10,584 points, is 0.4°C (1.0 F) or 2.92%, while for 68% operation the mean error is 0.6°C (1.4 F) or 3.48%. While the average agreement is very good, the maximum local error is large at a few points suggesting that more observations are needed to capture these effects.

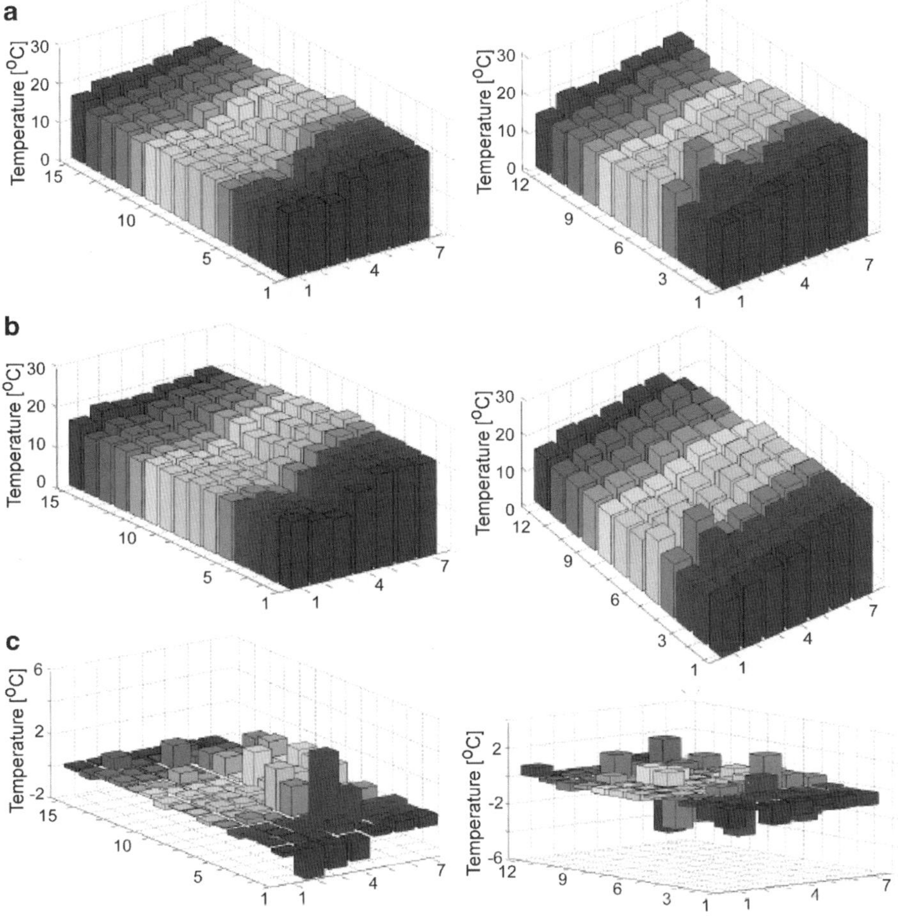


**Fig. 7.27** (**a**) Measured temperatures, (**b**) POD generated temperatures, and (**c**) temperature errors for 84% of ACU operation at the inlets of racks R1–R5 (*first column*) and racks R10–R12 (*second column*). (Figure reprinted, with permission, from [36]. © 2009 ASME)

## 7.3 Conclusions

In this chapter, we reviewed different techniques for making relevant measurements for characterizing the environmental conditions in a DC. System implications were discussed, in particular how design choices of the measurement system (e.g., measurement frequency, numbers of sensors, placement of sensors) depend on the supporting modeling method. Next, we reviewed different modeling technologies and discussed how the measurements can be leveraged as inputs for subsequent modeling. We presented in more detail two modeling approaches: a simplified physics-based model (Laplacian model), where the measurements define the boundaries and a reduced order modeling approach using POD, where three-dimensional thermal assessment data is used to compute POD modes and coefficients. Case studies for the different approaches were presented.

**Acknowledgments** We acknowledge contributions from the world-wide MMT team including Andrew Stepanchuk, Alan Claassen, James Lacey, Hongfei Li, Srinivas Yarlanki, Hans-Dieter Wehle, Raymond Lloyd, Tom Keller, Levente Klein, Michael Schappert, Fernando Jiminez, and many more colleagues. We also thank Jon Lenchner, Jeff Kephart, Raja Das, Tom Sarasin, and Wayne Riley from the Tivoli/Maximo team. This work was partially supported by the Department of Energy (Grant Number DE-EE0002897). Portions of this chapter reprinted, with permission, from [19]: H. F. Hamann, V. López, and A. Stepanchuk, *Thermal zones for more efficient data center energy management*, In Proceedings of ITherm2010, © 2010 IEEE; [21]: V. López and H. F. Hamann, *Measurement-based modeling for data centers*, In Proceedings of ITherm2010, © 2010 IEEE; and [45]: V. López and H. F. Hamann, *Heat transfer modeling in data centers*, Int. J. Heat Mass Transfer, **54** (2011), © 2011 Elsevier.

## References

1. Belady C, Rawson A, Pfleuger J, Cader T (2008) Green grid data center power efficiency metrics: PUE and DCIE. The Green Grid – White Paper 6
2. American Society of Heating, Refrigerating and Air-Conditioning Engineers (2008) 2008 ASHRAE environmental guidelines for datacom equipment—expanding the recommended environmental envelope, 2nd ed. http://tc99.ashraetcs.org/documents/ASHRAE_Extended_Environmental_Envelope_Final_Aug_1_2008.pdf. Last accessed on 27 Dec 2011
3. Walker IS, Wray CP, Dickerhoff DJ, Sherman MH (2001) Evaluation of flow hood measurements for residential register flows. LBNL White Paper 47382
4. Radmehr A, Schmidt RR, Karki KC, Patankar SV (2005) Distributed leakage flow in raised-floor data centers. In: Proceedings of IPACK2005, IPACK 2005–73273
5. Karki KC, Radmehr A, Patankar SV (2003) Use of computational fluid dynamics for calculating flow rates through perforated tiles in raised-floor data centers. Int J HVAC Res 9:153
6. Idelchik IE (1994) Handbook of hydraulic resistance. CRC, Florida
7. Hamann HF, Schappert M, Iyengar M, van Kessel T, Claassen A (2008) Methods and techniques for measuring and improving data center best practices. *n:* 11th intersociety conference on thermomechanical phenomena in electronic systems, Orlando, p 1146
8. Joshi Y (2010) Role of thermal engineering in improving data center energy efficiency. Workshop on thermal management in telecommunication systems and data centers, Richardson
9. Shepard D (1968) A two-dimensional interpolation function for irregularly-spaced data. Proceedings of the 1968 ACM national conference, p 517
10. Zhou Y, Xhionghui L, Zhong X, Klein L, Schappert MA, Hamann HF (2010) A tele-operative RMMT system facilitating the management of cooling and energy in data centers, Tianjin, China
11. Callen HB (1985) Thermodynamics and an introduction to thermostatistics. New Delhi, Wiley
12. Hamann HF, Lacey J, O'Boyle M, Schmidt RR, Iyengar M (2008) Rapid three dimensional thermal characterization of large-scale computing facilities. IEEE Trans Comp Pack Techn 31:444
13. Hamann HF, van Kessel TG, Iyengar M, Chung J-Y, Hirt W, Schappert MA, Claassen A, Cook JM, Min W, Amemiya Y, Lopez V, Lacey JA, O'Boyle M (2009) Uncovering energy efficiency opportunities in data centers. IBM J Res Dev 53:10:1–10:12
14. Awtrey D (1997) Transmitting data and power over a one-wire bus. Sensors 14(2):48

15. Pouchard LC, Poole S, Lothian J (2009) Open standards for sensor information processing. Oakland Ridge National Laboratory White Paper ORNL/TM-2009/145. http://www.csm.ornl.gov/~7lp/publis/Sensor_TR_2009.pdf. Last accessed on 27 Dec 2011
16. Javvin Technologies (2004) Network protocol handbook, 2nd edn. http://books.google.com/books?id=D_GrQa2ZcLwC&printsec=frontcover&source=gbs_ge_summary_r&cad=0#v=onepage&q&f=false. Last accessed on 27 Dec 2011
17. VanGilder JW, Shrivastava SK (2006) Real-time prediction of rack-cooling performance. ASHRAE Trans 112:151
18. VanGilder JW, Shrivastava SK (2007) Capture index: an airflow-based rack cooling performance metric. ASHRAE Trans 113:126
19. Hamann HF, López V, Stepanchuk A (2010) Thermal zones for more efficient data center energy management. In: 12th intersociety conference on thermomechanical phenomena in electronic systems, Las Vegas, p 1
20. Hamann HF, Iyengar MK, van Kessel TG (2009) Techniques for thermal modeling of data centers to improve energy efficiency. US Patent Application 2009/0326879A1
21. López V, Hamann HF (2010) Measurement-based modeling for data centers. In: 12th intersociety conference on thermomechanical phenomena in electronic systems. Las Vegas, p 1
22. Toulouse M, Doljac G, Carey V, Bash C (2009) Exploration of a potential-flow-based compact model of air-flow transport in data centers. In: ASME International Mechanical Engineering Congress & Exposition, Technical Publication
23. Bergman S, Schiffer M (1953) Kernel functions and elliptic differential equations in mathematical physics. Dover Publications (2005); unabridged republication of the work originally published by Academic, New York
24. Landau LD, Lifshitz EM (1959) Fluid mechanics, volume 6 of course of theoretical physics. Pergamon Press (1959); translated from the Russian by Sykes JB, Reid WH
25. Heath MT (2002) Scientific computing: an introductory survey. McGraw-Hill, Boston
26. Johnson C (1987) Numerical solution of partial differential equations by the finite element method. Cambridge University Press, Cambridge
27. Hughes TJR (1987) The finite element method: linear static and dynamic finite element analysis, Dover Publications (2000); reprint of the original publication from Prentice Hall
28. Kernighan BW, Ritchie DM (1988) The C programming language. Prentice Hall, Englewood Cliffs
29. FlexPDE. A multi-physics finite element solution environment for partial differential equations. http://www.pdesolutions.com/. Last accessed on 27 Dec 2011
30. OpenFOAM. The open source CFD toolbox. http://www.openfoam.com/. Last accessed on 27 Dec 2011
31. Alberty J, Carstensen C, Funken SA (1999) Remarks around 50 lines of Matlab: short finite element implementation. Numer Algorithms 20:117
32. Anderson E, Bai A, Bischof C, Blackford S, Demmel J, Dongarra J, Du Croz J, Greenbaum A, Hammarling S, McKenney A, Sorensen D (1999) LAPACK users' guide, SIAM. http://www.netlib.org/clapack
33. Davis TA (2004) Algorithm 832: UMFPACK, an unsymmetric-pattern multifrontal method. ACM Trans Math Software 30:196. http://www.cise.ufl.edu/research/sparse/umfpack/
34. Tschudi W, Mills E, Greenberg S, Rumsey P (2006) Measuring and managing data-center energy use. HPAC Engineering 45
35. Bonilla CJ, Ferrer E, Bash C (2007) Thermal zone mapping: data center design and assessment automated visualization tool for thermal metric analysis. IMAPS ATW
36. Samadiani E, Joshi Y, Hamann HF, Iyengar MK, Kamalsy S, Lacey J (2009) Reduced order thermal modeling of data centers via distributed sensor data. In: Proceedings of IPACK2009, InterPACK 2009-89187
37. Brooks AN, Hughes TJR (1982) Streamline upwind/Petrov-Galerkin formulations for convection dominated flows with particular emphasis on the incompressible Navier–Stokes equations. Comp Methods Appl Mech Eng 32:199

38. Samadiani E, Joshi Y (2010) Multi-parameter model reduction in multi-scale convective systems. Int J Heat Mass Transfer 53:2193
39. Samadiani E, Joshi Y (2010) Proper orthogonal decomposition for reduced order thermal modeling of air cooled data centers. ASME Trans J Heat Transfer 132:0714021
40. Holmes P, Lumley JL, Berkooz G (1996) Turbulence, coherent structures, dynamical systems and symmetry. Cambridge University Press, Cambridge
41. Ravindran SS (2002) Adaptive reduced-order controllers for a thermal flow using proper orthogonal decomposition. SIAM J Sci Comput 23:1924
42. Rolander N (2005) An approach for the robust design of air cooled data center server cabinets. MS thesis, G.W. School of Mechanical Engineering, Georgia Institute of Technology, Atlanta
43. Rambo J, Joshi Y (2007) Reduced-order modeling of turbulent forced convection with parametric conditions. Int J Heat Mass Transfer 539:50
44. Rambo JD (2006) Reduced-order modeling of multiscale turbulent convection: application to data center thermal management. PhD Dissertation, in Mechanical Engineering, MS thesis, G.W. School of Mechanical Engineering, Georgia Institute of Technology, Atlanta
45. López V, Hamann HF (2011) Heat transfer modeling in data centers. Int J Heat Mass Transfer 54:5306–5318. Published online 15 Sept 2011. doi:10.1016/j.ijheatmasstransfer.2011.08.012

# Chapter 8
# Numerical Modeling of Data Center Clusters

**Bahgat Sammakia, Siddharth Bhopte, and Mahmoud Ibrahim**

**Abstract** This chapter deals with the numerical modeling of data centers. The chapter presents an overview of the fundamental equations governing the conservation of mass, energy, and momentum, with an emphasis on the most widely used numerical approaches used for discretizing the equations and solving them. The specific simplifications and assumptions that are typically used in modeling data centers are reviewed. Turbulent modeling is covered in some detail, with an emphasis on the suitability of different models for data centers. A review of recent numerical studies of data centers is presented and compared to available measurements and characterization studies. Results for different air cooling protocols are presented and ranked according to their overall performance. A detailed discussion of the impact of blockages in the plenum, due to wiring and cooling water pipes, is presented and general design guidelines are made pertaining to placement of such blockages. Specific attention is given to the modeling of data centers during dynamic fluctuations in power, airflow, and temperature. This is of particular relevance for the establishment of dynamic self-regulating data centers that may be optimized to operate at the lowest possible energy level while they are meeting specific performance metrics. A case is made for verified reduced order modeling of dynamic data centers. Such an approach may be the most suitable and pragmatic one to achieve real-time holistic models that are capable of predicting and optimizing the overall performance of complex data centers.

B. Sammakia (✉) • S. Bhopte • M. Ibrahim
Small Scale System Integration & Packaging Center, Binghamton University—State University of New York, Binghamton, NY 13902, USA
e-mail: bahgat@binghamton.edu; Sbhopte@binghamton.edu; mibrahi1@binghamton.edu

Y. Joshi and P. Kumar (eds.), *Energy Efficient Thermal Management of Data Centers*, DOI 10.1007/978-1-4419-7124-1_8,

## 8.1 Introduction

As discussed in earlier Chaps. 1 and 2, one of the most popular configurations from thermal management standpoint is the raised-floor data center with alternating hot- and cold-aisle (HACA) arrangement. The reader may refer to Fig. 1.15 for details on the HACA arrangement. Rapidly increasing heat fluxes and volumetric heat generation rates of the computer servers has led to very high flow rates of cooling air from the computer room air conditioning (CRAC) units resulting in turbulent flow regimes with large variability in velocity magnitude. Due to this complex nature of flow inside the data center computational fluid dynamics (CFD), heat transfer is usually required to investigate thermal performance of data center.

In this chapter, system level numerical modeling of coupled heat transfer and fluid flow is discussed. Impact of model complexity on numerical prediction of data center performance is discussed in detail. As we have already emphasized in Chap. 2, this chapter addresses the problem of under-floor blockages in a data center using CFD modeling techniques. Detailed comparisons between numerical and experimental results are presented.

## 8.2 Overview of Fundamental Equations

The basic equations governing the conservation of mass, momentum, and energy are shown below in (8.1)–(8.3). Interested reader may refer to a standard fluid mechanics book such as [1] for details regarding the derivation of the governing equations. Solutions to these equations provide information of flow, pressure, and temperature distributions in the data center. These complex nonlinear equations require the use of numerical techniques. The numerical technique typically reduces the differential equations to simple algebraic ones which can be easily solved for a finite number of grid points in the flow field.

Conservation of mass:

$$-\frac{1}{\rho}\frac{D\rho}{Dt} = \frac{\partial u_i}{\partial x_i}. \tag{8.1}$$

Conservation of momentum:

$$\rho\frac{Du_i}{Dt} = -\frac{\partial P}{\partial x_i} + \rho g_i + \frac{\partial}{\partial x_j}\left[\mu\left(\frac{\partial u_i}{\partial x_j} + \frac{\partial u_j}{\partial x_i}\right) - \frac{2}{3}\mu\left(\frac{\partial u_i}{\partial x_i}\right)\delta_{ij}\right]. \tag{8.2}$$

Conservation of energy:

$$\begin{aligned}\rho\frac{De}{Dt} = &-P\frac{\partial u_i}{\partial x_i} + \frac{\partial}{\partial x_j}\left[k\left(\frac{\partial T}{\partial x_j}\right)\right] + \left[\mu\left(\frac{\partial u_i}{\partial x_j} + \frac{\partial u_j}{\partial x_i}\right) - \frac{2}{3}\mu\left(\frac{\partial u_i}{\partial x_i}\right)\delta_{ij}\right]\\ &\times\left[\frac{1}{2}\mu\left(\frac{\partial u_i}{\partial x_j} + \frac{\partial u_j}{\partial x_i}\right)\right].\end{aligned} \tag{8.3}$$

### 8.2.1 Overview of Turbulence Modeling

Turbulence can be defined as the continual mixing of adjacent fluid layers with different mean velocities, which is a more effective mean of transferring momentum than viscous stresses. The instantaneous value of any flow variable can be decomposed into a mean component and a fluctuating component. The Reynolds averaging procedure is conducted by decomposing the different flow variables into a mean component and a fluctuating component. Velocity, pressure, and temperature can be decomposed as shown in (8.4)–(8.6).

$$\tilde{u}_i = U_i\,(\text{mean}) + u_i'\,(\text{fluctuating}), \tag{8.4}$$

$$\tilde{p} = P\,(\text{mean}) + p\,(\text{fluctuating}), \tag{8.5}$$

$$\tilde{T} = \overline{T}\,(\text{mean}) + T'\,(\text{fluctuating}). \tag{8.6}$$

And by definition, the mean of the fluctuations is zero: $\overline{u}' = \overline{p} = \overline{T}' = 0$.

Although the averages of individual fluctuations (e.g., for the velocities $u'$ or $v'$) are zero, the average of a product (e.g., $u'$, $v'$) is not and may lead to a significant net flux. Taking the average of these components and applying them to the mass continuity equation, momentum equations, and energy equation yields the Reynolds-averaged Navier–Stokes (RANS) equations. Assuming incompressible flow of a Newtonian fluid and ignoring the viscous dissipation term in the energy equation, and using the Boussinesq approximation of buoyancy, the RANS equations can be written as shown in (8.7)–(8.10).

Mean continuity equation:

$$\frac{\partial U_i}{\partial X_i} = 0, \tag{8.7}$$

$$\frac{\partial u'_i}{\partial X_i} = 0. \tag{8.8}$$

Mean momentum equation:

$$\frac{\partial U_i}{\partial t} + U_j \frac{\partial U_i}{\partial x_j} + \frac{\partial}{\partial x_j}(\overline{u}_i \overline{u}_j) = -\frac{1}{\rho_0}\frac{\partial P}{\partial x_i} - g[1 - \beta(\overline{T} - T_0)]\delta_{i3} + \upsilon \frac{\partial^2 U_i}{\partial x_j \partial x_j}. \tag{8.9}$$

Mean energy equation:

$$\frac{\partial \overline{T}}{\partial t} + U_j \frac{\partial \overline{T}}{\partial t} + \frac{\partial}{\partial x_j}(\overline{u}_j \overline{T}') = \kappa \frac{\partial^2 \overline{T}}{\partial x_j{}^2}. \tag{8.10}$$

The RANS equations listed above can be used to account for turbulence using different assumptions.

In a simple shear flow, the total stress is:

$$\tau = \mu \frac{\partial \overline{u}}{\partial y} - \rho \overline{u'v'}. \tag{8.11}$$

In fully turbulent flow, turbulent stress is usually substantially bigger than viscous stress. Stresses and kinetic energy due to velocity fluctuations:

Normal stresses: $u'^2$, $v'^2$, $w'^2$
Shear stresses: $v'w'$, $w'u'$, $u'v'$
Kinetic energy: $k = \frac{1}{2}(\overline{u'^2} + \overline{v'^2} + \overline{w'^2})$
Turbulence intensity:

$$\frac{\text{Root} - \text{mean} - \text{square} - \text{fluctuation}}{\text{Mean velocity}} = \frac{u'_{\text{rms}}}{U} = \frac{\sqrt{\left(\frac{2}{3}\right)k}}{U}. \tag{8.12}$$

### 8.2.2 Eddy-Viscosity Models

The mean shear stress has both viscous and turbulent parts. In simple shear:

$$\tau = \mu \frac{\partial \overline{u}}{\partial y} - \rho \overline{u'v'}. \tag{8.13}$$

The most popular type of turbulence model based on RANS equations is an eddy-viscosity model (EVM) which assumes proportionality between turbulent stress and mean velocity gradient similar to that between viscous stress and velocity gradient. In simple shear:

$$-\rho \overline{u'v'} = \mu_t \frac{\partial U}{\partial y}, \tag{8.14}$$

where $\mu_t$ is called an eddy viscosity or turbulent viscosity. The kinematic eddy viscosity is

$$\nu_t = \frac{\mu_t}{\rho}. \tag{8.15}$$

The total mean shear stress in (8.13) above is then written as:

$$\tau = \mu_{\text{eff}} \frac{\partial U}{\partial y}, \tag{8.16}$$

where the effective viscosity $\mu_{\text{eff}}$ is the sum of molecular and turbulent viscosities:

$$\mu_{\text{eff}} = \mu + \mu_{\text{t}}. \tag{8.17}$$

With the eddy-viscosity hypothesis, closure of the mean-flow equations depends only on the specification of $\mu_{\text{t}}$, a property of the turbulent flow.

The kinematic eddy viscosity $\nu_{\text{t}} = \mu_{\text{t}}/\rho$ has dimensions of [velocity] × [length], which suggests that it can be modeled as

$$\nu_t = u_0 l_0. \tag{8.18}$$

On physical grounds, $u_0$ should be a velocity scale reflecting the magnitude of turbulent fluctuations and $l_0$, a length scale characteristic of the size of turbulent eddies. Today, common practice is to solve transport equations for one or more turbulent quantities (usually $k$ + one other) from which $\mu_{\text{t}}$ can be derived on dimensional grounds. The following classification of EVMs is based on the number of transport equations.

Zero-equation models

- Constant EVMs.
- Mixing-length models: $l_0$ specified algebraically; $u_0$ from mean-flow gradients.

One-equation models

- $l_0$ specified algebraically; transport equation is used to derive $u_0$.

Two-equation models

- Transport equations for quantities from which $u_0$ and $l_0$ can be derived.

Below is a list of the RANS-based turbulent models that employ the linear eddy-viscosity hypothesis:

1. Algebraic models
   (a) Cebeci–Smith model
   (b) Baldwin–Lomax model
   (c) Johnson–King model
   (d) A roughness-dependent model
2. One-equation models
   (a) Prandtl's one-equation model
   (b) Baldwin–Barth model
   (c) Spalart–Allmaras model

3. Two-equation models:

   (a) $k$–$\varepsilon$ models

      - Standard $k$–$\varepsilon$ model
      - Realizable $k$–$\varepsilon$ model
      - RNG $k$–$\varepsilon$ model
      - Near-wall treatment

   (b) $k$–$\omega$ models

      - Wilcox's $k$–$\omega$ model
      - Wilcox's modified $k$–$\omega$ model
      - SST $k$–$\omega$ model
      - Near-wall treatment

   (c) Realizability issues

      - Kato–Lunder modification
      - Durbin's realizability constraint
      - Yap correction
      - Realizability and Schwarz' inequality

Other RANS-based turbulent models that employ the nonlinear-eddy-viscosity hypothesis are shown below:

1. Explicit nonlinear constitutive relation

   (a) Cubic $k$–$\varepsilon$
   (b) EARSM

2. $v^2 - f$ models

   (a) $\overline{v}^2 - f$ model
   (b) $\xi - f$ model

Finally, another RANS-based turbulent model is the Reynolds stress model (RSM); however, it does not use the eddy-viscosity approach, but rather computes the Reynolds stresses directly.

Other turbulent models that are not based on the RANS equations but rather based on large eddy simulation (LES) are listed below:

1. Smagorinsky–Lilly model
2. Dynamic subgrid-scale model
3. RNG-LES model
4. Wall-adapting local eddy-viscosity (WALE) model
5. Kinetic energy subgrid-scale model
6. Near-wall treatment for LES models

### 8.2.3 The k–ε Model

The $k$–ε model is the most common turbulence model in use today, especially for numerical modeling of data centers. $k$ is the kinetic energy and ε is the rate of dissipation of energy. It has been favored for industrial applications due to its low computational expense and generally better numerical stability. It is a two-equation EVM with the following specification:

$$\mu_{\mathrm{t}} = \rho \upsilon_{\mathrm{t}}, \tag{8.19}$$

where

$$v_{\mathrm{t}} = C_{\mu} \frac{k^2}{\varepsilon}, \tag{8.20}$$

$C_{\mu}$ is a constant (with a typical value of 0.09), $k$ and ε are determined by solving (8.21)–(8.22).

$$\frac{\partial \rho \overline{U}_i k}{\partial X_j} = \frac{\partial}{\partial X_j}\left(\left(\mu + \frac{\mu_{\mathrm{t}}}{\sigma_{\mathrm{k}}}\right)\frac{\partial k}{\partial X_i}\right) + P + G - \rho\varepsilon, \tag{8.21}$$

$$\frac{\partial \rho \overline{U}_i \varepsilon}{\partial X_j} = \frac{\partial}{\partial X_j}\left(\left(\mu + \frac{\mu_{\mathrm{t}}}{\sigma_{\mathrm{k}}}\right)\frac{\partial k}{\partial x_i}\right) + C_1 \frac{\varepsilon}{k}(P + C_3 G) - C_2 \rho \frac{\varepsilon^2}{k}, \tag{8.22}$$

where $P$ is the shear production which is defined as:

$$P = \mu_{\mathrm{eff}} \frac{\partial \overline{U}_i}{\partial x_j}\left(\frac{\partial \overline{U}_i}{\partial X_j} + \frac{\partial u'_i}{\partial x_i}\right), \tag{8.23}$$

$G$ is the production of turbulence kinetic energy due to buoyancy, and is given by the following equation:

$$G = \frac{\mu_{\mathrm{eff}}}{\sigma_{\mathrm{T}}} \beta \frac{\partial T}{\partial x_i}. \tag{8.24}$$

And the model constants are:

$$C_{\mu} = 0.09, \quad C_1 = 1.44, \quad C_2 = 1.92, \quad C_3 = 1.0. \tag{8.25}$$

Using the $k$–ε model, the RANS momentum equation in (8.9) becomes:

$$\frac{\partial}{\partial x_j}(\rho \overline{U}_i \overline{U}_j) = -\frac{\partial P}{\partial x_i} + \frac{\partial}{\partial x_j}\left(\mu \frac{\partial \overline{U}_i}{\partial x_j} - \rho \overline{u}'_i \overline{u}'_j\right) + g_i(\rho - \rho_0), \tag{8.26}$$

where

$$-\rho u_i u_j = \mu_t \left( \frac{\overline{U}_i}{\partial x_j} + \frac{\overline{U}_j}{\partial x_i} \right) - \frac{2}{3} \rho k \delta_{ij}. \tag{8.27}$$

The RANS energy equation in (8.10) becomes:

$$\frac{\partial}{\partial t}(\rho e) + \frac{\partial}{\partial x_i}\left[\overline{U}_i(\rho e + p)\right] = -\frac{\partial}{\partial X_j}\left( k_{\text{eff}} \frac{\partial T}{\partial X_j} + \overline{U}_i (\tau_{ij})_{\text{eff}} \right) + S_h, \tag{8.28}$$

where $e$ is the total energy, $(\tau_{ij})_{\text{eff}}$ is the deviatoric stress tensor defined in (8.29), and $k_{\text{eff}}$ is the effective thermal conductivity defined in (8.30).

$$(\tau_{ij})_{\text{eff}} = \mu_{\text{eff}} \left( \frac{\overline{U}_i}{\partial x_j} + \frac{\overline{U}_j}{\partial X_i} \right) - \frac{2}{3} \mu_{\text{eff}} \frac{\overline{U}_i}{\partial X_i} \delta_{ij}, \tag{8.29}$$

$$k_{\text{eff}} = k + \frac{c_p \mu_t}{\text{Pr}_t}, \tag{8.30}$$

where $k$, in this case, is the thermal conductivity. The default value of the turbulent Prandtl number is usually around 0.85.

## 8.3 Literature Review on Numerical Modeling of Data Centers

This section provides a detailed review of literature regarding numerical modeling of data centers. Other literature reviews are made available by Schmidt and Shaukatullah [2], who looked at the thermal management of computer data centers and telecommunication rooms. Great similarities were noted in the challenges faced by both of them, and two main areas of focus were highlighted to maintain the enhancement of the air cooling technologies: the integration of energy efficiency and improving room ventilation for cooling clusters of concentrated heat loads. Rambo and Joshi [3] also provided an informative literature review. This chapter focuses on different areas to aid the reader in better understanding the state of the art in each. The sections are broken as follows:

1. Numerical modeling
   (a) Raised-floor airflow supply
   (b) Data center dimensions, rack layout, and power distribution
   (c) Alternative airflow supply and return schemes
   (d) Energy efficiency and thermal performance metrics
   (e) Rack-level thermal analysis
   (f) Dynamic thermal management of data centers

2. Experimental measurements and validation
3. Reduced order models and prediction models
4. Energy efficiency
5. Water cooling

1. Numerical modeling

   (a) Raised-floor airflow supply
   Many of the legacy data centers utilize the raised-floor configuration for chilled air supply. The use of the under-floor supply poses many difficulties, such as how to ensure providing the correct airflow rate through a specific tile to cool particular clusters of racks. Considerable focus in the literature was oriented toward this topic. Kang et al. [4] introduced a computational model to predict tile flow rates in data centers. The model assumed uniform pressure in the plenum, which is valid if the air velocity in the plenum is not significant enough to introduce variations in pressure. Kang et al. [5] studied the flow underneath the floor and its effect on the distribution of air flow from the perforated tiles.

   In another study, Karki et al. [6] introduced a 3D numerical model using a software package Tileflow [7]. The novelty of their model is that it is concerned only with the space underneath the plenum to predict tile airflow rates. The model was valid for any plenum height, as opposed to the study by Schmidt et al. [8], which was only restricted to a small plenum depth of 0.284 m high. Karki and Patankar [9] and Patankar and Karki [10] examined different techniques in controlling the airflow distribution in a raised-floor data center. Three main techniques were considered: changing plenum height, changing open area of perforated tiles, and installing thin partitions to guide the air. The results showed significant variations in airflow distribution when changing plenum height or tile open areas; however, the thin partitions provided the best flexibility in controlling the airflow distribution. Rambo et al. [11] performed experimental measurements on a raised-floor plenum data center with two CRACs. The focus of the study was to investigate the perforated tile flow rate, especially tiles closest to the CRACs. Results for large plenum depths of ~1 m showed very few cases of reversed flow for certain CRAC operation scenarios, unlike the conclusions drawn from Schmidt et al. [8, 12] for shallower plenums where reverse flow is expected for perforated tiles near an operating CRAC. Kumar and Joshi [13] conducted an experimental study to look closely the path lines of air in the cold aisle for different tile flow rates. Results showed that increasing the perforated tile flow rate is not necessarily the best way to provide cooling for the portion of a rack.

   In another study concerned with under-floor plenum flow, Radmehr et al. [14] conducted experimental and modeling study to quantify the distributed leakage flow in a data center. The authors provided a procedure to measure the leakage flow through gaps between tiles and estimated leakage flow in a conventional data center to be around 5–15%. A computational model was used to predict leakage flow and was found to be in good agreement with experimental results. Karki et al. [15] introduced two methods of calculating distributed air leakage in raised-floor data centers. The first method used a CFD model and showed good agreement with the results presented in Radmehr et al. [14]. The second method was based on the assumption that the pressure underneath the plenum is uniform. Despite the simplicity of the model, it still showed good agreement with the results from the first approach.

The latest work on airflow rates through perforated tiles was done by Abdelmaksoud et al. [16] who presented an experimental and computational investigation on the effect of detailed tile geometry on the velocity distribution across the tiles in a data center cell. Also in another paper, Abdelmaksoud et al. [17] presented a study investigating the effects of buoyancy and proposed different models for tile flow and rack exhaust to conserve both mass and momentum.

(b) Data center dimensions, rack layout, and power distribution

Considering only raised-floor data centers, many differences are present between data center facilities, such as dimensions, ceiling heights, plenum height, rack and CRAC layouts, etc. However, one could make general observations regarding different parameters in a data center and understand the effect of varying them. Many studies sought optimizing the data center by looking at these parameters and some of this literature is included below.

Schmidt [18] conducted a series of CFD studies that examine the cooling of racks configured in a cold-aisle/hot-aisle layout. The studies looked at the effect of server heat load, tile flow rate, CRAC unit location, and room height on the inlet temperatures of data processing equipment. Different configurations were examined, and the general conclusions showed that all of the parameters having a substantial effect on the inlet temperatures of servers. The key result was that when the chilled air from the perforated tiles is drawn in by the lower portion of the racks, the air in upper portions is drawn in from the rest of the room, possibly from the exhaust of the same rack. A novel idea was examined by Schmidt and Cruz [19] by looking at the effects of introducing some of the chilled air into the hot aisle and its effect on the rack inlet temperatures. The thought of cooling the hot exhaust air will result in lower inlet temperatures overall, especially for the top portions of the racks that are starved. It was found that for every case, it is better to concentrate the chilled air in the cold aisle only. In Schmidt and Cruz [20], the effect of placing high-powered racks among low-powered racks on the inlet temperatures of the racks was investigated. The one key result from this study was that for very low tile flows, the inlet temperatures were similar for both low- and high-powered racks. This is a significant conclusion since less tile flow rate is required to keep both low- and high-powered racks at acceptable inlet temperatures.

Schmidt and Cruz [21] focused on the effect of inlet rack air temperatures when adjacent racks are removed. Best improvements to the rack inlet air temperature were noted when one rack was removed nearby a high-powered rack. Also in clusters of high-powered racks, removing every other rack showed improvements to the rack inlet air temperatures. Schmidt and Cruz [22] investigated the effect of reducing rack flow rates on the inlet temperatures of the rack. Reducing the rack flow rate while keeping the heat load fixed resulted in increased air temperatures exiting the rack, and higher inlet temperatures were noted. Schmidt [23] presented, for the first time, a complete thermal profile of a high-power-density data center. A methodology was provided on how to thermally characterize a data center, to aid in comparisons and future designs. The effect of maldistributed flow from perforated tiles on rack inlet temperatures was examined by Schmidt and Cruz [24]. For lower tile flow rates, it was noted that the more maldistributed the flow, the lower the rack inlet temperatures, where for higher tile flow the maldistribution did not have such effect. Patel et al. [25] investigated the effect of changing the data center layout on the provisioning of CRAC units through the use of CFD modeling and created guidelines for effective infrastructure design. Using different room layouts with a different number of CRACs, they investigated the provisioning of each CRAC unit for a fixed heat load scenario. They examined the energy savings that can be achieved given that the CRAC units can vary in capacity according to their percentage of mal-provisioning. The energy savings reached 35%; however, they varied depending on the rack heat loads and the severity of mal-provisioning of CRAC units. The study discussed how highly dynamic the data center environment is and that the use of variable capacity CRAC units is vital. Gondipalli et al. [26] discussed a numerical case study on cold-aisle containment methods using doors and roofs. The numerical study was further expanded by optimizing the designs of doors and roofs having slits [27].

(c) Alternative airflow supply and return schemes
The ventilation configuration in data centers is known to greatly affect its efficiency from a cooling perspective. Optimizing the air supply and return configurations has seen extensive research efforts and recommendations. Nakao et al. [28] numerically investigated four different airflow configurations: under-floor air supply/overhead return, under-floor air supply/horizontal return, overhead air supply/under-floor return, and overhead air supply/horizontal return. Under-floor air supply/overhead return was found to be the best configuration, providing the lowest average inlet air temperature at the least required supply airflow rate. Similarly, Noh et al. [29] used CFD modeling to compare three different configurations and found the same conclusions as in [28]. Patel et al. [30] developed a CFD model of a data center facility with overhear supply and return. Results were compared to experimental measurements and reported errors between 7 and 12%. Shrivastava et al. [31] also used a numerical CFD model to investigate seven different airflow supply and return ventilation schemes. Further CFD modeling was used by Herrlin and Belady [32], and Schmidt and Iyengar [33] compared under-floor supply to overhead supply. Rambo and Joshi [34, 35]

also looked at different airflow supply and return configurations along with different CRAC placements with respect to racks.

Given the various cooling schemes considered, the general outcome shows that under-floor air supply and overhead return is the best scheme to adopt for maximum efficiency. The performance of the data center has been characterized on the basis of average and maximum mean region rack inlet air temperature. Also the worst cooling scheme is using overhead airflow supply and floor return [28, 31].

(d) Energy efficiency and thermal performance metrics

Evaluating data center efficiency and thermal performance in most cases is accomplished by looking at server inlet air temperatures. Although this is of great importance in ensuring the correct environment for operation and reliability of the servers, it does not provide any indication of the efficiency of the data center or the extent of the presence of flow distribution complications, such as infiltration and recirculation. Hence, it is recognized that metrics are needed to assess the data center environment and aid in comparing data centers with different configurations. Sharma et al. [36] and Sharma and Bash [37] proposed for the first time two-dimensionless parameters, namely, supply heat index (SHI) and return heat index (RHI), which evaluate the thermal performance of data centers. The SHI quantifies the infiltration of heat into the cold aisles and ideally is desired to be zero (no infiltration). The RHI quantifies the mixing of rack outlet air with cooler air from the cold aisles, before returning to the CRAC. Ideally, the RHI is desired to be one. The study quantified the indices through CFD modeling. Effects of geometric changes on the indices were investigated for raised-floor infrastructure with room return and ceiling return configurations. Geometric variations focused on cold-aisle width, hot-aisle width, and ceiling height, which all proved to be significant. Escobar and sharma [38] introduced a nondimensional parameter $G$ to evaluate data center layout and its cooling performance through the SHI. In discussing the complexity involved in cooling data centers, specifically due to the rise in compute loads, Schmidt et al. [39] introduced the Beta index which measures the extent of increase in inlet temperature due to recirculation. Norota et al. [40] used the statistical definition of the Mahalanobis generalized distance to describe the nonuniformity in rack thermal performance.

A foundational data center efficiency metric called the power usage effectiveness (PUE) was introduced in [41] and is currently widely used for assessing data center efficiency. This metric is defined as the ratio of the total facility power in a data center to the power of the IT equipment. The total facility power is the power delivered to operate the data center, which includes the power for operating the IT equipment and the cooling infrastructure. The authors show how the metric can be used to capture the actual cost of operation for a watt of IT equipment power. The PUE metric was used in [42] to assess 13 enterprise data centers with variable designs, showing an average PUE of 2.1 and stressed the importance for data center managers in using PUE as a measure of benchmark data center efficiency. The authors also proposed a new metric, compute power efficiency (CPE), which gives insight onto what percentage of the total facility power is actually used to compute. The study showed that for a data center with a PUE of 2.0, the CPE is 10%, meaning that only 1 W for every 10 W of utility power is actually used for computing. This indicated the importance of optimizing such a metric along with the PUE.

One of the latest contributions is by Tozer and Salim [43] who introduced a number of metrics that evaluate different data center phenomenon: bypass ratio, negative pressure ratio, recirculation ratio, and balance ratio. Different data center configurations and practices were compared using a recirculation and bypass chart and they showed the significance of such a chart in the evaluation of data centers.

(e) Rack-level thermal analysis

Other research attempts focused specifically on server racks. Zhang et al. [44] examined the effect of modeling the rack details on the numerical results of the data center cell and did not find great variations with a more detailed rack. Rambo and Joshi [45] presented a parametric study based on a multiscale model in which each rack was divided into a series of submodels with the possibility of varying their power and airflow. In another study, Rambo and Joshi [46] investigated computationally the optimal arrangement of servers and power dissipation profile within a forced air cooled rack and reported the best case to be the one where power increases vertically within the rack. Herrlin [47] proposed a method and a metric for analyzing how effective a rack is in cooling data centers. The proposed metric is named as rack cooling index (RCI) and was tested for different data center configurations, showing how it could be used to provide useful information about effective cooling of racks.

Rolander et al. [48, 49] combined the reduced order model introduced by Rambo and Joshi [50] with robust design principles to obtain thermally efficient server cabinet configurations. Results showed that redistributing the heat load in the correct manner allows current cabinets to dissipate 50% more power without the need of increase in cooling supply.

(f) Dynamic thermal management of data centers

Most of the literature reviewed above is focused only on fixed configurations of data centers with fixed CRAC airflow supply rate and temperature, fixed server power dissipations and fan speeds. Each investigation represents a steady-state scenario. However, in reality, a data center environment is highly dynamic. Server power dissipations vary according to the workload, accordingly server fan speeds change. CRAC supply airflow rate and temperature are bound to change to respond to forming hot spots in data centers. These changes require dynamic analysis for safer, more efficient, and effective data center management. One of the first works discussing transient data in a data center was presented by Stahl and Belady [51], where they collected empirically at different heat loads to compare with theoretical calculations.

Beitelmal and Patel [52] performed transient simulations to study the impact of CRAC failure on the temperature variations within the data center. The study highlighted the mal-provisioning of CRAC units when failure occurs, along with inlet temperatures reaching unacceptable levels within 80s of failure. It is important to point out the study that did not account for the thermal mass of rack units or CRAC units, which strongly affect the transient analysis as shown by Ibrahim et al. [53]. Ibrahim et al. [54] developed a transient model to look at time-varying server power and CRAC supply airflow rate. A generic profile was used to vary the power and airflow, and a number of case studies showed the variation of inlet temperature with time. Sharma et al. [55] highlighted the importance of dynamically managing data centers and discussed the ways of allocating workloads for more uniform temperature distributions within a data center.

Patel et al. [56] proposed the idea of smart cooling of data centers, to reach energy savings of 50%. They looked at fully controlling the data center environment through attributes such as distributed sensing, variable air conditioning, data aggregation, and a control system that links all the attributes together. The authors also investigated "smart tiles," which are variable opening plenum tiles to aid in more flexible control of the cooling distribution. Energy savings are obtained through directing cooling resources where and when needed. In addition, the redistribution of workloads, whether within a data center or globally through a computing grid, is proposed for the most energy efficient scenario. Bash et al. [57] adopted this idea of smart cooling to dynamically control the thermal environment of a particular data center. The data center is located in Hewlett-Packard Laboratories in Palo Alto and is designed for providing a production information technology (IT) environment while being used as a test lab for research purposes. The data center adopts the under-floor cooling design with room and ceiling return. A distributed sensor network was placed to manipulate CRAC supply, which was regulated using proportional–integral–derivative controllers (PID controllers). Experimental results were presented showing the variation of CRAC fan speed and CRAC supply temperature with time and were used to compare conventional control methods to the proposed. A promising 58% reduction in energy consumption was shown in one of the test cases.

2. Experimental measurements and validation studies

   Given the complexity of real data centers, numerical modeling may not fully capture their real thermal environment. Many investigators have compared experimental measurements with CFD results for validation and to justify dependence on such models in conducting parametric studies. Boucher et al. [58] focused on controlling the cooling resources in a data center. Their work included experimental studies of varying CRAC temperatures, fan speeds, and openings of vent tiles to control data center cooling. Rack inlet temperatures were found to vary linearly with CRAC supply temperatures, which lead the authors to conclude that buoyancy effects are small. This conclusion however is based on constant rack powers and a fixed rack location respective to the CRACs. Ibrahim et al. [53] show how buoyancy effects may be significant in a data center. Boucher et al. [58] also found the fan speed to have an unpredictable effect on the SHI, indicating that it is not a practical method of control. Changing the vent tiles opening did not impact the global flow patterns in the data center. However, it locally affected the rack air inlet temperatures, indicating that it is another practical method of control. The authors also investigated the relationship between CRAC fan speed and power consumption and found that an optimal CRAC fan speed can be utilized to consume the lowest power and remain within the constraints of the thermal management of the data center.

Schmidt et al. [8] compared experimental data with a computational model based on depth-averaged equations for the velocity and pressure distributions under the plenum. Given the simplicity of the model used, results are considered to be in OK agreement. In general, the model was able to capture the tiles flow rate pattern; however, in many cases, the CFM value modeled varied significantly from the experimental measurements. Iyengar et al. [59] compared results for a small data center test cell composed of one CRAC unit and one rack and obtained an average absolute error of 3°C, between experimental and CFD results. A series of studies followed, using the same data center test cell. Bhopte et al. [60–62] discussed the detrimental impact of under-floor blockages in data centers. Based on detailed numerical study, broad guidelines on managing under-floor blockages were presented. The established guidelines were experimentally validated on a different data center layout [63]. Cruz et al. [64] compared experimental data for three different layouts of perforated tiles to a CFD model with seven turbulence models. In another paper, Cruz et al. [65] compared experimental data to numerical modeling using eight different turbulence models and a laminar flow model. For a different data center, Samadiani et al. [66] used a numerical model to quantify the effect of plenum pipes and perforated tiles openness on the total CRAC air flow rates and its distribution. The numerical model was validated against the experimental results obtained by Rambo et al. [11], giving an average error of 10–13%.

Schmidt et al. [12] conducted an experimental study involving different configurations of tiles and a variation of turning CRACs on and off. The room used was 6.06 m × 20 m in size and had two CRAC units. The location of the open tiles was changed for different cases investigated. It was shown that the airflow distribution is strongly influenced by the number of CRACs in operation. Reverse flow was observed at tiles operating near the CRAC units due to the high horizontal air velocity in the under-floor region, causing air to bypass the tiles closest to the CRACs. Also in the same study, two of the experimental scenarios investigated were compared to a CFD model and were found to be in good agreement, where the tile flow rates were within 3 $m^3/s$. In a later study, Schmidt et al. [67] examined a data center housing a high performance supercomputer cluster and compared measurements to CFD results. The cluster had a peak performance of 77.8 TFlops/s and employed more than 12,000 processors. Comparisons did not show a good agreement and the authors suggested that it is due to inaccuracies in the CFD model such as the need for a finer grid or due to the physical nature of the turbulent mixing not accurately accounted for in the model.

3. Reduced order models and prediction models
Achieving a dynamically controlled data center for optimized energy efficiency requires real-time predictions and decisions based on changing attributes in the data center such as varying power dissipations and airflow supply. The ideal situation is to be capable of predicting the temperature and airflow distribution given a specific power distribution scenario and use the cooling resources to efficiently cool the room. In attempting to move away from CFD modeling due to its computational expense, Karki and Patankar [68] tried to simplify the data center problem by looking at a one-dimensional model of the plenum and compared to an analytical solution showing excellent agreement. The model is restricted to certain conditions, however could be a valuable tool for quick observations. Lopez and Hamann [69] used real-time sensor data as boundary inputs into a simplified, physics-based model using potential flow theory. Hamann et al. [70] further used the model to define thermal zones within a data center and provided case studies to explain how CRAC units can be utilized efficiently by using the model as a management tool. Potential flow theory was also previously used in data center applications by VanGilder and Shrivastava [71] and Toulouse et al. [72]. Samadiani et al. [73] discussed the requirements for an ideal thermal design of a data center to face the increasing trends in power dissipation, which would consist of a multiscale thermal solution.

   Rambo and Joshi [50] and Rambo [74] used proper orthogonal decomposition (POD) to create approximate solutions of steady, multiparameter RANS simulations. The model was demonstrated by looking at a single rack with ten servers. The authors predicted an order of $10^4$ reduction in model size and results within 5% of true solutions. Also Rambo and Joshi [75] introduced the idea of unit cells architecture by successively increasing the number of racks in a row. This was investigated to provide a common basis in comparing the thermal performance of various cooling schemes. It was found that using seven racks in a row is sufficient to model high-power density data centers. Somani and Joshi [76] presented an algorithm named ambient intelligence based load management (AILM) which improves the data center heat dissipation capacity. The algorithm is trained using inlet temperature of racks, and it distributes workloads according to the thermal environment in a data center and numerical results showed 50% enhancements in the heat dissipation capacity.

   Shrivastava et al. [77] used design of experiment (DOE) technique to screen the significant factors impacting the data center's thermal performance. Detailed comparison between the experimental and numerical results for the rack inlet temperatures in a 68.75-$m^2$ (7,400 $ft^2$) data center is presented in Shrivastava et al. [78]. The development of a software tool that estimates cooling performance of cluster of racks placed in a common cold aisle in a raised-floor data center in real time was discussed by Shrivastava et al. [79, 80]. A fundamental assumption within the algorithm of the tool is that computation of air flow

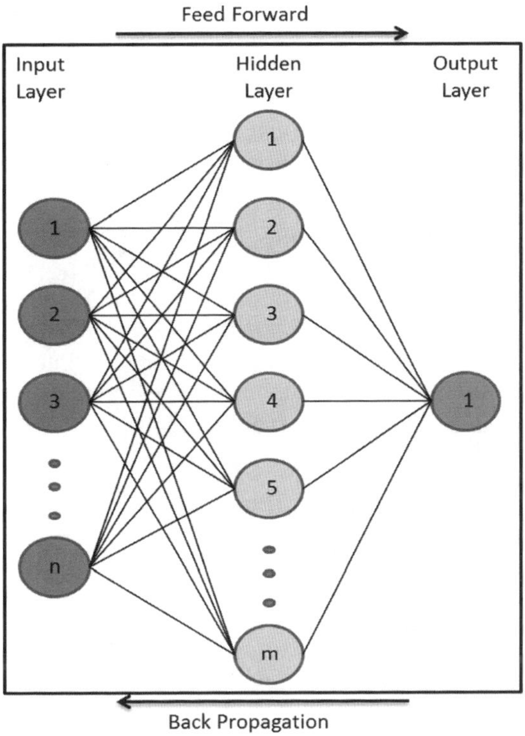


**Fig. 8.1** Example neural network topology [82]

patterns inside the cold aisle can be decoupled from room environment. This partially decoupled analysis (PDA) requires a single cluster of equipment consisting of two equal length rows of equipment, which makes the aisle to be analyzed as a simple rectangular shape. When the aisle is cut-out of the larger environment, it is easier to analyze and obtaining boundary conditions for the different faces of the rectangle is simple. If the correct boundary conditions are available, then the solution for the rectangle will be accurate. The boundary conditions are determined by the empirical analysis of full CFD simulations. PDA simulations are shown to yield solution in the range of 10–30 s [81]. Using another approach, Shrivastava et al. [82] discussed the neural network (NN) model to predict CI. A NN typically consists of an input layer, hidden layers, and an output layer. Figure 8.1 shows a graphical representation of the three-layer NN model adopted for the data center application. A NN requires a number of training sets, where each training set provides an input and a corresponding output value. These training sets are used to develop the hidden layers, which are later used to predict output values, given a certain input. This analytical approach using artificial intelligence was used to predict cooling performance of data centers with accuracy within 3.8% for the set of example scenarios considered. This approach was further used to develop genetic algorithm (GA) to optimize cooling performance of cluster of equipment composed of two approximately equal lengths of rows of racks and coolers bounding a common cold aisle [83].

In a recent study, Marwah et al. [84] compared four different techniques in the prediction of thermal anomalies using experimental data obtained over a period of 3 months. The authors define a thermal anomaly as temperature exceeding a specific threshold for a specific interval of time at a particular rack sensor location. The naïve Bayesian classifier was shown to perform best out of the four techniques considered, giving an 18% success rate in predicting thermal abnormalities at an average time of 12 min. The naïve Bayesian classifier works in the following manner: given a set of input data a number of classes are assigned where different data fall into. The naïve Bayesian classifier learns the conditional probability of each class and predicts the presence of a thermal anomaly based on that probability.

4. Energy efficiency
With the increasing power dissipations of computer equipment, reducing energy consumption in data centers has become a major challenge. Sharma et al. [55] suggested two workload-redistribution approaches to improve robustness and energy efficiency of the cooling resources. First approach is row-wise thermal control which is implemented by calculating the power dissipation of each rack using local air temperature measurements in a row and redistributing the workload accordingly for a more uniform temperature distribution. The second approach is a regional approach which follows the same concept as row-wise thermal control, however, handling larger regions instead of row-by-row. Iyengar et al. [85] also investigated reducing energy usage in data centers through the control of CRAC units. Three methods of control were studied: reducing CRAC fan speed, systematic shut down of CRACs, and increasing the refrigeration chiller plant set point temperature. Results showed that reducing the fan speed have the most effect on energy savings, while increasing the refrigeration chiller plant set point temperature was not very efficient. Breem et al. [86] and Walsh et al. [87] created a model that represents the cooling infrastructure involved in a data center from the rack level to the cooling tower. They performed a number of parametric studies to look for improving data center efficiency and found that tight temperature control at the chip level yields the best efficiency.

A number of recent publications by the American Society for Heating Refrigeration and Air-Conditioning (ASHRAE) have focused on ways to improve the cooling infrastructure. For instance, Scofield and Weaver [88] discussed the possibility of using air handling units to cool data center facilities, rather than the use of CRAC units. Air handling units are normally placed in a mechanical room away from where the location of cooling is required while the CRAC units are designed to be placed inside the data center facility. The authors present a number of advantages some of which include the use of wet bulb economizers and use of outside air at cool environments to boost the energy efficiency. Munther [89] also recommended the use of outside air for lower energy consumption and provides benchmarks in classifying data center from energy efficiency prospective. The use of a re-circulating air conditioning by evaporation (RACE) unit for better cooling efficiency is shown to possibly lower data center cooling costs by 75% [90]. Also Judge et al. [91] provided ten energy-saving strategies for data centers that vary from power supply cable types to dynamic cooling.

From a different perspective, Shah et al. [92, 93] and McAllister et al. [94] looked at exergy-based analyses of data centers to evaluate its energy efficiency. Also Bash et al. [95] introduced a new technique for the evaluation of the performance and efficiency of data centers called damage boundaries. The authors define four regions of operation within any data center, which may aid data center operators in optimally provisioning their data center.

5. Water-cooled data centers
   To augment heat dissipation capabilities of racks beyond approximately 20 kW, water cooling is being explored in high-density data centers. Schmidt et al. [96] introduced a water-cooled heat exchanger attached to the rear door of a rack (RDHx). Results showed that the heat exchanger reduces hot air recirculation and that extracting the heat at the point of generations is much more effective than CRACs distributed around the data center. Modeling studies that illustrate the efficiency of the RDHx are described in Schmidt et al. [97]. Also a vapor compression refrigeration version of the RDHx is described in Tsukamoto et al. [98]. Kelkar et al. [99] developed a computational model for the analysis of two-phase pumped-loop cooling systems and investigated the use of R134A as a coolant instead of water. Kumari et al. [100] discussed the possibility of introducing mist in cooling server racks through modeling. Schmidt et al. [101] introduced the open side car heat exchanger, designed to reduce the effect of hot air recirculation and remove the heat load of a rack of up to 35 kW. It is similar to the RDHx introduced in [96], however, is said to have roughly twice the heat removal capability. More water cooling is expected in data centers in the near future, and ultimately, a 100% water-cooled data center is expected to have the highest efficiency yet.

## 8.4 CFD Modeling of Data Centers

A CFD code solves incompressible Navier Stokes equation with $k$–$\varepsilon$ turbulence model and energy equation to compute flow and temperature distributions within data center. Detailed description of these equations is found in Sect. 8.2. Karki et al. [6] have discussed a detailed computational modeling methodology used in commercial CFD code Tileflow™.

### 8.4.1 Boundary Conditions

*Walls*: At the walls of plenum, data center room, and other solid surfaces, the usual no-slip boundary condition is used. For wall shear stress and turbulence quantities, standard wall function treatment is employed.

*Inflow through CRACs*: The flow exiting out of the CRAC units is treated as inflow into the plenum. All the dependent variables are assumed to be known at the inflow. The inlet vertical velocity is deduced from the CRAC flow rate and the outlet open area. The horizontal velocity components are assumed to be zero.

*Flow through perforated tiles*: The flow through the perforated tiles is calculated using the pressure drop equations specified in Sect. 2.2 of Chap. 2.

There are several commercially available CFD codes which can efficiently solve the problem of data center thermal management. Some of the popular codes are Flovent/Flotherm(TM, Mentor Graphics), Tileflow (TM, Innovative Research Inc.), 6-Sigma (TM, Future Facilities Inc.), Coolsim (TM, Applied Math Modeling), etc.

All these codes discretize the coupled partially differential equations using finite volume method and solve them iteratively. For example, Tileflow™ uses *additive correction multigrid method* to solve the coupled continuity and momentum equations. Multigrid methods recognize that low-frequency components of error appear as high frequencies on a coarser grid and can thus be smoothed by applying solver on a coarser grid. These methods solve the problem on a series of successively coarser grids so that all frequency components are reduced at comparable rates [9].

In the following section, numerical case studies addressing some of the key problems of data center thermal management are presented. These case studies have been solved using commercial CFD code Flotherm™.

### 8.4.2 Prominent Numerical Modeling Approaches

In this section, key features of heat transfer and fluid flow in raised-floor data center clusters are discussed. Figure 8.2 shows the layout of data center considered. The data center has 20 server racks, each dissipating 32 kW of power with 1.37 $m^3$/s (2,905 CFM) air flow rate through each rack (20°C temperature rise). Four CRAC units supply cooling air to the data center.

1. Simple model or data center model without plenum
   Figure 8.3 shows a simple raised-floor data center where the cold aisles are modeled using flow sources delivering the chilled air at specified flow rate and temperature. CRAC units are modeled as devices which extract the hot exhaust from the room at specified flow rate. To supply 1.37 $m^3$/s (2,905 CFM) of cold air to each of the 20 racks, the CRAC units must supply a total of 58,100 CFM (20 × 2,905) (27.42 $m^3$/s) of cold air to the data center. However, CRAC units may not supply 100% of required cold air. In this case study, 80% CRAC supply

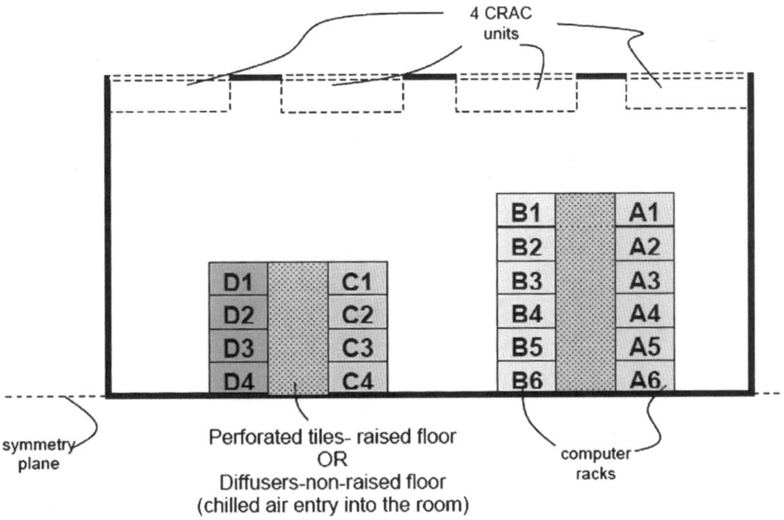


**Fig. 8.2** Plan view of the data center layout considered to present numerical modeling case studies

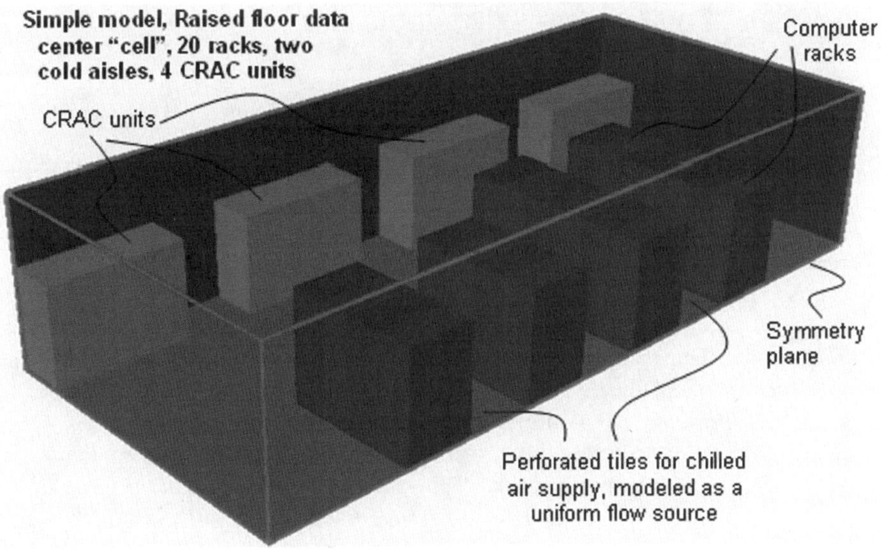


**Fig. 8.3** Simple model with flow sources representing the perforated tiles

is considered, which means only 80% of required cooling air is supplied to the room (58,100 $\times$ 0.8 = 46,480 CFM).

As plenum is not included in the solution domain, this model reduces the number of grid cells significantly, thereby reducing the computational time.

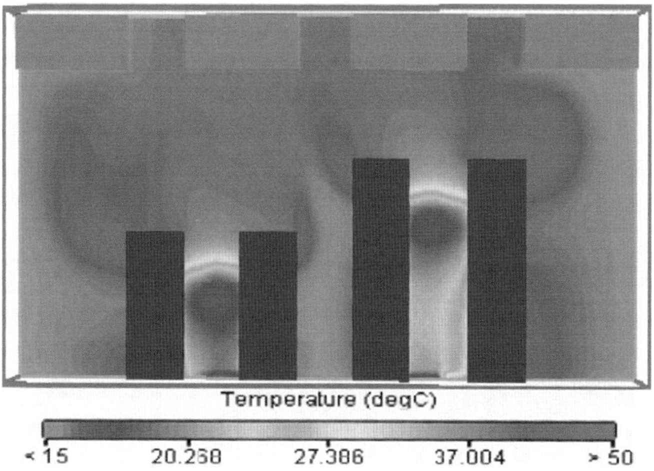


**Fig. 8.4** Temperature contours for a simple model at a height of 1.5 m above the raised floor

The flow is modeled as turbulent using the $k$–$\varepsilon$ model. Figure 8.4 shows temperature contours at a height of 1.5 m above the raised floor. This model is a good approximation for evaluating thermal performance of a raised-floor data center with uniform tile flow rates, or where individual tile flow rates are experimentally known. However, if the tile flow rates in the data center considered are expected to be highly nonuniform, this model may not predict its thermal performance accurately.

2. Detailed model or data center model with plenum:
Figure 8.5 shows a detailed raised-floor data center model in which the CRAC units supply the chilled air to the hollow raised floor and the air is introduced into the room via perforated tiles. Tile flow rates are computed as a function of pressure drop through the individual tiles, which ultimately is governed by various under-floor parameters such as plenum depth, tile resistance, CRAC supply flow rates, etc. This model incorporates the under-floor effects and produces results which are more accurate and enable better thermal design decisions.

In this approach, CRAC units are modeled as both flow supply and extraction devices. In the detailed model considered, 80% of CRAC supply is considered with 0.3-m deep plenum and 50% open tiles. An 80% of air supply equates to 46,480 CFM, which is divided evenly among the 4 CRAC units, where they each introduce air into the plenum, and air is introduced into the room through the floor tiles. Along with predicting thermal performance, this model can predict the air flow patterns below and above the raised floor. Figure 8.6 shows velocity vectors in a vertical plane, 305 mm in front of the rack row A. The vectors clearly indicate very high tile flow rates near rack A6 and warm air re-circulating over the entire height of rack A1. Detailed modeling approach, due to the addition of several details, increases the number of grid cells and computational time significantly.

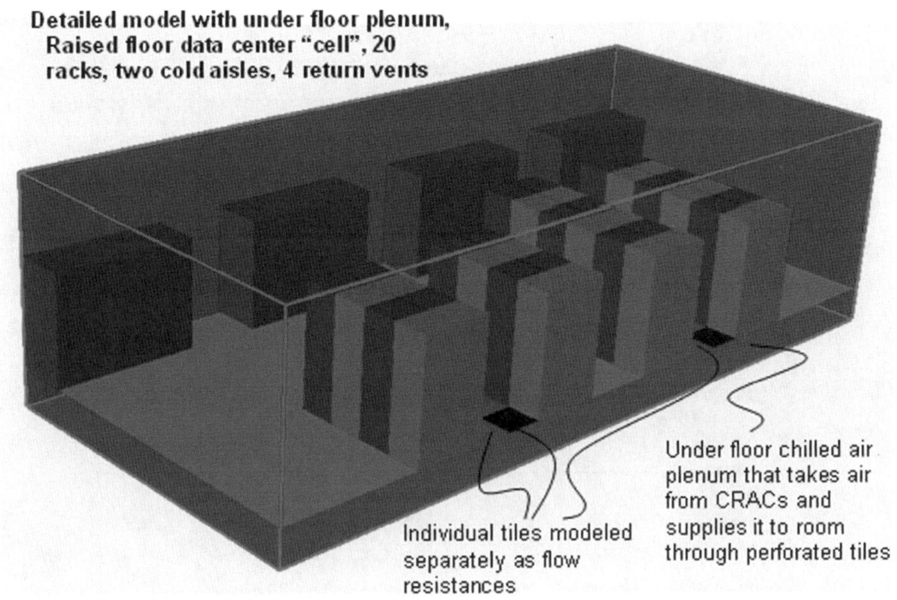


**Fig. 8.5** Detailed model with plenum and floor tiles

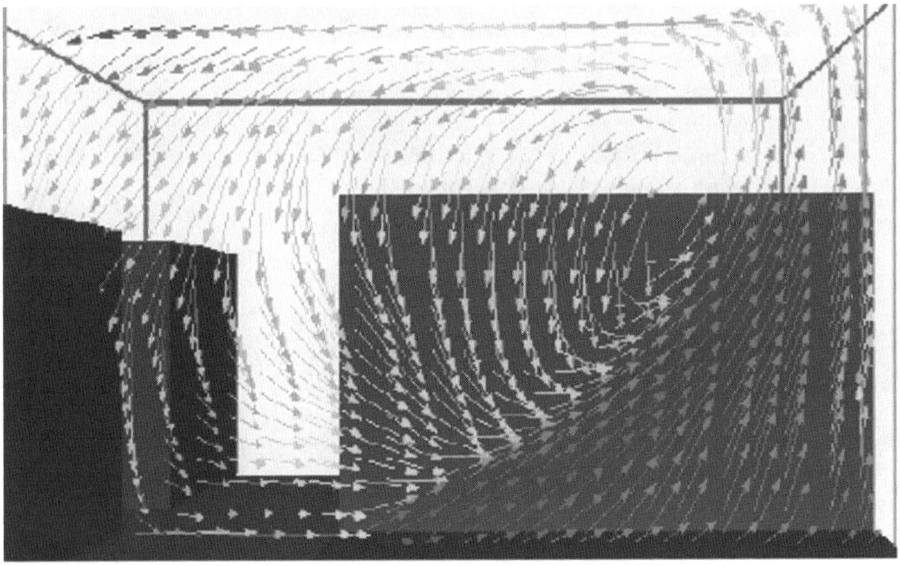

**Fig. 8.6** Velocity vectors predicted by detailed model shown 305 mm in front of rack row A

3. Detailed model with CRAC blower characteristic curve
   Detailed modeling approach discussed above assumes the CRAC unit as a fixed flow rate device supplying the cold air and extracting the warm air. As we have discussed earlier in Chap. 2, the CRAC flow rate is governed by the blower and

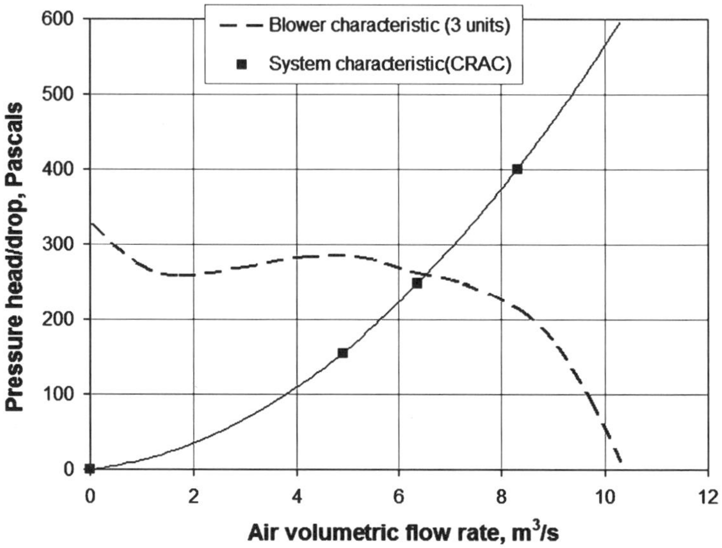


**Fig. 8.7** Blower and system characteristic curve as specified by the vendor

system characteristic curve as specified by the vendor. Figure 8.7 shows typical blower and system characteristic curves for a data center. As stated earlier in Chap. 2, the operating point is at the intersection of the two curves, at the air flow rate of 6.37 $m^3/s$ for the pressure drop of 248.84 Pa. This operating point corresponds only to internal resistance of CRAC unit due to heat exchanger coils, etc. (included in numerical model only as a planar resistance). Perforated tiles and plenum also add resistance to flow, so operating range of CRAC units in our data center is approximately between 4 and 6 $m^3/s$.

Parametric study with three tile resistances (25% open, 50% open, and 75% open) and three plenum depths (0.15, 0.3, and 0.6 m) is discussed [60]. For all the possible combinations of input parameters, nine cases were solved and analyzed. Figure 8.8 shows the effect of plenum depth and tile resistance on total CRAC supply. Both plenum depth and tile resistance have significant impact on total CRAC supply rate. For 25% tiles, by increasing the plenum depth from 0.15 to 0.6 m, the total CRAC supply rate increased by ~10%. For 50% tiles, by increasing the plenum depth from 0.15 to 0.6 m, the total CRAC supply increased by ~17%. For 75% tiles, by increasing the plenum depth from 0.15 to 0.6 m, the total CRAC supply rate increased by ~20%. Hence, CRAC characteristic curves are numerically shown to have a significant impact on data center air supply.

4. Detailed model with CRAC thermal characteristic curve
   CRAC units are usually modeled as fixed flow devices which supply cold air into the room through the plenum at a specified flow rate and temperature (e.g., 15°C), regardless of the temperature of the extracted hot exhaust. However, they are heat exchangers, with the hot exhaust transferring heat to cold fluid flowing

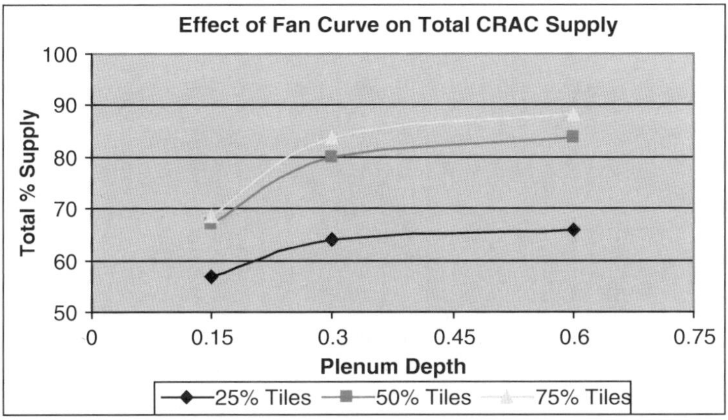


**Fig. 8.8** Effect of plenum depth and tile resistance on total CRAC supply

in coils inside the CRAC. The supply air temperature is therefore a function of the hot exhaust and cold fluid inlet temperatures and flow rates. Figure 8.9 shows thermal characteristic curve for CRAC unit, as specified by vendor.

The case investigated in this section is a 0.3 m plenum and 50% open tiles. For the modeling of the CRAC unit, the cold fluid is water at inlet temperature of 7.2°C and flow rate of 1.35E−12 $m^3/s$ (89.2 gpm) as specified by the vendor. The heat exchanger is modeled in a counter-flow configuration. The thermal conductance, a measure of how well heat can be transferred between the extracted air and the cold fluid in the coils, was 6,200 W/K and 7,100 W/K corresponding to 80% and 100% supply, respectively, which is obtained using the thermal conductance curve provided by the vendor in Fig. 8.9.

Figure 8.10 shows the effect of modeling of the CRAC thermal characteristic curve on the supply temperatures. It is clear that accounting for this effect is important as the supply temperatures are higher, reaching temperatures of 25°C compared to 15°C. The higher supply temperature will evidently affect the inlet temperatures into the racks which in some cases may reach unacceptable values.

### *8.4.3 Transient Detailed Data Center Model*

As shown in the review of literature section, most of the analyses conducted in the past have been concerned only with steady state. Fixed flow rates and rack powers are used, and the code is solved to obtain the steady-state air temperatures within the data center. Various changes can occur in an operating data center, such as the rack power, the CRAC supply flow rate, and the supply air temperature. Understanding the effect of these changes as a function of time is of great importance.

Transient analysis of data centers was conducted by Beitelmal et al. [52] to study the impact of CRAC failure on the temperature variations within the data center.

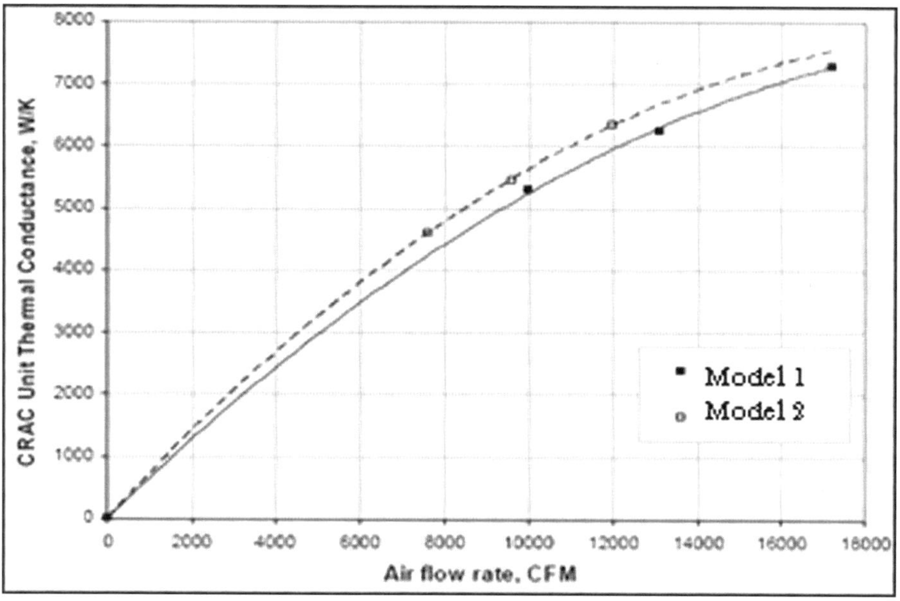


**Fig. 8.9** CRAC thermal conductance curve specified by the vendor

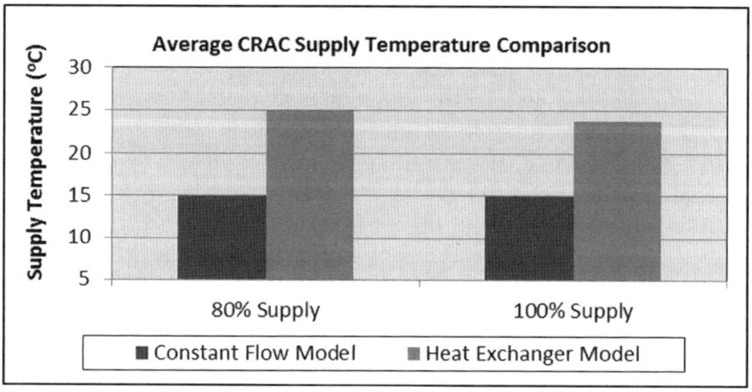


**Fig. 8.10** Effect of CRAC thermal characteristic curve on supply temperature

They modeled a data center with a total area of 267 $m^2$, which contained 66 server racks and 6 CRAC units. With the failure of one CRAC unit, the inlet air temperatures in some regions reached unacceptable levels as high as 40°C only within 80 s of failure. Recently, Ibrahim et al. [54] developed a numerical model to investigate different transient scenarios, and highlighted their importance. The effect of thermal mass on transient modeling was briefly introduced by Ibrahim et al. [53]. Some of this work is discussed here as an introduction to transient modeling of data centers.

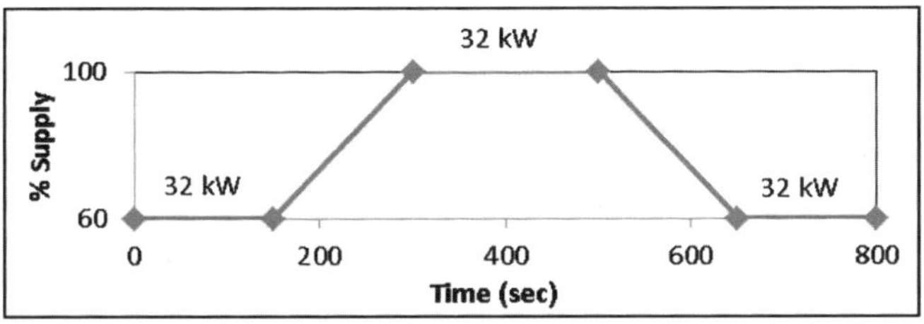


**Fig. 8.11** CRAC airflow supply profile with time and constant rack power

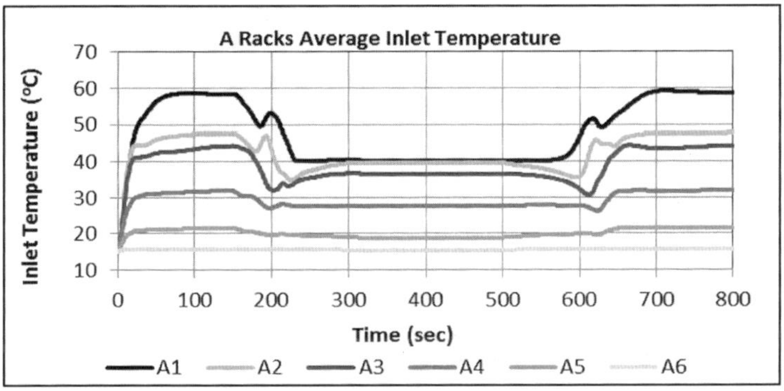


**Fig. 8.12** Average inlet air temperature (°C) of row A racks vs. time

#### 8.4.3.1 Case Study: Varying CRAC Airflow; Constant Rack Power

This case study is intended to give a general preview of the temperature and flow distribution changes in the room with time-dependent power and CRAC airflow, however, without accounting for the thermal mass of the server racks. Figure 8.11 shows the CRAC airflow supply profile with time, while power is kept constant at 32 kW.

Figure 8.12 shows the average inlet temperature with time for the racks of row A. Intuitively, the lower CRAC air supply rates cause higher air inlet temperature values. An interesting behavior is observed when the CRAC air supply is ramped up and down, where fluctuations in the inlet temperatures take place, especially for row A racks. These fluctuations are explained by the changing airflow pattern in the room, where recirculation zones are changing with time.

Figure 8.13 shows plots of the velocity vectors in the room at various times with a zoomed in plot of the temperature variation of rack A1. We can note from the figure the change in the recirculation zones around rack A1, which follow the fluctuations in inlet temperatures.

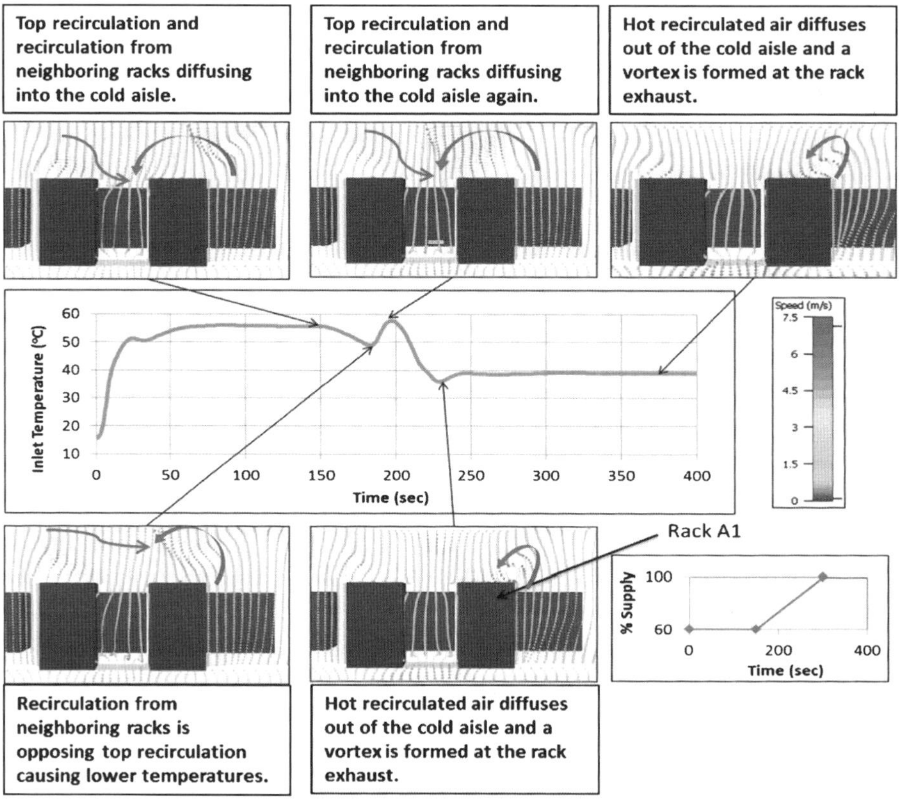


**Fig. 8.13** Velocity profiles near rack A1 for various times

## 8.4.4 *Effect of Thermal Characteristics of Electronic Enclosures*

The thermal mass of a body is its capacity to store heat. It is also known as thermal capacity or heat capacity. Typically, the symbol used to refer to thermal mass is $C_{\mathrm{th}}$ and it is measured in the units of J/°C. The general equation relating heat energy to thermal mass is:

$$Q = C_{\mathrm{th}}\Delta T, \tag{8.31}$$

where $Q$ is the heat energy transferred and $\Delta T$ is the change in temperature.

When looking at a data center, thermal mass is present in any body within the facility, including CRAC units, server racks, servers, tiles, etc. Accounting for thermal mass does not affect the overall steady state of any given scenario in a data center, only the time it takes to reach that steady state. In the case of transient

**Table 8.1** Server thermo-physical properties and thermal mass

| | Conductivity (W/m K) | Density (kg/m$^3$) | Specific heat (J/kg K) | Thermal mass per server (J/°C) |
|---|---|---|---|---|
| Copper | 385 | 8,930 | 385 | 377,497 |
| Steel | 55.4 | 7,850 | 490 | 422,345 |
| Mix (50% copper/ 50% steel) | 220 | 8,390 | 438 | 403,495 |
| 10% of Mix | 220 | 839 | 43.8 | 4,035 |
| 100% of Mix | 220 | 8,390 | 438 | 403,495 |

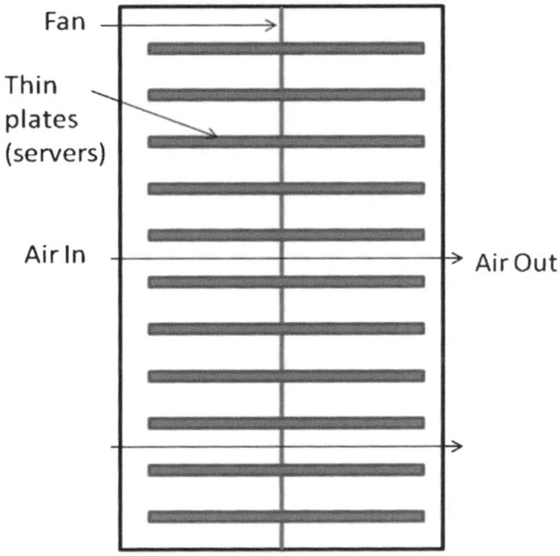


**Fig. 8.14** Schematic side view of the server rack

analysis, it becomes crucial to account for thermal mass, as it controls the rate of temperature rise, and the data center response to airflow changes. This section introduces a straightforward way of modeling thermal mass of servers only, which can be expanded to account for thermal mass of any bodies within a data center facility. It is intended only to give general observations on how thermal mass may be very crucial in modeling data centers.

Let us assume specifications of a typical commercial 2U server (89.08 mm in height) with a mass of 27.22 kg and the dimensions of width = 44.54 cm and depth = 69.98 cm. To get an estimate of the material properties of the server, we make a crude approximation assuming it is made of 50% copper and 50% steel and we name this material as "Mix" material. The material properties are averaged and the corresponding density, specific heat, heat capacity, and conductivity are used. Table 8.1 shows the details of the material properties.

Each server is modeled as a thin plate with the same width and depth of a typical 2U server. However, the height is adjusted to obtain the same weight of 27.22 kg, given a Mix material density of 8,390 kg/m$^3$. Knowing that a server rack used in data centers nowadays is expected to hold approximately twenty 2U servers, 20 thin plates are used in the model. Figure 8.14 shows a schematic view of the modeled rack with servers and a fan in the middle.

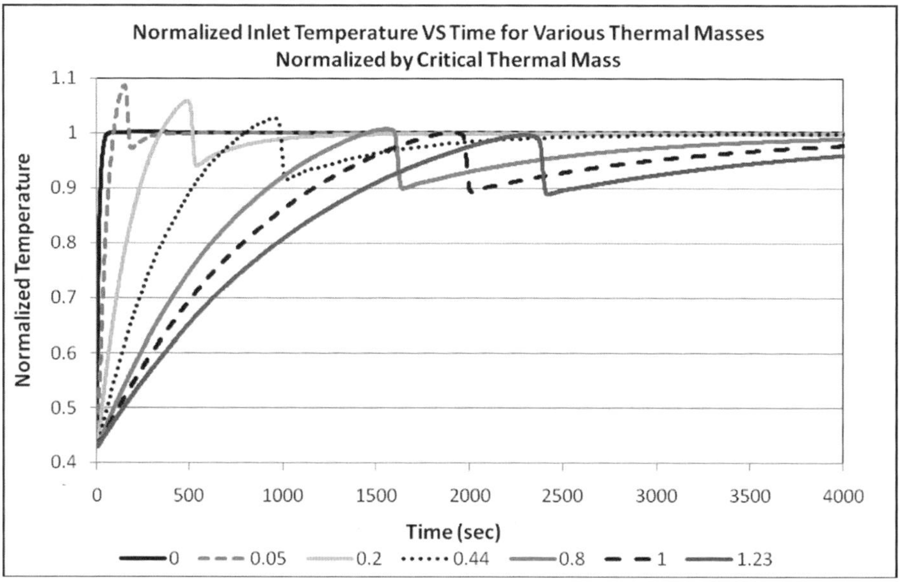


**Fig. 8.15** Normalized inlet temperature vs. time for various thermal masses normalized by the critical thermal mass ($Q^*_{th}$)

The same data center room in Fig. 4.3 is used, with the servers set to 1.6 kW of power dissipation, totaling to 32 kW per rack. An 80% of airflow supply at 15°C was distributed evenly among the four CRACs. The data center was initially at 15°C and established airflow. At time zero, the power in the servers is applied and the model is left to reach steady state. A total of 11 transient runs were conducted for various percentages of the Mix material, ranging from 0% (no thermal mass) to a 100%, with increments of 10%. The thermal mass per server $Q_{th}$ was calculated as the mass of the server times the specific heat capacity of the corresponding percentage of material.

Figure 8.15 shows the average temperature of rack A1 normalized by the steady-state temperature, for the various percentages of the Mix material. It is noted from the figure that there are overshoots beyond the steady-state temperature for a number of thermal mass percentages. For clarity, a critical thermal mass $Q^*_{th}$ is said to be the thermal mass at which no overshoots occur. Any $Q_{th}$ higher than $Q^*_{th}$ will not exhibit any overshoots in temperatures. The different $Q_{th}$ investigated in Fig. 8.15 are normalized by the $Q^*_{th}$. Various conclusions can be drawn from the presented results. In many data centers, operation managers tend to put a few servers per rack, and not fill up the racks for overheating purposes, however, the results presented here suggest that this could raise the possibility of temperature overshoots, as having fewer servers means lower thermal mass. The results also indicate that it takes a long time for a data center to reach steady state due to high thermal mass, which suggests that supplying fewer or more percentage of airflow to the data center facility may have a slow effect on the overall environment.

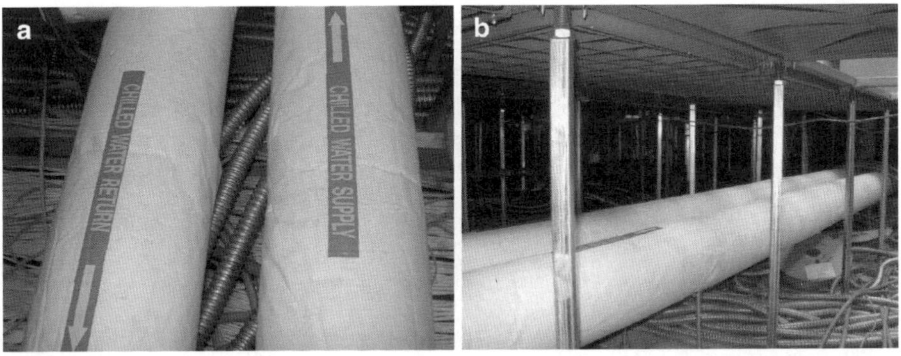


**Fig. 8.16** (**a**) Magnified view of pipes that circulate coolant to and from A/C units. (**b**) Chiller pipes running under the floor tiles

Case studies presented above show that numerical modeling is a powerful technique that helps an engineer to understand and solve the problem of data center thermal management. This tool is further used to investigate and address the problem of under-floor blockages in data centers.

## *8.4.5 Effects of Under-Floor Blockages*

### 8.4.5.1 Under-Floor Blockages in Data Centers

Raised-floor data centers offer considerable flexibility in placing the servers above the raised floor. The under-floor plenum serves as a distribution chamber for the cooling air. Without the need for ducting, cold air can be delivered to any location simply by replacing solid floor tile by a perforated tile. Plenums are also used to route piping that circulate coolant from the CRAC units and the chillers, and cables that supply power and network connections to the servers.

Figure 8.16a, b shows chiller pipes under the raised floor. These chiller pipes supply coolant to the CRAC units (usually water at ~7°C) and re-circulate the warm water back to the outside chillers or cooling tower. Often, inadequate consideration is given while laying these large diameter chiller pipes. These blockages impede the cold air stream coming from the discharge of the CRAC units and yield highly complex flow patterns, resulting in maldistributed tile flow rates, which can create hot spots above the raised floor. Hence, in this section, effect of under-floor blockages on overall data center performance is discussed in detail. Figure 8.17 shows examples of cables and wires lying under the raised floor. Such cluster of wires if lying directly under the perforated tiles can reduce their flow rates significantly and can impact the thermal performance of corresponding and neighboring server racks.

**Fig. 8.17** Examples of cluster of cables and wires lying carelessly under the rack

#### 8.4.5.2 Numerical Modeling of Under-Floor Blockages

To model under-floor blockages, data center shown in Fig. 8.2 is considered. Perforated tiles are 40% open and the plenum is 0.45 m deep. CRAC units are modeled as per the specifications from the CRAC manufacturer. For this numerical model, CRAC units supply a total of 48,800 CFM or ~85% of the required cooling air.

In the numerical model, inlet temperatures are monitored ~300 mm in front of the racks. There are six monitor points over the entire height of server racks. Monitor point 1 is 333 mm above the raised floor and monitor point 6 is 1,750 mm above the raised floor. Hence monitor point A1–6 represents air inlet temperature of rack A1 at a height of 1,750 mm above the raised floor. References have shown that due to lack of cooling air near the edge racks (A1, B1, C1, and D1), recirculation cells are formed over their entire height. Since rack A1 has the highest inlet air temperatures, over its entire height, temperature at monitor points A1–1 (333 mm above raised floor) and A1–6 (1,750 mm above the raised floor) are used to present the comparison between numerical cases. For the baseline case, average inlet temperature of rack A1 (averaged over six monitor points) is 36.7°C. Inlet temperature at A1–1 is 35.3°C and at A1–6 is 36.7°C.

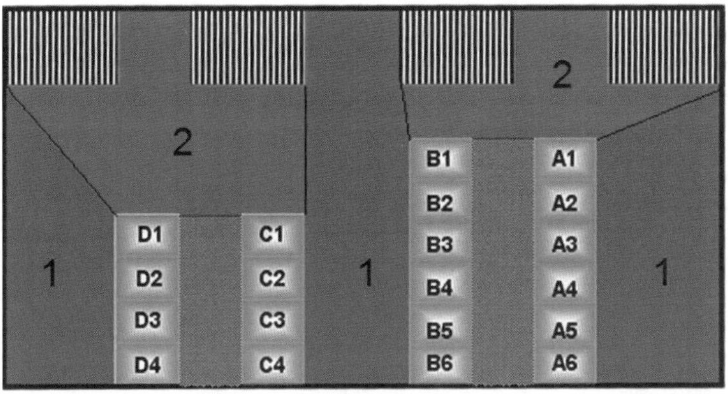


**Fig. 8.18** Plan view of the data center highlighting critical and safe flow paths

#### 8.4.5.3 Concept of Critical and Safe Flow Paths

Bhopte et al., based on parametric study of blockage size and locations under plenum, have defined the concept of safe and critical paths. Figure 8.18 highlights safe and critical paths under plenum for the defined data center configuration.

Zone 1 in Fig. 8.18 is considered the "*Safe*" path which is defined as the plenum region which does not directly come in the path between CRAC and cold aisle. Zone 2 is considered the "*Critical*" path which is defined as the plenum region which lies directly in the path between CRAC and cold aisle. Using this concept of safe and critical paths, guidelines are presented on defining safe chiller pipe layouts and on managing randomly lying blockages such as cables, wires, tubes.

#### 8.4.5.4 Guidelines for Installing Chiller Pipes

As mentioned previously, plenums are also used to route chiller pipes. These pipes, if installed in a critical path, may reduce the CRAC flow rates considerably. Such pipes if run under perforated tiles can correspondingly reduce their flow rates and can create additional hot spots above the raised floor. Hence in this section, piping patterns, closer to real world, are investigated to come up with a layout or pattern which will be least adverse on data center performance. In some cases, such safe piping patterns may even have a positive effect on data center performance.

Chiller pipes, run under the plenum either parallel (termed as parallel piping pattern) or perpendicular (termed as perpendicular piping pattern) to CRAC flow. Figures 8.19 and 8.20 show parallel and perpendicular pipes in critical flow paths. The following numerical study is presented to investigate the impact of chiller pipes on tile flow distribution and rack inlet temperatures. Two perpendicular pipes are assumed to be running under the plenum. Two parametric locations for the pipes are considered in Fig. 8.21. First case has pipes in the critical flow path, while the

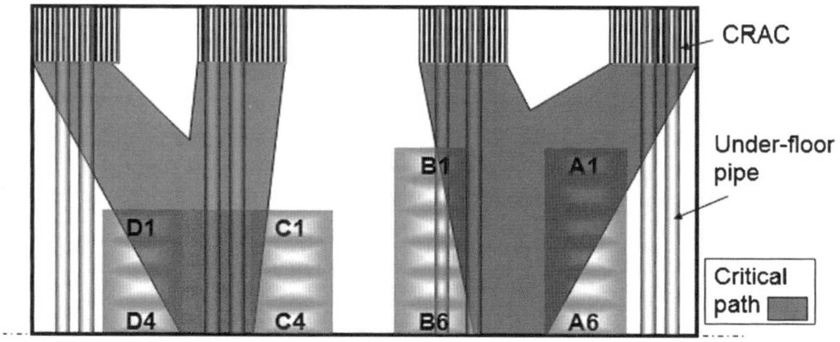


**Fig. 8.19** Parallel pipes in critical flow path

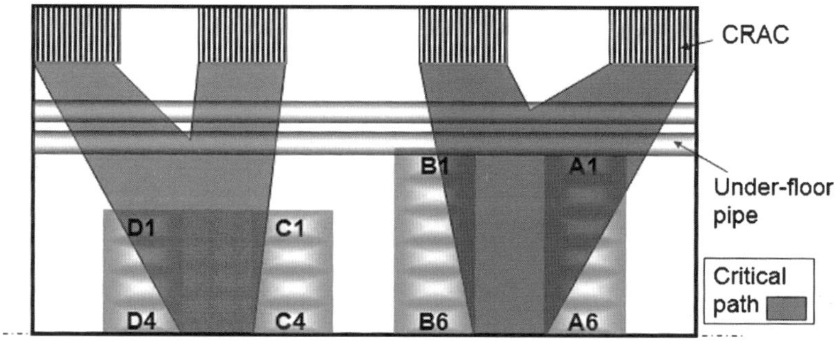


**Fig. 8.20** Perpendicular pipes in critical flow path

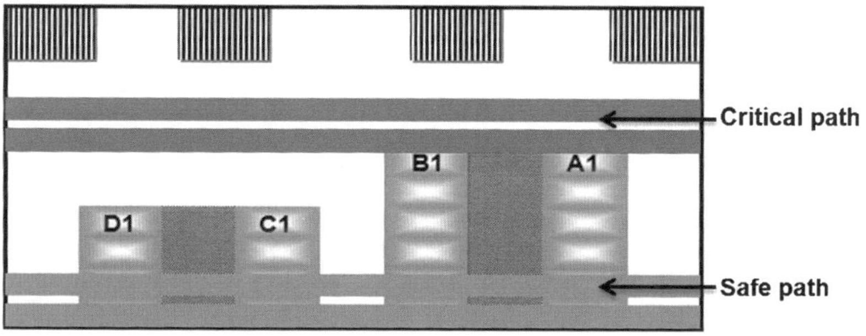


**Fig. 8.21** Two parametric locations for one pair of perpendicular chiller pipes

second case has pipes in the safe flow path. Pipe diameters are very close to typical data center pipes with a diameter of 0.15 m.

Detailed parametric study on sizes and locations of parallel and perpendicular pipes can be found in references [63]. In general, perpendicular piping pattern has a more detrimental impact on data center performance than parallel piping pattern and should be avoided. Putting the perpendicular pipes in the vicinity of the CRAC

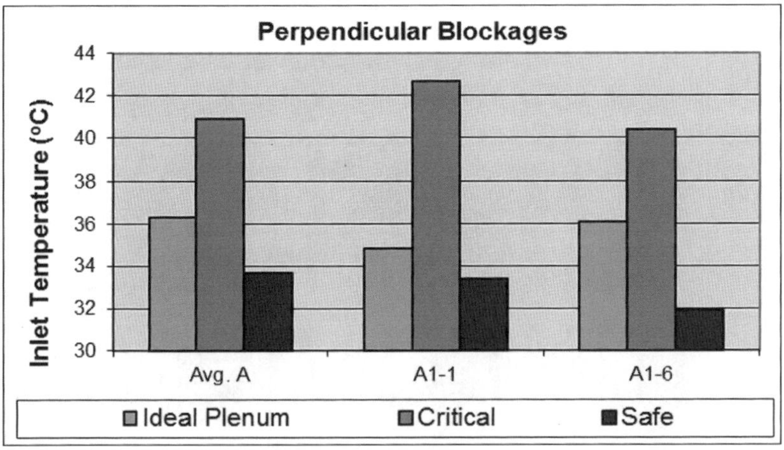


**Fig. 8.22** Thermal performance comparison for perpendicular piping pattern

units considerably increases the local pressure, because of which CRAC units supply lower flow rate.

Thermal comparison between ideal plenum with no blockages and with blockages in two parametric locations is presented in Fig. 8.22. Putting the pipes in critical path reduces the total CRAC flow rate by ~14%. Due to the significantly reduced CRAC flow rates, over heating of servers is possible. On the other hand, pipes in safe path reduce the total CRAC flow rates by ~1%. Bhopte et al. discussed a numerical case study showing much higher tile flow rates for the tiles near symmetry wall (A6, B6, C4, and D4) compared to edge tiles (A1, B1, C1, and D1). Due to this maldistribution of air, higher air inlet temperatures exist for edge server racks. Routing the chiller pipes under tiles A6, B6, C4, and D4 reduces their flow rates (refer to Fig. 8.23). Correspondingly, flow rates of tiles A1, B1, C1, and D1 are increased, thereby reducing their inlet air temperatures significantly without compromising inlet air temperatures of racks A6, B6, C4, and D4. Hence pipes in safe locations are shown to improve the data center performance even with 1% reduction in CRAC flow rate. This reiterates the fact that blockages, if installed with prior consideration, may help in improving the data center performance without changing the cold air supply.

Based on the case study presented, safe parallel and perpendicular piping patterns are proposed for data center layouts. Figures 8.24 and 8.25 show the safe piping patterns for the data center layout considered.

Figure 8.26 is an example of complex data center room layout showing highlighted critical paths and possible safe piping patterns. This example shows that by using the guidelines on installing the chiller pipes, a data center facility engineer can easily decide on routing the chiller pipes in safe flow paths without doing the thermal analyses.

Blockages (refer to Fig. 8.17) such as server rack cables, optical cables, power supply cables randomly lie under the plenum and make the air flow more

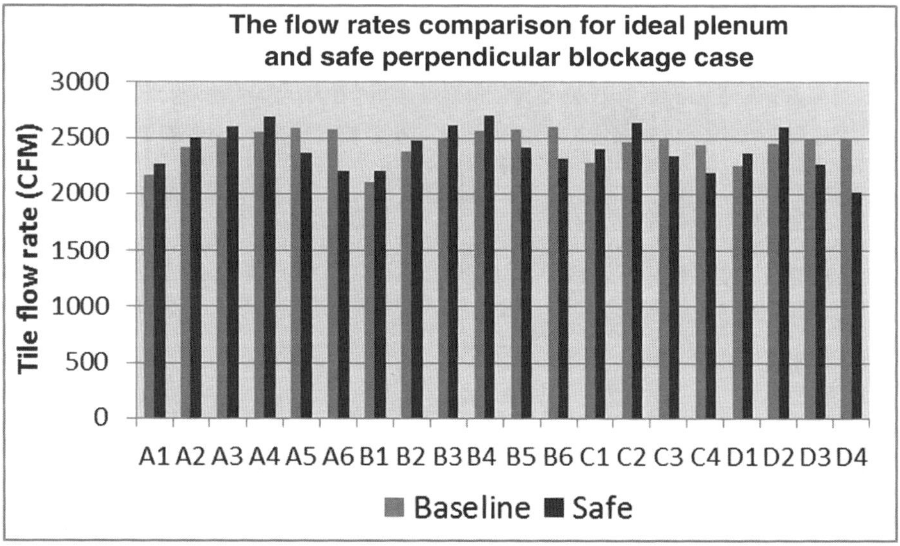


**Fig. 8.23** Tile flow rates for baseline case and safe perpendicular blockage case

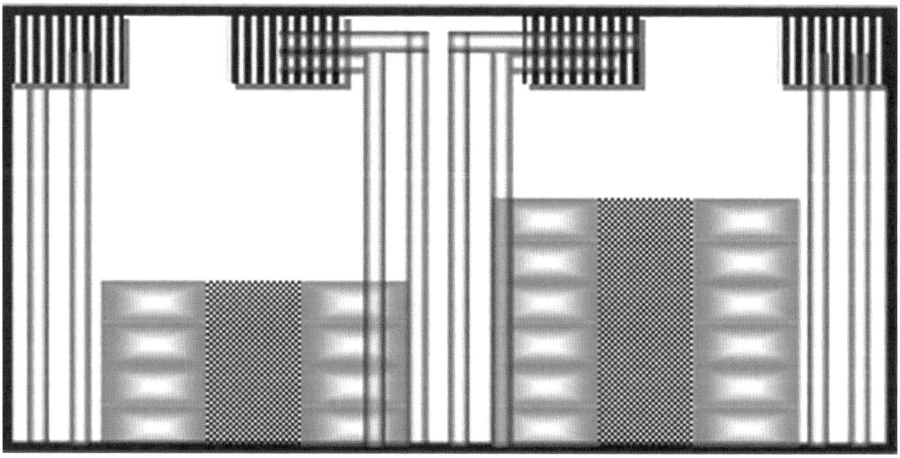

**Fig. 8.24** Recommended parallel chiller piping pattern

unpredictable. If they lie under any perforated tile, they may offer substantial resistance to flow and reduce the corresponding tile flow rate and possibly even the CRAC flow rate. Bhopte et al. have presented a numerical case study where 100% blocked regions in safe path were shown to have no adverse impact on under-floor blockages. Such regions can potentially be used to stack such cables and wires. Overall, guidelines on managing under-floor blockages are summarized in Fig. 8.27.

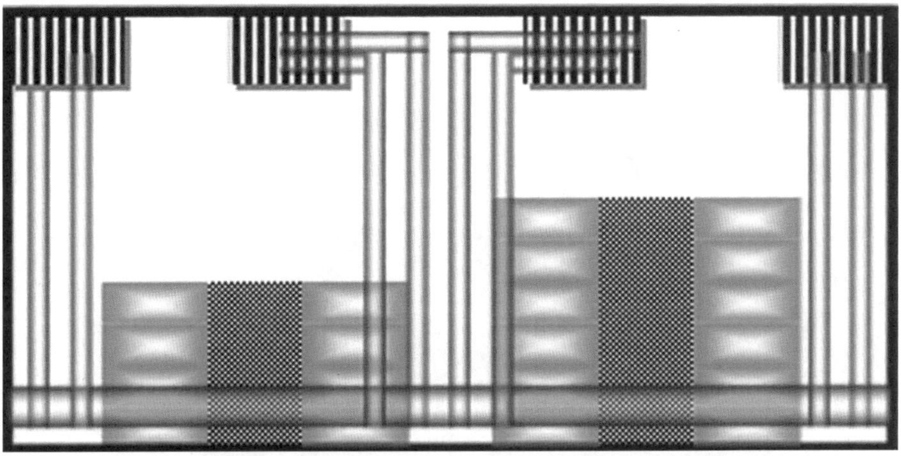

Fig. 8.25 Recommended perpendicular chiller piping pattern

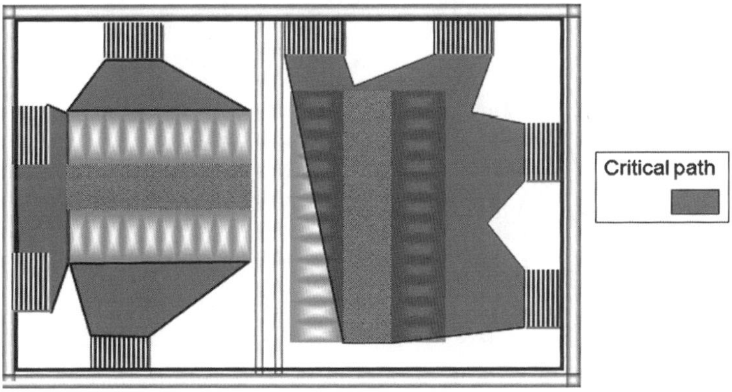


Fig. 8.26 Example of a complex data center room with highlighted critical paths and possible safe piping pattern

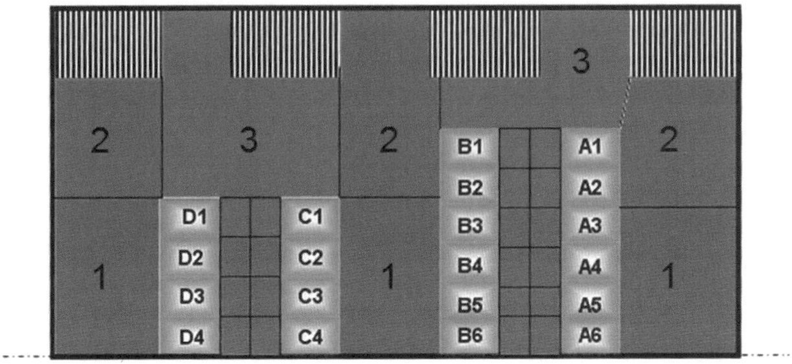


Fig. 8.27 Plenum color code demonstrating broad guidelines on managing under floor blockages for improved data center performance

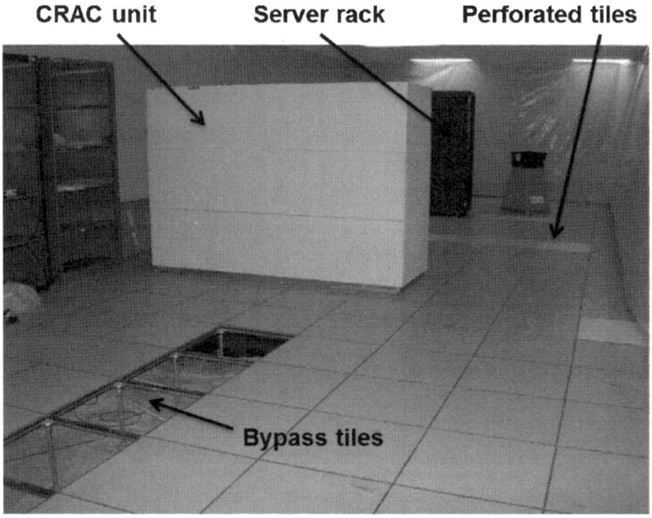


**Fig. 8.28** Test facility at Poughkeepsie, New York

Zone 1 refers to the zone of safe path where potentially 100% blocked regions can be created. Such area should be used to stack unruly cables, wires, etc. Zone 2 refers to a less critical path. This path can be used for routing chiller pipes, ducts, and tubes. These blockages placed here will not have that adverse impact as they would have, if they are lying in critical path. Zone 3 refers to the most critical path. This zone must always remain blockage free for optimal thermal performance.

Numerically developed guidelines are experimentally validated on a different data center layout. The measurements were performed in a raised-floor data center at the IBM facility in Poughkeepsie, NY. The raised-floor height from the subfloor to the bottom of tiles is 16.5 in. The size of the test area is 50 $\times$ 18 ft. over a 25 $\times$ 9 tiles region. One CRAC unit provides airflow to the test area. Experimentally determined flow rate of the CRAC unit in this data center was 10,200 CFM. The test area also has one thermal simulator, with perforated tiles in front of it. Figure 8.28 shows a photograph of the test facility showing open tiles, A/C unit, perforated tiles, and the thermal simulator.

In this study, thermal performance of the simulator is not addressed. A combination of 16 and 30% open perforated tiles are placed along layout. Chilled water pipes under the raised floor are included in the numerical model. Effects of perimeter and distributed leakage flow from the plenum and tiles are included in the model. Complete method of mass balance used for the numerical models can be found in reference [63]. Figure 8.29 shows the plan view of the layout with the existing blockages (shown in gray).

To experimentally verify the established guidelines for the placement of underfloor structures (piping, etc.), additional blockage is installed alternately in critical flow paths (between CRAC unit and tiles R1–R6) and safe flow path (behind tiles

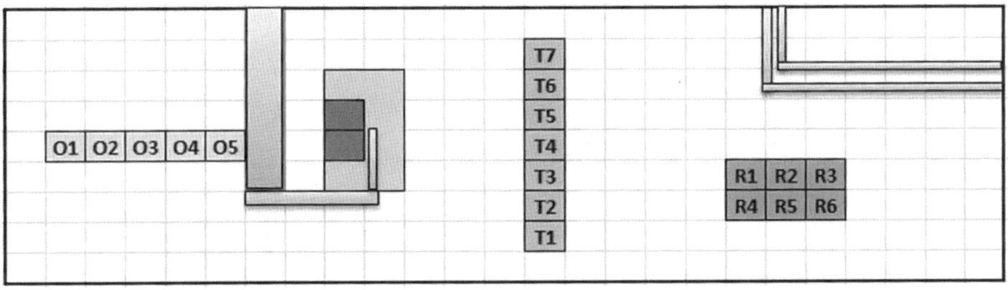


**Fig. 8.29** Layout with six 16% open (R1–R6) supply tiles. Seven 30% open tiles (T1–T7) are located between CRAC and supply tiles. Behind CRAC are five fully open tiles (O1–O5)

**Fig. 8.30** Flexible plastic pipes, similar to CRAC pipes, are additionally installed alternately in critical and safe flow paths

R1–R6). These flexible plastic pipes (refer to Fig. 8.30) are similar to CRAC pipes and blocked approximately 80% plenum height.

Figure 8.31 summarizes two scenarios considered in one figure. Scenario (a) has additional pipes behind the tiles T1–T7. Scenario (b) has additional pipes near the right wall. However, the two lines on the figure are just to show the locations of the two scenarios. It must not be confused with putting additional blockages at those locations at the same time.

For the scenarios mentioned above, air flow rates coming out of each of the 18 tiles (O1–O5, T1–T7, and R1–R6) are measured, and the results are compared with numerical predictions in Figs. 8.32 and 8.33. Overall there is a fairly good agreement for all the three cases presented above. This confirms that a careful numerical model can help in understanding and solving many problems of data center air delivery.

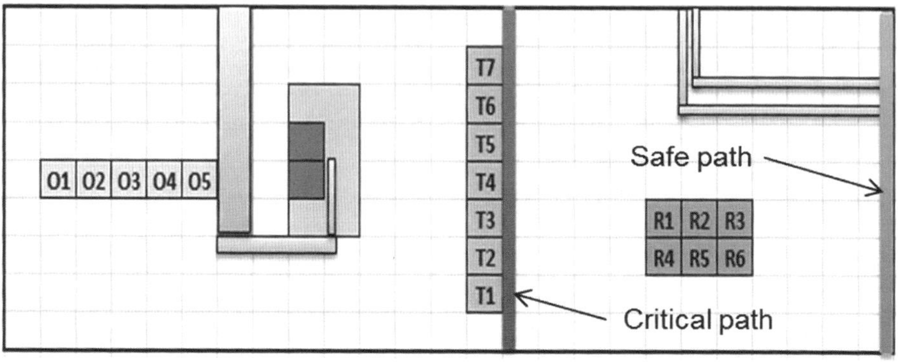


**Fig. 8.31** Layout showing additional pipe blockages in critical path and safe path

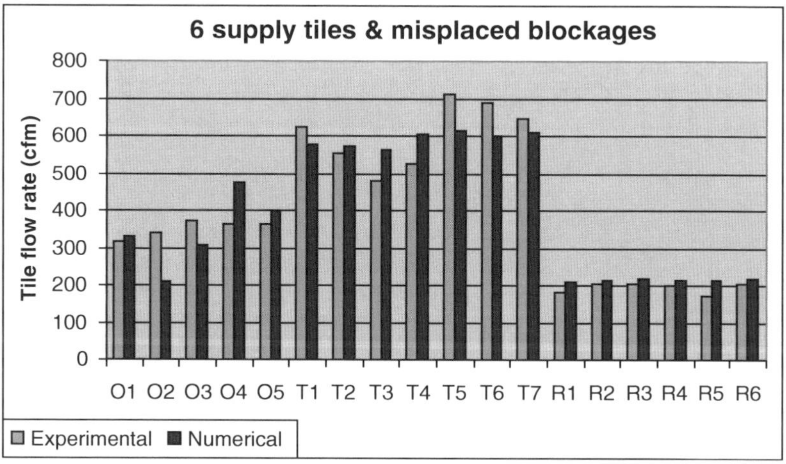


**Fig. 8.32** Comparison between experimental and numerical results for layout 1 with additional blockage in critical flow path

On average, the agreement between experimental and numerical results is around 12%. Additional blockage has significantly adverse impact on the air flow rate of the supply tiles (R1–R6). Average experimental tile flow rate for the six supply tiles (R1–R6) is 2.7E−11 $m^3/s$ (~240 CFM). This flow rate became 2.2E−11 $m^3/s$ (~195 CFM) when the blockage was put in critical flow path (under tiles T1–T7), with a reduction of ~19%. Due to this reduction, an increase in flow rate for tiles T1–T7 seen. In a fully operational data center, if flow rate of supply tiles is reduced by 19%, over heating of servers is inevitable. On the other hand, when the blockage was installed in safe flow path, the average supply tile flow rate becomes 2.63E−11 $m^3/s$ (~232 CFM) with a reduction of just 3%. Hence blockages in safe path are shown to have minimal impact on flow rate

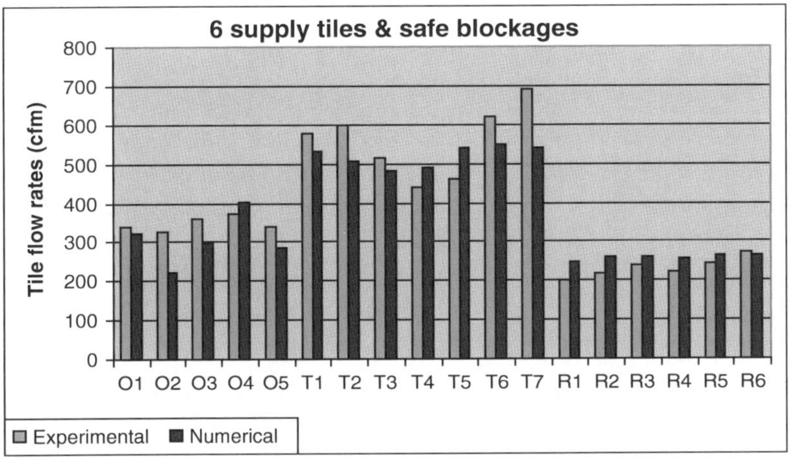


**Fig. 8.33** Comparison between experimental and numerical results for layout 1 with additional blockage in safe flow path

through supply tiles. Therefore, numerical modeling is shown to successfully address the problem of adverse effect under-floor blockages on data center performance.

## 8.5 Conclusions and Future Projections

As pointed out in earlier chapters, data center power consumption is already very high and is continuing to increase at a fairly robust level. This trend will continue over time due to the ever increasing demand for applications that drive data processing, storage, and retrieval. The trends are not sustainable from many perspectives, including cost and notably environmental impact. It should also be noted that the efficiency in terms of total energy consumed compared to energy consumed to actually process information is relatively very low for a variety of reasons discussed in detail in other chapters. This is particularly true for legacy data centers which may have been designed originally to operate at a specific efficiency and mode, but are repeatedly upgraded and changed over time as equipment gets obsolete. In many cases, the data centers are operating at an unknown efficiency and may be very wasteful of energy.

It is therefore imperative to strive to operate data centers as dynamic systems that are operated according to best known design principles in order to approach optimal energy consumption at any specified performance level. The system efficiency is directly tied to the workload distribution in the data center. Running the data center at low workload levels is known to result in low efficiency. This is generally true for all data centers whether they are legacy or new. There are also circumstances wherein performance and reliability are the main concern of the user. However, it should always be the objective of the operator to run the data center at near optimal conditions for the specified performance set by the users. This is however easier said than done. The complexity of a data center operation stems from its size, inherent multiscale nature, the variations in power consumption due to workload variations, the challenges in collecting and processing useful data (such as temperature, humidity, air velocity, and pressure) during operations, and quite often the hybrid nature of multicooling modes used.

The objective of running data centers at near optimal energy efficiency will require real-time models and physical measurements on a continuous basis. The models must be dynamic, predictive, and verified and must be holistic combining workload distributions with energy efficiency and thermal response. Accurate predictions of optimal workload distribution, energy efficiency, and thermal response are required. The data centers must also have a control system that is programmable and predictable. The control system will allow the operator (or operating system) to select the right workload distribution along with the correct thermal management solution at all times. Numerical methods will play a major role in building data centers that are optimally designed. As shown earlier in this chapter, it is possible to accurately predict temperature and velocity distributions in a data center using full physics-based numerical solutions. However, such solutions are time-consuming and cannot be run in real time. It is therefore necessary to develop solutions that are reduced order or compact models that are capable of producing fairly accurate solutions in a fraction of the time. It is also possible to use neural networking algorithms that are trained using full-scale, physics-based solutions.

Using approximate solutions, compact solutions or neural network solutions in order to simulate the thermal response of a data center operating under dynamic conditions in real time carry the potential risk of errors in the analysis. Such errors are inevitable because the number of variables to be considered in a full-scale data center is large. The number of permutations required in order to fully train a neural network is unattainable. It is therefore necessary to couple the real-time models to dynamic live sensing of key variables during the operation of the data center. In most cases, the real-time models will predict the data center accurately and will allow near optimal performance from an energy standpoint. If the models deviate from the measured temperatures, then the operating system would revert, temporarily, to a conservative predetermined cooling mode until the models recover and are back to tracking temperatures with sufficient accuracy.

## References

1. Kundu PK, Cohen IM (2008) Fluid mechanics. Elsevier, Burlington
2. Schmidt R, Shaukatullah H (2003) Computer and telecommunications equipment room cooling: a review of literature. IEEE Trans Compon Packag Technol 26:89–98

3. Rambo J, Joshi Y (2007) Modeling of data center airflow and heat transfer: state of the art and future trends. Distrib Parallel Databases 21(2):193–225
4. Kang S, Schmidt R, Kelkar KM, Radmehr A, Patankar S (2001) A methodology for the design of perforated tiles in raised floor data centers using computational flow analysis. IEEE Trans Compon Packag Technol 24(2):177–183
5. Kang S, Schmidt R, Kelkar K, Patankar S (2001) A methodology for the design of perforated tiles in raised floor data centers using computational flow analysis. IEEE-CPMT J 24:177–183
6. Karki K, Radmehr A, Patankar S (2003) Use of computational fluid dynamics for calculating flow rates through perforated tiles in raised-floor data centers. Int J Heat Vent Air-Conditioning Refrig Res 9(2):153–166
7. Innovative Research, Inc. (2010) TileFlow: a simulation tool for airflow distribution in raised-floor data centers. Tile Flow Software, Innovative Research Inc, 3025 Harbor Lane N., Suite 300, Plymouth, MN 55447, USA
8. Schmidt R, Karki K, Kelkar K, Radmehr A, Patankar S (2001) Measurements and predictions of the flow distribution through perforated tiles in raised-floor data centers. In: Proceedings of the Pacific Rim/ASME international electronic packaging technical conference and exhibition (IPACK'01), Kauai, Hawaii
9. Karki K, Patankar S (2003) Techniques for controlling airflow distribution in raised-floor data centers. In: Proceedings of the Pacific Rim/ASME international electronic packaging technical conference and exhibition (IPACK'03), Maui, Hawaii
10. Patankar S, Karki K (2004) Distribution of cooling airflow in a raised-floor data center. ASHRAE Trans 110(2):629–635
11. Rambo J, Nelson G, Joshi Y (2007) Airflow distribution through perforated tiles in close proximity to computer room air conditioning units. ASHRAE Trans 113(2):124–135
12. Schmidt R, Karki K, Patankar S (2004) Raised-floor data center: perforated tile flow rates for various tile layouts. In: Proceedings of the eighth intersociety conference on thermal and thermo-mechanical phenomena in electronic systems (ITHERM), Las Vegas, NV
13. Kumar P, Joshi Y (2010) Experimental investigation on the effect of perforated tile air jet velocity on server air distribution in a high density data center. In: Proceedings of the inter society conference on thermal phenomena (ITherm), Las Vegas, NV
14. Radmehr A, Schmidt R, Karki K, Patankar S (2005) Distributed leakage flow in raised-floor data centers. In: Proceedings of the Pacific Rim/ASME international electronic packaging technical conference and exhibition (IPACK'05), San Francisco, CA
15. Karki K, Radmehr A, Patankar S (2007) Prediction of distributed air leakage in raised-floor data centers. ASHRAE Trans 113(1):219–226
16. Abdelmaksoud W, Khalifa HE, Dang T, Iyengar M, Schmidt R (2010) Experimental and computational study of perforated floor tile in data centers. In: Proceedings of the inter society conference on thermal phenomena (ITherm), Las Vegas, NV
17. Abdelmaksoud W, Khalifa HE, Dang T, Iyengar M, Schmidt R (2010) Improved CFD modeling of a small data center test cell. In: Proceedings of the inter society conference on thermal phenomena (ITherm), Las Vegas, NV
18. Schmidt R (2001) Effect of data center characteristics on data processing equipment inlet temperatures, Advances in Electronic Packaging 2001. In: Proceedings of the Pacific Rim/ASME international electronic packaging technical conference and exhibition (IPACK'01), Kauai, Hawaii, vol 2. pp 1097–1106
19. Schmidt R, Cruz E (2002) Raised floor computer data center: effect on rack inlet temperatures of chilled air exiting both the hot and cold aisles. In: Proceedings of the eighth intersociety conference on thermal and thermo-mechanical phenomena in electronic systems (ITHERM), San Diego, CA, pp 580–594
20. Schmidt R, Cruz E (2002) Raised floor computer data center: effect on rack inlet temperatures when high powered racks are situated amongst lower powered racks. In: IMECE conference, New Orleans, LA, pp 297–309

21. Schmidt R, Cruz E (2003) Raised floor computer data center: effect on rack inlet temperatures when adjacent racks are removed. In: Proceedings of the Pacific Rim/ASME international electronic packaging technical conference and exhibition (IPACK'03), Maui, Hawaii, pp 481–493
22. Schmidt R, Cruz E (2003) Raised floor computer data center: effect on rack inlet temperatures when rack flowrates are reduced. In: Proceedings of the Pacific Rim/ASME international electronic packaging technical conference and exhibition (IPACK'03), Maui, Hawaii, pp 495–508
23. Schmidt R (2004) Thermal profile of a high density data center-methodology to thermally characterize a data center. In: Proceedings of the ASHRAE Nashville conference, pp 604–611.
24. Schmidt R, Cruz E (2004) Cluster of high-powered racks within a raised-floor computer data center: effect of perforated tile flow distribution on rack inlet air temperatures. ASME J Electron Packag 126(24):510–518
25. Patel CD, Sharma RK, Bash CE, Beitelmal A (2002) Thermal considerations in cooling large scale high computer density data centers. In: Proceedings of the eighth intersociety conference on thermal and thermo-mechanical phenomena in electronic systems (ITHERM), San Diego, CA
26. Gondipalli S, Bhopte S, Sammakia B, Iyengar M, Schmidt R (2008) Effect of isolating cold aisles on rack inlet temperatures. In: IEEE ITHERM, Orlando, FL
27. Gondipalli S, Bhopte S, Sammakia B, Murray B, Iyengar M, Schmidt R (2009) Optimization of cold aisle isolation designs for a data center with roofs and doors using slits. ASME InterPACK, San Francisco, CA
28. Nakao M, Hayama H, Nishioka M (1991) Which cooling air supply system is better for a high heat density room: underfloor or overhead. In: Proceedings of the international telecommunications energy conference (INTELEC), Kyoto, Japan, pp 393–400
29. Noh H, Song K, Chun SK (1998) The cooling characteristic on the air supply and return flow system in the telecommunication cabinet room. In: Proceedings of the international telecommunications energy conference (INTELEC), San Francisco, CA, pp 777–784
30. Patel CD, Bash CE, Belady C, Stahl L, Sullivan D (2001) Computational fluid dynamics modeling of high compute density data centers to assure system inlet air specification. In: Proceedings of the Pacific Rim/ASME international electronic packaging technical conference and exhibition (InterPACK), Kauai, HI
31. Shrivastava SK, Sammakia B, Schmidt R, Iyengar M (2005) Comparative analysis of different data center airflow management configurations. In: Proceedings of InterPACK, 17–22 July, San Francisco, CA
32. Herrlin MK, Belady C (2006) Gravity-assisted air mixing in data centers and how it affects the rack cooling effectiveness. In: Proceedings of the tenth intersociety conference on thermal and thermo-mechanical phenomena in electronic systems (ITHERM), San Diego, CA, pp 434–438
33. Schmidt R, Iyengar M (2007) Comparison between underfloor supply and overhead supply ventilation designs for data center high-density clusters. ASHRAE Trans 113(1):115–125
34. Rambo J, Joshi Y (2006) Convective transport processes in data centers. Num Heat Transf A Appl 49(10):923–945
35. Rambo J, Joshi Y (2006) Thermal modeling of technology infrastructure facilities: a case study of data centers. In: Minkowycz WJ, Sparrow EM, Murthy JY (eds) Handbook of numerical heat transfer, vol II. Taylor and Francis, New York, pp 821–849
36. Sharma RK, Bash CE, Patel CD (2002) Dimensionless parameters for evaluation of thermal design and performance of large-scale data centers. In: Proceedings of the eighth ASME/AIAA joint thermophysics and heat transfer conference, St. Louis, MO
37. Sharma RK, Bash CE (2002) Dimensionless parameters for energy efficient data center design. In: Proceedings of the IMAPS advanced technology workshop on thermal management (THERM ATW), Palo Alto, CA

38. Escobar S, Sharma R (2008) Data center characteristic temperature signatures and SHI correlation to nondimensional parameters. In: Proceedings of the ninth intersociety conference on thermal and thermo-mechanical phenomena in electronic systems (ITHERM), Orlando, FL, pp 1203–1209
39. Schmidt R, Cruz E, Iyengar M (2005) Challenges of data center thermal management. IBM J Res Develop 49:709–723
40. Norota M, Hayama H, Enai M, Mori T, Kishita M (2003) Research on efficiency of air conditioning system for data center. Presented at INTELEC'03 – 25th International telecommunications energy conference, Yokohama, Japan, pp 147–151
41. Malone C, Belady C (2006) Metrics to characterize data center & IT equipment energy use. In: Proceedings of the 2006 digital power forum, Richardson, TX
42. Belady C, Malone C (2007) Metrics and infrastructure model to evaluate data center efficiency. In: Proceedings of the Pacific Rim/ASME international electronic packaging technical conference and exhibition (InterPACK), Vancouver, British Columbia, Canada
43. Tozer R, Salim M (2010) Data center air management metrics – practical approach. In: Proceedings of the inter society conference on thermal phenomena (ITherm), Las Vegas, NV
44. Zhang X, VanGilder J, Iyengar M, Schmidt RR (2008) Effect of rack modeling detail on the numerical results of a data center test cell. In: Proceedings of the inter society conference on thermal phenomena (ITherm), 28–31 May, Orlando, FL
45. Rambo J, Joshi Y (2003) Multi-scale modeling of high power density data centers. In: Proceedings of the Pacific Rim/ASME international electronic packaging technical conference and exhibition (IPACK'03), Maui, Hawaii, USA
46. Rambo J, Joshi Y (2005) Thermal performance metrics for arranging forced air cooled servers in a data processing cabinet. ASME J Electron Packag 127:452–459
47. Herrlin M (2005) Rack cooling effectiveness in data centers and telecom central offices: the Rack cooling index (RCI). ASHRAE Trans 111(2):725–731
48. Rolander N, Rambo J, Joshi Y, Mistree F (2005) Towards sustainable design of data centers: addressing the lifecycle mismatch problem. Presented at IPACK'05 – international electronic packaging technical conference and exhibition, San Francisco, CA
49. Rolander N, Rambo J, Joshi Y, Mistree F, Allen JK (2006) Robust design of turbulent convective systems using the proper orthogonal decomposition. ASME J 128:844–855
50. Rambo J, Joshi Y (2005) Reduced order modeling of steady turbulent flows using the POD. Presented at ASME summer heat transfer conference, San Francisco, CA
51. Stahl L, Belady C (2001) Designing an alternative to conventional room cooling. In: Proceedings of international telecommunications energy conference, pp 109–115
52. Beitelmal A, Patel CD (2004) Thermo-fluids provisioning of a high performance high density data center. Technical Report No. HPL-2004-146(R.1), Hewlett Packard Laboratories, Palo Alto CA
53. Ibrahim M, Bhopte S, Sammakia B, Iyengar M, Schmidt R (2010) Effect of thermal characteristics of electronic enclosures on dynamic data center performance. In: IMECE conference, Vancouver, Canada
54. Ibrahim M, Gondipalli S, Bhopte S, Sammakia B, Murray B, Ghose K, Iyengar M, Schmidt R (2010) Numerical modeling approach to dynamic data center cooling. In: Proceedings of the intersociety conference on thermal phenomena (ITHERM), Las Vegas, USA
55. Sharma RK, Bash CE, Patel CD, Friedrich RJ, Chase JS (2003) Balance of power: dynamic thermal management of internet data centers. Technical Report No. HPL-2003-5, Hewlett Packard Laboratories, Palo Alto, CA
56. Patel CD, Bash CE, Sharma RK, Beitelmal A, Friedrich RJ (2003) Smart cooling of data centers. In: Proceedings of the Pacific Rim/ASME international electronic packaging technical conference and exhibition (InterPACK), Kauai, HI
57. Bash C, Patel C, Sharma K (2006) Dynamic thermal management of air cooled data centers. In: Proceedings of the tenth intersociety conference on thermal and thermo-mechanical phenomena in electronic systems (ITHERM), San Diego, CA, pp 445–452

58. Boucher TD, Auslander DM, Bash CE, Federspiel CC, Patel CD (2004) Viability of dynamic cooling control in a data center environment. In: Proceedings of the ninth intersociety conference on thermal and thermo-mechanical phenomena in electronic systems (ITHERM), San Diego, CA
59. Iyengar M, Schmidt R, Hamann H, VanGilder J (2007) Comparison between numerical and experimental temperature distributions in a small data center test cell. In: Proceedings of the Pacific Rim/ASME international electronic packaging technical conference and exhibition (IPACK'07), Vancouver, Canada, pp 819–826
60. Bhopte S, Sammakia B, Iyengar M, Schmidt R, Agonafer D (2006) Effect of under floor blockages on data center performance. In: Proceeding of IEEE ITHERM, San Diego, CA
61. Bhopte S, Sammakia B, Iyengar M, Schmidt R, Agonafer D (2007) Numerical modeling of data center clusters – impact of model complexity. In: Proccedings of ASME IMECE, Chicago, IL
62. Bhopte S, Sammakia B, Iyengar M, Schmidt R (2007) Guidelines on managing under floor blockages for improved data center performance. In: Proceedings of ASME IMECE, Chicago, IL
63. Bhopte S, Sammakia B, Iyengar M, Schmidt R (2007) Experimental investigation of the impact of under floor blockages on flow distribution in a data center cell. In: Proceedings of the Pacific Rim/ASME international electronic packaging technical conference (InterPACK), Vancouver, Canada
64. Cruz E, Joshi Y, Iyengar M, Schmidt R (2009) Comparison of numerical modeling to experimental data in a small data center test cell. In: Proceedings of the Pacific Rim/ASME international electronic packaging technical conference (InterPACK), July, Paper number IPACK2009-89306
65. Cruz E, Joshi Y, Iyengar M, Schmidt R (2009) Comparison of numerical modeling to experimental data in a small, low power data center test cell. In: Proceedings of ASME IMECE conference, November, Paper number IMECE2009-12860
66. Samadiani E, Rambo J, Joshi Y (2010) Numerical modeling of perforated tile flow distribution in a raised-floor data center. J Electron Packag 132(2):021002 (8 pp)
67. Schmidt R, Iyengar M, Caricari J (2010) Data center housing high performance supercomputer cluster: above floor thermal measurements compared to CFD analysis. J Electron Packag 132(2):021009 (8 pp)
68. Karki K, Patankar S (2006) Airflow distribution through perforated tiles in raised-floor data centers. Build Environ 41(6):734–744
69. Lopez V, Hamann HF (2010) Measurement-based modeling for data centers. In: Proceedings of the inter society conference on thermal phenomena (ITherm), Las Vegas, NV
70. Hamann HF, Lopez V, Stepanchuk A (2010) Thermal zones for more efficient data center energy management. In: Proceedings of the inter society conference on thermal phenomena (ITherm), Las Vegas, NV
71. VanGilder JW, Shrivastava SK (2006) Real-time prediction of rack-cooling performance. ASHRAE Trans 112(2):151–162
72. Toulouse M, Doljac G, Bash C (2009) Exploration of a potential-flow-based compact model of air-flow transport in data centers. In: ASME International Mechanical Engineering Congress & Exposition, Technical Publication
73. Samadiani E, Joshi Y, Mistree F (2008) The thermal design of a next generation data center: a conceptual exposition. J Electron Packag 134(4):041104 (8 pp)
74. Rambo J (2006) Reduced-order modeling of multiscale turbulent convection: application to data center thermal management. Ph.D. Thesis, Department of Mechanical Engineering, Georgia Institute of Technology
75. Rambo J, Joshi Y (2003) Physical models in data center airflow simulations. In: Proceedings of IMECE'03 – ASME international mechanical engineering congress and R&D exposition, Washington, DC

76. Somani A, Joshi Y (2009) Data center cooling optimization – ambient intelligence based load management (AILM). In: Proceedings of the ASME heat transfer summer conference, San Francisco, CA
77. Shrivastava SK, Sammakia B, Schmidt R, Iyengar M (2005) Significance level of factors for different airflow management configurations of data centers. In: Proceedings of ASME IMECE, 5–11 Nov, Orlando, FL
78. Shrivastava S, Iyengar M, Sammakia B, Schmidt R, VanGilder JW (2009) Experimental-numerical comparison for a high-density data center: hot spot heat fluxes in excess of 500 W/ft$^2$. IEEE Trans Compon Packag Technol 32(1):166–172
79. Shrivastava SK, VanGilder JW, Sammakia B (2006) A statistical prediction of cold aisle end airflow boundary conditions. In: Proceedings of IEEE ITHERM 2006, San Diego, CA
80. Shrivastava SK, VanGilder JW, Sammakia B (2007) Prediction of cold aisle end airflow boundary conditions using regression modeling. IEEE Trans Compon Packag Technol 30(4):866–874
81. VanGilder JW, Zhang X, Shrivastava SK (2007) Partially decoupled aisle method for estimating rack cooling performance in near real time. In: Proceedings of ASME InterPACK 2007, 8–12 July, Vancouver, BC, Canada
82. Shrivastava SK, VanGilder JW, Sammakia BG (2007) Data center cooling prediction using artificial neural network. In: Proceedings of ASME InterPack, 8–12 July, Vancouver, BC, Canada
83. Shrivastava SK, VanGilder JW, Sammakia BG (2008) Optimization of cluster cooling performance of data centers. In: Proceedings of IEEE ITHERM, Orlando, FL
84. Marwah M, Sharma R, Bash C (2010) Thermal anomaly prediction in data centers. In: Proceedings of the inter society conference on thermal phenomena (ITherm), Las Vegas, NV
85. Iyengar M, Schmidt R, Caricari J (2010) Reducing energy usage in data canters through control of room air conditioning units. In: Proceedings of the inter society conference on thermal phenomena (ITherm), Las Vegas, NV
86. Breem T, Walsh E, Punch J, Shah A, Bash C (2010) From chip to cooling tower data center modeling: part I influence of server inlet temperature and temperature rise across cabinet. In: Proceedings of the inter society conference on thermal phenomena (ITherm), Las Vegas, NV
87. Walsh E, Breem T, Punch J, Shah A, Bash C (2010) From chip to cooling tower data center modeling: part II influence of chip temperature control philosophy. In: Proceedings of the inter society conference on thermal phenomena (ITherm), Las Vegas, NV
88. Scofield C, Weaver T (2008) Data center cooling: using wet-bulb economizers. ASHRAE J 50(8):52–54, 56–58
89. Munther S (2009) Energy in data centers: benchmarking and lessons learned. Eng Syst 26(4):24–32
90. Scofield C, Weaver T, Dunnavant K, Fisher M (2009) Reduce data center cooling cost by 75%. Eng Syst 24(6):34–41
91. Judge J, Pouchet J, Ekbote A, Dixit S (2008) Reducing data center energy consumption. ASHRAE J 50(11):14–26
92. Shah A, Carey V, Bash C, Patel C (2006) An exergy-based figure-of-merit for electronic packages. ASME J Electron Packag 128(4):360–369
93. Shah A, Carey V, Bash C, Patel C (2008) Exergy analysis of data center thermal management systems. ASME J Heat Transf 130(2):021401 (1–9)
94. McAllister S, Carey V, Shah A, Bash C, Patel C (2008) Strategies for effective use of exergy-based modeling of data center thermal managements systems. Microelectron J 39(7):1023–1029
95. Bash C, Shih R, Shah A, Patel C (2010) Data center damage boundaries. In: Proceedings of the inter society conference on thermal phenomena (ITherm), Las Vegas, NV
96. Schmidt R, Chu RC, Elssworth M, Iyengar M, Porter D, Kamath V, Lehmann B (2005) Maintaining datacom rack inlet air temperatures with water cooled heat exchangers. In: Proceedings of the Pacific Rim/ASME international electronic packaging technical conference and exhibition (IPACK'05), San Francisco, vol 2, pp 905–914

97. Schmidt R, Iyengar M (2009) Server rack rear door heat exchanger and the new ASHRAE recommended environmental guidelines. In: Proceedings of the Pacific Rim/ASME international electronic packaging technical conference (InterPACK), Paper number IPACK2009-89212.
98. Tsukamoto T, Takayoshi J, Schmidt R, Iyengar M (2009) Refrigeration heat exchanger systems for server rack cooling in data centers. In: Proceedings of the Pacific Rim/ASME international electronic packaging technical conference (InterPACK), Paper number IPACK2009-89258.
99. Kelkar K, Patankar S, Kang S, Iyengar M, Schmidt R (2010) Computational method for generalized analysis of pumped two-phase cooling systems and its applications to a system used in data-center environments. In: Proceedings of the inter society conference on thermal phenomena (ITherm), Las Vegas, NV
100. Kumari N, Bahadur V, Hodes M, Salamon T, Lyons A, Kolodner P, Garimella S (2009) Numerical modeling of mist-cooled high power components in cabinets. In: Proceedings of the Pacific Rim/ASME international electronic packaging technical conference and exhibition (IPACK'09), San Francisco, CA
101. Schmidt R, Iyengar M, Porter D, Weber G, Graybill D, Steffes J (2010) Open side car heat exchanger that removes entire server heat load without and added fan power. In: Proceedings of the inter society conference on thermal phenomena (ITherm), Las Vegas, NV

# Chapter 9
# Exergy Analysis of Data Center Thermal Management Systems*

**Amip J. Shah, Van P. Carey, Cullen E. Bash, Chandrakant D. Patel, and Ratnesh K. Sharma**

**Abstract** Data center thermal management systems exist to maintain the computer equipment within acceptable operating temperatures. As power densities have increased in data centers, however, the energy used by the cooling infrastructure has become a matter of growing concern. Most existing data center thermal management metrics provide information about either the energy efficiency or the thermal state of the data center. There is a gap around a metric that fuses information about each of these goals into a single measure. This chapter addresses this limitation through an exergy analysis of the data center thermal management system. The approach recognizes that the mixing of hot and cold streams in the data center airspace, which is often a primary driver of thermal inefficiency in the data center, is an irreversible process and must therefore lead to the destruction of exergy. Experimental validation in a test data center confirms that such an exergy-based characterization in the cold aisle reflects the same recirculation trends as suggested by traditional temperature-based metrics. Further, by extending the exergy-based model to include irreversibilities from other components of the thermal architecture, it becomes possible to quantify the amount of available energy supplied to the cooling system which is being utilized for thermal management purposes. The energy efficiency of the entire data

---

**Credit*: Portions of this chapter (particularly related to Sects. 9.2, 9.3.1, 9.3.3, and 9.3.4) are reproduced, with permission, from the work of Shah A, Carey V, Patel C, Bash C (2008) Exergy analysis of data center thermal management systems. J Heat Transfer 130(2), Article No. 021401, © American Society of Mechanical Engineers.

A.J. Shah (✉) • C.E. Bash • C.D. Patel • R.K. Sharma
Hewlett Packard Laboratories, 1501 Page Mill Road, M/S 1183, Palo Alto, CA 94304-1126, USA
e-mail: amip.shah@hp.com; cullen.bash@hp.com; chandrakant.patel@hp.com; ratnesh.sharma@hp.com

V.P. Carey
Department of Mechanical Engineering, University of California, Berkeley, CA 94720-1740, USA
e-mail: vcarey@me.berkeley.edu

Y. Joshi and P. Kumar (eds.), *Energy Efficient Thermal Management of Data Centers*, DOI 10.1007/978-1-4419-7124-1_9,

center cooling system can then be collapsed into the single metric of net exergy consumption. When evaluated against a ground state of the external ambience, this metric enables an estimate of how much of the energy emitted into the environment could potentially be harnessed in the form of useful work. The insights availed from the above analysis include a wide range of considerations, such as the viability of workload placement within the data center; the appropriateness of airside economization as well as containment; the potential benefits of reusing waste heat from the data center; as well as the potential to install additional compute capacity without needing to increase the data center cooling capacity. In addition, the analysis provides insight about how local thermal management inefficiencies in the data center can be mitigated. The chapter concludes by suggesting that the proposed exergy-based approach can provide a foundation upon which the data center cooling system can be simultaneously evaluated for thermal manageability and energy efficiency.

## Nomenclature

| | |
|---|---|
| $A$ | Availability or available energy (exergy) [J] |
| $\bar{A}$ | Area [$m^2$] |
| $C$ | Coefficient matrix (9.67) |
| $C_p$ | Specific heat at constant pressure [J/kg K] |
| CFD | Computational fluid dynamics |
| cfm | Cubic feet per minute |
| const | Arbitrary constant value |
| COP | Coefficient of performance defined as the heat removed by a cooling system normalized to the work input to the cooling system for removal of heat |
| CRAC | Computer room air conditioning unit |
| $E$ | Energy stored in system [J] |
| $e$ | Energy transported per unit massflow [J/kg] |
| $f$ | Function |
| $F$ | Matrix defining flow conditions at boundaries (9.67) |
| $g$ | Acceleration due to gravity [$m/s^2$] |
| $H$ | Enthalpy of a system [J] |
| HVAC | Heating, Ventilation, and Air-Conditioning system |
| $h$ | Specific enthalpy (enthalpy per unit mass) [J/kg] |
| $\bar{h}$ | Convection coefficient [$W/m^2K$] |
| $i$ | Counter in summation |
| $J$ | Joule unit of energy |
| KE | Kinetic energy [J] |
| $M$ | Mass stored in system [kg] |
| Ma | Mach number |
| $m$ | Mass flow [kg] |
| $N$ | Number especially as a limit for the counter in summation |
| $n$ | Number especially as a limit for the counter in summation |

| | |
|---|---|
| $\hat{n}$ | Normal (unit) vector in a direction perpendicular to the plane of consideration |
| $P$ | Pressure [Pa or N/m$^2$] |
| PE | Potential energy [J] |
| $Q$ | Amount of heat transferred [J] |
| $q$ | Heat dissipation [J] |
| $R$ | Universal gas constant [J/kg-K] |
| RHI | Return heat index (9.49) non-dimensional measure of recirculation |
| $S$ | Entropy in system [J/K] |
| $s$ | Specific entropy (entropy per unit mass) [J/kg K] |
| SHI | Supply heat index (9.48) non-dimensional measure of recirculation |
| SOR | Successive over relaxation |
| $T$ | Absolute temperature [K] |
| $t$ | Time [s] |
| $U$ | Internal energy [J] |
| $u$ | Component of velocity nominally in the direction of $x$-axis [m/s] |
| $\forall$ | Volume [m$^3$] |
| $V$ | Velocity [m/s] |
| $v$ | Component of velocity nominally in the direction of $y$-axis [m/s] |
| $W$ | Amount of work transferred [J] nominally in the form of electricity or mechanical work |
| $w$ | Component of velocity nominally in the direction of $z$-axis [m/s] |
| $x$ | Distance in the direction of $x$-axis [m] |
| $y$ | Distance in the direction of $y$-axis [m] |
| $z$ | Height or distance in the direction of $z$-axis [m] |
| $\alpha$ | Thermal diffusivity defined as the ratio of thermal conductivity to the volumetric thermal capacity ($\rho C_p$) |
| $\beta$ | Nondimensional measure of recirculation (9.50) at the rack inlet |
| $\Delta$ | Change in value or state |
| $\delta$ | Incremental (infinitesimally small) change |
| $\Phi$ | Exergy of a closed system [J] |
| $\varphi$ | Potential flow function (9.60, 9.61) |
| $\eta$ | Efficiency |
| $\rho$ | Density [kg/m$^3$] |
| $\omega$ | Relaxation factor for Gauss–Seidel iteration (9.70) |
| $\Psi$ | Stream exergy for flow through an open system [J] |
| $\psi$ | Specific stream exergy (stream exergy per unit mass) [J/kg] |

## Subscripts

| | |
|---|---|
| 0 | Ground state |
| 1 | State of a system (nominally initial state) or arbitrary index value (e.g., in summation) |

| | |
|---|---|
| 2 | State of a system (nominally final state) or arbitrary index value (e.g., in summation) |
| II | Related to second law (such as second-law efficiency) |
| a | Ambient |
| airspace | related to the airspace within a data center |
| b | At the boundary of a control volume nominally the boundary of a cell in a finite volume mesh |
| C | Carnot |
| CRAC | Related to the CRAC units in a data center |
| cv | Control volume |
| cycle | Integrated over an entire cycle i.e., final and initial states are identical |
| d | Destroyed or irreversibly consumed |
| f | Related to the faces (surfaces) of a cell within a finite volume mesh |
| gen | Generated |
| H | Related to high-temperature reservoir (source) |
| i | Inlet or counter in summation, often for variables relating to the *x*-direction |
| in | Inlet or flowing into a given control volume |
| j | Counter in summation often for variables relating to the *y*-direction |
| k | Counter in summation often for variables relating to the *z*-direction |
| KE | Related to the transfer of kinetic energy |
| L | Related to low-temperature reservoir (sink) |
| max | (Theoretical) maximum value |
| o | Outlet |
| out | Outlet or flowing out of a given control volume |
| P | Related to the processor within a computer system |
| PE | Related to the transfer of potential energy |
| Q | Related to the transport of heat |
| sup | Supply state |
| rack | Related to the computer racks in a data center |
| rec | Recoverable |
| ret | Return state |
| rev | Reversible |
| t, th | thermal |
| W | Related to work |

## Superscripts

| | |
|---|---|
| . | Rate (i.e., per unit time) |
| $'''$ | Per unit volume |
| $\rightarrow$ | Vector |

## 9.1 Background

### *9.1.1 Review: The First Law of Thermodynamics*

Most people are familiar with the concept of energy conservation: energy is neither created nor destroyed, merely transformed from one form to another. This is the main embodiment of the first law of thermodynamics: when a net transfer of energy to (or from) a system, the energy of the system must increase (or decrease) by exactly that amount.

For closed "control mass" thermomechanical systems (i.e., those across which no flow of mass takes place), energy transfer to or from the system typically occurs in the form of the transport of heat or work. Then, as shown in Fig. 9.1a, application of energy conservation as dictated by the first law of thermodynamics yields the following energy balance for closed systems:

$$\Delta E = Q - W. \tag{9.1}$$

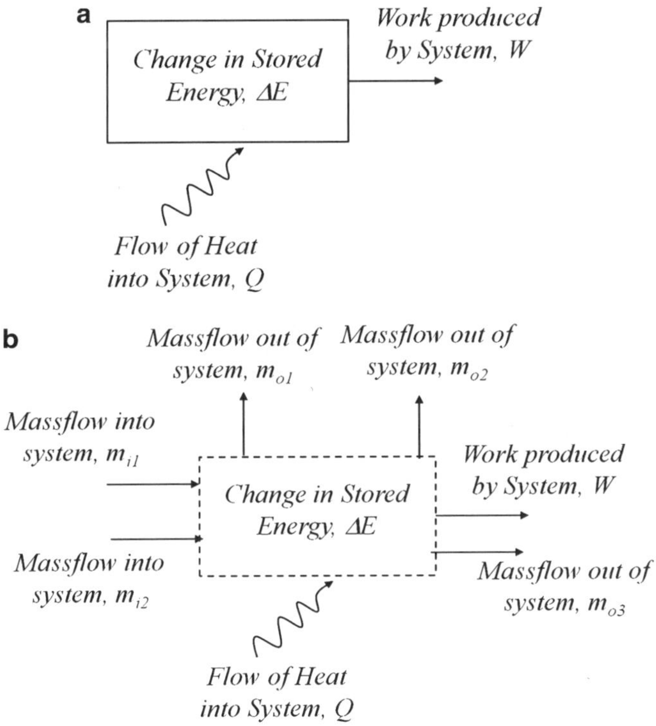


**Fig. 9.1** Energy balance from the first law of thermodynamics. (**a**) Closed (control mass) system; (**b**) Open (control volume) system

Note that the sign convention adopted in this work is that the flow of energy into the system is considered positive, while the flow of energy out of the system is considered negative. Thus, in (9.1), heat is assumed to be flowing into the system, while work is being done by the system (and that this performance of work requires energy to leave the system).

Nominally, the change in energy of a closed system over a finite time interval can be considered in terms of its constituent elements:

$$\Delta E = \Delta U + \Delta KE + \Delta PE. \tag{9.2}$$

On an instantaneous basis, the energy balance is obtained in terms of rates of energy transfer:

$$\frac{\mathrm{d}E}{\mathrm{d}t} = \dot{Q} - \dot{W} = \frac{\mathrm{d}U}{\mathrm{d}t} + \frac{\mathrm{d(KE)}}{\mathrm{d}t} + \frac{\mathrm{d(PE)}}{\mathrm{d}t}. \tag{9.3}$$

For systems at steady state, there will be no change in the energy of the system:

$$\dot{Q} - \dot{W} = \frac{\mathrm{d}U}{\mathrm{d}t} + \frac{\mathrm{d(KE)}}{\mathrm{d}t} + \frac{\mathrm{d(PE)}}{\mathrm{d}t} \approx 0. \tag{9.4}$$

Formulations of the above for open systems (which have mass flowing across the boundary of the system) will be similar, but must include energy carried into and out of the system in the form of enthalpy accompanying the mass flow. Then, as shown in Fig. 9.1b, energy conservation yields:

$$\dot{Q} - \dot{W} + \sum_i \dot{m}_i e_i - \sum_o \dot{m}_o e_o = \frac{\mathrm{d}E}{\mathrm{d}t}. \tag{9.5}$$

where $e$ is the energy transferred per unit mass flow, given by:

$$e = h + \frac{V^2}{2} + gz \approx \left(\frac{U}{m} + P\mathcal{V}\right) + \frac{V^2}{2} + gz. \tag{9.6}$$

For an open system at steady state with negligible changes in potential and kinetic energy, (9.5) reduces to:

$$\dot{Q} - \dot{W} + \sum_i \dot{m}_i h_i - \sum_o \dot{m}_o h_o = 0. \tag{9.7}$$

Equation (9.7) is the simplified form of the first law of thermodynamics for open systems at steady state, and will be used quite often in subsequent sections of this chapter.

The same concepts of conservation can also be extended to the notion of mass flow. That is, for any system:

$$\Delta M = \sum_i m_i - \sum_o m_o. \tag{9.8}$$

$$\frac{\mathrm{d}M}{\mathrm{d}t} = \sum_i \dot{m}_i - \sum_o \dot{m}_o. \tag{9.9}$$

The above equations are true in general, for all types of systems. The interested reader is referred to standard textbooks on thermodynamics [1–8] for a more detailed introduction to the field of thermodynamics.

### *9.1.2 Review: The Second Law of Thermodynamics*

The first law of thermodynamics governs the *quantity* of energy transfer that may occur across a system. The second law of thermodynamics is a reflection of how the *quality* of energy transferred across a system may change.

For example, consider the energy transport within a typical computer system. Energy is input to the system in the form of electricity, while energy is dissipated from the system in the form of heat. From the first law of thermodynamics discussed above, because there is no change in the energy storage within the system, the net amount of heat dissipated must equal the amount of electricity supplied. Thus, the heat transmitted from the system has the same quantity of energy as that required to run the computer. Suppose another computer system is connected downstream, so that all the heat dissipated from the first system were to be captured by this second system, and the net energy input to this second system matches the net energy input to the first system. Would it be possible to theoretically design a computer system that operates exclusively on this dissipated heat? The answer is no, and the reason lies within the second law of thermodynamics.

Formally, the Kelvin–Planck statement of the second law suggests that it is impossible for any device that operates on a cycle to receive heat from a single reservoir and produce a net amount of work. Thus, as shown in Fig. 9.2a, even an idealized heat engine drawing heat from a reservoir cannot operate at a thermal efficiency of 100%. Some heat loss must occur to the surrounding environment, which effectively reduces the thermal efficiency by a corresponding amount. Alternatively, the Clausius statement of the second law indicates that it is impossible to construct a device that operates in a cycle and produces no effect other than the transfer of heat from a lower temperature body to a higher temperature body. In other words, as shown in Fig. 9.2b, heat transfer will not occur on its own from a cold body to a warm body; such transport must be driven through external work. For example, for a refrigerator to function as intended (i.e., transfer heat away from

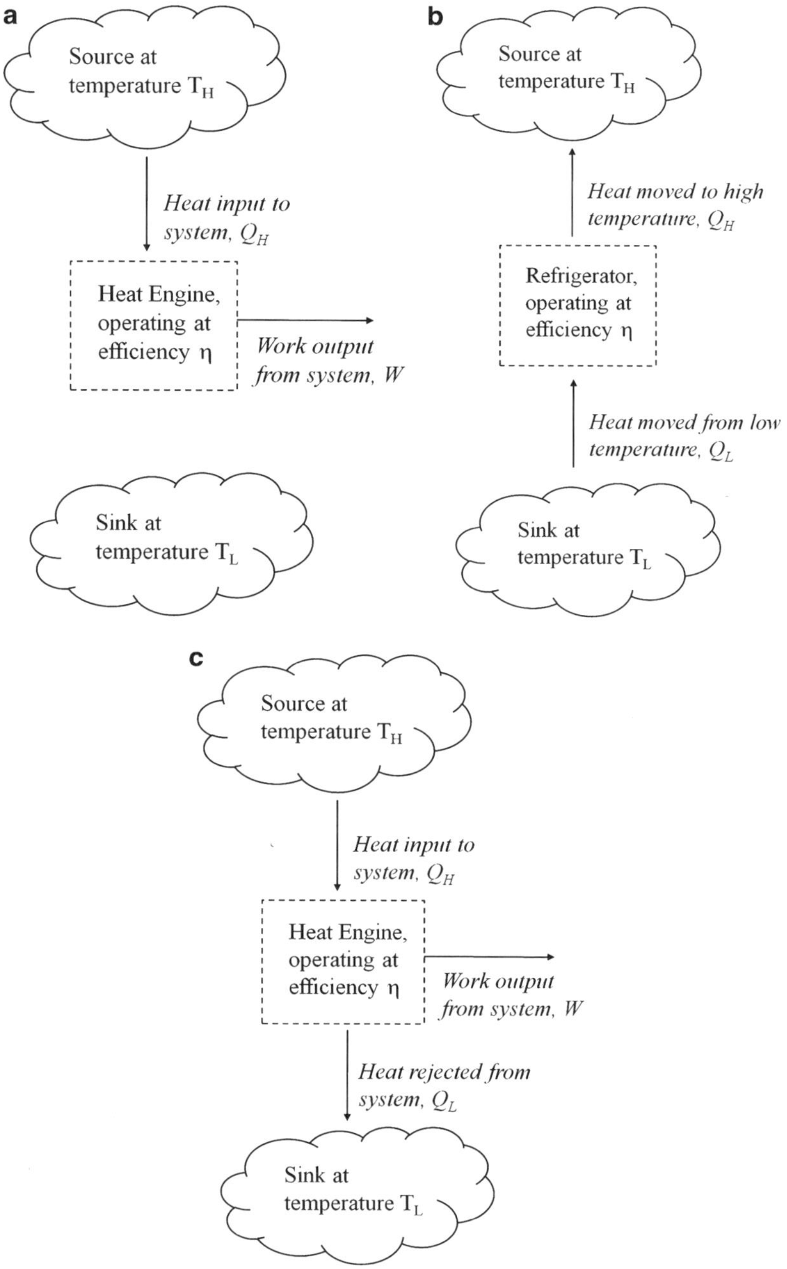


**Fig. 9.2** Examples of the second law of thermodynamics. (**a**) A system which violates the Kelvin–Planck statement of the second law, because heat input cannot be entirely converted to work in a cyclic device (i.e., some heat must also be rejected to the environment). (**b**) A system which violates the Clausius statement of the second law, because heat cannot be moved from a lower temperature to a higher temperature without an external input of work. (**c**) A heat engine that complies with the requirements of the second law

its low-temperature contents into a higher temperature surrounding), external power must be supplied to a compressor.

The above examples emphasize that in any real system, the change of energy from one form to another (e.g., from work to heat) will generally be accompanied by a change in the amount of energy that is available to do useful work. That is, while energy is entirely converted (and conserved) across different forms in accordance with the first law of thermodynamics, the second law of thermodynamics indicates that the amount of available energy will not be conserved; in fact, available energy will be continually destroyed due to irreversibilities in physical systems. Overcoming this irreversible destruction of available energy to ensure that the system can maintain its level of performance will require a continuous input of availability from a source external to the system, in an amount at least proportional to the amount of exergy that was destroyed.

To illustrate this applicability of the second law, consider the heat engine shown in Fig. 9.2c which draws input in the amount $Q_H$ from a high-temperature reservoir and produces a net work output in the amount of $W$. From the Kelvin–Planck statement of the second law, $W$ is necessarily lower than $Q_H$, which implies that—using a simple first-law energy balance—a non-zero amount of heat must be rejected to the lower temperature environment. (It is worth noting that the impossibility of reversing the system, viz. trying to transport heat from the lower temperature environment to the higher temperature reservoir without any net input of work, is simply a representation of the Clausius statement of the second law.)

Let the amount of heat rejected by the work-producing engine be $Q_L$. Then, at steady state, the first law (9.1) provides:

$$Q_H - Q_L - W = 0. \tag{9.10}$$

The thermal efficiency of the heat engine can be defined as the useful output (work) from the system, relative to the input (cost) to the system. That is:

$$\eta_{th} = \frac{W}{Q_H}. \tag{9.11}$$

Combining (9.10) and (9.11):

$$\eta_{th} = 1 - \frac{Q_L}{Q_H}. \tag{9.12}$$

Classical thermodynamics provides that the maximum efficiency attainable by such a heat engine will be achieved when:

$$\frac{Q_L}{Q_H} = \frac{f(T_L)}{f(T_H)}. \tag{9.13}$$

For convenience, the above functional relationships for temperature are typically arbitrarily defined linearly in terms of the Kelvin temperature scale, so that the maximum efficiency attainable by a heat engine occurs when:

$$\frac{Q_L}{Q_H} = \frac{T_L}{T_H}. \tag{9.14}$$

Substituting into (9.12), the resulting efficiency (known as the Carnot efficiency) will be:

$$\eta_C = 1 - \frac{T_L}{T_H}. \tag{9.15}$$

Combining (9.11) and (9.15), the maximum amount of work which such a heat engine might produce is obtained as:

$$W_{max} = \left(1 - \frac{T_L}{T_H}\right) Q_H. \tag{9.16}$$

A Carnot engine—defined as a heat engine which operates at the above efficiency—is an idealized engine that produces the above (maximum) amount of theoretical work. Such an idealized Carnot engine is also *reversible*. That is, if work in the amount of $W_{max}$ were to be provided as input to a refrigeration cycle operating at the Carnot efficiency, then the maximum amount of heat which could be removed from the low-temperature reservoir would be $Q_L$, and a minimum undesirable output (waste) of heat in the amount of $Q_H$ would occur. A real (nonidealized) engine would be unable to achieve such reversibility, because its operating efficiency would always be lower than the Carnot efficiency, due to losses incurred through friction and other effects. Hence, the maximum work which a real engine would produce would be less than $W_{max}$:

$$W < W_{max}. \tag{9.17}$$

If such a real engine operating below the Carnot efficiency were to be reversed, then reversing the work input to the system would be insufficient to drive the transport of heat in the amount of $Q_L$ across the reservoirs. Thus, to create a refrigeration cycle that moves heat in the amount of $Q_L$, work input larger than that produced by the heat engine would be required. For this reason, real (nonidealized) engines operating below the Carnot efficiency are considered as *irreversible*.

It follows from the above discussion, in particular (9.14), that for a reversible heat engine:

$$\frac{Q_L}{T_L} - \frac{Q_H}{T_H} = 0. \tag{9.18}$$

Based on the Clausius statement of the second law, (9.18) can be generalized for any cyclic system as follows:

$$\oint \frac{\delta Q}{T} \leq 0, \tag{9.19}$$

where the path of the integral traces the cycle. The equality in the above expression holds for reversible systems. If the system is irreversible, then the above ratio must always be less than zero (else the Clausius statement of the second law would be violated). This inequality provides a basis to define the *entropy generated* in a system as:

$$S_{\text{gen,cycle}} = -\oint \frac{\delta Q}{T}. \tag{9.20}$$

Thus, for a cyclic device, the entropy generated will be a measure of the irreversibility encountered within the system. The change in entropy of a system can then be given as:

$$\Delta S = S_2 - S_1 + S_{\text{gen}}. \tag{9.21}$$

Recalling from classical thermodynamics that the change in entropy of a system is defined as:

$$\mathrm{d}S = \left(\frac{\delta Q}{T}\right)_{\text{rev}}, \tag{9.22}$$

so that for an isothermal process where temperature is held constant at $T$, combining (9.21) and (9.22) yields:

$$\Delta S = \frac{Q}{T} + S_{\text{gen}}. \tag{9.23}$$

For reversible systems, $S_{\text{gen}} = 0$ and for irreversible systems, $S_{\text{gen}} > 0$. That is, any system where entropy is being generated will not be reversible, and will always operate at an efficiency below the maximum achievable efficiency.

For purposes of this work, rather than a more detailed exposition of the theory underlying the second law, a simplistic conceptualization is sufficient. The next section provides such a conceptualization. The interested reader is referred to standard textbooks on thermodynamics [1–12] for a more formal introduction to the second law.

### *9.1.3 Exergy, Irreversibility, and Useful Work*

A key concept of relevance to this work is the realization from the second law that a limit exists on the amount of useful work which may be extracted from a given system. This limit is independent of practical constructs such as design or operational inefficiencies. In other words, even for idealized systems such as Carnot engines, a limit will exist on the amount of useful work that can be extracted from the system. It follows, then, that not all energy supplied to a system will be available to do useful work. That is, in addition to conservation of the quantity of energy indicated by the first law, the second law provides the notion of *quality* of energy—a measure of how much energy is available to do useful work. In the example provided in Sect. 9.1.2 of two computer systems, where the first one is driven by electricity and dissipates heat which is captured by the second computer, the electricity supplied to the first computer system is mostly available for useful work. The heat dissipated by the first computer system will only be partially available for useful work. This amount of energy that is available for useful work is referred to as the *available energy* (*availability*) or *exergy (essergy)*.

Exergy is not a new concept, with the earliest references to exergy and available energy dating back more than 50 years [13]. In the USA, the concept was popularized in the context of "availability" in the 1940s while equivalent references to "exergy" came to be used in the 1950s. The latter term eventually gained global acceptance predominantly because it could be adapted without requiring translation [1]. It should be noted that while some authors define each term slightly differently (discussed later), both availability and exergy are used interchangeably in this text.

To understand the concept of exergy, it is useful to return to the theoretical framework of the second law of thermodynamics, which imposes a limitation on the amount of useful work that may be extracted from a given system at a particular state. For example, consider 1 J of heat held at 500°C relative to an ambient temperature of 25°C. According to the second law of thermodynamics, the maximum work from this source–sink system could be extracted in a Carnot engine operating at a thermodynamic efficiency of:

$$\eta_{\mathrm{C}} = 1 - \frac{T_{\mathrm{L}}}{T_{\mathrm{H}}} = 1 - \frac{(25+273)}{(500+273)} = 0.61. \tag{9.24}$$

Thus, the maximum work produced by the engine would be about 0.61 J. Note that this is a theoretical limit, and does not take into account any inefficiencies within the engine itself.

Consider now the same Joule of heat at a reduced temperature of 50°C relative to the same ambient reference temperature (25°C). The thermodynamic efficiency of the Carnot engine in this case would be:

$$\eta_{\mathrm{C}} = 1 - \frac{T_{\mathrm{L}}}{T_{\mathrm{H}}} = 1 - \frac{(25+273)}{(50+273)} = 0.08. \tag{9.25}$$

Thus, if the system is cooled to 50°C, the available work decreases to 0.08 J. In both cases, the quantity of energy input is the same; however, there is clearly a difference in utility or usefulness of the energy. This "quality" of energy is captured by the metric of exergy, which is a quantitative measure of the amount of useful work that may be extracted from a given system relative to an arbitrarily chosen reference state (also known as the "ground" state). Systems with higher utility have higher exergy content; in the above example, the exergy of the 500°C system would be 0.61 J, while the exergy of the 50°C system would be 0.08 J.

An exergy analysis is also beneficial when considered within the context of *exergy destruction.* According to the first law of thermodynamics, energy is always conserved; however, the same is not true for exergy. In the above example, as the system is cooled from 500 to 50°C, the available work reduces from 0.61 to 0.08 J (so the reduction in availability between the two states is 0.53 J). Because the cooling process is *irreversible*—without any changes in the surrounding or an additional work input, it is thermodynamically impossible for the system to warm back up to 500°C—potentially available work in the amount of 0.53 J can no longer be recovered. This exergy has been destroyed, and the usefulness of the system correspondingly decreases. Exergy is *not* conserved; an irreversible process will necessarily destroy exergy.

Note that the metric of exergy destruction scales with the extent of irreversibility encountered in the system. In the above example, if the system were to be cooled to 25°C, then a temperature drop from 500 to 25°C would represent a larger irreversibility compared to a temperature drop from 50 to 25°C. Therefore, the exergy destruction would be higher for the case with a 500°C source. Reversible processes (where no exergy is destroyed) are the most thermodynamically efficient; systems with lower exergy destruction are always more thermodynamically efficient.

In addition to exergy destruction, sometimes exergy may be *lost* to the surrounding due to inefficiencies in system design. Unlike exergy destruction, which is theoretical in nature and driven from the second law, *exergy loss* is a practical consideration that can be minimized through improved engineering. That is, for ideal systems, the exergy lost to the surrounding may effectively be zero. However, in real systems, both exergy loss and exergy destruction effectively reduce the energy available to do useful work. In this text, for convenience, this sum of exergy loss and exergy destruction is collectively referred to as the *exergy consumption.*

It is also important to note the dependence of exergy on the ground (reference) state. In the above example, the ground state considered was the ambient temperature of 25°C. Suppose the ambient temperature was actually 50°C, and this temperature was chosen as the ground state. Then the exergy values would be:

$$\text{At } 500°\text{C}: \quad \eta_C = 1 - \frac{T_L}{T_H} = 1 - \frac{(50+273)}{(500+273)} = 0.58. \tag{9.26}$$

$$\text{At } 50°\text{C}: \quad \eta_C = 1 - \frac{T_L}{T_H} = 1 - \frac{(50+273)}{(50+273)} = 0. \tag{9.27}$$

Thus, the exergy content of a system will vary with the definition of the ground state. Therefore, while comparing the available work from two different systems, it is necessary to ensure that the ground state is identical in both cases. While the choice of ground state is somewhat arbitrary, it is customary to choose an environmentally benign state as the appropriate ground state. Under this definition, the exergy content of a system will actually be the amount of work potential available prior to returning the system to its environmentally benign ground state. A positive exergy value indicates work is available from the system; a negative exergy value indicates work must be input to the system to restore the ground state. (Some authors choose to represent available energy as the useful work which can be derived from the system with respect to the surrounding, while defining exergy as the useful work which can be derived from the system with respect to the ground state. In this work, because the ground state is almost always defined to be the surroundings of the system, both available energy and exergy are used interchangeably.)

Returning now to the example provided in Sect. 9.1.2 of two computer systems, where the first one is driven by electricity and dissipates heat which is captured by the second computer, an exergy analysis can be considered. As discussed, the electricity supplied to the first computer system is in a highly ordered state (low entropy generation during transport), and will mostly be available for useful work. Thus, the first computer is being supplied with high exergy. The conversion from electricity to heat, however, is an irreversible process—entropy is generated during this conversion from a highly ordered state (electricity) to a lower ordered state (heat), and from the second law, even with an idealized Carnot engine, the amount of work which can be recovered from the transport of heat is limited by the temperature difference between the heat source and heat sink. The heat can never be converted back into electrical work with 100% efficiency. Exergy is destroyed in an amount proportional to this irreversibility, which limits the amount of useful work which can be extracted from the heat delivered to the second computer. That is, even though the energy delivered to both computers is the same, not enough of the energy supplied is *available* to the second computer: the amount of *exergy* delivered to the second computer is much lower than what was delivered to the first computer. This is why the second computer will be unable to function exclusively on the heat recovered from the first computer. Beyond identifying why the second computer cannot function exclusively on heat, the above exergy analysis also provides important information regarding what *can* be achieved with the heat dissipated by electronic systems. For example, it is clear that the quality of the heat dissipated is insufficient to drive highly ordered electronic componentry. However, perhaps this heat is of sufficient quality where it may be viable to serve alternative applications—space heating, biodigesters, certain HVAC functions—adjacent to the data center. Such collocation opportunities become particularly attractive in the context of data centers, where the rate of heat dissipation is sufficiently large that economies-of-scale may begin to favorably drive the reuse of waste heat. The viability of an exergy analysis for providing insight into these types of opportunities is discussed subsequently in this chapter.

### 9.1.4 Exergy Analysis of Thermal Systems

To formalize the exergy analysis conceptualized above, consider the (maximum) amount of useful work which may be extracted from a given system relative to a surrounding ground state (denoted by the subscript 0). It is customary to consider a quiescent ground state ($V_0 = 0$) at no elevation ($z_0 = 0$). Then, for a system involving the transport of heat from a temperature $T$ to a surrounding held at temperature $T_0$, the maximum theoretical work which could be availed would be when the heat is inputted to a Carnot heat engine. From (9.16), the available energy associated with heat transfer $Q$ would then be:

$$A_Q = \left(1 - \frac{T_0}{T}\right) Q. \tag{9.28}$$

Similarly, if the system were producing net work in the amount $W$, the maximum useful work available would be the net output from the system. That is:

$$A_W = W. \tag{9.29}$$

Thus, work produced from a system is entirely available as exergy. Similarly, kinetic energy and potential energy are entirely available to do useful work:

$$A_{\text{KE}} = \frac{MV^2}{2}. \tag{9.30}$$

$$A_{\text{PE}} = Mgz. \tag{9.31}$$

However, in the case of internal energy, the maximum work potential is limited by the entropy generated internal to the system. To illustrate this, consider a closed system producing work $W$ and dissipating heat in the amount of $Q$ to the environment. Then, assuming no change in kinetic and potential energy, an energy balance from the first law provides (9.1):

$$\Delta U = Q - W. \tag{9.32}$$

But, the second law of thermodynamics necessitates that entropy be generated along with the loss of heat $Q$ to the surrounding at temperature $T_0$:

$$S_{\text{gen}} = \Delta S - \frac{Q}{T_0}. \tag{9.33}$$

Combining (9.32) and (9.33), the work available relative to the ground state of the system will be:

$$W = T_0 \Delta S - T_0 S_{\text{gen}} - \Delta U = (U - U_0) - T_0 (S - S_0) - T_0 S_{\text{gen}}. \tag{9.34}$$

Equation (9.34) confirms that, as expected, the maximum work will be achieved if the system is reversible, viz. when $S_{\text{gen}} = 0$. The relationship between exergy destruction and entropy generation is more formally described through the Guoy-Stodola theorem, which suggests that the loss of availability is proportional to the generation of entropy. For a more thorough discussion on the topic, the interested reader is referred to Bejan [3].

For a closed system, the potential to accomplish boundary work by expanding the boundaries of the system also exists. This boundary ($P\mathcal{V}$) work which occurs between the system and its surroundings will be entirely available for work. Then, adding this term to (9.34) and neglecting the potential to do work through any changes in the chemical constituency of the system (i.e., ignoring chemical exergy), the exergy of a closed system will be:

$$\Phi = (U - U_0) - T_0(S - S_0) + P_0(\mathcal{V} - \mathcal{V}_0). \tag{9.35}$$

Similarly, for an open (control volume) system, the system does not have the potential to accomplish any boundary work but the potential to do work through a transport of enthalpy (flow work) exists. In addition, due to the possible change in velocity and elevation of the flow going through the system, the opportunity to do work from transferring kinetic energy and potential energy can also be availed in an open system. Using the same arguments as above, the exergy associated with mass flowing through a control system (i.e., stream exergy) can be obtained as:

$$\Psi = (H - H_0) - T_0(S - S_0) + \frac{V^2}{2} + gz. \tag{9.36}$$

To illustrate the use of the above formulations, consider an open system with massflow $m$ that dissipates heat $Q$ to the surrounding. No work is done by the system, and the system is chemically stable (i.e., no change in chemical exergy). Then, for such a system, the potential to do work exists in both the massflow (i.e., stream exergy) as well as extracting work from the dissipation of heat. The total exergy of the system would then be:

$$A = \Psi + A_{\text{Q}} = m\left[h - h_0 - T_0(s - s_0) + \frac{V^2}{2} + gz\right] + \left(1 - \frac{T_0}{T}\right)Q. \tag{9.37}$$

Equation (9.37) represents the maximum amount of work which can be done by the system.

Now, consider a scenario where a system cools down from an initial temperature $T_1$ to a final temperature $T_2$. As discussed earlier, this change in temperature is an irreversible process which must correspond to a destruction of exergy. To quantify this destruction of exergy, one can simply consider the difference in work potential

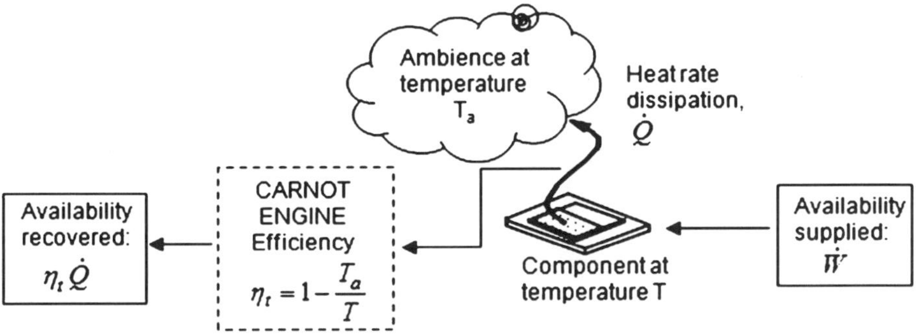


**Fig. 9.3** Exergy analysis of a component in a system operating at temperature $T$ dissipating heat $\dot{Q}$ to an ambient at temperature $T_a$. The exergy consumption will be the difference between the availability supplied and availability recovered

between the initial and final states of the system. Assuming no change in velocity or elevation of the system, the change in exergy of the system will be:

$$\Delta A = A_1 - A_2 = m[(h_1 - h_2) - T_0(s_1 - s_2)] + \left[\left(1 - \frac{T_0}{T_1}\right) - \left(1 - \frac{T_0}{T_2}\right)\right] Q. \quad (9.38)$$

This change in exergy of the system is irreversible, so that the potential of the system to do work has been decreased. Recovering the destroyed exergy will require an external input in an amount greater than or equal to $\Delta A$. Note that if the final state of the system is the ground state, then (9.38) will simply indicate that the entire availability of the system was destroyed. That is, a system at the ground state has no capacity to do useful work.

### *9.1.5 Exergy Analysis of Electronic Systems*

To extend the above example of exergy analysis to a computer system, consider a sample component of a computer system as shown in Fig. 9.3. The component draws electricity in the amount $\dot{W}$ at temperature $T$. For simplicity, a single (average) temperature is assumed over the surface of the component. For the uniform temperature case shown in Fig. 9.3, all the electricity input to the component gets dissipated as heat to the ambient in accordance with the first law of thermodynamics. That is,

$$\dot{Q} \approx \dot{W}. \quad (9.39)$$

Theoretically, some work could be extracted if this dissipated heat were input to a heat engine. From (9.16), the maximum power output available from the above Carnot engine would be:

$$\dot{W}_{\text{out}} = \left(1 - \frac{T_{\text{a}}}{T}\right)\dot{Q}. \tag{9.40}$$

By definition, this maximum recoverable work is equal to the exergy content (quality) of the heat dissipated by the component at temperature $T$ operating against an ambient $T_{\text{a}}$. Then, (9.39) and (9.40) suggest:

$$\dot{A}_{\text{out}} = \left(1 - \frac{T_{\text{a}}}{T}\right)\dot{Q} = \left(1 - \frac{T_{\text{a}}}{T}\right)\dot{W}. \tag{9.41}$$

Note that this quality is lower than the exergy supplied to the component, since almost all the electricity supplied to the component is available for useful work. That is,

$$\dot{A}_{\text{in}} = \dot{W}. \tag{9.42}$$

At steady state ($\text{d}A/\text{d}t = 0$), the rate of exergy consumption in the component will be:

$$\dot{A}_{\text{d}} = \dot{A}_{\text{in}} - \dot{A}_{\text{out}} = \dot{W} - \left(1 - \frac{T_{\text{a}}}{T}\right)\dot{W} \tag{9.43}$$

or

$$\dot{A}_{\text{d}} = \left(\frac{T_{\text{a}}}{T}\right)\dot{W}. \tag{9.44}$$

Equation (9.44) suggests that the rate of exergy consumption of the component is proportional to the electricity input of the component and inversely proportional to the operating temperature of the component. This may seem somewhat counter-intuitive, since it would appear that maintaining a higher component temperature could in fact reduce the exergy consumption. However, such an analysis would be incorrect because the component temperature and electricity input are coupled (i.e., for a constant ambient state, a higher component temperature could only result if the electricity input—and consequent heat dissipation—were adequately increased). To decouple this dependence, it is useful to also consider the convective heat transfer from the component to the surrounding:

$$\dot{Q} = \bar{h}\bar{A}(T - T_{\text{a}}). \tag{9.45}$$

Combining (9.39) and (9.45),

$$\dot{W} = \bar{h}\bar{A}(T - T_{\mathrm{a}}). \tag{9.46}$$

Substituting into (9.44),

$$\dot{A}_{\mathrm{d}} = \bar{h}\bar{A}T_{\mathrm{a}}\left(1 - \frac{T_{\mathrm{a}}}{T}\right). \tag{9.47}$$

Equation (9.47) provides the desired relationship between component temperature and exergy consumption for a computer system dissipating uniform heat. For a fixed convection coefficient $\bar{h}$, higher component temperatures will then lead to higher exergy consumption as expected. Note, however, that direct dissipation of heat to the ambient has been assumed in the above formulation, and that any exergy consumption associated with air handling has been ignored in the present formulation. These assumptions are revisited in subsequent sections of this chapter.

### 9.1.6 Prior Work

Several researchers have previously utilized a second law approach to assess and improve the performance of various thermal management systems, including electronics cooling systems. Much of the early work related to second law optimization was focused around the operation of large-scale complex systems, such as power plants [14–19] and other systems focused around energy conversion [20–32]. By quantifying the major sources of irreversibility in the operation of these systems and understanding the second-law efficiencies, it became possible to prioritize the relative opportunities for improvement across the system, including the choice and design of machinery; operating conditions for the working fluids; equipment and process parameter setpoints; etc. The successful application of second-law analysis to optimizing transport in energy conversion systems in turn gave rise to a new slew of industrial applications. These included manufacturing processes [33–38], chemical systems [39–45], heating and air-conditioning applications [46–49], and so on.

More recently, several studies related to the applicability of a second-law analysis for electronics thermal management can be found in the literature. The earliest applications of significance were related to optimizing the placement of heat-dissipating components within airflow [50–54]. For example, Bejan and Ledezma [50] modeled the airflow through an electronics system as flow in a rectangular duct over a heat-dissipating roughened plate. Using entropy generation minimization, the optimal placement of the board within the system could be identified. Ogiso [53] proposed the evaluation of different thermal management systems for electronics using a single metric based on the second law, which explored trade-offs between increasing the rate of heat transfer from the system versus increasing the work required by the cooling solution. Ngao et al. [54] took a similar approach in comparing the efficiency and performance of different types of thermal management systems used for electronics cooling. Simultaneously, recognizing that most electronic components dissipating high heat fluxes consist of a heat sink with active airflow, a large body of work emerged related to the application of the second law for the optimization of heat sink design [55–67]. This included work on optimizing the fin characteristics, such as shape; material; dimensions; spacing; etc. for maximizing the heat transfer rate at minimum cost, both operationally and in terms of manufacturing of the heat sink.

More recently, Carey and Shah [68] have assessed the energy efficiency of multiple generations of computer processors through an exergy analysis. Of interest particularly in this work is the applicability of exergy analysis for assessing the energy efficiency of large-scale computer infrastructures, such as data centers [69–74]. The earliest work in such application of the second law was that of Shah et al. [69], who applied the concept to identify recirculation patterns in the data center. Subsequently, Shah et al. extended the approach for considering energy efficiency of data center air-conditioning units [70], evaluating IT workload placement within the data center [71, 72], as well as electronic packaging considerations [74, 75].

Increasingly, the bounds of second-law analysis are extending beyond the optimization of traditional thermofluidic systems. For example, thermoeconomics [76–80] has been a topic of growing interest in the community due to its ability to merge operational idealizations from the second law with a detailed accounting of resource consumption across multiple components. In addition, the applicability of exergy analysis for resource accounting, environmental assessments, and evaluation of sustainability has become increasingly commonplace [81–93] across a wide variety of systems, within and beyond the IT domain.

The present work, given the topic of this book, is focused around the application of a second-law analysis for gaining insights into the energy efficiency of data center thermal management systems. Building upon the original work of Shah et al. [69], an approach to evaluating the data center cooling infrastructure in terms of exergy consumption is detailed. The approach is applied to the test case of a data center, and validated by comparison to experimental data. The chapter suggests that a second-law analysis of data center thermal management systems can be a viable approach to identifying and eliminating inefficiencies in the data center cooling infrastructure. Methods to extend this work for a more thorough analysis are discussed at the conclusion of the chapter.

## 9.2 Second Law View of Data Centers

### *9.2.1 Review of Data Center Thermal Management Systems*

As with all computer equipment, the electrical power supplied to the racks gets dissipated into the data center in the form of heat. As seen in Chap. 1, to cool the data center, cold air is supplied from the CRAC unit. The best practice is to divide

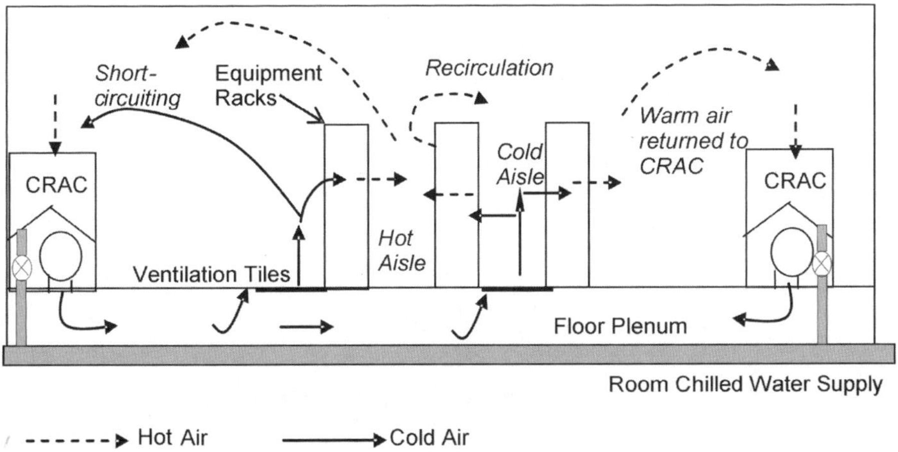


**Fig. 9.4** Typical thermal architecture in a raised-floor data center. Best practice dictates dividing the room into hot aisles and cold aisles on either side of a row of racks

the physical airspace of the data center into cold and hot aisles, so that cold air enters the room from the CRAC, gets heated up in the rack, and then is removed from the room in the hot aisle [94, 95]. Such a layout, discussed in detail in Chap. 1, is schematically represented in Fig. 9.4. (Some equipment manufacturers supply cold air from the top and/or sides of the equipment and exhaust it from the top (or bottom) [96–98], but the vast majority of data center cooling systems consist of a raised floor as described above.) The warm air from the hot aisle is removed from the computing environment either via a room return or ceiling return mechanism. This warm air is then returned to the CRAC unit, which is refrigerated by a chilled water stream or other thermal work input mechanisms before being resupplied to the data center environment at the desired supply temperature.

Over the last decade, the power dissipated per unit area of a chip has increased by an order of magnitude; microprocessor heat dissipation has similarly gone up by a factor of ten [99]. A typical data center today operating a thousand 10 kW racks over an area of 2,700 $m^2$ (30,000 $ft^2$) might require about 5 MW of power for cooling. At such high heat loads in data centers, the inefficiencies of both thermal work and flow work become significant [100]. Without major improvements in the cooling system, a data center housing five thousand 10 kW racks over a 9,200-$m^2$ (100,000 $ft^2$) area might require an additional 20 MW of electricity for cooling. At \$100/MWh, the total cost of operating the data center would be around \$44 million per year in rack operation and an additional \$18 million per year in cooling [101–103]. Clearly, data center thermal performance is of the utmost interest, both for purposes of environmental energy conservation and potential economic savings.

## 9.2.2 *Challenges in Data Center Thermal Management*

A data center characterized by the high density of deployment of computer racks outlined above can in itself be viewed as a computer, where the walls are like the enclosure and the racks are like electronic devices dissipating heat [104]. Unfortunately, the following problems are encountered in data centers with traditional cooling system design:

### 9.2.2.1 Mixing of Hot and Cold Airstreams

Many existing data center architectures provide no method of control to prevent the mixing of cold and warm airstreams in current data center thermal management systems. For example, cold supply from the CRAC may pass through an equipment rack and pick up dissipated heat, thereby becoming a warm air stream. (Note that in this text, the air handling units in the data center are broadly described as CRAC units with the intent of representing both direct expansion (DX) air-conditioning units as well as air handling units where the air is cooled by chilled water flowing through a heat exchanger.)

But as first illustrated by Patel et al. [105] and Schmidt [106] using computational fluid dynamics (CFD), such a warm stream may then return to a neighboring rack and subsequently nullify the effect of cooling from vent tiles. To avoid the resulting increase in inlet temperatures to the computer rack units (which could lead to system failure), data center operators typically enable additional cooling capacity to compensate for recirculation effects. However, this results in a waste of energy through short-circuiting, which refers to the direct passage of cold air from vent tiles or CRAC units to a system outlet. No thermal control is accomplished through the input of such streams, and the significant amount of energy spent in the CRAC units—meant for cooling of the system—is essentially wasted.

Increasingly, data centers are incorporating aisle containment to prevent such mixing and short-circuiting. However, this is not yet pervasively implemented in existing facilities. Moreover, not all facilities *require* aisle containment—for example, if recirculation could be minimized in the traditional open environments through proper airflow management, the costs of installing containment could be avoided altogether. Lastly, while containment accomplishes the thermal management goals of the data center by avoiding recirculation, it is not always clear that containment will necessarily result in energy efficiency gains that provide the requisite payback. A modeling approach that quantifies the energy savings achievable through elimination of airflow inefficiencies is lacking.

### 9.2.2.2 Lack of Information About Local Conditions

Most data centers operate on single-input, single-output environmental controls. A single sensor is placed at an arbitrary location, usually near the CRAC supply or return, and the amount of cooling is governed by the readings on this sensor. Although cost is the dominant factor for such design, it must be noted that even if a multitude of sensors were to be used at the CRAC, accurate information about local conditions in the data center would be difficult to obtain. For example, owing to the uneven distribution of high heat loads across a data center floor, local hotspots are quite common. Alternatively, it is foreseeable that individual sensors placed inside the room may sense only the exceedingly high temperature of such a hotspot, thereby calling for vast but unnecessary amounts of cooling. Conversely, a specific piece of equipment in the data center may become excessively hot and catalyze a shutdown if not detected. Thus, sole reliance on temperature sensing at a single CRAC location for cooling control can lead to operational problems and low thermal efficiency.

#### 9.2.2.3 Malprovisioning of Cooling Resources

The problems of mixing streams and local hotspots are further accentuated by the rudimentary approach taken to the design of data center cooling systems. Typically, the air conditioners or CRAC units are provisioned arithmetically based on a worst-case summation of all heat loads in the data center. The return air temperature to the CRAC units is then set at a fixed level during operation, with the goal of uniformly cooling the entire data center to a desired low temperature. Unfortunately, such an approach can only work if the hot return air from the racks is fluidly separated from the cold supply air. Without such separation, failure to prevent mixing may cause the designed capacity from the CRAC units to be insufficient for the prevention of local hotspots. To overcome such a situation, data center operators often simply run CRAC units at supply temperatures that are too low and fan speeds that are too high. But measurements such as those by Mitchell-Jackson et al. [107, 108] show that data centers rarely operate at full capacity. Thus, when the CRAC units are operated for maximum cooling, much of the cooling potential goes unused. In other cases, the data center operator believes that the available cooling capacity is being fully utilized only at partial compute loads and ends up installing unnecessary additional capacity into the infrastructure. The redundant and inefficient nature of traditional data center cooling systems can be a tremendous source of wastage.

#### 9.2.2.4 Inadequate Metrics to Optimize Overall Efficiency

As discussed subsequently, most existing data center metrics are focused on either the thermal manageability (such as metrics that quantify recirculation in the data center) or energy efficiency (such as metrics that quantify the fraction of energy supplied that goes towards satisfying computational workload). Increasingly, these competing goals are of equal importance to the data center operator. But the disjoint nature of existing data center metrics causes holistic optimization of the entire facility to be a difficult task. In addition, certain areas of peripheral relevance—such as quantifying opportunities for energy scavenging in the data center—are almost entirely ignored by existing metrics. Approaches that can address the multiple goals associated with data center thermal management are required.

#### 9.2.2.5 Workload Placed Without Consideration of Environmental Conditions

Computing workload is not equally distributed throughout data centers. Moreover, the temperatures prevalent in different areas of a data center are rarely uniform. Little, if any, consideration is given to the correlation between the above two situations. Quite often, heavy workloads are allotted to racks where temperatures are already high; such assignments greatly reduce the sensory and cooling capabilities of thermal management systems [109, 110]. Improved control algorithms are necessary to prevent wastage of energy by correlating computational workload with cooling equipment [111–114]. Moreover, the amount of energy spent on cooling is also a function of the conditions *outside* the data center. Computing is quickly becoming a global activity: future visions of the industry suggest a worldwide network of data centers that dynamically allocate resources to provide trillions of services to billions of users [115]. Such a distribution has great ramifications for thermal management: for example, cooling requirements tend to be much higher in the summer months than in the winter months; similarly, greater cooling is required for data centers in locations where the ambient temperature is high (e.g., a data center operating in Phoenix, Arizona in the summer daytime is likely to consume more energy for cooling than a data center operating in New Delhi, India at nighttime) [116]. It becomes extremely difficult to compare the performance of data centers operating in different geographical locations using traditional DC thermal management metrics.

#### 9.2.2.6 Summary and Related Work

To summarize, the following problems are encountered in current data center thermal management techniques:

- Recirculation and short-circuiting result in great wastage of energy. It is difficult to prevent these phenomena, because exact locations where mixing takes place cannot be pinpointed using traditional thermal analysis or measurement techniques.
- Local hotspots cast doubt on the accuracy and reliability of feedback from thermal sensors. Conservative designs based on single-input, single-output conditions often cause either unnecessary shutdown of the entire system or unneeded increase in cooling.
- Internal and external environmental conditions are not evaluated during normal data center operation or assessment of thermodynamic efficiency.

Previous researchers have suggested numerous solutions to address various components of the above issues. These solutions have included localized fixes such as control of recirculation through blanking panels [117], refrigerant-assisted spot-cooling [118], or use of liquid heat exchangers on rack doors [119]. Other solutions have explored containment between the hot and cold aisles [120, 121], distributed sensing throughout the data center [122–125], and dynamic control of the air-conditioning units based on sensed temperature data [126]. However, not all data centers incorporate these types of solutions; moreover, most of these solutions are targeted at addressing specific inefficiencies in the infrastructure (such as eliminating recirculation or improved provisioning of the CRAC units). A holistic approach or metrology for exploring and evaluating inefficiencies in the cooling infrastructure is lacking. As a result, many data centers continue to reflect several inefficiencies in the cooling infrastructure—particularly in the airspace surrounding the rack units.

Given the importance of thermofluidic phenomena within the computer room, much recent research has focused on characterizing airflow patterns in data centers. Early studies, such as those by Kang et al. [127], Patel et al. [105], and Schmidt [106], were based on CFD models of the data center airspace. Subsequent studies have explored in greater depth the impact of CRAC handling on data center airflow. For example, VanGilder and Lee [128] and Schmidt et al. [129] have explored flow-management techniques via control of subplenum configurations and vent tile layout. Karki et al. [130, 131] have performed various numerical simulations to characterize the relationship between airflow and data center physical characteristics, while VanGilder and Schmidt [132] and Radmehr et al. [133] have characterized the flow-through vent tiles in raised-floor data centers via computational modeling [132] and experimental measurement [133] for different CRAC settings. The supply air distribution resulting from a single CRAC unit was numerically modeled by Rambo and Joshi [134], who also presented computational models of subrack thermal responses to various CRAC responses.

In addition to the above studies which focus almost exclusively on the fluidic aspects of recirculation, several researchers have also explored the impact of other parameters on data center thermal management. For example, Patel et al. [135] and Bash and Forman [136] have studied the impacts of asymmetry in data center rack layout on cooling effectiveness, while Bhopte et al. [137] and Schmidt and Iyengar [138] have investigated and attempted to optimize the data center layout for various design parameters such as plenum height, ceiling height, and row length. Similarly, the impact of rack parameters such as location, loading, and local fan flowrates on the resulting rack inlet temperatures has been parametrically assessed in several studies [139–145]. For nonraised-floor data center cooling configurations (such as room-supply/ceiling-return or ceiling-supply/ceiling-return, etc.), variations in the rack inlet air temperature have been modeled by Shrivastava et al. [97] and Iyengar et al. [98], who found significantly different patterns of recirculation for different data center architectures. The physics behind inefficiencies such as recirculation and short-circuiting have been examined in greater detail by Bash et al. [100], who in conjunction with Sharma et al. [146, 147] proposed the development of high-level control policies to enable real-time feedback and control of data center thermal architecture. To enable such policies, the following nondimensional parameters were proposed based on rack inlet and exhaust temperature [146, 147]:

$$\mathrm{SHI} = \frac{\delta Q}{Q + \delta Q} = \frac{\text{Enthalpy rise in cold aisle due to infiltration of warm air}}{\text{Total enthalpy rise at rack exhaust}}, \quad (9.48)$$

$$\mathrm{RHI} = \frac{Q}{Q + \delta Q} = \frac{\text{Total enthalpy rise in data center airspace}}{\text{Total enthalpy rise at rack exhaust}}, \quad (9.49)$$

where SHI is the *supply heat index* and RHI is the *return heat index* of the data center. Note that the sum of SHI and RHI is always unity, and both parameters are scaled to vary between 0 and 1. Ideally, a case of no recirculation corresponds to SHI $= 0$, which will simultaneously lead to a value of RHI $= 1$.

Numerical simulations by Sharma et al. [146] have demonstrated that larger amounts of recirculation do indeed lead to higher SHI values. The effects of geometric layout, rack height, and row length can all be captured indirectly in the changing values of SHI. These metrics can be useful for real-time feedback regarding data center airspace behavior, particularly because the normalized non-dimensional nature enables scalability across the rack, row, and room levels.

While the high-level estimates of mixing provided by the SHI are useful for purposes of cooling control and thermal management, Schmidt et al. [148, 149] have argued that SHI–RHI by itself is insufficient because a data center with favorable global profiles (low SHI and high RHI) can still be subjected to local hotspots that hinder adequate thermal management. To address this concern, the following metric was proposed [148, 149]:

$$\beta = \frac{\Delta T_{\mathrm{in}}}{\Delta T_{\mathrm{rack}}} = \frac{T_{\mathrm{rack,in}} - T_{\mathrm{sup}}}{T_{\mathrm{rack,out}} - T_{\mathrm{rack,in}}}. \quad (9.50)$$

A value of $\beta = 0$ indicates that no hot air is recirculated to the front of the rack, while a value of $\beta = 1$ implies insufficient cold air supply so that air from the hot aisle is directly recirculated back to the inlet of the rack. A value of $\beta > 1$ suggests that a local self-heating loop exists which is causing the air to be heated even beyond the caloric heat gain due to heat addition from the rack. It would appear that rack-level approximations of SHI and RHI can perform the same function as $\beta$, and therefore use of either metric would suffice for purposes of thermal management. Subrack approximations of recirculation would require estimates of SHI or $\beta$ at corresponding granularities.

While useful for estimating thermal manageability of a given data center configuration, metrics such as SHI, $\beta$, the Rack Cooling Index [150], etc. only provide limited information regarding the energy efficiency of the system. For example, it is difficult to ascertain whether local hotspots are formed primarily due to poor CRAC performance, suboptimal equipment layout, unfavorable airflow patterns, etc. Other energy efficiency indicators for the data center have been proposed elsewhere [151–153], but these metrics provide no information about recirculation phenomena and corresponding impacts on the thermal manageability of data centers. Thus, there remains a gap with regards to an appropriate metric that fuses information about thermal manageability and energy efficiency while providing insight into methods of reducing recirculation in the data center airspace. This chapter proposes such a metric using the thermodynamic concept of exergy. The next section discusses, conceptually, how such an approach might fill the metrology gap with regards to energy-efficient data center thermal management.

### *9.2.3 Irreversibilities in Data Center Architectures*

As discussed previously, the exergy of a system is a representation of the amount of useful work that can be obtained from a given quantity of energy. For example, the electricity supplied to the computer equipment is a very high quality of energy, as it can almost entirely be converted into work. Therefore, the electricity supplied to the data center has a high exergy value. On the other hand, heat is a relatively lower quality of energy, owing to the Carnot limitation on the amount of work which can be extracted from a heat source. So, the heat rejected by the racks inside the data center has a lower exergy value. Alternatively, the conversion of electricity to heat is an irreversible process—without any external input, the work available from the electricity can never be recovered once the conversion to heat has taken place. That is, exergy has been destroyed during the conversion of electricity to heat, so that the exergy content of the heat source will be lower than the exergy content of the electricity source.

Similarly, within the context of data center thermal management, there will be a difference in the exergy value of the airflow depending on the temperature and velocity in the data center. For example, relative to an ambient reference state, more work is theoretically available from a high-temperature exhaust stream in the hot aisle than a low-temperature input stream in the cold aisle. Or, the exergy value of a cold aisle with no recirculation will be lower than the exergy value of a cold aisle with recirculation where the average temperature is higher. Equivalently, the mixing of hot and cold air in the data center is an irreversible process—the streams cannot be separated into their original hot and cold aisles without any external work input—and therefore, any occurrence of recirculation will also be a source of exergy consumption. Following the discussion of Sect. 9.2.2, this mixing of airstreams is an example of a common inefficiency within the data center cooling infrastructure. Thus, it appears that the metric of exergy consumption could provide information about local inefficiencies in the data center. In addition, following the discussion of prior work in Sect. 9.1.6, an exergy analysis should be able to provide information about the thermodynamic efficiency of the data center as a whole (including the CRAC and rack units). In this sense, it may be possible to utilize exergy consumption as a common metric to quantify both local and global thermodynamic efficiencies—something that is not possible using existing metrics within the data center.

For such a metric of energy efficiency to be useful, however, the metric would also need to provide the necessary information regarding data center thermal manageability. For most raised-floor air-cooled data centers, as long as the supply temperature delivered by the CRAC units is below the allowable inlet temperature of the computer systems, questions around thermal manageability will only be dependent on the amount of recirculation: if there is no recirculation, then there will be no concerns about the possibility of over-heating the computer systems; while if recirculation is significant, it is likely that the data center thermal management system will prove inadequate. But owing to the irreversibility of the recirculation process, there is reason to believe that a map of exergy consumption throughout the data center airspace can pinpoint locations of recirculation, thus providing an indication of data center thermal manageability. In addition, such a diagram could also provide insights into the relative local energy efficiency across the room. For example, the opportunity to trickle charge batteries in wireless sensors at locations of high available energy could be identified; or, alternatively, the potential savings which can be availed from eliminating inefficiencies in the data center can be quantified. The next section discusses how such a localized exergy consumption map could be created for a given data center. The following section provides insights obtained from combining knowledge about localized inefficiencies with an understanding of the global thermodynamic efficiency of the data center, both derived from a second-law analysis.

## 9.3 Exergy Analysis of Data Center Thermal Management Systems

From a thermal management perspective, three components of the data center infrastructure are crucial: the sources of heat (computing racks), the medium of heat removal (air), and the air-conditioners removing the heat from the medium (CRAC units). The total exergy consumption in the data center will be the sum of the exergy consumption in these individual components:

$$\dot{A}_{\mathrm{d}} = \dot{A}_{\mathrm{d_{rack}}} + \dot{A}_{\mathrm{d_{CRAC}}} + \dot{A}_{\mathrm{d_{airspace}}}. \tag{9.51}$$

The exergy consumption in each of these components will now be evaluated in further detail.

## 9.3.1 Modeling Approach

### 9.3.1.1 Exergy Consumption in Rack Units

For simplicity, these are treated as a single computer unit dissipating heat at a uniform temperature. This assumption will be valid if:

1. All the servers in the rack are homogeneous. For racks with nonhomogeneous servers, the exergy consumption occurring local to each server must be considered individually, and the total exergy consumption of the rack can then be calculated as the sum of the exergy consumption in these individual servers.
2. The load in the rack is uniformly distributed across the servers. For nonuniform loading, the effects of turbulence and buoyancy inside the rack can be significant. For example, Rolander et al. [154] show how an optimal arrangement of servers inside a rack can allow 50% higher loads compared to an arbitrary suboptimal configuration meeting the same thermal management criteria, while Rambo and Joshi [155] show how different server arrangements inside a single rack will lead to different thermal characteristics inside the cabinet. Inclusion of such considerations in the exergy model for racks could be achieved by approximating the entropy generation rate related to the airflow through the cabinet. After adding this to the entropy generated due to dissipation of heat in the cabinet, the Guoy-Stodola theorem could be invoked to predict the total rack exergy consumption.
3. The thermofluidic operating conditions (processor temperature, air flowrate, etc.) internal to each server are the same. This assumption will be valid if all the servers have the same package-level thermal solution, which is likely to be the case for a homogeneously loaded rack per (1) and (2) above.

Subject to the above assumptions, from a second law perspective, the rack can be assumed to behave like a single heat-dissipating component in a larger compute infrastructure. Then, as shown by Carey and Shah [68], the predominant exergy consumption within the rack units will occur due to the conversion of high-quality electrical energy to low-quality thermal energy. Assuming that the electrical energy is entirely available for useful work, and that the heat dissipation equals the amount of electricity to the package, the exergy consumption of the racks can be calculated as [68]:

$$\dot{A}_{d_{\mathrm{rack}}} = \dot{W} - \left(1 - \frac{T_0}{T_{\mathrm{P}}}\right)\dot{Q} \approx \left(\frac{T_0}{T_{\mathrm{P}}}\right)\dot{Q}, \tag{9.52}$$

where $T_{\mathrm{P}}$ is the temperature at which the heat is dissipated, in this case assumed to be the processor temperature. To be precise, the heat is actually transferred to the airflow after it has been dissipated through the package-level thermal solution. Shah et al. [74] have presented such a package-level exergy consumption model which takes into account the heat spreader, thermal interface material, and heat sink. At the data center level, however, the added exergy consumption caused by the package-level thermal solution will be small (usually less than 1%), and is therefore neglected.

Equation (9.52) is a simplified version of the exergy consumption in the rack units for the idealized case characterized by the assumptions (1)–(3) above. For this example, a constant processor temperature of 85°C was assumed in all the racks. If desired, subrack thermal models could be incorporated into the data center numerical model for a more accurate temperature estimate.

#### 9.3.1.2 Exergy Consumption in CRAC Units

The CRAC units can be modeled as simple air-handling units, for which a basic second-law analysis yields the following exergy consumption:

$$\dot{A}_{d_{\mathrm{CRAC}}} = \dot{m}_{\mathrm{CRAC}}\left(\psi_{\mathrm{ret}} - \psi_{\mathrm{sup}}\right) - \dot{W}_{\mathrm{CRAC}}, \tag{9.53}$$

where $\dot{W}_{\mathrm{CRAC}}$ is the electrical work supplied to the CRAC units. In all air-handling units, this will include the work required in the blower. In addition, for DX CRAC units, this may include the work done by the compressor, in which case the $\dot{W}_{\mathrm{CRAC}}$ term can also be rewritten in terms of the COP of the CRAC unit and the heat load on the CRAC unit $\left(\dot{W}_{\mathrm{CRAC}} = \dot{W}_{\mathrm{blowers}} + \dot{Q}_{\mathrm{CRAC}}/\mathrm{COP}\right)$. However, in comparing two different data centers, it is important to keep the boundary conditions of the control volume identical—so that if one data center has a DX unit and the compressor work is included in the control volume, then the chiller from a second data center with air-handling units should also be included in the control volume.

#### 9.3.1.3 Exergy Consumption in Airspace

As discussed earlier, the mixing of hot and cold air is an irreversible process. Therefore, exergy must be destroyed in this process. If the exact locations of exergy consumption in the data center airspace could be pinpointed, then it becomes possible to physically identify locations within the data center where mixing is particularly prominent. Solutions to address this mixing could then be developed as required.

To computationally estimate the extent of recirculation in the data center airspace by measuring the exergy consumption across the airspace, a finite volume approach is used by dividing the physical volume of the room into a large mesh of smaller volumes, so that by evaluating the exergy consumption of each cell against a common ground (reference) state, the total exergy consumption of the airspace can be computed as the sum of the individual cell exergy consumption:

$$\dot{A}_{d_{\mathrm{airspace}}} = \sum_{i=1}^{N} \dot{A}_{d_i}, \tag{9.54}$$

where $N$ is the total number of cells in the room. The same rules used to define the grid for numerical thermofluidic analysis are used to govern the choice of mesh in the current analysis, i.e., the number, uniformity, and size of the cells chosen for the analysis depend on the thermal variability encountered within the system for a given data center setting. It is essential to choose a cell size that is small enough to capture field variations at length scales of room subregion dimensions. Beyond this, cell sizes should be chosen as a compromise between the required level of granularity (or the degree of accuracy necessary) and the desired speed of computation.

It should be noted that the total exergy consumption estimate of (9.54) is proposed as a global metric of mixing in the airspace. The magnitude and location at which recirculation occurs will be represented by the exergy consumption within a given cell of the grid, which requires knowledge about the exergy content of the cell (relative to the ground state). For a given control volume, the specific exergy of the system can be defined with respect to the ground state as:

$$\psi = (h - h_0) + \frac{V^2}{2} + g(z - z_0) - T_0(s - s_0). \tag{9.55}$$

The subscript "0" corresponds to the ground (reference) state against which the system is operating. It is important that a common ground state be maintained throughout the second-law analysis, otherwise the summation of (9.54) will not be valid.

The rate of exergy consumed in each individual cell can then be calculated by performing an exergy balance for the individual cell:

$$\frac{\mathrm{d}A}{\mathrm{d}t} = \sum_j \left(1 - \frac{T_0}{T_j}\right)\dot{Q}_j - \left(\dot{W}_{\mathrm{cv}} - P_0 \frac{\mathrm{d}\mathcal{V}_{\mathrm{cv}}}{\mathrm{d}t}\right) + \sum_{\mathrm{in}} \dot{m}_{\mathrm{i}}\psi_{\mathrm{i}} - \sum_{\mathrm{out}} \dot{m}_{\mathrm{o}}\psi_{\mathrm{o}} - \dot{A}_{\mathrm{d}}, \tag{9.56}$$

where $j$ is the boundary across which heat $\dot{Q}_j$ is added at a temperature $T_j$. Note that the first term in (9.56) relates to the loss of exergy due to boundary heat transfer from high (processor) to low (ambient) temperature in the airspace of the data center, and is different from the exergy consumption during heat dissipation in the rack itself (which is due to conversion of high-quality electrical work to low-quality heat at the processor temperature). Additionally, the walls of the racks are assumed to be insulated, so that the heat addition will only occur as the air flows through the racks from inlet to outlet. Lastly, we account for any electrical power input to the system within the component exergy consumption, so that no work is done within or across the control volume of each cell. The system is assumed to operate at steady state, and no changes occur to the control volume boundaries, so that:

$$\dot{A}_{\mathrm{d}} = \sum_j \left(1 - \frac{T_0}{T_j}\right)\dot{Q}_j + \sum_{\mathrm{in}} \dot{m}_{\mathrm{i}}\psi_{\mathrm{i}} - \sum_{\mathrm{out}} \dot{m}_{\mathrm{o}}\psi_{\mathrm{o}}. \tag{9.57}$$

Substituting for the specific exergy from (9.55),

$$\dot{A}_{\mathrm{d}} = \sum_j \left(1 - \frac{T_0}{T_j}\right)\dot{Q}_j + \sum_{\mathrm{in}} \dot{m}_{\mathrm{i}}\left[(h - h_0) + \frac{V^2}{2} + g(z - z_0) - T_0(s - s_0)\right]_{\mathrm{in}}$$
$$- \sum_{\mathrm{out}} \dot{m}_{\mathrm{o}}\left[(h - h_0) + \frac{V^2}{2} + g(z - z_0) - T_0(s - s_0)\right]_{\mathrm{out}}. \qquad (9.58)$$

Thus, for a cell at a given location, if the temperature and flow (pressure and velocity) properties at that point are known in addition to the heat input, then the exergy consumption for the air flowing through the cell can be calculated per (9.58). Typical state-of-the-art thermofluidic analysis already involves a determination of the temperature and flow throughout the data center airspace; thus, only one additional calculation is required to gain the desired information about exergy consumption in the system. Applying (9.58) to each cell of the numerical grid will provide a map of exergy consumption throughout the data center and enable identification of local recirculation effects. The measure of total airspace exergy consumption per (9.54) provides a single metric that can be used to evaluate the overall recirculation patterns in the cold aisle or the entire data center.

Having defined the exergy consumption in the individual components, (9.51)–(9.54) can be combined to evaluate the total exergy consumption in the data center.

#### 9.3.1.4 Summary: Data Center Exergy Analysis

To summarize the formulations developed in this section, the total exergy consumption in the data center can be calculated as follows:

$$\dot{A}_{\mathrm{d}} = \dot{A}_{\mathrm{d_{racks}}} + \dot{A}_{\mathrm{d_{CRACs}}} + \dot{A}_{\mathrm{d_{airspace}}} \qquad \text{[per (9.51)]}$$

Where $\dot{A}_{\mathrm{d_{racks}}}$ is predicted in Sect. 9.3.1.1 [per (9.52)]

$\dot{A}_{\mathrm{d_{CRACs}}}$ is predicted in Sect. 9.3.1.2 [per (9.53)]

and $\dot{A}_{\mathrm{d_{airspace}}}$ is predicted in Sect. 9.3.1.3 [per (9.54)]

The exergy consumption for the racks and CRAC units can be estimated based on input conditions regarding heat load, processor temperatures, and CRAC flow, while the exergy consumption in the airspace can be estimated numerically using a finite volume analysis in the airspace. Note also that the difference between the exergy supplied to the data center and the exergy consumed in the data center will yield the amount of available energy which can be (theoretically) recovered from the data center. This can be quantified in terms of the second-law efficiency of the data center thermal infrastructure, which is nominally defined as the exergy

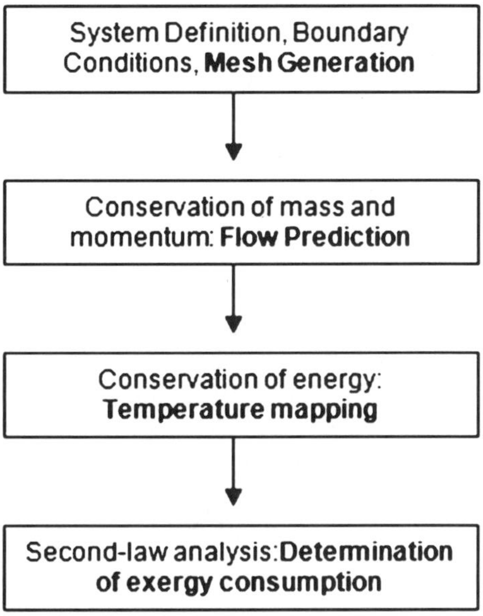


**Fig. 9.5** Flowchart for determining exergy consumption in data center airspace. The algorithm is applied to each individual cell and then solved at the system-wide level

contained in the outflows normalized by the exergy contained in the inflows to the data center:

$$\eta_{\mathrm{II}} = \frac{\dot{A}_{\mathrm{out}}}{\dot{A}_{\mathrm{in}}} = \frac{\dot{A}_{\mathrm{sup}} - \dot{A}_{\mathrm{d}}}{\dot{A}_{\mathrm{sup}}} = 1 - \frac{\dot{A}_{\mathrm{d}}}{\dot{A}_{\mathrm{sup}}}. \tag{9.59}$$

Note that the maximum theoretical value for the second-law efficiency—which would occur when the system is reversible—is 100%. Values below this maximum represent the extent of irreversibility in the system. Thus, the above metric of (9.59) is basically a representation of how close the data center is operating to its thermodynamically "ideal" state.

### *9.3.2 Computational Considerations*

Figure 9.5 summarizes the sequence of steps required to progress from system definition to exergy consumption. It is important to note that each step must be implemented in succession, but otherwise is independent. Thus, for example, the flow prediction and temperature mapping functions could be implemented separately using commercial CFD software and the exergy consumption could then be implemented as a postprocessing function for the fields predicted using off-the-shelf CFD software. For completeness, we illustrate both approaches in this text. The below description provides a grounds-up approach, where all steps must be implemented afresh. A case study demonstrating the results from an exergy analysis is presented in Sect. 9.3.3, wherein the flow and temperature prediction has been implemented using commercial CFD software.

#### 9.3.2.1 System Definition

The first step in developing the numerical model for exergy consumption is setting the system boundaries. In this case, the physical contour of data center comprises the boundary of the control volume. For purposes of simplicity, the plenum and ceiling are not included in the present model. The approach could easily be conceptually extended if desired; some of the additional considerations to be kept in mind while modeling the plenum and ceiling vents have been discussed by Karki et al. [130], Schmidt et al. [156], and Kang et al. [127].

For the thermal architecture of the room, the following assumptions are made in the system definition of the present model:

1. The location and magnitude of input flows are known. These could be measured at each perforated vent tile using a flow-hood or specified from plenum models of CFD simulations.
2. The location and magnitude of the return flows (ceiling or room return) to each CRAC are known from steady-state conditions (i.e., the total flow returned to the CRAC must match the flow supplied from each CRAC).
3. The location and magnitude of the heat load throughout the room are known. That is, the depth, width, length, and height of each row of racks are specified along with the heat load on each individual rack.
4. The flowrate through each of the rack units is known based on the fan curve of each system within the rack.

#### 9.3.2.2 Flow Modeling

Since the flow involves air at low temperatures and low Mach number, the flow is assumed to be incompressible, inviscid, and irrotational (i.e., potential flow). This assumption will be poor near the walls of the room or racks where a boundary layer of finite thickness may be formed, but the assumed model will generally be accurate elsewhere. It should be noted that this potential flow approach is simply used for illustrative purposes and to ensure expeditiousness of the model; and the accuracy of such an approach in terms of predicting flow patterns themselves is somewhat limited. McAllister et al. [157] and Shah et al. [73] discuss these limitations in more detail, suggesting that a hybrid model which includes viscous models near the racks and a potential flow model elsewhere in the room may be one approach to achieving higher accuracy, although identifying the optimal flow model to couple with exergy-based analysis in the data center remains an area of open research. Nonetheless, as discussed in Sect. 9.3.4, one of the benefits of an exergy-based approach is that for purposes of pinpointing inefficiencies in the data center infrastructure, exact numerical solutions are not necessary. Qualitative results based on trends from higher error, lower resolution models of the data center can still be useful into gaining insights as to how the data center may be changed from an energy efficiency standpoint. In this sense, using coarse models to identify an approximate design space which leads to optimal energy efficiency followed by a higher resolution, viscous-based model to ensure thermal manageability may be an appropriate compromise. Future research may explore this area of hybrid models in more detail.

In general, for potential flow at steady state, the conservation of mass and conservation of momentum equations can be reduced to obtain the Laplace equation [158–160]:

$$\frac{\partial^2 \phi}{\partial x^2} + \frac{\partial^2 \phi}{\partial y^2} + \frac{\partial^2 \phi}{\partial z^2} = 0, \tag{9.60}$$

where $\phi$ is the potential function specified as:

$$\vec{V} = \vec{\nabla}\phi \quad \left(\text{i.e., } u = \frac{\partial \phi}{\partial x}, v = \frac{\partial \phi}{\partial y}, w = \frac{\partial \phi}{\partial z}\right). \tag{9.61}$$

Because (9.60) is elliptic in nature, Dirichlet boundary conditions are required for the problem to be well posed, i.e., $\phi$ must be specified along each boundary of the system. This is straightforward for the data center, since the velocity components along the system boundary are known per assumption (1) and (2) earlier. The 18 boundary conditions (three components of velocity along each face of the data center boundary) have been summarized in Fig. 9.6.

A general closed-form analytical solution of (9.60) is difficult to derive owing to the discretized nonharmonic nature of the boundary conditions. Therefore, it is useful to instead numerically determine an approximate solution for the flow in each cell of the mesh. This can be accomplished when the cell sizes are sufficiently small that the exact differentials of (9.61) can be approximated as a difference term as follows:

$$\left(\frac{\partial \phi}{\partial x}\right)_{i,j,k} \approx \frac{\phi_{i+1,j,k} - \phi_{i,j,k}}{\Delta x}. \tag{9.62}$$

The above approximation is of the order $O(\Delta x)$. This approximation is only one of many methods in which the differential equation could be reduced to a difference equation. For example, the above is a first-order "forward difference" approximation (because the differential at $i,j,k$ has been approximated with the next $i + 1$ cell in the mesh). A "backward-difference" approximation could similarly be formulated using the $i - 1$ cell as follows:

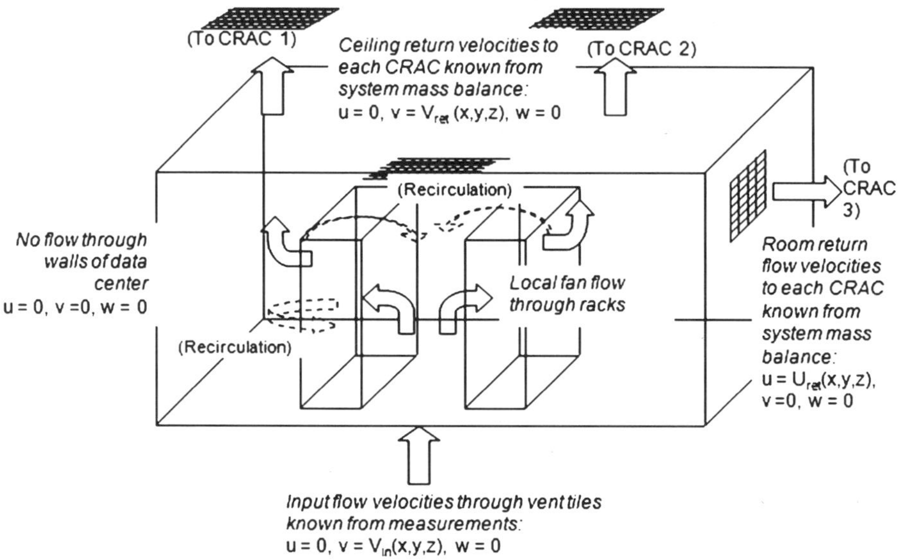


**Fig. 9.6** Summary of flow model and boundary conditions used to solve flow conditions in the data center

$$\left(\frac{\partial \phi}{\partial x}\right)_{i,j,k} \approx \frac{\phi_{i,j,k} - \phi_{i-1,j,k}}{\Delta x}. \tag{9.63}$$

This approximation is also of the order $O(\Delta x)$.

A higher order approximation could also be considered by averaging the slope of the forward- and backward-difference methods. In this case, a "central difference" approximation is obtained as follows:

$$\left(\frac{\partial \phi}{\partial x}\right)_{i,j,k} \approx \frac{\phi_{i+1,j,k} - \phi_{i-1,j,k}}{2\Delta x}. \tag{9.64}$$

The approximation of (9.64) can be shown to be of the order of $O(\Delta x^2)$.

Similarly, an approximation for the second derivative could be obtained as follows:

$$\left(\frac{\partial^2 \phi}{\partial x^2}\right)_{i,j,k} = \left(\frac{\partial}{\partial x}\left[\frac{\partial \phi}{\partial x}\right]\right)_{i,j,k} \approx \frac{\phi_{i+1,j,k} + \phi_{i-1,j,k} - 2\phi_{i,j,k}}{(\Delta x)^2}. \tag{9.65}$$

The above approximation is of the order $O(\Delta x^2)$. Further details regarding higher-order approximation schemes as well as stability criteria and error estimations can be found in a standard textbook on numerical or computational methods (e.g., [161–167]). In the present model, the approximations of (9.64) and (9.65) are used.

Discretizing the differential Laplace equation of (9.60) as an algebraic difference equation,

$$\frac{\phi_{i+1,j,k}+\phi_{i-1,j,k}-2\phi_{i,j,k}}{(\Delta x)^2}+\frac{\phi_{i,j+1,k}+\phi_{i,j-1,k}-2\phi_{i,j,k}}{(\Delta y)^2}+\frac{\phi_{i,j,k+1}+\phi_{i,j,k-1}-2\phi_{i,j,k}}{(\Delta z)^2}=0. \tag{9.66}$$

Equation (9.66) can readily be applied to each cell of the mesh, with the appropriate adjustments being made at each of the boundaries per the conditions of Fig. 9.6. Having done this, a set of simultaneous equations is obtained that can be expressed as:

$$[C][\phi] = [F], \tag{9.67}$$

where for a system of $n$ total cells, $[C]$ is a square $n \times n$ coefficient matrix, $[\phi]$ is an $n \times 1$ column vector containing the unknown variables $(\phi_1, \phi_2, \phi_3 \ldots \phi_n)$, and $[F]$ is an $n \times 1$ column vector containing the prescribed boundary conditions. From (9.66), it can be inferred that each row of $[C]$ will consist of seven nonzero entries; thus, the matrix $[C]$ will be quite sparse. In general, $[C]$ will be a block-diagonal matrix. Equation (9.67) can be readily solved using a variety of matrix inversion methods, including Gaussian elimination or triangular decomposition. However, such techniques generally do not take advantage of the sparseness of the matrix $[C]$. Therefore, it is common to use an iterative method for solving (9.67). These methods usually consist of guessing an initial solution $\phi^{(0)}$and then repeating the calculations across the entire mesh till a solution which agrees with all the specified boundary conditions is found. The simplest iteration scheme is the Jacobi method, which is found from (9.66) by solving for the $\phi$ term with the largest coefficient as follows:

$$\phi_{i,j,k}^{(n+1)}=\frac{\left\{\dfrac{\phi_{i+1,j,k}^{(n)}+\phi_{i-1,j,k}^{(n)}}{(\Delta x)^2}+\dfrac{\phi_{i,j+1,k}^{(n)}+\phi_{i,j-1,k}^{(n)}}{(\Delta y)^2}+\dfrac{\phi_{i,j,k+1}^{(n)}+\phi_{i,j,k-1}^{(n)}}{(\Delta z)^2}\right\}}{\left\{\dfrac{2}{(\Delta x)^2}+\dfrac{2}{(\Delta y)^2}+\dfrac{2}{(\Delta z)^2}\right\}}, \tag{9.68}$$

where the superscript denotes the iteration level. A slightly more sophisticated form of the iteration can be developed by utilizing the most recent values of the computation on the right-hand side, i.e.,

$$\phi_{i,j,k}^{(n+1)}=\frac{\left\{\dfrac{\phi_{i+1,j,k}^{(n)}+\phi_{i-1,j,k}^{(n+1)}}{(\Delta x)^2}+\dfrac{\phi_{i,j+1,k}^{(n)}+\phi_{i,j-1,k}^{(n+1)}}{(\Delta y)^2}+\dfrac{\phi_{i,j,k+1}^{(n)}+\phi_{i,j,k-1}^{(n+1)}}{(\Delta z)^2}\right\}}{\left\{\dfrac{2}{(\Delta x)^2}+\dfrac{2}{(\Delta y)^2}+\dfrac{2}{(\Delta z)^2}\right\}}. \tag{9.69}$$

The iteration approach of (9.69) is known as the Gauss–Seidel method [168–171].

From a computational standpoint, the Gauss–Seidel method is inherently more efficient than the Jacobi method because the latter requires the continuous storage of two matrices, while the former can be accomplished with only one matrix (since the new values can directly be rewritten into the matrix at each iteration). Techniques to further accelerate convergence in the Gauss–Seidel method have also been developed by introducing a relaxation factor $\omega$ as follows:

$$\phi_{i,j,k}^{(n+1)} = \frac{\left\{\frac{\phi_{i+1,j,k}^{(n)} + \phi_{i-1,j,k}^{(n+1)}}{(\Delta x)^2} + \frac{\phi_{i,j+1,k}^{(n)} + \phi_{i,j-1,k}^{(n+1)}}{(\Delta y)^2} + \frac{\phi_{i,j,k+1}^{(n)} + \phi_{i,j,k-1}^{(n+1)}}{(\Delta z)^2}\right\}}{\left\{\frac{2}{(\Delta x)^2} + \frac{2}{(\Delta y)^2} + \frac{2}{(\Delta z)^2}\right\}} + (1-\omega)\phi_{i,j,k}^{(n+1)}. \tag{9.70}$$

Iterative convergence is achieved for choices of $0 < \omega < 2$ [where $\omega = 1$ corresponds to the Gauss–Seidel iteration of (9.69)]. Values of $0 < \omega < 1$ are considered successive under relaxation (SUR), while values of $1 < \omega < 2$ are considered successive over relaxation (SOR). The appropriate choice for $\omega$ depends on the grid size, number of cells, and convergence criteria. If $\omega$ is chosen such that convergence is reached in a minimum number of iterations, the choice of $\omega$ is said to correspond to an optimized SOR.

In the present model, the SOR iterative method is chosen to solve (9.67). The value of $\omega$ was chosen based on the data center characteristics, grid size, and convergence criteria. For most cases, a tolerance of $10^{-6}$ was required for convergence, i.e., the convergence criteria were established as:

$$\left\| \phi_{i,j,k}^{(n+1)} - \phi_{i,j,k}^{(n)} \right\| \leq 10^{-6} \tag{9.71}$$

Having solved the given mathematical problem of (9.67) for the boundary conditions of Fig. 9.6, the velocities at each point on the grid can be calculated by discretizing (9.64) as follows:

$$u_{i,j,k} \approx \frac{\phi_{i+1,j,k} - \phi_{i-1,j,k}}{\Delta x} \tag{9.72}$$

$$v_{i,j,k} \approx \frac{\phi_{i,j+1,k} - \phi_{i,j-1,k}}{\Delta y}, \tag{9.73}$$

$$w_{i,j,k} \approx \frac{\phi_{i,j,k+1} - \phi_{i,j,k-1}}{\Delta z}. \tag{9.74}$$

Thus, the flow velocities at each point of the grid can be mapped. If the pressure distribution is desired, this can be back-calculated from the Bernoulli equation for potential flow:

$$P + \frac{1}{2}\rho\left[u^2 + v^2 + w^2\right] = \text{const}, \tag{9.75}$$

where the constant is determined based on the boundary conditions of the problem.

#### 9.3.2.3 Temperature Approximations

Once the velocities and pressure distribution of the system are known, the temperature distribution of the airspace can be approximated from the conservation of energy equation. For an ideal gas with constant thermal properties and no viscous dissipation (because Ma $\ll$ 1), the conservation of energy equation at steady state yields:

$$u\frac{\partial T}{\partial x} + v\frac{\partial T}{\partial y} + w\frac{\partial T}{\partial z} - \alpha\left(\frac{\partial^2 T}{\partial x^2} + \frac{\partial^2 T}{\partial y^2} + \frac{\partial^2 T}{\partial z^2}\right) - \frac{\dot{q}_{\text{gen}}'''}{\rho C_{\text{p}}} = 0, \tag{9.76}$$

where $\dot{q}_{\text{gen}}'''$ is the heat generated per unit volume of the cell.

The above equation is a parabolic equation, so Dirichlet or mixed boundary conditions are required at each boundary for the problem to be well posed. In this case, the walls of the data center are assumed to be well-insulated, while the temperatures of the input streams are assumed to be known. The temperature of the return streams can be determined based on an energy balance for the entire system. Figure 9.7 summarizes these boundary conditions.

Discretizing (9.76) using a central-difference scheme similar to (9.64) and (9.65),

$$\begin{aligned}
&\left[u_{i,j,k}\left(\frac{T_{i+1,j,k} - T_{i-1,j,k}}{\Delta x}\right) + v_{i,j,k}\left(\frac{T_{i,j+1,k} - T_{i,j-1,k}}{\Delta y}\right) + w_{i,j,k}\left(\frac{T_{i,j,k+1} - T_{i,j,k-1}}{\Delta z}\right)\right] \\
&- \alpha\left[\left(\frac{T_{i+1,j,k} + T_{i-1,j,k} - 2T_{i,j,k}}{(\Delta x)^2}\right) + \left(\frac{T_{i,j+1,k} + T_{i,j-1,k} - 2T_{i,j,k}}{(\Delta y)^2}\right)\right. \\
&\left. + \left(\frac{T_{i,j,k+1} + T_{i,j,k-1} - 2T_{i,j,k}}{(\Delta z)^2}\right)\right] = \frac{\left(\dot{q}_{\text{gen}}\right)_{i,j,k}}{\rho C_{\text{p}}(\Delta x \Delta y \Delta z)}.
\end{aligned} \tag{9.77}$$

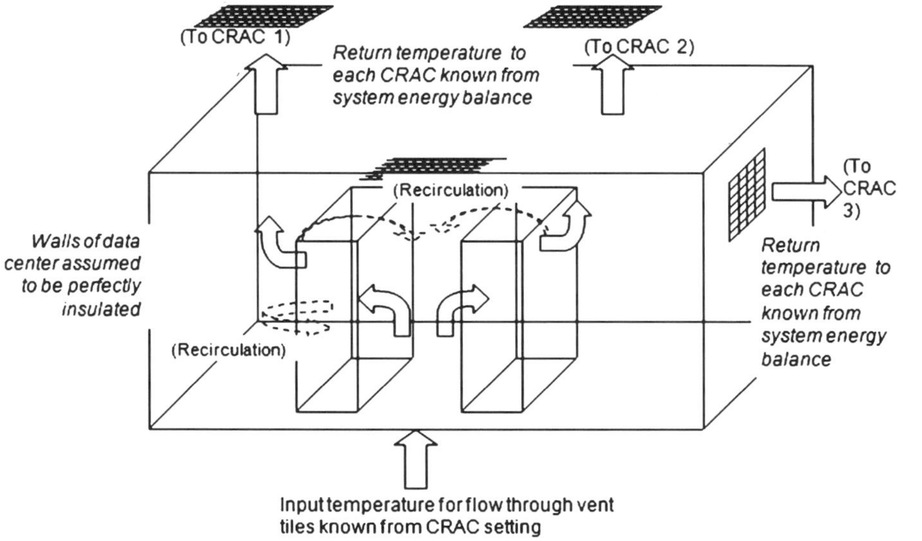


**Fig. 9.7** Summary of boundary conditions used to solve for temperature in the data center airspace

Using a Gauss–Seidel iterative scheme with SOR to solve,

$$T_{i,j,k}^{(n+1)} = \frac{1}{\frac{2}{\alpha}\left[\frac{1}{(\Delta x)^2}+\frac{1}{(\Delta y)^2}+\frac{1}{(\Delta z)^2}\right]}\left\{\begin{array}{l} \frac{(\dot{q}_{\mathrm{gen}})_{i,j,k}}{\rho C_{\mathrm{p}}(\Delta x\Delta y\Delta z)} - \left[\frac{u_{i,j,k}}{\Delta x} - \frac{\alpha}{(\Delta x)^2}\right]T_{i+1,j,k}^{(n)} \\ - \left[\frac{v_{i,j,k}}{\Delta y} - \frac{\alpha}{(\Delta y)^2}\right]T_{i,j+1,k}^{(n)} - \left[\frac{w_{i,j,k}}{\Delta z} - \frac{\alpha}{(\Delta z)^2}\right]T_{i,j,k+1}^{(n)} \\ + \left[\frac{u_{i,j,k}}{\Delta x} - \frac{\alpha}{(\Delta x)^2}\right]T_{i-1,j,k}^{(n)} + \left[\frac{v_{i,j,k}}{\Delta y} - \frac{\alpha}{(\Delta y)^2}\right]T_{i,j-1,k}^{(n)} \\ + \left[\frac{w_{i,j,k}}{\Delta z} - \frac{\alpha}{(\Delta z)^2}\right]T_{i,j,k-1}^{(n)} \end{array}\right\}$$

$$+ (1-\omega)T_{i,j,k}^{(n)}. \tag{9.78}$$

Again, the value of $\omega$ was chosen based on the data center characteristics, grid size, and convergence criteria. For most cases, a tolerance of $10^{-3}$ was required for convergence, i.e., the convergence criteria were established as:

$$\left\| T_{i,j,k}^{(n+1)} - T_{i,j,k}^{(n)} \right\| \leq 10^{-3}. \tag{9.79}$$

Thus, the temperatures in the system at each $(i,j,k)$ location can be determined.

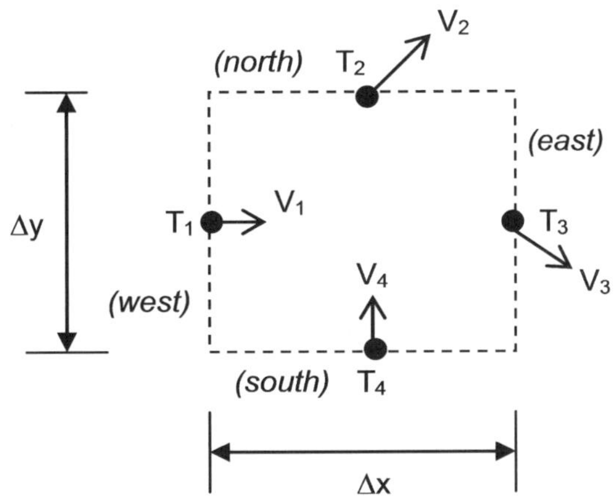


**Fig. 9.8** Simplification of mesh as control volumes to enable calculation of exergy consumption. Each face of the control volume is assumed to correspond to the temperature and flow of that node in the mesh, thus providing a map of exergy transport throughout the system

#### 9.3.2.4 Exergy Consumption Calculations

Knowing the temperature and flow in each cell of the mesh, the exergy consumption in each cell can be calculated per (9.58). Because the exergy consumption in the airspace is related mostly to the flow of air, it is useful to consider the transport of exergy *across* a cell in the mesh (rather than the exergy inside the cell of a mesh). In other words, consider a control volume drawn between adjacent individual nodes of the mesh such that the discretized values calculated for flow and temperature are valid at the cell boundaries. Such an approximation is shown in Fig. 9.8 for two dimensions; the extension to three dimensions that includes a front and back face is straightforward.

Using such an approximation, the exergy transport across each control volume in the mesh becomes apparent. In the example of Fig. 9.8, exergy is transported into the control volume at the west (W) and south (S) faces at temperatures $T_1$ and $T_4$, respectively. Similarly, exergy leaves the control volume at the north (N) and east (E) faces at temperatures $T_2$ and $T_3$, respectively. The velocities associated with the flow can be calculated by the norm of the velocity vectors at each component, that is, at the $j$th node:

$$V_j = \sqrt{u_j{}^2 + v_j{}^2 + w_j{}^2}. \tag{9.80}$$

Since constant density is assumed, the massflow corresponding to each flow will be:

$$\dot{m}_j = \pm \rho \bar{A}_j V_j, \tag{9.81}$$

where the area $\bar{A}_j$ is the cross-sectional area corresponding to the face across which the flow occurs. For example, for the north or south faces, the area will be $\bar{A}_{\mathrm{N}} = \bar{A}_{\mathrm{S}} = \Delta x \Delta z$, while for the east or west faces, the area will be $\bar{A}_{\mathrm{W}} = \bar{A}_{\mathrm{E}} = \Delta y \Delta z$.

In three dimensions, the area corresponding to the front or back faces will be $\bar{A}_{\mathrm{F}} = \bar{A}_{\mathrm{B}} = \Delta x \Delta y$. Net flows going into the cell at each face are positive, while flows going out of the cell are considered negative. The sign could be determined by inspection or mathematically as follows:

$$\mathrm{sgn}(\dot{m}_j) = \mathrm{sgn}(-\vec{V}_j \cdot \hat{n}_j), \tag{9.82}$$

where $\hat{n}_j$ is a normal vector perpendicular to the face $j$ but pointing away from the center of the cell. That is, if $\vec{V}_j$ and $\hat{n}_j$ are in opposite directions, then the flow will be into the cell, so $\dot{m}_j$ will be positive.

Because the flow into or outside of the cell has inherently been captured by the sign of $\dot{m}_j$, (9.58) can be collapsed into a simplified form as follows:

$$\dot{A}_{\mathrm{d}} = \sum_j \left(1 - \frac{T_0}{T_j}\right)\dot{Q}_j + \sum_j \dot{m}_j \left[(h_j - h_0) + \frac{V_j^2}{2} + g(z_j - z_0) - T_0(s_j - s_0)\right]. \tag{9.83}$$

Assume that the heat input from the racks occurs at an average temperature of the cell boundaries $T_{\mathrm{b}}$, that is:

$$T_{\mathrm{b}} = \sum_{j=1}^{N_{\mathrm{f}}} \left(\frac{T_j}{N_{\mathrm{f}}}\right), \tag{9.84}$$

where $N_{\mathrm{f}}$ is the total number of faces of each cell ($N_{\mathrm{f}} = 4$ for two dimensions, $N_{\mathrm{f}} = 6$ for three dimensions). The approximation of (9.84) will be valid if the cells are small enough such that the rack heat diffuses equally to all the boundaries of the cell.

Combining (9.83) and (9.84), the exergy consumption in the $j$th cell is obtained as:

$$\dot{A}_{\mathrm{d}_j} = \left(1 - \frac{T_0}{T_{\mathrm{b}}}\right)\dot{Q}_j + \sum_j \dot{m}_j \left[(h_j - h_0) + \frac{V_j^2}{2} + g(z_j - z_0) - T_0(s_j - s_0)\right]. \tag{9.85}$$

Approximating the enthalpy and entropy for air using the Gibbs' formulation for an ideal gas with constant specific heat,

$$\dot{A}_{\mathrm{d}_j} = \left(1 - \frac{T_0}{T_{\mathrm{b}}}\right)\dot{Q}_j + \sum_j \dot{m}_j \left[C_{\mathrm{p}}(T_j - T_0) - T_0\left(C_{\mathrm{p}}\ln\frac{T_j}{T_0} - R\ln\frac{P_j}{P_0}\right) + \frac{V_j^2}{2} + g(z_j - z_0)\right]. \tag{9.86}$$

Equation (9.86) is the desired equation to calculate the exergy consumption in the $j$th cell of the mesh. Since the ground state is maintained common across each cell of the mesh, the total exergy consumption in the airspace will be the sum of all the individual cell exergy consumption:

$$\dot{A}_{\mathrm{d_{airspace}}} = \sum_{j=1}^{N} \dot{A}_{d_j} \tag{9.87}$$

where $N$ is the total number of cells in the room.

### 9.3.3 Case Study

For demonstration purposes, the model was applied to the data center shown in Fig. 9.9. Ceiling return mechanism was considered for all experimental and modeling cases, with the locations of return vents noted in Fig. 9.9. Air is supplied from two 105-kW (30-ton) Liebert FH600C-AAEI CRAC units, each of which has a maximum air flowrate of 7.9 $m^3/s$ (17,100 cfm) per manufacturer's specifications. The air enters the room either through low-flow vent tiles [which have dampeners that restrict maximum throughput to 0.3 $m^3/s$ (750 cfm)] or high-flow vent tiles with no dampeners. The exact magnitude of flow through each tile will be a function of plenum pressure.

Heat input is provided from two rows of computing racks. The heat dissipation from each rack is specified using rack power measurements or from manufacturer's specifications. The relevant heat loads are summarized in Fig. 9.9. A base case scenario for the model was run with all of the CRAC units operating at 60% fan speed at a supply temperature of 16°C for a ground state corresponding to the supply temperature.

For convenience, because the focus of this work is around modeling the exergy consumption within the data center, off-the-shelf CFD software is utilized for solving the momentum and energy transfer within the data center. This enables ensuring that the results from an exergy-based model are being compared to a model that has fairly wide acceptance within the industry. Specifically, the commercial CFD code Flovent™ [172] was used to numerically solve for the flow and temperature properties using an LVEL k–ε model across a nonuniform grid of 569,772 cells. Grid independence and convergence criteria of subunity residuals were verified per the work of Patel et al. [105]. The model converged in roughly 6 h on an HP Proliant ML370 machine. For the established grid, the temperature and flow values were then postprocessed per the procedure outlined in Sect. 9.3.2 to calculate the exergy consumption values in each cell. (It should be noted that results from this model were then also compared to those obtained from a lower resolution potential flow model of the data center, and we find that the qualitative conclusions drawn from each model are quite similar. For purposes of conciseness, we only discuss the results from the former, and refer the reader to the work by McAllister et al. [157] for a more detailed discussion of the latter.)

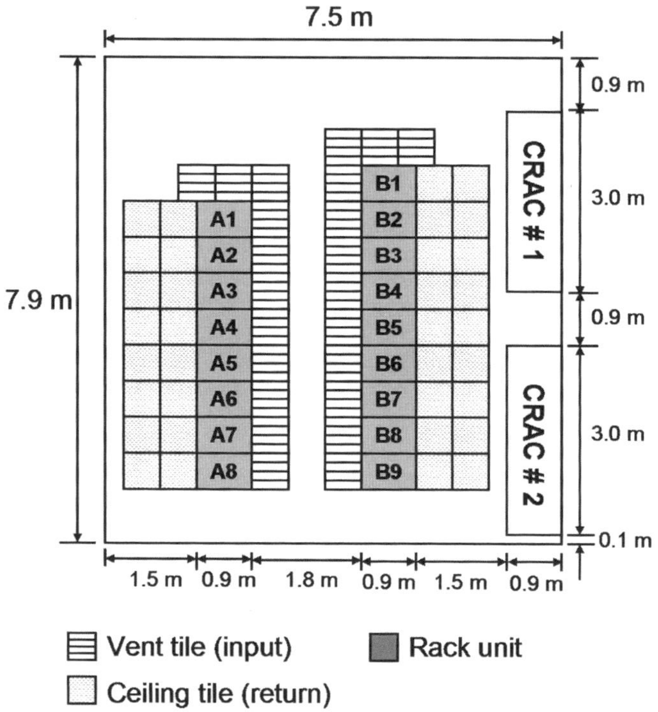


| Rack | Power (kW) |
|---|---|
| A1 | 2.0 |
| A2 | 2.6 |
| A3 | 2.6 |
| A4 | 2.6 |
| A5 | 2.6 |
| A6 | 3.2 |
| A7 | 3.2 |
| A8 | 0.0 |

| Rack | Power (kW) |
|---|---|
| B1 | 9.9 |
| B2 | 9.9 |
| B3 | 6.6 |
| B4 | 6.6 |
| B5 | 6.6 |
| B6 | 6.6 |
| B7 | 6.6 |
| B8 | 6.6 |
| B9 | 0.0 |

**Fig. 9.9** System layout and rack loading for test data center (*top view*). Each floor tile has an area of 0.6 m × 0.6 m (2 ft × 2 ft). The floor-to-ceiling height is 2.7 m (9 ft), while the depth of the under-floor plenum is 0.6 m (2 ft) and the height of the ceiling return plenum is 1.2 m (4 ft)

## 9.3.4 *Results*

### 9.3.4.1 Airspace Inefficiencies

Results from the model for temperature and exergy consumption are shown in Figs. 9.10a and b, respectively, at heights of 0.9 m (3 ft), 1.5 m (5 ft), 2.1 m (7 ft) and 2.7 m (9 ft) from the floor. At a height of 0.9 m (3 ft), the temperature distribution is as desired, with the hot aisles being slightly warmer than the cold aisles. No recirculation effects are noticed in the cold aisles. The same trend continues at a height of 1.5 m (5 ft), with the hot aisle being slightly warmer due to the mixing of exhausts from the servers loaded in the bottom and middle of the racks. At a height of 2.1 m (7 ft), the recirculation effects are clearly noticeable in the middle of the rows. The temperature map at 2.7 m (9 ft) suggests that this is due to recirculation over the top of the rack, although some recirculation around the end of the rows is also observable.

The exergy consumption maps of Fig. 9.10b suggest similar trends of recirculation. At a height of 0.9 m (3 ft), some exergy consumption is observed in the cold aisles due to nonuniform temperature of the supply air exiting from the plenum. But the magnitude of this exergy consumption is small (less than 50 W), and does not significantly impact the temperature of the cold aisle. Further, the effects of

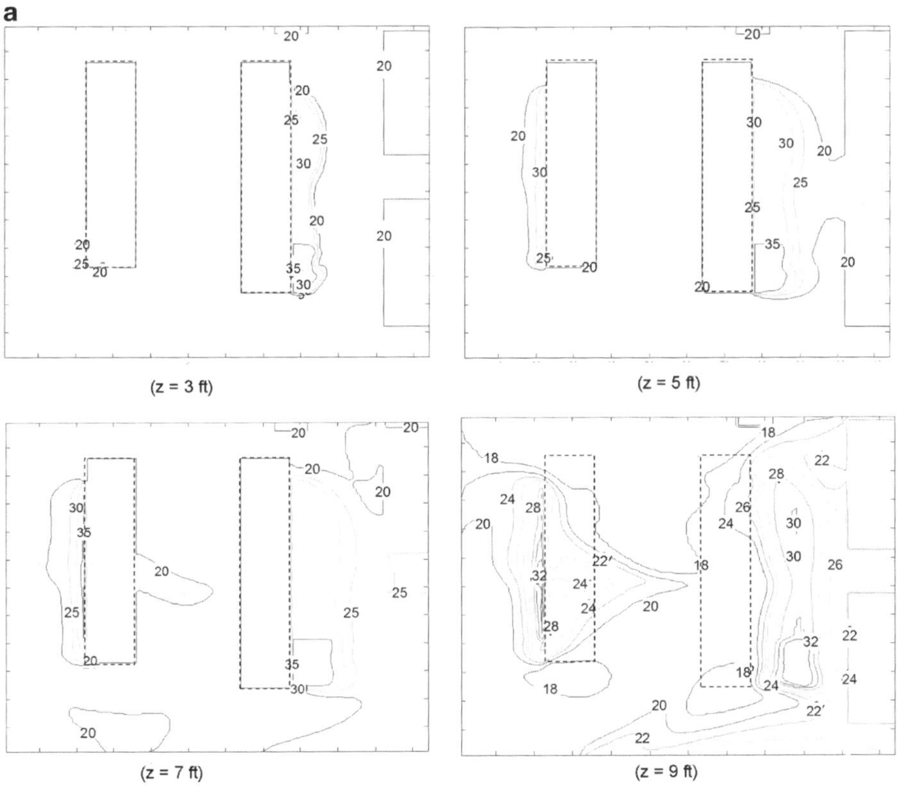


**Fig. 9.10** Results from numerical model (*top view*). Each map is a top view of the data center airspace at heights of 0.9 m (3 ft), 1.5 m (5 ft), 2.1 m (7 ft), 2.7 m (9 ft). (**a**) Temperature predictions [in °C] from model, and (**b**) Map of exergy consumption rate [in W] throughout data center. *Dotted lines* show the rack locations. The regions of exergy consumption correspond well to hotspots or areas of recirculation estimated from the temperature map

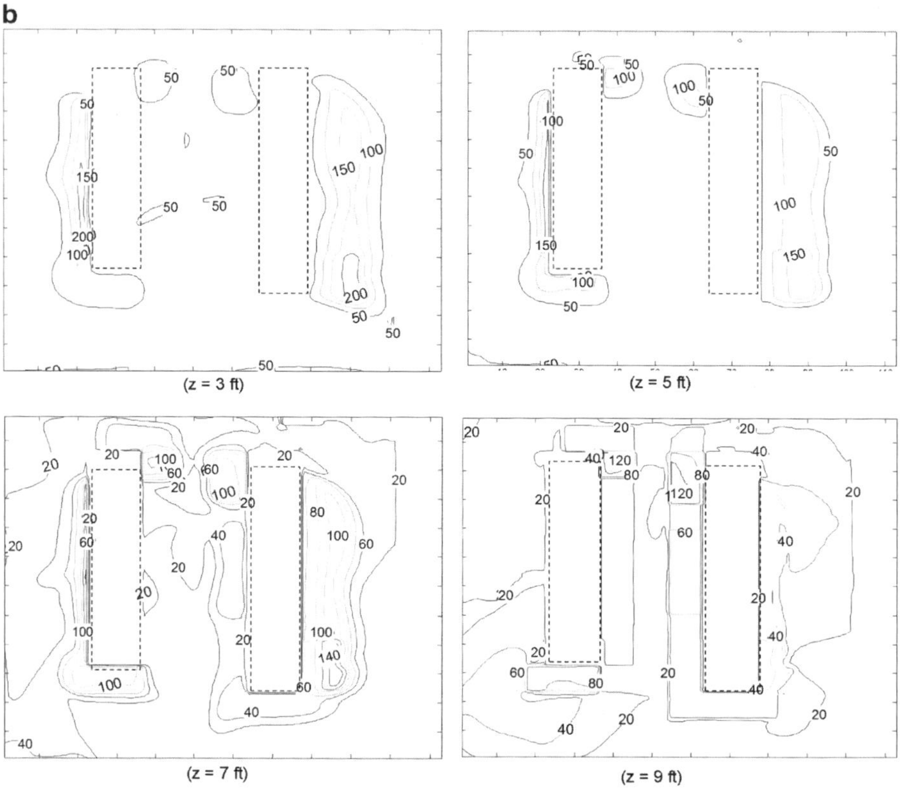


**Fig. 9.10** (continued)

mixing are clearly observable in the hot aisles, where exergy is destroyed due to outlet temperature gradients caused by nonuniform heat loading of the racks. (The impact of uniformity of rack heat load distribution on energy efficiency, as well as the relevance of this map for equipment placement, is discussed shortly.) At a height of 1.5 m (5 ft), the supply air through the vent tiles is well mixed and therefore devoid of significant temperature gradients. As a result, almost no exergy consumption is noticed in the cold aisle. The effects of recirculation are clearly evident at a height of 2.1 m (7 ft), where mixing in the cold aisle is noticeable in the middle of the rows A and B. This corresponds to the observations from Fig. 9.10a, where higher rack inlet temperatures were noticed in these locations. Lastly, at a height of 2.7 m (9 ft), it is interesting to note that the exergy consumption is actually quite low even though the temperatures are high. This is because most of the high-temperature air in the hot aisles is removed through the ceiling vents, and although the temperatures in the cold aisle may be higher, there is not much mixing taking place at these locations. Some effects of recirculation are evident around the end of the rows, which is the cause of the higher temperatures observed in these locations in Fig. 9.10a.

Thus, there is good qualitative agreement between the temperature-based and exergy-based predictions of recirculation patterns from the numerical model. The exergy-based predictions of recirculation agree with the temperature-based assessments, although the exergy consumption model fuses information regarding flow and temperature into a single map to provide insight regarding the locations of mixing. To verify the numerical accuracy of these qualitative assessments, the predictions from the model were compared against actual measurements in the data center. Note that identifying the locations of mixing within the data center is only one aspect of the second-law analysis; additional applications are described later in this chapter.

In performing experimental validation, we recognize that the exergy-based data center model discussed above essentially performs three computational operations in sequential order:

- Flow approximation (based on conservation of mass and momentum),
- Temperature estimation (based on energy balances),
- Determination of system exergy consumption (based on flow and temperature maps).

Thus, to estimate the accuracy of the exergy-based model, it is necessary to first verify the variability of the flow and temperature inputs provided to the exergy model. The error introduced during the exergy consumption calculations can then be estimated through standard statistical analysis [173–177].

The input flow conditions through the vent tiles were measured using a flow hood (estimated to be accurate to within 5% of actual values). Measurements were also made with a hand-held anemometer, empirically estimated to be within 20% of calibrated values, at several locations in the room. Hand-held thermometers and type-K thermocouples (accurate to within $\pm 0.1^\circ C$ per manufacturer's specifications) are used to obtain temperature readings at the data collection points. The points of data collection were chosen to allow for assessment of model performance at different key locations of the system, including rack inlet, rack outlet, CRAC return, and potential hotspots in the hot aisles. Additional measurements were also made in the cold aisles to assist in the identification of locations of recirculation or short-circuiting suggested by the model, yielding a total of 34 flow measurements and 102 temperature measurements throughout the data center.

Figure 9.11a shows a summary comparison of actual measurements and predicted values from the model. The mean difference between predicted and measured flow data is 18%, while the mean difference for temperature data is 13%. This is considered an acceptable level of accuracy for a first approximation. If further accuracy is desired, then finer meshes or more accurate flow modeling techniques should be considered. For example, the turbulence parameters in the k–ε model may need to be adjusted to suit the environment being modeled, or a full data center model that includes flow through the servers as well as the plenums may yield a more accurate pressure map within the room.

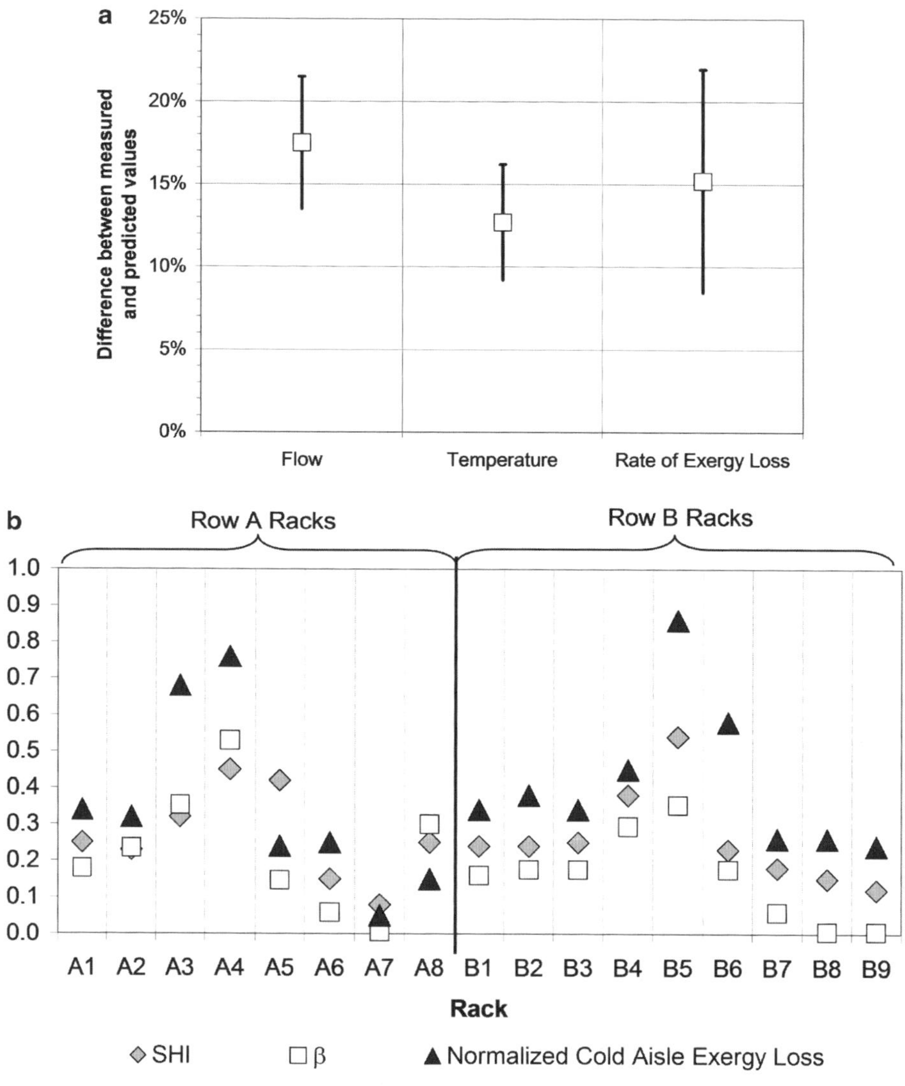


**Fig. 9.11** Experimental validation of model. (**a**) Comparison of predicted and measured values. (**b**) Comparison of rack-level estimates given by different metrics

Because exergy consumption is calculated in the model based on system temperature and flow values, and since both of these quantities have already been validated, it is expected that predictions for exergy consumption will also be within the same limits of accuracy. Nonetheless, to obtain a quantitative estimate of model accuracy with regards to prediction of exergy consumption, an indirect estimate of the actual exergy consumption in the physical airspace was obtained from the following equation:

$$\dot{A}_{d_{\text{airspace}}} = \dot{A}_{d_{\text{airspace,aisles}}} + \dot{A}_{d_{\text{airspace,CRAC}}} + \dot{A}_{d_{\text{airspace,racks}}} = \dot{A}_{\text{airspace,sup}} - \dot{A}_{\text{airspace,rec}}. \quad (9.88)$$

The logic implicit to the expression of (9.88) is that any difference between the exergy supplied to the data center and the exergy recoverable from the data center must be due to the destruction of exergy in the air going through the data center (i.e., if no exergy were destroyed, then in the absence of useful work, all of the exergy supplied should be recoverable). This exergy consumption has been assumed to arise from three major sources:

- The exergy consumption for the air in the hot and cold aisles. For simplicity of nomenclature, it is assumed that the space above the rack units is part of the hot aisle, while any space between two rows of rack units comprises a cold aisle. This exergy consumption is mostly due to mixing, and cannot be measured directly in the data center.
- The exergy consumption as the air flows through the rack units, which can be estimated based on a caloric heat balance across the rack using rack power measurements and inlet–outlet temperature and flow measurements.
- The exergy consumption as the air is refrigerated in the CRAC units, which can be measured based on the fan speed and COP of the CRAC unit.

All the known quantities can be substituted into (9.88), and an experimental estimate of the exergy consumption for the airflow through the hot and cold aisles can then be obtained (this is the term predicted by the model which needs to be experimentally verified). It should be noted that small errors in the exergy consumption computed using (9.88) are to be expected, since some phenomena (such as exergy consumption due to heat escape through walls or air leaks) will not be accounted for. However, the magnitude of these losses will be small, and as shown in Fig. 9.11a, the estimated values based on measurements compare quite well to those predicted by the model. The uncertainty in estimating the difference between predicted and actual exergy consumption values stems mostly from the uncertainty in temperature and flow measurements as well as the inaccuracy of flow and temperature predictions discussed earlier. Predictions of airspace exergy consumption with higher accuracy would require the development of more accurate models for predicting temperature and flow.

The above validation is essentially a confirmation of the accuracy of the numerical values calculated by the model. It would also be desirable to obtain some types of validation regarding the viability of using airspace exergy consumption as a metric to estimate recirculation effects in the data center. The existing state-of-the-art utilizes the metrics of SHI (9.48) and $\beta$ (9.50) to measure recirculation. Both of these are temperature-based measures of the infiltration of warm air into the cold aisle, and thus only represent thermal effects of mixing in the cold aisle (while the airspace exergy consumption calculated in (9.86) includes thermal and fluidic mixing in both the hot and cold aisles). Therefore, to ensure a valid comparison, the exergy consumption model was rerun for the data center of Fig. 9.6, and the total cold aisle thermal exergy consumption was determined.

To normalize these values, the exergy consumption estimate was divided by the exergy supplied to the cold aisle [from (9.88)], assuming the ground state to be the external ambient. This normalized expression of exergy consumption is scaled so that a value of 0 corresponds to no recirculation while a value of 1 would be a case where all the warm air from the hot aisle gets recirculated to the cold aisle. This exergy-based estimate of recirculation is then compared to rack-level averages of SHI and $\beta$ obtained by temperature measurements in the actual data center. Figure 9.11b summarizes the results of this evaluation. Although the magnitude of each metric on the normalized scale differs—which is to be expected, because each metric is defined differently—the trends of recirculation predicted by all the metrics are very similar. The highest SHI and $\beta$ values, which represent those racks with the highest inlet temperatures, also correspond to the racks with the highest exergy consumption. Additionally, the exergy-based metric is more sensitive to recirculation than the temperature-based metrics.

To summarize, for the sample data center shown in Fig. 9.9, it has been shown that:

- Recirculation trends suggested by temperature maps (Fig. 9.10a) or exergy consumption maps (Fig. 9.10b) are qualitatively similar,
- Numerical error between estimated and predicted values are comparable for temperature and exergy consumption (Fig. 9.11a),
- Quantitative estimates of recirculation from the exergy-based metrics agree with other temperature-based metrics (Fig. 9.11b).

These observations confirm that an exergy-based approach can provide, at a minimum, the same type of information regarding data center thermal manageability as the more traditional temperature-based metrics. In addition, because the exergy maps provide granular data regarding thermofluidic phenomena in the data center (in the same manner as temperature and velocity fields), the exergy analysis can also be used to address local inefficiencies in the data center thermal management system. For example, locations of high exergy consumption in the exergy maps of Fig. 9.10b pinpoint where thermal inefficiencies may need to be mitigated through the use of localized solutions, such as containment or blanking panels. This is an important finding, as it suggests the viability of using exergy as a platform upon which integrated control of thermal management and energy efficiency within the data center may be achieved. For example, due to heterogeneity within the data center infrastructure, certain racks are often "easier" to cool than others. Placing computational workload on racks with specific thermal characteristics (rather than allocating workloads without regards to thermal considerations) has been shown to have the potential for significant energy savings [136]. Existing metrics to quantify the thermal characteristics of different racks are based on the temperature distribution within the data center. The analysis presented suggests that exergy consumption could be an alternative metric for such ranking of rack units. While harder to quantify in practice, an exergy-based metric could potentially provide a more physically meaningful way to represent the differences resulting from workload placement across racks with heterogeneous thermal characteristics. As illustration, by measuring the reduction in thermal irreversibility availed by shutting down unused racks after workload consolidation, it may become possible to directly quantify the energy savings of workload management. By contrast, in existing approaches, only the most efficient rack locations are identified without necessarily gaining insight into the extent of savings that might be achievable from actively managing or migrating workloads.

Similarly, instead of workload placement, the exergy maps of Fig. 9.10b can help guide the choice of rack placement within a data center. For example, if new equipment—or higher density equipment than the previous generation of equipment—is being situated in the data center, then placing this equipment in the locations with the lowest exergy consumption would have the smallest impact in terms of overall data center efficiency. On the other hand, situating the equipment in a location with high exergy consumption will lead to further irreversibilities and a lower second-law efficiency.

Another perspective that the above analysis potentially enables relates to the use of outside air for cooling. Airside economization is a growing trend in data center thermal management, since this can reduce capital and operational costs in the data center by eliminating the need for a chiller. Effectively, outside air is a viable method to cool the data center when there is agreement between the desired supply side (cold aisle) temperature and the temperature of the external ambient outside the data center. But, in such a scenario, the only difference between the supply-side airstream and the external ambient is the velocity of the air. Thermally, the air is in an identical state in the cold aisle and outside the data center. That is, the air in the cold aisle has no capacity to do thermal work—it is at the ground state—and thus, there is no thermal irreversibility between the outside air and the supply air. In such an internally reversible system, the air being supplied from the CRAC unit would have the same cooling capacity as the air outside the data center. No gains are derived by doing additional work in between the ground state and the point at which air is supplied to the cold aisle, which motivates the opportunity to use airside economization and providing outside air directly into the cold aisle. Such a metric that defines the difference between the desired supply state and the ambient ground state could also be used for making judgments around global workload placement in networks with a multiplicity of data centers are available for hosting a given workload. Effectively, from the perspective of energy-efficient thermal management, the location where the difference in exergy content between the desired supply-side state and the ground state is smallest will be the preferred processing location. Note that additional constraints related to considerations such as computational performance, distance of the data center from its core user base, security, and differences in the infrastructure within each data center would also need to be taken into account while making such placement decisions.

Thus, there is reason to believe that the above exergy analysis provides insights into opportunities for enhancing the data center energy efficiency. The next section explores such opportunities beyond just the airspace of the data center.

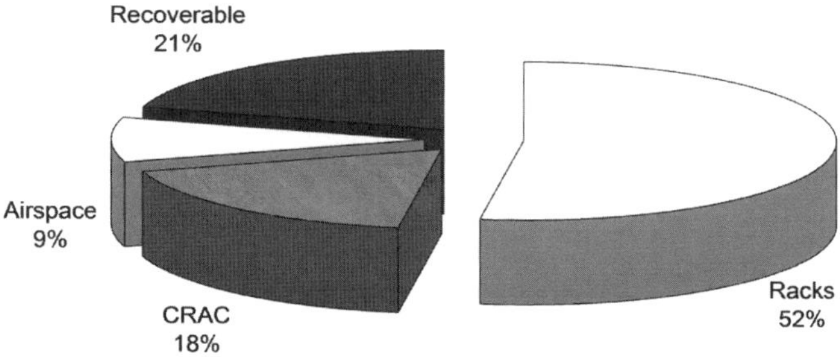


**Fig. 9.12** Exergy-based assessment of the entire data center. The majority of exergy consumption occurs in the rack and CRAC units, although significant amount of exergy is also consumed in the airspace. Nearly 21% of the supplied exergy is vented to the exhaust

#### 9.3.4.2 Beyond the Airspace

Figure 9.12 shows the results for the sample data center analyzed earlier beyond the airspace, in terms of total exergy consumption across the thermal management infrastructure. The maximum exergy consumption occurs in the rack units, owing to the irreversibility of converting electricity to heat. Exergy consumption in the CRAC units mostly stems from the pressure drop and heat removal as the air flows from the return to supply vents. The airspace exergy consumption is mostly due to recirculation and flow irreversibilities.

The insights from Fig. 9.12 are somewhat novel in terms of data center energy efficiency.

First, the airspace—which has been such a focus of recent research—is the smallest source of inefficiency, although it is also the easiest to address (and arguably the one which can be most influenced by the data center operator after the remaining equipment has been installed). A key benefit of the type of analysis presented here is that a theoretical limit can be provided around the extent of savings possible through the implementation of such efficiency measures. To illustrate this limit, consider that the exergy consumed in the airspace is effectively comprised of two components: the fluidic irreversibility due to friction, viscous dissipation, and other irreversibilities; and the thermal irreversibility due to recirculation and mixing of air at different temperatures. If the airspace had no mixing, then the thermal irreversibility in the cold aisle should approach zero—or, put differently, the delivery of air from the CRAC units to the rack units would be reversible. That is, if the CRAC units were to be inverted from refrigeration units to heat engines, then work should be available in an amount proportional to the efficiency of a Carnot engine operating between the supply and return temperature difference. For a reversible system, this is equivalently the minimum theoretical amount of work that would be required to drive the CRAC units in refrigeration mode for delivering the amount of cooling capacity required. As discussed in Sect. 9.1.3, the presence of irreversibilities means that additional work must be supplied to the CRAC units in order to deliver the same cooling capacity as an air-conditioner operating at the Carnot efficiency. For example, exergy consumption at the rate of 5 kW due to thermal irreversibilities in the airspace implies that, at a minimum, work must be continually supplied to the CRAC units at a minimum rate of 5 kW. (In practice, the amount of work required to compensate for the irreversibilities would be even higher due to inefficiencies within the CRAC unit related to converting electricity into kinetic energy of the airstreams.) In this sense, a lower bound on the added energy (exergy) costs of operating a data center with inefficiencies in the airspace can be obtained from the above analysis.

Secondly, there is scope for improvement in all the thermal management components of the data center, from the racks to the CRAC units, although the greatest long-term gain would arise from a type of disruptive improvement in rack power handling. The ability to process information in a manner that retains the quality of the electrical energy supplied to the rack units would be beneficial to the data center efficiency. But such a computational mechanism is currently not available, and may not be physically feasible within the scope of existing semiconductor technology that drives information systems.

Lastly, and perhaps most importantly, nearly 21% of the exergy being supplied is simply being vented through the exhaust. For example, it has recently been suggested [178] that co-locating a data center with a biomass-driven energy source could be synergistic, as feeding the waste heat from the data center into the anaerobic digester processing the farm waste could improve the rate of methane production, which in turn might reduce the need for data center grid power. Returning to the example from Sect. 9.1.2 of two computers placed downstream—one being fed high-quality electricity and the other merely capturing the low-quality waste heat from the first computer—it becomes possible to envision the "data center as a computer" [104], such that high-quality electricity is being fed into the data center and waste heat is being ejected downstream of the data center. But rather than situate another data center downstream (which would not be feasible), the opportunity to collocate another energy conversion system with the capability of utilizing low-grade waste heat can have significant ramifications on the ability to design data centers where the supply side and demand side of the facility have been holistically integrated.

Even if practically reusing the waste energy from the data center in a cost-feasible manner were challenging, the theoretical potential is interesting to notice. For example, even if energy reuse is not possible, the above analysis suggests that additional equipment with an exergy consumption of 21% can theoretically be installed in the data center simply by optimizing the existing thermal architecture. This additional exergy could be installed, as an example, in terms of additional compute power or redundant CRAC capacity. It should be noted that because the majority of the exergy supplied to the racks would be destroyed during conversion of electricity to heat, the actual compute capacity corresponding to this exergy supply would be lower, i.e., the above observation does not imply that it would be possible to obtain 21% additional compute capacity, but rather than the compute capacity for the given thermal architecture could be increased by a factor of$(0.21)\left(\eta_{\mathrm{II}_{\mathrm{racks}}}\right)$. For the example data center analyzed above, this corresponds to roughly 2.8 kW of additional compute capacity which could be installed without needing extra cooling capacity. It should also be noted that the recirculation patterns of the system would likely change upon installation of the additional compute capacity, and a corresponding increase in airspace exergy consumption may therefore offset some of the 2.8-kW availability. Additionally, a higher amount of recirculation may change the thermal manageability of the system. Therefore, a full thermal assessment—or equivalently, an assessment of the cold aisle exergy consumption corresponding to the higher rack power—should be repeated to confirm the viability of installing this additional compute power in the data center. Nonetheless, the exergy analysis presented above provides an upper bound against which the appropriate thermal management criteria can be evaluated.

## 9.4 Summary and Future Work

This chapter began with an introduction to the second law of thermodynamics, focusing around its applicability for the assessment of data center thermal management systems. A conceptual approach was outlined from a second-law perspective, and a finite-volume computational model to predict data center exergy consumption was demonstrated. Flow and temperature measurements from an actual data center were used to validate the numerical predictions of the model, and the viability of using exergy consumption to predict recirculation patterns in data centers was verified by comparing rack-level estimates with existing metrics such as SHI and $\beta$. Finally, the exergy consumption of the airspace obtained from the model was combined with exergy consumption predictions of the CRAC and rack units to obtain a net prediction of total data center exergy consumption. Insights into the relative energy efficiencies of the different data center components were obtained from this analysis. These insights included a wide range of considerations, including the viability of workload placement within the data center; the appropriateness of airside economization and/or containment; the potential benefits of reusing waste heat from the data center; as well as the potential to install additional compute capacity without necessarily needing to increase the data center cooling capacity.

Thus, by considering the exergy consumption among the different data center thermal management components (viz. cold aisle airspace, hot aisle airspace, CRAC units, and rack units), this chapter has provided a metric to simultaneously evaluate the thermal manageability, energy efficiency, and potential for energy scavenging of the data center. The cold aisle exergy consumption is a suitable metric for evaluating the temperature rise of rack inlet temperatures due to recirculation, while the net data center exergy consumption provides a measure of energy efficiency across the data center thermal infrastructure. The exergy content of the data center exhaust is an indicator of the theoretical viability of reusing waste heat from the data center. Furthermore, the two components of the overall exergy consumption metric—namely, exergy destruction, and exergy loss—provide guidance around how to prioritize efforts for reducing energy use in data centers. For example, in cases where the exergy destruction dominates the exergy consumption—such as the conversion of electricity to heat in the computer racks, or the recirculation of hot rack exhaust to the intake—it appears that a basic technology redesign (such as exploring nonsilicon technology) or avoiding the irreversible phenomena altogether (e.g., by containing hot aisles) might be most advantageous. On the other hand, in cases where the exergy loss dominates the exergy consumption—such as the waste heat emitted to the environment, or inefficiencies in heat exchanger design within the CRAC units—a redesign of the system based on a second-law optimization may lead to the greatest returns. It is important to note, however, that localized changes may not always be advantageous owing to the coupled nature of the data center infrastructure. For example, changes in air handling within the data center in turn may affect the temperatures within the system. In all scenarios, therefore, the goal behind data center design remains the same: minimizing the total exergy consumption across the entire infrastructure. The subtle difference highlighted above relates to prioritizing design alternatives for purposes of achieving this goal. Additional work is required to validate this intuition.

Historically, for large thermomechanical systems such as power plants, a gradual evolution of the detail and the breadth of second-law analysis can be observed in the literature. A similar trend may be anticipated for the second-law analysis of data centers. This work only presents a conceptual foundation of how key thermomechanical concepts may be translated into the data center and IT industries. Moving forward, a variety of opportunities are available to extend such an analysis. This may include the applicability of thermoeconomics, resource accounting, or even extended exergy accounting to evaluate the sustainability of data center thermal management systems. Even within its present scope, there are numerous opportunities to leverage the insights gained from a second-law analysis for purposes of improving the data center energy efficiency. For example, this work stops short of evaluating trade-offs that may arise across the (often competing) goals of effective data center thermal management and maximum infrastructure energy efficiency. Similarly, this work only considers the aggregate exergy consumption across each component. In the future, decomposing the sources of exergy consumption—particularly in terms of thermal versus fluidic exergy consumption—would enable a data center operator to better understand the trade-offs between increasing flow work (i.e., changing the flowrate of air) or increasing thermal work (e.g., changing the temperature setpoint) in the data center. As illustration, Shah et al. [179, 180] found the applicability of such an exergy-based model could instruct optimization of workload placement within the data center such that computational load distribution were assigned to those servers, which would result in the lowest irreversibility. This approach was shown to improve the energy efficiency by about 22% from a baseline case where all CRAC units were homogeneously operated.

The above analyses could also have implications for the design and control of IT systems. In current architectures, the focus is to remove heat from the system as rapidly as possible. But, the exergy analysis presented in this work suggests that a longer term incentive may exist to consider coupling the work-consuming devices (IT systems, which consume exergy) with the work-producing systems (power sources, which supply exergy) to take advantage of synergies, such as the temperature at which heat is dissipated. One example previously mentioned was the opportunity to collocate data center facilities with alternative energy systems that have use for low-grade waste heat, such as cogeneration systems or waste processing plants [178]. Another illustration is the opportunity of using such exergy-based metrics for workload placement decisions, both within the data center and also across data centers. Eventually, such supply–demand integration would lay the foundation for more sustainable data center design.

In summary, this work provides a foundation for future analyses by demonstrating the viability of using exergy-based metrics for the concurrent assessment of the thermal manageability and energy efficiency of data centers. Future work may begin by investigating the optimal compromise between thermal manageability and energy efficiency considerations in the data center, beyond which additional dimensions of analysis may gradually be added within and across data centers. Eventually, as the data center becomes the computer, a second-law analysis of data centers can provide the opportunity to identify opportunities for more efficient energy conversion within the data center, but also in terms of integrating energy conversion sources on both the supply and the demand side of the data center.

## References


1. Çengel YA, Boles MA (2001) Thermodynamics: an engineering approach, 4th edn. McGraw-Hill, Hightstown, NJ
2. Moran MJ, Shapiro HN (2004) Fundamentals of engineering thermodynamics, 5th edn. Wiley, New York, NY
3. Bejan A (1997) Advanced engineering thermodynamics, 2nd edn. Wiley, New York, NY
4. Sonntag RE, Borgnakke C (2000) Introduction to engineering thermodynamics, 2nd edn. Wiley, New York, NY
5. Reynolds WC (1977) Engineering thermodynamics. McGraw-Hill, Hightstown, NJ
6. Turns SR (2006) Thermodynamics: concepts and applications. Cambridge University Press, Cambridge, UK
7. Tester JW, Modell M (1996) Thermodynamics and its applications. Prentice-Hall, Englewood Cliffs, NJ
8. Jones JB, Dugan RE (1995) Engineering thermodynamics. Prentice-Hall, Englewood Cliffs, NJ
9. Moran MJ (1982) Availability analysis: a guide to efficient energy Use. Prentice-Hall, Englewood Cliffs, NJ
10. Bejan AD (1982) Entropy generation through heat and fluid flow. Wiley, New York
11. Bejan A, Tsatsaronis G, Moran M (1996) Thermal design and optimization. Wiley, New York, NY
12. Gyftopoulos EP, Beretta GP (2005) Thermodynamics: foundations and applications. Dover, Mineola, NY
13. Keenan JH (1951) Availability and irreversibility in thermodynamics. Br J Appl Phys 2:183–192
14. Kotas TJ (1995) The exergy analysis of thermal plant analysis. Krieger, Malabar, FL
15. Verkhivker GP, Kosoy BV (2001) On the exergy analysis of power plants. Energy Convers Manag 42(18):2053–2059

16. Rosen MA (2001) Energy- and exergy-based comparison of coal-fired and nuclear steam power plants. Exergy 1(3):180–192
17. Koroneos C, Haritakis I, Michaloglou K, Moussipolos N (2004) Exergy analysis for power plant alternative designs—parts I and II. Energy Sources 26:1277–1295
18. Ameri M, Ahmadi P, Khanmohammadi S (2008) Exergy analysis of a 420 MW combined cycle power plant. Int J Energy Res 32(2):175–183
19. Sengupta S, Datta A, Duttagupta S (2007) Exergy analysis of a coal-based 210 MW thermal power plant. Int J Energy Res 31(1):14–28
20. Tsatsaronis G, Park M-H (2002) On avoidable and unavoidable exergy destructions and investment costs in thermal systems. Energy Convers Manag 43(9–12):1259–1270
21. Bejan A, Mamut E (1999) Thermodynamic optimization of complex energy systems. Springer, New York, NY
22. Hepbasli A, Akdemir O (2004) Energy and exergy analysis of a ground source (geothermal) heat pump system. Energy Convers Manag 45(5):737–753
23. Balli O, Aras H, Hepbasli A (2008) Exergoeconomic analysis of a combined heat and power system. Int J Energy Res 32(4):273–289
24. Taniguchi H, Mouri K, Nakahara T, Arai N (2005) Exergy analysis on combustion and energy conversion processes. Energy 3(2–4):111–117
25. Rosen MA, Dincer I (2003) Exergy methods for assessing and comparing thermal storage systems. Int J Energy Res 27(4):415–430
26. Wright SE, Rosen MA (2004) Exergetic efficiencies and the exergy content of terrestrial solar radiation. J Sol Energ Eng 126(1):673–676
27. Caton JA (2000) On the destruction of availability (exergy) due to combustion processes—with specific application to internal combustion engines. Energy 25(11):1097–1117
28. Rosen MA (1995) Energy and exergy analyses of electrolytic hydrogen production. Int J Hydrogen Energ 20(7):547–553
29. Hotz N, Senn SM, Poulikakos D (2006) Exergy analysis of a solid oxide fuel cell micropowerplant. Proceedings of the 13th international heat transfer conference (IHTC-13), Sydney, Australia
30. Chan SH, Low CF, Ding OL (2002) Energy and exergy analysis of simple solid-oxide fuel-cell power systems. J Power Sources 103(2):188–200
31. Hussain MM, Baschuk JJ, Li X, Dincer I (2005) Thermodynamic analysis of a PEM fuel cell power system. Int J Thermal Sci 44(9):903–911
32. Bavarsad PG (2007) Energy and exergy analysis of internal reforming solid oxide fuel cell-Gas turbine hybrid system. Int J Hydrogen Energ 32(17):4591–4599
33. Szargut J, Morris DR, Steward FR (1988) Exergy analysis of thermal, chemical and metallurgical processes. Hemisphere, New York, NY
34. Wall G (1988) Exergy flows in industrial processes. Energy 13(2):197–208
35. Brodyansky VM, Le Goff P, Sorin MV (1994) The efficiency of industrial processes: exergy analysis and optimization. Elsevier, Amsterdam (The Netherlands)
36. Creyts JC, Carey VP (1997) Use of extended exergy analysis as a tool for assessment of the environmental impact of industrial processes. Proceedings of the international mechanical engineering conference and exposition, Dallas, TX, AES-37, pp. 129–137
37. Creyts JC, Carey VP (1999) Use of extended exergy analysis to evaluate the environmental performance of machining processes. Proceedings of the institution of mechanical engineers, vol. 213(4), Part E: Journal of Process Mechanical Engineering, pp 247–264, doi: 10.1243/0954408991529861
38. Sato N (2004) Chemical energy and exergy: an introduction to chemical thermodynamics for engineers. Elsevier, San Diego, CA
39. Morris D, Steward F, Szargut J (1994) Technological assessment of chemical metallurgical processes. Can Metall Q 33(4):289–295
40. Dincer I, Hussain MM, Al-Zaharnah I (2003) Energy and exergy use in industrial sector of Saudi Arabia. Proceedings of the institution of mechanical engineers, vol. 217(5), Part A: Journal of Power and Energy, pp 481–492, doi: 10.1243/095765003322407539

41. Szargut J, Morris DR (1990) Cumulative exergy losses associated with the production of lead metal. Int J Energy Res 14(6):605–616
42. Gong M (2004) Exergy analysis of a pulp and paper mill. Int J Energy Res 29(1):79–83
43. Dewulf J, Van Langenhove H (2004) Thermodynamic optimization of the life cycle of plastics by exergy analysis. Int J Energy Res 28(11):969–976
44. Uche J, Serra L, Valero A (2008) Exergy costs and inefficiency diagnosis of a dual-purpose power and desalination plant. J Energ Resource Technol 128(3):186–193
45. Gutowski T, Dahmus J, Thiriez A, Branham M, Jones A (2007) A thermodynamic characterization of manufacturing processes. Proceedings of the IEEE international symposium on electronics and the environment, Orlando, FL
46. Chengqin R, Nianping L, Guanga T (2002) Principles of exergy analysis in HVAC and evaluation of evaporative cooling schemes. Build Environ 37(11):1045–1055
47. Dikic A, Akbulut A (2008) Exergetic performance evaluation of heat pump systems having various heat sources. Int J Energy Res 32(14):1279–1926
48. Paoletti S, Rispoli F, Sciubba E (1989) Calculation of exergetic losses in compact heat exchanger passages. Proc ASME Adv Energ Syst 10–2:21–29
49. Yumruta R, Kunduz M, Kanolu M (2002) Exergy analysis of vapor compression refrigeration systems. Exergy 2(4):266–272
50. Bejan A, Ledezma G (1996) Thermodynamic optimization of cooling techniques for electronic packages. Int J Heat Mass Transf 39(6):1213–1221
51. Bejan A, Morega AM, Lee SW, Kim SJ (1993) The cooling of a heat generating board inside a parallel plate channel. Int J Heat Mass Transf 14:170–176
52. Bejan A (1997) Constructal-theory network of conducting paths for cooling a heat generating volume. Int J Heat Mass Transf 40(4):799–811
53. Ogiso K (2001) Assessment of overall cooling performance in thermal design of electronics based on thermodynamics. J Heat Transfer 123(5):999–1005
54. Ndao S, Peles Y, Jensen MK (2009) Multi-objective thermal design optimization and comparative analysis of electronics cooling technologies. Int J Heat Mass Transf 52:4317–4326
55. Culham JR, Muzychka YS (2001) Optimization of plate Fin heat sinks using entropy generation minimization. IEEE Trans Compon Packag Technol 24(2):159–165
56. Shuja SZ (2002) Optimal Fin geometry based on exergoeconomic analysis for a Pin-Fin array with application to electronics cooling. Exergy 2(4):248–258
57. Shih CJ, Liu GC (2004) Optimal design methodology of plate-fin heat sinks for electronic cooling using entropy generation strategy. IEEE Trans Compon Packag Technol 27(3): 551–559
58. Zhou J-H, Yang C-X, Zhang L-N (2008) Minimizing the entropy generation rate of the plate-finned heat sinks using computational fluid dynamics and combined optimization. Appl Therm Eng 29(8–9):1872–1879
59. Bar-Cohen A, Iyengar M, Kraus AD (2003) Design of optimum plate-Fin natural convective heat sinks. ASME J Electron Packag 125(2):208–216
60. Iyengar M, Bar-Cohen A (2001) Design for manufacturability of SISE parallel plate forced convection heat sinks. IEEE Trans Compon Packag Technol 24(2):150–158
61. Khan WA, Culham JR, Yovanovich MM (2004) Optimization of pin-fin heat sinks using entropy generation minimization. Proceedings of the 9th inter society conference on thermal and thermomechanical phenomena (ITHERM), pp 259–267, doi: 10.1109/ITHERM.2004.1319183
62. Khan WA, Culham JR, Yovanovich MM (2008) Optimization of pin-fin heat sinks in bypass flow using entropy generation minimization. J Electron Packag 130(3), Article 031010, doi: 10.1115/1.2965209
63. Bejan AD (1996) Entropy generation minimization. Wiley, New York, NY, pp 104–109
64. Kern DQ, Kraus AD (1972) Extended surface heat transfer. McGraw-Hill, New York, NY
65. Kraus AD, Aziz A, Welty J (2001) Extended surface heat transfer. Wiley, New York, NY

66. Khan WA, Culham JR, Yovanovich MM (2009) Optimization of microchannel heat sinks using entropy generation minimization methods. IEEE Trans Compon Packag Technol 32(2): 243–251
67. Bar-Cohen A, Iyengar M (2002) Design and optimization of Air-cooled heat sinks for sustainable development. IEEE Trans Compon Packag Technol 25(4):584–591
68. Carey VP, Shah AJ (2006) The exergy cost of information processing: a comparison of computer-based technologies and biological systems. J Electron Packag 128(4):346–352
69. Shah AJ, Carey VP, Bash CE, Patel CD (2003) Exergy analysis of data center thermal management systems. Paper IMECE2003-42527. Proceedings of the ASME international mechanical engineering congress and exposition, Washington, DC
70. Shah AJ, Carey VP, Bash CE, Patel CD (2004) An exergy-based control strategy for computer room air-conditioning units in data center. Paper IMECE2004-61384, Proceedings of the 2004 international mechanical engineering congress and exposition, Anaheim, CA
71. Shah AJ, Carey VP, Bash CE, Patel CD (2005) Exergy-based optimization strategies for multi-component data center thermal management: part I, analysis. Paper IPACK2005-73137. Proceedings of the pacific rim/ASME international electronic packaging technical conference and exhibition (InterPACK), San Francisco, CA
72. Shah AJ, Carey VP, Bash CE, Patel CD (2005) Exergy-based optimization strategies for multi-component data center thermal management: part II, application and validation. Paper IPACK2005-73138. Proceedings of the pacific rim/ASME international electronic packaging technical conference and exhibition (InterPACK), San Francisco, CA
73. Shah A, Carey V, Patel C, Bash C (2008) Exergy analysis of data center thermal management systems. J Heat Transfer 130(2), Article No. 021401, doi: 10.1115/1.2787024
74. Shah AJ, Carey VP, Bash CE, Patel CD (2008) An exergy-based figure-of-merit for electronic packages. J Electron Packag 128(4):360–369
75. Sciubba E (2005) Exergo-economics: thermodynamic foundation for a more rational resource use. Int J Energy Res 29(7):613–636
76. Sciubba E (2001) Beyond thermoeconomics? The concept of extended exergy accounting and its application to the analysis and design of thermal systems. Exergy 1(2):68–84
77. Sayed-El YM (2003) The thermoeconomics of energy conversions. Pergamon, Oxford, UK
78. Tsatsaronis G (2002) Application of thermoeconomics to the design and synthesis of energy plants. In: Grangopoulos CA (ed) Exergy, energy system analysis, and optimization, encyclopaedia of life support systems. EOLSS Publishers, UK (website: www.eolss.net) pp 160–172, ISBN 978-1-84826-614-8
79. Valero A (1995) Thermoeconomics: the meeting point of thermodynamics, economics and ecology. In: Sciubba E, Moran M (eds) Second law analysis of energy systems: towards the 21st century. Circus, Rome, pp 293–305
80. Ayres RU, Ayres LW, Warr B (2002) Exergy, power and work in the US economy, 1900–1998. Energy 28(3):219–273
81. Dincer I, Rosen M (2007) Exergy: energy environment and sustainable development. Elsevier, Elsevier, NY
82. Rosen MA, Dincer I (1997) On exergy and environmental impact. Int J Energy Res 21(7): 643–654
83. Zvolinschi A et al (2007) Exergy sustainability indicators as a tool in industrial ecology. J Ind Ecol 11(4):85–98
84. Hammond GP (2004) Engineering sustainability: thermodynamics, energy systems, and the environment. Int J Energy Res 28(7):613–639
85. Zevenhoven R, Kavaliauskaite I (2004) Mineral carbonation for long-term $CO_2$ storage: an exergy analysis. Int J Thermodyn 7(1):23–31
86. Cornelissen RL (1997) Thermodynamics and sustainable development: the use of exergy analysis and the reduction of irreversibility. PhD Dissertation, University of Twente, The Netherlands

87. Connelly L, Koshland CP (2001) Exergy and industrial ecology—part I: an exergy-based definition of consumption and a thermodynamic interpretation of ecosystem evolution. Exergy 1(3):146–165
88. Connelly L, Koshland CP (2001) Exergy and industrial ecology—part II: a non-dimensional analysis of means to reduce resource depletion. Exergy 1(4):234–255
89. Gong M, Wall G (2001) On exergy and sustainable development—part I: conditions and concepts. Exergy 1(3):128–145
90. Gong M, Wall G (2001) On exergy and sustainable development—part II: indicators and methods. Exergy 1(4):217–233
91. Balocco C et al (2004) Using exergy to analyze the sustainability of an urban area. Ecol Econ 48(2):231–244
92. Hannemann C, et al. (2008) Lifetime exergy consumption of an enterprise server. Proceedings of the IEEE international symposium on electronics and the environment, San Francisco, CA
93. Haseli Y, Dincer I, Naterer GF (2008) Unified approach to exergy efficiency environmental impact and sustainable development for standard thermodynamic cycles. Int J Green Energ 5:105–119
94. Sullivan RF (2003) Alternating cold and hot aisles provides more reliable cooling for server farms. White Paper by The Uptime Institute, Santa Fe, NM
95. APC (2003) Avoidable mistakes that compromise cooling performance in data centers and network rooms. White Paper # 49 by American Power Conversion, Washington, DC
96. ASHRAE (2004) Thermal guidelines for data processing environments. Atlanta, GA
97. Shrivastava S, Sammakia B, Schmidt R, Iyengar M (2005) Comparative analysis of different data center airflow management configurations. Proceedings of the ASME international electronic packaging technical conference and exhibition (InterPACK), San Francisco, CA
98. Iyengar M, Schmidt R, Sharma A, McVicker G, Shrivastava S, Sri-Jayantha S, Amemiya Y, Dang H, Chainer T, Sammakia B (2005) Thermal characterization of non-raised floor air cooled data centers using numerical modeling. Proceedings of the pacific rim/ASME international electronic packaging technical conference and exhibition (InterPACK), San Francisco, CA
99. Patel CD (2000) Enabling pumped liquid loop cooling: justification and the key technology and cost barriers. International Conference on High-Density Interconnect and Systems Packaging, Denver, CO
100. Bash CE, Patel C, Sharma RK (2003) Efficient thermal management of data centers—immediate and long-term research needs. Int J HVAC&R Res 9(2):137–152
101. Patel CD, Bash CE, Sharma R, Beitelmal A, Malone CG (2005) Smart chip, system and data center enabled by advanced flexible cooling resources. Proceedings of the IEEE semiconductor thermal management and measurement symposium (SEMITHERM), San Jose, CA
102. Patel CD, Shah AJ (2005) Cost model for planning, development and operation of a data center. Technical Report HPL-2005-107R1, Hewlett Packard Laboratories, Palo Alto, CA
103. Koomey J, Brill K, Turner P, Stanley J, Taylor B (2008) A simple model for determining true total cost of ownership for data centers. White Paper No. TUI 3011C Version 2.1, The Uptime Institute, Santa Fe, NM.
104. Patel CD, Friedrich R (2002) Towards planetary scale computing—technical challenges for next generation Internet computing. In: Joshi YK, Garimella SV (eds) Thermal challenges in next generation electronic systems. Millpress, Rotterdam, The Netherlands
105. Patel CD, Bash CE, Belady C, Stahl L, Sullivan D (2001) Computational fluid dynamics modeling of high compute density data centers to assure system inlet air specifications. Proceedings of the pacific rim/ASME international electronic packaging technical conference and exhibition (InterPACK), Kauai, HI
106. Schmidt R (2001) Effect of data center characteristics on data processing equipment inlet temperatures. Proceedings of the pacific rim/ASME international electronic packaging technical conference and exhibition (InterPACK), Kauai, HI

107. Mitchell-Jackson J, Koomey JG, Nordman B, Blazek M (2001) Data center power requirements: measurements from silicon valley. Energy 28(8):837–850
108. Mitchell-Jackson JD (2001) Energy needs in an internet economy: a closer look at data centers. M.S. Thesis (Energy and Resources Group), University of California, Berkeley, CA
109. Patel CD, Bash CE, Sharma R, Beitelmal M, Friedrich R (2003) Smart cooling of data centers. Proceedings of the international electronic packaging technical conference and exhibition (InterPACK'03), Maui, HI
110. Patel CD (2003) A vision of energy aware computing—from chips to data centers. Proceedings of the international symposium on micromechanical engineering, Tsuchiura, Japan
111. Tolia N, WangZ, Marwah M, Bash C, Ranganathan P, Zhu X (2009) Zephyr: a unified predictive approach to improve server cooling and power efficiency. Proceedings of the ASME/pacific rim technical conference and exhibition on packaging and the integration of electronic and photonic systems, MEMS, and NEMS (InterPACK), San Francisco, CA
112. Sharma R, Bash C, Patel C, Friedrich R, Chase J (2005) Balance of power: dynamic thermal management for internet data centers. IEEE Internet Comput 9(1):42–49
113. Moore J, Chase J, Ranganathan P, Sharma R (2005) Making scheduling "cool": temperature-aware workload placement in data centers. Usenix annual technical conference, Anaheim, CA
114. Moore J, Chase J, Farkas K, Ranganathan P (2005) Data center workload monitoring, analysis, and emulation. Proceedings of the eighth workshop on evaluation using commercial workloads (CAECW-8), San Francisco, CA
115. Fiorina C (2000) Transforming companies, transforming countries. Keynote address at the Japanese Chamber of Commerce and Industry, New York, NY
116. Patel C, Sharma R, Bash C, Graupner S (2003) Energy aware grid: global workload placement based on energy efficiency. Proceedings of the ASME international mechanical engineering congress and exposition (IMECE), Washington DC.
117. Wang D (2004) A passive solution to a difficult data center problem. Proceedings of the intersociety conference on thermal and thermomechanical phenomena (ITHERM), San Diego, CA, pp 586–592
118. Heydari A, Sabounchi P (2004) Refrigeration assisted spot cooling of a high heat density data center. Proceedings of the intersociety conference on thermal and thermomechanical phenomena (ITHERM), San Diego, CA, pp 601–606
119. Schmidt R, Chu R, Ellsworth M, Iyengar M, Porter D, Kamath V, Lehmann B (2005) Maintaining datacom rack inlet air temperatures with water cooled heat exchangers. Proceedings of the ASME international electronic packaging technical conference and exhibition (InterPACK), San Francisco, CA
120. Hewlett Packard (2009) Improving data center efficiency—incorporating air stream containment. White Paper. http://h10134.www1.hp.com/insights/whitepapers/downloads/air_stream_containment.pdf. Accessed 14 Dec 2009
121. Gondipalli S, Sammakia B, Bhopte S, Schmidt R, Iyengar MK, Murray B (2009) Optimization of cold aisle isolation designs for a data center with roofs and doors using slits. Proceedings of the ASME international electronic packaging technical conference and exhibition (InterPACK), San Francisco, CA
122. Marwah M, Sharma R, Patel C, Shih R, Bhatia V, Mekanapurath M, Velayudhan S (2009) Data analysis, visualization and knowledge discovery in sustainable data centers. Proceedings of the ACM compute conference, Bangalore, India
123. Samadiani E, Joshi Y, Hamann H, Iyengar MK, Kamalsy S, Lacey J (2009) Reduced order thermal modeling of data centers via distributed sensor data. Proceedings of the ASME international electronic packaging technical conference and exhibition (InterPACK), San Francisco, CA

124. Sharma R, Marwah M, Lugo W (2009) Application of data analytics to heat transfer phenomena for optimal design and operation of complex systems. Proceedings of the ASME summer heat transfer conference, San Francisco, CA
125. Marwah M, Sharma R, Lugo W (2009) Autonomous detection of thermal anomalies in data centers. Proceedings of the ASME international electronic packaging technical conference and exhibition (InterPACK), San Francisco, CA
126. Bash CE, Patel CD, Sharma RK (2006) Dynamic thermal management of an air-cooled data center. Proceedings of the intersociety conference on thermal and thermomechanical phenomena (ITHERM), San Diego, CA
127. Kang S, Schmidt RR, Kelkar KM, Radmehr A, Patankar SV (2000) A methodology for the design of perforated tiles in a raised floor data center using computational flow analysis. Proceedings of the eighth intersociety conference on thermal and thermomechanical phenomena in electronic systems (ITHERM), Las Vegas, NV
128. VanGilder JW, Lee T (2003) A hybrid flow network-CFD method for achieving any desired flow partitioning through floor tiles of a raised-floor data center. Proceedings of the international electronic packaging technical conference and exhibition (InterPACK), Maui, HI
129. Schmidt R, Karki K, Patankar S (2004) Raised-floor data center: perforated tile flow rates for various tile layouts. Proceedings of the ninth intersociety conference on thermal and thermomechanical phenomena in electronic systems (ITHERM), Las Vegas, NV
130. Karki KC, Radmehr A, Patankar SV (2003) Use of computational fluid dynamics for calculating flow rates through perforated tiles in raised-floor data centers. Int J Heat Ventil Air-Condition Refrig Res 9(2):153–166
131. Karki KC, Patankar SV, Radmehr A (2003) Techniques for controlling airflow distribution in raised floor data centers. Proceedings of the pacific rim/ASME international electronic packaging technical conference and exhibition (InterPACK), Maui, HI
132. VanGilder JW, Schmidt RR (2005) Airflow uniformity through perforated tiles in a raised floor data center. Proceedings of the pacific rim/ASME international electronic packaging technical conference and exhibition (InterPACK), San Francisco, CA
133. Radmehr A, Schmidt RR, Karki KC, Patankar SV (2005) Distributed leakage flow in raised floor data centers. Proceedings of the pacific rim/ASME international electronic packaging technical conference and exhibition (InterPACK), San Francisco, CA
134. Rambo JD, Joshi YK (2004) Supply air distribution from a single air handling unit in a raised floor plenum data center. Proceedings of the joint indian society of heat and mass transfer/american society of mechanical engineers heat and mass transfer conference (ISHMT/ASME), Kalpakkam, India
135. Patel CD, Sharma RK, Bash CE, Beitelmal A (2002) Thermal considerations in cooling large scale high computer density data centers. Proceedings of the eighth intersociety conference on thermal and thermomechanical phenomena in electronic systems (ITHERM), San Diego, CA
136. Bash C, Forman G (2007) Cool job allocation: measuring the power savings of placing jobs at cooling-efficient locations in the data center. Proceedings of the Usenix annual technical conference, Santa Clara, CA
137. Bhopte S, Agonafer D, Schmidt R, Sammakia B (2005) Optimization of data center room layout to minimize rack inlet air temperature. Proceedings of the pacific rim/ASME international electronic packaging technical conference and exhibition (InterPACK), San Francisco, CA
138. Schmidt R, Iyengar M (2005) Effect of data center layout on rack inlet air temperatures. Proceedings of the pacific rim/ASME international electronic packaging technical conference and exhibition (InterPACK), San Francisco, CA
139. Schmidt R, Cruz E (2002) Raised floor computer data center: effect on rack inlet temperatures of chilled air exiting both the hot and cold aisles. Proceedings of the intersociety conference on thermal and thermomechanical phenomena (ITHERM), San Diego, CA, pp 580–594
140. Schmidt R, Cruz E (2002) Raised floor computer data center: effect on rack inlet temperatures when high powered racks are situated amongst lower powered racks. Proceedings of the

ASME international mechanical engineering congress and exposition (IMECE), New Orleans, LA
141. Schmidt R, Cruz E (2003) Cluster of high powered racks within a raised floor computer data center: effect of perforated tile flow distribution on rack inlet air temperatures. Proceedings of the ASME international mechanical engineering congress and exposition (IMECE), Washington, DC, pp 245–262
142. Schmidt R, Cruz E (2003) Raised floor computer data center: effect on rack inlet temperatures when adjacent racks are removed. Proceedings of the pacific rim/ASME international electronic packaging technical conference and exhibition (InterPACK), Maui, HI
143. VanGilder JW, Shrivastava SK (2007) Capture index: an airflow-based rack cooling performance metric. ASHRAE Transact 113(1):126–136
144. Herrlin MK (2008) Airflow and cooling performance of data centers: two performance metrics. ASHRAE Transact 114(2):182–187
145. Bedekar V, Karajgikar S, Agonafer D, Iyengar M, Schmidt R (2006) Effect of CRAC location on fixed rack layout of a data center. Proceeding of the 9th intersociety conference on thermal and thermomechanical phenomena (ITHERM), San Diego, CA
146. Sharma RK, Bash CE, Patel CD (2002) Dimensionless parameters for evaluation of thermal design and performance of large-scale data centers. Proceedings of the eighth ASME/AIAA joint thermophysics and heat transfer conference, St. Louis, MO
147. Sharma RK, Bash CE (2002) Dimensionless parameters for energy-efficient data center design. Proceedings of the IMAPS advanced technology workshop on thermal management (THERM ATW), Palo Alto, CA
148. Schmidt RR, Cruz EE, Iyengar MK (2005) Challenges of data center thermal management. IBM J Res Dev 49(4/5):709–723
149. Schmidt R, Iyengar M, Chu R (2005) Meeting data center temperature requirements. ASHRAE J 47(4):44–49
150. Herrlin MK (2005) Rack cooling effectiveness in data centers and telecom central offices: the rack cooling index (RCI). ASHRAE Transact 111(2):725–731
151. Aebischer B, Eubank H, Tschudi W (2004) Energy efficiency indicators for data centers. International conference on improving energy efficiency in commercial buildings, Frankfurt, Germany
152. Norota M, Hayama H, Enai M, Kishita M (2003) Research on efficiency of air conditioning system for data center. Proceedings of the IEEE international telecommunications energy conference (INTELEC), Yokohama, Japan, pp. 147–151
153. The Green Grid (2007) Green grid metrics: describing datacenter power efficiency. Technical Committee White Paper. http://www.thegreengrid.org. Accessed 14 June 2010
154. Rolander N, Rambo J, Joshi Y, Mistree Y (2005) Robust design of air cooled server cabinets for thermal efficiency. Proceedings of the ASME international electronic packaging technical conference and exhibition (InterPACK), San Francisco, CA
155. Rambo J, Joshi Y (2005) Thermal performance metrics for arranging forced Air cooled servers in a data processing cabinet. J Electron Packag 127(4):452–459
156. Schmidt RR, Karki KC, Kelkar KM, Radmehr A, Patankar SV (2001) Measurements and predictions of the flow distribution through perforated tiles in raised-floor data centers. Proceedings of the pacific rim/ASME international electronic packaging technical conference and exhibition (InterPACK), Kauai, HI
157. McAllister S, Carey VP, Shah AJ, Bash CE, Patel CD (2008) Strategies for effective Use of Exergy-based modeling of data center thermal management systems. Microelectron J 39(7): 1023–1029
158. Munson BR, Young DF, Okiishi TH (1998) Fundamentals of fluid mechanics, 3rd edn. Wiley, New York, NY
159. White EM (1987) Fluid mechanics. McGraw-Hill, New York, NY
160. Batchelor GJ (1967) An introduction to fluid dynamics. Cambridge University Press, Cambridge, UK

161. Patankar SV (1980) Numerical heat transfer and fluid flow. Taylor & Francis, New York, NY
162. Versteeg H, Malasekra W (1996) An introduction to computational fluid dynamics: the finite volume method approach. Prentice-Hall, Englewood Cliffs, NJ
163. Tannehill JC, Anderson DA, Pletcher RH (1997) Computational fluid mechanics and heat transfer. Taylor and Francis, Washington, DC
164. Jaluria Y, Torrance KE (2003) Computational heat transfer (2nd edition). Taylor and Francis, New York, NY
165. Kotake S, Hijikata K (1993) Numerical simulations of heat transfer and fluid flow on a personal computer. Elsevier Science, Amsterdam (The Netherlands)
166. Ketkar SP (1999) Numerical thermal analysis. ASME Press, New York, NY
167. Cebeci T, Bradshaw P (1988) Physical and computational aspects of convective heat transfer. Springer, New York, NY
168. Kahan W (1958) Gauss-seidel methods of solving large systems of linear equations. Ph.D. Thesis, University of Toronto, Toronto, Canada
169. Varga R (1962) Matrix iterative analysis. Prentice-Hall, Englewood Cliffs, NJ
170. Young D (1971) Iterative solutions of large linear systems. Academic, New York, NY
171. Hageman L, Young D (1981) Applied iterative methods. Academic, New York, NY
172. Mentor Graphics (2010) FloVENT: Built Environment HVAC CFD Software. http://www.mentor.com/products/mechanical/products/flovent. Accessed 26 June 2010
173. Moffat RJ (1988) Describing the uncertainties in experimental results. Exp Therm Fluid Sci 1:3–7
174. Sen A, Srivastava M (1997) Regression analysis: theory, methods and applications. Springer, New York, NY
175. Pope AJ (1976) The statistics of residuals and the detection of outliers. NOAA Technical Report, Washington, DC
176. Draper NR, Smith H (1998) Applied regression analysis. Wiley, New York, NY
177. Box GEP, Hunter WG, Hunter SJ, Hunter WG (1978) Statistics for experimenters: an introduction to design, data analysis and model building. Wiley, New York, NY
178. Sharma R, Christian T, Arlitt M, Bash C, Patel C (2010) Design of farm waste-driven supply side infrastructure. Proceedings of the 4th ASME international conference on energy sustainability, Phoenix, AZ
179. Shah AJ, Carey VP, Bash CE, Patel CD (2005) Exergy-based optimization strategies for multi-component data center thermal management: part I, analysis. Proceedings of the pacific rim/ASME international electronic packaging technical conference and exhibition (IPACK2005-73138), San Francisco, CA
180. Shah AJ, Carey VP, Bash CE, Patel CD (2005) Exergy-based optimization strategies for multi-component data center thermal management: part II, application and validation. Proceedings of the pacific rim/ASME international electronic packaging technical conference and exhibition (IPACK2005-73138), San Francisco, CA

# Chapter 10
# Reduced Order Modeling Based Energy Efficient and Adaptable Design

**Emad Samadiani**

**Abstract** In this chapter, the sustainable and reliable operations of the electronic equipment in data centers are shown to be possible through a reduced order modeling based design. First, the literature on simulation-based design of data centers using computational fluid dynamics/heat transfer (CFD/HT) and low-dimensional modeling are reviewed. Then, two recent proper orthogonal decomposition (POD) based reduced order thermal modeling methods are explained to simulate multiparameter-dependent temperature field in multiscale thermal/fluid systems such as data centers. The methods result in average error norm of ~6% for different sets of design parameters, while they can be up to ~250 times faster than CFD/HT simulation in an iterative optimization technique for a sample data center cell. The POD-based modeling approach is applied along with multiobjective design principles to systematically achieve an energy efficient, adaptable, and robust thermal management system for data centers. The framework allows for intelligent dynamic changes in the rack heat loads, required cooling airflow rates, and supply air temperature based on the actual momentary center heat loads, rather than planned occupancy, to extend the limits of air cooling and/or increase energy efficiency. This optimization has shown energy consumption reduction by 12–46% in a data center cell.

## 10.1 Thermal Modeling Based Design of Data Centers

The multiscale nature of data centers spanning length scales from the chip to the facility level is shown in Fig. 10.1. As introduced in Chap. 1, a state-of-the-art approach for air cooling using an under-floor plenum is shown in Fig. 10.2.

E. Samadiani (✉)
G.W. Woodruff School of Mechanical Engineering, Georgia Institute of Technology, Atlanta, GA 30332, USA
e-mail: esamadia@binghamton.edu

Y. Joshi and P. Kumar (eds.), *Energy Efficient Thermal Management of Data Centers*, DOI 10.1007/978-1-4419-7124-1_10,

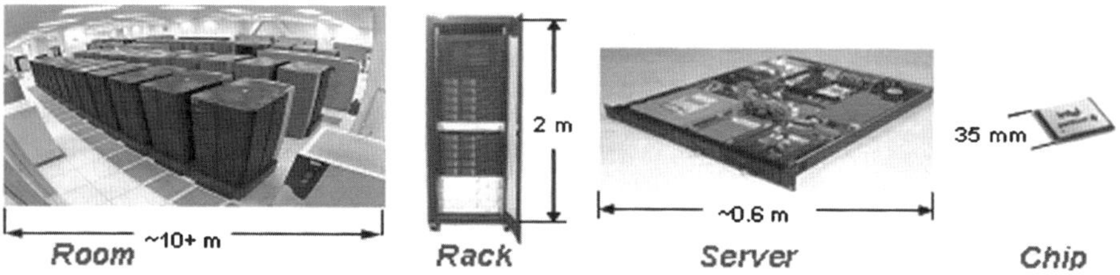


**Fig. 10.1** Data center and its multiscale nature

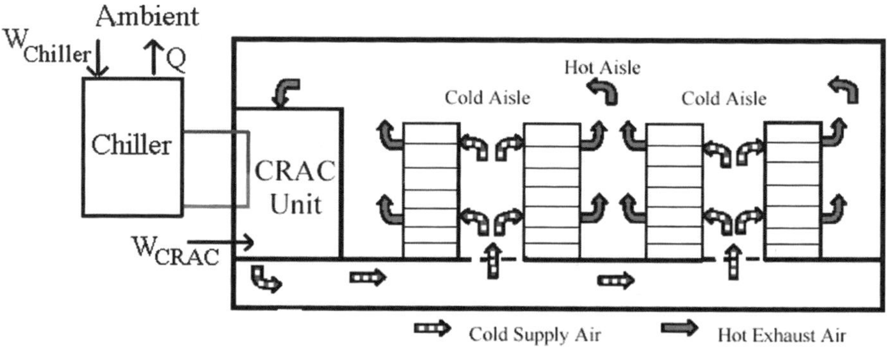


**Fig. 10.2** State-of-the-art cooling system in data centers

The computer room air conditioning (CRAC) units themselves are cooled by a chilled water loop, transferring the data center heat load to an outdoor chiller and ultimately to the ambient. The chip level determines the overall rate of the heat generation in the data center, while the CRAC units at the facility level are responsible in providing the cooling solution to maintain the chip temperatures in a safe range [1].

As seen in Chap. 1, the increasing energy usage of data centers [2–4] is a key concern regarding their sustained growth. A significant fraction of this goes towards powering and heat removal in the facilities [5]. As such, energy-efficient design of the cooling systems is essential for containing operating costs and promoting sustainability. Through better design and preventing over-provisioning, it should be possible to reduce energy consumption by the cooling systems. Predicting the flow and specially temperature fields inside data centers in terms of the involved design parameters is necessary for an energy efficient and reliable cooling system design [1]. The modeling approaches in the literature for flow and temperature prediction in data centers are reviewed in [1] and discussed in more detail in Chaps. 7 and 8. A brief summary, pertinent to their use in adaptable design, is provided below in Sect. 10.1.1.

### 10.1.1 Computational Fluid Dynamics/Heat Transfer Modeling

As seen in Chap. 8, the airflow inside data centers is often turbulent. Also, buoyancy effects can usually be neglected [6]. The Reynolds-averaged Navier–Stokes equations (RANS), introduced in Chap. 8, are commonly used to simulate the turbulent mean flow in air-cooled data centers, by modeling the effect of turbulence on the mean flow as a spatially dependent effective viscosity:

$$\nabla u = 0, \tag{10.1}$$

$$u\nabla u - \nabla(\upsilon_{\text{eff}}\nabla u) + \frac{1}{\rho}\nabla p = 0. \tag{10.2}$$

Also, the mean energy equation with effective thermal conductivity can be used to compute the temperature field. The mean energy equation, ignoring viscous dissipation, from Chap. 4, is:

$$\rho c_{\text{p}} u \nabla T - \nabla(k_{\text{eff}}\nabla T) = q. \tag{10.3}$$

As detailed in Chap. 8, several researchers have simulated the airflow and temperature fields in data centers [2, 3, 7–15]. Optimization [16–18] and design [19–23] incorporating different parameters involved in these systems have also been performed. Early applications of computational fluid dynamics/heat transfer (CFD/HT) modeling to data centers are reported in [8, 21, 24, 25]. Schmidt et al. [25] compared experimental measurements through raised-floor data center perforated tiles with two-dimensional computational models. Their experimental validation shows fair overall agreement with mean tile flow rates, with large individual prediction errors. Van Gilder and Schmidt [13] parametrically studied plenum airflow for various data center footprints, tile arrangements, tile porosity, and plenum depth. Initial studies to determine the airflow rates from the perforated tiles [12, 13, 25–27] have modeled the plenum only and do not simulate the effect of the airflow inside the computer room on the perforated tile flow distribution. Samadiani and Joshi [28] have shown that modeling the computer room, CRAC units, and/or the plenum pipes could change the tile flow distribution by up to 60% for the facility with 25% open perforated tiles and up to 135% for the facility with 56% open perforated tiles [28].

Numerical thermal modeling has been used for geometrical optimization of plenum depth, facility ceiling height and cold-aisle spacing for a single set of CRAC flow rates, and uniform rack flow and power dissipation [18]. A unit cell architecture of a raised-floor plenum data center is formulated in [9] by considering the asymptotic flow distribution in the cold aisles with increasing number of racks in a row. The results indicated that for high flow rate racks, a “unit cell” containing four rows of seven racks adequately models the hot-aisle/cold-aisle configuration in a “long” row of racks [9].

In [7, 11, 24, 29], researchers have either modeled individual racks as black-boxes with prescribed flow rate and temperature rise, or with fixed flow rate and uniform heat generation. A procedure to model individual servers within each rack was developed in [8]. Rambo and Joshi [8] developed a multiscale model of typical air-cooled data centers using commercial finite volume software. In their work, each rack is modeled as a series of submodels designed to mimic the behavior of a server in a data center. Rambo and Joshi [30] performed a parametric numerical study of various air supply and return schemes, coupled with various orientations of the racks and the CRAC units, to identify the causes of recirculation and nonuniformity in thermal performance throughout the data center.

The multiscale nature of data centers needs to be considered in the numerical modeling. Also, as suggested in [15], the future state of the art of thermal management in data centers will include a combination of cooling solutions at different scales. This increases the need to have a multiscale model for thermal phenomena happening at all important scales. The multiscale model of a representative data center in [8, 30] consists of ~1,500,000 grid cells and needs more than 2,400 iterations to obtain a converged solution. This model took about 8 h to converge on a 2.8 GHz Xeon with 2 GB memory [30]. Also, it should be noted that this model is still a significant departure from reality because it does not include finer details at the server and chip level. In light of this, a comprehensive CFD/HT multiscale model of operational data centers, which may contain thousands of racks, seems infeasible due to limits on available computing. A compact or low-dimensional model which could run much faster, while including the influence of all important scale parameters with sufficient fidelity is essential, especially for iterative, optimization-based design methods. A comprehensive review of literature on data center numerical modeling with a study on the necessity of compact airflow/thermal modeling for data centers has been done in [31].

### *10.1.2 Low-Dimensional Modeling of Data Centers*

Aside from CFD/HT, simulation methods based on heuristic approaches have also been explored [32–39] to predict the air temperature at discrete points, such as server inlets/outlets, for a new heat load distribution among the data center racks or servers. In [32–35], machine learning techniques based on the input from several deployed sensors are used to understand the relation between workload and internal and ambient temperatures. These methods require a large number of data points for interpolation and usually need a lengthy calibration for each data center of interest before they can be used for simulation. In [39], a threefold latent variable model, using structural-equation method (SEM) and errors-in-variables (EV) parameterization, is proposed to generate a surrogate model for maximum rack inlet temperatures in a nonraised-floor data center in terms of nine design variables. The data center model in [39] has four rows with six racks for the first two rows and four for the last two. They simulated the data center for 148 configuration runs using the commercial software Flotherm. However, they just monitored the temperature at five points for each of the 20 rack positions, resulting in a total of 100 points. The surrogate model has been used for determining practical values of the configuration variables of the data center to meet some physical and usage requirements.

In [36] , the rate of heat transferred by the airflow recirculation is described by a cross-interference coefficient matrix, which shows how much of the heat transferred by the air exiting from the outlet of each server contributes to the inlet of every other server. Having obtained this matrix through a calibration process for a specific data center, an abstract heat flow model is developed to predict the temperatures at the server inlets/outlets vs. server power consumption. In [37, 38], a coefficient matrix is assembled through a calibration process to provide an estimate of the sensitivity of each server inlet temperature to every other server heat load unit step change, for a given CRAC velocity. So an ambient intelligence-based load management (AILM) approach is designed to determine the maximum possible heat loads of each server to meet the corresponding thermal constraint within a given air velocity.

The mentioned works above simulate the effects of the system parameters on the temperature field in data centers based on some heuristic approaches. These methods can predict the air temperatures only at discrete points, such as server inlets/outlets. As mentioned in Chap. 7, reduced order thermal modeling of data centers is becoming increasingly important in improving operational efficiencies of data centers. The proper orthogonal decomposition (POD) [40, 41] is a physics-based approach which can predict the temperature fields in data centers much faster than CFD/HT simulations. This technique and representative results from [40, 41] are explained in Sect. 10.2, with a view towards its use in data center design, the focus of this chapter.

## 10.2 POD for Data Center Reduced Order Modeling

In Sect. 10.2.1, the basic POD technique and POD based reduced order modeling approaches are explained [1]. In Sect. 10.2.2, the recent work [40] on using POD and Galerkin projection for data centers is explained while the recent study [41] on using POD and energy balance for data centers is explained in Sect. 10.2.3.

### *10.2.1 Review of POD*

POD, also known as the Karhunen–Loeve decomposition, is a statistical technique which has several properties that make it well suited for low-dimensional modeling of turbulent flows. [6, 23, 42, 43]. First, it has been shown experimentally that low-dimensional models using POD can well address the role of coherent structures in turbulence generation [43]. Second, it captures more of the dominant dynamics for a given number of modes than any other linear decomposition [43]. Finally, the

Fig. 10.3 General POD based reduced order modeling approach [1]

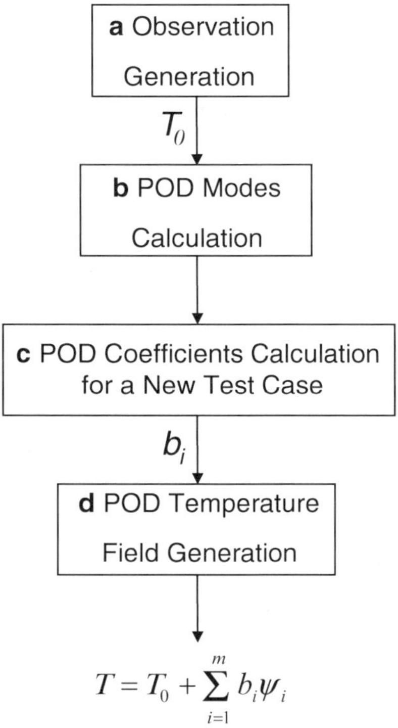


empirical determination of the basis functions makes this method ideal for nonlinear problems. A review of the POD method and its application for turbulence modeling has been done in [43].

In the POD-based model reduction technique, a set of data are expanded on empirically determined basis functions for modal decomposition. It can be used to numerically predict the temperature field more rapidly than full-field simulations. The temperature field is expanded into basis functions or POD modes:

$$T = T_0 + \sum_{i=1}^{m} b_i \psi_i. \tag{10.4}$$

The general algorithm to generate a POD based reduced order thermal modeling in a system is illustrated in Fig. 10.3 and is explained in the following:

(a) *Observation generation*: In the first step, the design variables of the system are changed $n$-times and the temperature field for the entire domain is obtained by CFD/HT simulations or detailed experimental measurements for each case. These thermal fields are called observations or snapshots. An element of the reference temperature field, $T_0$ in (10.4), is typically considered as the average of the all observed data for a field point.
(b) *POD modes, $\psi_i$, calculation*: The POD modes of a thermal system, $\psi_i$, can be calculated from observations. In (10.4), $m$ is the number of retained POD modes

in the decomposition which can be 1 up to $n - 1$, where $n$ is the number of observations. Using the method of snapshots, each POD mode can be expressed as a linear combination of the linearly independent observations [43]:

$$\psi_i = \sum_{k=1}^{n} a_k (T_{\text{obs},k} - T_0), \tag{10.5}$$

where $T_{\text{obs}}$ is a matrix of which each column, $T_{\text{obs},i}$, includes a complete temperature field data from an observation. The weight coefficients, $a_k$, in (10.5) are obtained by solving the following $n \times n$ eigenvalue problem:

$$\sum_{k=1}^{n} R(i,k) a_k = \lambda a_i\,; \quad i = 1, \dots, n, \tag{10.6}$$

where $R = (T_{\text{obs}} - T_0)^* \otimes (T_{\text{obs}} - T_0)/n$ [6, 23, 42, 43]. For a given set of observations, $n$ eigenvalues, $\lambda_i$, and their relevant eigenvectors are obtained from (10.6). Each eigenvector includes the weight coefficients, $a_k$, of the relative POD mode in (10.5), so $n$ POD modes are finally calculated. The energy captured by each POD mode in the system is proportional to the relevant eigenvalue. The eigenvalues are sorted in a descending order, so the first few POD modes in (10.4) capture larger energy compared with the later modes.

(c) *POD coefficients, $b_i$, calculation for a new test case*: This key step is where the POD can be used to create a reduced-order thermal/fluid model as a function of the system design variables. Generally, there are three methods to calculate the POD coefficients $b_i$ for a new test case with a new set of design variables:

- *Galerkin projection of the system POD modes onto the governing equations*: This results in a set of coupled nonlinear ordinary differential equations (ODEs) in time for transient systems, or a set of algebraic equations for steady-state systems, to be solved for the POD coefficients. This method has been used to create reduced-order models of transient temperature fields in terms of mostly one parameter, such as Reynolds/Raleigh number [44–51]. The previous investigations have been either for prototypical flows (such as flow around a cylinder) or for simple geometries such as channel flow where inhomogeneous boundary conditions are easily homogenized by the inclusion of a source function in the decomposition.
- *Interpolation among modal coefficients*: In steady state, the POD coefficients at a new set of design variables can be obtained by an interpolation between the weight coefficients at the observed variables to match a desired new variable value [51, 52]. In this approach, the coefficients used to reconstruct an observed field $T_{\text{obs},k}$ are found first by projecting each of the POD modes onto the observation in turn:

$$b_{i,\text{obs}} = (T_{\text{obs},k} - T_0)\psi_i, \quad i = 1, \dots, m. \tag{10.7}$$

This can be computed for all observations within the ensemble $T_{\text{obs}}$. The complete coefficient matrix $B \in \Re^{m \times n}$, in which each column is the coefficient vector to reconstruct the corresponding observation from the ensemble $T_{\text{obs}}$, can be more efficiently computed as:

$$B = \psi^{+} \otimes (T_{\text{obs}} - T_0), \tag{10.8}$$

where $(.)^{+}$ is the Moore–Penrose pseudo-inverse giving the least squares solution [53]. Once $b_{i,\text{obs}}$ has been found for all observations, each of which represents the solution under a specified combination of design variables, the POD coefficients $b_i$ for a new set of design variables are calculated through the interpolation of the coefficients $b_{i,\text{obs}}$ between the corresponding observations. In other words, rather than directly interpolating between observations, interpolation is performed in the POD mode space using the coefficients $b_{i,\text{obs}}$. For systems with one design variable, this interpolation can be done through linear, or the slightly more accurate piecewise cubic spline interpolation between coefficients. This method has been applied only for a system with one parameter and simple geometry such as cavity flow [51, 52]. However, the approach can be extended to more complex systems with multiple design variables using higher order multidimensional interpolation approaches, such as krieging or multivariate adaptive regression splines (MARS) [23].

– *Flux matching process*: In the flux matching process [6, 42], the coefficients $b_i$ are obtained by applying (10.4) to some locally specified region, such as system boundaries to match the known mass or heat fluxes. Although the flux matching process has been used to develop reduced-order models of the flow behavior in complex steady-state systems successfully [6, 23, 54, 55], it has been applied only for thermal modeling of a simple 2D geometry of a channel with two iso-heat flux blocks [42, 55], with no consideration of complex 3D geometry. Nie and Joshi [56] have presented a POD based reduced order modeling of steady turbulent convection in connected domains with the application for a 3D electronic rack. They developed a POD-based modeling for each component separately and then subsequently combined the models together using boundary profile based flux matching. Their method is only applicable to systems consisting of a series of nested subdomains. Also, they applied their methods for a case study, where the thermal parameters, which are chip heat generation rates, existed only in one subdomain, making the temperature distribution in other domains almost uniform. So, matching of the subdomains' temperature fields was much easier than matching the flow and pressure fields in [56].

(d) *POD temperature field generation*: With calculated $T_0$, $\psi_i$, and $b_i$ for a new set of design variables, the corresponding temperature field for the test case can be generated inside the entire domain from (10.4) for different numbers of used POD modes, $m$ [1].

## 10.2.2 POD and Galerkin Projection for Data Center Temperature Modeling

A recently developed POD based reduced order modeling approach for temperature field calculation in multiscale convective systems such as data centers is presented and applied for a data center cell in [40]. The approach is applicable for systems where the temperature field at selected scales drives the thermal design decision. The energy equation is solved only at these dominant scales via system POD modes and Galerkin projection to obtain a more accurate zoomed prediction at these scales, instead of the entire domain. The effects of the phenomena at other scales are modeled through simple energy balance equations and known heat flux and temperature matching, as well as appropriate matching conditions at the scale interfaces.

### 10.2.2.1 Review of the Method

The POD-based method for the thermal modeling of multiscale systems has been illustrated in Fig. 10.4. The reduced-order model is developed assuming the same POD temperature equation for the entire domain:

$$T = T_0 + \sum_{i=1}^{m} b_i \psi_i. \tag{10.9}$$

So, the first and second steps in Fig. 10.4 are similar to the basic POD technique, as explained in Sect. 10.2.1 and Fig. 10.3. The difference is in the key step of the POD technique, where the POD coefficients, $b_i$, must be calculated. In this method, after the POD thermal modes have been calculated for the entire domain, the required algebraic equations to calculate the POD coefficients are obtained separately by focusing on different scales of the system.

In any multiscale thermal/fluid system, there are often one or few important length scales dominating the thermal performance of the entire system, and driving thermal design decisions. For instance, the temperature field at the rack scale usually drives the thermal decisions for designing a cooling system in a data center.

At the dominant scales, the governing energy equation is solved via POD modes and Galerkin projection to obtain a more accurate prediction at these scales compared with the entire domain. Considering each dominant scale as the computational domain as seen in Fig. 10.5, the mean energy equation, ignoring viscous dissipation, is:

$$\rho c_{\mathrm{p}} u \nabla T - \nabla(k_{\mathrm{eff}} \nabla T) = q. \tag{10.10}$$

In (10.10), $q_{\mathrm{Domain}}$ is the domain volumetric heat generation. In Galerkin projection, the governing equation, (10.10), is projected into the space spanned by POD modes.

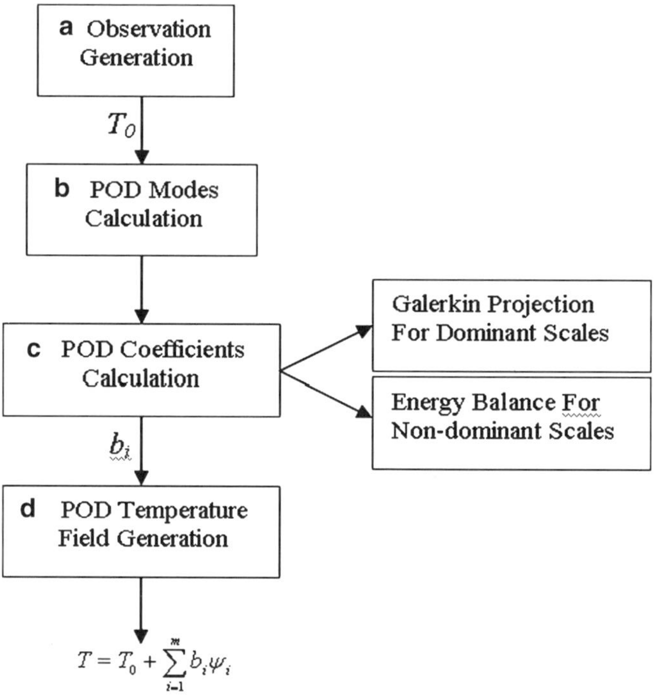


**Fig. 10.4** POD- and Galerkin projection-based thermal modeling method [40]

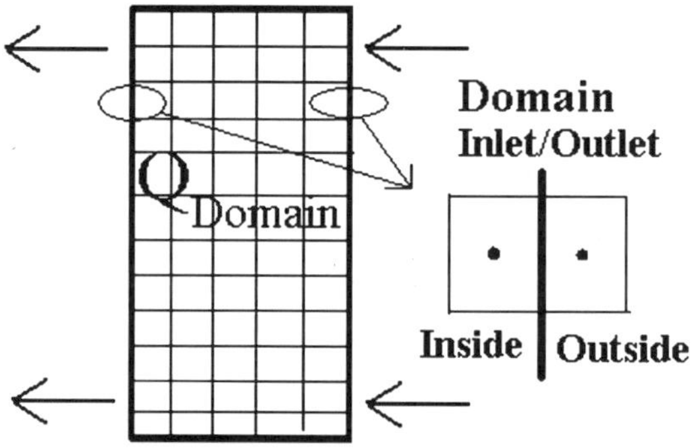


**Fig. 10.5** Dominant scale as the computational domain [40]

Using (10.9) as the temperature field, Galerkin projection results in a set of linear algebraic equations:

$$\int_{\Omega} \psi_i(\rho c_{\mathrm{p}} u \nabla(T_0 + \sum_{l=1}^{m} b_l\psi_l) - \nabla(k_{\mathrm{eff}} \nabla(T_0 + \sum_{l=1}^{m} b_l\psi_l)) - q_{\mathrm{Domain}})\,\mathrm{d}x\,\mathrm{d}y\,\mathrm{d}z = 0, \quad i = 1, 2, ..., m. \tag{10.11}$$

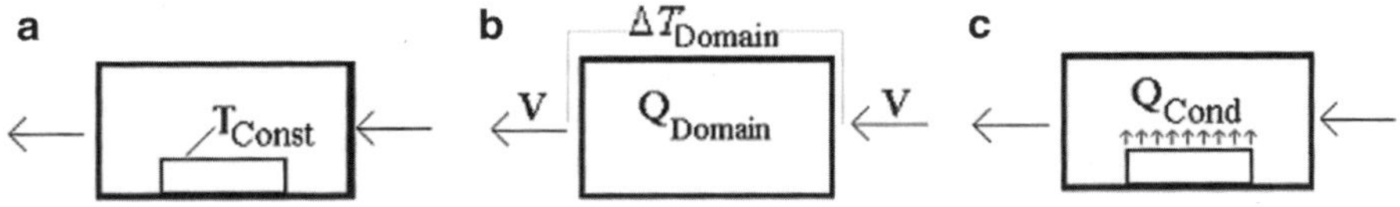


**Fig. 10.6** Nondominant scale simplifications [40]

In (10.11), $m$ is the number of used POD modes, which can change from 1 up to $n-1$ where $n$ is the number of observations. So we get $m$ algebraic equations for each dominant scale, if $m$ modes are retained in the linear decomposition of temperature field into POD modes. Also, the effect of phenomena at other scales on the dominant scale modeling is considered as boundary conditions at the dominant domain inlet/outlets. Since the reference temperature field, $T_0$ ,and the POD modes are known from the previous steps at the nodes inside and outside of the dominant domain boundary, the following equations can be used as required boundary conditions while integrating (10.11) by parts on the domain of Fig. 10.5:

$$T_{\mathrm{B.C.}} = \frac{T_{\mathrm{Inside}} + T_{\mathrm{Outside}}}{2} = \frac{T_{0,\mathrm{Inside}} + \sum_i b_i \psi_{i,\mathrm{Inside}} + T_{0,\mathrm{Outside}} + \sum_i b_i \psi_{i,\mathrm{Outside}}}{2}, \tag{10.12}$$

$$\frac{\partial T}{\partial x}_{\mathrm{B.C.}} = \frac{T_{\mathrm{Outside}} - T_{\mathrm{inside}}}{\Delta x} = \frac{T_{0,\mathrm{Outside}} + \sum_i b_i \psi_{i,\mathrm{Outside}} - T_{0,\mathrm{Inside}} - \sum_i b_i \psi_{i,\mathrm{Inside}}}{\Delta x}. \tag{10.13}$$

On the other hand, at nondominant scales, the algebraic equations are obtained simply through energy balance equations, heat flux matching, and/or surface temperature matching. Although simple, heat flux matching has been used as an effective way to calculate the POD coefficients [42]. Generally, the nondominant domains can be simplified in three ways, as illustrated in Fig. 10.6. For case (a) in Fig. 10.6, the fluid temperature at a specific surface of the domain is kept at a known constant value of $T_{\mathrm{Const}}$. From (10.9), we get

$$T_{\mathrm{Const}} = \bar{T}_0 + \sum_{i=1}^{m} b_i \bar{\psi}_{i,\mathrm{Surf}}, \tag{10.14}$$

where $\bar{\psi}_{i,\mathrm{Surf}}$ and $\bar{T}_0$ are the average values of the temperature POD modes and temperature reference on the surface with a constant temperature. We get one algebraic equation for each constant temperature surface of the domain.

For domains like case (b) in Fig. 10.6, one equation is obtained to satisfy the conservation of the energy across the domain. Applying the total energy balance across the inlet and outlet surfaces of the domain results in:

$$Q_{\mathrm{Domain}} = \rho V A c_{\mathrm{p}} \Delta T_{\mathrm{Domain}}. \tag{10.15}$$

By separating the known and unknown variables and substituting (10.9) in (10.15), we obtain:

$$\begin{aligned}\frac{Q_{\mathrm{Domain}}}{VA\rho c_{\mathrm{p}}} &= \Delta T_{\mathrm{Domain}} \\ &= \bar{T}_{0,\mathrm{Domian\,Outlet}} - \bar{T}_{0,\mathrm{Domain\,Inlet}} \\ &\quad + \sum_{i=1}^{m} b_i(\bar{\psi}_{i,\mathrm{Domain\,Outlet}} - \bar{\psi}_{i,\mathrm{Domain\,Inlet}}),\end{aligned} \tag{10.16}$$

where $\bar{\psi}_{i,\mathrm{Domain\,Inlet}}$ and $\bar{\psi}_{i,\mathrm{Domain\,Outlet}}$ are the average values of the temperature POD modes on the inlet and outlet surfaces of the domain of case (b) in Fig. 10.6, respectively. Also, $\bar{T}_{0,\mathrm{Domain\,Inlet}}$ and $\bar{T}_{0,\mathrm{Domain\,Outlet}}$ are the average values of the reference temperature, $T_0$, on the inlet and outlet surfaces of the domain, respectively.

For domains like case (c) in Fig. 10.6, one equation is obtained by matching the heat flux at the surface with a constant heat flux.

$$q''_{\mathrm{Cond}} = -k\frac{\partial T}{\partial n}|_{\mathrm{wall}}. \tag{10.17}$$

Since the flux function involves a gradient, substituting the POD temperature of (10.9) in (10.17) may produce large errors. To address this issue, a modal heat conduction function, $F_{i,\mathrm{ModalCond}}$, is defined in the POD space. All $m$ modal heat conduction functions can be calculated together by [42]:

$$F_{\mathrm{ModalCond}} = Q_{\mathrm{CondObs}} \otimes (T_{\mathrm{obs}} - T_0)^{+} \otimes \psi, \tag{10.18}$$

where $Q_{\mathrm{CondObs}}$, a $1 \times m$ matrix, includes $m$ observation surface heat inputs and (.) + is the Moore–Penrose pseudo-inverse giving the least squares solution. This definition results in the following algebraic equation for this case [42]:

$$Q_{\mathrm{Cond}} = \sum_{i=1}^{m} b_i F_{i,\mathrm{ModalCond}}. \tag{10.19}$$

All equations at dominant and nondominant scales are subsequently solved together to obtain a single set of POD coefficients, assuming the same POD temperature equation for the entire domain. With calculated $T_0$ , $\psi_i$, and $b_i$ for a new combination of design variables, the corresponding temperature field

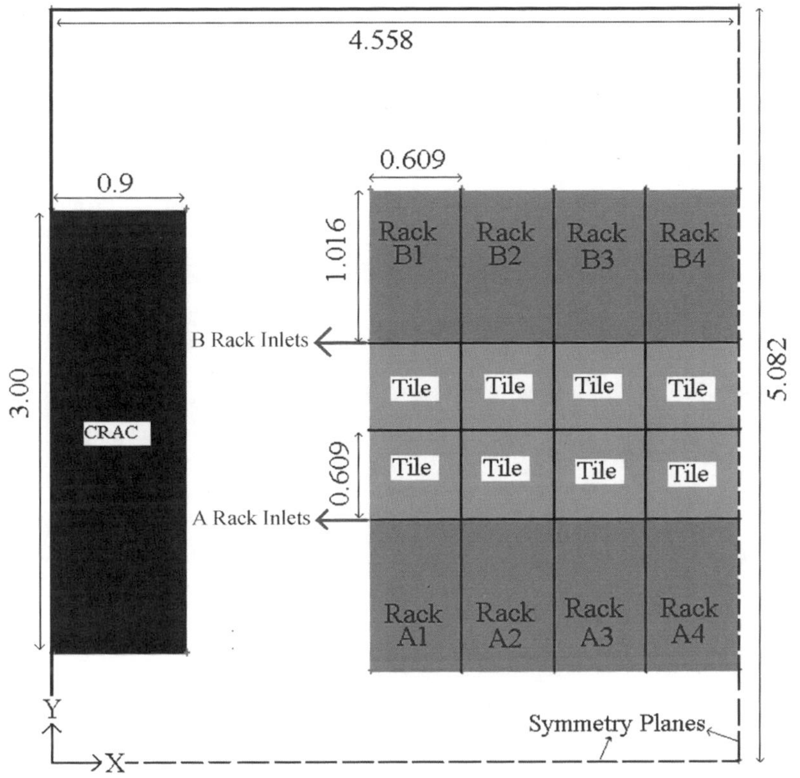


**Fig. 10.7** Data center cell top view; dimensions in meter. Only one quarter of the cell is shown due to symmetry [40]

for the test case can be generated inside the entire domain from (10.9) for different numbers of used POD modes, $m$ [40].

We should note that to solve (10.11) using Galerkin projection, the flow field and effective thermal conductivity at the dominant scales are required. The average of the velocity and effective thermal conductivity fields between two neighboring observations of each test case are used instead of the exact values in (10.11). Also, a POD based reduced order velocity model inside the domain can be obtained using flux matching process [6, 42] and used for greater accuracy. In the next section, the method outlined above is applied to an air-cooled data center cell.

#### 10.2.2.2 Application to a Data Center Cell

The POD-based method illustrated in Fig. 10.4 is applied to a data center cell shown in Fig. 10.7 to simulate the temperature field as a function of CRAC unit air delivery velocity and rack heat loads. Each CRAC unit takes in hot return air from the room and discharges cold air into a subfloor plenum for delivery to the data center through perforated tiles. Since the air temperature field at the rack scale drives the design of a cooling system in a data center, the turbulent energy equation is solved at the rack domain, see Fig. 10.5, via POD modes and Galerkin projection. Also, the effect of room scale phenomena, such as room level air recirculation, on the rack scale modeling is considered as boundary conditions at the rack inlet/outlets in Galerkin projection. So, (10.11–10.13) are used to obtain $m$ algebraic equations for each rack. At the data center scale, a simple energy balance is applied across the CRAC unit, case (b) in Fig. 10.6 and (10.16). Also, the temperature field at the perforated tile surfaces is kept fixed at the known constant air discharge temperature, case (a) in Fig. 10.6 and (10.14). Ultimately, $(m \times N_{\mathrm{racks}} + 1 + 1)$ equations are obtained to solve for the $m$ POD mode coefficients, $b_i$. All the mentioned equations are solved together using least square approach to obtain a single set of POD coefficients, assuming the same POD temperature equation for the entire domain.

To construct a POD based reduced order model of the temperature field, the rack heat loads and CRAC airflow rate are considered to change between 500 W–30 kW and 0.94 (2,000 CFM)–25.45 $m^3/s$ (54,000 CFM), respectively. To reduce the number of design variables for illustration purposes, we assume that corresponding racks in each column have the same heat load. This leads to five design variables for the data center cell of Fig. 10.7:

1. Inlet air velocity of CRAC unit, $V_{\mathrm{in}}$
2. Heat load of racks A1 and B1, $Q_1$
3. Heat load of racks A2 and B2, $Q_2$
4. Heat load of racks A3 and B3, $Q_3$
5. Heat load of racks A4 and B4, $Q_4$

To obtain the required observations for the POD algorithm, one-fourth of the representative data center and the plenum are simulated using CFD/HT code, Fluent v. 6.1. The geometry of this section of the data center is shown in Fig. 10.7. Each rack is modeled as a volumetric heat source with six representative fans at its exit and a lumped pressure jump at its inlet. The CRAC unit is modeled with a constant inlet and exit velocity, discharging the cooling air into the plenum at 15°C.

The CRAC velocity and rack heat loads are varied to generate 19 observed temperature fields for the data center example. The design variables for these observations are collected in Table 10.1. The rack inlet air temperatures are usually used for thermal design of data centers. The contours of the average of all 19 observations, $T_0$ in (10.9), at the inlets of racks A1 through A4 and B1 through B4 of the data center cell in Fig. 10.7 are shown in Fig. 10.8a, b, respectively. It is seen that the dominant hot spots for the data center cell occur at the middle and top of the first rack. The energy captured by each POD mode in the system is proportional to the relevant eigenvalue in (10.6). The energy percentage captured by each POD mode is plotted vs. the mode number in Fig. 10.9. The magnitude of the eigenvalue and the energy captured by each mode decreases sharply with the index of POD modes. The modes with largest eigenvalues take the shape of large-scale smooth structures, e.g., see Fig. 10.10a, b. The modes with large index numbers include small-scale

**Table 10.1** Design parameters for the observations [40]

| Observation # | 1 | 2 | 3 | 4 | 5 | 6 | 7 | 8 | 9 | 10 | 11 | 12 | 13 | 14 | 15 | 16 | 17 | 18 | 19 |
|---|---|---|---|---|---|---|---|---|---|---|---|---|---|---|---|---|---|---|---|
| VCRACin (m/s) | 0.4 | 0.6 | 0.8 | 1 | 1.4 | 1.9 | 2.1 | 2.31 | 2.4 | 2.6 | 3.5 | 4 | 4.8 | 5 | 6 | 7 | 8 | 9 | 9.4 |
| RackAB 1 (kW) | 1 | 2 | 4 | 5 | 6 | 4 | 12 | 15 | 30 | 21 | 14 | 10 | 30 | 21 | 20 | 18 | 17 | 27 | 30 |
| RackAB 2 (kW) | 1 | 3 | 3 | 5 | 7 | 10 | 8 | 15 | 5 | 11 | 22 | 15 | 30 | 21 | 20 | 18 | 13 | 26 | 30 |
| RackAB 3 (kW) | 1 | 2 | 1 | 5 | 8 | 12 | 19 | 15 | 5 | 7 | 20 | 20 | 28 | 21 | 25 | 29 | 24 | 26 | 30 |
| RackAB 4 (kW) | 1 | 1 | 5 | 5 | 9 | 16 | 9 | 15 | 20 | 6 | 14 | 30 | 23 | 21 | 25 | 29 | 18 | 26 | 30 |

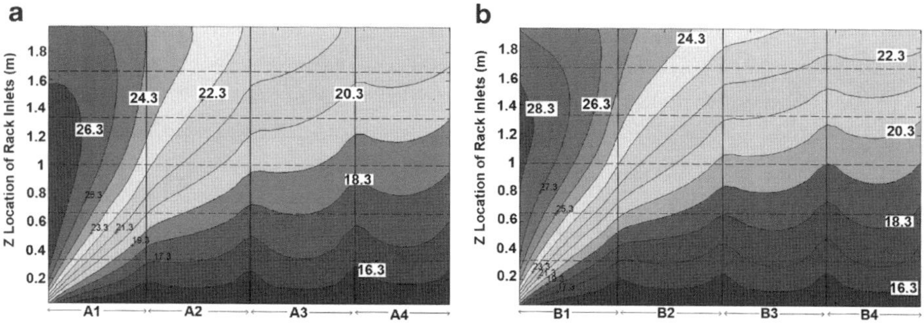


**Fig. 10.8** Reference air temperature contours (°C) at the racks inlets [40]

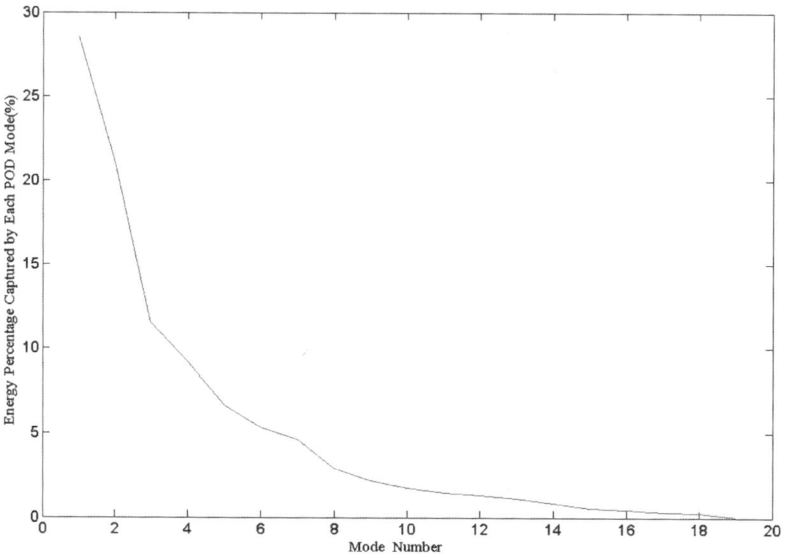


**Fig. 10.9** Energy percentage captured by each POD mode vs. the mode number [40]

structures, such as the temperature boundary layer, e.g., Fig. 10.10c, d. Figure 10.10 shows the contours of the first two and last two POD modes at the inlet surfaces of racks A1, A2, A3, and A4 of the data center. To solve (10.11) obtained from Galerkin projection, the velocity field inside the racks and at its boundaries is required. Here, we use the CFD solution to verify the presented POD-based algorithm.

The presented algorithm is used to generate temperature field for several new combinations of the design variables. For this case, there are $8 \times 18 + 1 + 1 = 146$ equations to be solved for the 18 unknown POD coefficients, using least square approach. POD coefficients of different modes, $b_i$, for 5 arbitrary test cases are shown in Fig. 10.11 when all 18 modes are retained in the decomposition of (10.9).

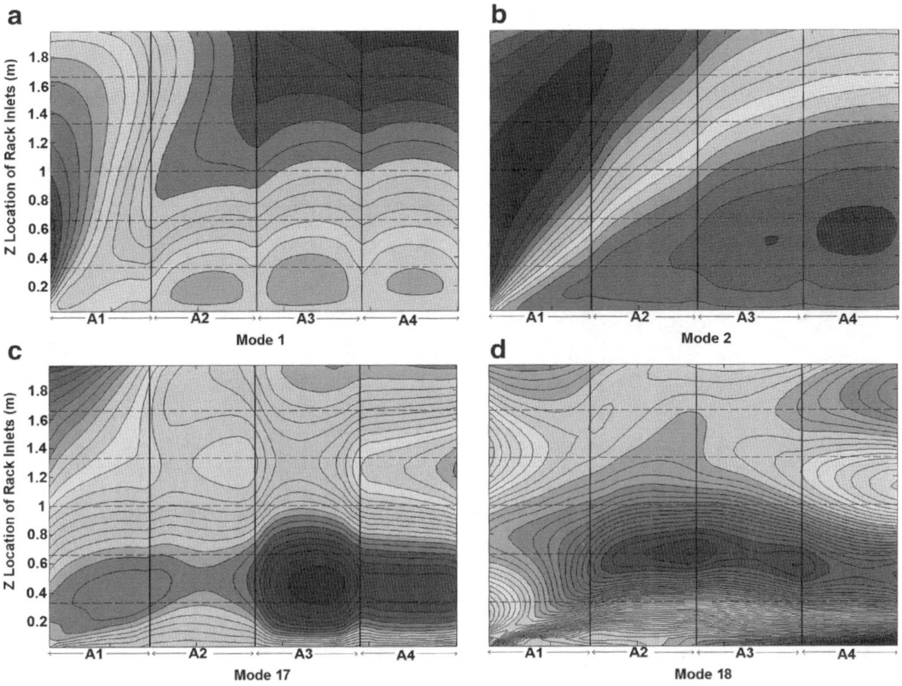


**Fig. 10.10** Contours of the first two and last two POD modes at the racks inlet surfaces for racks A1–A4 [40]

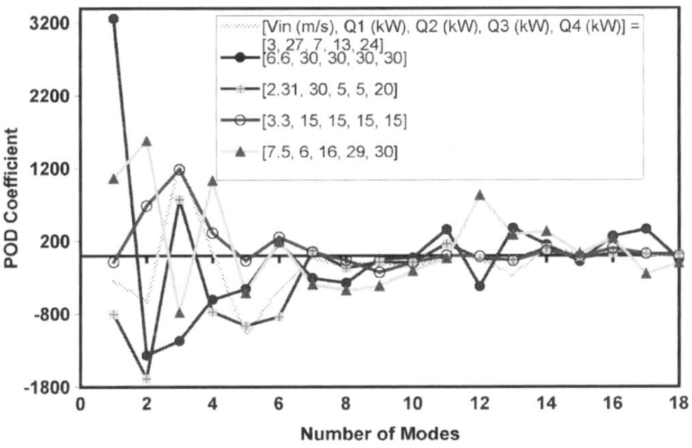


**Fig. 10.11** POD coefficients of different modes for five test cases which are specified in the legend [40]

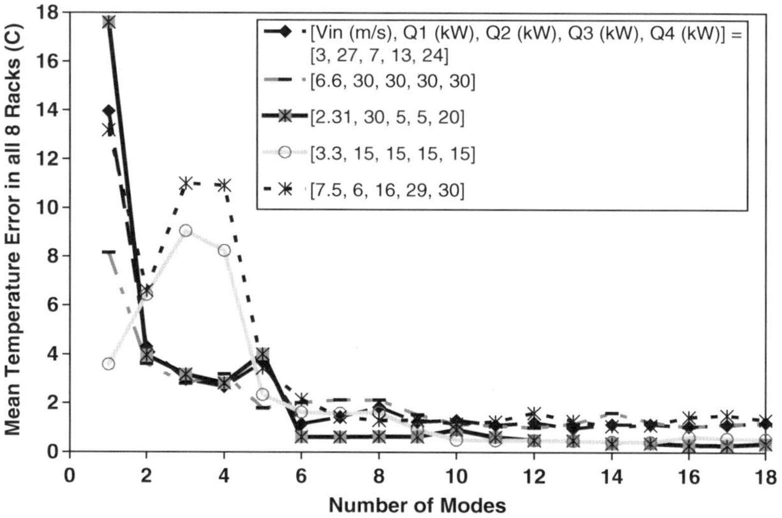


**Fig. 10.12** Mean temperature error (°C) within eight racks and at their boundaries vs. used mode numbers for five test cases which are specified in the legend [40]

It is seen that the value of POD coefficients decreases for modes with higher index and lower energy content. So the first few terms in the decomposition of (10.9) are dominant.

To study the fidelity of the POD method, the POD temperature values within the racks and at their boundaries are compared with full numerical simulations. A mean error, $\bar{T}_{\text{error}}(x, y, z)$ (°C), is calculated by taking an average of the absolute values of the temperature difference between POD and full numerical predictions for all points:

$$T_{\text{error}}(x, y, z) = |T_{\text{POD}}(x, y, z) - T_{\text{Fluent}}(x, y, z)|, \qquad (10.20)$$

$$\bar{T}_{\text{error}}(x, y, z) = \frac{\sum_{i=1}^{N_{\text{nodes}}} T_{\text{error}}(x, y, z)}{N_{\text{nodes}}}. \qquad (10.21)$$

$N_{\text{nodes}}$ is the number of nodes/points in the domain, 114,000 at the rack scale. The mean error at the rack scale is plotted for five different cases in Fig. 10.12 when the number of used POD modes changes from 1 to 18. The converged mean error at the rack scale for these cases is less than 1.3°C or 6%. As shown in Fig. 10.12, the local temperatures at the rack scale converge after ~14 modes.

To see if the POD method can predict the air temperatures at the rack inlets accurately for use in design decisions, the full-field predictions, POD simulations, and the POD temperature error are shown in Fig. 10.13 for racks A1 through A4 for two test cases. The average error is less than 1°C, while the maximum local error is ~2.5°C for some small regions. Considering that the error in deployed sensor measurements can be around 1°C, the POD-based method can be used effectively

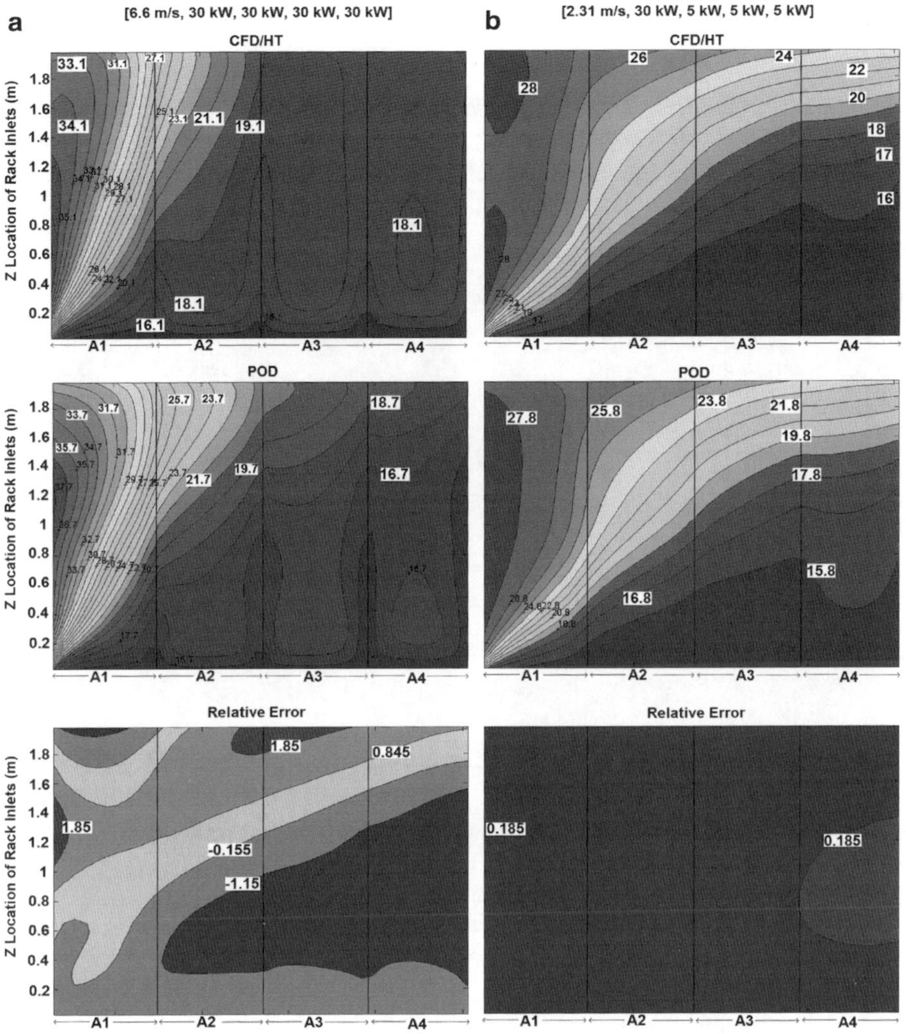


**Fig. 10.13** Contours of CFD/HT temperature, POD temperature, and relative error (°C) at racks inlets for two test cases. Relevant test case is mentioned at the top of each contour plot [40]

in solving data center thermal design problems. The mean error, the standard deviation, and the Euclidean L2 norm of the POD temperature error at all 114,000 points of the rack scale for the 15 test cases are tabulated in Table 10.2. The standard deviation and the error norm for each test case are defined by:

$$T_{\text{Std}} = \left( \frac{\sum_{i=1}^{N_{\text{nodes}}} \left( T_{\text{error}}(x, y, z) - \bar{T}_{\text{error}}(x, y, z) \right)^2}{N_{\text{nodes}} - 1} \right)^{1/2}, \tag{10.22}$$

**Table 10.2** POD temperature error at rack scale and whole domain for 15 test cases [40]

| | | Rack scale (114,000 points) | | | Whole data center (383,826 points) | | |
|---|---|---|---|---|---|---|---|
| Case # | [$V_{in}$ (m/s), $Q_1$ (kW), $Q_2$ (kW), $Q_3$ (kW), $Q_4$ (kW)] | Error norm (%) | Mean error (°C) | Standard deviation (°C) | Error norm (%) | Mean error (°C) | Standard deviation (°C) |
| 1 | [7, 5, 6, 16, 29, 30] | 6.35 | 1.32 | 1.16 | 19.00 | 3.06 | 2.98 |
| 2 | [2, 21, 21, 21, 21] | 3.73 | 0.96 | 0.89 | 30.10 | 8.29 | 5.89 |
| 3 | [2.31, 5, 5, 20, 30] | 6.59 | 1.61 | 1.02 | 21.07 | 4.39 | 3.94 |
| 4 | [3, 27, 7, 13, 24] | 6.08 | 1.21 | 0.99 | 12.75 | 2.32 | 2.19 |
| 5 | [2.31, 30, 5, 5, 20] | 1.80 | 0.35 | 0.32 | 15.80 | 2.89 | 2.89 |
| 6 | [3.3, 15, 15, 15, 15] | 3.51 | 0.54 | 0.60 | 10.73 | 1.64 | 1.76 |
| 7 | [6.6, 30, 30, 30, 30] | 4.48 | 1.18 | 0.96 | 23.04 | 4.53 | 4.45 |
| 8 | [5.5, 14, 23, 3, 19] | 7.39 | 1.35 | 1.25 | 17.74 | 2.78 | 2.70 |
| 9 | [1.5, 2, 30, 1, 3] | 8.89 | 1.98 | 1.36 | 18.51 | 3.39 | 3.27 |
| 10 | [4, 30, 29, 9, 28] | 10.12 | 2.25 | 1.87 | 19.30 | 3.82 | 3.40 |
| 11 | [7.8, 4, 8, 19, 3] | 9.11 | 1.40 | 1.17 | 35.67 | 5.39 | 3.36 |
| 12 | [1.155, 2.5, 2.5, 10, 15] | 8.66 | 2.29 | 1.89 | 30.32 | 6.31 | 6.58 |
| 13 | [4.62, 30, 30, 30] | 3.93 | 1.00 | 0.91 | 27.69 | 6.04 | 5.79 |
| 14 | [6.5, 14, 29, 22, 28] | 6.69 | 1.56 | 1.23 | 16.41 | 2.94 | 2.77 |
| 15 | [1.4, 6, 7, 8, 9] | 6.23 | 1.33 | 1.13 | 18.06 | 3.44 | 3.12 |
| Average | | 6.24 | 1.36 | 1.12 | 21.08 | 4.08 | 3.67 |

$$\text{Error Norm} = \frac{||T_{\text{error}}(x,y,z)||}{||T_{\text{Fluent}}(x,y,z)||} \times 100\%$$
$$= \left( \frac{\sum_{i=1}^{N_{\text{nodes}}} T_{\text{error}}{}^2(x,y,z)}{\sum_{i=1}^{N_{\text{nodes}}} T_{\text{Fluent}}{}^2(x,y,z)} \right)^{1/2} \times 100\%. \qquad (10.23)$$

In the error norm, the values of temperature are in degree Celsius. As seen in Table 10.2, the mean error varies from 0.35 to 2.29°C, while the average is 1.36°C, and the average standard deviation 1.12°C. Also, the error norm changes from 1.8 to 10.1%, while the average is 6.2%. These values confirm that the presented POD method is accurate enough at the rack scale to use for design purposes.

Although the suggested algorithm mainly focused on the rack scale to predict temperatures at the rack inlet/outlets and inside the racks, it would be interesting to see the POD temperature prediction for the entire data center domain. A very accurate representation of the temperature field at the room scale is not expected, since only a total energy balance and a perforated tile temperature match were used to simulate the details at the room level. The mean error, the standard deviation, and the Euclidean L2 norm of the POD temperature error at all 383,826 points of the entire domain are tabulated in Table 10.2 for all 15 cases considered before. The mean error changes from 1.64 up to 6.31°C, while the average is 4.08°C. The average of all standard deviations is 3.67°C. Also, the error norm changes from 10.7 up to 35.7% while the average is 21.1%.

All these values confirm that the presented POD method is not accurate enough at the room scale.

Regarding the method efficiency, the POD-based algorithm generates the temperature field for a new test case with different CRAC velocity and rack heat loads in 12 min, while the CFD/HT simulation done by Fluent takes ~2 h for the same test case on the same computing platform (a desktop computer with Xeon™ CPU, 2.8-GHz and 2.75 GB of RAM). Also, the most time-consuming part of the method, integrating the velocity terms in (10.11) over the domain, can be done once for all observed CRAC velocities, if the method is to be used for many simulations. It takes ~38 min to calculate these terms. After that, the algorithm is ready to obtain the POD temperature field for each new test case in only 4 s. So, if we assume that 100 additional runs beyond the initial observations are needed to find an optimal thermal design in data centers, the CFD/HT model by Fluent takes ~200 h (~8 days) to find the design solution, while the POD algorithm can do it in ~(38 + 7 = 45) min which is ~250 times faster. This confirms the ability of the presented method to provide a quick and accurate enough thermal modeling of a multiscale thermal/fluid system in order to design around several input parameters.

Although the presented method in [40] provides a quick and reasonably accurate thermal modeling of air-cooled data centers for design purposes, the approach is applicable only for systems where the temperature field at selected scales, called dominant scales, drives the thermal design decision. Accordingly, the generated temperature field based on this method at scales other than dominant scales is not very accurate. Also, the method requires the fluid flow solution at these dominant scales for integration of the energy equation via system POD modes and Galerkin projection [1].

### *10.2.3 POD and Energy Balance for Data Center Temperature Modeling*

Samadiani and Joshi [41] have presented a simpler POD-based method to generate a reduced-order thermal modeling of complex systems such as air-cooled data centers. The method and its application to a data center cell are reviewed in Sects. 10.2.3.1 and 10.2.3.2.

#### 10.2.3.1 Review of the Method

The POD-based method has been illustrated in Fig. 10.14. The first and second steps in Fig. 10.14 are similar to the previous method explained in Sect. 10.2.2 and Fig. 10.4. The difference is where the POD coefficients, $b_i$, must be calculated. In this method, the algebraic equations to be solved for POD coefficients in (10.9) are obtained simply through energy balance equations, heat flux matching, and/or

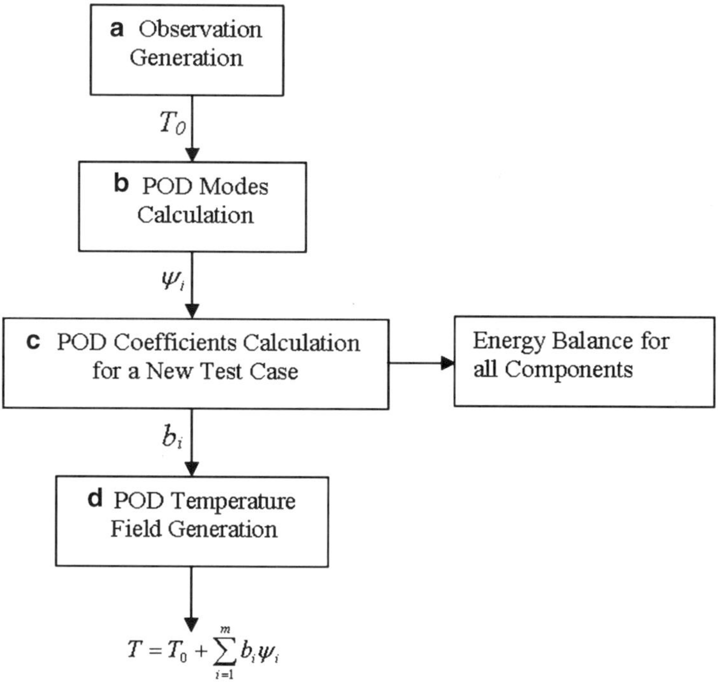


**Fig. 10.14** POD- and energy balance-based thermal modeling method [41]

surface temperature matching for all components of the system regardless of being dominant or not.

As explained in Sect. 10.2.2.1, the key convective phenomena at each component or subsystem of the main system are illustrated in Fig. 10.6. The corresponding equations for the three cases in Fig. 10.6 are (10.14), (10.16), and (10.19), respectively. After the algebraic equations have been obtained for all components of the system, they are solved together to find the associated POD coefficients for a new set of design variables. Sometimes, especially if the method is used for thermal modeling of real-world systems, the inlet velocity and/or heat load in case (b) of Fig. 10.6 is not known, but the temperature difference across the domain is measured and known instead. In this case, (10.16) can be still used to find the appropriate POD coefficients associated with the measured temperature difference.

We should note that the number of obtained algebraic equations, $s$, in this method, can be less, equal, or more than the number of available POD modes, $n-1$. $n$ is the number of observations. Since we need at least the same number of equations as the number of unknown POD coefficients to avoid an underdetermined system of equations, the maximum possible number of POD modes to use, $m$ in $T = T_0 + \sum_{i=1}^{m} b_i \psi_i$, is limited by the number of available equations, $s$, in addition to the number of available modes, $n-1$. Accordingly, $m$ can be 1 up to $\min(n-1, s)$ in this method. On the other hand, the number of available equations is limited by the number of convective components and available thermal information for the components in the system. This brings a limitation to the presented method whose effect on the results for a data center cell is studied in [41] and briefly discussed in Sect. 10.2.3.2 as well. In the next section, the method outlined above is applied to an air-cooled data center cell.

#### 10.2.3.2 Application to a Data Center Cell

The method in Fig. 10.14 is applied to the data center cell shown in Fig. 10.7 with the five design variables mentioned in Sect. 10.2.2.2. The difference with the case study in Sect. 10.2.2.2 is that each server within a rack here is modeled as a separate volumetric heat source. In Sect. 10.2.2.2, each rack was modeled as a uniform volumetric heat source.

The CRAC velocity and rack heat loads are varied to generate 21 observed temperature fields throughout the data center cell. The design variables for these observations are collected in Table 10.3. To obtain the appropriate algebraic equations to calculate the POD coefficients for a test case with new design variables, (10.16) associated with case (b) in Fig. 10.6 is applied to each server in the data center. Having applied (10.16) to all servers, $N_{\text{servers}}$ equations are obtained; $N_{\text{servers}}$ is 48 here. Similarly, energy balance equation, (10.16), is applied for the CRAC unit with known total heat load of the data center and CRAC inlet velocity for the new test case. Also, the temperature field at the perforated tile surfaces is kept fixed at the known constant air discharge temperature by applying (10.14) for case (a) in Fig. 10.6. Ultimately, ($N_{\text{servers}} + 1 + 1 = 50$) equations are obtained to solve for 20 POD mode coefficients. All the obtained equations are solved together using least square approach to obtain a single set of POD coefficients for a new set of design variables.

POD coefficients associated with different modes, $b_i$, are shown in Fig. 10.15 for four arbitrary test cases, which are distinct from the observations. These coefficients have been obtained when all 20 modes are retained in the POD reconstruction in (10.9). It is seen that the value of POD coefficients decreases for modes with higher index and lower energy content. Also, the last mode coefficients are almost zero. So, the first few terms in the decomposition of (10.9) are dominant. Also, the changes in the POD coefficients after using ~10 modes are much less than the coefficient changes in the initial part of the graph in Fig. 10.15. It seems that the solution has been converged after ~10 modes.

To study the convergence of the solution with the number of used POD modes and also to examine the fidelity of the POD method, the POD temperatures are compared with full CFD/HT simulations. A mean error, $\bar{T}_{\text{error}}(x, y, z)$ (°C), is calculated using (10.21). $N_{\text{nodes}}$ in (10.21) is 431,120 for the studied data center cell. The mean error is plotted for the four test cases in Fig. 10.16 when the number of used POD modes changes from 1 to 20. As shown in Fig. 10.16, the local temperatures converge after ~10 modes. This is consistent with the relative flattening in the POD coefficient changes after ~10 modes in Fig. 10.15, as discussed

**Table 10.3** Design variables for the observations [41]

| Observation # | 1 | 2 | 3 | 4 | 5 | 6 | 7 | 8 | 9 | 10 | 11 | 12 | 13 | 14 | 15 | 16 | 17 | 18 | 19 | 20 | 21 |
|---|---|---|---|---|---|---|---|---|---|---|---|---|---|---|---|---|---|---|---|---|---|
| $V_{in}$ (m/s) | 0.4 | 0.6 | 0.8 | 1 | 1.2 | 1.4 | 1.6 | 1.9 | 2.1 | 2.4 | 2.6 | 3 | 3.25 | 3.5 | 3.75 | 4 | 4.25 | 4.5 | 5 | 5.5 | 6 |
| $Q_1$ (kW) | 1 | 2 | 4 | 5 | 7 | 6 | 11 | 4 | 12 | 30 | 21 | 17 | 4 | 14 | 15 | 10 | 10 | 21 | 21 | 29 | 20 |
| $Q_2$ (kW) | 1 | 3 | 3 | 5 | 4 | 7 | 11 | 10 | 8 | 5 | 11 | 6 | 7 | 22 | 16 | 15 | 10 | 16 | 21 | 28 | 20 |
| $Q_3$ (kW) | 1 | 2 | 1 | 5 | 8 | 8 | 11 | 12 | 19 | 5 | 7 | 22 | 10 | 20 | 16 | 20 | 30 | 27 | 21 | 16 | 25 |
| $Q_4$ (kW) | 1 | 1 | 5 | 5 | 9 | 9 | 11 | 16 | 9 | 20 | 6 | 20 | 8 | 14 | 25 | 30 | 30 | 23 | 21 | 26 | 25 |

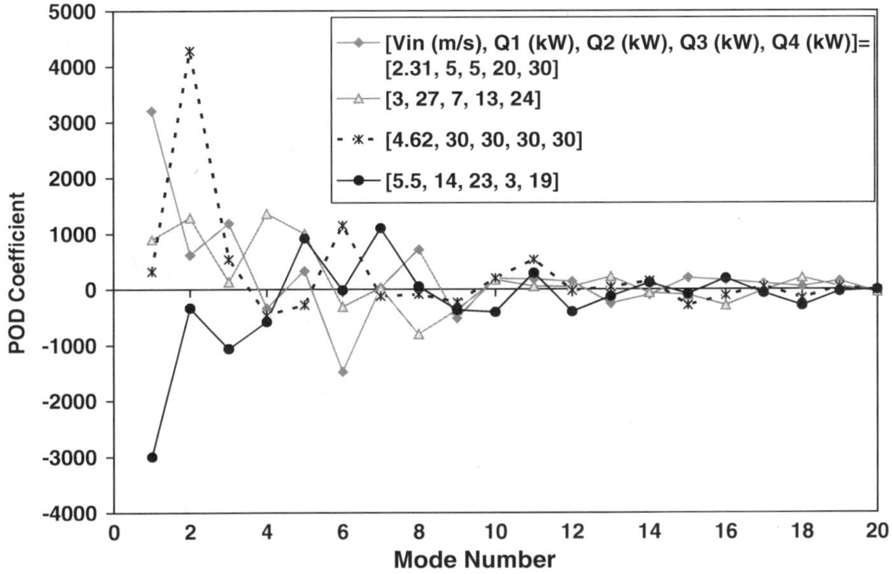


**Fig. 10.15** POD coefficients of the associated modes for four test cases, when all 20 modes are used in the POD reconstruction [41]

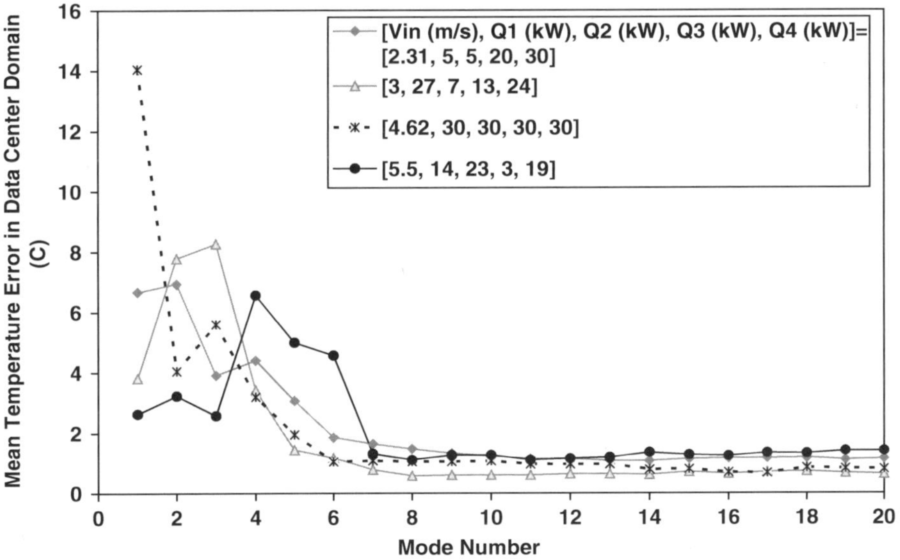


**Fig. 10.16** Effect of the number of retained POD modes on the mean POD temperature error (°C) for the entire data center for four test cases [41]

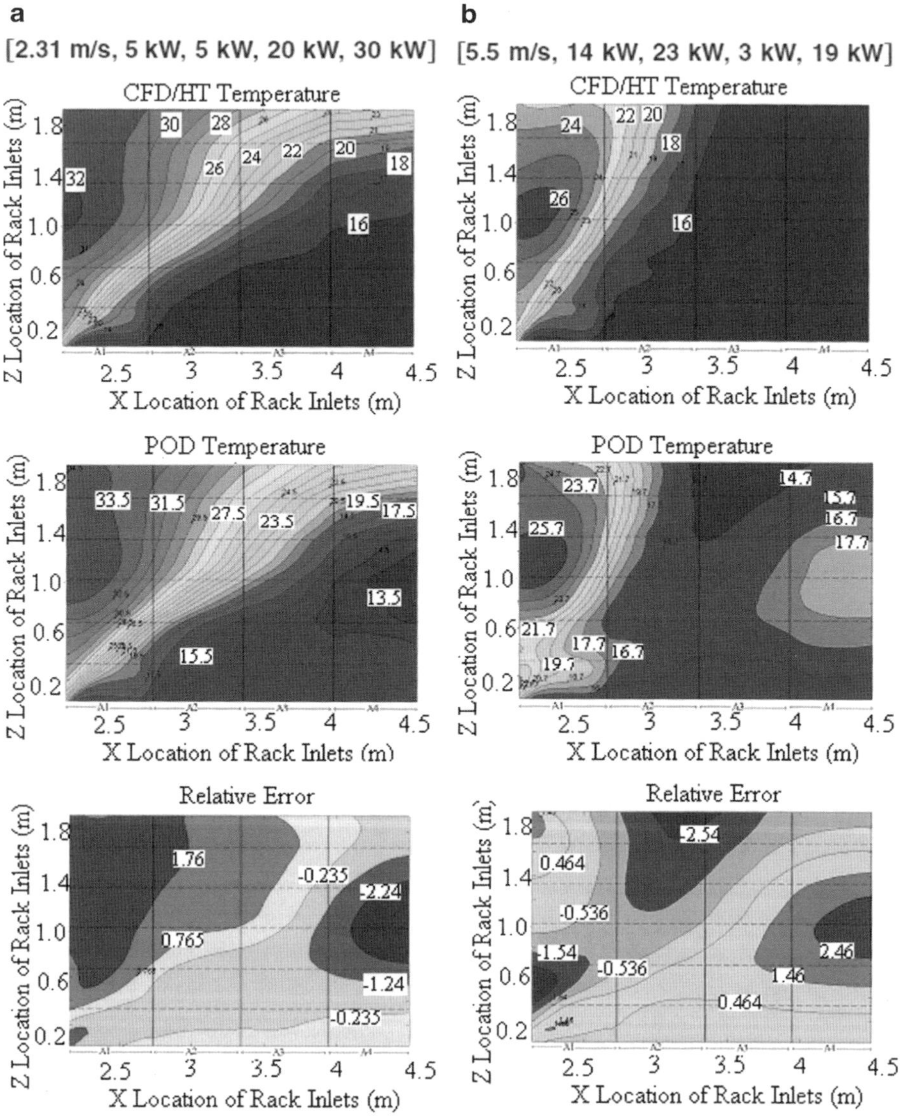


**Fig. 10.17** Contours of CFD/HT temperature, POD temperature, and relative error (°C) at racks inlets for two test cases. Relevant test case is mentioned at the top of each contour plot [41]

above. Also, the converged mean error for the entire domain for these cases is less than 1.4°C or 7.2%, as seen in Fig. 10.16.

To see if the POD method can predict the air temperatures at the rack inlets accurately for use in design decisions, the full-field predictions, POD simulations, and the POD temperature error are shown in Fig. 10.17 for racks A1 through A4 for two test cases. The average error is less than 1.5°C, while the maximum local error

**Table 10.4** POD temperature error and its standard deviation compared with CFD/HT solution for six test cases [41]

| Case # | [$V_{in}$ (m/s), $Q_1$ (kW), $Q_2$ (kW), $Q_3$ (kW), $Q_4$ (kW)] | Mean relative error (%) | Mean error (°C) | Standard deviation (°C) |
|---|---|---|---|---|
| 1 | [1.5, 2, 30, 1, 3] | 8.35 | 2.13 | 2.02 |
| 2 | [2.31, 5, 5, 20, 30] | 3.59 | 1.14 | 1.39 |
| 3 | [3, 27, 7, 13, 24] | 2.39 | 0.63 | 0.76 |
| 4 | [4, 30, 29, 9, 28] | 4.77 | 1.31 | 1.99 |
| 5 | [4, 30, 29, 9, 28] | 3.05 | 0.80 | 0.94 |
| 6 | [4.62, 30, 30, 30, 30] | 7.23 | 1.41 | 1.63 |
| Average | [5.5, 14, 23, 3, 19] | 4.90 | 1.24 | 1.46 |

is ~2.5°C for some small regions. Since the uncertainty in deployed sensor measurements can be around 1°C, the POD-based method can be used effectively in solving data center thermal design problems. The mean error (10.21), the standard deviation in the error (10.22), and the mean relative error (10.23) of the POD temperature field at all 431,120 points of the domain for six test cases, which are distinct from the observations, are tabulated in Table 10.4.

As seen in Table 10.4, the mean error for the six test cases varies from 0.63°C or 2.4% to 2.13°C or 8.4%. The average in the mean absolute and relative error for all cases is 1.24°C and 4.9% while the average standard deviation is 1.46°C. These values confirm that the presented POD method is reasonably accurate at the entire data center cell.

Regarding the computational speed of the POD-based method, it should be noted that the POD-based thermal model has only 20 DOF, representing a five order of magnitude decrease compared to the CFD/HT model. The CFD/HT simulation done by Fluent takes ~2 h to obtain the temperature field for a new test case on a desktop computer with Xeon™ CPU, 2.8-GHz and 2.75 GB of RAM. However, it takes only ~48 s to obtain the POD temperature field for the same test case on the same computing platform, which is ~150 times faster.

It was shown that the POD-based method can predict a new temperature field in the entire data center cell of Table 10.4 with an average error of 5% if the temperature differences across the 48 servers are given as known information, in addition to the CRAC air velocity and discharge temperature. One interesting question is how the POD solution and error change if lesser thermal information about the components such as the server temperature differences is supplied. It is shown [41] that the POD results remain accurate for the data center example even if the given thermal information at the component boundaries decreases by 67%, if we use the required dominant POD modes to capture the most important phenomena of the system. In fact, the method could predict the air temperatures at all 431,120 points in the data center cell with an average error of 6.2% even with knowing the temperature differences for only two servers per rack, which makes the method very appropriate for operational data centers.

As mentioned in Sect. 10.2.3.1, the maximum possible number of used POD modes is limited by the number of available algebraic equations in the presented method. This number is limited to the number of interior convective components or subsystems like the ones in Fig. 10.6, for which we can use energy balance equations, heat flux matching, and/or surface temperature matching. The effect of the number of these components in the main system on the POD solution is studied in [41] and the following statements are concluded about the presented method:

1. A converged and accurate temperature field in a complex system is generated only if the number of components and given thermal information is much higher than the number of available modes and the number of required dominant modes to capture the main physics of the system.
2. The solution starts to diverge when the number of equations, which is equal to the number of components or given thermal information, decreases and becomes closer to the number of available POD modes.

This limitation would not typically cause a problem in thermal model reduction of operational data centers with several housed servers if enough numbers of servers have thermal sensors at their inlet/outlet [41].

## 10.3 Adaptable and Robust Energy Efficient Design of Data Centers

The POD based reduced order modeling methods explained in Sects. 10.2.2 and 10.2.3 can be used for simulation-based optimization and design of data centers. Samadiani et al. [57] presented a POD-based design method to achieve an adaptable and robust energy efficient data center. The method and its application to a data center cell are reviewed in Sect. 10.3.

### *10.3.1 Simulation-Based Design Method for Energy-Efficient Data Centers*

In the traditional design of an air-cooled data center, the required CRAC airflow rate is calculated based on an acceptable temperature difference across the servers, $\Delta T$, typically 11°C (20°F):

$$V_{\text{in}} A_{\text{CRAC}} = \frac{Q_{\text{total}}}{\beta \rho c_{\text{p}} \Delta T}, \tag{10.24}$$

where $V_{\text{in}}$ is the average velocity of the supply air from the CRAC unit's discharge surface, ACRAC. Also, $Q_{\text{total}}$ is the data center heat load. Coefficient $\beta > 1$ is a rule-of-thumb parameter which accounts for the air recirculation effect on the temperature field in the data centers with cold-/hot-aisle arrangement. The CRAC supply temperature is fixed for intended data center heat load, while the work/heat load among the servers and racks are distributed randomly. While the heat generated by the electronic equipment in data centers is increasing year after year due to demands

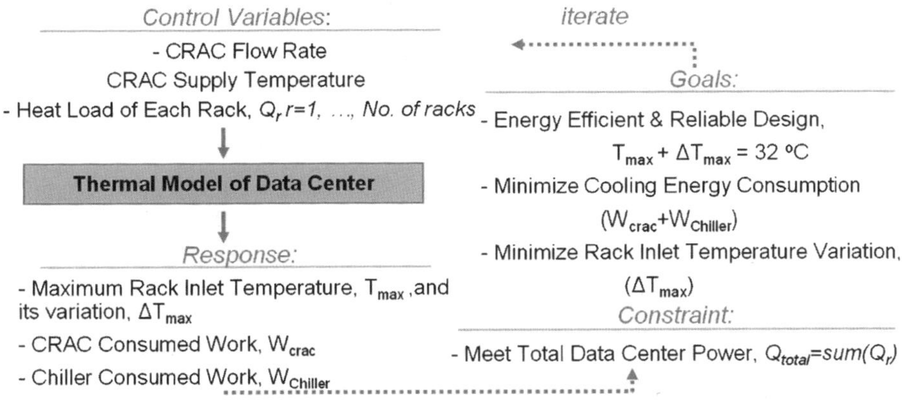


**Fig. 10.18** Adaptable robust design in energy-efficient data centers [57]

for higher processor speeds and miniaturization, sustainable and reliable operation of data centers is possible through energy-efficient and adaptable design of cooling systems for containing operating costs and promoting sustainability. In this regard, the main design requirements of the air-cooled systems can be classified into [58]:

(a) *Operating temperature limits*: The cooling system must keep the operating temperature in a specified range, below a typical value of 85°C for silicon components. To analyze the thermal performance of the typical air-cooling systems in data centers, a corresponding criterion may be to maintain the inlet cooling air temperature to the servers, considering the possible changes in the system parameters, under 32°C [59].
(b) *Energy efficiency*: Total cooling energy consumption of the data center, which is the summation of CRAC and chiller consumed work, $W_{\text{Cooling}} = W_{\text{CRAC}} + W_{\text{Chiller}}$, determines the operating cost of the data center cooling system, which should be minimized. Also, the cooling system should be designed to remove the actual data center heat load, rather than planned occupancy, to have an energy-efficient design which is neither under-cooled nor overcooled.
(c) *Robustness*: This requires maintaining the energy efficiency, effectiveness, reliability, and performance stability of the equipment, in spite of large uncertainty and variability. The typical variability sources are variations of CRAC supply airflow rate and rack heat loads.
(d) *Adaptability*: This allows additional flexibility to adjust and adapt to future technology, changes in environment and changes in customer demands. Air-cooled systems should be designed to be adaptable to these changes through, for example, intelligent rack heat load re-allocation and changes in supply temperature and airflow rates of CRAC units to handle the lifecycle mismatch between the IT equipment and facility thermal management systems.

Considering the requirements of an air-cooled data center, the design problem can be summarized as in Fig. 10.18. This diagram shows a design methodology to handle the actual momentary total heat load with minimum cooling energy consumption and

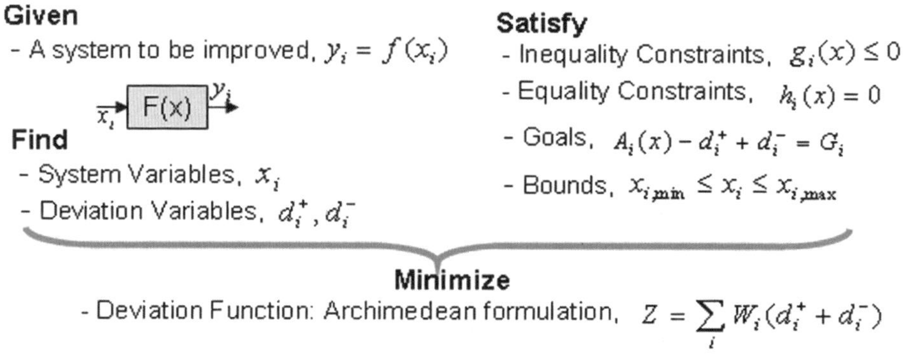


**Fig. 10.19** cDSP structure [57]

maximum efficiency, through adaptable changes in the rack heat load allocation, and CRAC supply airflow rate and temperature. The method also seeks minimum variation in the rack inlet temperature due to changes in CRAC supply airflow rate and temperature and rack heat loads. The heat load re-allocation can be implemented through physical re-location of the hardware and/or by distributing the processing tasks among the servers through virtualization technology [22]. The airflow rate of CRAC units can be varied using variable frequency drive motors. Also, the CRAC supply temperature can be easily changed in operational data centers.

The simulation-based approach to solve the design problem in Fig. 10.18 is based on the integration of three tools: (1) the POD based reduced order thermal multiscale modeling [40], explained in Sect. 10.2.2, (2) the compromise decision support problem (cDSP) [5], and (3) robust design principles [6]. The cDSP and robust design principles are explained in Sects. 10.3.2 and 10.3.3, respectively. In Sect. 10.3.4, the design method is demonstrated through application for an air-cooled data center cell.

### *10.3.2 Compromise Decision Support Problem*

The compromise decision support problem (cDSP) [60] provides a modular, adaptable, and computationally efficient mathematical framework for solving design problems with multiple objectives and constraints, making the cDSP very suitable for designing adaptable robust systems. Its structure, based on the Archimedean, or weighted sum, formulation is illustrated in Fig. 10.19. The conceptual basis of the cDSP is to minimize the difference between what is desired, the target $G_i$, and what can be achieved, $A_i(x)$. The difference between these values is called the deviation variable, $d_i$. The key benefit of the cDSP is that the designer preferences over different goals can be applied by easily weighing the coefficients, $W_i$, of the deviation variables $d_i$ associated with each goal. A simulation model, $f(x_i)$, relating the objective function to control variables is needed in the cDSP framework. For the presented design method, this model is obtained using the POD-based thermal modeling.

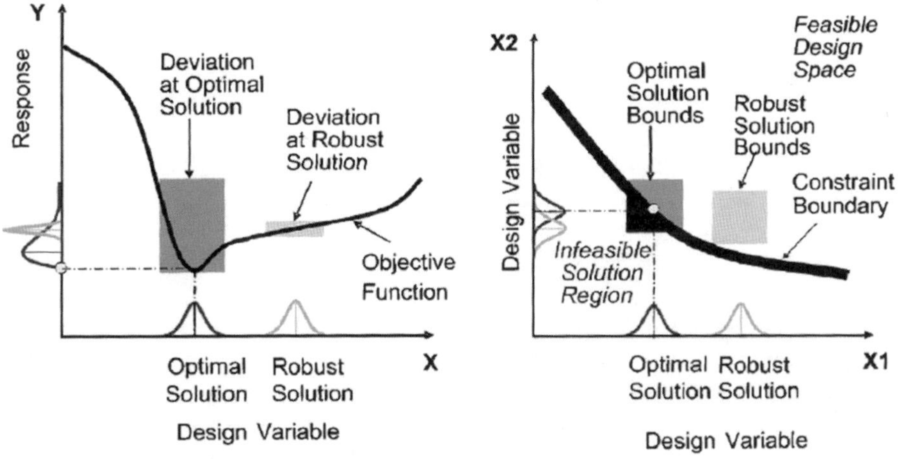


**Fig. 10.20** Type II robust design: (**a**) goals and (**b**) constraints representation [54]

## *10.3.3 Robust Design*

In typical optimization approaches for design, only the mean response is moved to a target while the effects of the variation in the system parameters or design variables on the performance evaluation are ignored. By accounting for variation, the results obtained by robust design techniques are effective regardless of changing noise factors, uncontrollable parameters (Type I), and/or design variables (Type II) [54]. The difference between having an optimal solution and robust solution for design goals and constraints is illustrated in Fig. 10.20. Considering the design goals, a robust solution happens in a flat region with minimal variability of the response, while an optimal solution happens at the lowest or highest value of the response regardless of the response variability. The trade-off between finding the robust or optimal solution is based on the level of variation of each design variable and the designer's preferences, which could be implemented through the cDSP. Considering the constraints, a robust solution always happens in the feasible design space despite the variable changes, while an optimal solution might fail to satisfy the constraints due to the changes in the design variables.

## *10.3.4 Application to a Data Center Cell*

A case study is defined first in Sect. 10.3.4.1. Then, a reduced-order thermal modeling of the data center using the POD-based method is developed in Sect. 10.3.4.2. The design problem is formulated using the cDSP in Sect. 10.3.4.3. Finally, the results and discussion are presented in Sect. 10.3.4.4.

#### 10.3.4.1 Case Study

An adaptable robust and energy-efficient design of an air-cooled data center cell which will be used for the next 10 years is considered. The data center cell is the same as in Sect. 10.2.2.2. One-fourth of the data center cell is shown in Fig. 10.7. The data center will house 1,033 W/m$^2$ (96 W/ft$^2$) for the first year of the operation, which is equal to 10% utilization of the full capacity. New IT equipment is integrated into the data center annually so that the data center will cope with 10,355 W/m$^2$ (962 W/ft$^2$) during the tenth year of the operation at 100% utilization.

As explained in the following, we do not directly consider the CRAC supply temperature as a control variable in iterative optimization. Also, to reduce the number of design variables for illustration purposes, we assume that corresponding racks in each column of the data center cell have the same heat load. This leads to five design variables, $x_i$, for the data center cell of Fig. 10.7:

1. Inlet air velocity of CRAC unit, $V_{in}$
2. Heat load of racks A1 and B1, $Q_1$
3. Heat load of racks A2 and B2, $Q_2$
4. Heat load of racks A3 and B3, $Q_3$
5. Heat load of racks A4 and B4, $Q_4$

The rack heat loads and CRAC air velocity are considered to change between 500–30 kW and 0.35–9.4 m/s, respectively. This causes the CRAC airflow rate to change from 0.94 (2,000 CFM) to 25.45 (54,000 CFM). Depending on the thermal capacity of the CRAC heat exchanger, the CRAC unit must provide a minimum flow rate to be able to remove a given data center heat load. The typical relationship between the heat removal capacity of a CRAC unit as a function of airflow rate is shown in Fig. 10.21. Accordingly, the following relationship is used to calculate the minimum required CRAC velocity (in m/s) as a function of total heat load (in kW):

$$V_{\text{in, minimum}} = \frac{-0.01 + \sqrt{(0.0001) - (28E - 8)Q_{\text{total}}}}{-0.0008}. \tag{10.25}$$

For a given data center heat load and initial CRAC supply temperature (15°C in this study) , the maximum inlet cooling air temperature to the servers, considering the possible changes in the system parameters, is designed to be equal to 32°C in order to have reliable, and neither overcooled nor under-cooled data center based on ASHRAE standards [59]. For this purpose, the initial data center air temperatures, obtained at the CRAC initial supply temperature, are increased by a specific value $(32 - (T_{\max} + \Delta T_{\max}))$.

$$T_{\text{Supply, new}} = T_{\text{Supply, initial}} + (32 - (T_{\max} + \Delta T_{\max})). \tag{10.26}$$

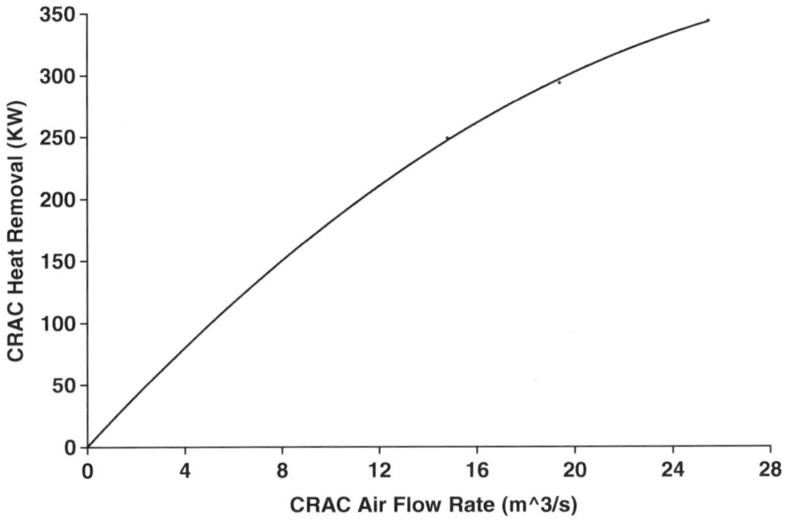


**Fig. 10.21** CRAC heat removal capacity as a function of airflow rate [57]

Considering the new CRAC supply temperature, the chiller work is calculated by:

$$W_{\text{Chiller}} = \frac{Q}{\text{COP}}, \tag{10.27}$$

where the coefficient of performance (COP) of the chiller-CRAC loop as a function of CRAC supply temperature can be calculated by modeling the chiller performance and the CRAC heat exchanger. This relationship for a water-chilled CRAC unit in a Hewlett-Packard (HP) Utility Data Center [61] is shown in Fig. 10.22. At higher supply temperatures, the COP increases and the chiller consumes less energy to remove a given center heat load. This relationship from [61] is also used here for COP calculation.

$$\text{COP} = 0.0068T_{\text{Supply}}^2 + 0.0008T_{\text{Supply}} + 0.458. \tag{10.28}$$

In addition to the chiller work, the work consumed by the CRAC blower motor should be calculated. The consumed work by CRAC is usually a linear function of the airflow rate. The following equation, obtained from the available data for typical CRAC units, is used in this study to calculate $W_{\text{CRAC}}$:

$$W_{\text{CRAC}}(\text{kW}) = 2.7\,V_{\text{in}}\ (\text{m/s}). \tag{10.29}$$

At each iteration of the design problem in Fig. 10.18, the chiller and CRAC work are calculated as explained above. Then, iteration continues to find optimal and robust values of the control variables to minimize the cooling energy consumption

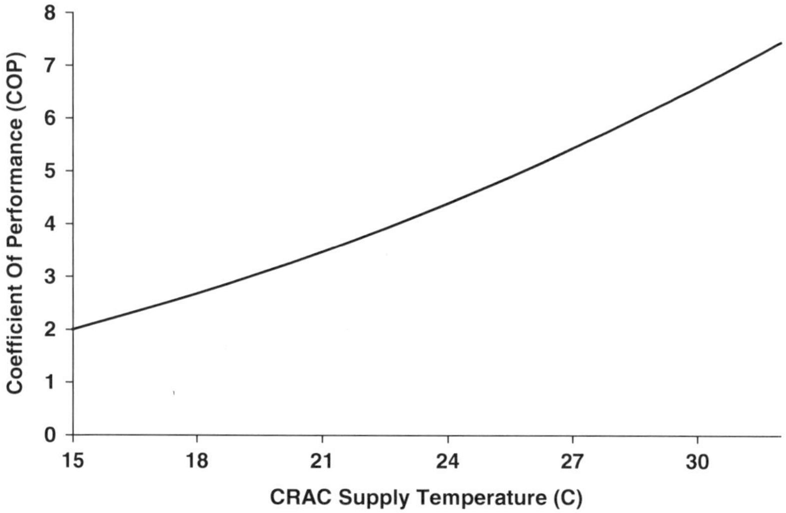


**Fig. 10.22** COP of a chilled water loop in the HP Utility Data Center [61]

and rack inlet temperature variation. In order to model the temperature field inside the data center and obtain $T_{\mathrm{max}}$, the POD-based algorithm explained in Sect. 10.2.2 is used. Also, for Type II robust design, the variability of the response, $\Delta T_{\mathrm{max}}$, is calculated by Taylor expansions of the system response, considering the worst variation scenario in the design variables:

$$\Delta T_{\mathrm{max}} = \sum_{i=1}^{5} \left| \frac{\delta T_{\mathrm{max}}}{\delta x_i} \Delta x_i \right|. \quad (10.30)$$

The derivatives are computed using first-order difference technique since no closed form solution exists.

To summarize, the following steps are taken to solve the design problem in Fig. 10.18 for the data center example:

1. Find the maximum rack inlet temperature $T_{\mathrm{max}}$ as a function of control variables of CRAC velocity and racks heat loads:
   (a) For known CRAC initial supply temperature (15°C), generate observations by varying the control variables, i.e., CRAC velocity and racks heat loads.
   (b) Obtain POD modes and coefficients for a new set of control variables.
   (c) Find $T_{\mathrm{max}}$.
   (d) Calculate $T_{\mathrm{supply,\ new}}$, $W_{\mathrm{Chiller}}$, $W_{\mathrm{crac}}$, $\Delta T_{\mathrm{max}}$ using (10.26)–(10.30)
2. Using cDSP and designer's preferences over the goals, $W_i$, formulate and solve the design problem for a given data center heat load at different years, $Q_{\mathrm{total}}$.

3. Save the results for energy efficient and robust operation of the data center at each year:

   (a) Optimal/robust CRAC flow rate and racks heat loads
   (b) Optimal/robust CRAC supply temperature

Following the listed steps, the POD-based algorithm is applied to the data center cell in Sect. 10.3.4.2. Then, the design problem is formulated by cDSP in Sect. 10.3.4.3. The design problem is solved and results for the example are discussed in Sect. 10.3.4.4.

#### 10.3.4.2 POD-Based Thermal Modeling of the Data Center Cell

Obtaining a POD thermal modeling of the data center cell is explained in Sect. 10.2.2.2. In this section, only the comparison between the maximum rack inlet temperature, $T_{max}$, obtained by POD and CFD/HT is presented since $T_{max}$ drives the thermal design decision as discussed before and shown in Fig. 10.18. More discussion about the POD method application and results for this case study can be found in [57].

The maximum rack inlet temperatures obtained by POD are compared with full CFD/HT solutions in [57] for 41 arbitrary test cases, of which 36 are distinct from the observations. The average of the error for all test cases is 1.3°C or 4.6% while the error for few test cases, especially for the cases out of the range of observed temperature fields or near the extreme limits, is higher than 2.5°C. To be used within the iterative robust design problem of Fig. 10.18, the compact model of the data center must predict the effect of the total center heat load, rack heat load allocation, and the CRAC velocity on $T_{max}$ accurately.

In Fig. 10.23, the $T_{max}$ obtained by POD and CFD/HT simulations are compared when the total heat load of the data center varies from 24 kW (10% utilization) to 240 kW (100% utilization). For these results, the CRAC velocity is fixed at 9.4 m/s and the total center heat load is distributed uniformly among all eight racks. The maximum temperature increases with the center heat load linearly. As seen in the figure, the POD predicts the effect of the total heat load on $T_{max}$ accurately.

The effect of the racks heat load distribution is shown in Fig. 10.24 when the CRAC velocity is fixed at 2.31 m/s and the total center heat load is 120 kW but with different distributions among eight racks. It is interestingly seen that a simple work load distribution change among the racks can decrease the maximum temperature at the rack inlets as much as 10°C, which could be translated to significant energy saving in the chiller work. This shows there are opportunities to save energy in air-cooled data centers through intelligent workload re-allocation if an efficient design method is applied. As seen in the figure, the POD-based method predicts this trend accurately.

The effect of the CRAC velocity on the $T_{max}$ obtained by POD is shown in Fig. 10.25 for six different center heat loads with a uniform distribution among the racks.

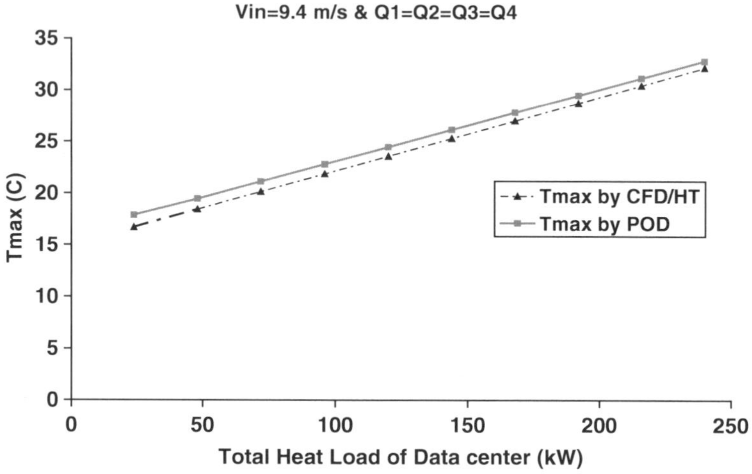


**Fig. 10.23** $T_{max}$ obtained by POD and CFD/HT for CRAC velocity of 9.4 m/s and uniform distribution of the data center heat load [57]

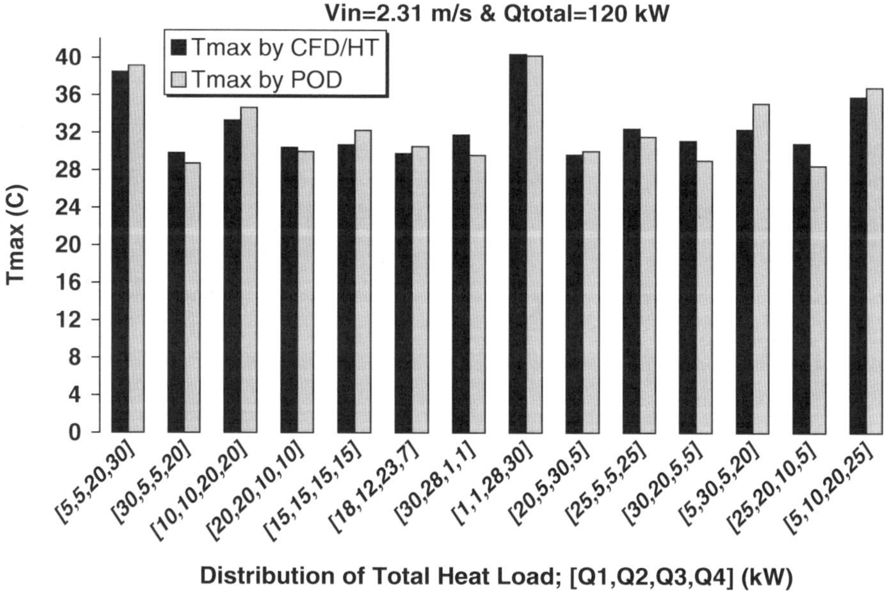


**Fig. 10.24** $T_{max}$ obtained by POD and CFD/HT for CRAC velocity of 2.31 m/s and different distributions of the data center heat load, 120 kW [57]

Also, $T_{max}$ is obtained by CFD/HT simulation for some limited velocities when $Q_{total} = 120$, 168, and 240 kW as shown in this figure. It is seen that the trend and values of $T_{max}$ obtained by POD are in a good agreement with CFD/HT simulations. The design constraint ($T_{max} = 32°C$) is shown in this figure as well.

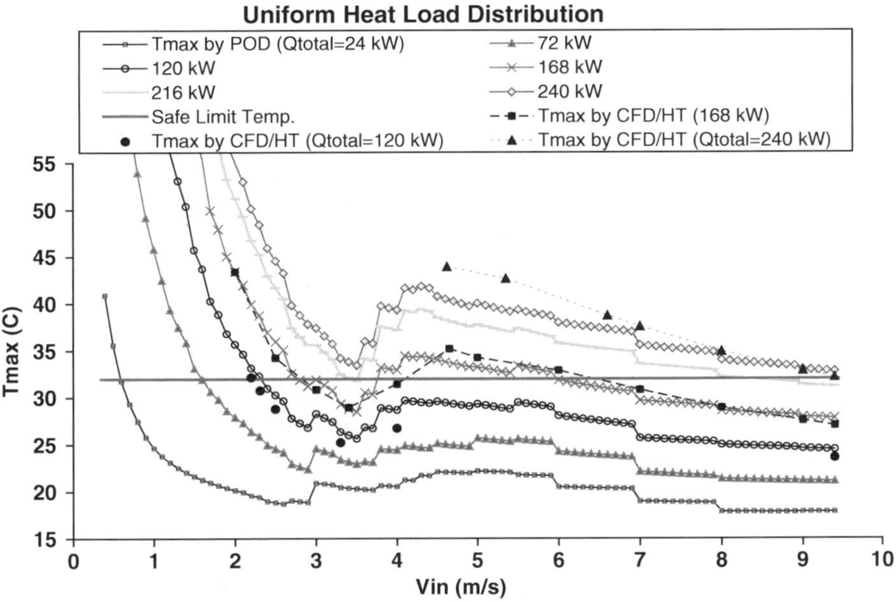


**Fig. 10.25** $T_{\max}$ obtained by POD and CFD/HT vs. CRAC velocity for different data center heat loads with uniform distributions among racks [57]

As seen in Fig. 10.25, the change trend of the maximum rack inlet temperature with CRAC velocity is highly nonlinear, having at least one local minimum and maximum for each total heat load. For example, when $Q_{\text{total}} = 168$ kW, the maximum inlet temperature decreases from ~43°C gradually to reach ~29°C, a local minimum, by increasing the inlet velocity from 2 to ~3.4 m/s. Increasing the CRAC velocity more than ~3.4 m/s increases the temperature unexpectedly to reach a local maximum, ~35°C, at ~4.6 m/s. Afterwards, $T_{\max}$ decreases linearly by increasing the velocity. As seen in the figure, the POD-based method predicts this trend accurately. This trend shows that the thermal management of data centers cannot be done by simply increasing the cooling airflow rate of CRAC units. As seen in Fig. 10.25, increasing the CRAC velocity by 129% from 2.8 to 6.4 m/s requires 129% more power, but does not change the maximum rack inlet temperature or the thermal performance of the center when $Q_{\text{total}} = 168$ kW. This confirms that there are opportunities to save energy in air-cooled data centers through intelligent changes in CRAC velocity if an efficient design method is applied.

The reason for the nonlinear changes in $T_{\max}$ with CRAC velocity in Fig. 10.25 is the change of the air recirculation pattern in the data center around 3.8 m/s. The operating point of the server fans is ~1.07 m/s or equivalently ~0.21 $m^3$/s (455 CFM). So, the required airflow rate for each rack is ~1.28 $m^3$/s (2,730 CFM) and for all eight racks is ~10.26 $m^3$/s (21,840 CFM). If we assume the CRAC airflow rate is distributed uniformly among all eight perforated tiles in Fig. 10.7, the CRAC unit needs to provide at least 10.26 $m^3/s$ (21,840 CFM), to match the required rack airflow rates. This is equal to providing the velocity of 3.8 m/s by the CRAC unit. Below this limit, air recirculation from the hot aisle to the cold aisle will provide the rest of the required rack flow rate. On the other hand, at velocities much above this limit, the extra rate of flow provided by the CRAC unit will re-circulate mainly between the CRAC and the closest racks, i.e., racks A1 and B1 in Fig. 10.7. We see an unexpected increase in $T_{max}$ as a result of increasing the CRAC velocity in some regions in Fig. 10.25, where a transition in the air recirculation pattern occurs inside the center. This trend can start at CRAC velocities as low as ~3 m/s and end at velocities as large as ~5 m/s depending on the total heat load and its distribution.

Accurate and computationally efficient prediction of the effects of the CRAC velocity, total data center heat load, and rack heat load allocation on $T_{max}$ by the POD-based method makes it a suitable tool to be used within the cDSP in order to design around the five design variables. The cDSP for this example is constructed in the next section.

#### 10.3.4.3 CDSP for the Data Center Cell Design

Using the cDSP construct in Fig. 10.19, the cDSP for an adaptable robust and energy-efficient design of the data center cell shown in Fig. 10.7 is constructed. The mathematical formulation of the cDSP is shown in Table 10.5 while each section of it is explained in the following.

Given

The POD-based method is used to calculate $T_{max}$ as a function of control variables. The initial CRAC supply temperature is 15°C while the new CRAC supply temperature after each iteration is obtained from (10.26). The total cooling energy consumption, $W_{Cooling}$, is the summation of the CRAC work, calculated from (10.29), and the chiller work, calculated from (10.28) and (10.27) at the new supply temperature. The variation of the control variables is determined by literature review and experience. For a more accurate representation, manufacturers' or experimental statistical data can also be used if available. As stated before, the given total data center power increases annually by 10% from 24 to 240 kW during 10 years of operation. Since at 100% utilization, when $Q_{total} = 240$ kW, the heat loads of all racks must be 30 kW and there is no space for energy-efficient design, this is not considered in the cDSP.

The target cooling energy is the minimum possible energy consumption of the data center for a given $Q_{total}$. The minimum of $W_{crac}$ in (10.29) is obtained if the minimum possible inlet velocity, (10.25), is provided by the CRAC unit. The chiller work is minimal if there is no air recirculation in the center and so the supply temperature is equal to $T_{max}$ and equal to 32°C. Using (10.28), the maximum COP

**Table 10.5** Mathematical formulation of cDSP for the adaptable robust and energy-efficient design of the data center cell [57]

*Given*

- Response model of maximum rack inlet temperature, $T_{\max}$, new CRAC supply temperature, and total cooling energy as a function of
  $x_1, x_2, x_3, x_4, x_5 = V_{\text{in}}(\text{m/s}),\ Q_1\ (\text{kW}),\ Q_2\ (\text{kW}),\ Q_3\ (\text{kW}), Q_4\ (\text{kW})$
- $T_{\text{Supply,iniial}} = 15\,°\text{C}$
- $\Delta V_{\text{in}}(\text{m/s})$ and $\Delta Q_i\ (\text{kW})\quad i = 1, \ldots, 4$
- Total data center power, $Q_{\text{total}} = 24, 48, 72, 96, 120, 144, 168, 192, 216$ kW
- Target cooling energy consumption,

$$G_{\text{CoolingEnergy}}(\text{kW}) = 2.7\, V_{\text{in,minimum}}\ (\text{m/s}) + \frac{Q_{\text{total}}(\text{kW})}{7.45} \tag{10.31}$$

$$\text{where } V_{\text{in, minimum}}(\text{m/s}) = \frac{-0.01 + \sqrt{(0.0001) - (28E - 8)Q_{\text{total}}(\text{kW})}}{-0.0008} \tag{10.25}$$

- Target for total maximum possible variation, $\text{Max}(\Delta T_{\max})(°\text{C})$

*Find*

- The values of control factors:
  $x_1$, CRAC inlet velocity, $V_{\text{in}}$
  $x_2$, Heat load of racks A1 and B1, $Q_1$; $x_3$, Heat load of racks A2 and B2, $Q_2$
  $x_4$, Heat load of racks A3 nad B3, $Q_3$; $x_5$, Heat load of racks A4 and B4, $Q_4$
- The values of deviation variables: $d_i^+,\ d_i^-\ \ i = 1, 2$

*Satisfy*

- *The constraints*:
  - The maximum rack inlet temperature cannot exceed 32°C

$$T_{\max} + \sum_{i=1}^{n} \left| \frac{\delta T_{\max}}{\delta x_i} \Delta x_i \right| \leq 32°\text{C} \tag{10.32}$$

  - CRAC inlet velocity must be higher than the minimum required CRAC velocity to cope with the total center heat load:

$$V_{\text{in,minimum}} \leq x_1 \tag{10.33}$$

  - The total heat load of the center must equal $Q_{\text{total}}$

$$2x_2 + 2x_3 + 2x_4 + 2x_5 = Q_{\text{total}} \tag{10.34}$$

- *The goals*:
  - Minimize cooling energy consumption for a neither overcooled nor under-cooled design:

$$T_{\text{max,new}} + \sum_{i=1}^{n} \left| \frac{\delta T_{\max}}{\delta x_i} \Delta x_i \right| = 32 \tag{10.35}$$

$$\frac{G_{\text{CoolingEnergy}}}{W_{\text{CRAC}} + W_{\text{Chiller}}} + d_1 = 1 \tag{10.36}$$

  - Minimize variation of $T_{\max}$

$$\frac{\sum_{i=1}^{n} |(\delta T_{\max}/\delta x_i)\Delta x_i|}{\text{Max}(\Delta T_{\max})} - d_2 = 0 \tag{10.37}$$

- *The bounds*:
  $0.4\ \text{m/s} \leq x_1 \leq 9.4\ \text{m/s}$
  $0.5\ \text{kW} \leq x_i 30\ \text{kW}\quad i = 2, \ldots, 5$
  $d_i \geq 0\ \ i = 1, 2$

*Minimize*

- The Archimedean objective function:

$$f = \sum_{i=1}^{2} W_i d_i\,,\ \text{with}\ \sum_{i=1}^{2} W_i = 1\,,\quad W_i \geq 0\,,\ \ i = 1, 2 \tag{10.38}$$

of the center can be 7.45 and accordingly the minimum chiller work is calculated from (10.27).

The variability of the response, $\Delta T_{max}$, is calculated by (10.30) and the maximum possible value of this variation, $\mathrm{Max}(\Delta T_{max})$, is obtained by searching the domain for different design variables.

### Find

The design variables and the associated deviations from the target values are the parameters to be found.

### Satisfy

There are three constraints for this problem. The maximum temperature at the rack inlet considering the worst scenario, $T_{max} + \Delta T_{max}$, must be less than the limit, 32°C, in (10.32). Also, the CRAC velocity must be at least equal to the required velocity based on the CRAC capacity and the center heat load, calculated from (10.33). The final constraint is keeping the total data center heat load at the given $Q_{total}$, (10.34).

There are two goals associated with (1) minimization of the cooling energy consumption for an energy-efficient design and (2) minimization of the variation of $T_{max}$ for a robust operation. The maximum air temperature at the rack inlets for the worst possible changes is designed as 32 C in order to have a not overcooled design, (10.35), while the cooling energy is minimized to reach the target, (10.36). Deviation variable $d_1$ in (10.36) represents the overachievement of the goal since the minimum possible energy consumption of the data center for a given $Q_{total}$ has been considered as the target in (10.31). In fact, $d_1$ shows how much larger the cooling energy consumption of the center is than the possible minimum. Also, the variation of $T_{max}$ respect to the changes in the design variables is minimized to reach zero, (10.37). Deviation variable $d_2$ in (10.37) represents the overachievement of the goal since the minimum possible $T_{max}$ variation is zero. In fact, $d_2$ shows how larger the system response variation is than zero, considering the worst possible instability, $\mathrm{Max}(\Delta T_{max})$, as the comparison reference.

### Minimize

Both $d_1$ and $d_2$ need to be minimized, ideally zero, to reach the associated goals. In the cDSP formulation, the designer's preferences over the goals are applied through weighting each deviation variable. The total deviation function, defined by (10.38), is minimized to calculate the control variables.

#### 10.3.4.4 Results and Discussion

Two different scenarios are studied here. In the first scenario, robustness is not considered and optimal solution is obtained by minimizing the cooling energy consumption function. In the second scenario, the effects of variations in the control variables and robustness on the solution are studied through solving the cDSP in Table 10.5. The optimal solution and the minimum of the objective function in (10.38) for solving the cDSP are found through a pattern search [62] using the MATLAB Genetic Algorithm and Direct Search Toolbox.

Optimal vs. Baseline Design

First, we assume there is no variation in the design parameters, i.e., $\Delta V_{\text{in}} = \Delta Q_i = 0$. The optimal solutions are obtained to have an adaptable and energy-efficient data center for 10 years. Five design variables, one CRAC flow rate and four racks heat loads, along with new CRAC supply temperature are found to change each year to cope with the annual total data center work load increase to guarantee the data center remains reliable and energy efficient, neither overcooled nor under-cooled. The obtained variables and cooling energy consumptions for 9 years are shown in Table 10.6. Also, the adaptable energy-efficient design is compared with the traditional design in Table 10.6 and Fig. 10.26. In the traditional design, the required CRAC airflow rate is calculated using (10.24) with recirculation effect of $\beta = 1.15$, while the total data center heat load each year is distributed randomly among all racks, as listed in Table 10.6, to represent today's data centers.

The CRAC supply temperature is fixed in the traditional design, while the adaptable design results in new higher supply temperatures for each year to avoid overcooling the center and have the maximum rack inlet temperature equal to the limit, $T_{\text{max, new}} = 32°\text{C}$, according to (10.26). As shown in the table and figure, the traditional design of the data center cell fails to meet the reliability constraint after 2 years, i.e., $T_{\text{max}} > 32°\text{C}$, while the adaptable design method application guarantees that the IT equipment operation remains safe for all years. In addition, through adaptable intelligent changes in the rack heat loads and CRAC unit flow rate and supply temperature, the energy consumption and cost of powering the required cooling systems are always kept minimal. As shown in the table and figure, the adaptable design consumes 12–46% less energy than the traditionally designed cooling system during different years of the operation.

As shown in Table 10.6, optimal value of the CRAC velocity and the distribution of the heat load among the racks are different for different years, depending on the total center heat load. To minimize the cooling energy consumption, the summation of CRAC and chiller work needs to be minimized. CRAC work is minimum at lower CRAC velocities, $V_{\text{in}}$, based on (10.29), while the chiller work becomes minimum at lower $T_{\text{max}}$ based on (10.26)–(10.28) and Fig. 10.22. Considering the nonlinear change of $T_{\text{max}}$ with $V_{\text{in}}$ shown in Fig. 10.25, lowering $V_{\text{in}}$ in a specific

**Table 10.6** Adaptable optimal design vs. baseline/traditional design [57]

| | Year # | First | Second | Third | Fourth | Fifth | Sixth | Seventh | Eighth | Ninth |
|---|---|---|---|---|---|---|---|---|---|---|
| | Data center utilization (%) | 10 | 20 | 30 | 40 | 50 | 60 | 70 | 80 | 90 |
| | Data center heat load (kW) | 24 | 48 | 72 | 96 | 120 | 144 | 168 | 192 | 216 |
| Traditional/baseline design | $V_{in}$ (m/s) | 0.76 | 1.51 | 2.27 | 3.03 | 3.79 | 4.54 | 5.30 | 6.06 | 6.82 |
| | $Q_1$ (kW) | 2 | 8.4 | 4.8 | 7.5 | 14.8 | 9.9 | 4.8 | 25.7 | 27.3 |
| | $Q_2$ (kW) | 6.1 | 5.1 | 3.4 | 13.5 | 11.7 | 15.1 | 29.7 | 23.2 | 28.5 |
| | $Q_3$ (kW) | 2.8 | 7.1 | 2.1 | 18 | 14.8 | 17.7 | 28 | 29.7 | 22.8 |
| | $Q_4$ (kW) | 1.1 | 3.4 | 25.7 | 9 | 18.7 | 29.3 | 21.5 | 17.4 | 29.4 |
| | $T_{max}$ | 27.74 | 26.08 | 34.41 | 26.32 | 28.06 | 31.85 | 39.09 | 33.19 | 35.56 |
| | Supply temperature (°C) | 15 | 15 | 15 | 15 | 15 | 15 | 15 | 15 | 15 |
| | $W_{CRAC}$ (kW) | 2.05 | 4.09 | 6.14 | 8.18 | 10.23 | 12.27 | 14.32 | 16.36 | 18.41 |
| | $W_{Chiller}$ (kW) | 12.00 | 24.00 | 36.00 | 48.00 | 60.00 | 72.00 | 84.00 | 96.00 | 108.00 |
| | $W_{Cooling,\ total}$ (kW) | 14.05 | 28.09 | 42.14 | 56.18 | 70.23 | 84.27 | 98.32 | 112.36 | 126.41 |
| | PUE | 1.59 | 1.59 | 1.59 | 1.59 | 1.59 | 1.59 | 1.59 | 1.59 | 1.59 |
| Adaptable optimal design | $V_{in}$ (m/s) | 1.12 | 1.93 | 3.00 | 3.00 | 3.71 | 3.24 | 3.39 | 9.40 | 9.38 |
| | $Q_1$ (kW) | 8.35 | 12.9 | 3.8 | 9.5 | 2.725 | 11.1 | 9.3 | 11.9 | 18 |
| | $Q_2$ (kW) | 1.05 | 4.5 | 29.8 | 29.9 | 15.625 | 29.9 | 29.7 | 24.7 | 30 |
| | $Q_3$ (kW) | 1.55 | 3.3 | 1 | 2.3 | 29.925 | 13.9 | 26.9 | 29.9 | 30 |
| | $Q_4$ (kW) | 1.05 | 3.3 | 1.4 | 6.3 | 11.725 | 17.1 | 18.1 | 29.5 | 30 |
| | $T_{max}$ | 21.48 | 21.90 | 21.29 | 23.33 | 23.72 | 26.55 | 27.14 | 27.40 | 29.69 |
| | New supply temperature (°C) | 25.52 | 25.10 | 25.71 | 23.67 | 23.28 | 20.45 | 19.86 | 19.60 | 17.31 |
| | $W_{CRAC}$ (kW) | 3.03 | 5.22 | 8.10 | 8.10 | 10.01 | 8.74 | 9.16 | 25.38 | 25.33 |
| | $W_{Chiller}$(kW) | 4.89 | 10.08 | 14.48 | 22.40 | 28.84 | 43.41 | 53.22 | 62.22 | 86.04 |
| | $W_{Cooling,\ total}$ (kW) | 7.92 | 15.30 | 22.58 | 30.50 | 38.85 | 52.15 | 62.39 | 87.60 | 111.37 |
| | PUE | 1.33 | 1.32 | 1.31 | 1.32 | 1.32 | 1.36 | 1.37 | 1.46 | 1.52 |
| | Energy saving (%) | 43.6 | 45.5 | 46.4 | 45.7 | 44.7 | 38.1 | 36.5 | 22.0 | 11.9 |

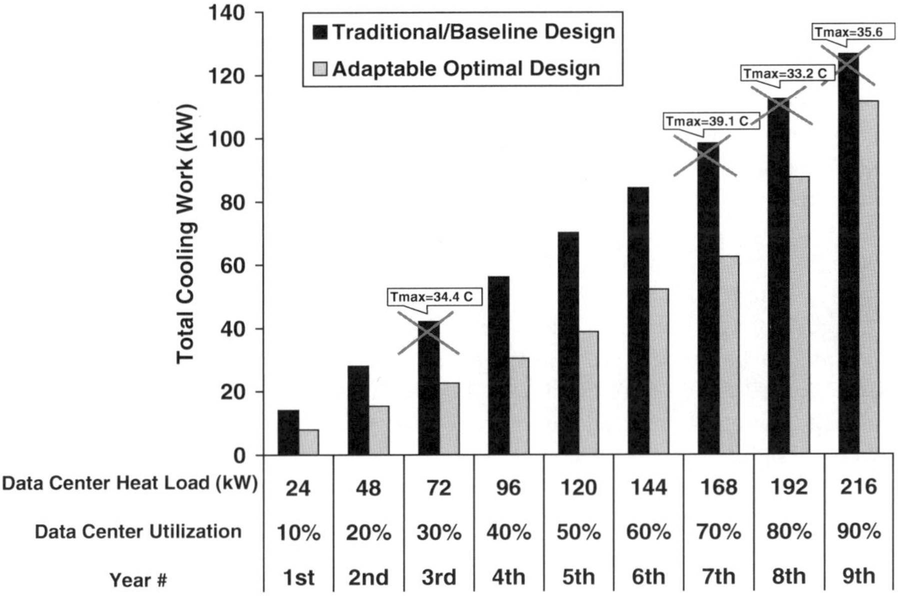


**Fig. 10.26** Total cooling energy consumption of adaptable and traditional designs for 9 years. Cross signs show that the reliability requirement has failed, i.e., $T_{\max} > 32°C$, by the traditional design at years# 3, 7, 8, and 9 [57]

range, roughly between 3 and 5 m/s depending on the total heat load and its distribution, will unexpectedly reduce $T_{\max}$ and also the chiller work in addition to reducing the CRAC work. This is the reason that the design solution has resulted in an optimal $V_{\text{in}}$ between 3 and 3.7 m/s in the third year through the seventh year of the operation, as shown in Table 10.6. When the data center heat load increases from 72 kW in the third year to 168 kW in the seventh year, the optimal CRAC velocity changes slightly between 3 and 3.7 m/s and even decreases in the sixth year. On the other hand, in the first and second years, the optimal velocities happen at 1.12 and 1.93 m/s. In these cases, the reduction in the CRAC work as a result of having lower velocities than 3 m/s has been larger than the reduction in the chiller work as a result of having a minimum $T_{\max}$ at ~3 m/s. In the eighth and ninth years, the competition between the effects of $V_{\text{in}}$ and $T_{\max}$ on the CRAC and chiller work will result in an optimal velocity around the upper bound of $V_{\text{in}}$, 9.4 m/s. This happens since $T_{\max}$ is very close to the limit of 32°C at velocities ~3–4 m/s and so there is almost no gain on the chiller work compared with the gain at higher velocities close to the upper bound, as seen in Fig. 10.25.

Although the optimal heat load distribution in a data center depends on the details of the center flow/temperature fields at different CRAC velocities and heat load distributions, being obtained through the exact solution of the governing equations, some general conclusions can be obtained based on the results of Table 10.6 and the air recirculation pattern around the racks in Fig. 10.7.

As explained at the end of Sect. 10.3.4.2, the air recirculation pattern of the center changes at roughly 3.8 m/s. As seen in Table 10.6, in the first and second years with the optimal velocities of 1.12 and 1.93 m/s, when the provided flow rate by the CRAC unit is much lower than the required flow rate by the racks, the best place to put most of the heat is the first racks, A1 and B1 in Fig. 10.7. In these years, the air recirculation happens largely from the tops of the racks and slightly from the sides of the first racks. On the other hand, in the third, fourth, sixth, and seventh years, when the transition in the recirculation pattern is happening at the associated optimal velocities, the best racks to put most of the total heat on are the second racks, A2 and B2. Conversely, in the fifth, eighth, and ninth years, when the optimal $V_{in}$ is 3.71, 9.4, 9.38 m/s and the CRAC flow rate is at least equal to the required flow rate, the best racks to put most of the total heat on is the third racks, A3 and B3, while the worst racks to put heat on is the first racks, A1 and B1. In these years, the air recirculation happens largely between the CRAC and the closest racks, i.e., racks A1 and B1.

## Robust vs. Optimal Design

In the first scenario, an energy-efficient data center was designed to have the maximum rack inlet temperature, $T_{max,new}$, equal to the limit 32°C in each year. Although it is the most energy-efficient configuration, small changes in the system parameters can cause $T_{max,new}$ to increase and IT equipment operation to fail. For illustration purpose, we consider the operation of the data center cell in the fifth year when $Q_{total} = 120$ kW. If we assume each control variable can vary by ±5% during operation, i.e., $\Delta V_{in} = \pm 0.05\, V_{in}$ and $\Delta Q_i = \pm 0.05\, Q_i \;\; i = 1, ..., 4$, the variation in $T_{max}$, $\Delta T_{max}$ in (10.30), can be as high as 3.68°C. This maximum possible variation, $\mathrm{Max}(\Delta T_{max})$, obtained by searching the domain, happens at $\{x_1, x_2, x_3, x_4, x_5\} = \{2.707$ m/s, 1 kW, 29.8 kW, 28.2 kW, 1 kW$\}$. As seen in Table 10.6, the design variables, $\{x_1, x_2, x_3, x_4, x_5\}$, at the optimal energy-efficient configuration are $\{3.71$ m/s, 1 kW, 29.8 kW, 28.2 kW, 1 kW$\}$. In this configuration, $T_{max}$ variation, calculated by (10.30), is 2.31°C and the maximum temperature at the rack inlets can reach 34.31°C because of the small variations in the control variables, making the IT equipment not satisfy the reliability design constraint.

To handle this issue, robustness in design constraints as illustrated in Fig. 10.20b must be considered in solving the design problem. For this purpose, the maximum air temperature at the rack inlets for the worst scenario, $T_{max} + \Delta T_{max}$, is considered in the associated constraints in (10.32) and (10.35) of the cDSP in Table 10.5, while the weighing coefficient associated with the robustness in the goals, minimizing $T_{max}$ variation, is zero, i.e., $W_1 = 1$ and $W_2 = 0$. The cDSP is solved for the total heat load, $Q_{total} = 120$ kW. The new values of the control variables, energy consumption, and $T_{max}$ variation are compared with the results of the optimal energy-efficient design in Table 10.7, see Case #1 and Case #2. As seen in Table 10.7, considering robustness in constraints guarantees that the maximum possible rack inlet temperature remains 32°C despite the changes in the design

**Table 10.7** Pareto frontier; robust design vs. optimal design specifications in the fifth year with $Q_{total}$ = 120 kW [57]

| Case # | 1. Optimal energy-efficient design | 2. Energy-efficient design with robustness in constraints | 3. Energy-efficient design with robustness in constraints and goals | 4. Energy-efficient design with robustness in constraints and goals | 5. Energy-efficient design with robustness in constraints and goals | 6. Energy-efficient design with robustness in constraints and goals |
|---|---|---|---|---|---|---|
| Energy efficiency weighing coefficient, $W_1$ | 1 | 1 | 0.75 | 0.5 | 0.25 | 0 |
| Robustness, minimizing $T_{max}$ variation, weighing coefficient, $W_2$ | 0 | 0 | 0.25 | 0.5 | 0.75 | 1 |
| $V_{in}$ (m/s) | 3.707 | 3.507 | 3.507 | 8.307 | 9.4 | 9.4 |
| $Q_1$ (kW) | 2.725 | 2.1 | 1.3 | 7.6 | 9 | 9 |
| $Q_2$ (kW) | 15.625 | 22.1 | 22.5 | 14 | 11.2 | 11.2 |
| $Q_3$ (kW) | 29.925 | 26.5 | 28.7 | 14 | 22.6 | 22.6 |
| $Q_4$ (kW) | 11.725 | 9.3 | 7.5 | 24.4 | 17.2 | 17.2 |
| $T_{max}$ (°C) | 23.72 | 24.90 | 24.83 | 24.22 | 22.95 | 22.95 |
| New supply temperature (°C) | 23.28 | 21.69 | 21.79 | 22.24 | 23.60 | 23.60 |
| $W_{CRAC}$ (kW) | 10.01 | 9.47 | 9.47 | 22.43 | 25.38 | 25.38 |
| $W_{Chiller}$ (kW) | 28.84 | 32.66 | 32.39 | 31.26 | 28.14 | 28.14 |
| $W_{total}$ (kW) | 38.85 | 42.13 | 41.86 | 53.69 | 53.52 | 53.52 |
| Maximum variation in $T_{max}$ (°C) | 2.31 | 0.7066 | 0.6059 | 0.5426 | 0.4583 | 0.4583 |
| New $T_{max}$ plus the worst possible variation (°C) | 34.31 | 32 | 32 | 32 | 32 | 32 |
| Reduction in variability (%) compared with optimal design | – | 69.41 | 73.77 | 76.51 | 80.16 | 80.16 |
| Increase in $W_{total}$ (%) compared with optimal design | – | 8.44 | 7.76 | 38.20 | 37.76 | 37.76 |

variables. Also, comparing Case #2 and Case #1 in Table 10.7 shows that the variation in $T_{max}$ is reduced by 69.4% from 2.31 to 0.71°C. This might be unexpected since the weighing coefficients associated with the minimization of $\Delta T_{max}$ are zero in this scenario, i.e., $W_1 = 1$ and $W_2 = 0$ in (10.38). This reduction happens because lower value of $\Delta T_{max}$ results in a higher new supply temperature, based on (10.26), and as a result, lower chiller work. So, satisfying only the first goal in the cDSP, (10.35), indirectly considers $\Delta T_{max}$ minimization and the second goal, (10.37), somewhat as well. This is one example showing the linear weighting system used in the cDSP does not accurately translate the designer's preferences over different goals in complex design problems. Generally, a Pareto frontier [63] should be developed between two extreme solution points in order to investigate the trade-offs between robust and optimal solutions in designing highly nonlinear complex systems such as data centers.

The Pareto frontier is made through changing the weights in (10.38) in the cDSP to determine the design specifications as the goal changes from an optimal solution, when $W_1 = 1$ and $W_2 = 0$, to a robust solution, when $W_1 = 0$ and $W_2 = 1$. Six different cases with the associated weighing coefficients and design specifications are shown in Table 10.7 for the data center cell with 120 kW total heat load. Case #1 in the table is the optimal design, when $T_{max}$ has the highest variation, 2.31°C, in the Pareto frontier but the energy consumption is minimal, $W_{total} = 38.85$ kW. While the first case denies the design constraint and IT equipment reliability limit with small changes in the design variables, Case #2 considers the robustness in constraints, as explained above. This results in a 69.41% reduction in $\Delta T_{max}$ but 8.44% increase in the energy consumption. To have a more stable IT operation, the variation in $T_{max}$ should be reduced. This is done by increasing the associated weighing coefficient, $W_2$, in (10.38), and reducing the energy efficiency weighing coefficient. Cases #3 through #6 in Table 10.7 show the design specifications as the data center design becomes more robust. As the weighing coefficients change linearly through the frontier, nonlinear changes in the $T_{max}$ variation and energy consumption, $W_{total}$, are observed. The last two cases have the same design specifications with the lowest $\Delta T_{max}$, 0.46°C (80.16% reduction compared with the optimal solution), and so the most stability and robustness in the operation. But, they consume 37.8% more energy than the optimal solution. If a data center is loosely controlled or needs a high level of reliability and stability, the last case should be selected as the final solution. However, Case #3, when $W_1 = 0.75$ and $W_2 = 0.25$, results in a better balance between energy efficiency and robustness; it brings 73.77% reduction in $\Delta T_{max}$ but only 7.76% increase in $W_{total}$ compared with the optimal solution, as seen in Table 10.7. Overall, the Pareto frontier in Table 10.7 gives designers a much greater amount of information and freedom in configuring the data center for their desired goals over a single application of the weighted sum approach.

Although the Pareto frontier results in Table 10.7 have been obtained for the dynamic rack heat load allocation, their general trend can be validated through investigating a graphical illustration of $T_{max}$ vs. the CRAC velocity for the uniform distribution of the total heat load $Q_{total} = 120$ kW, shown in Fig. 10.25.

Comparing the obtained CRAC velocity and $T_{max}$ for different weighing coefficients in Table 10.7 with the associated graph for $Q_{total} = 120$ kW in Fig. 10.25 shows an agreement in the trend of the results when changing from optimal to robust solution. For more optimal and more energy-efficient solutions, i. e., Case #1, 2, and 3 in Table 10.7, the obtained CRAC velocity is ~3.5 m/s. As seen in Fig. 10.25, $T_{max}$ is almost a minimum at ~3.5 m/s and so the chiller and data center operation will be relatively efficient in these cases. On the other hand, in order to have more robustness in the equipment operation, we should look for the flat regions, where $T_{max}$ slightly changes with changes in the variables.

As seen in Fig. 10.25, variation of $T_{max}$ with the CRAC velocity becomes weak after ~8 m/s, while $T_{max}$ is also relatively small. This is the reason that the solution for Case #4, when both robustness and energy efficiency are important for designer, happens at 8.3 m/s, as shown in Table 10.7. Also, as seen in Fig. 10.25, $T_{max}$ reaches its minimum variation at upper bound of 9.4 m/s, where the most robust solutions, Cases #5 and 6 in Table 10.7, happen as well. Small discrepancy of the results in Table 10.7 with the trend of $T_{max}$ variation in Fig. 10.25 is due to the effect of the dynamic heat load allocation. The nonlinear behavior of the data center energy efficiency and robustness with the associated weighing coefficients in the cDSP confirms the necessity of obtaining a Pareto frontier for complex nonlinear systems.

## 10.4 Conclusion

Developing adaptable energy-efficient air-cooled data centers that are readily adaptable to changes through continuous improvement of an existing base is necessary in today's global market. CFD/HT-based methods are too time-consuming and costly for readily adaptable simulation-based designs. In this chapter, two recently developed deterministic POD based reduced order modeling approaches for prediction of temperature field in data centers are presented. The methods are demonstrated through application to similar data center cells. The method results in average error norm of ~6% for different sets of design parameters, while they can be up to ~250 times faster than CFD/HT simulation in an iterative optimization technique. Also, a simulation-based design approach is reviewed to bring adaptability and robustness in energy-efficient data centers. The approach is centered on the integration of three constructs: (a) POD-based multiscale modeling, (b) cDSP, and (c) robust design. The results for a data center case study show a 12–46% reduction in the energy consumption of the center in addition to being adjustable to the newer IT equipment and higher heat loads compared with a traditional design. Also, compared with an optimal solution, a robust solution can reduce the variability in the thermal response by 73.8% with only 7.8% increase in the center energy consumption.

The presented reduced-order thermal modeling and design approaches were focused on the rack and room level in an air-cooled data center. The details at the server level such as chip numbers, dimensions, and server power distribution can be modeled separately and connected to the already developed POD-based modeling. Developing such a multiscale reduced-order modeling approach can enable designing an adaptable energy-efficient air-cooled data center with more freedom in exploring several design parameters at different scales. Such a multiscale design can increase the energy efficiency in data centers substantially in addition to being able to model and design next-generation multiscale solutions integrating air, liquid, two phase, etc. cooling systems for future high heat load data centers.

## References

1. Samadiani E, Joshi Y (2010) Reduced order thermal modeling of data centers via proper orthogonal decomposition – a review. Int J Numer Methods Heat Fluid Flow 20(5):529–550
2. Report to Congress on Server and Data Center Energy Efficiency Public Law 109–431 (2007, August 2) U.S. Environmental Protection Agency ENERGY STARProgram
3. Greenberg S et al (2006) Best practices for data centers: lessons learned from benchmarking 22 data centers. In: ACEEE summer study on energy efficiency in buildings in asilomar, CA. pp 76–87
4. ASHRAE (2005) Datacom equipment power trends and cooling applications. American Society of Heating, Refrigeration and Air-Conditioning Engineers Atlanta, Atlanta
5. Belady CL (2007) In the data center, power and cooling costs more than the it equipment it supports. Electronics Cooling 13(1):24–27
6. Rambo JD (2006) Reduced-order modeling of multiscale turbulent convection: application to data center thermal management. In: Mechanical engineering. Georgia Institute of Technology, Atlanta
7. Patel CD et al (2002) Thermal considerations in cooling of large scale high compute density data centers. In: ITHERM 2002 – Eight intersociety conference on thermal and thermomechanical phenomena in electronic systems, San Diego, CA, pp 767–776
8. Rambo J, Joshi Y (2003) Multi-scale modeling of high power density data centers. In: ASME InterPACK03. ASME, Kauai, Hawaii
9. Rambo J, Joshi Y (2003) Physical models in data center airflow simulations. In: IMECE-03 – ASME international mechanical engineering congress and R&D exposition. IMECE03-41381, Washington, DC
10. Shrivastava S, et al (2005) Comparative analysis of different data center airflow management configurations. In: ASME InterPACK05. ASME, San Francisco, CA
11. Iyengar M et al (2005) Thermal characterization of non-raised floor air cooled data centers using numerical modeling. In: ASME InterPACK05. ASME, San Francisco, CA
12. Schmidt R, Karki KC, Patankar SV (2004) Raised-floor data center: perforated tile flow rates for various tile layouts. In: ITHERM 2004 – Ninth intersociety conference on thermal and thermomechanical phenomena in electronic systems, Las Vegas, NV
13. VanGilder JW, Schmidt R (2005) Airflow uniformity through perforated tiles in a raised-floor data center. In: ASME InterPACK. ASME, San Francisco, CA, IPACK2005-73375
14. Lawrence Berkeley National Laboratory and Rumsey Engineers. Data center energy benchmarking case study. http://datacenters.lbl.gov/
15. Samadiani E, Joshi Y, Mistree F (2007) The thermal design of a next generation data center: a conceptual exposition. In: Thermal issues in emerging technologies, ThETA 1, Cairo, Egypt
16. Shah A et al (2005) Exergy-based optimization strategies for multi-component data center thermal management: Part I, analysis. In: ASME InterPACK05. ASME, San Francisco, CA

17. Shah A et al (2005) Exergy-based optimization strategies for multi-component data center thermal management: Part II, application and validation. In: ASME InterPACK05. ASME, San Francisco, CA
18. Bhopte S et al (2005) Optimization of data center room layout to minimize rack inlet air temperature. In: ASME InterPACK. ASME, San Francisco, CA, IPACK2005-73027
19. Schmit R, Iyengar M (2005) Effect of data center layout on rack inlet air temperatures. In: ASME InterPACK 05. ASME, San Francisco, CA
20. Sharma RK, Bash CE, Patel CD (2002) Dimensionless parameters for the evaluation of thermal design and performance of large-scale data centers. In: The eighth AIAA/ASME joint thermophysics and heat transfer conference, St. Louis
21. Kang S et al (2000) A methodology for the design of perforated tiles in a raised floor data center using computational flow analysis. In: ITHERM 2000 – Intersociety conference on thermal and thermomechanical phenomena in electronic systems, Las Vegas, NV
22. Boucher TD et al (2004) Viability of dynamic cooling control in a data center environment. In: International society conference on thermal phenomena, Las Vegas, NV
23. Rolander N (2005) An approach for the robust design of air cooled data center server cabinets. MS thesis, G.W. School of Mechanical Engineering, Georgia Institute of Technology, Atlanta
24. Patel C et al (2001) Computational fluid dynamics modeling of high compute density data centers to assure system inlet air specifications. In: ASME IPACK'01. ASME, Kauai, Hawaii
25. Schmidt R et al (2001) Measurements and predictions of the flow distribution through perforated tiles in raised floor data centers. In: ASME InterPACK01, Kauai, Hawaii
26. Radmehr A et al (2005) Distributed leakage flow in raised-floor data centers. In: ASME InterPACK. ASME, San Francisco, CA, IPACK2005-73273
27. Karki KC, Radmehr A, Patankar SV (2003) Use of computational fluid dynamics for calculating flow rates through perforated tiles in raised-floor data centers. HVAC and R Res 9(2):153–166
28. Samadiani E, Rambo J, Joshi Y (2007) Numerical modeling of perforated tile flow distribution in a raised-floor data center. In: ASME InterPACK '07, Vancouver, British Columbia, Canada
29. Schmidt R, Cruz E (2003) Cluster of high powered racks within a raised floor computer data center: effects of perforated tiles flow distribution on rack inlet air temperature. In: IMECE-03 – ASME international mechanical engineering congress and R&D exposition, Washington, DC
30. Rambo J, Joshi Y (2006) Thermal modeling of technology infrastructure facilities: a case study of data centers. In: Minkowycz WJ, Sparrow EM, Murthy JY (eds) The handbook of numerical heat transfer. Taylor and Francis, New York, pp 821–850
31. Rambo J, Joshi Y (2007) Modeling of data center airflow and heat transfer: State of the art and future trends. Distributed and Parallel Databases 21:193–225
32. Moore J et al (2005) Making scheduling cool: temperature-aware workload placement in data centers. In: Usenix technical conference, Anaheim, CA
33. Karlsson M, Karamanolis C, Zhu X (2004) Triage: performance isolation and differentiation for storage systems. In: The twelfth international workshop on quality of service, Montreal, Canada
34. Sharma RK et al (2003) Balance of power: dynamic thermal management for internet data centers. Whitepaper issued by Hewlett Packard Laboratories
35. Moore J et al (2005) Data center workload monitoring, analysis, and emulation. In: Eighth workshop on computer architecture evaluation using commercial workloads, San Francisco, CA
36. Tang Q et al (2006) Sensor-based fast thermal evaluation model for energy-efficient high-performance datacenters. In: Fourth international conference on intelligent sensing and information processing (ICISIP), Bangalore, India
37. Nathuji R et al (2008) CoolIT: coordinating facility and IT management for efficient datacenters. In: HotPower '08; workshop on power aware computing and systems, San Diego, CA
38. Somani A, Joshi Y (2008) Ambient intelligence based load management, USPTO Provisional Patent, GTRC ID No. 4524

39. Qian Z (2006) Computer experiments: design, modeling, and integration. School of Industrial and Systems Engineering, Georgia Institute of Technology, Atlanta
40. Samadiani E, Joshi Y (2010) Multi-parameter model reduction in multi-scale convective systems. Int J Heat Mass Transfer 53:2193–2205
41. Samadiani E, Joshi Y (2010) Proper orthogonal decomposition for reduced order thermal modeling of air cooled data centers. J Heat Trans 132(7):0714021–07140214
42. Rambo J, Joshi Y (2007) Reduced-order modeling of turbulent forced convection with parametric conditions. Int J Heat Mass Trans 50(3–4):539–551
43. Holmes P, Lumley JL, Berkooz G (1996) Turbulence, coherent structures, dynamical systems and symmetry. Cambridge University Press, Great Britain
44. Ravindran SS (2002) Adaptive reduced-order controllers for a thermal flow using proper orthogonal decomposition. SIAM J Sci Comput 23(6):1924–1942
45. Park HM, Cho DH (1996) The use of the Karhunen-Loeve decomposition for the modeling of distributed parameter systems. Chem Eng Sci 51(1):81–98
46. Park HM, Cho DH (1996) Low dimensional modeling of flow reactors. Int J Heat Mass Trans 39(16):3311–3323
47. Sirovich L, Park HM (1990) Turbulent thermal convection in a finite domain: Part I. Theory. Phys Fluids 2(9):1649–1658
48. Sirovich L, Park HM (1990) Turbulent thermal convection in a finite domain: Part II. Numerical results. Phys Fluids 2(9):1659–1668
49. Tarman IH, Sirovich L (1998) Extensions of Karhunen-Loeve based approximations of complicated phenomena. Comput Methods Appl Mech Eng 155:359–368
50. Park HM, Li WJ (2002) Boundary optimal control of natural convection by means of mode reduction. J Dyn Syst Meas Cont 124:47–54
51. Ding P et al (2008) A fast and efficient method for predicting fluid flow and heat transfer problems. ASME J Heat Trans 130(3):032502
52. Ly HV, Tran HT (2001) Modeling and control of physical processes using proper orthogonal decomposition. Math Comput Model 33:223–236
53. Strang G (1988) Linear algebra and its applications. Thomson Learning, Singapore
54. Rolander N et al (2006) Robust design of turbulent convective systems using the proper orthogonal decomposition. J Mech Des 128(Special Issue Robust and Risk Based Design):844–855
55. Nie Q, Joshi Y (2008) Multiscale thermal modeling methodology for thermoelectrically cooled electronic cabinets. Numer Heat Transf A Appl 53(3):225–248
56. Nie Q, Joshi Y (2008) Reduced order modeling and experimental validation of steady turbulent convection in connected domains. Int J Heat Mass Trans 51(25–26):6063–6076
57. Samadiani E et al (2010) Adaptable robust design of multi-scale convective systems applied to energy efficient data centers. Numer Heat Trans A Appl 57(2):69–100
58. Samadiani E, Joshi Y, Mistree F (2008) The thermal design of a next generation DataCenter: a conceptual exposition. J Electron Packag 130(4):0411041–0411048
59. ASHRAE (2004) Thermal guidelines for data processing environments. American Society of Heating, Refrigeration, and Air-Conditioning Engineers, New York
60. Mistree F, Hughes OF, Bras B (1993) The compromise decision support problem and the adaptive linear programming algorithm. In: Kamat MP (ed) Structural optimization: status and promise. AIAA, Washington, DC, pp 247–286
61. Moore J et al (2005) Making scheduling cool: temperature-aware workload placement in data centers. In: USENIX annual technical conference, Anaheim, CA, pp 61–75
62. Lewis RM, Torczon V (2000) Pattern search methods for linearly constrained minimization. SIAM J Optim 10(3):917–941
63. Steuer RE (1986) Multiple criteria optimization, theory computations and applications. Wiley, New York

# Chapter 11
# Statistical Methods for Data Center Thermal Management

**Ying Hung, Peter Z.G. Qian, and C.F. Jeff Wu**

**Abstract** A data center is an integrated IT system housing multiple-unit servers intended for providing various application services. A significant portion of the costs associated with operating and maintaining a data center is used for heat removal. A growing trend in the IT industry is to use computer experiments to study thermal properties of data centers because the corresponding physical experimentation can be time consuming, costly, or even infeasible. This chapter presents useful statistical methods for the design and analysis of data center computer experiments.

## 11.1 Introduction

Modern data centers cost millions of dollars to operate and maintain each year. A significant portion of these costs is used for heat removal. Therefore, efficient cooling mechanism of a data center has become a major challenge. An objective for the thermal management study is to model the thermal distribution in data center, and the final goal is to design a data center with an efficient heat-removal mechanism [1].

To create an energy-efficient data center, studies based on physical experimentation can be time consuming, costly, or even infeasible. The advent of modern computers

Y. Hung (✉)
Department of Statistics and Biostatistics, Rutgers, The State University of New Jersey, Piscataway, NJ, USA
e-mail: yhung@stat.rutgers.edu

P.Z.G. Qian
Department of Statistics, University of Wisconsin-Madison, Madison, WI, USA
e-mail: peterq@stat.wisc.edu

C.F.J. Wu
School of Industrial and Systems Engineering, Georgia Institute of Technology, Atlanta, GA, USA
e-mail: jeff.wu@isye.gatech.edu

Y. Joshi and P. Kumar (eds.), *Energy Efficient Thermal Management of Data Centers*, DOI 10.1007/978-1-4419-7124-1_11,

and the advancement of computer-aided design methodologies have given rise to another mode of experimentation known as computer experiments. Computer experiments based on finite elements analysis or network simulations have been widely used as economical alternatives. On the statistical side, design and analysis for computer experiments have received a lot of attention in the past decades [2, 3].

The discussions in this chapter focus on the statistical design and analysis of data center computer experiments. Existing statistical designs for conducting data center computer experiments are reviewed in Sect. 11.2.1. A recent development of statistical designs is introduced in Sect. 11.2.2 which takes into account the irregular shape of experimental regions in data center applications. Based on the outputs of data center computer simulations, statistical emulators can be built. Modeling and analysis techniques for building such emulators are reviewed in Sect. 11.3. These techniques are useful in enhancing the prediction accuracy and identifying important variables in the study of data center thermal management. Conclusions and remarks are discussed in Sect. 11.4.

## 11.2 Designs for Data Center Computer Experiments

### *11.2.1 Space-Filling Designs*

Computer experiments are widely used for the design and optimization of complex systems, including the data center thermal management study. In data center computer experiments, instead of physical experimentation, mathematical models describing the temperature distribution are developed using engineering/physical laws and solved on computers through numerical methods such as the finite element analysis. Because deterministic models are used for experiments, the output of a data center computer experiment is not subject to random variation. That is, the *same* inputs produce exactly the *same* outputs. Such a deterministic property makes the design of data center computer experiments different from that of physical experiments [4]. For example, replication is not required. In fact, it is desirable to avoid replicates when projecting the design onto a subset of factors. This is based on the effect sparsity principle [5], which stipulates that only a few out of the numerous factors in the system dominate its performance. Therefore, a good model can be fitted using only these few important factors. Consequently, when projecting the design onto these factors, replication is not required.

Latin hypercube designs (LHDs) are extensively used space-filling designs in data center computer experiments [6] that takes into account the deterministic property of the experimental outputs. A desirable property of an LHD is its one-dimensional balance, i.e., when an $n$-point design is projected onto any factor, there are $n$ different levels for that factor. Suppose the $n$ levels of a factor are denoted by $1, \ldots, n$. Table 11.1 illustrates an example of LHD with three factors in six design points. Figure 11.1 shows the projections of such a design onto any two factors.

**Table 11.1** A Latin hypercube in three factors

| $x_1$ | 1 | 2 | 3 | 4 | 5 | 6 |
|---|---|---|---|---|---|---|
| $x_2$ | 5 | 1 | 3 | 6 | 2 | 4 |
| $x_3$ | 3 | 4 | 1 | 2 | 6 | 5 |

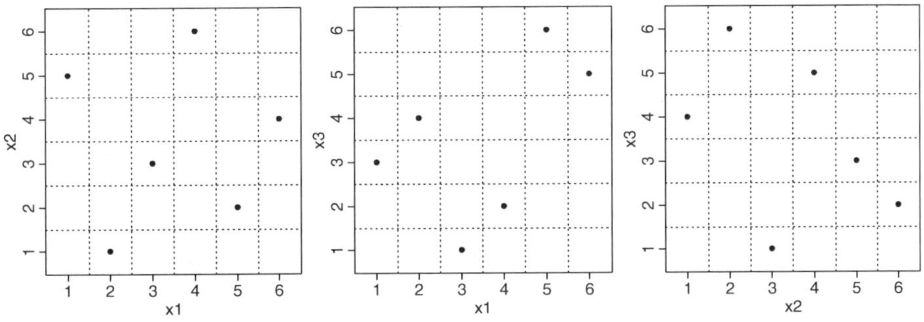

**Fig. 11.1** Projections of Latin hypercube design on two factors

In general, an $n$-run LHD can be generated using a random permutation of $\{1, \ldots, n\}$ for each factor. Each permutation leads to a different LHD. For $k$ factors, we can, thus, obtain $(n!)^k$ LHDs. A randomly generated LHD can have a systematic pattern: the variables may be highly correlated or the design may not have good space-filling properties. Therefore, there has been numerous work in the literature to avoid the above problems and obtain a "good" LHD. The idea is to find the best design by optimizing a criterion that describes a desirable property of the design. Iman and Conover [7], Owen [8], and Tang [9] proposed to find designs minimizing correlations among factors. Johnson et al. [10] and Morris and Mitchell [11] proposed to find the best LHD by maximizing the minimum distance between the points so that they can be spread out uniformly over the experimental region. Joseph and Hung [12] proposed a multiobjective criterion by combining the foregoing two performance measures and the resulting designs are shown to be good in terms of both the correlation and distance criteria. Other approaches of finding good optimal LHDs include Owen [13], Tang [14], Park [15], Ye [16], Ye et al. [17], and Jin et al. [18]. Nested LHDs for conducting multifidelity computer experiments have been proposed in Qian [19].

The standard framework for computer experiments assumes all input factors are quantitative [2, 3]. However, a data center computer experiment can involve qualitative factors like diffuser height and hot-air return-vent location [1]. Qian and Wu [20] proposes a general approach for constructing a new type of design, called *sliced space-filling design*, to accommodate both quantitative and qualitative factors. The basic idea of their approach is to construct a U design [14] based on a special orthogonal array for the quantitative factors and then partition the design into groups associated with different level combinations of the qualitative factors, each of which achieves uniformity in low dimensions.

A cornerstone of their approach is a special type of orthogonal array, called *sliced orthogonal array*. An orthogonal array of $n$ rows, $m$ columns, $s$ levels, and strength $t \geq 2$ is an $n \times m$ matrix with entries from a set of $s$ levels such that, for every $n \times t$ submatrix, all $s^t$ level combinations appear equally often [21]. Let $OA(n, m, s, t)$ denote such an array. Suppose $B$ is an $OA(n_1, k, s_1, t)$; the $n_1$ rows of $B$ can be partitioned into v subarrays each with $n_2$ rows, denoted by $B_i$; and there is a projection $\delta$ that collapses the $s_1$ levels of $B$ into $s_2$ levels with $s_1 > s_2$. Further, suppose $B_i$ is an $OA(n_2, k, s_2, t)$ after the $s_1$ levels of $B$ are collapsed according to the projection $\delta$. Then, $B$ is called a sliced orthogonal array [20].

Using a sliced orthogonal array $B$ defined above, construction of a sliced space-filling design consists of two steps. In step 1, $B$ is used to generate a U design $D$ [14] for the quantitative factors, where $D_i$ denotes the subset of points corresponding to $B_i$. In step 2, $D_i$'s are assigned to different level combinations of the qualitative factors. The array $D$ constitutes a sliced space-filling design. From the modeling aspect, such a design possess two attractive properties. First, for any qualitative factor level combination, the design points for the quantitative factors are space filling. Second, when collapsed over the qualitative factors, the whole design achieves uniformity in low-dimensions. Conceptually, these properties are related to those of sliced response surface designs with qualitative and quantitative factors [22, 23].

It is desirable to use a sliced space-filling design to conduct a data center computer experiment with qualitative and quantitative factors. The effect of the quantitative factors on the response may change from one level combination of the qualitative factors to another. The first property stated above ensures that, at any qualitative factor level combination, the values of the quantitative factors are spread evenly in the design space. If the effect of the qualitative factors on the response turns out to be small, because of the second property, the points of the collapsed design for the quantitative factors are uniformly distributed.

For illustration, Table 11.2 presents a sliced orthogonal array $B$, which can be divided into $B_{11}, B_{12}, B_{21}, B_{22}$, respectively, corresponding to runs 1–4, 9–12, 17–20, 25–28; runs 57–60, 49–52, 41–44, 33–36; 5–8, 13–16, 21–24, 29–32; and runs 61–64, 53–56, 45–48, 37–40. Each $B_{ij}$ becomes an orthogonal array with four levels after the eight symbols are collapsed as follows: $\{1, 2\} \rightarrow 1$, $\{3, 4\} \rightarrow 2$, $\{5, 6\} \rightarrow 3$, $\{7, 8\} \rightarrow 4$. We use $B$ to construct a U design $D$ for a data center computer experiment with five quantitative factors, $x_1, \ldots, x_5$. The design $D$ is partitioned into $D_{ij}$ with points corresponding to $B_{ij}$, $i, j = 1, 2$. Finally $D_{11}, D_{12}, D_{21}, D_{22}$ are assigned to the level combinations $(z_1, z_2) = (-, -)$, $(z_1, z_2) = (-, +)$, $(z_1, z_2) = (+, -)$ and $(z_2, z_2) = (+, +)$, respectively. Figures 11.2 and 11.3 present the bivariate projections of $D$ and $D_{11}$, where both designs achieve uniformity on $4 \times 4$ grids in two dimensions, and $D$ also achieves maximum uniformity in one dimension.

### 11.2.2 Probability-Based Latin Hypercube Design

A challenging design issue arises in the data center thermal study. To monitor and study the thermal distribution, computer experiments need to be conducted in sites uniformly over the experimental region such that the fitted thermal model can be

**Table 11.2** A 64 × 5 sliced orthogonal array

| Run # | $x_1$ | $x_2$ | $x_3$ | $x_4$ | $x_5$ | Run # | $x_1$ | $x_2$ | $x_3$ | $x_4$ | $x_5$ |
|---|---|---|---|---|---|---|---|---|---|---|---|
| 1 | 1 | 1 | 1 | 1 | 1 | 33 | 8 | 1 | 8 | 8 | 8 |
| 2 | 1 | 3 | 3 | 5 | 7 | 34 | 8 | 3 | 6 | 4 | 2 |
| 3 | 1 | 5 | 5 | 8 | 4 | 35 | 8 | 5 | 4 | 1 | 5 |
| 4 | 1 | 7 | 7 | 4 | 6 | 36 | 8 | 7 | 2 | 5 | 3 |
| 5 | 1 | 8 | 8 | 7 | 2 | 37 | 8 | 8 | 1 | 2 | 7 |
| 6 | 1 | 6 | 6 | 3 | 8 | 38 | 8 | 6 | 3 | 6 | 1 |
| 7 | 1 | 4 | 4 | 2 | 3 | 39 | 8 | 4 | 5 | 7 | 6 |
| 8 | 1 | 2 | 2 | 6 | 5 | 40 | 8 | 2 | 7 | 3 | 4 |
| 9 | 3 | 1 | 3 | 3 | 3 | 41 | 6 | 1 | 6 | 6 | 6 |
| 10 | 3 | 3 | 1 | 7 | 5 | 42 | 6 | 3 | 8 | 2 | 4 |
| 11 | 3 | 5 | 7 | 6 | 2 | 43 | 6 | 5 | 2 | 3 | 7 |
| 12 | 3 | 7 | 5 | 2 | 8 | 44 | 6 | 7 | 4 | 7 | 1 |
| 13 | 3 | 8 | 6 | 5 | 4 | 45 | 6 | 8 | 3 | 4 | 5 |
| 14 | 3 | 6 | 8 | 1 | 6 | 46 | 6 | 6 | 1 | 8 | 3 |
| 15 | 3 | 4 | 2 | 4 | 1 | 47 | 6 | 4 | 7 | 5 | 8 |
| 16 | 3 | 2 | 4 | 8 | 7 | 48 | 6 | 2 | 5 | 1 | 2 |
| 17 | 5 | 1 | 5 | 5 | 5 | 49 | 4 | 1 | 4 | 4 | 4 |
| 18 | 5 | 3 | 7 | 1 | 3 | 50 | 4 | 3 | 2 | 8 | 6 |
| 19 | 5 | 5 | 1 | 4 | 8 | 51 | 4 | 5 | 8 | 5 | 1 |
| 20 | 5 | 7 | 3 | 8 | 2 | 52 | 4 | 7 | 6 | 1 | 7 |
| 21 | 5 | 8 | 4 | 3 | 6 | 53 | 4 | 8 | 5 | 6 | 3 |
| 22 | 5 | 6 | 2 | 7 | 4 | 54 | 4 | 6 | 7 | 2 | 5 |
| 23 | 5 | 4 | 8 | 6 | 7 | 55 | 4 | 4 | 1 | 3 | 2 |
| 24 | 5 | 2 | 6 | 2 | 1 | 56 | 4 | 2 | 3 | 7 | 8 |
| 25 | 7 | 1 | 7 | 7 | 7 | 57 | 2 | 1 | 2 | 2 | 2 |
| 26 | 7 | 3 | 5 | 3 | 1 | 58 | 2 | 3 | 4 | 6 | 8 |
| 27 | 7 | 5 | 3 | 2 | 6 | 59 | 2 | 5 | 6 | 7 | 3 |
| 28 | 7 | 7 | 1 | 6 | 4 | 60 | 2 | 7 | 8 | 3 | 5 |
| 29 | 7 | 8 | 2 | 1 | 8 | 61 | 2 | 8 | 7 | 8 | 1 |
| 30 | 7 | 6 | 4 | 5 | 2 | 62 | 2 | 6 | 5 | 4 | 7 |
| 31 | 7 | 4 | 6 | 8 | 5 | 63 | 2 | 4 | 3 | 1 | 4 |
| 32 | 7 | 2 | 8 | 4 | 3 | 64 | 2 | 2 | 1 | 5 | 6 |

accurate. An obvious approach is to use existing space-filling designs, such as LHDs which were reviewed in Sect. 11.2.1. It has limitations because the experimental region here is *irregular*. First, data center may not be in a rectangular shape. This leads to irregular allocation of racks, where data center facilities are designed to be stored in. Second, even if the racks are located regularly in a rectangular region, the data center facilities can be stored irregularly because of the usage limitation. Such problem exists not only in the design of data center computer experiments but also in devising a sensor placement plan for physical experimentations.

To illustrate how the existing method can fail, consider the popular LHD. Its most important property is the one-dimensional balance. A naive way to apply LHDs is the following. Assume there are 6 design points and $x$, $y$, $z$ are three factors which stand for the 3-dimensional coordinates. One can first find an LHD with 6

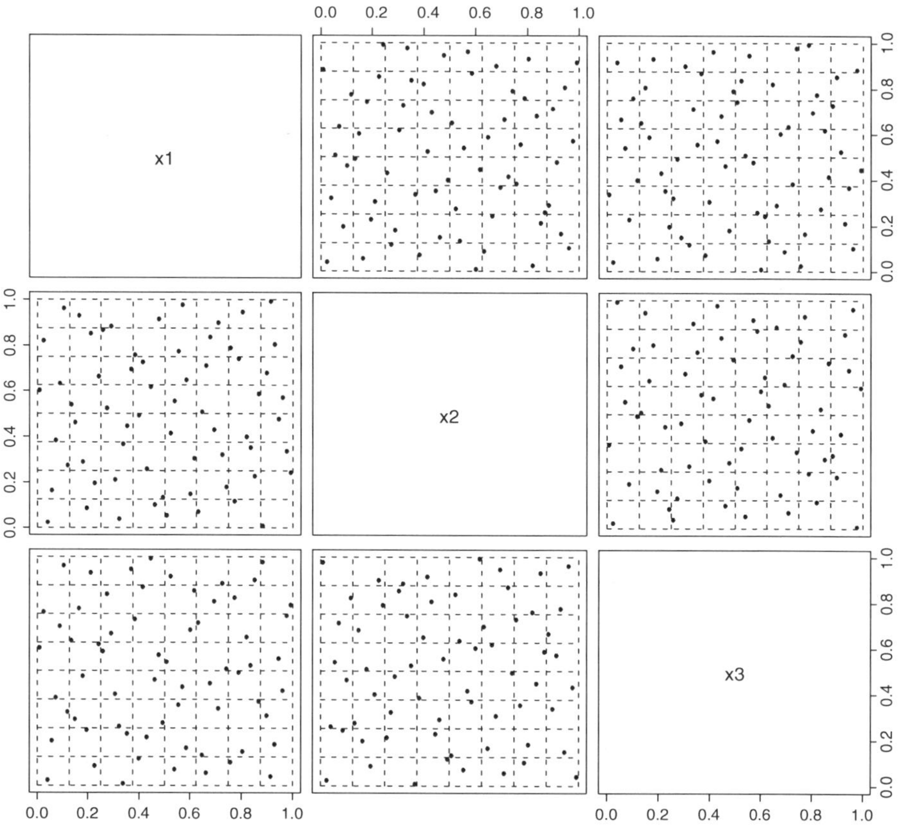


**Fig. 11.2** Bivariate projections of a 64 ×5 sliced space-filling design. Endoscope and instrument manipulators (©Biometrika 2009), reprinted with permission

runs and 3 factors, and adjust the design points to areas where the facilities are located. The left panel in Fig. 11.4 illustrates the design projected onto the $x$-, $y$-axis. By moving points in the white areas to the closest locations of the facility indicated by the gray areas, the adjusted design is given in the right panel. For irregular region, the design in the right panel of Fig. 11.4 can lead to a design in Fig. 11.5, which has three replicates on the $x$-axis and is, thus, undesirable. Hence, without the one-dimensional balance property, the design points cannot be located as uniformly as they should be.

Hung et al. [24] proposed a new class of space-filling designs for experiments with a specific type of irregular regions, where the desirable range of one factor depends on the level of another factor. For example, the range of the $x$-axis in Fig. 11.5 (gray areas) depends on the level of the $y$-axis. Such a region is called a *slid-rectangular region* and the space-filling designs for slid-rectangular regions are named probability-based LHDs.

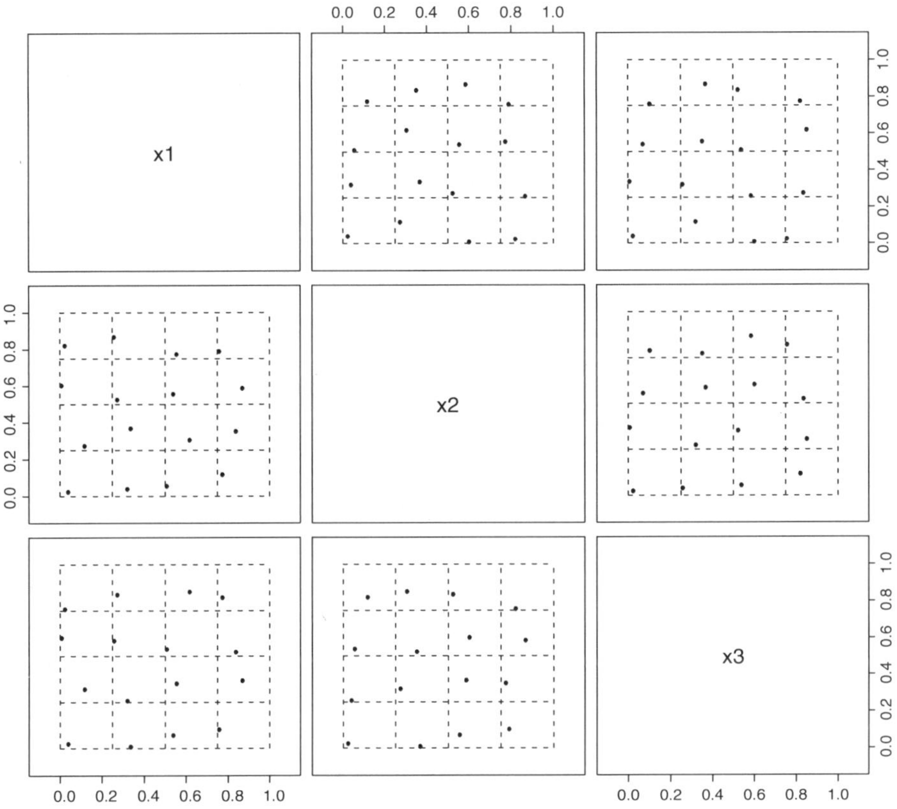


**Fig. 11.3** Bivariate projections of the slice associated with $(z_1, z_2) = (-1, -1)$ of the design in Fig. 11.2. Endoscope and instrument manipulators (© Biometrika 2009), reprinted with permission

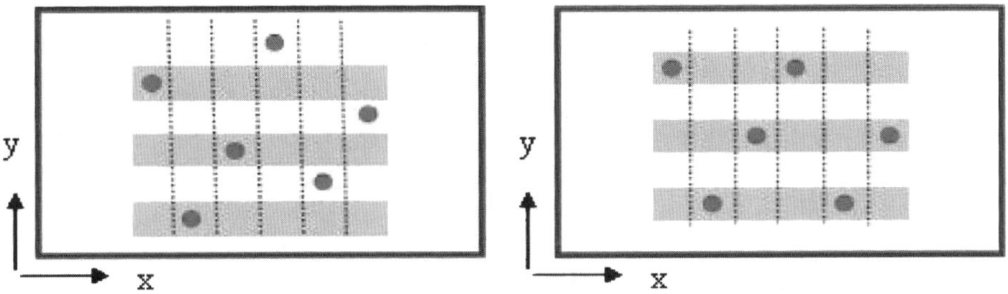


**Fig. 11.4** *Left*: 6-run Latin hypercube design. *Right*: adjusted Latin hypercube design

The basic idea of the probability-based LHDs is to take into account the irregularity in the design construction so that the design can still maintain the one-dimensional balance. To illustrate the construction of such a design, consider the region in Fig. 11.5 but assume that eight points are to be placed.

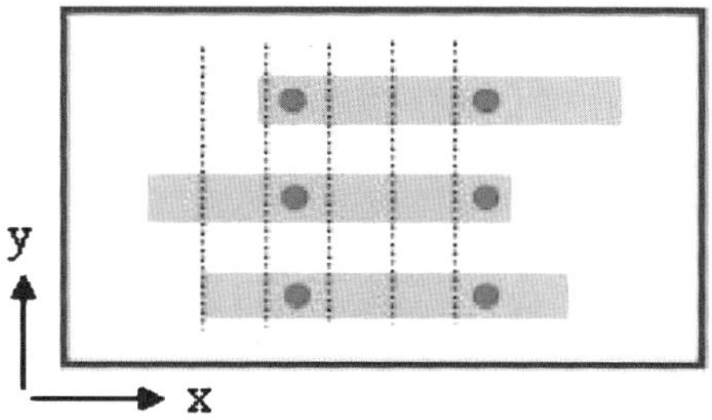


**Fig. 11.5** Latin hypercube design with slid-rectangular region

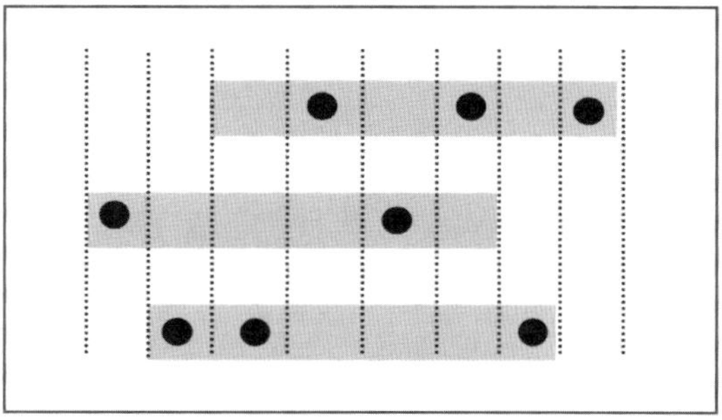

**Fig. 11.6** Probability-based Latin hypercube design

**Table 11.3** Design table for Fig. 11.6

| | | | | | | | | |
|---|---|---|---|---|---|---|---|---|
| $x$ | 1 | 2 | 3 | 4 | 5 | 6 | 7 | 8 |
| $y$ | 2 | 1 | 1 | 3 | 2 | 3 | 1 | 3 |

Denote them by $d_1, \ldots, d_8$, where $d_i = (x_i, y_i)$. One has to first divide the $x$-axis into eight mutually exclusive intervals and define $x_i = i$, for all $i = 1, \ldots, 8$. For choosing the $y$ values in $d_i$, the first step is to define the corresponding feasible range (i.e., set of levels), $C_i$, on the $y$-axis for each level of $x$. In Fig. 11.5, $C_1 = \{2\}$, $C_2 = \{1, 2\}$, etc. Next, for each level of $x$, assign the corresponding feasible $y_i$ with equal probability. For example, $pr(y_2 = 1) = pr(y_2 = 2) = 1/2$, and $\mathrm{pr}(y_2 = 3) = 0$ because $y_2 = 3$ is not included in the feasible set $C_2$. Figure 11.6 shows an example of probability-based LHDs generated by this procedure. A general construction procedure for probability-based Latin hypercube designs is given in Hung et al. [24] (see Table 11.3).

For a given number of runs and factors, probability-based Latin hypercube designs are not unique. A further modification is introduced to incorporate a proportional balance property, where the number of observations is proportional to the length of the interval. Such a design is named *balanced* probability-based LHD, which can be written as a modification of the probability-based LHD with the constraints. Table 11.4 lists such a design with three factors ($p = 3$) and 22 runs. For the slid-rectangular region, factor $x_2$ has five levels and the proportional lengths of $x_1$ at different levels of $x_2$ are 3 : 4 : 5 : 6 : 4. Clearly, both the proportional allocation property and the one-dimensional balance property hold for the first two

**Table 11.4** Design table for Fig. 11.7

| | | | | | | | | | | | | | | | | | | | | | | |
|---|---|---|---|---|---|---|---|---|---|---|---|---|---|---|---|---|---|---|---|---|---|---|
| $x_1$ | 1 | 2 | 3 | 4 | 5 | 6 | 7 | 8 | 9 | 10 | 11 | 12 | 13 | 14 | 15 | 16 | 17 | 18 | 19 | 20 | 21 | 22 |
| $x_2$ | 1 | 1 | 2 | 2 | 1 | 3 | 4 | 4 | 2 | 3 | 4 | 5 | 5 | 2 | 3 | 4 | 5 | 3 | 3 | 4 | 5 | 4 |
| $x_3$ | 18 | 1 | 8 | 14 | 20 | 4 | 11 | 17 | 7 | 21 | 3 | 15 | 10 | 19 | 5 | 13 | 22 | 9 | 16 | 2 | 12 | 6 |

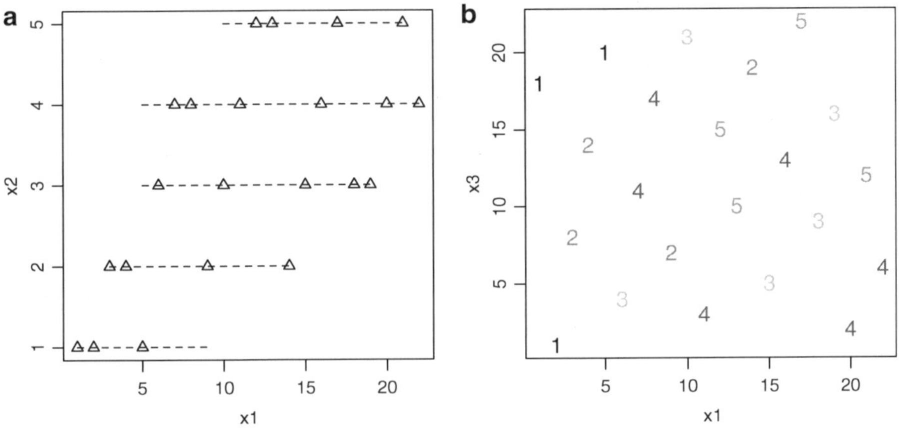


**Fig. 11.7** An example of balanced probability-based Latin hypercube design

factors. Detailed discussions on the drawing rules and the associated sampling probabilities can be found in Hung et al. [24].

Based on the computer experiments conducted using probability-based LHDs or its balanced version, an important follow-up question is how to design an adaptive plan that can efficiently increase the sampling effort. Such a plan is usually called adaptive sampling [25]. Thompson [26, 27] first developed adaptive sampling plans based on initial designs constructed for rectangular regions. These methods cannot be directly applied for slid-rectangular regions because the designs and related inferences are developed for independent and equal inclusion probabilities. It can be easily shown that the independence and equality are violated in probability-based LHDs because of the slid-rectangular shape of the regions. A new class of adaptive designs named *adaptive probability-based LHDs* is proposed by Hung [28] to conduct adaptive designs with slid-rectangular regions.

An adaptive probability-based LHD refers to an adaptive procedure in which the initial experiments are conducted according to a probability-based LHD (or a balanced version). The main idea for the adaptation is that whenever the response of a design point satisfies a given criterion, additional experiments are conducted in its neighborhood. The proposed approach is illustrated by a simulated data center example based on Hung [28]. Figure 11.8 is the two-dimensional layout of the data center. Sensors can be located in six rows (eight racks) with different lengths,

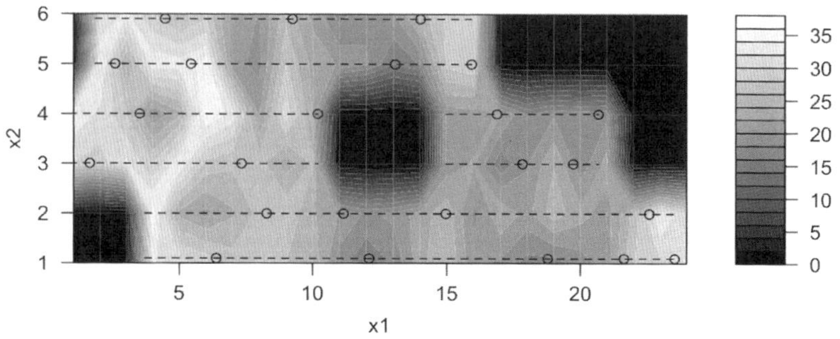


**Fig. 11.8** Two-dimensional layout of a data center

**Table 11.5** Comparison of estimators in the data center study

| | $\hat{\mu}$ | $\hat{\mu}^*$ | $\hat{\mu}_{SRS}$ |
|---|---|---|---|
| Mean | 24.20 | 24.20 | 24.21 |
| Variance | 0.52 | 0.46 | 0.77 |

denoted by the dashed lines. This layout results in a slid-rectangular experimental region which can be specified by $x_1$ and $x_2$. In this study, we fixed all sensors to the top of the rack because temperature increases theoretically with height and monitoring higher temperature is the main objective of this study. There are 24 sensors available for the initial design and, thus, a 24-run PLHD denoted by circles in Fig. 11.8 is considered. Sensors appear to be spread out uniformly over the experimental region according to this design. Based on these initial observations, three more sensors are added by the proposed adaptive sampling strategies with $\nu = 34°\text{C}$ and the estimated temperature is $\hat{\mu} = \hat{\mu}^* = 24.52°\text{C}$.

To study the advantage of the adaptive designs, simulations are conducted in this example to compare the performance of the adaptive designs with that of the nonadaptive simple random sample (without replacement) with sample size equal to the adaptive design. Simulations are performed based on a snapshot of the data center temperatures [28]. The temperature map is provided in Fig. 11.8 where the green areas indicate the nonexperimental regions. Using the snapshot data, detailed temperature observations are available for all 166 sites and, therefore, the comparison can be performed based on different choices of designs. The simulations consist of 3,000 iterations. A randomly generated 24-run PLHD is used for each iteration as an initial design and adaptive designs are conducted accordingly. The estimated mean temperatures and estimated variances are summarized in Table 11.5, where $\hat{\mu}$ and $\hat{\mu}^*$ are two unbiased estimators calculated based on the adaptive designs and $\hat{\mu}_{SRS}$ represents the estimator based on simple random sampling with sample size equal to the adaptive designs. As shown in the table, the improved unbiased estimator $\hat{\mu}^*$ based on adaptive designs has about 40% variance reduction compared to $\hat{\mu}_{SRS}$. Moreover, it provides about 12 % variance reduction compared to the original unbiased estimator $\hat{\mu}$. The average final sample size is 28.73 following the adaptive procedure.

## 11.3 Statistical Methods for Modeling Data Center Computer Experiments

Data center computer experiments, although cheaper than physical experimentations, can still be time consuming and expensive. An approach to reduce the computational time and cost is to perform optimization on a meta-model that approximates the original computer model. The meta-model can be obtained from the data by running a data center computer experiment on a sample of points in the region of interest.

Kriging is widely used for obtaining meta-models [2, 4, 29]. For examples, Pacheco et al. [30] uses kriging for the thermal design of wearable computers, Cappelleri et al. [31] uses kriging for the design of a variable thickness piezoelectric bimorph actuator. The popularity of kriging is due to its interpolating property which is desirable in deterministic data center computer experiments (i.e., no random error in the output) [4, 32].

A kriging model, known as *universal kriging*, can be stated as follows. Suppose the true function is $y(\boldsymbol{x})$, where $\boldsymbol{x} \in \mathbb{R}^p$. The core idea in kriging is to model this function as a realization from a stochastic process

$$Y(\boldsymbol{x}) = \mu(\boldsymbol{x}) + Z(\boldsymbol{x}), \tag{11.1}$$

where $\mu(\boldsymbol{x}) = \sum_{i=0}^{m} \mu_i f_i(\boldsymbol{x})$ and $Z(\boldsymbol{x})$ is a weak stationary stochastic process with mean 0 and covariance function $\sigma^2\psi$. The $f_i$'s are some known functions and $\mu_i$'s are unknown parameters. Usually $f_0(\boldsymbol{x}) = 1$. The covariance function is defined as $\text{cov}\{Y(\boldsymbol{x}+\boldsymbol{h}), Y(\boldsymbol{x})\} = \sigma^2\psi(\boldsymbol{h})$, where the correlation function $\psi(\text{h})$ is a positive semidefinite function with $\psi(0) = 1$ and $\psi(-\boldsymbol{h}) = \psi(\boldsymbol{h})$. In this formulation, $\mu(\boldsymbol{x})$ is used to capture some of the known trends, so that $Z(x)$ will be a stationary process. But, in reality, rarely will these trends be known and, thus, the following special case, known as *ordinary kriging*, is commonly used (see [33, 34]),

$$Y(\boldsymbol{x}) = \mu_0 + Z(\boldsymbol{x}). \tag{11.2}$$

The meta-model (or the predictor) can be obtained as follows. Suppose we evaluate the function at $n$ points $\{\boldsymbol{x}_1, \ldots, \boldsymbol{x}_n\}$ and let $\boldsymbol{y} = (y_1, \ldots, y_n)'$ be the corresponding function values. Then, the ordinary kriging predictor is given by

$$\hat{y}(\boldsymbol{x}) = \hat{\mu}_0 + \psi(\boldsymbol{x})'\Psi^{-1}(\boldsymbol{y} - \hat{\mu}_0\mathbf{1}_n), \tag{11.3}$$

where $1_n$ is a column of 1's with length $n$, $\psi(x)' = (\psi(\boldsymbol{x}-\boldsymbol{x}_1), \ldots, \psi(\boldsymbol{x}-\boldsymbol{x}_n))$, $\Psi$ is an $n \times n$ matrix with elements $\psi(\text{x}_i - \text{x}_j)$, and $\hat{\mu}_0 = \mathbf{1}_n'\Psi^{-1}y/\mathbf{1}_n'\Psi^{-1}\mathbf{1}_n$. It is the best linear unbiased predictor (BLUP), which minimizes the mean squared prediction error $E\hat{Y}(\boldsymbol{x}) - \boldsymbol{Y}(\boldsymbol{x})^2$ under (11.2).

It can be easily shown that the predictor in (11.3) is an *interpolating predictor*. To evaluate such a predictor, a correlation function has to be specified. A widely used correlation function in computer experiments is the power exponential product correlation function given by (see [2])

$$\psi(h) = \exp\left(-\sum_{j=1}^{p} \theta_j h_j^q\right), \quad 1 \leq q \leq 2, \tag{11.4}$$

where $q = 1$ and $q = 2$ correspond to the exponential and Gaussian correlation functions respectively. Other correlation functions, such as Matérn correlation functions [35], are also popular in practice. More discussions on the correlation functions can be found in Cressie [36]. Based on (11.4),the correlation parameters $\theta = (\theta_1, \ldots, \boldsymbol{\theta}_p)$ can be estimated using the maximum likelihood method assuming the Gaussian process for $Z(x)$ in (11.2). Other estimation methods such as the penalized likelihood method [37] can also be employed. Then the maximum likelihood estimate is given by

$$\hat{\boldsymbol{\theta}} = \arg_{\boldsymbol{\theta}} \min n \log \hat{\sigma}^2 + \log \det(\Psi), \tag{11.5}$$

where

$$\hat{\sigma}^2 = \frac{1}{n}(\boldsymbol{y} - \hat{\mu}_0 \mathbf{1}_n)' \Psi^{-1} (y - \hat{\mu}_0 \mathbf{1}_n).$$

These results can be easily extended to the universal kriging predictor.

Despite the prevalence of ordinary kriging in many applications, it was observed that the ordinary kriging prediction can be poor if there are some strong trends [38] and the prediction accuracy can be improved by selecting the important variables properly [39]. Take data center computer experiments as an example, there are many variable such as the rack power, diffuser flow rate, diffuser height, and hot-air return-vent location [40]. Among these variables, only some of them have significant contributions to the temperature distribution. In the literature, there are numerous methods for variable selection in computer experiments. Welch et al. [34] proposed an algorithm to screen important variables sequentially and Linkletter et al. [41] proposed a Bayesian variable selection procedure. These methods focus on identifying variables with significant impact on the process being studied, which is the main goal in the early stages of experimentation. This is referred to as *screening* in the experimental design literature [5]. The variable selection criteria is constructed based on the estimated correlation parameters $\boldsymbol{\theta}$, which can be numerically unstable [37, 42]. Therefore, the selected variables may not be reliable.

A recent approach called *blind kriging* [39], integrates a Bayesian forward selection procedure into the kriging model. Its objective is to enhance the prediction accuracy with the help of variable selection. This method performs variable selection through the mean function and the Gaussian process part is used to achieve interpolation. It effectively reduces the prediction error and demonstrates the advantages of combining variable selection techniques with kriging. Nevertheless, the iterative Bayesian estimation is computationally intensive. Moreover, studying the selection consistency is crucial because identifying the correct mean function is an important element to ameliorate the prediction accuracy. A new approach known as *penalized blind kriging* is proposed by Hung [43] to overcome the foregoing problems. The idea is to modify the blind kriging by incorporating a variable selection mechanism into kriging via the penalized likelihood functions. Thus, variable selection and modeling are achieved simultaneously.

Finally, we briefly discuss two other methods for modeling data center computer experiments. First, for situations with both quantitative and qualitative factors, new Gaussian process models with very flexible correlation structures [40, 44] should be used, instead of those given in (11.1) and (11.2). Second, a large data center computer model is often run with varying degrees of accuracy, resulting in computer experiments with different levels of fidelity. Data from multifidelity data center computer experiments can be modeled by using hierarchical Gaussian process models [45, 46, 47].

## 11.4 Conclusions and Remarks

Data center computer experiments have been widely used as economical proxies to physical experiments for efficient data center thermal management. This chapter reviews statistical methods for design and analysis of such experiments. Recent developments on designing experiments with quantitative and qualitative factors and experiments with irregular shape of regions are introduced. Interpolating models which take into account the deterministic property of computer experiments are discussed.

There are several issues deserve further investigation. Space-filling designs are popular in practice, but better designs may well exist. Sequential design seems particularly appropriate for expensive data center computer experiments. However, studies on constructing sequential design receive scant attention in the computer experiment literature. On the other hand, the existing techniques in modeling data center computer experiments can be computationally intensive when the number of observations is large. To overcome this difficulty, development of efficient modeling techniques is called for.

**Acknowledgments** This research of Hung is supported by NSF grants DMS 0905753 and CMMI 0927572. The research of Qian is supported by NSF grants DMS 1055214 and CMMI 0969616, and an IBM Faculty Award. The research of Wu is supported by ARO grant W911NF-08-1-0368 and NSF grant DMS 1007574.

## References


1. Schmidt RR, Cruz EE, Iyengar MK (2005) Challenges of data center thermal management. IBM J Res Dev 49:709–723
2. Santner TJ, Williams BJ, Notz WI (2003) The design and analysis of computer experiments. Springer, New York
3. Fang KT, Li R, Sudjianto A (2006) Design and modeling for computer experiments. CRC Press, New York
4. Sacks J, Welch WJ, Mitchell TJ, Wynn HP (1989) Design and analysis of computer experiments. Statist Sci 4:409–423
5. Wu CFJ, Hamada M (2009) Experiments: Planning, analysis, and parameter design optimization, 2nd edn. Wiley, New York
6. McKay MD, Beckman RJ, Conover WJ (1979) A comparison of three methods for selecting values of input variables in the analysis of output from a computer code. Technometrics 21:239–245
7. Iman RL, Conover WJ (1982) A distribution-free approach to inducing rank correlation among input variables. Commun Statist B—Simul Comput 11:311–334
8. Owen A (1994) Controlling correlations in Latin hypercube samples. J Am Statist Assoc 89:1517–1522
9. Tang B (1998) Selecting Latin hypercubes using correlation criteria. Statist Sinica 8:965–978
10. Johnson M, Moore L, Ylvisaker D (1990) Minimax and maximin distance design. J Statist Plan Infer 26:131–148
11. Morris MD, Mitchell TJ (1995) Exploratory designs for computer experiments. J Statist Plan Infer 43:381-402
12. Joseph VR, Hung Y (2008) Orthogonal-maximin Latin hypercube designs. Statist Sinica 18:171–186
13. Owen A (1992) Orthogonal arrays for computer experiments, integration and visualization. Statist Sinica 2:439–452
14. Tang B (1993) Orthogonal array-based Latin hypercubes. J Am Statist Assoc 88:1392–1397
15. Park JS (1994) Optimal Latin-hypercube designs for computer experiments. J Statist Plan Infer 39:95–111
16. Ye KQ (1998) Orthogonal column Latin hypercubes and their application in computer experiments. J Am Statist Assoc 93:1430–1439
17. Ye KQ, Li W, Sudjianto A (2000) Algorithmic construction of optimal symmetric Latin hypercube designs. J Statist Plan Infer 90:145–159
18. Jin R, Chen W, Sudjianto A (2005) An efficient algorithm for constructing optimal design of computer experiments. J Statist Plan Infer 134:268–287
19. Qian PZG (2009) Nested Latin hypercube designs. Biometrika 96:957–970
20. Qian PZG, Wu CFJ (2009) Sliced space-filling designs. Biometrika 96:945–956
21. Hedayat A, Sloane NJA, Stufken J (1999) Orthogonal arrays: theory and applications. Springer, New York
22. Draper NR, John JA (1988) Response-surface designs for quantitative and qualitaitve factors. Technometrics 30:423–428
23. Wu CFJ, Ding Y (1998) Construction of response surface designs for qualitative and quantitative factors. J Statist Plan Infer 71:331–348
24. Hung Y, Amemiya Y, Wu CFJ (2010) Probability-based Latin hypercube design. Biometrika 97:961–968
25. Thompson SK, Seber GAF (1996) Adaptive sampling. Wiley, New York
26. Thompson SK (1990) Adaptive cluster sampling. J Am Statist Assoc 85:1050–1059
27. Thompson SK (1991) Stratified adaptive cluster sampling. Biometrika 78:389–397
28. Hung Y (2011a) Adaptive probability-based Latin hypercube designs. J Am Statist Assoc 106:213–219

29. Jin R, Chen W, Simpson T (2001) Comparative studies of metamodeling techniques under multiple modeling criteria. J Struct Multidiscipl Optim 23:1–13
30. Pacheco JE, Amon CH, Finger S (2003) Bayesian surrogates applied to conceptual stages of the engineering design process. ASME J Mech Des 125:664–672
31. Cappelleri DJ, Frecker MI, Simpson TW, Snyder A (2002) Design of a PZT bimorph actuator using a metamodel-based approach. ASME J Mech Des 124:354–357
32. Laslett GM (1994) Kriging and splines: an empirical comparison of their predictive performance in some applications. J Am Statist Assoc 89:391–400
33. Currin C, Mitchell TJ, Morris MD, Ylvisaker D (1991) Bayesian prediction of deterministic functions, with applications to the design and analysis of computer experiments. J Am Statist Assoc 86:953–963
34. Welch WJ, Buck RJ, Sacks J, Wynn HP, Mitchell TJ, Morris MD (1992) Screening, predicting, and computer experiments. Technometrics 34:15–25
35. Matérn B (1986) Spatial variation, 2nd edn. Springer, New York
36. Cressie NA (1993) Statistics for spatial data. Wiley, New York
37. Li R, Sudjianto A (2005) Analysis of computer experiments using penalized likelihood in Gaussian kriging models. Technometrics 47:111–120
38. Martin JD, Simpson TW (2005) On the use of kriging models to approximate deterministic computer models. AIAA J 43:853–863
39. Joseph VR, Hung Y, Sudjianto A (2008) Blind kriging: A new method for developing metamodels. ASME J Mech Des 130:031102-1–8
40. Qian PZG, Wu H, Wu CFJ (2008) Gaussian process models for computer experiments with qualitative and quantitative factors. Technometrics 50:383–396
41. Linkletter CD, Bingham D, Hengartner N, Higdon D, Ye KQ (2006) Variable selection for Gaussian process models in computer experiments. Technometrics 48:478–490
42. Joseph VR (2006) Limit kriging. Technometrics 48:458–466
43. Hung Y (2011b) Penalized blind kriging in computer experiments. Statist Sinica 21:1171–1190
44. Han G, Santner TJ, Notz WI, Bartel DL (2009) Prediction for computer experiments having quantitative and qualitative input variables. Technometrics 51:278–288
45. Kennedy MC, O'Hagan A (2000) Predicting the output from a complex computer code when fast approximations are available. Biometrika 87:1–13
46. Qian Z, Seepersad C, Joseph R, Allen J, Wu CFJ (2006) Building surrogate models with detailed and approximate simulations. ASME J Mech Des 128:668–677
47. Qian PZG, Wu CFJ (2008) Bayesian hierarchical modeling for integrating low-accuracy and high-accuracy experiments. Technometrics 50:192–204

# Chapter 12
# Two-Phase On-Chip Cooling Systems for Green Data Centers

**John R. Thome, Jackson B. Marcinichen, and Jonathan A. Olivier**

**Abstract** Cooling of data centers is estimated to have an annual electricity cost of 1.4 billion dollars in the USA and 3.6 billion dollars worldwide. Currently, refrigerated air is the most widely used means of cooling data center's servers. According to recent articles published at the ASHRAE Winter Annual Meeting at Dallas, typically 40% or more of the refrigerated airflow bypasses the server racks in data centers. The cost of energy to operate a server for 4 years is now on the same order as the initial cost to purchase the server itself, meaning that the choice of future servers should be evaluated on their total 4-year cost, not just their initial cost. Based on the above issues, thermal designers of data centers and server manufacturers now seem to agree that there is an immediate need to improve the server cooling process, especially considering that modern data centers require the dissipation of 5–15 MW of heat, and the fact that 40–45% of the total energy consumed in a data center is for the cooling of servers. Thus, the manner in which servers are cooled and the potential of recovery of the dissipated heat are all more important, if one wishes to reduce the overall $CO_2$ footprint of the data center. Recent publications show the development of primarily four competing technologies for cooling chips: microchannel single-phase (water) flow, porous media flow, jet impingement cooling and microchannel two-phase flow. The first three technologies are characterized negatively for the relatively high pumping power to keep the temperature gradient in the fluid from inlet to outlet within acceptable limits, i.e., to minimize the axial temperature gradient along the chip and the associated differential expansion of the thermal interface material with the silicon created by it. Two-phase flow in microchannels, i.e., evaporation of dielectric refrigerants, is a promising solution, despite the higher complexity involved. The present chapter presents the thermo-hydrodynamic fundamentals of such a new green technology. Two potential cooling cycles making

---

J.R. Thome (✉) • J.B. Marcinichen • J.A. Olivier
Laboratory of Heat and Mass Transfer (LTCM), École Polytechnique Fédérale de Lausanne (EPFL), Station 9, CH-1015 Lausanne, Switzerland
e-mail: john.thome@epfl.ch; jackson.marcinichen@epfl.ch; jonathan.olivier@epfl.ch

Y. Joshi and P. Kumar (eds.), *Energy Efficient Thermal Management of Data Centers*, DOI 10.1007/978-1-4419-7124-1_12,

use of microchannel evaporators are also demonstrated. A case study was developed showing the main advantages of each cycle, and a comparison between single-phase (water and brine) and two-phase (HFC134a and HFO1234ze) cooling is given. Finally, an additional case study demonstrating a potential application for the waste heat of data centers is developed. The main aspects considered were reduction of $CO_2$ footprint, increase of efficiency (data centers and secondary application of waste heat), and economic gains.

## 12.1 Introduction

Cooling of data centers can represent up to 45% [1] of the total energy consumption using current cooling technologies (air cooling); see also Chap. 3. In the USA, this relates to an estimated 45 billion kW h usage by 2011, with an annual cost of $3.3 billion or $4 billion with the inclusion of a carbon tax. And this is just for cooling. This problem is aggravated by the current growth rate of data centers, being between 10 and 20% per annum. With the USA having an annual increase of total electrical generation of approximately 1.5%, data centers potentially will consume all of the electrical energy produced by 2030 if current growth rates remain! With cooling of data centers accounting for most of the non-IT energy usage, this is a logical aspect of data centers that needs to be addressed.

Most data centers make use of air-cooling technologies to ensure the correct running of the servers contained within. Air, however, is a very inefficient source of cooling due to its very low capacity for transporting heat and its low density, both which drive large power requirements to move it. The limits of air cooling are also being approached due to the performance increase in the microprocessors (CPUs) in the servers, which will have heat fluxes in the order of 100 W/cm$^2$ in the not too distant future. It was shown that air has a maximum heat removal capacity of about 37 W/cm$^2$ [2]. The problem is made worse with servers being more densely packed, such as blade centers with racks that will be generated in excess of 60 kW of heat, while today's data centers are designed for cooling capacities in the order of 10–15 kW per rack [3]. Hence, if data centers want to become green, other solutions to air cooling are required.

One long-term solution is to go to direct on-chip cooling. Recent publications show the development of primarily four competing technologies for on-chip cooling: microchannel single-phase flow, porous media flow, jet impingement, and microchannel two-phase flow [4]. Leonard and Philips [5] showed that the use of such new technology for cooling of chips could produce savings in energy consumption of over 60%. Agostini et al. [4] highlighted that the most promising of the four technologies was microchannel two-phase cooling, where Fig. 12.1 shows the heat sink thermal resistances for diverse cooling technologies as a function of the pumping power to the dissipated thermal power ratio. Using this criterion, the best heat sink solution should be that nearest the lower left corner because it represents the lowest thermal resistance at the lowest pumping power of the cooler

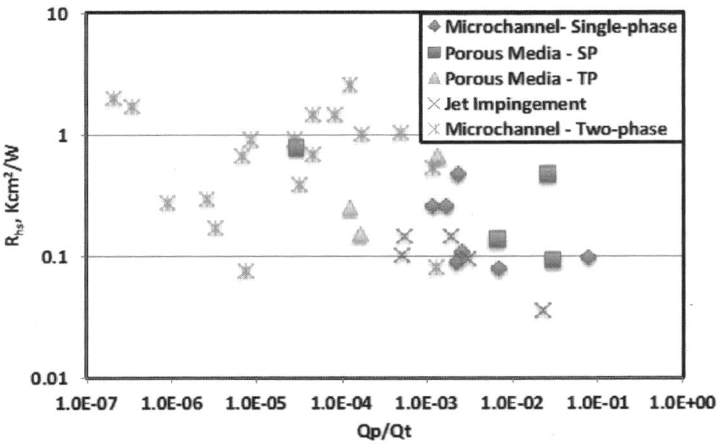


**Fig. 12.1** Thermal resistance of heat sinks for diverse cooling technologies as a function of the pump to the dissipated power ratio

(*attention*: not the pumping power of the entire cooling system). It is clear that two-phase microchannel cooling is the best performing technology in this confrontation from this perspective.

This chapter focuses on two-phase on-chip cooling technology within microchannels, highlighting some of the advantages it has over traditional air cooling as well as single-phase microchannel cooling. Some discussion regarding the fundamentals of two-phase flow within microchannels is given, which includes the description of mathematical prediction methods to determine critical parameters of such coolers. Further, the use of two-phase microchannels on a system level is described, showing the practical aspects of using such technology on a large scale. Finally, not mentioned in this introduction, the possibility to recover some of the waste energy generated by datacenters is discussed. Since this quantity is generally huge, recovering this energy could account for extra savings when distributed to a secondary application requiring its use. This is demonstrated in the last part of this chapter.

## 12.2 Two-Phase Flow and Refrigerants

When a pure substance is heated, its temperature rises as heat is added. This is referred to as *sensible* heat. The opposite is true when heat is removed; the temperature decreases. This happens when fluids or solids do not change from one phase to another. When instead this does happen, though, any heat added to or removed from the substance does not result in a change in temperature if the pressure remains constant. This is aptly referred to as *latent* heat, because the heat is hidden, as it cannot be measured.

Temperatures in substances change due to the increase or decrease in molecular interactions. The more the heat is added, the more interactions there are, with the result of a temperature increase. When a substance changes phase, as occurs during vaporization, energy is expended to break the intermolecular bonds. The molecular interaction is therefore not increased, explaining why the temperature does not increase. This is a very simplistic explanation, but is just intended to give a basic idea of phase change.

Every liquid has a tendency to turn into a vapor, which is associated with its pressure. *A liquid in a container empty of all gases but its own vapor comes to equilibrium with the vapor at a value of the pressure known as its saturated vapor pressure, or simply, vapor pressure for short* [6]. Therefore, a liquid will evaporate at a temperature which depends on the pressure. An example is water, which evaporates at 100°C at 1.0 atm pressure and 120°C at 2.0 atm.

So what is the main advantage of using a fluid's latent heat rather than its sensible heat? This can be shown best by means of an example. Electronic components should be cooled at low temperatures, say at a room temperature of 20°C for our example, while their maximum operating temperature should not exceed 85°C. When using the sensible heat of a liquid as a means of cooling, the maximum amount of energy available per kilogram of water is 273 kJ/kg (65 K $\times$ 4.2 kJ/kg). If, for an ideal case, the components were to be cooled by means of the fluid's latent heat at a temperature of 20°C, the amount of energy available per kilogram (latent heat of vaporization) of water would be 2,453.5 kJ/kg, almost ten times more! All this energy is absorbed while maintaining the saturation temperature at 20°C if the pressure is not varied! Figure 12.2 shows these values in a Mollier diagram or P-h diagram. This diagram is represented by the saturation lines (dash and dot dash), which bounds the two-phase region. To the left of this region, the fluid is fully liquid (subcooled with respect to the saturation temperature) and to the right it is fully vapor (superheated with respect to the saturation temperature). It can be seen that there is a huge difference between latent and sensible heat.

There is a caveat to this set-up though. For water to boil at 20°C requires that its pressure be dropped to below atmospheric pressure (to 0.023 atm), which has many impractical implications. One of these implications is that one's system would need to operate under vacuum, which is very difficult to maintain due to "undetectable" minor leaks of air into such systems. Operating at or above atmospheric pressures using water as the working fluid is also impractical for electronic cooling applications since its saturation temperature is then above 100°C. So, what alternative fluid can then be used?

This is where refrigerants come into play. Refrigerants have the important characteristic of having low saturation temperatures at atmospheric pressure. This implies that cooling systems can operate above atmospheric pressures while low operating temperatures can still be obtained for cooling. As an example, the most common working fluid used in most refrigeration systems is HFC134a, having a saturation temperature of −26°C at atmospheric pressure. So, following the example of the water, to cool electronics at a temperature of 20°C requires that the pressure in the system be 5.6 atm. The amount of energy per kilogram of HFC134a

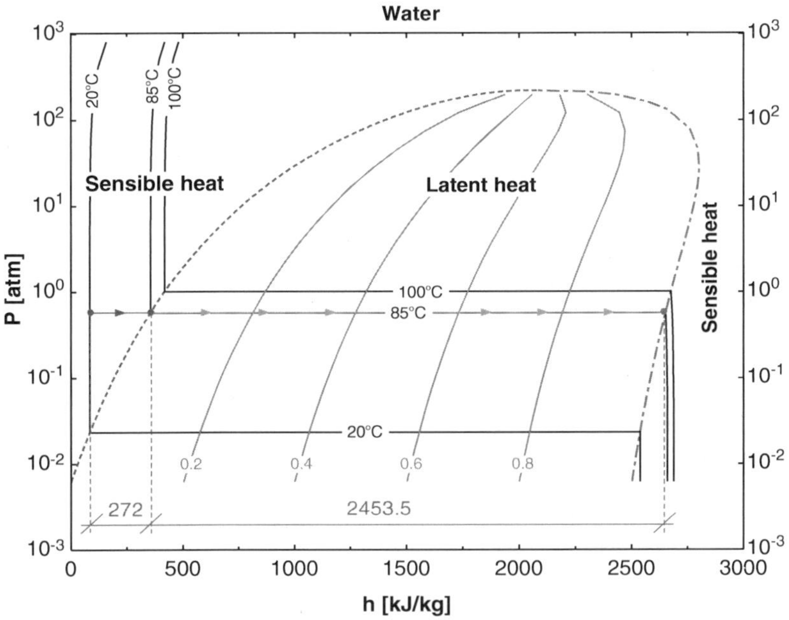


**Fig. 12.2** Mollier diagram for water

available for cooling by means of the fluid's latent heat is 182 kJ/kg. This is considerably lower than the latent heat of water and even lower than the sensible heat. One would tend to dismiss refrigerants as a cooling option due to this fact; however, it should be remembered that, unlike the sensible heat of water having a temperature gradient of 65°C, the temperature gradient using refrigerants is zero if the pressure remains constant.

This leads to another important difference in two-phase cooling compared to single-phase (water or air) cooling. For the latter, the temperature of the fluid rises as its flows through the cooling channels to take away the heat dissipated by the electronics, while however the temperature profile is not significantly affected by the pressure drop of the flow (only relatively small variations in specific heat and density). For a two-phase evaporation process, the pressure drop from inlet to outlet reduces the local saturation pressure along the coolant flow path; hence, the local saturation temperature (fluid temperature) falls from inlet to outlet. This can be deduced from inspection of Fig. 12.2 above for the 100 and 85°C saturation lines within the two-phase *dome* for a flow experiencing such a pressure drop. Thus, a two-phase coolant actually gets *colder* as it flows through an on-chip multi-microchannel evaporator. This of course means that accurate prediction of two-phase pressure drops is an intrinsic part of two-phase cooling technology.

The use of synthetic refrigerants has had a long and successful history since their introduction in the 1920s. The first synthetic refrigerant went under the trade name *Freon* of Dupont, which was a chlorofluorocarbon (CFC). However, environmental concerns have seen the phase-out of such refrigerants. The introduction of the Montreal Protocol in 1987 [7] imposed the phase-out of all CFC refrigerants by 1996 and 2010 for non-Article 5 and Article 5 countries, respectively. Transition hydrochlorofluorocarbon (HCFC) refrigerants are to be completely phased out by 2020 and 2040 for the respective countries. These two refrigerant types have been identified as depleting the ozone layer, hence their phase-out.

With global warming becoming an ever bigger concern, the Kyoto Protocol [8] sets binding targets for greenhouse gas emissions based on calculated equivalents of carbon dioxide, methane, nitrous oxide, hydrofluorocarbons (HFC), perfluorocarbons (PFC), and sulfur hexafluoride. The European Parliament has also set the timing for banning fluorochemical refrigerants having global warming potentials (GWPs) exceeding 150 for automotive air conditioners for new model vehicles effective from 2011 and for all other new vehicles in 2017 [9].

These protocols have led to the development of the so-called fourth-generation refrigerants having a zero-ozone depleting potential (ODP) and a very low GWP. Two of these refrigerants are HFO1234yf and HFO1234ze, seen as replacement refrigerants for the widely used HFC134a. The first is primarily targeted for automotive air-conditioning systems while the second is targeted for electronic cooling applications. A selection of refrigerants and their basic properties are listed in Table 12.1. Also listed is HFC236fa, mainly used as a fire suppressant, although it has also found its use in naval applications. This is the replacement refrigerant for CFC114 and is mainly used because of its relatively low pressure when compared to the other refrigerants, its low toxicity, and inflammability. HCFC123 was a very promising refrigerant due to its high refrigerating cycle thermal efficiency [10], which translates to a much smaller $CO_2$ footprint due to lower compressor power requirements; however, due to the fact that it has a negligible but non-zero ODP, it is set for phase-out and has been abandoned in nearly all chiller applications.

Reviewing the refrigerants in Table 12.1, HFC134a is the current front-runner for application in two-phase on-chip cooling systems. However, at an operating saturation temperature of 60°C, its pressure is rather high (16.8 bar); on the other hand, the piping is small in diameter and can easily handle much, much higher pressures (automotive air-conditioning systems with HFC134a are normally designed to withstand 100 bars or more with well understood fabrication techniques while also losing less than 40 g per year to meet European standards [14]). Both versions of HFO1234 would be good lower pressure alternatives, once they become available in larger quantities. They, however, each have a minor flammability classification by ASHRAE and one must wait to see what impact that has on their future application to electronics cooling. Of the remaining viable fluids, HFC245fa is an excellent low-pressure solution, except that during transportation of such two-phase cooling systems its pressure will fall below atmospheric in cold climates. Thus, HFC236fa would be more appropriate, but still falls into vacuum for subzero conditions during transport.

Dielectric fluids, such as FC-72 and others in that category, are also possible candidates as two-phase working fluids. They primarily have four important adverse characteristics besides their numerous positive ones: (1) they fall into

**Table 12.1** Basic fluid and environmental properties at a saturation temperature of 25°C

| | HFC134a | HFO1234yf[a] | HFO1234ze[a] | HCFC123 | HFC236fa | HFC245fa | CFC114 |
|---|---|---|---|---|---|---|---|
| GWP[b] | 1,320 | 4 | 6 | 76 | 9,650 | 1,020 | 9,880 |
| ODP | 0 | 0 | 0 | 0.012 | 0 | 0 | 0.94 |
| Atmospheric lifetime (years) | 14 | 0.030[c] | 0.038[d] | 1.3 | 240 | 7.6 | 300 |
| Boiling point at 1 bar (°C) | −26.1 | −29 | −19 | 27.5 | −1.4 | 15.1 | 3.6 |
| Triple point (°C) | −103.3 | −150.4 | −150.4 | −107 | −93.6 | −102 | −92.5 |
| $\rho_l$ (kg/m$^3$) | 1,207 | 194 | 1,180 | 1,464 | 1,360 | 1,339 | 1,518 |
| $\rho_v$ (kg/m$^3$) | 32.4 | 37.6 | – | 5.9 | 18.4 | 8.6 | 7.8 |
| $h_{lv}$ (kg/m$^3$) | 177.8 | 149[e] | 195[e] | 171.4 | 145.9 | 190.3 | 135.9 |

[a]Classified fluid. All properties obtained from sources available in the public domain
[b]100-year integration time horizon (ITH), $CO_2 = 1$
[c]Nielsen et al. [11]
[d]Søndergaard et al. [12]
[e]Estimated from Brown et al. [13]

vacuum at temperatures below about 50–60°C, (2) they have larger viscosities than the above refrigerants and thus prose pressure drop/pumping power penalties, (3) they have much lower critical heat fluxes (CHF) than the above refrigerants, and (4) they tend to absorb noncondensable gases which is particularly detrimental to the system's condenser performance. Even so, good design may make these a viable alternative.

## 12.3 Multi-microchannel Evaporator

I. *Microchannel fundamentals*

Two-phase flow phenomena on large-scale systems have been well studied for almost a century. A detailed discussion, including design correlations, on two-phase flow and heat transfer in macroscale tubes can be found in the free Web-book *Wolverine Engineering Databook III* of Thome [15]. For those desiring a more detailed review of two-phase flow, boiling and condensation in microchannels than presented below, one can refer to Chaps. 1, 20, and 21 of [15] to see several hundred embedded two-phase videos. Furthermore, Chap. 4 covers single-phase laminar flows and heat transfer, including prediction methods applicable to noncircular microchannels.

While initial studies in the literature reported significant size effects on friction factors and heat transfer coefficients in very small channels in single-phase flows, more accurate recent tests and analysis done with very smooth internal channels have shown that macroscale methods for single-phase flows work well at least down to diameters of 5–10 μm [16]. This is not the case for macroscale two-phase flow methods, which usually do not work very well when compared to data for channels below about 2.0 mm diameter. Thus, it is inappropriate (unreliable) to extrapolate macroscale research results and methods to the microscale. Furthermore, many of the controlling phenomena and heat transfer mechanisms change when passing from macroscale two-phase flow and heat transfer to the microscale. For example, surface tension (capillary) forces become much stronger as the channel size diminishes while gravitational forces are weakened. Or, for example, many two-phase microchannel flows have laminar liquid Reynolds falling in the laminar regime, whereas most macroscale methods are based on turbulent flow data. Furthermore, the nucleate boiling contribution of flow boiling is more or less suppressed in microchannels. Therefore, it is not physically sensible to "refit" macroscale methods to microscale data simply by fitting new empirical constants since the underlying physics has substantially changed, and thus dedicated research is required to investigate these microscale two-phase flows and develop new models and methods to describe them.

As a first view, Fig. 12.3 depicts the buoyancy effect on elongated bubbles flowing in 2.0, 0.790, and 0.509 mm horizontal circular channels taken in the LTCM lab [17]. In the 2.0-mm channel, the difference in liquid film thickness at the top compared to that at the bottom of the bubble is still very noticeable.

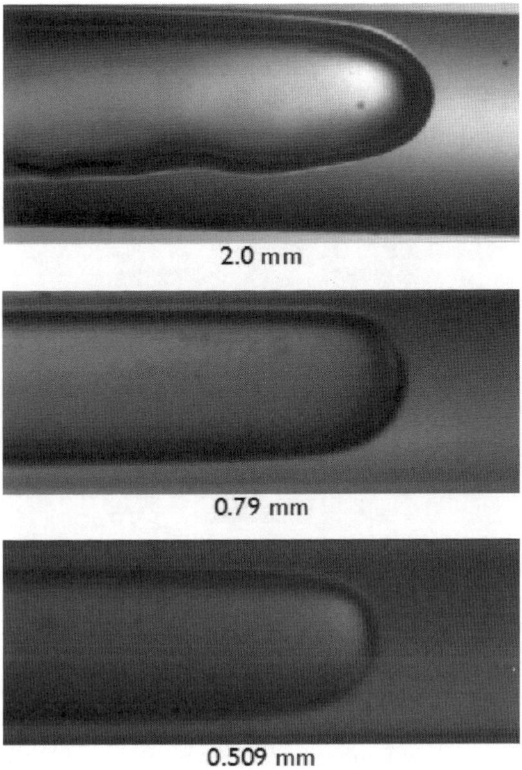


**Fig. 12.3** Video images of slug (*elongated bubble*) flow in 2.0-, 0.8-, and 0.5-mm horizontal channels with HFC134a

Since the local heat transfer is primarily by conduction across such thin laminar liquid films, the film thickness is the main resistance to heat transfer and thus cooling, and therefore its variation around the perimeter is an important thermal issue. Similarly, the film thickness in the 0.790-mm channel is still not uniform above and below the bubble. Instead, in the 0.509-mm channel, the film is now quite uniform. Interpreting these images and other images, one also ascertains that stratified types of flows (all vapor at top with liquid at the bottom) disappear in small horizontal channels. This transition is thus perhaps an indication of the lower boundary of macroscale two-phase flow, in this case occurring for a diameter somewhat greater than 2.0 mm. The upper boundary of microscale two-phase flow may be interpreted as the point at which the effect of gravity on the liquid–vapor interface shape becomes insignificant, such that the uniformly flowing bubble in the 0.509-mm channel is thus a microscale flow, with the transition occurring at about this diameter at the present test conditions.

Heat transfer and pressure drop mechanisms are strongly affected by the type of flow patterns present in the channels, which need to be determined to develop prediction methods. With the aid of high-speed videography (videos up to 120,000 digital images per second) and laser photo diode signals, it is possible to determine these regimes. Figure 12.4 shows a typical flow pattern map for

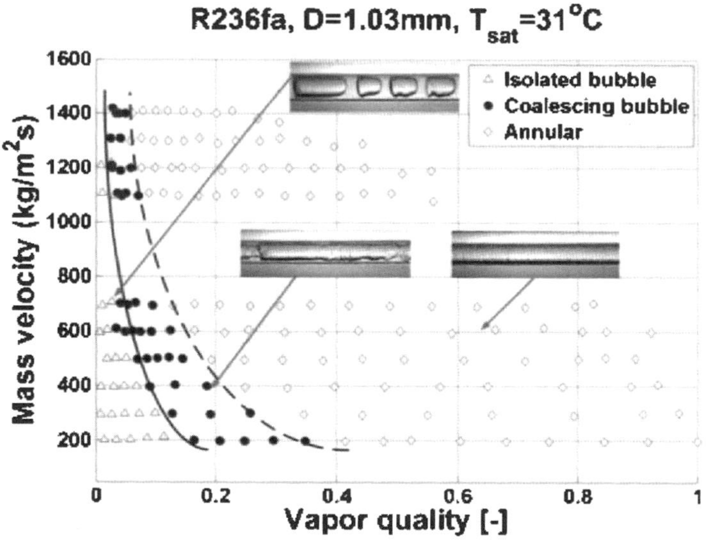


**Fig. 12.4** Three flow regimes shown on a flow pattern map [18]

HFC236fa inside a channel having a diameter of 1.03 mm obtained by Revellin et al. [17]. The flow patterns are identified as isolated bubbles (mostly bubbles shorter in length than the channel diameter), coalescing bubbles (mostly bubbles much longer than the channel diameter and coalescing due to their different axial velocities), and annular flow (characterized by a thin liquid film on the channel perimeter with a high-speed vapor core flow).

II. *Heat transfer*

Many heat transfer correlations have been developed to predict heat transfer coefficients within microchannels, such as that of Lazarek and Black [19], Tran et al. [20], Zhang et al. [21], Kandlikar and Balasubramanian [22], to name a few. Most of these are purely empirical correlations, mostly based on macroscale methods refitted to predict their small channel diameter data, and hence do not incorporate specific microscale mechanisms within them. In fact, the macroscale methods were initially developed to predict nucleate boiling phenomena and convective boiling heat transfer within macroscale tubes, while the first phenomena, except for the onset of boiling, is more or less not encountered in microchannel boiling.

Jacobi and Thome [23] proposed the first theoretically based, elongated bubble (slug) flow boiling model for microchannels, modeling the thin film evaporation of the liquid film trapped between these bubbles and the channel wall, and also accounting for the liquid-phase convection in the liquid slugs between the bubbles. The focus of their study was to demonstrate that the thin film evaporation mechanism was the principal heat transfer mechanism controlling heat transfer in slug flows in microchannels, not nucleate boiling as assumed by others in extrapolation of macroscale ideology to the microscale.

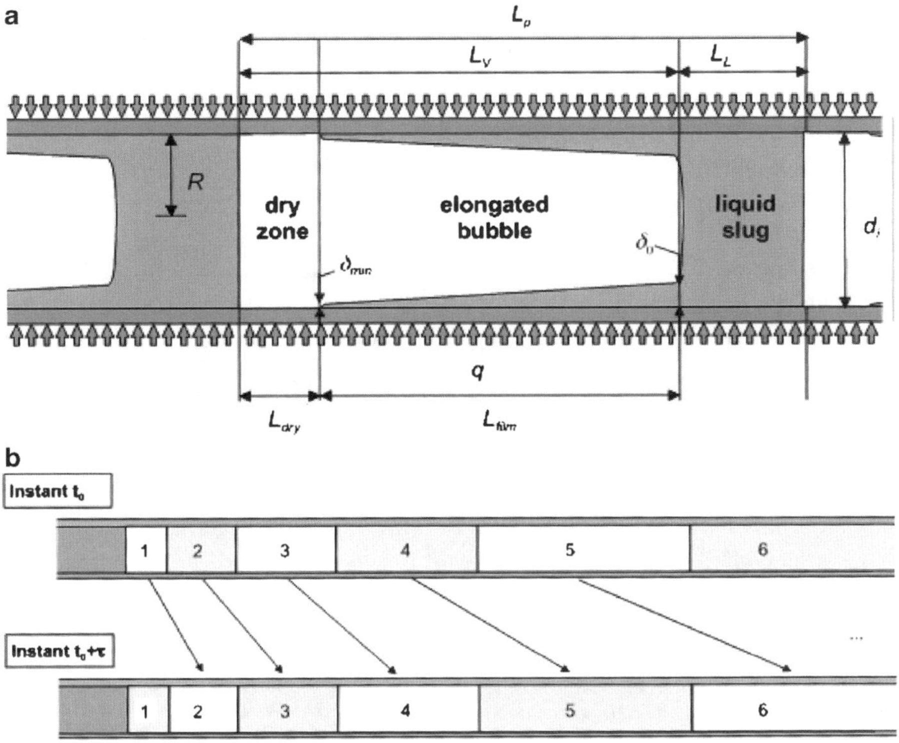


**Fig. 12.5** Three-zone heat transfer model of Thome et al. [24] for elongated bubble flow regime in microchannels. *Top*: Diagram illustrating a triplet composed of a liquid slug, an elongated bubble, and a vapor slug; *bottom*: Bubble tracking of a triplet with passage of a new bubble at time intervals of $\tau$

Following this, the three-zone model of Thome et al. [24] was proposed to cover also the intermittent dry-out regime following the elongated bubbles. This method was developed specifically for microchannel slug flows, but they also applied to annular flows since no reliable flow pattern map to determine the transition was available at that time.

Figure 12.5 shows a representation of the three-zone model [24] where $L_P$ is the total length of the pair or triplet (liquid slug, bubble, and rewettable dry zone), $L_L$ is the length of the liquid slug, $L_V$ is the length of the bubble including the length of the dry wall of the vapor slug $L_{dry}$, and $L_{film}$ is the length of the liquid film trapped by the bubble. The internal radius and the diameter of the tube are $R$ and $d_i$ while $\delta_o$ and $\delta_{min}$ are the thicknesses of the liquid film trapped between the elongated bubble and the channel wall at its formation and at dryout of the film (only when dryout occurs). The evolution of successive bubbles is shown in the lower diagram. The local vapor quality, heat flux, microchannel internal diameter, mass flow rate, and fluid physical properties at the local saturation pressure are input parameters to the model to predict the above parameters as well as the frequency of the bubbles, transient onset of local dryout, etc. The three-zone model predicts both the local instantaneous and time-averaged heat transfer coefficient at a fixed location along a microchannel during evaporation of a succession of elongated bubbles (with frequencies of passage as high as 900 Hz). The elongated bubbles are assumed to nucleate and quickly grow to the channel size upstream such that successive elongated bubbles are formed that are confined by the channel and grow in axial length, trapping a thin film of liquid between the bubble and the inner tube wall as they flow along the channel. The thickness of this film plays an important role in heat transfer. At a fixed location, the process is assumed to proceed as follows: (1) a liquid slug passes, (2) an elongated bubble passes (whose evaporating liquid film is formed from liquid removed from the liquid slug), and (3) a vapor slug passes if the thin evaporating film of the bubble dries out before the arrival of the next liquid slug. The cycle then repeats itself upon arrival of the next liquid slug at a frequency $f$ ($=1/\tau$). Thus, a liquid slug and an elongated bubble pair or a liquid slug, an elongated bubble, and a vapor slug triplet pass this fixed point at a frequency $f$ that is a function of the formation and coalescence rate of the bubbles upstream.

The complete development of the local heat transfer coefficient based on the three-zone model is found in Thome et al. [24] or in Chap. 20 of the free online *Databook III* [15]. This model has been compared to a large database from laboratories around the world for various refrigerants with good results [25], especially when substituting the measured channel roughness for the empirical value of the dryout thickness $\delta_{\min}$ as shown by Ong and Thome [26] for stainless steel microchannels and by Agostini et al. [27] and Costa-Patry et al. [28] for multi-microchannels made in silicon. The model is best limited to applications to the coalescing bubble regime of flow boiling, whose flow pattern boundaries can be predicted using the diabatic flow pattern map of Ong and Thome [29] and shown earlier in Fig. 12.4. Annular flow heat transfer data are instead best predicted using the new 1-D turbulence model of Cioncolini and Thome [30]. The isolated bubble flow regime usually only persists for the first few percent of vapor quality and can tentatively be predicted using the value of the three-zone model at the isolated bubble-to-coalescing bubble transition.

III. *Pressure drop*

The two principles approaches to predict frictional pressure gradients in microscale two-phase flow are the homogeneous and the separated flow models. The homogeneous model assumes that the two-phase fluid behaves as a single-phase fluid but pseudoproperties are used for the density and viscosity that are weighted relative to the vapor and liquid flow fractions. Instead, the separated flow model considers that the phases are artificially segregated into two separate streams, one liquid and one vapor, and interact through their common interface. For more information regarding these models, refer to ref. [15].

Various studies have shown that homogeneous model is a relatively good first choice for calculating two-phase pressure drops in microchannels. Based on their database of 913 data points from nine independent studies, Ribatskie et al. [31] found that the homogeneous model, using the homogeneous viscosity expression of Cicchitti et al. [32], predicted 54.3% of the data within ±30% with a mean absolute error of 61.6%. This was the highest success rate within ±30% of all 12 macroscale and microscale methods compared to their database. The macroscale method of Müller-Steinhagen and Heck [33] came in second with 53.1% of the data points within ±30% with a mean absolute error of 31.3%, and was regarded as the better method. Meanwhile, the Mishima and Hibiki [34] small channel correlation had the second best mean absolute error in that study (37.4%) with 47.7% of the database within ±30%.

Revellin and Thome [35] obtained 2,210 two-phase pressure drop data points for 0.790- and 0.509-mm glass channels at the exit of a microevaporator for HFC134a and HFC245fa, deducing the drop in pressure from the drop in saturation temperature measured using thermocouples, so as not to disturb the flow with pressure taps. Only a few data were obtained in the laminar regime ($Re < 2{,}000$) whilst no method in the literature was able to predict their large database located in the intermediate range ($2{,}000 < Re < 8{,}000$). For their data in the turbulent regime (1,200 points for $Re > 8{,}000$), the McAdams et al. [36] homogeneous model was not accurate, whereas the homogeneous model using the viscosity expression of Cicchitti et al. [32] again worked better (predicting 52% of the database within ±20% while the McAdams expression only achieved 6%), although the Cicchitti et al. expression still tended to systematically under predict the turbulent flow data. On the other hand, the separated flow correlation of Müller-Steinhagen and Heck [33] worked best, considering again only the turbulent database, and predicted 62% of the database within ±20%.

In summary, the prediction of two-phase pressure drops of air–water and refrigerants using the homogeneous model is only approximately reliable, and then apparently only applicable for turbulent microchannel two-phase flows. Of these methods, the Cicchitti et al. [32] viscosity expression works best. At least one macroscale method, that of Müller-Steinhagen and Heck [33], gives better results, although the desired level of accuracy and reliability for engineering design is still quite elusive.

More recently, Cioncolini et al. [37], from a database of 3,908 data points, have proposed a two-phase friction factor correlation for the annular flow regime valid for macro- and microchannels. This correlation was derived by means of a dimensional analysis and is based on a single dimensionless parameter as follows:

$$f_{\mathrm{tp}} = 0.172 We_{\mathrm{c}}^{-0.372},$$

where $We_{\mathrm{c}}$ is the gas core Weber number, which is based on the vapor core velocity, density, and diameter:

$$We_{\mathrm{c}} = \frac{\rho_{\mathrm{c}} V_{\mathrm{c}}^2 d_{\mathrm{c}}}{\sigma}.$$

This, in turn, is based on the fraction of liquid entrained as droplets in the vapor core that influences the two-phase core density $\rho_c$, the core velocity $V_c$, and the core diameter $d_c$ (see their publication for details on how to calculate these parameters) and $\sigma$ is the surface tension. In annular two-phase flow, the core flow can be considered as a spray interacting with the liquid film, which is shear-driven by the core flow and characterized by surface tension waves appearing at its surface. The tips of such waves are atomized by the core flow, giving rise to the entrainment. This correlation was found to predict the macrochannel data with a mean absolute error of 13.1%, predicting 70% of the data to within 15%. This correlation also predicted the microchannel data with fair accuracy, having a mean absolute error of 28.8%. The accuracy was improved by the introduction of the liquid film Reynolds number $Re_L$, such that the two-phase friction factor becomes

$$f_{tp} = 0.0196 We_c^{-0.372} Re_L^{0.318},$$

where $Re_L$ is also based on the fraction of liquid entrained as droplets in the vapor core. This correlation predicts almost all the microchannel data of the authors' database to within ±30% with a mean absolute error of 13.1%.

Currently, there is only one microchannel two-phase pressure drop model that attempts to model the physics of the slug flow. Garimella et al. [38] have proposed such a model for slug (elongated bubble or intermittent) flows, based on their work with air–water and condensation of refrigerants. Basically, they divided the flow into two zones, one for the elongated bubbles and one for liquid slugs between them, similar to the microchannel elongated bubble flow boiling model of Jacobi and Thome [23]. They then developed frictional pressure gradients expressions for the two types of flow and methods to predict the relative fraction of time and the length that each occupied in a cycle, thus coming up with a frictional pressure drop model based on the frequency of the bubbles. Readers are referred to their paper for complete details. While their method worked well for their own database, Revellin and Thome [35] found that this method only predicted about 20% of their database to within ±20% for HFC134a and HFC245fa in 0.5- and 0.8-mm channels. On the other hand, their model should only specifically be applied to slug flow data and such methods will become more useful as two-phase flow pattern maps for microchannels become more accurate and reliable, thus making possible comprehensive flow regime methods.

IV. *Critical heat flux*

For high heat flux cooling applications using multi-microchannel cooling channels, the CHF in saturated flow boiling conditions is a very important operational limit. CHF signifies the maximum heat flux that can be dissipated at the particular operating conditions by the evaporating fluid. Surpassing CHF means that the heated wall becomes completely and irrevocably dry, and is associated with a very rapid and sharp increase in the wall temperature due to the replacement of liquid by vapor adjacent to the heat transfer surface. Only a brief summary is presented below.

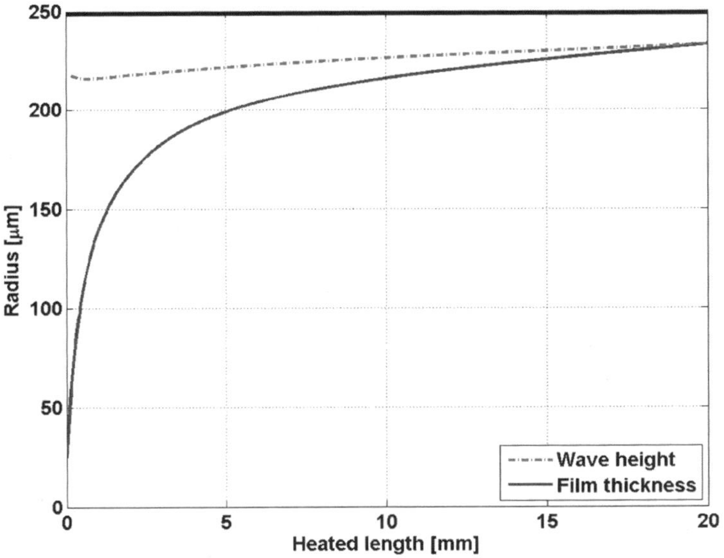


**Fig. 12.6** Revellin and Thome [39] CHF model showing the annular film thickness variation along the channel plotted versus the wave height

Revellin and Thome [39] proposed the first partially theoretically based model for predicting CHF in microchannels. Their model is based on the premise that CHF is reached when local dryout occurs during evaporation in annular flow at the location where the height of the interfacial waves matches that of the annular film's mean thickness. To implement the model, they first solve one-dimensionally the conservation of mass, momentum, and energy equations assuming annular flow to determine the variation of the annular liquid film thickness $\delta$ along the channel. Then, based on the slip ratio given by the velocities of the two phases (liquid and vapor) and a Kelvin–Helmholtz critical wavelength criterion (assuming the height of the wave scales proportional to the critical wavelength), the wave height was predicted with the following empirical expression:

$$\Delta\delta = 0.15\left(\frac{u_{\mathrm{V}}}{u_{\mathrm{L}}}\right)^{-\frac{3}{7}}\left(\frac{g(\rho_{\mathrm{L}} - \rho_{\mathrm{V}})(d_{\mathrm{i}}/2)^2}{\sigma}\right)^{-\frac{1}{7}}$$

where $u_{\mathrm{V}}$, $u_{\mathrm{L}}$, $g$, and $d_{\mathrm{i}}$ are, respectively, the mean velocities of the vapor and liquid phases, the acceleration due to gravity, and the internal channel diameter.

Then, when $\delta$ equals $\Delta\delta$ at the outlet of the microchannel, CHF is reached. Refer to Fig. 12.6 for a simulation. The leading constant and two exponents were determined with a database including three fluids (HFC134a, HFC245fa, and CFC113) and three circular channel diameters (0.509, 0.790, and 3.15 mm) taken from the CHF data of Wojtan et al. [40] and data from the Argonne

Laboratory by Lazarek and Black [41]. Their model also satisfactorily predicted the Purdue CFC113 data of Bowers and Mudawar [42] for circular multi-microchannels with diameters of 0.510 and 2.54 mm of 10 mm length. Furthermore, taking the channel width as the characteristic dimension to use as the diameter in their 1-D model, they were also able to predict the Purdue rectangular multi-microchannel data of Qu and Mudawar [43] for water. Altogether, 90% of the database was predicted within ±20%. As noted above, this model also accurately predicted the HFC236fa multi-microchannel data of Agostini et al. [44]. In addition, this model also predicts CHF data of $CO_2$ in microchannels from three additional independent studies as well as other fluids.

Notably, the above 1-D numerical method can also be applied to nonuniform wall heat flux boundary conditions when solving for the annular liquid film profile. Hence, it can tentatively simulate the effects of single or multiple hot-spots, their size and location, etc. or it can use, as an input, a CPU chip's heat dissipation map to investigate if the locally high heat fluxes will locally trip CHF.

Regarding simpler empirical methods, Ong [45] has more recently updated the CHF correlation developed by Wojtan et al. [40] for a wider range of operating parameters and fluid properties. This correlation is given as

$$q_{\mathrm{CHF}} = 0.12Gh_{\mathrm{LV}}\left(\frac{\mu_{\mathrm{LO}}}{\mu_{\mathrm{VO}}}\right)^{0.183}\left(\frac{\rho_{\mathrm{VO}}}{\rho_{\mathrm{LO}}}\right)^{0.062}We_{\mathrm{LO}}^{-0.141}\left(\frac{L_{\mathrm{ev}}}{d_{\mathrm{i}}}\right)^{-0.7}\left(\frac{d_{\mathrm{i}}}{d_{\mathrm{th}}}\right)^{0.11},$$

where the threshold $d_{\mathrm{th}}$ is defined as

$$d_{\mathrm{th}} = \frac{1}{\mathrm{Co}}\sqrt{\frac{\sigma}{g(\rho_{\mathrm{LO}} - \rho_{\mathrm{VO}})}}$$

and the confinement number is $\mathrm{Co} = 0.5$. $We_{\mathrm{LO}}$ is the liquid-only Weber number defined as

$$We_{\mathrm{LO}} = \frac{G^2 L_{\mathrm{ev}}}{\sigma\rho_{\mathrm{LO}}}$$

$G$ is the mass velocity per unit cross-sectional tube area of the fluid, $\sigma$ the fluid surface tension, $\rho_{\mathrm{LO}}$ and $\rho_{\mathrm{VO}}$ the liquid-only and vapor-only densities, $\mu_{\mathrm{LO}}$ and $\mu_{\mathrm{VO}}$ the liquid-only and vapor-only viscosities, $h_{\mathrm{LV}}$ the latent heat of vaporization, $L_{\mathrm{ev}}$ the heated length of the channel, and $d_{\mathrm{i}}$ the inner channel diameter. This correlation is valid for $0.35<d_{\mathrm{i}}<3.04$, $84<G<3,736$, $7<We_{\mathrm{LO}}<201,232$, $14.4<\mu_{\mathrm{LO}}/\mu_{\mathrm{VO}}<53.1$, $0.024<\rho_{\mathrm{VO}}/\rho_{\mathrm{LO}}<0.036$ and $22.7<L_{\mathrm{ev}}/d_i<177.6$.

This correlation predicted 94.4% of the data on which it was based and that of Wojtan et al. [40] for single microchannels within ±30% with a mean absolute error of 13.6%. It also predicted 100% of the split flow multi-microchannel data of Mauro et al. [46] and Agostini et al. [44] to within

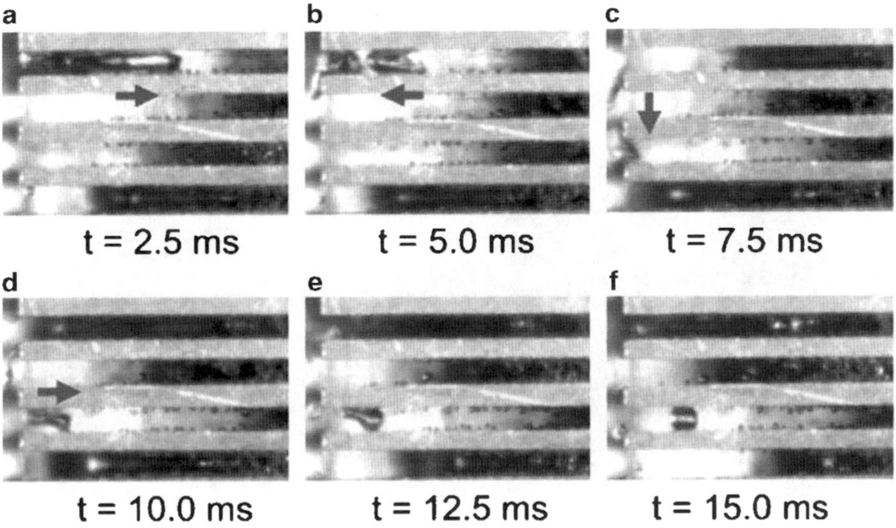


**Fig. 12.7** Dramatic effect that maldistribution can have on the heat transfer process [47]

±30% with a mean absolute error of 20.7 and 15.6%, while also predicting 91.9% of the once-through flow multi-microchannel data of Park [47] to within ±30% with a mean absolute error of 16.1%. Notably, like any empirical CHF correlation, it is only applicable to uniform heat flux boundary conditions.

V. *Flow distribution and flow stability*

Multi-microchannel flow boiling test sections can suffer from flow maldistribution among the numerous parallel channels, two-phase flow instabilities, and even backflow effects. The flow may in fact flow back into the inlet header and some channels may become prematurely dry from too low of an inlet liquid flow rate.

Figure 12.7 shows a sequence of video images to demonstrate back flow and parallel channel instability in multi-microchannel test section (something to be avoided). A slug bubble was observed at the inlet of the topmost channel in (a). If the flow in the channel is pushed upstream by a bubble growth downstream, the bubble goes back into the inlet plenum in (b), as there is no restriction at the channel inlet of the channel to prevent this. This reverse flow bubble quickly moves to one of the adjacent channels (c), and breaks down into smaller parts before entering these channels (d). Depending on its location, the inserted bubble becomes stagnant, (e) and (f), before moving forwards or backwards again.

Using microinlet orifices can completely prevent backflow, flow instabilities, and maldistribution. Figure 12.8 shows the maldistribution effect when no inlet orifice is used, with a large dry zone being visible in the top right corner. A CHF of only 115 W/cm$^2$ was achieved.

Figure 12.9 shows that maldistribution is avoided when making use of microinlet orifices at the entrance of each channel (created by placement of an

**Fig. 12.8** Without orifice by Park [47]

**Fig. 12.9** With orifice by Park [47]

insert in the entrance plenum), with heat fluxes in excess of 350 W/cm$^2$ being obtainable…this is equivalent to cooling of thirty-five thousand 100-W light bulbs per meter squared surface area! Hence, microscale flow boiling can dissipate very high heat fluxes as long as proper attention is paid to obtain

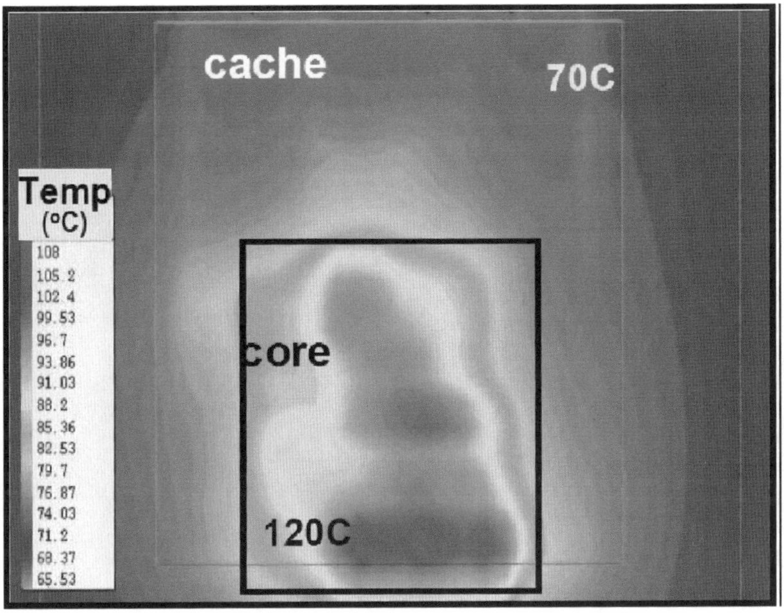


**Fig. 12.10** Temperature map of a typical chip [50]

good flow distribution and stable flow by the use of microorifices. Such orifices often make up about 1/3 of the pressure drop of the microevaporator cooling element, according to LTCM lab simulations; however, this pressure drop penalty is not all that significant when compared to the total pressure drop of the entire two-phase cooling loop, a topic discussed later in this chapter.

Figure 12.9 also depicts some flashing of the nearly saturated inlet liquid into a small fraction of vapor, visible at the left of the image in some channels. This flashing process is also an important feature of the inlet orifices as this takes slightly subcooled inlet liquid directly into the flow boiling regime without passing through the onset of nucleate boiling, the latter which requires a large wall superheat (with respect to the saturation temperature) to activate the boiling process and hence is to be avoided in electronics cooling because of the temperature overshoot involved.

VI. *Hot-spot management*

Nonuniform power dissipation across a chip leads to local hot-spots, resulting in elevated temperature gradients across the silicon die. These hot-spots could result in the degradation of reliability and performance of the chip [48], with a complete thermal breakdown of the chip also being possible. The reliability of a chip decreases by 10% for every 2°C rise in temperature [49]. An example of a temperature map showing the results of nonuniform power dissipation is given in Fig. 12.10. In this example, the temperature gradient is approximately 50°C.

From the numerous experimental data obtained in the last decade, various trends regarding microchannel two-phase flow boiling have been observed by

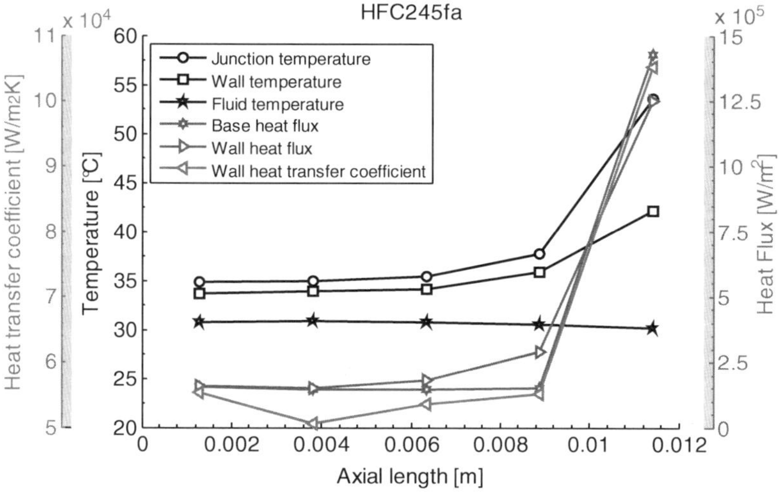


**Fig. 12.11** Thermal profile of a pseudochip [53]

Agostini and Thome [51]. One of these trends is that the local heat transfer coefficient for microchannel flow boiling at low to intermediate vapor qualities increases proportionally with the heat flux, essentially in the isolated bubble and coalescing bubble flow regimes but not in the annular flow regime as noted by Ong and Thome [26]. In general, it has been found that it approximately increases as $\alpha \propto q^{0.7}$. More recent results by Costa-Patry et al. [52], focusing on cooling of hot-spots of a pseudochip with 35 local heaters and temperature sensors cooled with a silicon multi-microchannel evaporator, have shown that this proportionality is closer to $q^{0.4}$, with conduction (heat spreading) effects within the evaporator being the main differentiating factor. One such result is shown in Fig. 12.11, which is the thermal profile of their pseudochip being cooled by a two-phase refrigerant evaporating in 135 parallel microchannels of 87 μm width engraved in opposite face of the silicon die. For a hot-spot heat flux in the outlet row 5 being ten times higher than the base heat flux in rows 1–4 (with fluid inlet at the left and exit at the right), the hot-spot heat transfer coefficient was measured to be two times higher with consequently a wall superheat of only 4.5 times higher [53]. *Hence, two-phase cooling has been proven experimentally to have a built-in passive hot-spot cooling mechanism, unlike single-phase cooling.*

Figures 12.12 and 12.13 show the results of simulations using a multipurpose internal LTCM lab code considering the presence of three hot-spots on a chip of 20 mm length. Base and hot-spot heat fluxes of 50 and 200 W/cm$^2$ were considered, respectively. HFC134a, water and 50% water–ethylene glycol were evaluated as working fluids with microchannels of 1.7 mm height and 0.17 mm width, fins 0.17 mm thick and a base of 1 mm thickness to the junction for a copper cooling element. The results, as expected, show an increase of heat transfer coefficient when using HFC134a two-phase flow boiling, resulting in a

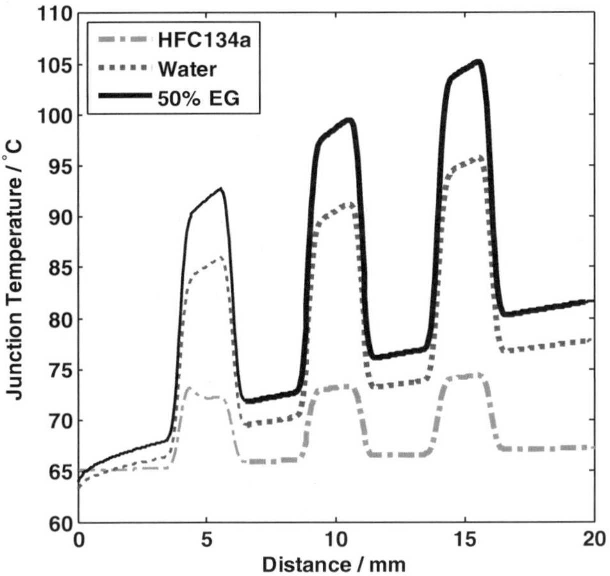


**Fig. 12.12** Junction temperature for nonuniform heat flux

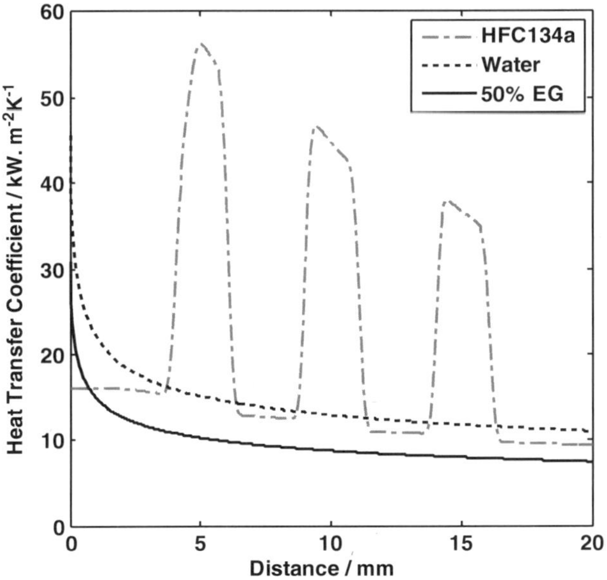


**Fig. 12.13** Heat transfer coefficient for nonuniform heat flux

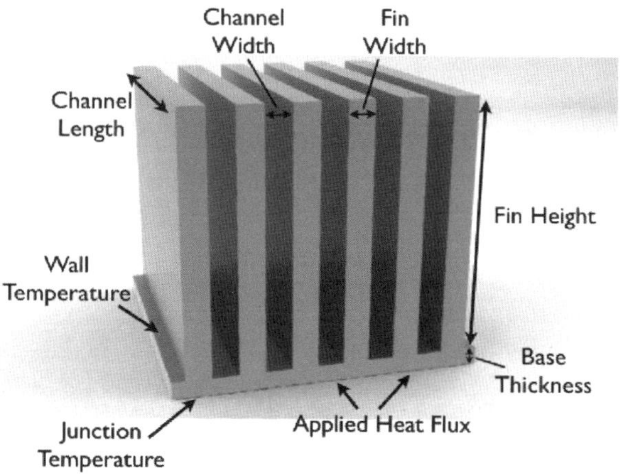


**Fig. 12.14** Schematic of multi-microchannel cooler

much lower increase of junction temperature than for the other two working fluids, which do not have an effect of the hot-spot on their heat transfer coefficient (except for the change in their physical properties with the locally rising liquid temperature).

VII. *Multi-microchannel evaporator/cooler simulations*

By using the latest two-phase correlations, discussed earlier, simulations can be performed to determine the performance of microevaporators for different operating situations (heat fluxes, fluids, fluid temperatures, etc.). Further, initial design estimations can also be determined by using such simulations, giving invaluable information to the thermal designer. This section shows how such a design tool (developed within the LTCM lab) can be used to make decisions regarding the type of coolant fluid, their thermodynamic conditions, and geometric characteristics. Aspects such as junction temperature, junction temperature uniformity, and pressure drop are discussed.

In the following simulations, heat fluxes will be varied from 30 to 150 W/cm$^2$ and mass fluxes from 300 to 1,000 kg/m$^2$ s (mass flux refers to the mass flow rate of coolant in kg/s divided by the cross-sectional area of the flow in m$^2$). It should be noted that an Intel Xeon chip generates a heat flux on the order of 50 W/cm$^2$. Nine fluids were simulated and compared, seven of which were refrigerants, the other two being pure water and a 50% water–ethylene glycol (EG) mixture. Ethylene glycol is of interest since it is added to water where environments are such that freezing becomes a concern.

The multi-microchannel cooler (MMC), a schematic of which is given in Fig. 12.14, consists of an integral piece of copper plate having fins 1.7 mm high, channel width and fin thickness of 0.17 mm, and a base thickness of 1 mm. The multi-microchannel will cool a chip having a footprint of 13.5 mm × 18.5 mm, a typical size for standard or future microprocessors. The flow configuration used for the MMC is a one inlet, two outlet configuration, i.e., a so-called split flow configuration. Flow enters at the center of the cooler and leaves at

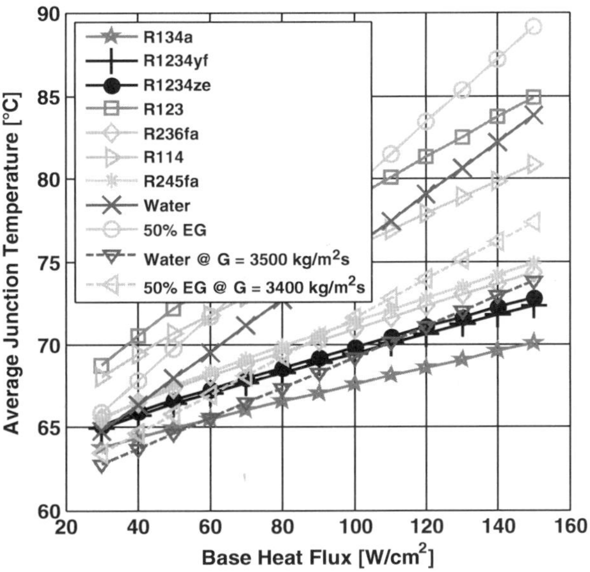


**Fig. 12.15** Junction temperature as a function of the base heat flux

the two opposite sides. This has the advantage of a much lower pressure drop and much higher CHF than a one inlet/one outlet (once-through) design. Furthermore, for the whole process to be energy efficient, the fluid temperatures entering the chip's cooler are set to 60°C. This has the advantage that no refrigeration chiller unit is required, with the excess heat being easily exchanged with ambient temperatures or being captured for use where this type and quality of heat is required by a secondary process.

*Junction temperature and junction temperature uniformity*

Figure 12.15 shows the length-averaged temperature of the junction of the MMC (viz. Fig. 12.14), which is calculated from the wall temperature at the base of the fins plus the conductive temperature difference across the copper die of 1 mm thickness. For the refrigerants, the average junction temperature for all the mass fluxes (an average for the range from 300 to 1,000 kg/m$^2$ s) was used since their values did not change significantly (0.1–2°C). For water and EG, the junction temperature variations at mass fluxes of 300 kg/m$^2$ s (for both) and 3,500 and 3,400 kg/m$^2$ s, respectively, are given. The diagram shows that the junction temperature is below the 85°C limit imposed on microprocessors for all fluids, except for EG at heat fluxes greater than 125 W/cm$^2$. These temperatures are directly related to the heat transfer coefficients. HFC134a has the lowest junction temperature over the range of heat fluxes, followed by HFO1234yf and HFO1234ze. The junction temperatures of water and EG are only comparable when their mass flow rates are increased to values of almost an order of magnitude greater than for the refrigerants.

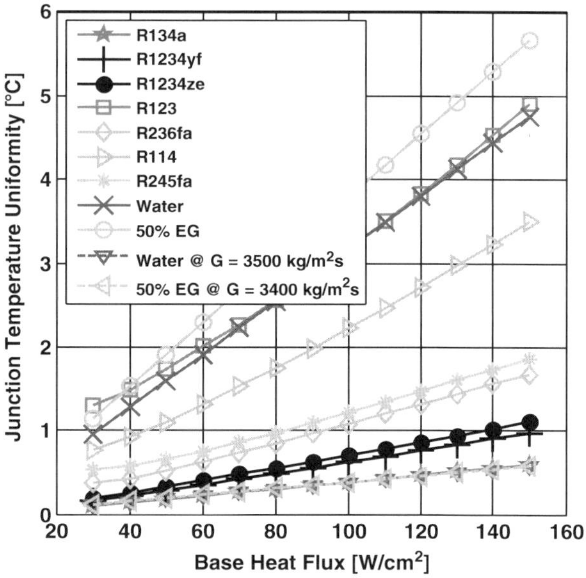


**Fig. 12.16** Junction temperature uniformity as a function of the base heat flux for a mass flux of 500 kg/m$^2$ s

Another thermal design criterion to consider is the uniformity of the junction (chip) temperature. This is an important aspect with regard to the cooling of integrated circuits and electronics, as too high a temperature gradient along their base surface will create an adverse nonuniform thermal stress. This could lead to the chip or electronics being damaged (silicon is very brittle) and also a breakdown of the thermal interface material.

The temperature uniformity can be expressed by taking the standard deviation of the temperatures at the junction surface along the length of the chip, calculated for a specific mass flux and all heat fluxes. The standard deviations of all the fluids are given in Fig. 12.16 as the junction temperature uniformity. The overall trend is a decrease in uniformity with an increase in heat flux. Once again, HFC134a has the best temperature uniformity (the lowest curve), with temperature variations of less than 1°C at the maximum base heat flux, while HCFC123 is consistently the worst refrigerant. The water–ethylene glycol mixture is worse than HCFC123, with water being only slightly better than HCFC123. However, by increasing the mass flux of water and EG to 3,500 and 3,400 kg/m$^2$ s, respectively (dashed lines in Fig. 12.16), their curves fall on top of that of HFC134a. This requires a considerable flow rate and will have a huge impact on pumping power requirements. After HFC134a, the new refrigerants HFO1234yf and HFO1234ze perform the best, with temperature variations always being 1.1°C or less along the length of the chip.

*Pressure drop*

Figure 12.17 shows the pressure drops for different fluids as a function of the base heat flux. For refrigerants, there is an increase in pressure drop with an

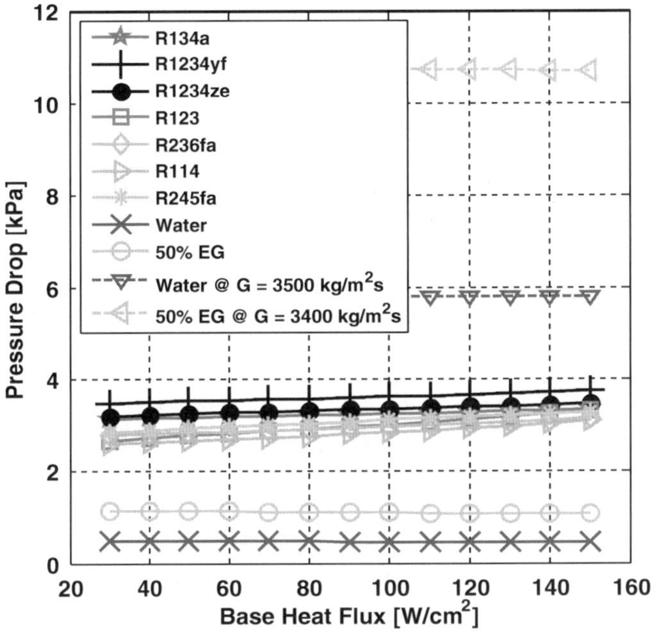


**Fig. 12.17** Pressure drop as a function of the base heat flux for a mass flux of 500 kg/m$^2$ s

increase in base heat flux. This is due to higher outlet vapor qualities being reached, where pressure gradients are greater, for higher heat fluxes. For single-phase water and EG, the opposite is seen where the pressure drop decreases slightly for an increase in heat flux. This is due to a decrease in fluid viscosity as the temperatures of these fluids are increased.

Pressure drops for the water can be much higher or much lower than for the refrigerants, depending on the allowable junction temperature nonuniformity (Fig. 12.16). Keeping the nonuniformity to a safe value of less than 2°C, one notices that the pressure drops of water and EG are, respectively, three and seven times larger than for most of the refrigerants. It should be remembered that the MMCs using refrigerants have an extra pressure loss at the inlet due to the use of an orifice to force good fluid flow distribution (this is included in the present calculations), which aids also in stabilizing the flow. This orifice represents about 40–60% of the total pressure drop in the present simulations and could be reduced to half of this by verifying flow stability with some actual test data. The pressure drop across the orifice is also used to flash a small fraction of the entering refrigerant, thus taking the heat transfer process directly into two-phase flow and avoiding both a single-phase cooling zone and the onset-of-nucleate boiling temperature overshoot.

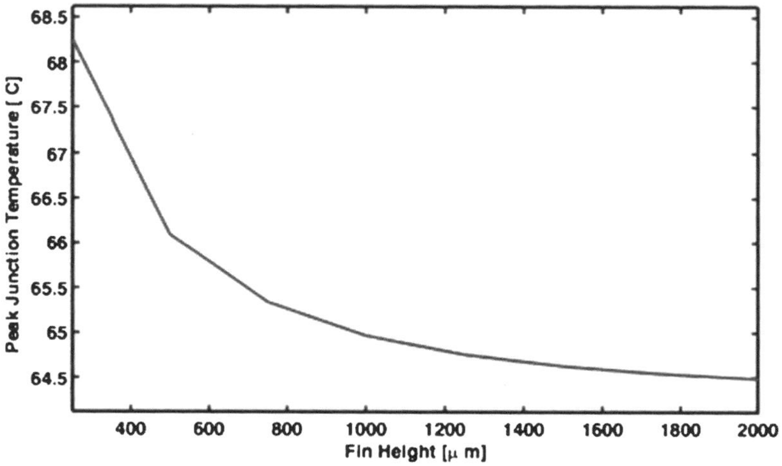


**Fig. 12.18** Effect of fin height on maximum junction temperature

It can be concluded, from the above simulations, that a two-phase refrigerant has a clear thermal advantage compared to single-phase water or a water–ethylene glycol mixture flow. Temperatures are much more uniform for lower flow rates, with single-phase water and EG only matching this when their flow rates are almost seven times higher than that of two-phase flow. Junction temperature increase with heat flux is also much lower for refrigerants, which is advantageous when considering microprocessor cycling or hot-spots.

*Geometric effects*

Geometric characteristics can be analyzed by making use of the simulation code. As an example, the fin height and channel width will be varied below while the fin width will remain fixed. This will show the optimal design for a given mass flux. Two parameters will be considered for the optimal design: the maximum junction temperature and the CHF. Figures 12.18 and 12.19 show these parameters as a function of the fin height.

It is seen that the CHF increases as the fin height increases while, at the same time, the maximum junction temperature decreases. The fin height of 1.7 mm presented in the previous simulations above was chosen on the basis of these results.

Figures 12.20 and 12.21 show the CHF and maximum junction temperatures as a function of the channel width. The choice of a channel width of 170 μm is not too bad when the maximum junction temperature is considered, as any further increase in width would not bring about much further reduction in temperature. For the maximum heat flux, however, a channel width of 100 μm might have been better, although this would be at the expense of additional pressure losses.

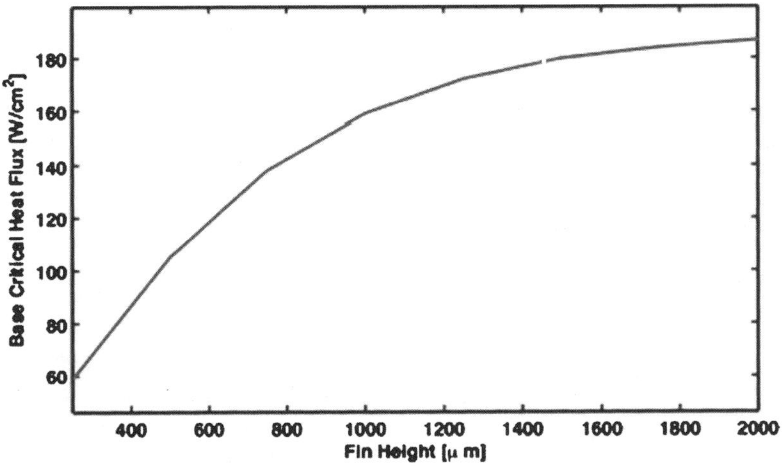


**Fig. 12.19** Effect of fin height on critical heat flux

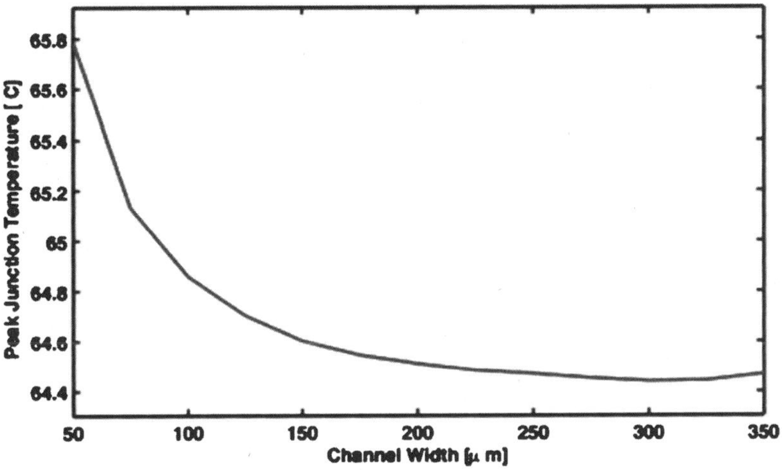


**Fig. 12.20** Effect of channel width on maximum junction temperature

VIII. *Remarks*

It was proven experimentally that, using a refrigerant evaporating at 60°C, microprocessors could be kept well below their 85°C limit, while removing heat fluxes in excess of 180 W/cm$^2$ or more [54]. Two-phase flow is also ideally suited for cooling of electronic hot-spots (local heat fluxes on CPUs up to 400 W/cm$^2$ or more) as heat transfer coefficients (thermal resistances) naturally increase (decrease) over hot-spot locations in the flow boiling process described earlier. This has the implications that electronics can have a more uniform temperature, implying that problems associated with

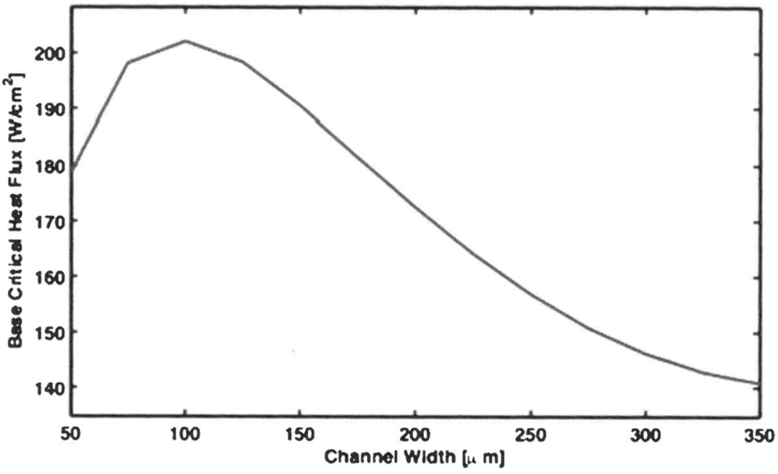


**Fig. 12.21** Effect of channel width on critical heat flux

adverse temperature gradients are greatly diminished and hence higher clock speeds can be utilized. The power requirements for removing the heat is also considerably lower than required for traditional air-cooling methods. This is due to the much larger heat carrying capacity refrigerants have over air, while also taking advantage of its latent heat.

Numerous other aspects of these microscale two-phase flows are under investigation in the LTCM lab, such as micro-PIV to characterize flow (with cavitation) through microorifices as small as 15–25 μm, transient aspects of the evaporation and bubble coalescence process, time-strip analysis of high-speed videos to discern information on the dry-out process and wave formation, flow pattern transition theory, etc. All these aspects are focused solely on the microevaporator. To see the actual effect on the cooling performance of a datacenter, such on-chip microevaporators need to be analyzed as part of a complete cooling cycle to determine their actual energetic characteristics.

## 12.4 Two-Phase MMC Cooling Cycle

Despite the numerous advantages water cooling might have with respect to air cooling, some negative aspects of making use of water as an on-chip cooling medium need to be addressed in a confrontation with two-phase on-chip refrigerant cooling [55]. In this section, the advantages two-phase refrigerant cooling has over single-phase water cooling for use with MMC are discussed on a system level. This does not mean that water on-chip cooling is not a good viable solution (*it is*), but it is the primary alternative cooling technology to two-phase on-chip cooling and hence the pros and cons should be discussed and some benchmark simulations

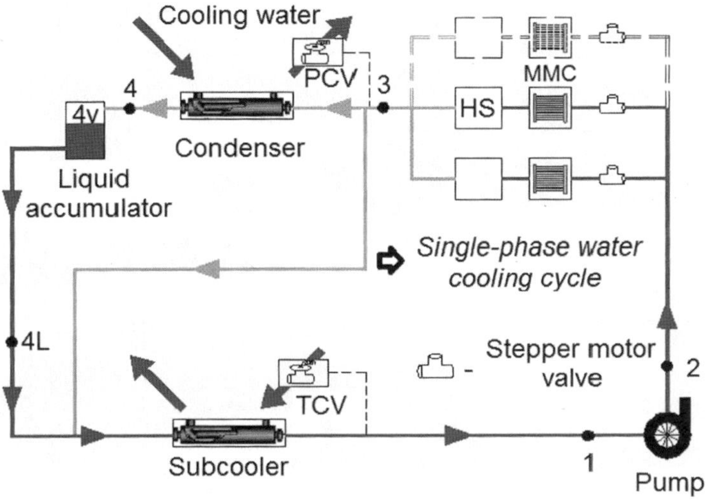


**Fig. 12.22** Schematic of the liquid pumping cooling cycle

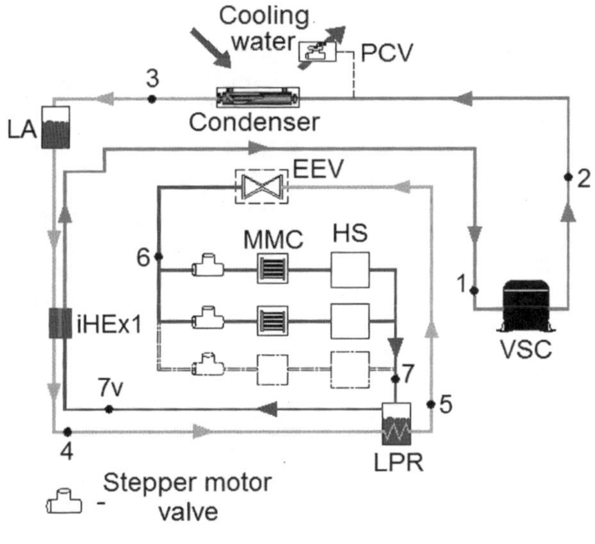


**Fig. 12.23** Schematic of the vapor compression cooling cycle

presented to see how these two solutions stand, and that is done here to help better understand the two-phase solution.

I. Cooling cycles
   Figures 12.22 and 12.23 depict potential two-phase cooling cycles, in which the cycle drivers are a liquid pump and a vapor compressor, respectively [56]. The goal is to control the chip temperature to a preestablished level by controlling

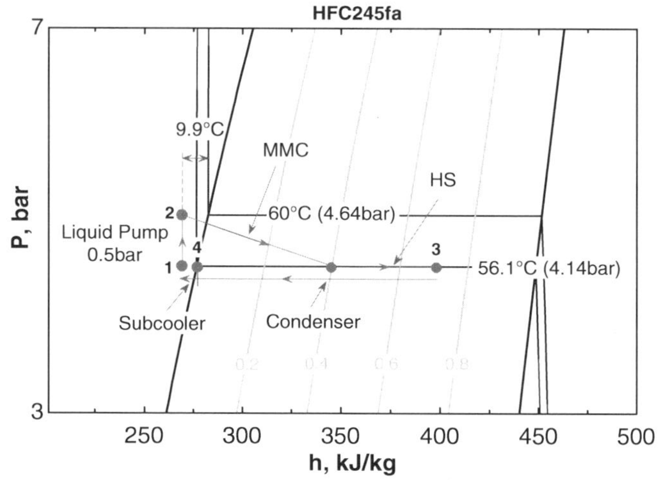


**Fig. 12.24** P-h diagram of the liquid pumping cooling cycle

the inlet conditions of the MMC (pressure, subcooling, and mass flow rate). It is imperative to keep the MMC outlet vapor quality below that of the critical vapor quality, which is associated with the CHF. Due to this exit vapor quality limitation (it is suggested not to surpass one-half of the critical vapor quality at the evaporator exit as a tentative safety margin), additional latent heat is available for further evaporation, which can be safely done in other low heat flux generating components, such as memory, DC/DC converters, etc.

Another parameter that must be controlled is the condensing pressure (condensing temperature). The aim is to recover the energy dissipated by the refrigerant in the condenser to heat buildings, residences, district heating, preheating boiler feedwater, etc. when that can be arranged and is viable.

The liquid pumping cooling cycle can be characterized in having low initial costs, a low vapor quality at the on-chip MMC outlet, a high overall efficiency, low maintenance costs, and a low condensing temperature. The heat spreader (HS) is for cooling of memory, etc., which are shown as "one" cooler here for simplicity purposes. This is a good operating option when the energy dissipated in the condenser is not recovered, typically during the summer season. However, the heat can still be recovered if there is an appropriate demand for low-quality heat (low exergy). On the other hand, the vapor compression cooling cycle can be characterized by a high condensing temperature (high heat recovery potential), a high range of controllability of the MMC inlet subcooling (characteristic of systems with variable speed compressors, VSC's, and electronic expansion valves, EEV's), a medium overall efficiency when compared with the liquid pumping cooling cycle. This is a good operating option when the energy dissipated in the condenser is recovered for other use, typically during the winter season when considering a district heating application (high exergy). Figures 12.24 and 12.25 show the Mollier diagram of the two cooling cycles

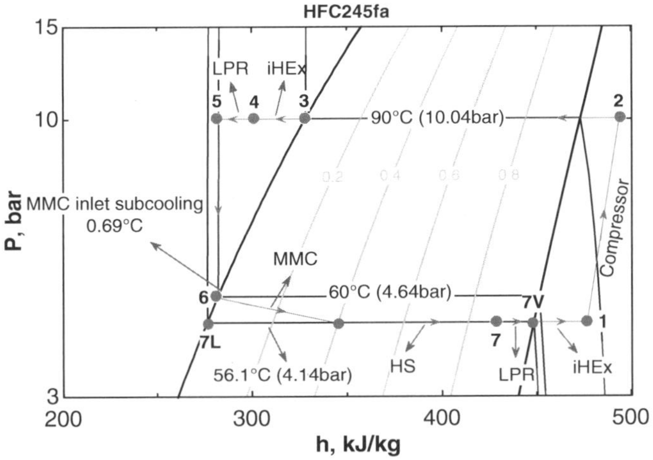


Fig. 12.25 P-h diagram of the vapor compression cooling cycle

Fig. 12.26 IBM blade with two microprocessors and heat generation capacity of 350 W [57]

for HFC245fa as the working fluid [56]. It can be noticed that the two cycles are differentiated mainly by the condensing temperatures with the evaporating temperature of the MMC set at 60°C (sufficient to keep the CPU below 85°C).

It is worth mentioning that the applicability of these cooling cycles is not restricted to only one microprocessor but can be applied to blade servers and clusters, which may have up to 70 blades per rack. Each blade, such as that of IBM shown in Fig. 12.26, can have two microprocessors with a heat generation capacity of 180 W. If the auxiliary electronics (memories, DC/DC, etc.) on the blade are included, the total heat generation per blade can be 350 W. Thus, the heat spreader (HS), shown earlier in Figs. 12.22 and 12.23, has the function to cool the auxiliary electronics, which can represent about 50% of the total heat load on the blade, but will have a larger surface area compared to the CPU and thus a lower heat flux.

Finally, when considering an entire rack, a very sizable heat load is generated, representing a good opportunity to recover the heat rejected. In this case, reuse of the heat removed from the blades for a secondary application will greatly reduce the $CO_2$ footprint of the system. For example, if we consider a data center with 50 racks packed with high-density blade servers (70 per rack), each blade dissipating 350 W, the total potential heat to be recovered will be 1.225 MW. Such a heat recovery system requires a secondary heat transfer fluid to pass through all the condensers (either water or a refrigerant) to transport the heat to its destination (to the environment or into a heat recovery system).

II. Case study: two-phase cooling cycle simulation

A detailed inhouse simulation code was developed in the LTCM lab to design and evaluate the performance of the liquid pumping and vapor compressor cooling cycles presented earlier, under steady-state operating conditions. The simulation takes into account the design of the condenser and subcooler, evaluates the performance of the MMC and various component coolers for a given heat load, and calculates the power consumption to drive the cooling cycle. The pressure drop of each component and piping are also calculated. Table 12.2 shows the principal methods used in the simulation. Most likely, this is the most detailed such code available to date and is based on extensive experience of the LTCM lab on single- and two-phase flows and heat transfer and numerous projects with lab sponsors.

The simulations consider the following input data: (1) the geometrical parameters of the MMCs and the heat exchangers, (2) the heat required to be absorbed by the MMCs and heat spreaders, (3) the evaporation temperature and subcooling at the MMC inlet, (4) the condensing temperature at the condenser inlet (only for the vapor compression cooling cycle), (5) the inlet and outlet water temperature of the secondary fluid for the condenser and the subcooler, and (6) the length, diameter, and direction of the pipes and elbows joining the components.

A comparison of five simulated cases considering different working fluids and cooling cycles is presented here. For all simulated cases, i.e., 1–4, which considers the liquid pumping cooling cycle, and 5, which considers the vapor compression cooling cycle, the design is such that the total cooling cycle pressure drop is about 1.5 bar. The design constraints on the condenser and the subcooler are that the pressure drops in the working and secondary fluids are, respectively, 0.1 and 1.5 bar. The difference among cases 1–4 is the working fluid used. They are, respectively, HFC134a, HFO1234ze, water, and brine (50% water–ethylene glycol mixture). For case 5, HFC134a is considered. The internal diameter of piping on the blade was considered to be 3 mm for all cases. This size is ideal for confined spaces as it can easily be bent to conform to the required profile of the system needing to be cooled. The analysis of results

**Table 12.2** Methods in the simulation

| Component | Type | Method |
|---|---|---|
| MMC | Multi-microchannel | Heat transfer coefficient by Thome et al. [24] |
| | | Pressure drop as suggested byRibatski et al. [31] |
| | | Critical heat flux by Revellin and Thome [39] |
| Condenser (tube-in-tube) | Inner tube: spiral μ-fin (single-phase flow) | Heat transfer coefficient by Meyer and Olivier [58a] |
| | | Pressure drop by Meyer and Olivier [58b] |
| | Inner tube: spiral μ-fin (two-phase flow) | Heat transfer coefficient by Cavallini et al. [59] |
| | | Pressure drop by Cavallini et al. [59] |
| | Annulus: smooth (single-phase flow) | Heat transfer coefficient by Dittus and Boelter [60] |
| | | Pressure drop by Blasius [61] |
| Subcooler (tube-in-tube) | Inner tube: ribbed (single-phase flow) | Heat transfer coefficient by Ravigururajan and Bergles [62] |
| | | Pressure drop by Ravigururajan and Bergles [62] |
| | Annulus: smooth (single-phase flow) | Heat transfer coefficient by Dittus and Boelter [60] |
| | | Pressure drop by Blasius [61] |
| Straight horizontal pipes | Adiabatic (single-phase flow) | Pressure drop by Blasius [61] |
| | Adiabatic (two-phase flow) | Pressure drop by Muller-Steinhagen and Heck [33] |
| Straight vertical pipes (upward) | Adiabatic (single-phase flow) | Pressure drop by Blasius [61] and Azzi et al. [63] |
| | Adiabatic (two-phase flow) | Pressure drop by Taitel et al. [64], Barnea et al. [65], Barnea [66], and Liu and Wang [67] |
| Straight vertical pipes (downward) | Adiabatic (single-phase flow) | Pressure drop by Blasius [61] and Azzi et al. [63] |
| | Adiabatic (two-phase flow) | Pressure drop by Barnea et al. [65], Barnea [66], and Perez-Tellez [68] |
| Elbow (horizontal) | Adiabatic (single-phase flow) | Pressure drop by Spedding et al. [69] |
| | Adiabatic (two-phase flow) | Pressure drop by Azzi et al. [70] |
| Elbow (vertical) | Adiabatic (single-phase flow) | Pressure drop by Spedding et al. [69] and Azzi et al. [63] |
| | Adiabatic (two-phase flow) | Pressure drop by Azzi et al. [63] |

are developed taking into account the thermal-hydraulic performance, the charge (mass) of working fluid, the volume of the heat exchangers, and the power consumption of the cooling cycles.

A blade center containing 16 blades is taken into consideration for the total heat load. Each blade, for example, as that shown in Fig. 12.26, presents two electronic systems in parallel, with each system being composed of one

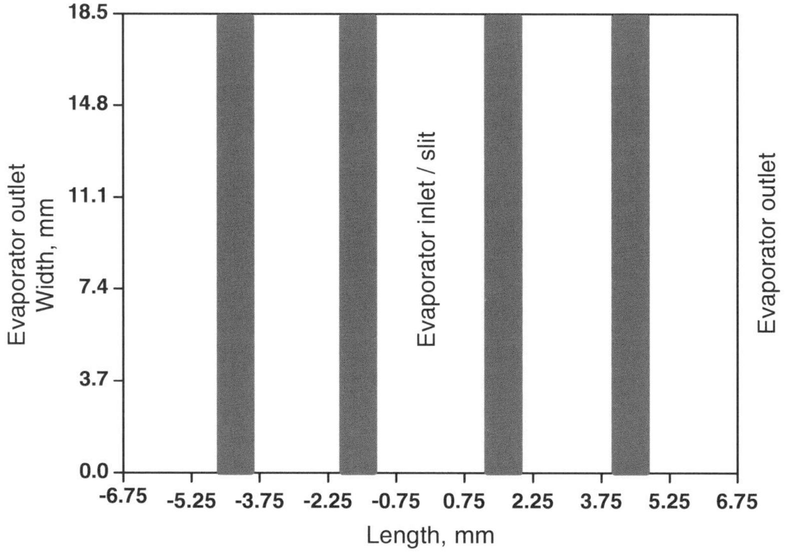


**Fig. 12.27** Hot-spots on the microprocessor

microprocessor (90 W of heat load for a footprint of 13.5 mm × 18.5 mm) and the auxiliary electronics (55.6 W of heat load). Four hot-spots, each with 18.5 mm width and 1 mm length, were considered per microprocessor (viz. Fig. 12.27). The microprocessors have a base heat flux of 17.9 W/cm$^2$ and each hot-spot having a value of 78.7 W/cm$^2$.

Table 12.3 shows the input parameters for the simulations. The other thermodynamic parameters required to determine the total energy balance of the cycles come from the linkage to the methods shown earlier in Table 12.2.

It is worth pointing out that the condenser and subcooler are tube-in-tube heat exchangers (double pipe) with spiral microfins on the internal surface of the inner tube with a smooth external surface, chosen to keep the present simulation simple. For both heat exchangers, the objective is to find its length, while the other geometrical parameters were held fixed. In the annulus of these exchangers, water is considered as the secondary fluid. The MMC had the same microchannel dimensions and split flow arrangement as in the previous section. It is important to mention that the required flow rate for the two-phase refrigerant cycles (cases 1, 2, and 5) is chosen such that the outlet vapor quality is 30%. This exit quality also ensures that the critical quality, calculated from the CHF, is never reached ($x_{\text{crit}} = 0.98$). This ensures that, for two-phase flow, the MMC will not burn out, thus giving it a large margin of safety. The flow at the entrance of the MMC is saturated liquid (0 K subcooling). The temperature uniformities obtained by the MMC simulations were 1.52 K for HFC134a and 1.86 K for HFO1234ze. These low values are due to the near ideal matching of the tandem

**Table 12.3** Input data

| Component | Working fluid | Input data |
|---|---|---|
| Auxiliary electronics | All of them | 55.6 W per half blade |
| MMC | HFC134a and HFO1234ze | Inlet evaporating temperature = 60°C |
| | | Inlet subcooling = 0°C |
| | | Outlet vapor quality = 30% |
| | | $Q_{MMC}$ = 90 W per MMC |
| | Water and brine | Inlet temperature = 60°C |
| | | Outlet temperature = 62°C |
| | | $Q_{MMC}$ = 90 W per MMC |
| Condenser (tube-in-tube) | HFC134a and HFO1234ze | Inlet condensing temperature = 95°C (case 5) |
| | Secondary fluid: water | Inlet temperature = 15°C |
| | | Outlet temperature = inlet condensing temperature—10 K |
| | | Outlet temperature = inlet condensing temperature—5 K (case 5) |
| Subcooler (tube-in-tube) | HFC134a and HFO1234ze | Inlet vapor quality = 0% |
| | Secondary fluid: water | Inlet temperature = 15°C |
| | | Outlet temperature = subcooler inlet temperature—10 K |
| Compressor | HFC134a | Isentropic compression |
| | | Condensing temperature = 95°C |
| Liquid pump | All of them | Isentropic pumping |

fall in the local flow boiling heat transfer coefficient and saturation temperature (pressure) along the MMC.

The flow rate for the single-phase cooling cycles (cases 3 and 4) is chosen so that the rise in water/brine temperature from inlet to outlet of the MMC is 2 K, which result in temperature uniformities on the chip of 2.53 and 3.13 K for water and brine as working fluids, respectively. The actual temperature rise could be more, depending on the server manufacturer's design specifications. Increasing this temperature difference will decrease the water/brine flow rate for its simulation, and hence also reduce its pressure drop and pumping power accordingly, but will induce greater temperature gradients along the chip.

Figures 12.28–12.31 and Table 12.4–12.5 show the simulation results for the five cases. The power consumption of the drivers is given in Fig. 12.28, with the pressure drop contribution of each component of the cooling cycles being given in Fig. 12.29 (those of the two-phase units include the pressure drop of the inlet orifices). The heat exchanger volume, the mass of working fluid, and the water mass flow rate of the secondary fluid are given in Fig. 12.30, while the shear stress exerted on the pipe walls and the average working fluid velocities, both at the liquid pump outlet, are given in Fig. 12.31. Table 12.4 shows the geometric parameters of piping used for each cycle to guarantee the design constraint of

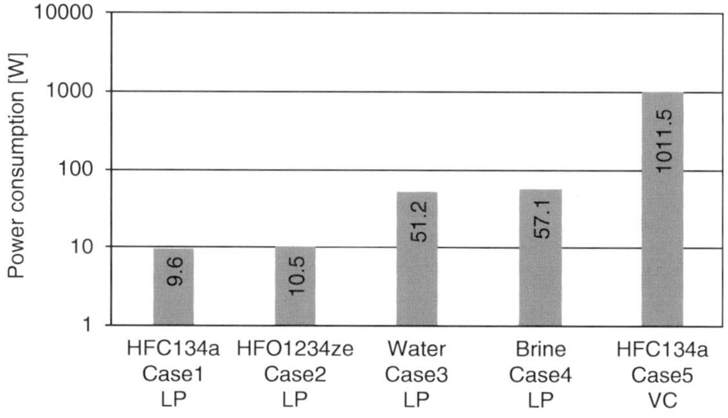


**Fig. 12.28** Power consumption of the drivers (*LP* liquid pump, *VC* vapor compressor)

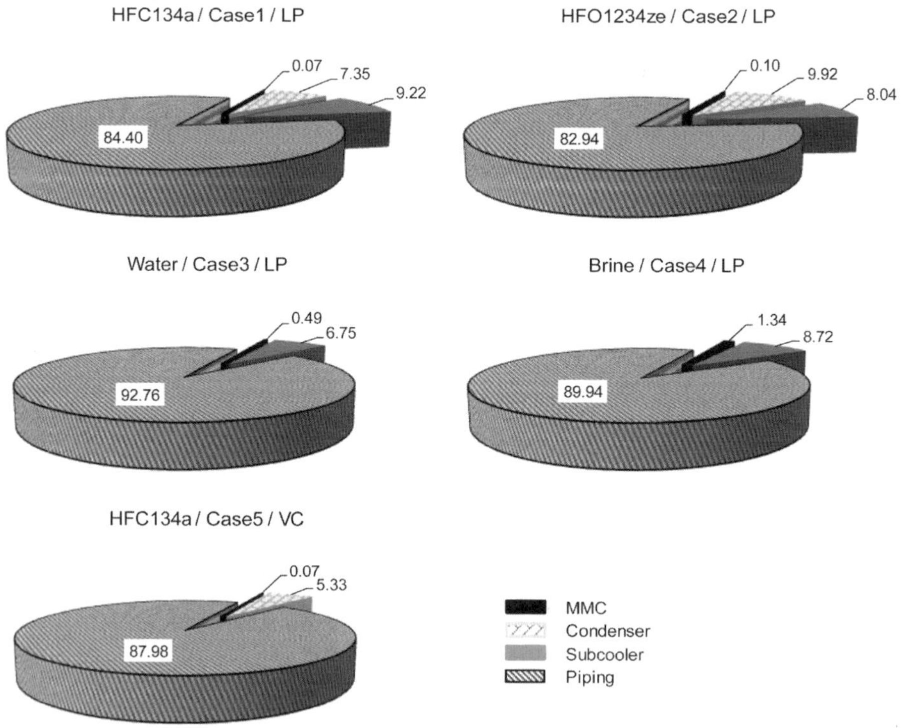


**Fig. 12.29** Percentage breakdown of pressure drop by component

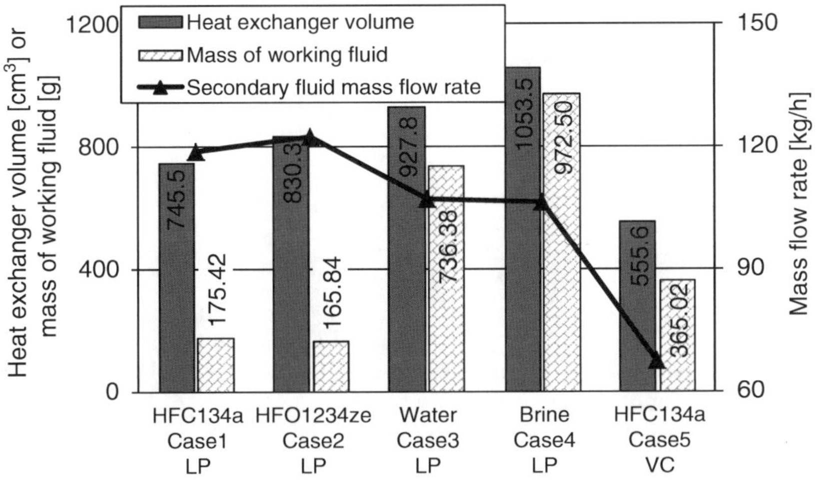


**Fig. 12.30** Heat exchanger volume, mass of working fluid, and secondary fluid mass flow rate

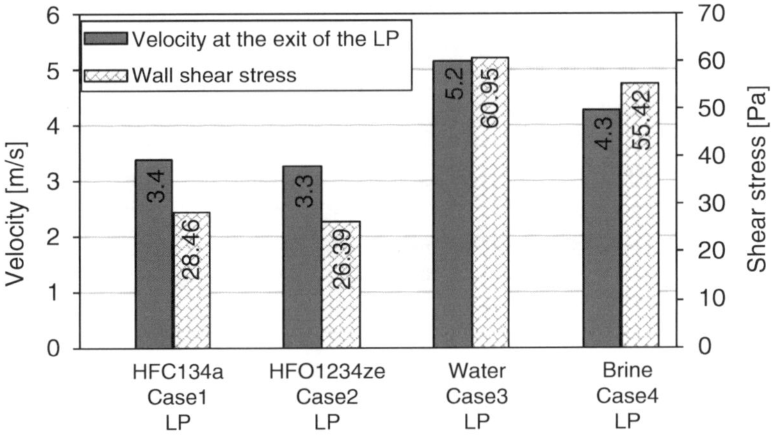


**Fig. 12.31** Shear stress on the pipe walls and working fluid velocities at the exit of the liquid pump

**Table 12.4** Cooling cycle piping designs and pressure drop on the blade

| Cases | 1 | 2 | 3 | 4 | 5 |
|---|---|---|---|---|---|
| Cooling cycle | LP | | | | VC |
| Working fluid | HFC134a | HFO1234ze | Water | Brine | HFC134a |
| Internal diameter of piping on the blade (mm) | 3.0 | 3.0 | 3.0 | 3.0 | 3.0 |
| Internal diameter of piping for single-phase flow (mm) | 4.9 | 5.1 | 9.2 | 10.6 | 5.2 |
| Internal diameter of piping for two-phase flow (mm) | 8.4 | 8.6 | – | – | 8.5 |
| Pressure drop in the piping on the blade (bar) | 0.04 | 0.05 | 0.25 | 0.42 | 0.04 |

**Table 12.5** Heat exchanger simulation results

| Cases | 1 | 2 | 3 | 4 | 5 |
|---|---|---|---|---|---|
| *Condenser* | | | | | |
| Outlet temperature of secondary fluid (°C) | **48.3** | **47.6** | – | – | **90.0** |
| Heat transfer rate (W) | 4195.0 | 4302.5 | – | – | 5670.7 |
| *Subcooler* | | | | | |
| Outlet temperature of secondary fluid (°C) | **48.0** | **47.0** | **52.8** | **52.8** | – |
| Heat transfer rate (W) | 473.8 | 367.2 | 4710.4 | 4716.3 | – |
| Total heat transfer rate (W) | **4668.8** | **4669.7** | **4710.4** | **4716.3** | **5670.7** |

1.5 bar for the total cooling cycle pressure drop. To put things into perspective, Table 12.4 also includes the pressure drop of the blade only. Finally, Table 12.5 shows the outlet temperature of secondary fluid and the heat transfer rate obtained in the heat exchangers, i.e., condenser and subcooler.

Figure 12.29 shows that, for all the cycles considered here, more than 80% of the pressure drop of the cycle is attributed to the piping. Only 5–18% of the contribution comes from the subcooler and condenser. It can be seen that the contribution of the microchannel coolers are negligible (and hence use of inlet orifices does not have much of an energetic penalty!).

Comparing case 1 with case 3, HFC134a and water liquid pumping cooling cycles, it is seen that when using water a larger internal diameter of piping (9.2 mm, viz. Table 12.4) for the system is required, while the overall power consumption is still 5.3 times higher than for the two-phase HFC134a refrigerant cycle (Fig. 12.28). The larger pipe diameter for case 3 is necessary to guarantee the 1.5 bar pressure drop design constraint, with the larger power consumption being a consequence of the much higher mass flow rate of water than refrigerant. The pressure drop on the blade for case 1 was six times lower than that of case 3.

As a point of interest, if the junction temperature uniformity of case 1 were imposed on case 3 (1.52 K), water flow rates will need to be increased to very high values (1,090 kg/h), which results in a pressure drop of 95 bar, just for the section on the blade. To limit this pressure drop to a manageable level (1 bar), the piping on the blade would need to be increased from 3 to 9.5 mm, which is highly impractical for the confines of the blade.

Another point to consider is that the internal diameter of the piping on the blade for case 1 can be decreased even further since its pressure drop is very low. By decreasing it to 2 mm, the pressure drop increases to 0.26 bar, which is on par with the value of case 3. The smaller diameter piping is inherently more flexible, making it easier to conform to the layout of the electronic components on the motherboard, which leads to less stress exerted on the blade, reduced weight (cost), and reduced refrigerant charge.

Case 2, which considers the new environmentally friendly working fluid HFO1234ze, showed a small increase in power consumption (1.1 times) and internal diameter of piping when compared with case 1. Therefore, the new refrigerant would be a suitable replacement for HFC134a, which will eventually be phased out as noted earlier in this chapter. The brine liquid pumping cooling cycle (case 4), when compared with case 3, presented a higher power consumption and internal diameter of piping. It is worth mentioning that the use of brine as working fluid is necessary for conditions of subzero ambient temperatures. Otherwise, one would plan to avoid the use of a brine and charge the water into the system after its installation on site.

Figure 12.30 show that the volume of the heat exchangers, when using single-phase water or brine as the working fluid (cases 3 and 4), is larger than when using two-phase refrigerant as the working fluid (cases 1, 2, and 5), which implies a higher investment cost. This is due to the overall thermal resistance of the heat exchangers for the single-phase cases being higher than that for the refrigerant two-phase cases. The charge or mass of working fluid required by the two-phase cooling cycles is also much lower than for the single-phase cooling cycles. The mass of working fluid for case 1 is 4.2 and 5.5 times lower than for cases 3 and 4, respectively. The diagram also shows that the required flow rate for the secondary fluid, needed to maintain the design constraints, is the lowest for case 5, i.e., vapor compression cooling cycle. This would lead to lower operating costs for the secondary fluid cycle.

Within the literature, it is stated that water velocities inside copper pipes should not exceed 1.2 m/s to avoid issues regarding erosion [71]. Some even suggest that velocities should be kept below 1 m/s for hot water [72]. This velocity can be represented by the shear stress exerted on the tube wall. It can then be used on a comparative basis with other fluids. It should be mentioned that shear stress is not the main reason for erosion. It also involves the interaction the fluid has with pipe connecting pieces (forming of local vortices), chemical stability of the working fluid, etc. However, the refrigerants used in this study are known to be very stable and highly compatible with copper and other materials, which would then justify this comparison. Figure 12.31 shows the shear stress on the pipe walls and the cross-sectional average working fluid velocities at the exit of the liquid pump for all the cases. The graph shows that for the single-phase cooling cycles (cases 3 and 4), the velocities and shear stresses are almost twice as high than those of the two-phase cooling cycles (cases 1 and 2).

Finally, the results for case 5 (HFC134a vapor compression cooling cycle) show a significant increase in pumping power compared to the other cases. However, what is of interest is the lower secondary fluid mass flow rate and heat exchanger volume (viz. Fig. 12.30) and the higher outlet temperature of secondary fluid (viz. Table 12.5, bold values), when compared with all the cases evaluated. This implies that a lower pumping power for the secondary fluid is required while material costs of the condenser is lower due to the lower volume and that there exists a higher potential for energy recovery (high exergy). As mentioned beforehand, this cycle would be a good operating option when the energy dissipated in the condenser is recovered for other uses, for example, into a district heating network. It is also worth mentioning that the higher heat transfer duty in the condenser (viz. Table 12.5, bold values) is associated with the additional work imparted by the compressor on the fluid. This subject will be explored in the following section.

## 12.5 Waste Heat Recovery

The main advantage of making use of on-chip cooling (water or two-phase) for the cooling of datacenters is that the heat gained from cooling the chips can be easily reused elsewhere. This is because the heat removal process is local to the chips, thus minimizing any losses to the environment, which is prone to traditional air-cooling systems. Second, the heat transfer coefficients of the microchannel cooling elements is very high, such that its low thermal resistance allows the coolant to enter at temperatures as high as 60–70°C and still maintain the chip's operating temperature below 85°C, which in turn allows the heat to be recovered at near the same temperature. The opportunity thus exists to reuse the datacenter's waste heat in secondary applications instead of being disposed of into the atmosphere. This not only has the potential to reduce datacenter energy costs, but also to reduce its carbon footprint and hence, its environmental impact. (The effect of being more efficient is a drop in $CO_2$ emissions, which, considering the taxation of such emissions, will bring about even further savings).

A potential secondary application to make use of the waste heat is a thermal power plant. This secondary application has the advantage that it requires the heat yearlong, unlike district heating, for example, which is dependent of seasonal changes. Typical coal thermal power plants have thermal efficiencies on the order of 33% and are one of the main causes of $CO_2$ emissions. Therefore, any improvement in efficiency will not only have less impact on the environment, but also bring about financial savings. A power plant making use of the heat from a datacenter could potentially improve its efficiency from 1 to 3%, depending on the temperature of this heat.

The difficult problem with recovering the waste heat of a datacenter is not in the quantity of heat available, but rather in the quality of heat. Currently, heat is being dumped into the atmosphere at temperatures of about 40°C when using traditional air-cooling methods. This is because chip temperatures are being cooled at 15–20°C. Due to the effective cooling of chips when using on-chip cooling, fluid approach temperatures of about 60°C can be realized, while removing high heat fluxes and keeping chip temperatures below 85°C [54]. These higher temperatures translate to a wider spectrum of applications that can make use of this energy while at the same time increasing its economical value.

Reusing heat at 60°C can easily be achieved by making use of a liquid pump to drive the refrigerant. Higher temperatures can, however, be achieved by making use of a vapor compression cycle, which can still cool chips with a saturation temperature near 60°C, but is capable of rejecting heat at temperatures in excess of 90°C. This, however, comes at the expense of additional energy expenditure of the compressor, which is much higher than that of the liquid pump cycle. A trade-off, therefore, exists between the energy consumed by the datacenter and the efficiency improvements gained by the secondary process, both being dependent on the quality of the cooling system's rejected heat.

### *12.5.1 Case Study: Datacenter and Power Utility*

Marcinichen et al. [56] characterized a vapor compression and a liquid pumping cooling cycle using on-chip cooling with multi-microchannel evaporators. According to them, the vapor compression cycle is characterized by a high condensing temperature on the order of 90°C (high heat recovery potential, i.e., high exergy) and a high range of controllability of the microevaporator inlet subcooling (characteristic of systems with a variable speed compressor and an electric expansion valve), but it has a high energy consumption. The liquid pumping cycle is characterized as having a lower condensing temperature on the order of 60°C (similar to the evaporating temperature), but its energy consumption is low.

The case study performed here considered that the data center uses on-chip cooling to cool the servers, with the heat recovered being redistributed to a power plant. The datacenter is modeled as IT equipment and a cooling cycle consisting of an evaporator (on-chip cooling elements), a condenser, a fluid driver (compressor or liquid pump), and an expansion valve in the case of a vapor compression system. The power plant was assumed to be a thermal Rankine cycle consisting of a boiler, a high- and low-pressure turbine, a condenser, a low-pressure and high-pressure feedwater pump, and a feedwater heater. The feedwater heater receives heat from steam tapped after the high-pressure turbine. The optimal pressure for tapping the steam is calculated to obtain maximum thermal efficiency. The datacenter waste heat is injected into the Rankine cycle after the condenser and prior to the feedwater heater. This is shown schematically in Fig. 12.32.

The operating conditions for both cycles are listed in Table 12.6. A first-law analysis is performed on the two cycles, showing their overall performances. The following simplifying assumptions are made:

- No pressure drop in components
- Isentropic compression
- Isentropic pumping
- Isenthalpic expansion
- 100% Exchanger efficiency

A. *Power plant*

Heat dissipated by the datacenter can be reused by a power plant. Since the waste heat of the datacenter is of low quality, it can only be inserted after the condenser of the power plant. This would then increase the temperature of the boiler water leaving the condenser (46°C typically) to a maximum temperature as defined by the condensing temperature of the datacenter cycle. Therefore, any additional heat added to the power plant's cycle will result in less fuel needing to be burned, thus saving fuel and reducing the $CO_2$ footprint of the power plant.

Figure 12.33 shows the efficiency improvement of such a power plant as a function of the datacenter cycle's condensing temperature, assuming there is an optimal match between the power plant's preheating duty and the heat duty recoverable from the datacenter. The figure shows that the higher the condensing temperature, the greater the efficiency improvements. The efficiency of the plant

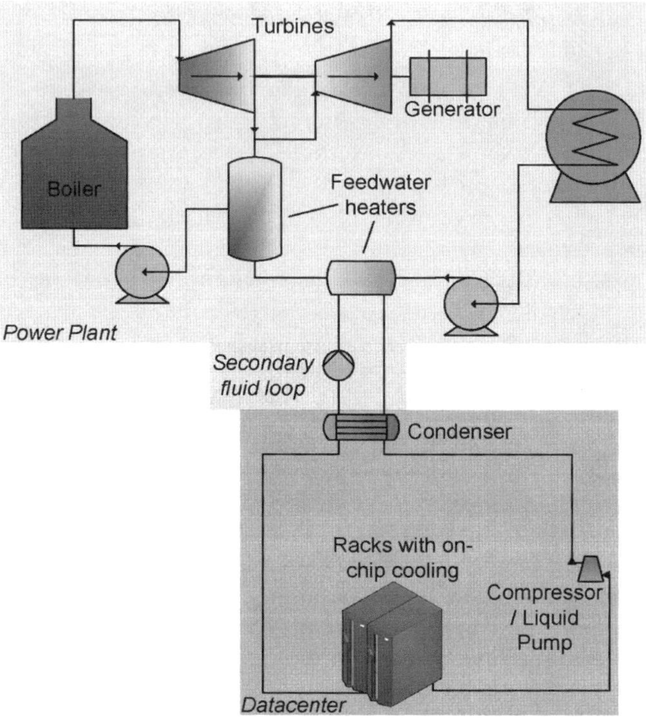


**Fig. 12.32** Datacenter integrated in a power utility

**Table 12.6** Operating conditions for the power utility and datacenter cooling cycle

| | Power utility | Datacenter cooling |
|---|---|---|
| Boiling/evaporation pressure | 16.55 MPa | 1.681 MPa |
| Turbine/compressor | | |
| *Inlet pressure* | 16.55 MPa | 1.681 MPa |
| *Inlet temperature* | 538°C | 70°C |
| Condensing temperature | 46°C | 60–100°C |

can be improved by up to 2.2% if the datacenter's waste heat is reused in the power plant. By using a liquid pumping cycle in the datacenter, condensing temperatures of 60°C can be reached since this would imply that the evaporating temperature on the chip is also about 60°C. For higher condensing temperatures, a vapor compression cycle would be required.

In terms of $CO_2$ footprint, Fig. 12.34 shows the reduction in the amount of $CO_2$ per kilowatt-hour output per year as a function of the datacenter condensing temperature. Also shown on the graph is the amount of $CO_2$ saved per kilowatt-hour output per year. These values assume that coal is used as the source of the power plant's energy and will be less for other types of fuel. Therefore, if a power plant with an output capacity of 500 MW were

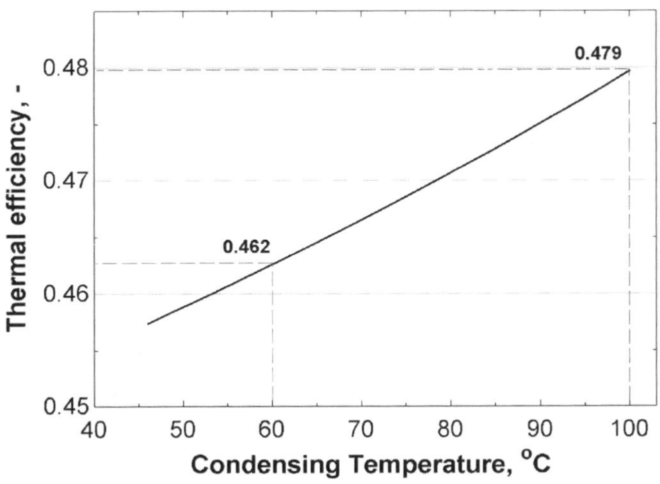


**Fig. 12.33** Thermal efficiency of power plant

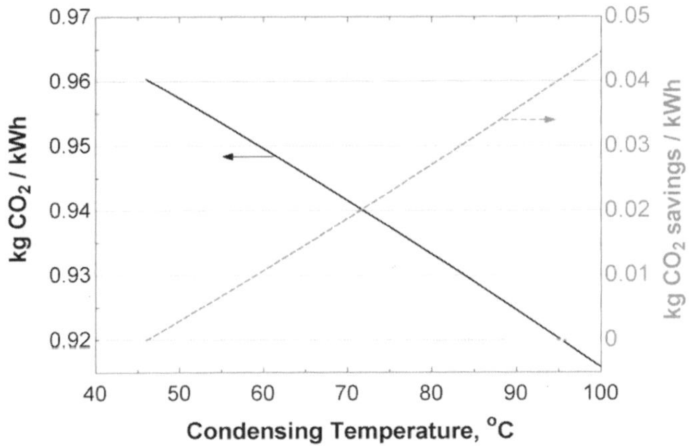


**Fig. 12.34** $CO_2$ footprint and savings per kilowatt-hour

considered (90 MW of waste heat recovery from the datacenter), the savings in $CO_2$ would be on the order of 195,000 tons of $CO_2$ per year. This could potentially save about \$3 million per year if a carbon tax of \$15 per ton of $CO_2$ [73] were considered.

B. *Datacenter*

Due to datacenter growth, we will likely see many datacenters in excess of 100,000 servers (viz. Fig. 12.35) in the future. Therefore, the simulations below will be based on a datacenter containing 100,000 servers, with each server assumed to be dissipating 300 W of Joule heating, which includes the main processor and auxiliary electronics (memories, DC–DC converters, hard drives, etc.). Figure 12.36 shows a graph of the total power supply required by

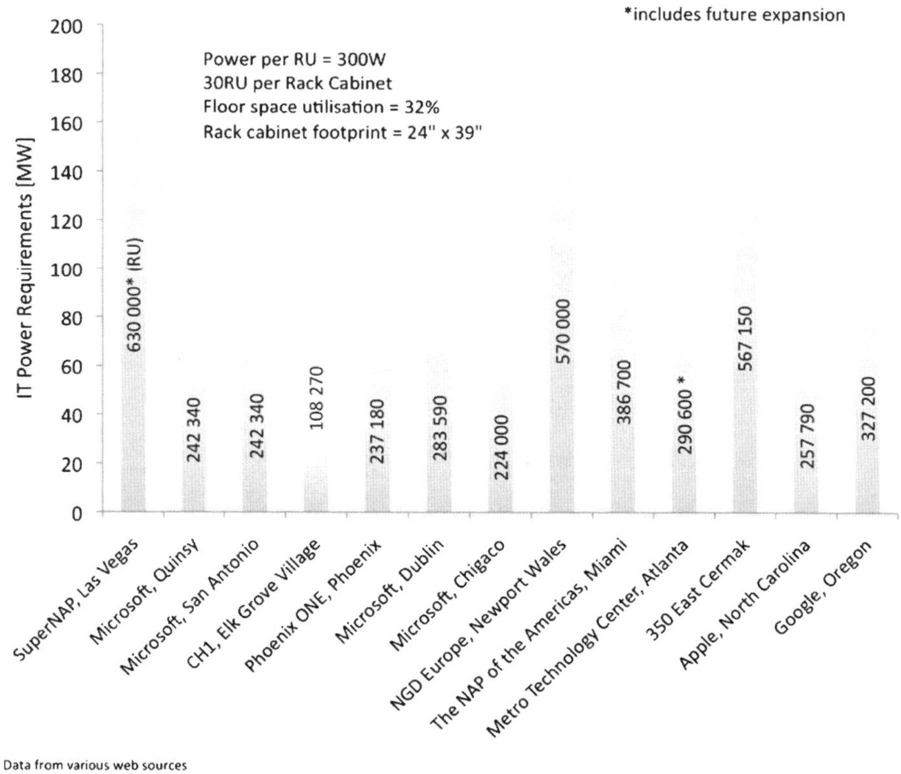


**Fig. 12.35** Datacenter size and information technology (IT) power requirements

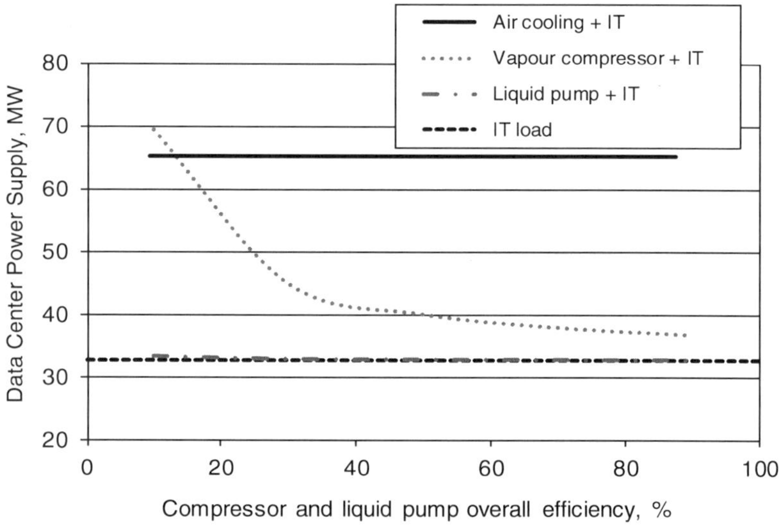


**Fig. 12.36** Datacenter power supply—IT and cooling system

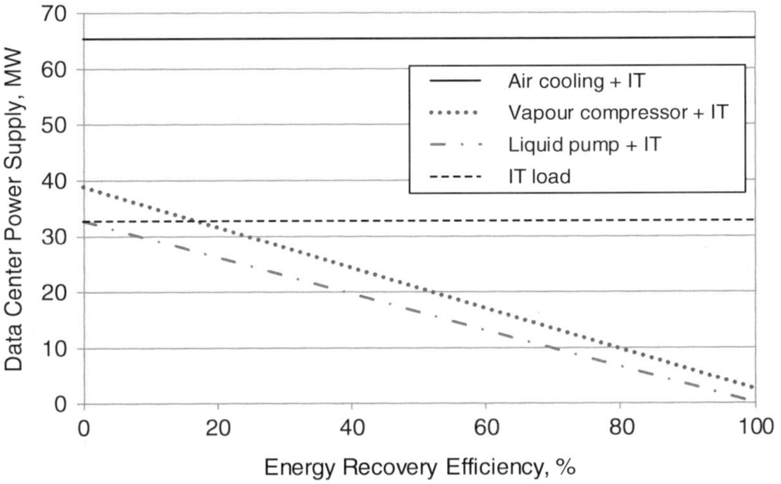


**Fig. 12.37** Datacenter power supply—IT and cooling system—Potential of energy recovery

the datacenter to operate the IT equipment and the cooling system. Included in the graph are two on-chip two-phase cooling methods using a vapor compression cycle and liquid pumping cycle. As a comparison, traditional air cooling is also simulated, where it is assumed that their power consumption is the same as that required to operate the IT equipment (Koomey [1] and Ishimine et al. [74]). The results are plotted as a function of the compressor or pump overall efficiencies (defined as the ratio between isentropic compression or pumping power and the electrical power for the drivers). All simulations were developed assuming an evaporating temperature of 60°C (on the chip), with condensing temperatures being 60 and 90°C for the liquid pumping and vapor compression cycles, respectively.

It is seen that the cycle using a vapor compressor is a strong function of the compressor's overall efficiency, up to a value of approximately 35% after which it becomes less dependent. Typically, compressors have an overall efficiency of about 60%. The liquid pumping cycle hardly shows any dependence on the pump efficiency. This is due to the fact that the power required to drive the pump is relatively very low [75]. The higher power consumption of the compressor is due to the energy required to increase the pressure from a saturation temperature of 60°C to 90°C. What is noticed, though, is that the datacenter's power requirements are reduced considerably for overall efficiencies above 15% (vapor compression cycle) when compared to traditional air cooling. This reduction is on the order of 50% when using a liquid pumping cycle and 41% for a vapor compression cycle with a compressor having an overall efficiency of 60%.

Further savings can be made when energy recovery is considered. Since on-chip cooling is used, recovering this energy would be a simple process by just incorporating a condenser, where the energy absorbed by the fluid from the server is transferred to a secondary fluid, like water, in the condenser. Figure 12.37 shows a plot of the datacenter power supply for the three types of cooling technologies as a function of the efficiency with which energy is recovered, where 100% efficiency implies that all the heat generated by the servers is recovered, while 0% means that none of the heat is recovered. Note that there is no change in power consumption for air cooling as it was assumed that the heat was not recovered, although this technology does exist, albeit not as effective as for on-chip cooling. The plots for the liquid pumping and vapor compression cycles assume that the pump and compressor has an overall efficiency of 100 and 60%, respectively, although the choice of efficiency for the pump is negligible (viz. Fig. 12.36). It should be noted that a 0 MW datacenter supply does not mean the datacenter requires no power, but rather that all the power received as electricity is sold as heat. The financial implications would show this since the value of the heat sold would be different from the electricity purchased and will be a function on the application to which the heat is sold.

Figure 12.37 shows that practically all the power purchased in the form of electricity can potentially be sold as heat. This is especially the case for a liquid pumping cycle due to its energy consumption being very low. The vapor compression cycle always has some unrecoverable heat due to inefficiencies, with this heat being lost to the environment. At this point in time, it would appear that the use of a liquid pumping cycle far outweighs use of a vapor compression cycle. However, what the graphs do not show is the quality of the heat being sold. The former can sell heat at a temperature of 60°C, while the latter can sell it at 90°C. The quality of heat is therefore important, not only due to the monetary value it adds, but also to the application to which it is sold. A restricted number of applications can use 60°C waste heat, with the limits becoming less as the temperature is increased.

C. *Carbon footprint*

For the calculation of the carbon footprint, only the contribution of the electricity used is considered. The effect of greenhouse gases (GHG) being formed by the manufacturing, transporting, storage, and disposal of the components of the datacenter, as well as the datacenter building, fall under a life cycle assessment analysis, which falls outside the scope of the current chapter. Furthermore, of the GHG, only $CO_2$ will be considered as it contributes to more than 75% of all the GHG and contributes the most to the greenhouse effect. Figure 12.38 shows the reduction of the quantity of $CO_2$ for the three cooling technologies as a function of the efficiency with which the energy is recovered. The quantity of $CO_2$ is calculated with the assumption that the datacenter purchases its electricity from a power plant running on coal and that it is selling waste heat back to the power plant, as discussed earlier. This graph, therefore, takes into consideration the efficiency increase of the power plant, since the amount of $CO_2$ released is a function of the power plant's efficiency, which in itself is a function of the efficiency with which energy is recovered. Since it is likely that the datacenter can be physically located at the power plant for this waste heat recovery system to be feasible, this also eliminates electrical transmission losses to off-site datacenters and provides an additional reduction in electrical consumption by the datacenter and a reduced $CO_2$ production.

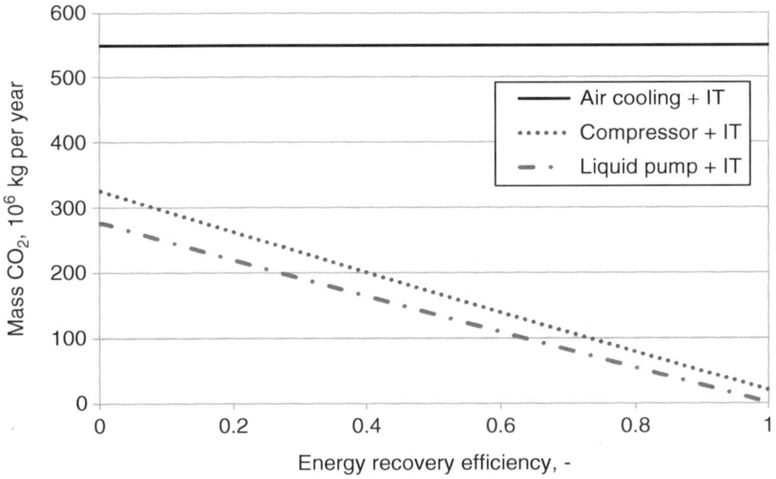


**Fig. 12.38** Reduction in $CO_2$ for three datacenter cooling technologies

The diagram shows that the datacenter could potentially have a zero carbon footprint regarding electricity usage when on-chip cooling using a liquid pump or vapor compression cycle is used. The use of a liquid pump or vapor compression cooling cycle without energy recovery (0% recovery efficiency), compared to traditional air cooling, reduces the carbon footprint of the datacenter considerably, with a reduction of 50% for the former and 40% for the latter. This reduction is improved further with energy recovery, with a potential reduction of almost 100 and 96% being achievable, at least in this simplified analysis.

As observed previously, Fig. 12.38 does not show the quality of the heat being recovered, with the value and applicability of this heat not being explicitly shown. This can be seen more clearly when applying the waste heat to the thermal power plant, which is shown in Fig. 12.39. This graph shows the $CO_2$ reduction of the datacenter due to energy recovery and the savings in $CO_2$ of the power plant due to efficiency improvements. Instead of plotting the carbon footprint as a function of the recovery efficiency, it is now plotted as a function of the condenser temperature, which is directly linked to the feedwater heater of the power plant. The effect is the same since a lower energy recovery efficiency would result in a lower temperature increase of the power plant's feedwater and, hence, a smaller increase in efficiency. The graph therefore shows the limit of each cooling cycle. Since the power plant's thermal efficiency improves with increase in condenser temperature, the amount of $CO_2$ saved by the power plant increases. However, when using a liquid pump cycle for the data center, only 25% of the total potential savings in $CO_2$ can be achieved, amounting to approximately 17,000 tons per year for a 173 MW power plant. By making use of a vapor compression cycle, however, the potential savings in $CO_2$ can reach as high as 70,000 tons per year. Therefore, although the liquid

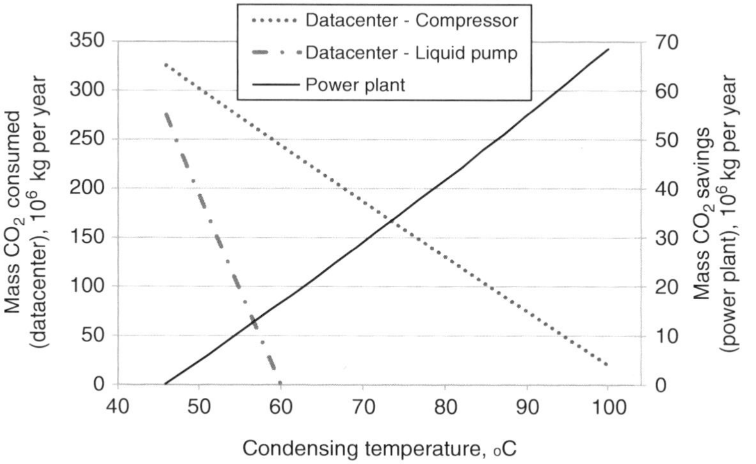


**Fig. 12.39** Carbon footprint of datacenter and $CO_2$ savings of power plant

pumping cycle was the better performing cooling cycle regarding energy usage and $CO_2$ reduction, due to the higher temperatures achievable by the vapor compression cycle, it has a larger impact on the secondary application making use of the waste heat.

Of course, depending on the point of view taken, it may be argued that due to the $CO_2$ savings of the power plant are all attributable to the datacenter, such that the datacenter could claim those savings are part of its own $CO_2$ reduction. This could potentially then lead to the datacenter having a negative carbon footprint regarding energy usage, which it can then use as a carbon offset, as shown in Fig. 12.40. This offset could be used to compensate for other $CO_2$ emitting processes in the datacenter, or could even be sold to other organizations via carbon trading. This should, however, be viewed as a tentative idea as regulations would determine whether this is possible or not. However, if this were the case, by making use of a vapor compression cycle, the offset could be 167% more than when using a liquid pumping cycle.

D. *Monetary savings*

Global warming is having a tremendous impact on the environment and on the livelihood of people regarding food production and natural resources, and thus on important industries and services, such as datacenters. To counter this, a carbon tax is expected to be introduced, with the aim of helping the environment by not only reducing carbon emissions by forcing people and organizations to become more energy efficient, but also by raising funds to be used for clean energy research. The tax will in some manner be levied on the carbon content of fuels, increasing the competitiveness of noncarbon technologies such as solar, wind, or nuclear energy sources. Therefore, organizations using electricity produced from the burning of fossil fuels will pay a higher tax than those produced from noncarbon burning fuels. The probability also exists for taxing the utility generating the electricity as

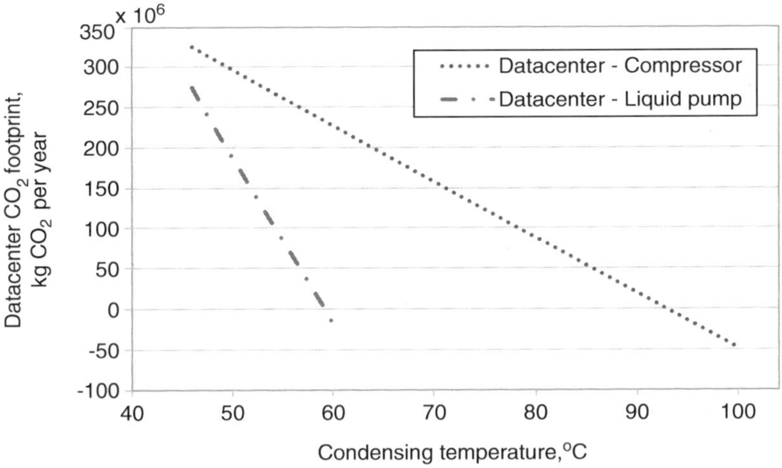


**Fig. 12.40** $CO_2$ footprint of a datacenter considering $CO_2$ reduction of power plant

an incentive to increase power plant efficiency. Carbon taxes have so far only been introduced in a few countries, with most European countries taking the lead, even though the way organizations are being taxed vary from country to country. In the USA, the introduction of a carbon tax has been made in California and the city of Boulder, Colorado, with taxes being on the order of 4 cents/ton of $CO_2$. European countries have been much more aggressive with taxes, with some countries, such as Sweden, applying taxes as high as \$100 per ton of $CO_2$ [76]. The Larson Bill [73] proposes to introduce a nationwide tax (in USA) of \$15/ton $CO_2$ starting in 2012, increasing by \$10/ton $CO_2$ every year. It also proposes to increase this increment to \$15/ton $CO_2$ after 5 years if the US emissions stray from the Environmental Protection Agency's (EPA) glide-path prediction, which proposes to cut emission to 80% that of 2005 levels by 2050.

A not unlikely potential price of \$30/ton $CO_2$ [77] would cost industries millions if efficiencies are not improved. Data centers are not exempt from these taxes, which will be introduced in the following years [78]. Figure 12.41 shows the potential savings made by a datacenter with 100,000 servers matched to a coal power plant with a size of 175 MW if heat were captured from the datacenter and sold to the power plant. The savings not only include that saved in energy costs by implementing a liquid pump or vapor compression cycle instead of a traditional air-cooling cycle, but also that saved in carbon tax. The savings of the power plant is in terms of fuel saved and the savings made in carbon tax. For fuel costs, a value of \$90/ton of coal was used.

For the datacenter, the graph shows that most savings are made if a liquid pump cycle is used, with the potential savings being in the order of \$45 million per year, while a vapor compression cycle would save in the order of \$40 million per year. The power plant, when recovering heat from a

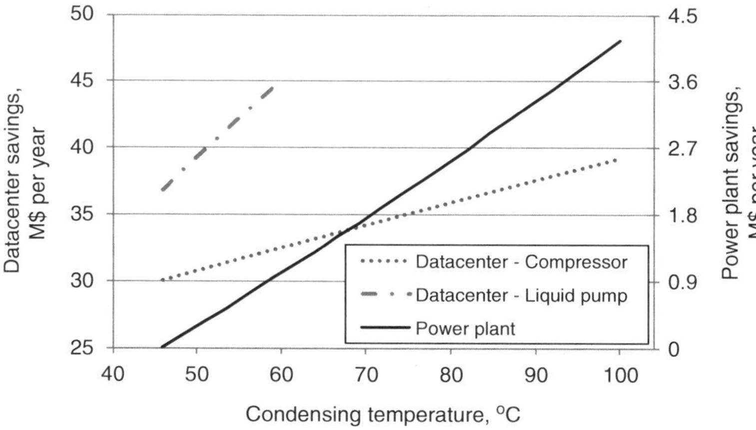


**Fig. 12.41** Datacenter and power plant savings

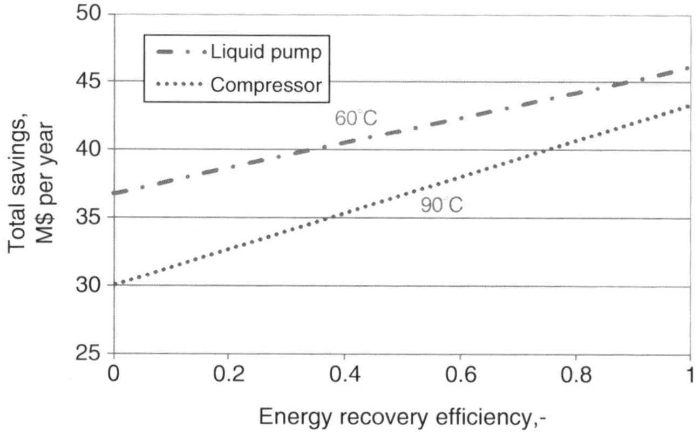


**Fig. 12.42** Total savings for liquid pumping and vapor compression cooling technologies

datacenter using a liquid pump cycle, will only save about \$1 million per year, with savings reaching almost \$4.5 million a year if a vapor compression cycle is used, which includes the additional electricity used to operate the compressor.

The overall savings are given in Fig. 12.42. This graph shows that the total savings, if a datacenter was to sell waste heat at 60°C with a liquid pump cycle, could be about \$46 million a year, while selling heat at 90°C with a vapor compression cycle could save a total of \$43 million per year. These are savings that a customer would potentially not have to pay. Even though it seems the clear solution is to use a liquid pump cycle within a datacenter, the incentive for a power plant to cooperate with a datacenter would be greater if a vapor compression cycle was used.

## 12.6 Summary

In this chapter, a general overview about multi-microchannel cooling technology was presented. Thermo-hydraulic aspects were considered highlighting the physical phenomena and the state-of-the art correlations available in the literature. The performance of a MMC was evaluated considering different working fluids, thermodynamic conditions, and geometries. The results showed the huge potential such a new technology has regarding the cooling of high heat flux components, capable of removing much more than 150 W/cm$^2$ at high evaporating pressures. This is an advantage when looking for waste heat recovery (high exergy). A case study was presented to demonstrate the application of such MMCs on a data center on-chip cooling system taking into account different drivers (liquid pump and vapor compressor) and working fluids. A comparison with the current cooling technology, i.e., air-cooling systems, showed an enormous reduction in energy consumption. In addition, comparing water single-phase flow and HFC134a two-phase flow on-chip cooling, a reduction in energy consumption in the order of five times for the latter was observed. Finally, an analysis on waste heat recovery was described considering the potential application in a coal power plant. The analysis considered aspects such as power plant and data center thermal performance improvements, reduction of carbon footprint and monetary savings. The huge quantity of energy coming forth from large datacenters were highlighted, further showing the huge opportunity available in applying on-chip multi-microchannel cooling to save and recovery energy.

Refrigerant on-chip cooling inherently has many more advantages over water on-chip cooling. This was not mentioned in this chapter, but is highlighted below.

1. Refrigerants are dielectric, making them intrinsically safe for electronic components.
2. Flow rate of two-phase refrigerant (kg/s) is low due to latent heat.
3. Chip temperatures are more uniform—longer chip life time.
4. Pumping power of refrigerant is low due to low flow rate.
5. Size of refrigerant pump is small due to low flow rate.
6. Refrigerant pipe diameters are small so flexible for easier installation.
7. Less copper required in two-phase cooling systems, so light in weight.
8. Refrigerants can be driven by a compressor to dissipate heat at higher temperature.
9. Refrigerants have a low freezing point, ideal for low temperature environments.
10. Compatibility of refrigerants with system materials is well known.
11. Refrigerants have no organic/fouling problems.
12. Refrigerants have no erosion problems.
13. Boiling refrigerants have higher (or similar) heat transfer coefficients to water.
14. Boiling has self-enhancement effect under hot-spots (heat transfer coefficient rises with heat flux).
15. Refrigerants have transient flow thermal storage capacity (latent heat).
16. Refrigerants have history of long, clean life in millions of air conditioners.

These are only some points mentioned and should be considered as open points for future discussions.

## References


1. Koomy JG (2007) Estimating regional power consumption by servers: a technical note. Lawrence Berkeley National Laboratory, Oakland, CA
2. Saini M, Webb RL (2003) Heat rejection limits of air cooled plane fin heat sinks for computer cooling. IEEE Trans Compon Packag Technol 26(1):71–79
3. Samadiani E, Joshi S, Mistree F (2008) The thermal design of a next generation data center: a conceptual exposition. J Electron Pack 130: 041104-1–041104-8
4. Agostini B, Fabbri M, Park JE, Wojtan L, Thome JR, Michel B (2007) State-of-the-art of high heat flux cooling technologies. Heat Transfer Eng 28:258–281
5. Leonard PL, Phillips AL (2005) The thermal bus opportunity – a quantum leap in data center cooling potential. In: ASHRAE transactions, Denver, CO, 2005
6. Gosney WB (1982) Principles of refrigeration, 1st edn. Cambridge University Press, Cambridge
7. Montreal, The Montreal protocol on substances that deplete the ozone layer. 2000, United Nations Environment Programme: Nairobi, Kenya
8. Kyoto, Kyoto protocol to the United Nations framework convention on climate change. 1998, United Nations: New York, USA
9. Calm J (2008) The next generation of refrigerants – historical review, considerations, and outlook. Int J Refrig 31:1123–1133
10. Calm J (2006) Comparative efficiencies and implications for greenhouse gas emissions of chiller refrigerants. Int J Refrig 29:833–841
11. Nielsen O, Javadi M, Sulbaek M, Hurley M, Wallington T, Singh R (2007) Atmospheric chemistry of $CF_3CF = CH_2$: kinetics and mechanisms of gas-phase reactions with CL atoms, OH radicals, and O3. Chem Phys Lett 439:18–22
12. Søndergaard R, Nielsen O, Hurley M, Wallington T, Singh R (2007) Atmospheric chemistry of trans-$CF_3CH = CHF$: kinetics of the gas-phase reactions with Cl atoms, OH radicals, and $O_3$. Chem Phys Lett 443:199–204
13. Brown J, Zilio C, Cavallini A (2010) Thermodynamic properties of eight fluorinated olefins. Int J Refrig 33:235–241
14. Directive 2006/40/EC of the European Parliament and of the Council of 17 May 2006 relating to emissions from air-conditioning systems in motor vehicles and amending Council Directive 70/156/EEC. 2006
15. Thome JR (2010) Wolverine engineering databook III. Available from: http://www.wlv.com
16. Morini GL (2006) Scaling effects for liquid flows in microchannels. Heat Transfer Eng 27:64–73
17. Revellin R, Dupont V, Ursenbacher T, Thome JR, Zun I (2006) Characterization of two-phase flows in microchannels: optical measurement technique and flow parameter results for R-134a in a 0.5 mm channel. Int J Multiphase Flow 32:755–774
18. Ong CL (2010) Macro-to-microchannel transition in two-phase flow and evaporation, In: Heat and Mass Transfer Laboratory (LTCM). École Polytechnique Fédérale de Lausanne: Lausanne
19. Lazarek GM, Black SH (1982) Evaporating heat transfer, pressure drop and critical heat flux in a small vertical tube with R-113. Int J Heat Mass Transfer 25:945–960
20. Tran TN, Wambsganss MW, France DM (1996) Small circular and rectangular channel boiling with two refrigerants. Int J Multiphase Flow 22:485–498
21. Zhang W, Hibiki T, Mishima K (2004) Correlation for flow boiling heat transfer in mini-channels. Int J Heat Mass Transfer 47:5749–5763
22. Kandlikar SG, Balasubramanian P (2004) An extension of the flow boiling correlation to transition, laminar, and deep laminar flows in minichannels and microchannels. Heat Transfer Eng 25:86–93
23. Jacobi AM, Thome JR (2002) Heat transfer model for evaporation of elongated bubble flows in microchannels. J Heat Transfer-Transact ASME 124(6):1131–1136

24. Thome JR, Dupont V, Jacobi AM (2004) Heat transfer model for evaporation in microchannels. Part I: presentation of the model. Int J Heat Mass Transfer 47:3375–3385
25. Dupont V, Thome JR, Jacobi AM (2004) Heat transfer model for evaporation in microchannels, part II: comparison with the database. Int J Heat Mass Transfer 47:3387–3401
26. Ong CL, Thome JR (2011) Macro-to-microchannel transition in two-phase flow, part 2: flow boiling heat transfer and critical heat flux. Exp Therm Fluid Sci 35(6):873–886
27. Agostini B, Thome JR, Fabbri M, Michel B, Calmi D, Kloter U (2008) High heat flux flow boiling in silicon multi-microchannels – part II: heat transfer characteristics of refrigerant R245fa. Int J Heat Mass Transfer 51(21–22):5415–5425
28. Costa-Patry E, Olivier JA, Michel B, Thome JR (2011) Two-phase flow of refrigerants in 85 μm-wide multi-microchannels, part II: heat transfer with 35 local heaters. Int J Heat Fluid Flow 32(2):464–476
29. Ong CL, Thome JR (2011) Macro-to-microchannel transition in two-phase flow: part 1 – two-phase flow patterns and film thickness measurements. Exp Therm Fluid Sci 35:37–47
30. Cioncolini A, Thome JR (2011) Algebraic turbulence modeling in adiabatic and evaporating annular flows. Int J Heat Fluid Flow, in review
31. Ribatski G, Wojtan L, Thome JR (2006) An analysis of experimental data and prediction methods for two-phase frictional pressure drop and flow boiling heat transfer in micro-scale channels. Exp Therm Fluid Sci 31:1–19
32. Cicchitti A, Lombardi C, Silvestri M, Soldaini G, Zavattarelli R (1960) Two-phase cooling experiments – pressure drop, heat transfer and burnout measurements. Energia Nucleare 7:407–425
33. Müller-Steinhagen H, Heck K (1986) A simple friction pressure drop correlation for two-phase flow in pipes. Chem Eng Process 20:297–308
34. Mishima K, Hibiki T (1996) Some characteristics of air-water two-phase flow in small diameter vertical tubes. Int J Multiphase Flow 22:703–712
35. Revellin R, Thome JR (2007) Adiabatic two-phase frictional pressure drops in microchannels. Exp Therm Fluid Sci 31(7):673–685
36. McAdams WH, Woods WK, Bryan RL (1942) Vaporization inside horizontal tubes – II – benzene-oil mixtures. Transaction of the ASME 64:193
37. Cioncolini A, Thome JR, Lombardi C (2009) Unified macro-to-microscale method to predict two-phase frictional pressure drops of annular flows. Int J Multiphase Flow 35:1138–1148
38. Garimella S, Killion JD, Coleman JW (2002) An experimental validated model for two-phase pressure drop in the intermittent flow for circular channel. J Fluid Eng 124:205–214
39. Revellin R, Thome JR (2008) An analytical model for the prediction of the critical heat flux in heated microchannels. Int J Heat Mass Transfer 51:1216–1225
40. Wojtan L, Revellin R, Thome JR (2006) Investigation of saturated critical heat flux in a single uniformly heated microchannel. Exp Therm Fluid Sci 30:765–774
41. Lazarek GM, Black SH (1982) Evaporating heat transfer, pressure drop and critical heat flux in a small vertical tube with R-113. Int J Heat Mass Transfer 25(7):945–960
42. Bowers MB, Mudawar I (1994) High flux boiling in low flow rate, low pressure drop mini-channel and micro-channel heat sinks. Int J Heat Mass Transfer 37:321–332
43. Qu W, Mudawar I (2004) Measurement and correlation of critical heat flux in two-phase micro-channel heat sink. Int J Heat Mass Transfer 47:2045–2059
44. Agostini B, Revellin R, Thome JR, Fabbri M, Michel B, Calmi D, Kloter U (2008) High heat flux flow boiling in silicon multi-microchannels: part III. Saturated critical heat flux of R236fa and two-phase pressure drops. Int J Heat Mass Transfer 41:5426–5442
45. Ong CL (2010) Macro-to-microchannel transition in two-phase flow and evaporation. Ph.D. Thesis, EPFL: Lausanne
46. Mauro AW, Thome JR, Toto D, Vanoli GP (2010) Saturated critical heat flux in a multi-microchannel heat sink fed by a split flow system. Exp Therm Fluid Sci 34:81–92
47. Park JE (2008) Critical heat flux in multi-microchannel copper elements with low pressure refrigerants. École Polytechnique Fédérale de Lausanne

48. Karajgikar S, Agonafer D, Ghose K, Sammakia B, Amon C, Refai-Ahmed G (2010) Multi-objective optimization to improve both thermal and device performance of a nonuniformly powered micro-architecture. J Electron Packag 132:81–87
49. Bar-Cohen A, Kraus A, Davidson S (1983) Thermal frontiers in the design and packaging of microelectronic equipment. Mech Eng 105(6):53–59
50. Borkar S, Karnik T, Narendra S, Tschanz J, Keshavarzi A, De V (2003) Parameter variations and impact on circuits and microarchitecture. In: DAC'03 40th annual design automation conference, 2003
51. Agostini B, Thome JR (2005) Comparison of an extended database for flow boiling heat transfer coefficient in multi-microchannels elements with the three-zone model. In: Proceedings of ECI international conference on heat transfer and fluid flow in microscale, Castelvecchio Pascoli, Italy, 2005
52. Costa-Patry E, Olivier JA, Paredes S, Thome JR (2010) Hot-spot self-cooling effects on two-phase flow of R245fa in 85 μm-wide multi-microchannels. In: 16th International workshop on thermal investigations of ICs and systems, Spain, 2010
53. Costa-Patry E, Nebuloni S, Olivier JA, Thome JR (2011) On-chip two-phase cooling with refrigerant in 85 μm-wide multi-microchannel evaporator under hot-spot conditions. IEEE Trans Compon Packag Technol. DOI: 10.1109/TCPMT.2011.2173572
54. Madhour Y, Olivier JA, Costa-Patry E, Paredes S, Michel B, Thome JR (2011) Flow boiling of R134a in a multi-microchannel heat sink with hotspot heaters for energy-efficient microelectronic CPU cooling applications. IEEE Trans Compon Packag and Manufac Technol 1(6): 873–883
55. Marcinichen JB, Olivier JA, Thome JR (2011) Reasons to use two-phase refrigerant cooling. In: ElectronicsCooling. Available from: http://www.electronics-cooling.com/
56. Marcinichen JB, Thome JR, Michel B (2010) Cooling of microprocessors with micro-evaporation: a novel two-phase cooling cycle. Int J Refrig 33(7):1264–1276
57. IBM BladeCenter QS22: overview. Available from: http://www-03.ibm.com/systems/bladecenter/hardware/servers/qs22/
58a. Meyer JP, Olivier JA (2010) Transitional flow inside enhanced tubes for fully developed and developing flow with different types of inlet disturbances: Part II–heat transfer. Int J Heat Mass Transfer 54(7–8):1587–1597
58b. Meyer JP, Olivier JA (2010) Transitional flow inside enhanced tubes for fully developed and developing flow with different types of inlet disturbances: Part I – Adiabatic pressure drops. Int J Heat Mass Transfer 54(7–8):1598–1607
59. Cavallini A, Del Col D, Doretti L, Longo GA, Rossetto L (2000) Heat transfer and pressure drop during condensation of refrigerants inside horizontal enhanced tubes. Int J Refrig 23:4–25
60. Dittus FW, Boelter LMK (1930) Publications on engineering, vol 2. University of California, Berkeley, p 443
61. Blasius H (1913) Das Ähnlichkeitsgesetz bei Reibungsvorgängen in Flussigkeiten. 1913: Fors-chg. Arb. Ing. -Wes., 131 Berlin
62. Ravigururajan TS, Bergles AE (1985) General correlations for pressure drop and heat transfer for single-phase turbulent flow in internally ribbed tubes. Augmentation of heat transfer in energy systems, vol. 52, pp 9–20
63. Azzi A, Alger USTHB, Friedel L (2005) Two-phase upward flow 90° bend pressure loss model. Forschung im Ingenieurwesen, vol. 69, pp 120–130
64. Taitel Y, Bornea D, Dukler AE (1980) Modelling flow pattern transitions for steady upward gas-liquid flow in vertical tubes. AIChE J 26(3):345–355
65. Barnea D, Shoham O, Taitel Y (1982) Flow pattern transition for vertical downward two phase flow. Chem Eng Sci 37(5):741–744
66. Barnea A (1986) Transition from annular flow and from dispersed bubble flow – unified models for the whole range of pipe inclinations. Int J Multiphase Flow 12(5):733–744
67. Liu D, Wang S (2008) Flow pattern and pressure drop of upward two-phase flow in vertical capillaries. Ind Eng Chem Res 47:243–255

68. Perez-Tellez C (2003) Improved bottomhole pressure control for underbalanced drilling operations. Ph.D. Thesis, Louisiana State University
69. Spedding PL, Benard E, McNally GM (2004) Fluid flow through 90 degree bends. Dev Chem Eng Mineral Process 12:107–128
70. Azzi A, Friedel L, Belaadi S (2000) Two-phase gas/liquid flow pressure loss in bends. Forschung im Ingenieurwesen, vol. 65, pp 309–318
71. ASHRAE (1997) Pipe sizing. In: ASHRAE Handbook – Fundamentals
72. Oliphant RJ (2003) Causes of copper corrosion in plumbing systems. A review of current knowledge, Bucks, UK
73. Larson JB (2009) America's Energy Security Trust Fund Act of 2009. H.R. 1337
74. Ishimine J, Ohba Y, Ikeda S, Suzuki M (2009) Improving IDC cooling and air conditioning efficiency. FUJITSU Sci Tech J 45:123–133
75. Marcinichen JB, Thome JR (2010) Refrigerated cooling of microprocessors with micro-evaporation/new novel two-phase cooling cycles: a green steady-state simulation code. In: 13th Brazilian congress of thermal sciences and engineering. Uberlândia, MG, Brazil, 2010
76. Wikipedia. Carbon tax, The Free Encyclopedia, from http://en.wikipedia.org/w/index.php?title=Carbon_tax&oldid=374709823. Retrieved 08:12, 26 July 2010
77. Nordhaus W (2008) A question of balance – weighing the options of global warming policies. Yale University Press, New Haven, CT
78. Mitchel RL (2010) Carbon tax could whack data centers. Computerworld, http://blogs.computerworld.com/16128/carbon_tax_could_whack_data_centers

# Chapter 13
# Emerging Data Center Thermal Management and Energy Efficiency Technologies

**Yogendra Joshi and Pramod Kumar**

**Abstract** This chapter introduces a number of emerging topics in data center design and operation. The use of ambient air, water, or ground for heat rejection is attractive for many facilities whenever the environmental conditions are conducive. Changes in equipment layout and real-time control of cooling and information technology (IT) resources also offer opportunities for savings in cooling energy costs. As rack level powers continue to increase for several equipment classes, there is increasing interest in hybrid liquid/air cooling and liquid cooling approaches. Safety issues arising due to elevated temperatures and ambient noise are also receiving increasing attention. Broadening of equipment operation temperature and humidity ranges is resulting in concerns for wiring corrosion. Need for on-demand, rapidly deployable, and expandable computing resources has resulted in rapid development of modular data centers.

As the number of data centers continues to increase, reduction in their fixed and recurring operational costs is of significant and growing interest. As discussed earlier, demonstrated approaches include the use of air- or water-side economizers (Chap. 1), improvements in air delivery and air flow management (Chap. 2), and revision of current standards to allow for higher rack inlet air temperatures and broader relative humidity range (Chap. 2). In this chapter, we first discuss a few additional approaches that have more recently been introduced and offer promise for further reductions in data center energy usage costs for cooling. Next, we discuss the growing trend towards higher cabinet powers and progress in liquid cooling-based approaches to handle these. We conclude with modular data centers, which have been introduced by several manufacturers.

Y. Joshi (✉) • P. Kumar
G.W. Woodruff School of Mechanical Engineering, Georgia Institute of Technology, Atlanta, GA 30332, USA
e-mail: yogendra.joshi@me.gatech.edu; pramod.kumar@me.gatech.edu

Y. Joshi and P. Kumar (eds.), *Energy Efficient Thermal Management of Data Centers*, DOI 10.1007/978-1-4419-7124-1_13,

## 13.1 Energy Usage Reduction Through Facility Architectural or Equipment Layout Changes

Following the introduction of energy usage metrics discussed in Chap. 6, users of large data centers have started to pay closer attention to reducing energy consumption. For air-cooled facilities with modest power dissipation per rack, typically well below 10 kW, availability of inexpensive real estate, cooler air temperatures, and naturally windy conditions may make it possible to utilize external air for cooling through most of the year.

### *13.1.1 Energy Efficient Facility Architecture: The Yahoo! "Chicken Coop" Facility*

Yahoo! commissioned a new facility in Lockport, New York, in 2010 that incorporates a number of novel features to take maximum advantage of available natural resources for cooling, with only modest use of external energy for supplemental cooling. As reported [1], the long narrow design, Fig. 13.1a, b, mimics a chicken coop to promote buoyancy-assisted air flow. Ambient air is drawn into the facility for cooling using a fan array, and the hot air is routed through tall cupolas vented at the top. Through a combination of the area's cooler climate, and prevailing winds from Lake Ontario, less than 1% of the buildings' energy consumption is needed to cool the 120-foot by 60-foot server buildings, holding 50,000 servers. The facility also benefits from low-cost hydropower from the New York Power Authority. During the brief summer period, the convective air cooling is augmented with evaporative cooling [2]. Evaporative coolers or swamp coolers, which are very common in residential air conditioning in hot and dry climates, involve forced air flow over a water saturated absorbent medium, as discussed in Chap. 2. Through the use of these energy efficiency advances, the quoted PUE value of the facility is reported around 1.08.

### *13.1.2 Energy Efficient Facility Architecture: The Facebook Open Compute Facility*

In April 2011, Facebook opened a new facility in Prineville, Oregon, with a reported PUE of 1.07 (http://www.facebook.com/note.php?note_id=10150144039563920). This was achieved through the adoption of a number of energy efficiency advances, described in the open literature under the Open Compute Project (http://opencompute.org/). A 480-V electrical distribution was used to reduce losses. Individual servers are packaged in a low-cost, "vanity-free" design, with reduced use of plastic. The servers are placed in triplet racks composed of three adjacent 42U

**Fig. 13.1** (**a**) The Yahoo! "chicken coup" data center facility. The central wing and two wings on each side are visible. The central cupola in each wing provides top venting of hot air. Photograph courtesy of Yahoo! (**b**) The Yahoo! "chicken coup" data center facility. A single wing is shown in the winter season, with air inlet vents at the bottom and exit vents at the cupola top. Backup utilities are seen in the foreground. Photograph courtesy of Yahoo!

**Fig. 13.2** The Facebook triplet rack architecture. Each 42U column houses 30 servers, with a total of 90 servers per triplet. Photograph courtesy Alan Brandt

columns, see Fig. 13.2. Each column houses 30 servers. Rack switches are located at the top of each triplet. A battery cabinet is located in-between two triplets in the data center aisle and provides backup power in case of AC power failure.

The Facebook Prineville facility uses a full air-side economizer system. The air distribution scheme is illustrated in Fig. 13.2. The outside air enters through double louvers and proceeds into the outside air intake corridor. It enters the evaporative cooling/humidification room equipped with a misting system. Subsequently, mist eliminators eliminate water carry over. The air then enters the supply fan room and is then introduced into the cold aisles in the data center space. The cold air enters the racks through the front and is discharged at the back of the racks into the contained hot aisle. The hot air then enters the return plenum. It is then returned to the filter room or exhausted out of the building, either by natural pressure gradient or through relief fans (Fig. 13.3).

In winter, the hot-aisle air is re-used to warm office spaces, as well as the incoming cooling air into the facility.

### *13.1.3 Alternate Rack Architectures*

As seen in Chaps. 1 and 2, the hot-aisle/cold-aisle (HACA) layout has been the most commonly utilized arrangement for air-cooled data center facilities. Improvements in cooling such as hot-aisle containment and cold-aisle containment have been discussed in Chap. 2. Another option that may provide improved usage of cooling

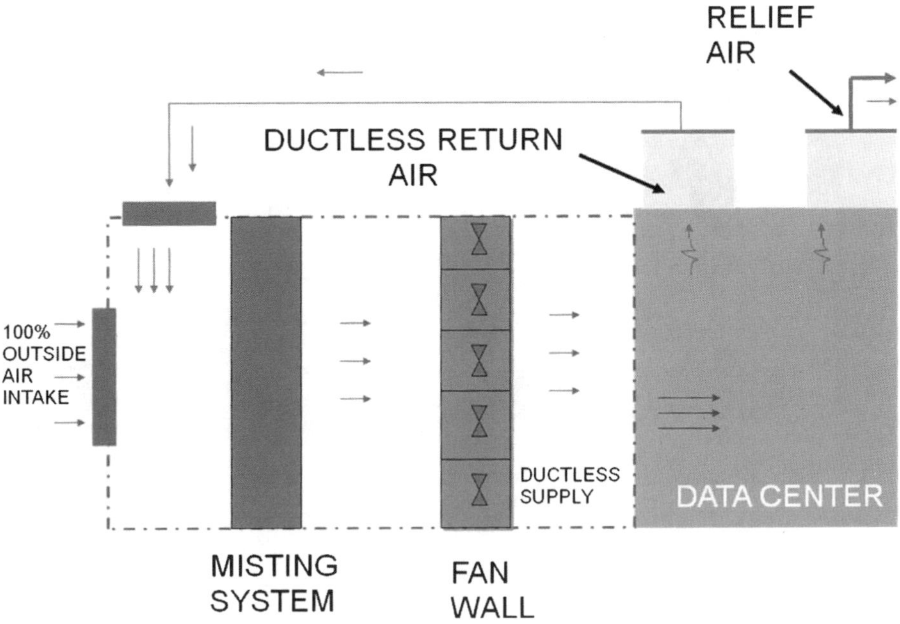


**Fig. 13.3** Air distribution in the Facebook Prineville, Oregon data center facility

resources but does not require additional hardware is alternate layouts. As an example, the S-Pod arrangement seen in Fig. 13.4 [3] refers to the layout of a data center space in which the racks are arranged in groups of near circular form—pods. The computer room air conditioning (CRAC) unit provides cooled air, which after passing through the under- or over-floor plenum, is released into the room through perforated tiles. As seen in Chaps. 2 and 8, the traditional HACA layout is subject to mixing of hot return air with the cold supply air and thus reducing the cooling potential of the cold air. It also forces some rack inlets to be devoid of direct supply of cold air altogether.

In the S-Pod layout, the perforated tiles are surrounded by racks on all four sides. The air coming out of the tiles is pulled into the racks with the help of fans and is exhausted outside the pod. Then it is returned to the CRAC unit for supply. The arrangement of these pods in the data center can be in-line or staggered. Figure 13.4 presents a staggered arrangement. It also presents a few possible schemes which can be employed for server access for maintenance. For the same floor space area, the HACA layout houses 44 racks while the S-Pod layout can accommodate 56 racks (27.3% more). The limitation in deciding the number of racks that can be housed is the minimum distance required between the CRAC and the nearest perforated tile, to avoid negative flow from the perforated tiles.

Due to the symmetric nature of the HACA arrangement, only one-fourth of the facility was simulated. The computational model chosen for simulations is shown in

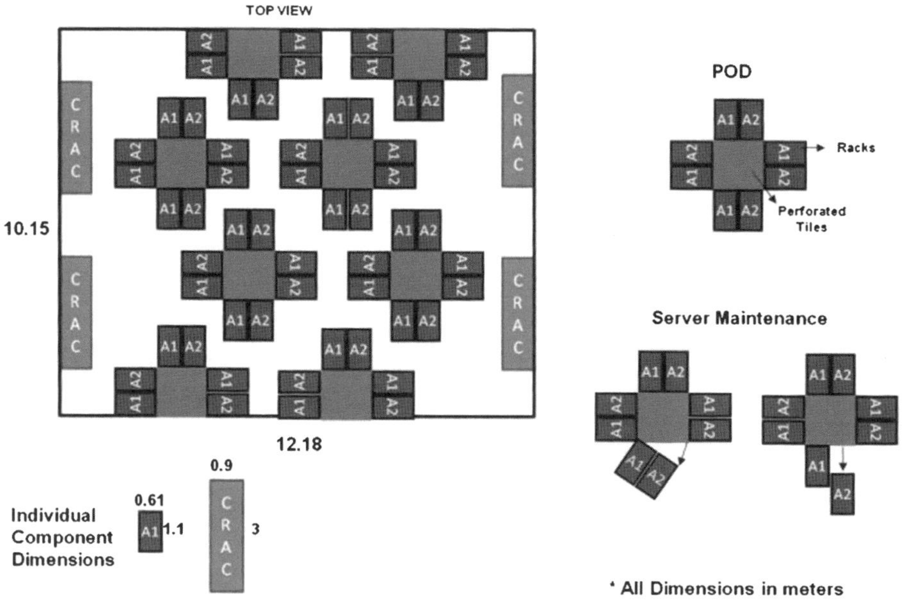


**Fig. 13.4** The S-POD layout as an alternate rack arrangement to the standard hot-aisle/cold-aisle [3]. Server maintenance can be performed as shown

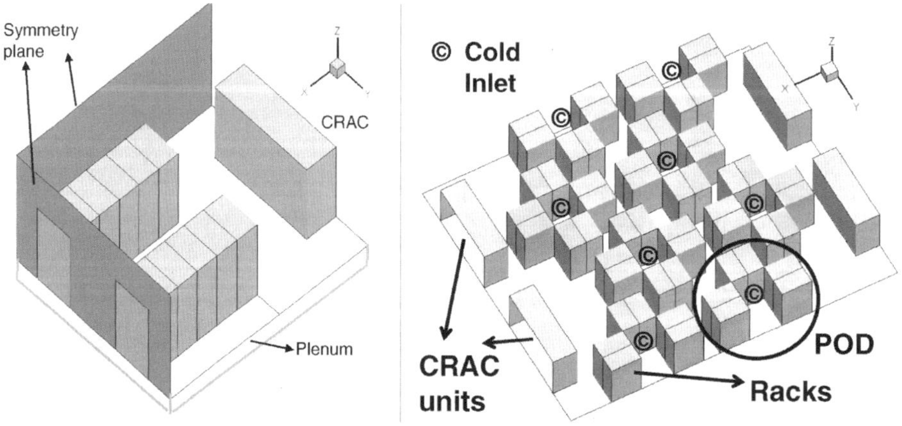


**Fig. 13.5** Computational model geometry for the S-POD layout and the standard hot-aisle/cold-aisle layout [3]

Fig. 13.5. The hot-aisle arrangement is such that it is opposite to the CRAC unit to minimize recirculation in HACA arrangement. A pod consists of four columns with two racks each and the space between them being the perforated tiles. The S-Pod layout consequently has 56 racks and 4 CRAC units. The racks have a height of 2 m, while the whole facility excluding the plenum is 3 m high. The racks are assigned a

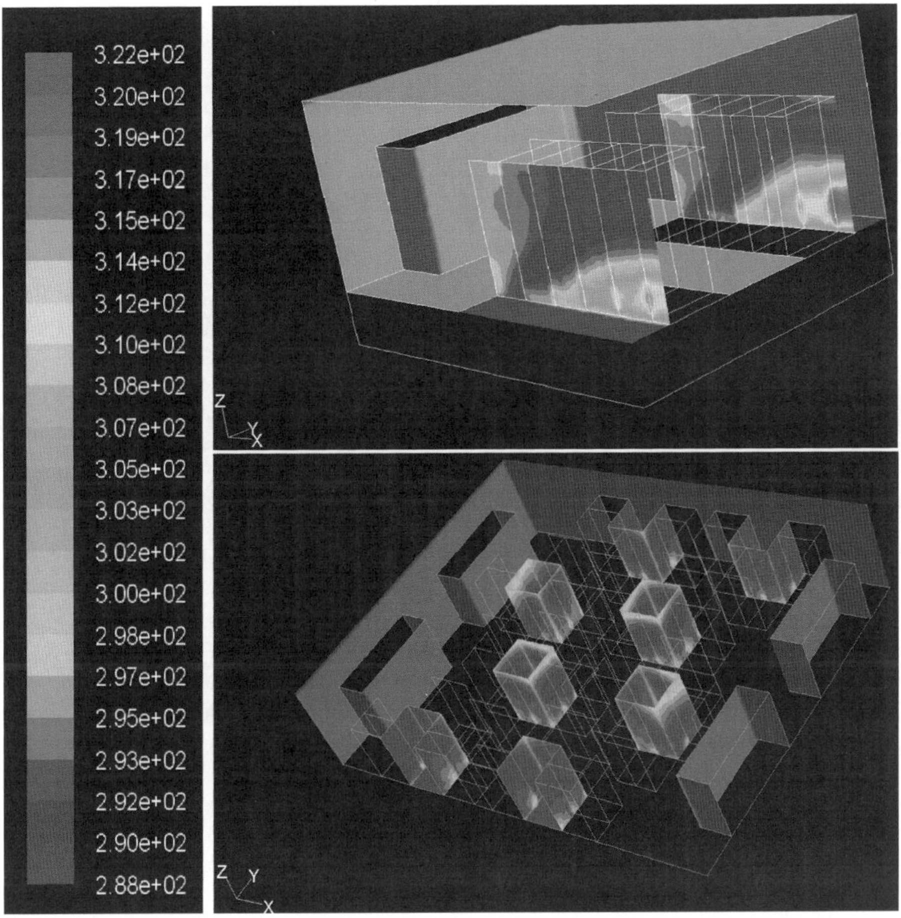


**Fig. 13.6** Computed air temperatures at rack inlets for the S-POD layout and hot-aisle/cold-aisle layout [3]. Cooler temperatures are found for the S-POD layout for all rack bottoms

uniform heat generation rate per volume, assumed to be the same for all racks for both the layouts. The plenum height is 0.86 m. The rack outlets are modeled as fans governed by a polynomial relationship between the pressure jump and velocity. A fixed flow rate boundary condition is given at the CRAC units with a supply air temperature of 288.15 K and velocities being 4, 7, and 10 m/s in three different cases.

Figure 13.6 shows the air temperature contours at rack inlets for the two layouts for CRAC velocity of 7 m/s. The volumetric heat generation rate for each rack is chosen to be 16,161.61 W/m$^3$ (21.7 kW per rack). In the HACA layout, the high air velocity deprives the lower parts of the racks from getting the required flow, thus raising their temperature. The maximum air temperature at rack inlets for HACA layout is 322 K, while for S-Pod case it is 302 K, for the allowed temperature limit

**Table 13.1** CRAC mass flow rate impact on rack inlet air flow rate for S-pod layout

| CRAC velocity (m/s) | Net CRAC mass flow rate (kg/s) | Net mass flow rate at rack inlets (kg/s) | Temperature profile remarks |
|---|---|---|---|
| 4 | 52.9 | 86.9 | Hot at top |
| 7 | 92.6 | 86.11 | Uniform |
| 10 | 132 | 85 | Hot at bottom |

of 305 K (32°C). Further calculations show that the maximum allowed HACA case heat generation limit is 15 kW per rack, while the S-Pod case can go to 22 kW. Thus, the S-Pod case can take up to 53% higher heat load per rack for this velocity and since they have 27.3% more number of racks, the net heat load capacity of the data center with S-Pod layout increases by 95%.

At lower CRAC velocities, the benefit reduces. The results for 4 m/s were obtained for 8,889 W/m$^3$ (11.93 kW per rack). In the HACA layout, it was noticed that the flow coming out of the perforated tiles lacks the momentum to reach the topmost levels of the rack. Thus, these levels get air re-circulated from the hot aisle. In the S-Pod case, the pods closest to the CRAC units and especially the racks closest to CRAC units have higher inlet temperatures, with maximum values of about 307 K. The maximum inlet air temperature for the HACA case increased up to 312 K, showing that the S-Pod layout can go up to 11 kW, whereas the HACA layout can handle only 8.6 kW of heat load per rack. Thus, S-Pod has 27.8% more heat dissipation capacity per rack and it has 27.3% more racks. Thus, the net heat dissipation capacity for the data center in S-Pod case increases around 63.6%.

As per Table 13.1, the supply mass flow rate from the CRAC units for the complete facility with an S-Pod layout increases linearly with velocity. The net mass flow rate required by the racks on the other hand decreases very slowly with increasing velocity because of the changing pressure differences affecting the fan velocity. It also shows that the 4-m/s case provides insufficient mass flow rate, with the coolant lacking momentum at the perforated tiles, thus leading to recirculation at top. While the 10-m/s case provides 55% extra mass flow rate, most of this just leaves the pod from the top, leading to wastage of cool air and thus energy. Also, because of the higher momentum of the flow coming out of the perforated tiles, the flow near the lower servers in the rack is highly directional in nature and is difficult to be sucked in. This leads to higher temperatures near the lower end. The 7-m/s case provides roughly the mass flow rate required by the racks, which reaches the top ends of the rack without abandoning the lower ends. Also, the average inlet temperature for 7-m/s case is lowest among the three velocities considered. This indicates that there exists an optimal velocity for this layout. As the heat dissipation increases, the return air temperature at CRAC inlets also increases. Since there is a cap on this temperature due to cooling limitations of the CRAC unit, one still needs to increase CRAC velocities to achieve higher heat dissipations. However, the efficiency of the data center reduces.

While the 7-m/s case provides almost the same mass flow rate as required by the racks and also yields the lowest maximum air inlet temperature for the racks, it is not the optimal velocity in every case. To determine the optimal velocity, one has to increase from lowest velocity to an operating point, which ensures maximum inlet temperature to be lower than a prescribed level, e.g., 305 K.

The computations showed that there is mixing from the upper side of the racks for both HACA and S-Pod arrangements for the 4-m/s air velocity. To avoid this in the S-Pod case, the cold air supply space above the raised floor inside the pod could be closed from the top using a physical barrier. In this case, it has to be ensured that the flow rate provided by each perforated tile is greater than or equal to the flow rate coming out of each pod to avoid creating negative pressures, in order to ensure adequate mass flow rate through the racks.

Based on the simulations, Somani et al. [3] provide the following reasons for why the S-Pod layout is better than the HACA layout:

1. The increase in the inlet temperatures of the racks closest to the CRAC unit in the HACA case for higher velocities such as 7 m/s indicates that side air recirculation exists. The inherent design of the S-Pod blocks the side air mixing if it is made leakage proof.
2. The number of racks per perforated tile (if same dimensions are maintained for both) is 2 for the S-Pod layout and 1 for the HACA layout. Thus, the pressure with which the flow comes out of the perforated tiles in S-Pod case is higher, which prevents top side recirculation from reaching the topmost servers in each rack.
3. The effective hot-aisle dimensions reduce in S-Pod case, even further with a staggered arrangement. Thus, more racks can be accommodated in the same space.
4. Higher heat dissipation is achieved in the S-Pod layout by increasing the effective inlet temperature from the room to the CRAC for a given velocity.

## 13.2 CRAC Fan Speed Control for Energy Efficiency Improvement

Typical control methods implemented in most data centers today have the CRAC fans operating at 100% of their maximum speed. These are paired with a P or PI (proportional or proportional integral) controller within the CRAC to regulate the flow of chilled water in the cooling coil, based on the supply and return temperatures on the water and air sides. As seen in Chap. 2, at higher tile air flow rates, the uniform distribution of cold supply air flow throughout the height of the rack is challenging. This difference in distribution can cause significant variations in server inlet air temperatures. Sundaralingam et al. [4] have implemented a server rack heat load based CRAC fan speed controller to provide the data center with only the necessary cooling, using an energy balance approach described through the study below.

### 13.2.1 An Experimental Case Study

The experiments to demonstrate feasibility of the control approach were performed in a facility with an area of 112 m$^2$ divided into two zones: (1) heterogeneous rack zone and (2) high-performance computational (HPC) facility zone. There are a total of 6 CRACs and 26 server racks to which power is supplied through power distribution units. All the CRACs have variable frequency drives (VFD) to allow control of cooling air. The experiments were conducted in the HPC facility zone shown in Fig. 13.7. The data center employs a HACA configuration. The heat flows across a chilled water-cooled rear door heat exchanger (RDHx) located at the exit of the racks to reduce its temperature prior to discharge into the hot aisle. We discuss the RDHx operation concept in Sect. 13.4. The cold aisle is located in the center of the computational zone shown in Fig. 13.7. A downflow CRAC unit was used, with VFD speeds ranging from 60% to 100%. The cooling coils within the CRAC and the RDHx are supplied with chilled water from the building chiller. The chillers have five independent controllers operating based on supply water temperature set points. A schematic of the distribution of chilled water from the building chiller is given in Fig. 13.8.

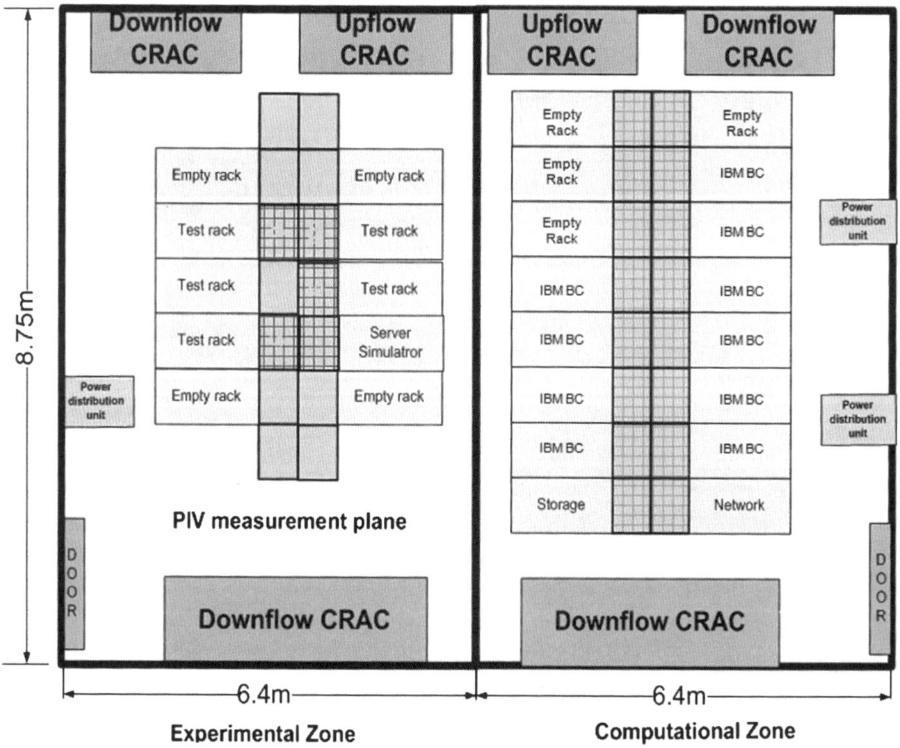


**Fig. 13.7** Experimental data center laboratory layout for the control of cooling resources [4]

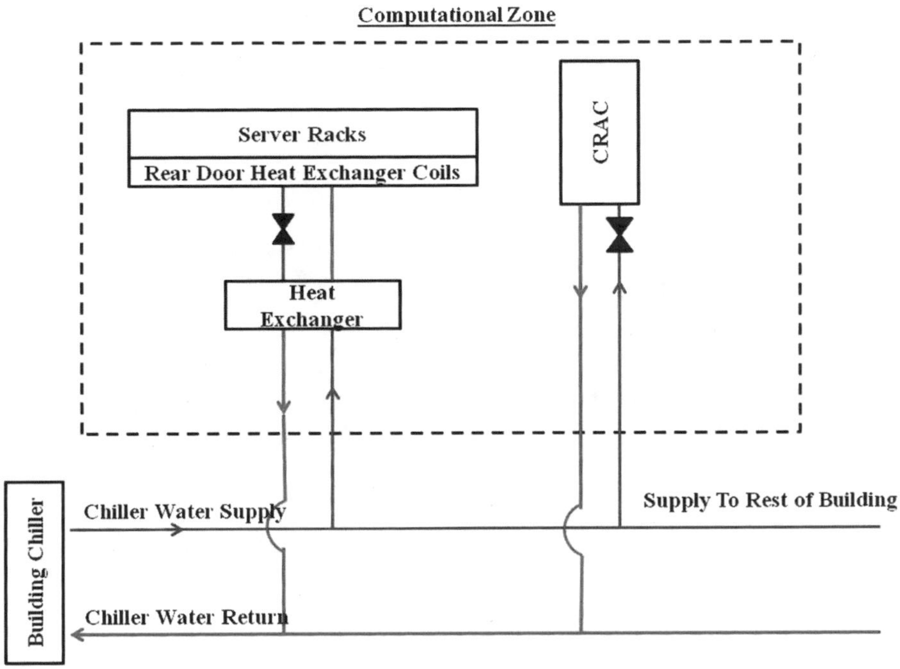


**Fig. 13.8** Chilled water distribution within data center facility [4]

A total of 60 IBM BladeCenter servers were housed in ten racks. Each rack with dimensions 0.58 m × 1.07 m × 2.13 m consists of 6 IBM BladeCenters, each consisting of 14 blades, each with 2 dual core processors, as seen in Fig. 13.9. The server management module provides access to various data points from its sensors. For this experiment, the temperature sensor on the front panel of the server is used. First, a set of experiments are conducted to determine the relationship of CRAC fan speed with the mass flow rate of air passing through the CRAC. Using an energy balance on the cooling coil and the hot air passing around it, correlations are derived as shown in (13.1). The parameters measured are the volume flow rate, $V_{\mathrm{w}}$, the difference between the return and supply water temperature, $\Delta T_{\mathrm{w}}$, and temperature difference between the return and supply air temperature across the CRAC, $\Delta T_{\mathrm{CRAC,a}}$.

$$m_{\mathrm{a}} = \frac{\rho_{\mathrm{w}} V_{\mathrm{w}} C_{p,\mathrm{w}} \Delta T_{\mathrm{w}}}{C_{p,\mathrm{a}} \Delta T_{\mathrm{CRAC,a}}}. \tag{13.1}$$

Runs for fan speeds 60–100% with increments of 10% are conducted and results are correlated using a linear fit given in (13.2). The total electrical load on the

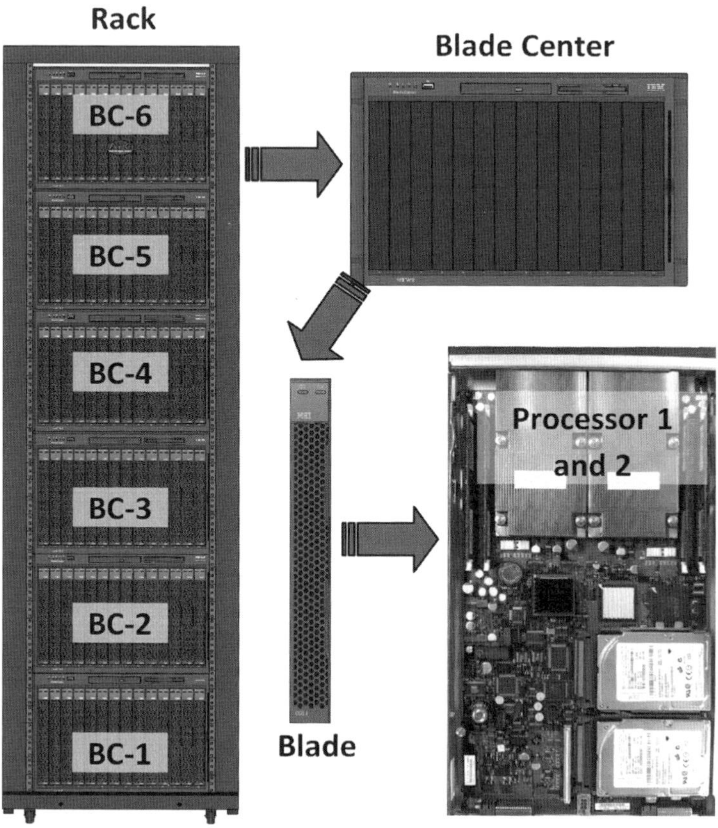


**Fig. 13.9** Blade center rack and dual processor server architecture [4]

servers throughout the experiment is kept nearly constant at 140 kW to reduce the number of variables.

$$m_a = Aw_{fan} + B. \tag{13.2}$$

Next, using the same data set, the ratio of electrical power, $P_{Rack}$, to heat removal rate by the CRAC, $H_{Removed}$, is determined experimentally using (13.3).

$$\eta = \frac{H_{Removed}}{P_{Rack}}. \tag{13.3}$$

The heat being removed was calculated using the mass flow rate and the temperature difference across the chilled water supply and return to the data center.

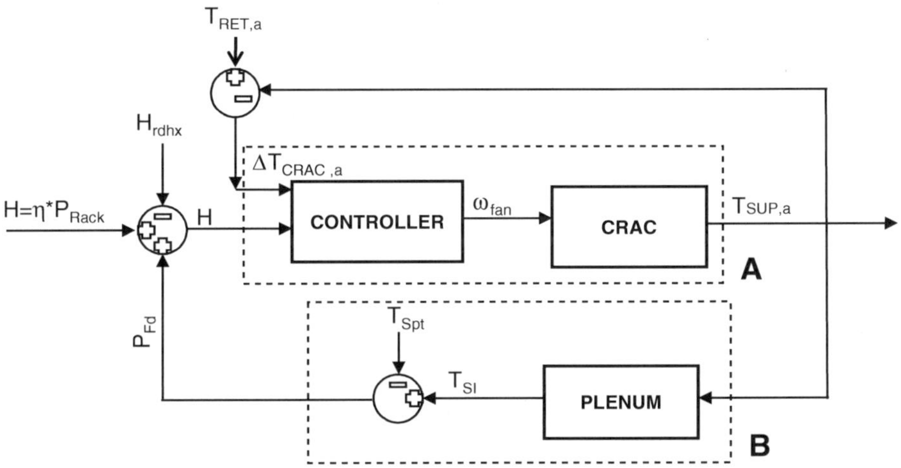


**Fig. 13.10** The control and feedback system for the blade center cooling [4]

Note that throughout this experiment, the RDHx is set to its maximum cooling capacity, which is roughly 60 kW. The constants determined from the steps above are used in developing the controller illustrated in Fig. 13.10.

There are two main parts of the control system described above: the controller and the feedback. The controller requires the estimated heat generation rate from power drawn for all the servers, which is measured through branch circuit monitoring system (BCMS) power monitoring equipment. In addition, the controller is supplied with the temperature difference between the supply and return air across the CRAC. Combining (13.1)–(13.3), (13.4) is used as the controller, where $H_{\mathrm{RdHx}}$ is the measured rate of heat removed by the RDHx, and $\alpha$ specifies the fraction of the heat rate removed through the CRAC. For this experiment, value for $\alpha$ of 0.8 was used. The CRAC fan speeds are controlled using this model and the temperature difference of the air across the CRAC after changing the fan speed is sent back as the input to the controller. $P_{\mathrm{Fd}}$ is the feedback power corresponding to the server inlet temperatures.

$$w_{\mathrm{fan}} = \frac{\alpha(\eta \times P_{\mathrm{Rack}} - H_{\mathrm{RdHx}} + P_{\mathrm{Fd}})}{C_{p,\mathrm{a}}\Delta T_{\mathrm{CRAC,a}}} - \frac{B}{A}. \tag{13.4}$$

The second part of the control system provides a feedback power, $P_{\mathrm{Fd}}$, based on the server air inlet temperatures, $T_{\mathrm{SI}}$. The feedback is described below using a simplifying assumption that all the air supplied by the CRAC is being drawn into the servers and using each server as discrete volume. See (13.5). This equation only applies to server inlet temperatures that exceed the maximum set point temperature, $T_{\mathrm{Spt}}$ of 28°C.

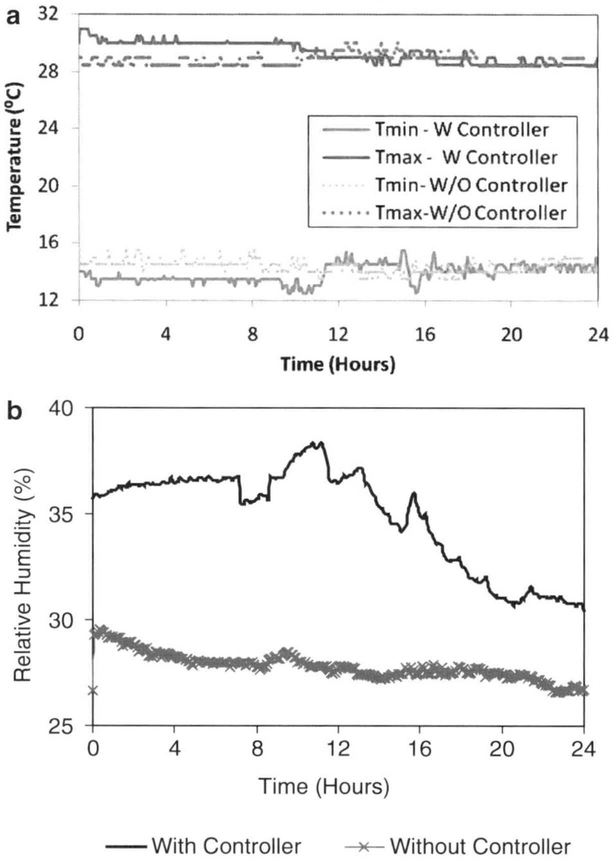


**Fig. 13.11** Minimum and maximum server inlet air temperatures (**a**) and relative humidity (**b**), without and with the controller

$$P_{\mathrm{Fd}} = \begin{cases} \sum \frac{1}{n} \dot{m}_{\mathrm{a}} C_{p,\mathrm{a}} (T_{\mathrm{SI}} - T_{\mathrm{Spt}}) \, \forall \, T_{\mathrm{SI}} > T_{\mathrm{Spt}} \\ 0 \, \forall \, T_{\mathrm{SI}} < T_{\mathrm{Spt}} \end{cases} \tag{13.5}$$

The data were recorded for every 5 min and the controller was programmed to sample every 2.5 min. There are two runs: 1) With controller implemented, and 2) the fan speed kept constant at 100%. The duration of each run was 1 day. Figure 13.11a shows a comparison of the minimum and maximum server inlet temperatures for every 5 min, with and without the controller. There is a reasonable agreement between both data sets. However, with the set point at 28°C, both sets of data had maximum server inlet temperature exceeding the set point limit of 28°C. The maximum temperature observed was 31.5°C for the controller and 30°C without the controller. For finer control of the server inlet air temperature, a better approach would be to manipulate the direction of air flow from the tiles. Figure 13.11b captures the variation of relative humidity throughout the experiment. The relative humidity of the space with the controller running is consistently higher than that at the constant fan speed. However, both runs were well within the recommended operating ranges for humidity.

### 13.2.2 Energy Savings and COP

It is expected that the COP of the chiller would be lower at lower CRAC fan speeds. However, there would also be some savings for the CRAC. Hence, it is important to determine how much extra power is required for the chiller, in comparison to the power reduction for the CRAC. For fan speed 60%, the CRAC draws 2.5 kW, and for fan speed 100%, 5 kW. By integrating both curves in Fig. 13.12 and computing the difference, the savings for a day is determined to be 30 kWh.

Next, effective COPs for the CRAC and Chiller are computed for both runs. The COP is computed using (13.6).

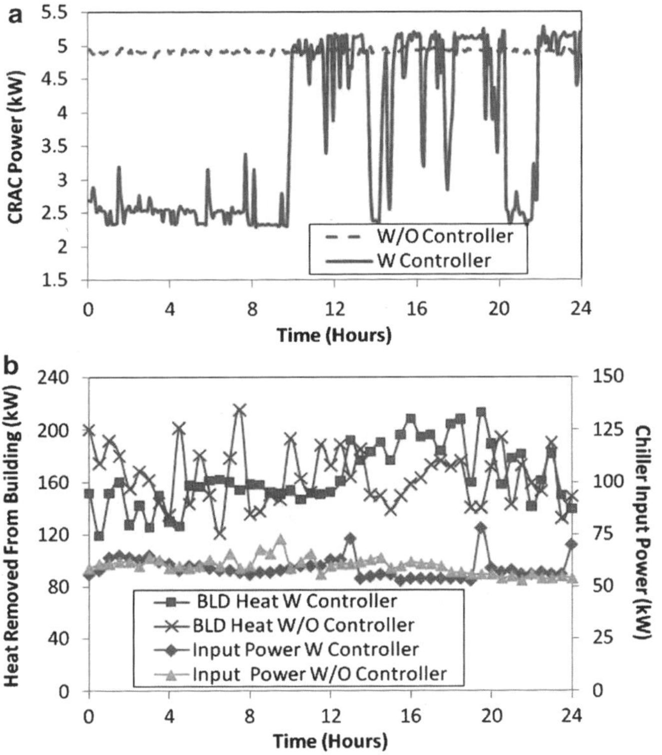


**Fig. 13.12** CRAC power consumption (**a**) and chiller and total building power consumption (**b**), without and with the controller

$$\mathrm{COP} = \frac{\text{Heat removed (kW)}}{\text{Total input power (kW)}}. \tag{13.6}$$

One approach to analyze the overall performance would be to calculate the ensemble COP of the entire data center. However, for this experiment, only the COPs of the CRAC and the building chiller were calculated using (13.6). The COP for the CRAC was computed using the ratio of the total power removed from the HPC zone of the data center by the CRAC to the total power of the CRAC unit for a 24-h period. The COP of the building chiller was calculated as the ratio of the total heat being removed from the building to the total power of the chiller compressors, pumps, and fans for a 24-h period. The data used to compute the averaged COP for a day is given in Fig. 13.12b. The average COP was computed as the ratio of the area under of the curve of the heat being removed from the building to the area under the curve of the input power. The values of COP can be used to compute the required chiller input power. To remove a constant heat load of 110 kW from the HPC zone for a period of 24 h, 950 kWh would be required by the setup with the controller, in comparison to 967 kWh of the setup without the controller. When combined with the savings from the CRAC, there is a savings of 47 kWh for the 24-h period. This is approximately a 6% saving in the total input power to the CRAC and the chiller, compared to the constant speed fan operation.

## 13.3 Holistic Consideration of Energy Consumption for IT and Facilities

In the IT space, workloads are managed using virtual machine (VM) migration capabilities to consolidate applications among servers in a manner that can reduce the power consumption of hardware, while meeting performance requirements. In the facilities cooling domain, a similarly useful mechanism that is being increasingly integrated into modern CRAC units is the ability to change the air flow rate into the data center. Management decisions made to maximize the efficiency of IT power usage can sometimes conflict with load allocations that allow for optimal efficiency of cooling components. For example, a heterogeneity-aware VM allocation policy may decide to migrate workloads to a set of servers that are optimal in terms of energy usage [5], but happen to be physically placed in a localized area of the data center facility. The heat generated by these resources may create a hot spot that requires the HVAC to utilize a higher air flow rate needed with an alternate allocation. Therefore, independent decisions made in these two domains often do not maximize overall data center efficiency.

There has been significant recent work focused on power management of IT resources. Methods have been developed to utilize capabilities such as processor voltage/frequency scaling for power profiles of processors and platforms. Storage resources have also provided a strong opportunity to reduce power and

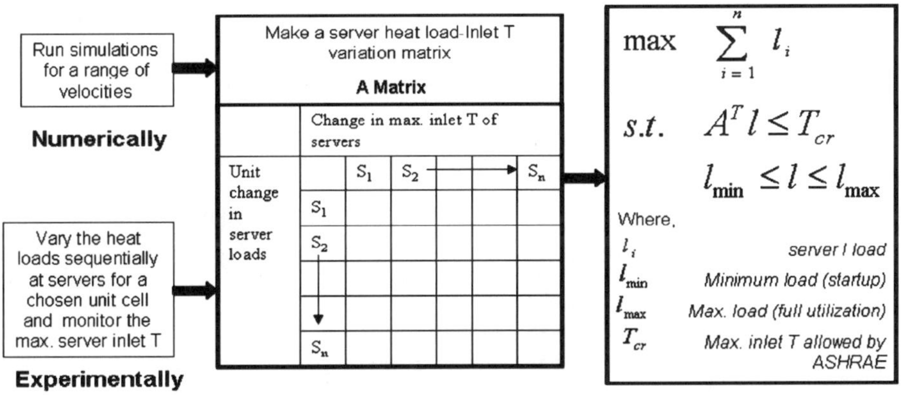


**Fig. 13.13** The ambient intelligence based load management (AILM) algorithm [10]

thermal usage in enterprise systems. At the data center level, power consumption can be reduced by turning servers off and bringing them online based on demand [6]. Other proposed data center management approaches have considered temperature-aware workload placement [7]. Emulation tools that estimate the thermal implications of power management can aid in the offline design of management policies as well [8].

Compared to such tools, and the heuristic-based thermal prediction approach proposed by Moore et al. [9], the ambient intelligence load management (AILM) approach [10] uses a temperature linearity concept when the flow field remains the same and applies a simplex-based optimization framework. The AILM approach is designed to manage a data center from a cooling-centric perspective, by determining within a given air velocity, the heat load limits at each server that must be observed to prevent thermal violations. The guiding principle of the AILM algorithm is the temperature field linearity concept when the flow conditions remain the same inside the room. It is assumed that temperature variations are not high enough to cause appreciable air density variation. Thermal radiation effects are also neglected. Thus, a change in the volumetric heat generation rate of server *i* (for the given facility) present in the room contributes towards the change in the inlet air temperature of server *j* (for the given facility). As a result, a data center can be calibrated based on how much a unit change in volumetric heat generation at server *i* can alter the inlet temperature of server *j*. An overview of the algorithm based upon this premise is provided in Fig. 13.13.

In order to use the AILM tool, it must be calibrated for a data center configuration. As the figure shows, the tool can use either experimental or simulated data. For a given set of supply velocities from the CRAC unit, a baseline is obtained based upon the minimum server power dissipation. For this condition, the maximum temperature at the inlet of each server is noted. Next, for a server, the power dissipated is increased by a unit amount and the system is left to reach steady state, without changing the CRAC velocity. The maximum

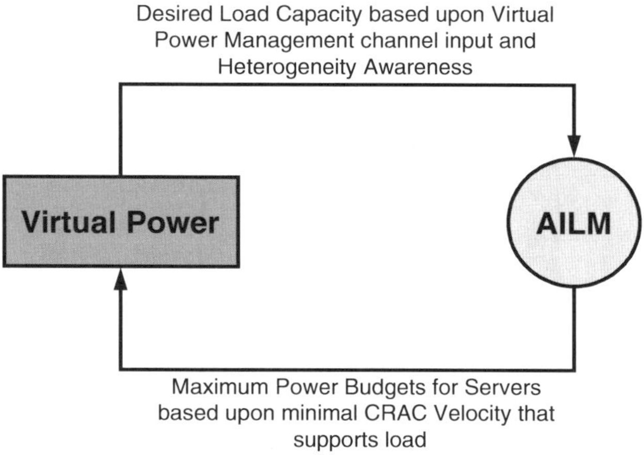


**Fig. 13.14** The CoolIT approach for coordinated management of resources [11]

temperatures at the server inlets are noted again, and the difference from the baseline case is recorded. This difference, when calculated for all the servers, provides an estimate of server *j*'s inlet air temperature sensitivity to server *i*'s heat load, for a given CRAC velocity. This process is repeated sequentially for all of the servers. The outcome is a $n \times n$ matrix of values, for a given CRAC velocity and for *n* servers. By using this method for each reasonable CRAC velocity, the output is the maximum power dissipation a data center's cooling system can handle. To calculate this and the respective power dissipation for each server, we optimize the server loads within the constraints of the maximum and minimum loads of servers and the maximum critical inlet server temperature. Once the above calibration is performed, AILM effectively provides for each velocity $Vx$ a total possible maximum load capacity, as well as a load vector that maps a power budget to each server. This interface information can be used by the IT management infrastructure to manage the data center in a coordinated fashion.

In the CoolIT approach [11], the IT management infrastructure strives to allocate and manage VMs in a manner that minimizes the power consumption of server resources. At the same time, the AILM thermal assessment capabilities determine whether implementing certain power limits on servers would allow for the load capacity being used, but would also provide for a reduced operating point for the CRAC. This results in a cyclical coordination relationship, as illustrated in Fig. 13.14.

As an example, consider an eight rack configuration with rows of four racks each and one cold aisle as shown in Fig. 13.15. The rack labels identify the type of rack (e.g., composed of type "A" or "B" hardware), assuming a heterogeneous equipment configuration. With a naming convention that assumes that racks across the cold aisle are of the same type, the setup depicted in Fig. 13.15 would be an "ABAB" configuration.

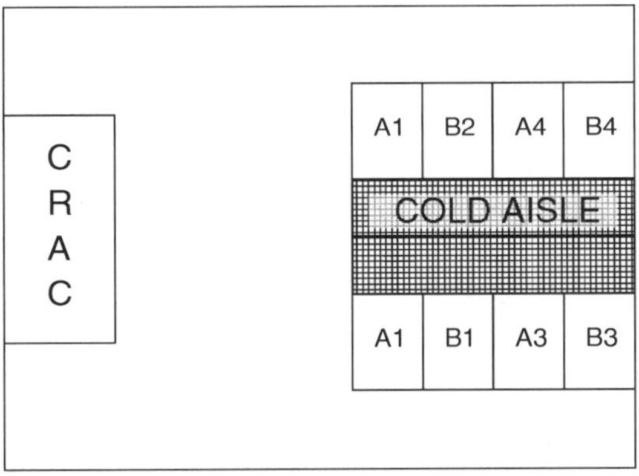


**Fig. 13.15** Application of the CoolIT approach to a data center cell [11]

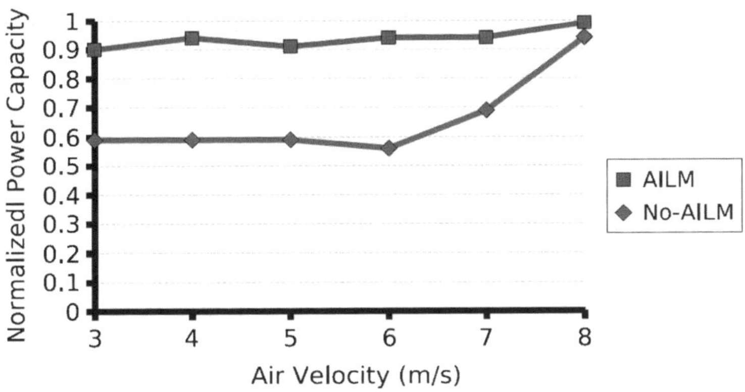


**Fig. 13.16** Application of the CoolIT approach to a data center cell [11]

Starting with a homogeneous data center configuration, we use AILM to determine the overall load capacity the data center can support for a given CRAC air velocity. Figure 13.16 provides the results normalized to the maximum possible load, comparing an "AILM" case to a "No-AILM" case.

The "AILM" case assumes that load is allocated based upon thermal effects captured by AILM to allow for intelligent load management within a given CRAC velocity. In the "No-AILM" case, load is distributed using a simple load balancing scheme where each server is provided equal workload up until the point that the thermal constraints are violated somewhere, thereby constituting a completely thermally unaware solution. As apparent from the figure, with "AILM" knowledge, across all air velocities, the IT system can operate to within 10% of full load. In the "No-AILM" case, however, there are limits of up to 40% for lower air velocities. This difference can be attributed to the fact that the AILM model is able to capture air recirculation effects [11]. These effects can cause significant hotspots at a small set of servers. By allowing an IT manager to be aware of these bottlenecks, close to

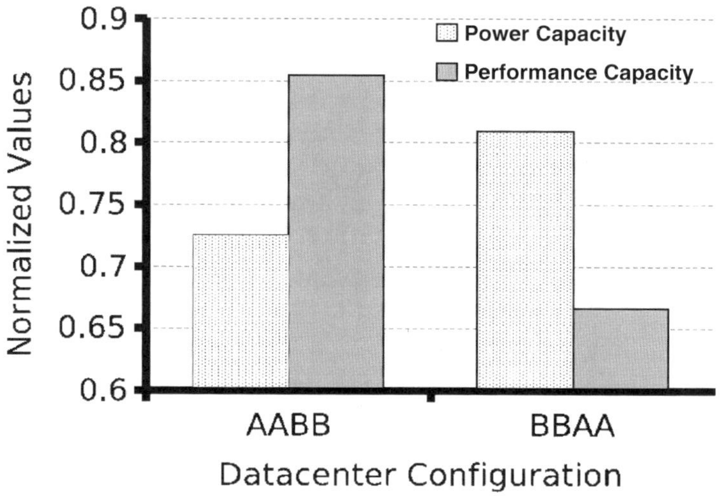


**Fig. 13.17** Power and performance tradeoffs for two rack configurations using the CoolIT approach [11]

maximum load can be supported by simply shutting these systems down. Therefore, in this example, with AILM we can actually utilize a low operating point for the CRAC, as long as we adhere to the set of per server power budgets provided by AILM.

Figure 13.17 presents the power and performance capacity at a particular air flow velocity (7 m/s), but with two different data center configurations. Here, the maximum load capacity that can be supported is higher with the "BBAA" configuration, by 8%. Without IT workload awareness, a physical configuration based upon this AILM result would dictate utilizing such a setup. However, as we observe from the associated performance capacity, the "AABB" configuration can achieve a maximum performance that is 18% higher. These simulation-based results again show that neither IT nor cooling governor alone should dictate management decisions, but, instead, coordination should be applied.

## 13.4 Advanced Cooling Approaches at the Server and Cabinet Levels for High-Performance Systems

A vast majority of servers are currently air-cooled, where heat sinks remove the heat from the processor chips and reject it to an outside air stream created through fans or blowers. Conventional heat sinks are either made of solid high thermal conductivity metal such as aluminum or copper. Newer generation heat sinks for high heat dissipation have a composite construction, which incorporates heat pipes to transport the heat to a fin array. As processor powers increase, the heat sink sizes and cooling air velocities have steadily increased, see Fig. 13.18, increasing both

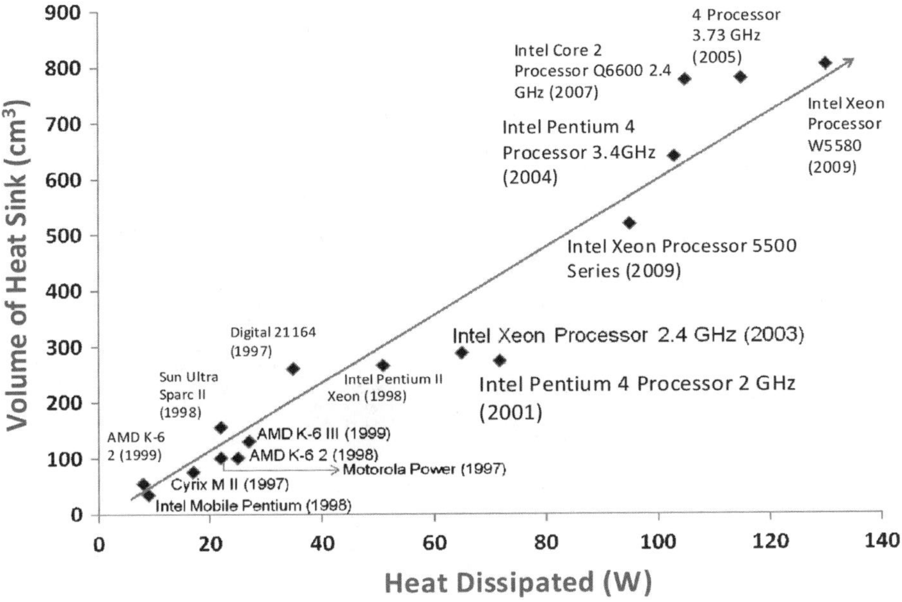


**Fig. 13.18** Increasing microprocessor powers and heat sink volumes

energy consumption and acoustic noise. The narrow form factor and high powers for servers such as blades require compact heat removal devices. This has prompted the exploration and use of liquid cooling for high-performance server applications. Several configurations utilizing liquids are possible.

### *13.4.1 Rack Level Liquid Cooling*

Liquid cooling could be utilized within a rack to cool the hot air from a server cabinet prior to discharge into a hot aisle. This could be done by circulating chilled water through a RDHx [12] or utilizing a circulating refrigerant that evaporates within the cabinet [13, 14]. These approaches are illustrated in Figs. 13.19 and 13.20. The circulating chilled water-cooled RDHxs are widely utilized for cabinets dissipating over ~20 kW. Ever higher cabinet powers are being encountered in applications such as high-performance computing. Use of fin-and-tube heat exchangers to cool internally circulating air in sealed electronic cabinets has been made for more than a decade for military electronics. For server cabinets, Schmidt et al. [12] and Webb and Nasir [15] proposed this concept.

Schmidt et al. [12] explored the use of chilled water-cooled RDHx in an air-cooled data center. This allows co-location of multiple high-density server cabinets in close proximity, while maintaining the existing layout of equipment racks. Based on computational modeling, they predicted a reduction of air inlet temperatures

Photograph courtesy: Coolcentric

Photograph courtesy: NASA/Goddard/Pat Izzo

**Fig. 13.19** Single-phase liquid-cooled rear door heat exchanger, displaying the finned tube rear door [12]

near the top of the cabinets, with RDHx. In conventional air-cooled cabinets, inlet air temperature rise from the bottom to top is a result of hot air recirculation from the hot to cold aisle. This is mitigated due to the reduced exit air temperatures due to the RDHxs. The RDHx was designed to attach to a 42-U rack, with a maximum depth of the door assembly of 143 mm and a mass of 31.7 kg for unassisted lift. Chilled water is delivered to the heat exchanger using flexible 1.9 cm (0.75 in.) supply and return hoses, connected to fluid supply and return manifolds on the heat exchanger via quick-connect couplings.

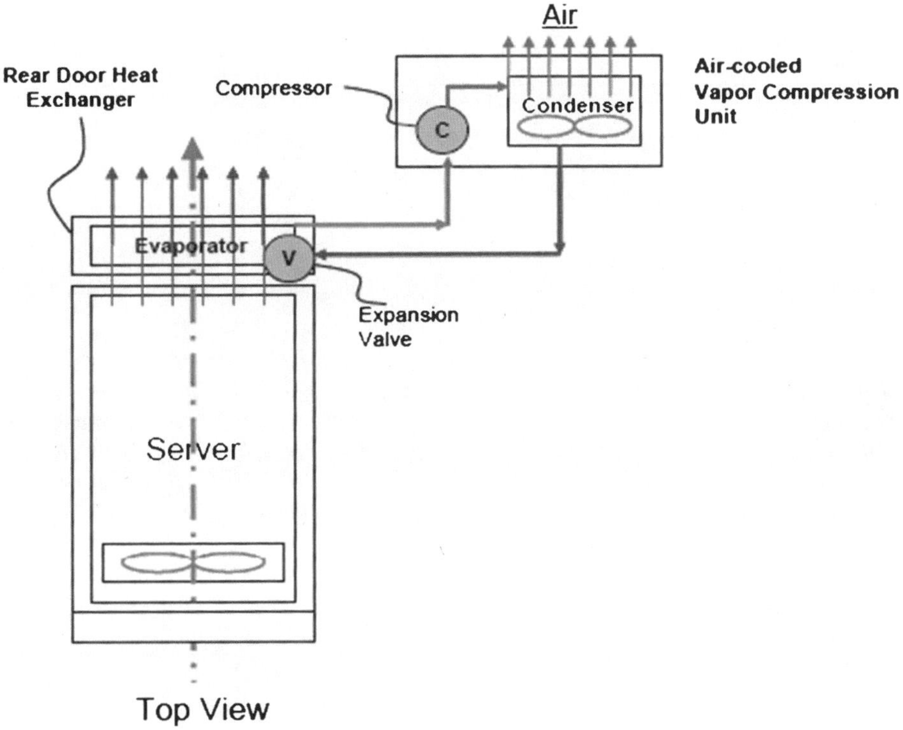


**Fig. 13.20** Schematic of a refrigerant-cooled evaporative rear door heat exchanger [17]

The heat transfer performance of the RDHx was characterized using the commonly employed effectiveness parameter:

$$\varepsilon = \frac{T_{ao} - T_{ai}}{T_{wi} - T_{ai}}, \tag{13.7}$$

where $T_{ai}$ and $T_{ao}$ are the inlet and outlet air temperatures, respectively, and $T_{wi}$ is the water inlet temperature. The pressure drop (in inches of water) across the heat exchanger as a function of the uniform inlet flow rate (in cubic feet per minute) was correlated as follows:

$$\Delta p = 1.52 \times 10^{-7}\ \mathrm{CFM}^{1.72} \tag{13.8}$$

At water flow rates of 0.693 kg/s, the values of ε ranged from 0.69 to 0.5 as air flow rate was increased from ~0.7 to ~1.58 kg/s. At a reduced water flow rate of 0.441 kg/s, the values of ε ranged from 0.64 to 0.43 for the same range of air flow rates.

Schmidt et al. [12] also performed a total cost of ownership analysis to assess the savings achievable through the use of RDHx. Their example considers a 48 IBM eServer BladeCenters installation. Each blade center dissipates 4 kW of

**Table 13.2** Total cost of ownership comparison (in k$) between conventional air-cooled and RDHx arrangements (adapted from [12])

| Cost item | Nature of cost | Conventional (k$) | RDHx (k$) |
|---|---|---|---|
| Electricity costs ($0.07/kWh) | Recurring annual | 118 | 118 |
| Facility lease/rental ($50–$150/ft$^2$) | Recurring annual | 27–82 | 15–46 |
| Facility cooling | Recurring annual | 49 | 46 |
| Facility construction ($800–$1,000/ft$^2$) | One time | 435–544 | 243–304 |
| Cooling hardware (including racks) | One time | 112 | 108 |
| | Total annual | 194–249 | 179–210 |
| | Total one time | 547–656 | 351–412 |

power, resulting in a total power dissipation of 192 kW. Each rack can accommodate up to six BladeCenters. They compare the use of conventional air-cooled racks cooled using chilled water-based CRACs, with RDHxs as an alternative for cooling. In the conventional installation, only 50% population of rack is considered in order to keep the power density manageable, requiring 16 equipment racks. The RDHx installation requires a total of nine racks, eight for the fully populated computing racks and one for housing the coolant distribution unit (CDU) that feeds chilled water to individual RDHxs. The estimated space required to accommodate this system with conventional cooling is 50.5 m$^2$ (544 ft$^2$) and with RDHxs it is 28.2 m$^2$ (304 ft$^2$). The cost of the two solutions is compared in Table 13.2.

The purchase costs of the IT equipment are not included in the above table, as these would be independent of how the equipment will be cooled. For the RDHx installation, the reduced space requirements are due to both fewer racks and one less CRAC. The reduction in the CRAC units occurs since the cooling requirements of 55 tons of sensible cooling are met entirely by CRACs in case of conventional cooling, whereas the RDHx arrangement removes approximately 55% of the cabinet load directly. The reduced facility footprint reduces both construction costs, as well as renting and leasing costs. Table 13.2 shows 40% lower one-time costs and 14% lower recurring annual costs for the RDHx installation compared to the conventional installation.

## *13.4.2 Effect of Tube Cross-Section Shape*

Current state-of-the-art fin tube RDHxs for rack cooling, Fig. 13.19, employ round tubes.

Webb and Nasir [15] investigated two automotive radiator inspired designs of RDHxs. They recommend brass tubes for chilled water flow (at ~18°C) since no corrosion inhibitor is needed. Performance of 9.5 mm diameter round copper tubes expanded on aluminum fins is compared with flat tube designs. They report superior heat transfer performance at low-pressure drop on the air side with one row of flat brass tubes of cross section 1.9 mm × 16 mm with 0.15 mm thick wall and copper

**Table 13.3** Performance of finned flat tube chilled water/air heat exchanger (adapted from [15])

| Design objective | Frontal air velocity (m/s) | Fin pitch (mm) | Q/ITD (W/K) | $P_{air}$ (kPa) | $P_{water}$ (kPa) |
|---|---|---|---|---|---|
| Low air DP | 0.9 | 1.69 | 427 | 0.00343 | 26.5 |
| Max $Q$ (kW) | 1.2 | 1.02 | 631 | 0.011 | 26.5 |

Where $Q$ is the total heat transfer rate and ITD the inlet temperature difference between the hot surface or chip and the inlet water temperature. Using a chip temperature of 85°C and chilled water inlet temperature of 18°C, the value of LTD is 67 K. For these conditions, the maximum heat removal rate is predicted to be approximately 42.2 kW. They point out that this can be further improved by changing the air- or water-side parameters. These calculations can be performed using well-established heat exchanger theory

louver fins 25–40-μm thick. Equivalent heat transfer performance is achieved with two rows of round tubes with 6.8 fins/cm and 44.0 mm deep unit. Performance computations for a rack cooling application with a 0.64-m$^2$ (1.6 m × 0.4 m) finned frontal area were carried out for two alternate design objectives of low air side pressure drop and maximum heat transfer rate. The results are summarized in Table 13.3 for 1.9 mm × 16 mm tube cross section, 8.45 mm fin height, and 40 μm fin thickness. The header length is 0.4 m and the water flow rate is 0.5 l/s in the tubes, with two water passes.

### 13.4.3 *Module Level Liquid Cooling*

Cold plates with internally circulating single-phase or phase change coolant could be brought in thermal contact with heat-generating components such as microprocessors to transfer heat. These are typically fabricated from metal and consist of internal coolant passages, allowing a large heat transfer surface area at low-pressure drop for effective thermal transport into the liquid. The circulation of the liquid could occur using a pump. Alternately, passive liquid circulation using natural convection could be utilized, forming a thermosyphon. Such liquid flow loops reject the heat to the ambient air or to another liquid stream.

Commercial implementation of module level liquid cooling of microprocessor chips has been made recently for high-performance computers. Campbell et al. [16] discussed liquid cooling for the IBM Power 575 Supercomputing node, which was introduced in 2008. The node, packaged in a dense 2U (88.9 mm) enclosure, contains 16 dual core processor modules, each cooled with a cold plate. A single rack can contain up to 14 nodes, with a total power dissipation of 72 kW in a single rack. Copper tubing connects four groups of four cold plates in series. Each group of four cold plates is connected to a common set of liquid supply and return headers. Each cold plate is designed to meet the cooling requirements of a single microprocessor module, with a total power dissipation of ~158 W, such that the average temperature of the core region does not exceed 65°C, for inlet water temperature of 28.8°C at the fourth module. The cold plate design point

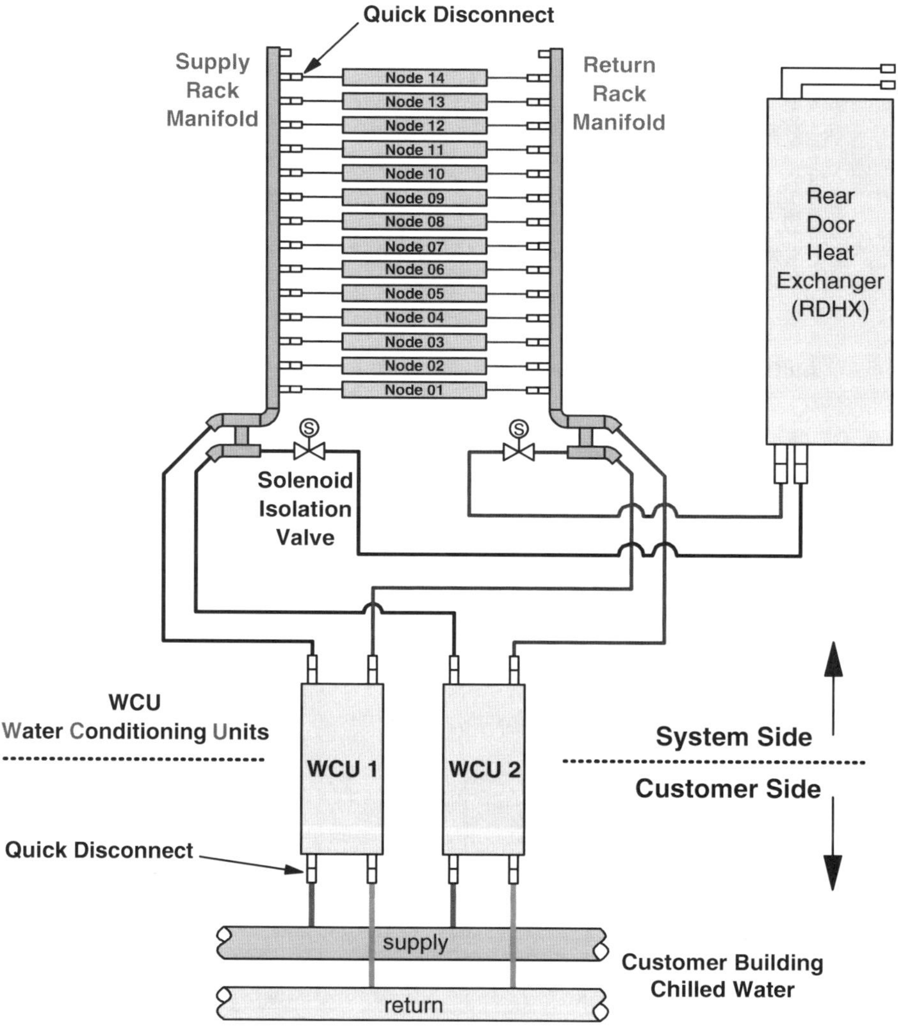


**Fig. 13.21** Chilled water distribution to the liquid-cooled computing nodes and the rear door heat exchanger in the IBM Power 575 supercomputer. Courtesy M. Ellsworth, IBM Corporation

corresponds to a flow rate of $1.32 \times 10^{-5}$ $m^3/s$ (0.21 gallons/min) and a pressure drop of 3.3 kPa (0.48 psi).

Cooling water is supplied to the nodes using the arrangement shown in Fig. 13.21, which also incorporates rack level RdHx-based cooling. Two modular water conditioning units (WCUs) provide the cooling water to the rack level manifolds, the nodes, as well as the RDHx. The WCU provides and controls the flow of coolant at specified temperature, pressure, and flow rate, and rejects heat to the building chilled water, via a plate heat exchanger. During normal operation, both WCUs are required to provide the flow and heat transfer needs for the maximum 14 node configuration. In case of failure of one WCU, the failed unit is shut down, and the flow path between the rack level manifolds and the RDHx is shut, which isolates the RDHx from the water cooling system. The remaining WCU is sufficient to meet the cooling requirements of the 14 computing nodes, without the RDHx.

Xu [17] presents an architecture and performance measurements for module level liquid cooling of high-power servers of 1U form factor. The processor heat is collected by a source plate and conducted to an array of heat pipe evaporators embedded within it. The condenser end of the heat pipe array is connected to a chilled water-cooled cold plate, for ultimate heat removal. The prototype had less than 15 mm total thickness. Thermal resistance of 0.09 K/W was achieved between the heat pipe evaporator base and the liquid inlet in the cold plate at a maximum power of 450 W at a water flow rate of 1 l/min.

### *13.4.4 Chip Level Liquid Cooling*

In order to get the maximum benefits from liquid cooling, which offers an order of magnitude or higher heat transfer coefficients than air cooling, the thermal resistances in the path from the chip to the coolant must be minimized. The best performance can be achieved by providing liquid cooling directly to the chip. Significant progress on this topic has been made during the past two decades. Heat transfer and pressure drop characteristics of single-phase and two-phase cooling at the chip level have been considered in detail in Chap. 12.

## 13.5 System Level Liquid Cooling Architectures

Liquid cooling can be implemented at one or more levels of the packaging architecture, as described above. Hanneman and Chu [18] have explored several system level architectures and compared their energy usage characteristics to standard air-cooled data centers. The specific architectures considered are: RDHx-based single-phase cooling with remote heat rejection to ambient air (Fig. 13.22), use of refrigerant-based heat exchangers at the rack level (Fig. 13.23), water-based individual module level cold plates (Fig. 13.24), and refrigerant-based individual module level cold plates (Fig. 13.25). They consider the scenario of accommodating 240 racks with powers increasing from 15 kW/rack to 30 kW/rack over two future generations, while keeping the floor area invariant at 1,113 m$^2$ (12,000 ft$^2$). This implies an increase in facility power density from 3,234 W/m$^2$ (300 W/ft$^2$) to 6,468 m$^2$ (600 W/ft$^2$) to be achieved through advanced cooling. Two locations, one in Texas and another in Minnesota, were considered, with a significantly higher heat rejection area or external condenser units needed for Texas due to the higher dry-bulb temperatures.

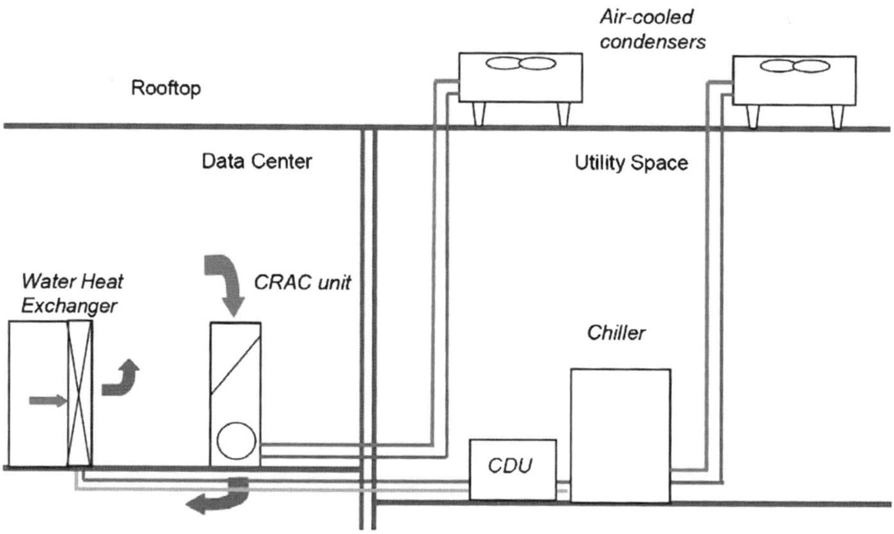


**Fig. 13.22** Rear door heat exchanger based cabinet cooling with remote heat dissipation (© 2007 ASME, reprinted with permission [18])

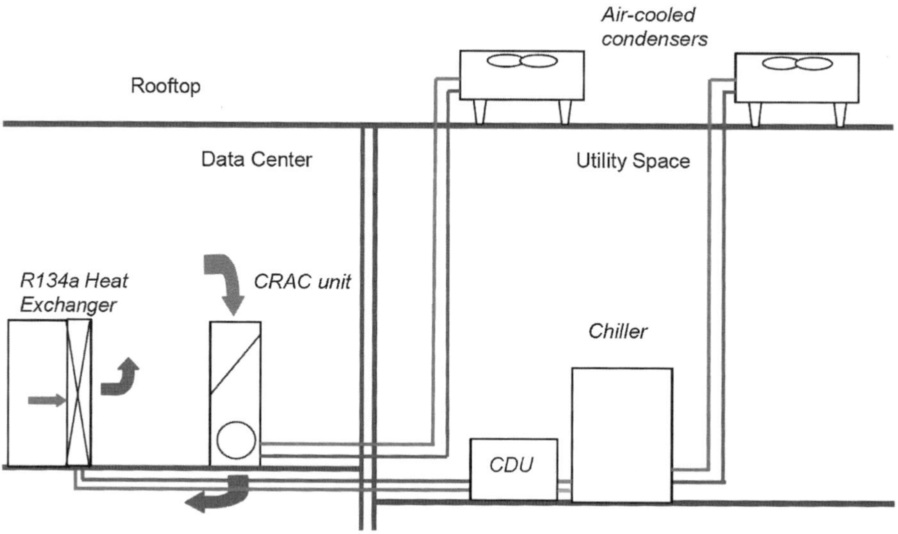


**Fig. 13.23** Rear door heat exchanger cooling with coolant delivers unit and remote heat dissipation (© 2007 ASME, reprinted with permission [18])

Hanneman and Chu [18] conclude from their analysis that the move to the 30-kW racks in their example would not be achievable with the air-cooled approach, without the construction of significant new data center space. The use of liquid cooling, however, would make such an expansion in capability easily feasible.

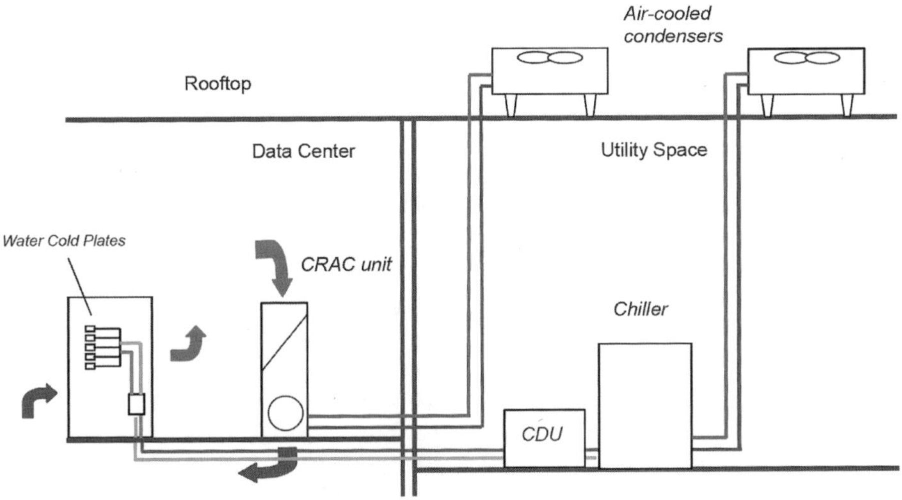


**Fig. 13.24** Module level single-phase liquid cooling using chilled water and remote heat dissipation (© 2007 ASME, reprinted with permission [18])

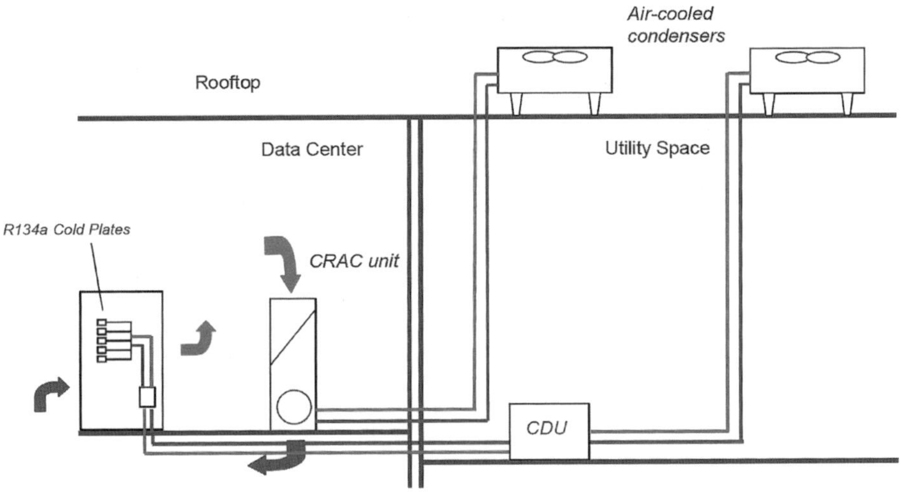


**Fig. 13.25** Module level refrigerant-based evaporative cooling, without a refrigerant chiller and with remote heat dissipation (© 2007 ASME, reprinted with permission [18])

For all four advanced cooling configurations, the analysis assumes that for each rack 15 kW is removed by CRAC supplied air flow, and the remaining 15 kW is removed by the liquid cooling solution. It is found for the assumed input data that the RDHx is the easiest to implement as a transitional approach from air cooling, since RDHx units can be easily integrated within cabinets. Also, the chilled water and CDU can usually be integrated within many existing facilities. The refrigerant-based rack level cooling reduces coolant flow rates compared to water, but requires a refrigerant chiller (Fig. 13.20) in order to meet the rack cooling air temperature requirements. This solution is found to offer only marginal advantages in capital and operational costs over the RDHx approach.

The module level cold plates using either water or refrigerant were able to easily meet the higher rack power requirements. The water-cooled module configuration required a chiller to supply water at temperatures as high as 38°C to achieve a maximum chip temperature of 100°C. The chiller was assumed to be the same as in the RDHx solution. The water flow rate per rack was estimated at $6.3 \times 10^{-4}$ $m^3/s$ (10 gallons/min), which is significant. The cooling power required was about 5% lower, and the capital and operating expenses for this solution were found to be comparable to the rack level liquid cooling approaches. For module level cooling using a refrigerant, the calculations were performed for R134a with an evaporative cold plate per server. The refrigerant vapor was condensed outside the data center. The saturation temperature was assumed to be 70°C, which provided maximum chip junction temperature below 100°C even for the hottest day in Texas. This could be accomplished without using a refrigerant chiller, as seen in Fig. 13.25. The refrigerant flow rates were found to be 25% or less than the water flow required in the module level water cooling case. Cooling power requirements were about 50% of other options for this case. The capital expense was about 33% lower, and operational expenses about 15% lower, assuming electricity costs of $0.10/kWh.

Ellsworth and Iyengar [19] performed an energy efficiency comparison of module level water-cooled servers, with air-cooled servers of comparable performance. The water-cooled system involved 4.7 GHz IBM Power 575 Supercomputing node-based racks, each dissipating 72 kW, with 80% of it going to the water. The fully air-cooled version of the same system developed by Hitachi operates at 3.5 GHz and dissipates 61 kW per rack. Three factors were cited as the key advantages of liquid cooling over air cooling. First, the 34% higher processor frequency provided a commensurate increase in performance, which would not be possible with an air-cooled server in the same 2U form factor. Second, the chip temperatures are about 20°C lower with liquid cooling, which results in lower leakage current in the chip transistors, and hence improved power utilization. Third, the cooling power consumption is reduced with liquid cooling due to the improved efficiency of the liquid delivery CDUs and RDHXs compared to CRACs.

The energy savings are analyzed by considering a 12-rack water-cooled cluster, and comparing its performance with a 16-rack air-cooled cluster. The control volume for the energy usage analysis was the facility, which included data center space conditioning devices such as CRACs, WCUs, building level pumps, chiller, and the cooling tower. The analyzed configurations are shown in Fig. 13.24. Water cooling with 20% power removal from the IT equipment by air and 80% by liquid was found to save 45% operational power at the facility level. Increased effectiveness of water cooling to remove 100% of the IT load increases the operational energy saving to 50%. Incorporation of water-side economizer can further reduce the power used to generate subambient chilled water in cold climates. Incorporation of water-side economizer for the CRACs reduced the cooling power consumption by 92% compared to the air-cooled facility. While these numbers will change depending upon the assumptions made in the analysis, the trends are reflective of the significant energy efficiency improvements that can be brought out by incorporating liquid cooling and careful usage of the ambient environment for cooling.

## 13.6 Thermal Storage

Storage of chilled water or ice is increasingly being utilized in newly constructed data center facilities to provide transient cooling. The most common reason for incorporating this is to provide additional cooling capability for several minutes in the event of a chiller shut down or failure. The most common reason for this may be a sag of 85% or below nominal voltage, or complete loss of utility power. As seen in Chap. 1, the most common scenario under such situations would be one where the uninterrupted power supply (UPS) would maintain the operation of servers, while the backup generators come on line to re-start the chillers and the CRAC units. For moderate-to-high data center IT loads, the air temperatures may exceed the maximum desirable value of 35°C (95°F) well within a minute. The use of chilled water or ice storage would allow handling this emergency situation and allow cooling of the IT equipment for several minutes.

Garday and Housley [20] describe the use of chilled water storage at an Intel regional hub data center facility. The conceptual layout of the chilled water storage is seen in Fig. 13.26. Two 24,000 gallon tanks containing water at 5.6°C (42°F)

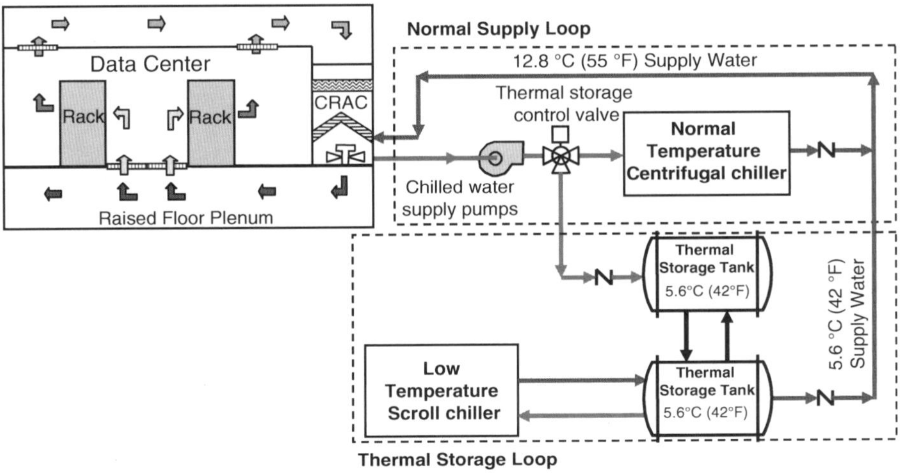


**Fig. 13.26** Dual tank-based thermal storage system [20]

allowed successful operation during an outage lasting several hours during 2006. Once the chillers stopped working, water from the storage tanks was added to the chilled water system to continue to maintain 12.8°C (55°F) water delivery temperature to the CRAC units. Chilled water pumps and CRAC fans were on UPS power for continued operation. The servers continued to operate for more than 15 min following the outage, due to the relatively small IT load on the system during the period. The cooling continued long enough afterwards to insure removal of the stored heat. The cost of the storage system was found to be significantly lower than putting the chiller on a UPS and standby generator.

Schmidt et al. [21] discuss the infrastructure design for the Power 6 575 high-performance cluster at the National Center for Atmospheric Research in Boulder, Colorado. Each of the 11 racks in this cluster generates 60 kW. Module level water cooling and RDHxs remove 80% of the heat generated by each rack, and the remainder is removed by the CRAC units. Two 1,500 gallon thermal storage tanks were employed. The storage system was designed so that the chilled water supply temperature to the Power 6 575 was not to exceed 17.8°C (64°F) for at least 10 min following a chiller failure. The tanks were made of carbon steel and externally insulated. Each tank was 145 cm (5′6″) in diameter and 2.13 m (7 ft) tall. Schmidt et al. [21] noted that considerable prior literature is available on the storage tank design due to the importance of solar energy storage, resulting from mismatch between supply and demand. The importance of stratification effects, which result in the settling of cooler denser liquid layers near the bottom, and warmer less dense layers near the top of the tank was noted. The importance of the aspect ratio of the tank, the ratio of height to diameter, was also pointed out. However, the design requirements for chilled water storage for data center cooling are quite different than solar energy storage, and more specific calculations are needed. Schmidt et al. [21] performed these for the specific dynamic conditions of the NCAR installation.

A thermosyphon-based storage concept was investigated by Wu et al. [22]. This system is best suited for cold climates and utilizes the one way nature of heat flow in a thermosyphon. As seen in Fig. 13.27, in the terrestrial gravity field if the device is oriented with the condenser above the evaporator, the working fluid within the thermosyphon will absorb the heat, resulting in a cooling effect. The hot working fluid will become lighter due to reduced density and by natural convection rise within the thermosyphon. It will reject the heat at the top in the condenser if the ambient temperature is lower, and sink towards the evaporator, completing the circulation pattern. If, on the other hand, as seen in Fig. 13.27, the ambient temperature is above the condenser temperature, or the evaporator is above the condenser, buoyant circulation of the working fluid would not be possible, and the thermal storage device will not work. Thus, the thermosyphon displays a thermal diode character.

Wu et al. [22] carried out performance calculations for a 8,800-kW data center consisting of racks cooled via module level water cooling. This heat is rejected to cold water supplied by an external chiller, along with a thermosyphon-based chilled water or ice storage. The thermosyphon tube is made of stainless steel and the fins are aluminum. The working fluid is R134a. The conceptual schematic is seen in Fig. 13.28. The water-based cold energy storage system at 25°C is designed for

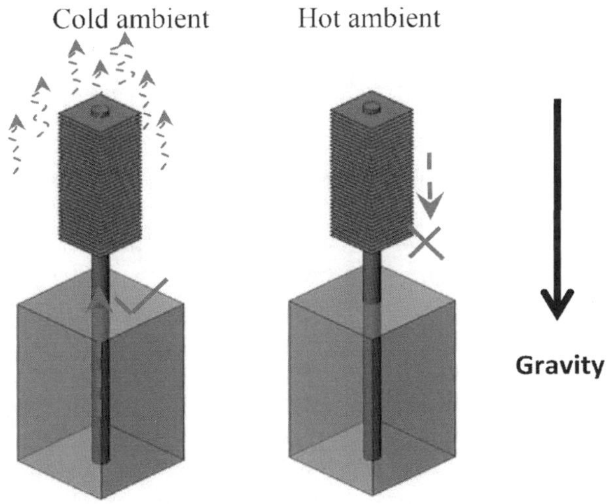


**Fig. 13.27** The thermal diode effect in a two-phase thermosyphon [22]. In the configuration on left, heat rejection to the ambient surroundings results in cooling of the working fluid which sinks down due to gravity. In the configuration on right, heat rejection to the surroundings is not possible and no circulation of working fluid within the thermosyphon exists

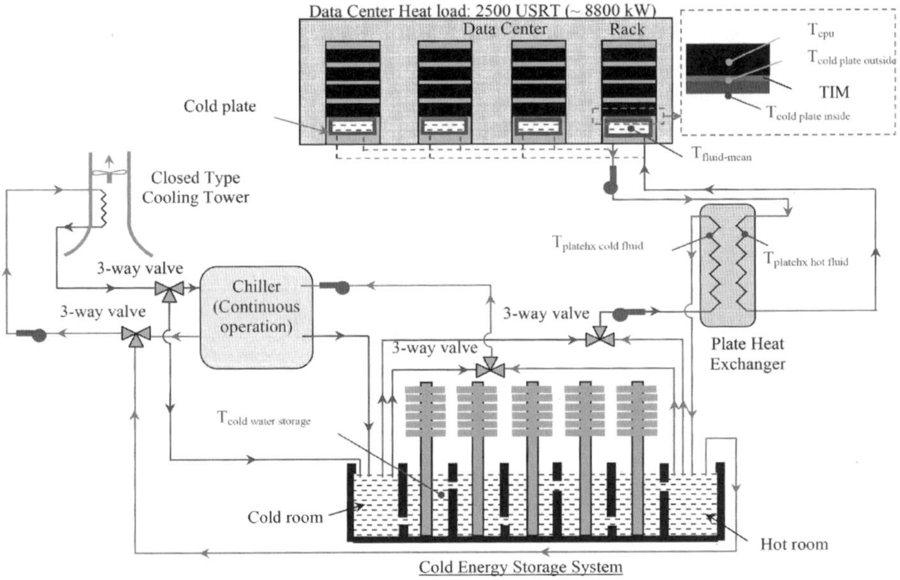


**Fig. 13.28** Thermosyphon-based chilled water/ice storage system for cold climates [22]

Poughkeepsie, New York, where the average annual temperature is 10°C and the average wind speed is 1.68 m/s. Depending upon the below freezing days per year, the storage system could redesigned to provide either chilled water or ice. Wu et al. [22] found that the heat pipe-based storage system is able to handle the entire data center heat removal requirements for winter. During fall and summer, partial use of the chiller is needed, and during summer months the chiller use is a requirement. An individual thermosyphon module was tested in Japan by Fujikura over a 25-day period, when the temperature was below 0°C. Approximately 113 kg of ice were formed, which agreed well with their model predictions.

## 13.7 Corrosion Management

Increasing trend towards utilizing ambient air for cooling to achieve gains in energy efficiency of data centers has raised concerns about the air quality within the data centers. This is particularly true in cities with poor air quality. Also, since 2008, ASHRAE in concert with the IT equipment and facilities vendors and operators has continually expanded inlet air temperature and humidity range guidelines for IT equipment operation. Both of these trends can potentially impact IT equipment failure rates due to increased corrosion. As discussed by Klein et al. [23], contamination risks of two kinds exist: gaseous and particulate based. Particulate contamination can be managed through appropriate filtering of the air. In case of gaseous contamination, it is recommended that copper and silver contamination rates be maintained below 300 Å/month.

The Instrumentation, Systems, and Automation Society [24] has identified four severity levels based on copper reactivity rates that could be applied to data center operating environments. The G1 level, below 300 Å/month, is classified as mild. The G2 level with moderate contamination levels is classified between 300 and 1,000 Å/month. The G3 level or harsh environment is classified as one with reactivity rates between 1,000 and 2,000 Å/month. The G4 level or severe environment corresponds to reactivity rates above 2,000 Å/month.

Klein et al. [23] present the development of a high-sensitivity real-time corrosion monitoring technology, which allows determination of spatial and temporal variation of corrosion rates within a data center. They advocate high-sensitivity corrosion rate measurements that can guide a corrosion management strategy. By placing corrosion sensors both inside and outside a data center, the facility operator can prevent outside air being used for cooling when pollution levels are high. Contamination monitoring is critical in areas which have shown higher corrosion rates than 300 Å/month, such as New York, California, and New Jersey in the USA. The corrosion sensor described in [23] is capable of measuring at the level of single atomic layer (~1 Å) and has a sensor lifetime of over 2 years. In their sensor, they monitor multiple film thicknesses of Cu and Ag simultaneously. Once the thinnest film is consumed by corrosion, the next higher thickness film is investigated, and subsequently the next higher thickness, and so on. The sensor consists of thin metal structure, on a glass or ceramic surface. Once the metal films are exposed to the corrosive environment, they transform into nonconductive corrosion products such as $Ag_2S$, $Cu_2S$, and $Cu_2O$. This results in a change in film thickness and an increase in film resistance that can be detected. This allows the determination of the loss in film thickness over a given time and hence the corrosion rate.

## 13.8 Safety and Health Issues

The safety and health of data center operators are increasingly become items of concern. A key area of focus has been the higher temperatures to which the operators are exposed to. This includes both higher air temperatures and hotter surfaces that are in thermal contact with the electronic components.

### *13.8.1 Air Temperatures*

As equipment power densities in data centers increase, air temperature variations within the facilities have the potential to be larger as well. In the standard HACA arrangement, hot air recirculation near the top servers in racks is well documented. This undesirable mixing of the exhaust air with the incoming cold air reduces the hot-aisle temperatures and the overall cooling efficiency. One way to address this is by utilizing hot-aisle containment, as described in Chap. 2. This increases the average air temperature within the hot aisle, since this air is physically prevented from leaving the hot aisle and mixing with the cold-aisle air. In a white paper, Fink [25] explores the effect of hot-aisle containment on the increase in air temperatures, which could potentially cause elevated heat-related stress in the operators.

Fink [25] points out that the commonly employed dry-bulb temperature is not a realistic measure of heat stress, as it does not take into account the effects of relative humidity, and any radiant heating. The wet bulb globe temperature (WGBT) is suggested as a more realistic measure for assessing heat stress effects, as also recommended by the Occupational Safety and Hazard Administration (OSHA). In the absence of solar irradiation, the WGBT is determined as follows:

$$\mathrm{WGBT} = 0.7\ \mathrm{NWB} + 0.7\ \mathrm{DBT}, \tag{13.9}$$

where NWB is the natural wet bulb temperature, which is a function of both the dry-bulb temperature and the relative humidity. This is the temperature that would be measured by a mercury thermometer, whose bulb was covered by a water saturated wick. Depending upon the room relative humidity, there will be evaporation from the wick and associated latent heat removal, which would result in a reduction of the dry-bulb temperature. This cooling effect would be equivalent to what would be achievable by a data center operator under identical conditions. The DBT is the dry-bulb temperature. Fink [25] has considered a computational fluid dynamics model of a 40-kW data center with 3 kW/rack power dissipation. The results indicate that the hot-aisle containment results in a slightly higher WGBT value of 23.1°C, compared to 22.6°C for the HACA arrangement. Both of these are within the OSHA maximum value of 30°C. As the author notes, these values are based on simulations and could vary drastically for higher rack powers and for larger facilities. However, they do bring out the need for assessment of safety implications of thermal design changes for improving energy efficiency.

### *13.8.2 Surface Touch Temperatures*

Surface temperatures are an important consideration in the design of electronic products. Increasing compactness and functionality of these has resulted in higher surface temperatures. Maximum surface temperature criteria have been based on burn thresholds for human skin in contact with hot surfaces. Roy [26] points out that the safety criteria provided by industry and government organizations have been commonly based on material temperatures and not skin contact temperatures. They also suffer from a number of other limitations. For example, the materials are classified into three groups: metals, ceramics/glasses, and plastics/insulators. Even though the material properties within these categories vary widely, further guidance is usually not forthcoming. Also, maximum allowable/recommended temperatures are typically based on qualitative contact criteria, such as "momentary" or "short duration." Finally, all standards are based on constant and uniform temperature of the contacting surface, which is typically not true in smaller portable products, which may also have localized hot spots.

Roy [26] recommends the following correlation for the estimation of maximum allowable surface temperature for the entire range of contact times:

$$T_{\max} = 41 + 2 \times 10^{-4}\left(\frac{t}{T}\right)^{-0.4} \exp\left[2.5 \times 10^{3} t_{\mathrm{sk}}\left(\frac{t}{T}\right)^{-\frac{1}{5}}\right] \times \left(\frac{1}{\sqrt{K\rho c}} + 0.00075\right) + 5F(t) \qquad (13.10)$$

where $T_{\max}$ is the maximum allowable temperature in °C and $t$ is the time in s. Also, $t_{\mathrm{sk}}$ is the skin thickness (90 μm), $k$ the thermal conductivity of the material in W/mK, $\rho$ the density in kg/m$^3$, and $c_p$ the specific heat in J/kg K. Also,

$$\text{For } t \geq 2\text{s} : T = 2.5 \text{ and } F(t) = \frac{1}{1 + \exp(-0.25(t-5))} - (1 - \exp(-0.01t^{\frac{1}{2}}))$$
$$\text{For } t = 1\text{s} : T = 1.0 \text{ and } F(t) = 1. \qquad (13.11)$$

The above equation covers casual accidental contacts, with $t = 1$ s. Also covered are short duration, momentary contacts with knobs with $t = 3$–$4$ s, continuous normal contacts with operating handles and knobs with $t = \sim 10$ s, and prolonged continuous contacts with $t = \sim 1$–$10$ min.

In cases where the object temperature varies with time, Roy [26] suggests using a cumulative skin damage model based on the above equation. The cumulative skin damage fraction $f_\mathrm{d}$ is calculated as:

$$f_\mathrm{d} = \int_0^{t(\mathrm{total})} \frac{\mathrm{d}t(T)}{t_{\max}(T)} \leq 1, \quad \text{for } T > 42^\circ\mathrm{C}, \tag{13.12}$$

where d$t$ is the actual time increment spent at the temperature $T$, where $t_{\max}$ is the maximum allowable time for that particular temperature calculated from the above equation.

Roy [26] finds that the current safety standards for touch times may be quite inaccurate and result in safety hazards for some of the newer higher $k$ materials, such as ceramics and filled polymers, as well as low-density materials, such as metal foams. For example, for plastics, the current standards suggest the maximum allowable temperature of 75–85°C for a short duration contact of ~3–4 s. However, based on the above equation, with $k = 20$ W/mK, $c_p = 900$ J/kg K, and $\rho = 1{,}840$ kg/m$^3$, the maximum allowable temperature should be ~61°C.

## 13.9 Acoustic Noise in Data Centers

Anyone who has set foot in a modern data center can attest to the noisy operation of these facilities. It is common for operators to wear ear protection if they are going to be inside the facility for a long time. In the last few years, some attention is beginning to be focused on this issue. Acoustic noise is becoming an increasing concern due to increasing server powers and compact form factors, requiring larger cooling air flow rates. Another contributing factor is the expansion of the operating envelopes due to energy conservation initiatives. ASHRAE has provided very timely updates on these since 2008, and the 2011 update [27] provides a very good summary of the acoustic noise concerns driven by increasing server air inlet temperatures. The workplace noise limits are set by the OSHA in the USA and by the EC Directives in Europe. The 2011 ASHRAE document [27] brings out the empirical fan law prediction of the sound power level being proportional to the fifth power of the rotational speed. The difference between two sound powers, $L_\mathrm{w}$, in decibels is given by:

$$L_\mathrm{w} = 10 \log_{10}\left(\frac{P_1}{P_0}\right), \tag{13.13}$$

where $P_1$ and $P_0$ are the sound powers. A 20% increase in speed would thus correspond to a ~3.9-dB increase in noise level. In 2008, the ASHRAE guidelines raised the recommended operating server inlet air temperature from 25 to 27°C. It is anticipated that a 3–5 dB increase in noise may result as a consequence. The 2011 update also presents estimates of expected increase in A-weighted sound power levels at several higher server inlet air temperatures above 25°C. The A-weighted scale (dBA) accounts for the sensitivity of human ear to certain frequencies. At 30°C, the expected increase in sound power emission is 4.7 dBA, at 35°C it increases to 6.4 dBA, at 40°C to 8.4 dBA, and at 45°C to 12.9 dBA. These numbers also depend on variables such as rack configuration, cooling scheme, and fan speed algorithms used. While these noise emission levels are instructive, the safety concerns center around the noise *imission* levels that employees and other support personnel may be exposed to. ASHRAE 2011 update advises data center managers to consult with acoustical or industrial hygiene experts for locally applicable regulations. The workplace noise levels of concern are expressed as A-weighted sound pressure levels, instead of A-weighted sound power used for noise emission. When the pressure level exceeds 85 dB(A), hearing conservation programs, including baseline audiometric testing, noise level monitoring, noise hazard signage, education, and training are needed. Beyond 87 dB(A) in Europe or 90 dB(A) in the USA, mandatory hearing protection and rotation of employees are necessary.

As stated in the 2011 update [27], modeling and predictions of server racks in the standard HACA configuration, with front to back air flow across racks, show typical levels of 85 dB(A), when each of the individual racks have measured power levels of about 84 dB. If this is considered, the starting value at 25°C inlet air temperature, then based on the above predictions, the sound pressure level at the center of the aisle would be expected to increase to 89.7 dB(A) at 30°C inlet, 91.4 dB(A) at 35°C inlet, 93.4 dB(A) at 40°C, and 97.9 dB(A) at 45°C inlet. These levels are well above the regulatory levels and would require mitigation.

Researchers and industry have recently started focusing on active approaches for noise monitoring and mitigation in data centers. Sommerfeldt et al. [28] explored the feasibility of utilizing active noise control to reduce data center noise. Passive acoustical absorption materials, such as foam or fiberglass panels, were considered inapplicable due to their flammable nature and also due to their release of particulates that could trigger early warning by fire detection systems. Active noise control uses microphones and loudspeakers to observe the noise environment and to emit a sound signal that cancels out the noise. Three methods were used to determine the feasibility of using active noise control in data centers. These included measured acoustical data, computational modeling, and device testing. The study concludes that active noise control of a large facility is not feasible due to the random nature of the noise and the high modal density of the data center. Modest attenuation in the range 2–4 dB was achievable in laboratory experiments for single noise source, indicating that active noise control may be feasible within the cold aisle. Noise canceling technology has been commercially implemented (www.silentium.com) as part of a six-server, high-performance computing "personal cluster" for deployment in an office environment, instead of a data center. Fan noise is reduced by canceling the original noise with an antinoise signal by up to 10 dB.

## 13.10 Modular Data Centers

A traditional bricks and mortar data center design and construction can take 18–24 months or longer. Modular or container data centers offer the appeal of rapid deployment, typically within 4–5 months, with the advantage of added flexibility. These were first promoted by and started to appear in large-scale production data centers operated by large IT companies within the past 5 years. A number of drivers may be behind this trend, including the ability to add capacity incrementally on an as needed basis, possibility of easy IT equipment refresh by swapping entire containers filled with IT equipment and possible gains in energy efficiency under certain conditions. Presumably due to perceived growing interest in such facilities, the number of suppliers of these units has sharply increased within the last 2–3 years. Some of the commercially available modular units are seen in Figs. 13.29 and 13.30. Figure 13.31 shows how multiple modular data centers can be combined to increase capacity.

Modular data centers may be of interest under a number of scenarios [29, 30]:

*Remote data center*: Capacity could be installed and operated at a location different from an existing data center. Drivers for such deployment may include lower energy costs, emerging markets, with limited critical data center design and build skills, and space challenges.

*Data center expansion*: An existing data center may be reaching its space, power, or cooling capacity. An expansion close to an existing facility can be accomplished through a modular addition. Specific situations requiring this may include need for

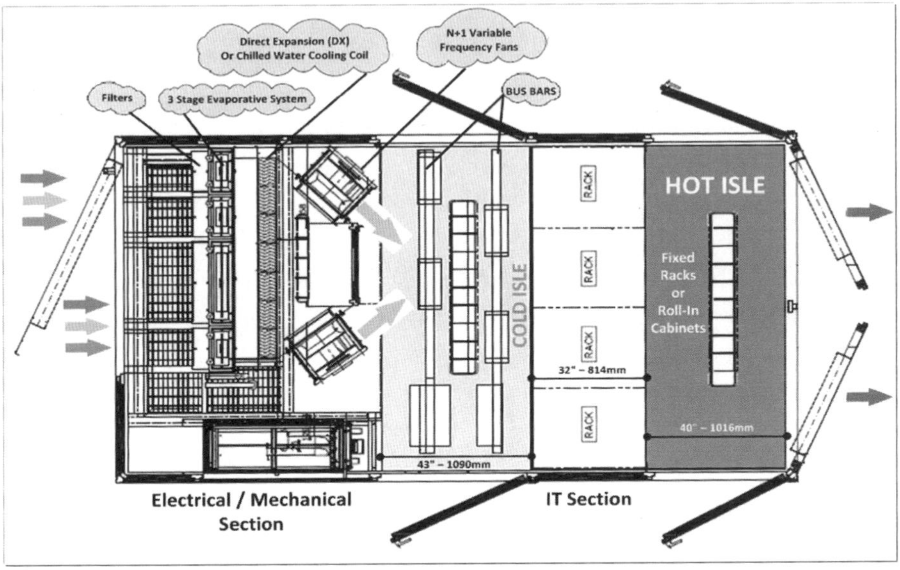


**Fig. 13.29** ICE cube air layout top view for a rack module. Courtesy of SGI

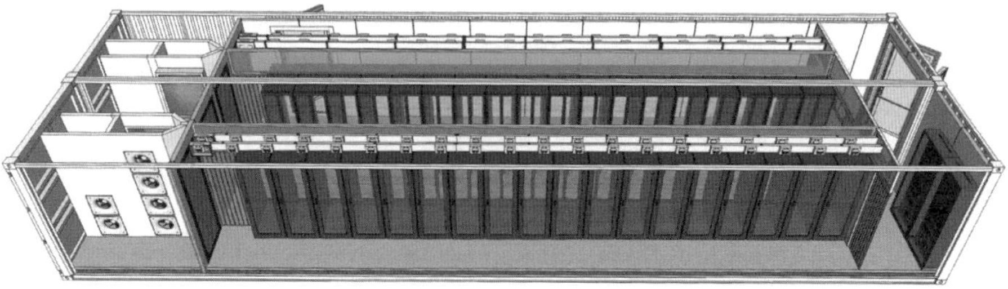

**Fig. 13.30** A single aisle multirack modular data center. Figure courtesy of IBM

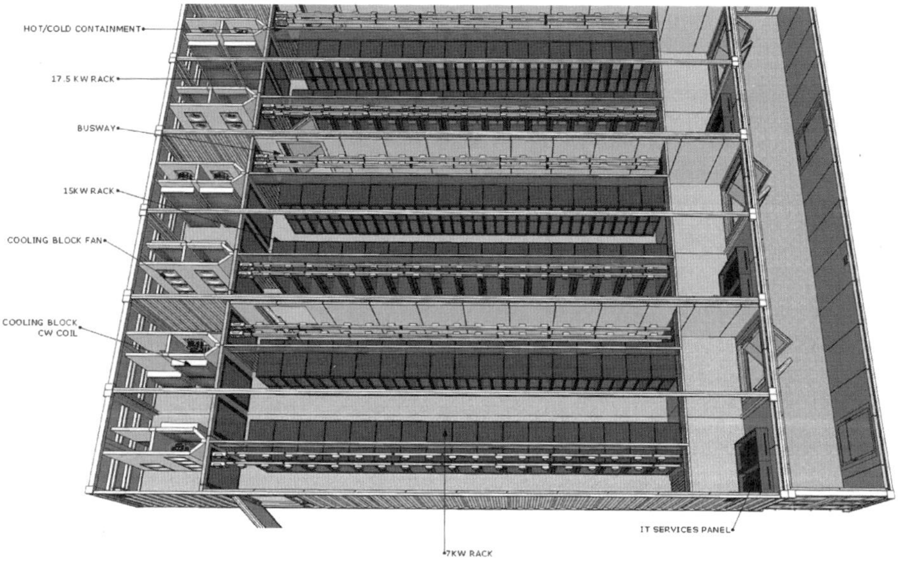


**Fig. 13.31** Expanding modular data center capacity to multiple aisles. Figure courtesy of IBM

high-density cooling solutions to complement a low-density facility, adding data center space without new construction, and lower real-estate costs.

*Temporary capacity during retrofit*: A modular data center may be a solution to bridge the gap while construction of a brick and mortar facility is ongoing. This may be needed to address increased needs for data processing during construction and coverage during high-demand business cycles.

*Mobile data center*: Modular solutions may be attractive when the facility users are on the move and may need the solution to move with them. Related examples where such facilities are attractive include emergency usage and disaster recovery or avoidance. Several other scenarios such as military needs, sporting events, and movie production may also need such facilities.

*Accelerating the growth of cloud computing*: Modular facilities could support high-performance computing in both private and public cloud environments. They could be located in low-cost areas for lower total cost of ownership.

In most applications, modular data centers are employed in outdoor environments. Containers are fully insulated and sealed to provide protection from external temperatures, humidity, smoke, dust, as well as shielded from electromagnetic interference (EMI) and radio frequency interference (RFI). Each module needs to provide proper security. Containers may be 2.4 m (8 ft) wide, 2.9 m (9.5 ft) high, and come in lengths ranging from 3.05 m (10 ft), 6.1 m (20 ft), to 16.2 m (53 ft). Operating environmental temperatures range from −46 to 66°C. The containers are rated to power density of 300 kW per 6.1-m container to 750 kW for 16.2-m container. Maximum rack power of 30 kW is listed [29]. Cooling options include in-row cooling, and overhead fan coil units, utilizing chilled water, or direct expansion (DX) units, and RDHxs utilizing chilled water for high-power racks.

### *13.10.1 Evolution in Capabilities*

Container data centers first came to market in 2007 with the "Project Blackbox" offering from Sun. Google received a patent in 2007 [31] for work initiated a few years earlier on modular data center. Currently, more than a dozen vendors offer modular data centers [30]. All modular units require external electrical power. The earlier modular units were based on chilled water or on-board refrigerant-based systems. Several configurations for on-board cooling have been utilized. Cooling coils could be located above, behind, or on the side of the equipment racks. For DX units, the compressors and condensers are located outside the container. Typical PUE values reported in the literature for these units are ~1.2 [30]. The common layout in these units is one of a single row of equipment racks running along the length of the container with complete or partial access to the front and back of the racks. This is similar to the HACA configuration in traditional facilities. Some 20-ft units feature a central aisle, with equipment racks arranged face to back on either side. Recently, several vendors have incorporated the use of outside air for cooling in modular data centers. These units offer improved energy efficiency with quoted PUE values as low as 1.02. These units are integrated with evaporative cooling or standard cooling units when conditions do not allow the use of the economizers. The rack equipment is typically arranged in a single pass through configuration.

## References


1. CNET News. Yahoo opens doors to self-cooled data center. http://news.cnet.com/8301-11128_3-20016849-54.html. Accessed 19 Sept 2010
2. Noteboom S (2010) Plenary Talk, ITherm 2010, Las Vegas, June 2010
3. Somani A, Gupta T, Joshi Y (2008) Scalable pods based cabinet arrangement and air delivery for energy efficient data center. In: International forum on heat transfer, Tokyo, 17–19 September 2008
4. Sundaralingam SV, Kumar P, Joshi Y (2011) Server heat load based CRAC fan controller paired with rear door heat exchanger. In: Proceedings of the ASME 2011 pacific rim technical conference & exposition on packaging and integration of electronic and photonic systems, InterPACK2011, 6–8 July 2011, Portland
5. Nathuji R, Isci C, Gorbatov E (2007) Exploiting platform heterogeneity for power efficient data centers. In: Proceedings of the IEEE international conference on autonomic computing (ICAC), June 2007
6. Chase J, Anderson D, Thakar P, Vahdat A, Doyle R (2001) Managing energy and server resources in hosting centers. In: Proceedings of the 18th symposium on operating systems principles (SOSP)
7. Moore J, Chase J, Ranganathan P, Sharma R (2005) Making scheduling cool: temperature-aware workload placement in data centers. In: Proceedings of the USENIX annual technical conference, June 2005
8. Heath T, Centeno AP, George P, Ramos L, Jaluria Y, Bianchini R (2006) Mercury and freon: temperature emulation and management in server systems. In: Proceedings of the international conference on architectural support for programming languages and operating systems (ASPLOS), October 2006
9. Moore J, Chase J, Ranganathan P (2006) Weatherman: automated, online, and predictive thermal mapping and management for data centers. In: Proceedings of the IEEE international conference on autonomic computing (ICAC), June 2006
10. Somani A, Joshi Y (2009) Data center cooling optimization–ambient intelligence based load management (AILM). In: Proceedings of the ASME 2009 heat transfer summer conference, HT2009, 19–23 July 2009, San Francisco
11. Nathuji R, Somani A, Schwan K, Joshi Y (2008) CoolIT: coordinating facility and IT management for efficient datacenters. In: The workshop on power aware computing and systems (HotPower), December 2008
12. Schmidt R, Kamath V, Chu RC, Lehman B, Ellsworth M, Iyengar M, Porter D (2005) Maintaining datacom rack inlet air temperatures with water cooled heat exchanger. In: ASME InterPACK '05, 17–22 July 2005, San Francisco
13. Heydari A, Gektin V (2005) Reliability challenges in design and operation of high heat powered processors. In: ASME InterPACK '05, 17–22 July 2005, San Francisco
14. Tsukamoto T, Takayoshi J, Schmidt RR, Iyengar MK (2009) Refrigeration heat exchanger systems for server rack cooling in data centers. In: InterPACK conference, IPACK2009, 19–23 July 2009, San Francisco
15. Webb RL, Nasir H (2005) Advanced technology for server cooling. In: Proceedings of IPACK2005, ASME InterPACK '05, 17–22 July, San Francisco
16. Campbell LA, Ellsworth MJ Jr, Sinha AK, Campbell LA, Ellsworth MJ Jr, Sinha AK (2009) Analysis and design of the IBM power 575 supercomputing node cold plate assembly. In: Proceedings of the ASME 2009 InterPACK conference IPACK2009, 19–23 July 2009, San Francisco
17. Xu G (2007) Evaluation of a liquid cooling concept for high power processors. In: 23 rd IEEE SEMI-THERM symposium
18. Hannemann R, Chu H (2007) Analysis of alternative data center cooling approaches. In: Proceedings of the ASME InterPACK conference, InterPACK 2007, 8–12 July, Vancouver

19. Ellsworth MJ Jr, Iyengar MK (2009) Energy efficiency analyses and comparison of air and water cooled high performance servers. In: InterPACK Conference, IPACK2009, 19–23 July 2009, San Francisco
20. Garday D, Housley J (2007) Thermal storage system provides emergency data center cooling, White Paper Intel Information Technology, Intel Corporation, September 2007
21. Schmidt R, New G, Ellsworth M, Iyengar M (2009) IBM's POWER6 high performance water cooled cluster at NCAR-infrastructure design. In: Proceedings of the ASME 2009 InterPACK conference IPACK2009, 19–23 July 2009, San Francisco
22. Wu XP, Mochizuki M, Mashiko K, Nguyen T, Wuttijumnong V, Cabsao G, Singh R, Akbarzadeh A (2010) Energy conservation approach for data center cooling using heat pipe based cold energy storage system. In: 26th IEEE SEMI-THERM Symposium, 2010
23. Klein LJ, Singh PJ, Schappert M, Griffel M, Hamann HF (2011) Corrosion management for data centers. In: 27th IEEE SEMI-THERM Symposium, 2011
24. ANSI/ISA-71.04-1985 (1986) Environmental conditions for process measurement and control systems: airborne contaminants, ANSI/ISA-the instrumentation, systems, and automation society, 3 February 1986
25. Fink J (2005) Impact of high density hot aisles on IT personnel work conditions. White paper #123, American Power Conversion
26. Roy SK (2011) An equation for estimating the maximum allowable surface temperatures of electronic equipment. In: 27th IEEE SEMI-THERM symposium
27. ASHRAE TC 9.9 (2011) Thermal guidelines for data processing environments – expanded data center classes and usage guidance. Whitepaper prepared by ASHRAE technical committee (TC) 9.9 Mission critical facilities, technology spaces, and electronic equipment
28. Sommerfeldt S, Maughan M, Daily J, Esplin J, Collins Z, Shaw M (2010) Acoustical analysis of active control in the server room of a C7 data centers colocation facility. Feasibility report accessed at http://www.c7dc.com/articles/C7-Active-Noise-Control.htm
29. IBM (2010) Portable modular data centers – add capacity where and when you need it. IBM Global Technology Services, Thought Leadership White Paper, December 2010
30. Bramfitt M, Coles H (2011) Modular/container data centers procurement guide: optimizing for energy efficiency and quick deployment. Department of Energy, Lawrence Berkley National Laboratory Report, 2011
31. Whitted WH, Aigner G (2007) Modular data center. US Patent 7,278,273, 9 Oct 2007

# Index





DOI 10.1007/978-1-4419-7124-1, © Springer Science+Business Media, LLC 2012

**E**

**R**

Printed by Printforce, the Netherlands